MOLECULAR NUTRITION

# 分子营养学

〔美〕J. 曾普尔尼　〔德〕H. 丹尼尔　编著

罗绪刚　吕　林　李爱科　主译

U0197547

科学出版社

北　京

图字:01-2005-2025 号

## 内 容 简 介

分子营养学是营养学领域中发展最快的一门科学。本书是目前国内已出版的唯一一本有关分子营养领域研究进展的中文译著。书中对国际一流专家发表的有关分子营养方面的最新和最重要的研究成果进行了综述。全书共分 24 章，详细叙述了分子营养学的研究方法、细胞内营养物质的动态平衡、细胞增殖和细胞凋亡、信号转导、基因表达和蛋白水解、核酸和分子水平事件的生理作用以及遗传修饰食物和食物过敏的分子机理等。

本书忠实原文，最大限度地反映了原书的风格与韵味。可作为营养学、动物营养学、生物化学、分子生物学及医药学专业等的教学、科研人员及研究生的参考书。



**图书在版编目(CIP)数据**

分子营养学（Molecular Nutrition）/（美）J. 曾普尔尼，（德）H. 丹尼尔编著. 罗绪刚等译. —北京：科学出版社，2008
ISBN 978-7-03-017371-3

Ⅰ. 分…　Ⅱ. ①曾…②丹…③罗…　Ⅲ. 细胞学：分子生物学：营养学　Ⅳ. Q2

中国版本图书馆 CIP 数据核字（2006）第 058451 号

责任编辑：王　静　李秀伟　沈晓晶　李　锋/责任校对：刘小梅
责任印制：赵　博/封面设计：耕者设计工作室

科 学 出 版 社 出版
北京东黄城根北街 16 号
邮政编码：100717
http://www.sciencep.com
北京厚诚则铭印刷科技有限公司印刷
科学出版社发行　各地新华书店经销
*
2008 年 5 月第　一　版　　开本：787×1092　1/16
2025 年 1 月第十次印刷　　印张：26 1/2
字数：614 000

**定价：98.00 元**

（如有印装质量问题，我社负责调换）

## 《分子营养学》参加翻译人员

**主　译**　罗绪刚　中国农业科学院北京畜牧兽医研究所矿物元素营养研究室与动物营养学国家重点实验室 研究员 博士研究生导师

吕　林　中国农业科学院北京畜牧兽医研究所矿物元素营养研究室与动物营养学国家重点实验室 副研究员 博士

李爱科　国家粮食局科学研究院 研究员 博士研究生导师

**翻译和审校人员**（按贡献大小排序）

余顺祥　黄金秀　黄艳玲　计　峰　李素芬
张铁鹰　张晓琳　于　昱　沈素芳　郝守峰
白世平　刘　彬

# 译 者 序

在过去20年中，新的强大的分子生物学技术的引入，使人类生物学知识迅猛增长。分子营养学是现代分子生物学理论与技术在营养学中应用而产生的一门新兴交叉边缘学科，也是营养学领域中发展最快的一门科学。它从分子水平上深入揭示营养物质在体内的确切代谢机制并用分子生物学技术评价营养需要量和食物营养价值等，是营养学的研究前沿和未来发展方向。

由J. 曾普尔尼和H. 丹尼尔编著的*Molecular Nutrition*，是目前国际上已出版的唯一一本有关分子营养学领域研究进展的英文著作。书中对国际一流专家发表的有关分子营养方面的最新和最重要的研究成果进行了综述。全书共分24章，详细叙述了分子营养学的研究方法、细胞内营养物质的动态平衡、细胞增殖和细胞凋亡、信号转导、基因表达和蛋白水解、核酸和分子水平事件的生理作用以及遗传修饰食物和食物过敏的分子机制等。本书论述精辟、阐述清晰、条理性强，且附有图表，每章后有概括的内容总结，并列出了参考文献，便于读者学习和查询。本书对于营养学、动物营养学、生物化学、分子生物学及医药学专业等的教学、科研人员及研究生而言，是一本适用的参考书。

本书的翻译和审校工作是由中国农业科学院北京畜牧兽医研究所矿物元素营养研究室与动物营养学国家重点实验室的研究人员和博士、硕士研究生们以及国家粮食局科学研究院饲料研究设计所的研究人员共同完成的。大家在繁重的工作之余，抽出时间翻译和审校本书，希望本书对我国分子营养学的发展起到很好的促进作用。

本书的翻译出版由国家自然科学基金重点项目（项目号：30530570）、中国农业科学院一级岗位杰出人才专项及国家973项目课题（课题号：2004CB117501）经费资助，特此致谢！

由于译者水平有限，不妥之处在所难免，敬请有关专家和读者批评指正。

译 者

2008年1月1日于北京

# 译者序

# 前　言

在过去的二三十年中，分子生物学为我们提供了许多有力的手段和技术，使得分子营养学作为营养研究中的一个新领域而诞生。分子营养学是在分子水平上研究营养物质的作用，如信号转导、基因表达和蛋白质的共价修饰的学科。利用分子技术获得的研究成果已引导我们进入了新的领域，此领域超越了如研究营养缺乏临床征兆特征的经典营养学的研究领域。那些在分子水平上研究营养物质作用效果的营养学家早已认识到这是一个多么令人振奋的领域。

在本书中，我们试图捕获分子营养学领域的一些兴奋点。我们很幸运能征募到在各自学科内起引领作用的章节作者，这些作者欣然地对分子营养学的不同方面进行了论述。但像这样的一本书不可能很全面，编辑更应从大量营养物质及其在分子水平上的更大量作用中进行选择。为了本书中的每一个专题，其他同样重要的专题可能已被忽略。作为编辑，我们对在本书中未见到他们所研究的营养物质的那些营养学家表示歉意。

我们试图涵盖分子营养学的广阔研究领域。本书前面两章由 J. Zhang 和 H. Daniel 撰写，内容包括基因组和后基因组方面的技术进展以及这些技术为未来营养研究带来的希望。接着，H. Daniel、D. B. McCormick 和 J. Zempleni 告诉我们营养物质怎样进入细胞、怎样到达胞内靶位置以及像细胞增殖这样的生理过程怎样影响营养物质的转运。该部分的最后一章由 J. C. Mathers 完成，他综述了营养物质在细胞凋亡过程中的作用。由 F. Foufelle 和 P. Ferré、M. S. Kilberg 等、U. Beisiegel 等、D. R. Soprano 和 K. J. Soprano、K. Dakshinamurti、A. Fischer 等、E. M. Schmelz、T. Badger 等和 D. Attaix 等撰写的一系列章节，用实例阐明了常量和微量营养成分在信号转导、基因表达和蛋白质水解中的作用。J. K. Christman、J. Zempleni、J. Kirkland 和 J. C. Spronck 综述了核酸和核酸结合蛋白质的营养依赖性修饰作用。

当然，分子营养学不是一门只服务于自我的科学。本书选择了关于脂蛋白装配（J. A. Higgins）、细胞中胆固醇代谢（J. Y. Lee *et al.*）、氧化应激（A. Taylor and M. Siegal）和免疫反应（P. Yaqoob and P. C. Calder）的调节等章节用以阐明分子水平的活动如何被溶入多器官及整体代谢途径之中。最后，J. S. Stanley、G. Bannon 和 S. L. Taylor 在遗传修饰食物与食物过敏的分子机制方面表达了各自的见解。

我们向参与本书各章编写的所有科学家表示感谢。这些学者在百忙之中抽出时间为本书贡献了高质量的章节。此外，我们还感谢 CABI 出版社的 Rebecca Stubbs 在本书准备期间对我们持续不断的支持。

Hannelore Daniel
Janos Zempleni

# 前 言

# 目　　录

# 第 1 部分
# 分子营养学研究的方法

# 1　基因组与后基因组

Ji Zhang[1,2]
（1 内布拉斯加大学医学中心病理与微生物系；
2 穆若-迈尔研究所人类分子遗传中心，
内布拉斯加大学医学中心，奥马哈，内布拉斯加，美国）

## 引　　言

近年来，基因组研究领域取得的巨大进展推动了人类健康与疾病的分子机制的精确测定，为增进健康、降低发病率和死亡率及预防疾病提供了巨大潜力。了解健康成年人对营养摄入的不同反应的分子机制将大大推动营养学的发展，因此，营养学家获取基因组研究的必要技术和资源的知识是相当重要的。最初，基因组学指的是对生物体基因组，即一整套基因和染色体进行作图、测序和分析的科学规则。目前，基因组学的研究重点已由基因组的结构分析（结构基因组学）向基因组的功能分析（功能基因组学）转移。结构基因组学的目的是构建生物体的高分辨遗传、物理和转录图谱，并最终测定其全部 DNA 序列。然而，功能基因组学代表着基因组研究的新阶段，是指在大量结构基因组学信息的基础上发展创新性技术。本章的第一部分将着重介绍结构基因组学中所使用的工具和试剂；第二部分将着重介绍 DNA 微阵列技术，即当今的功能基因组学的代表。

## 结构基因组学

### 基因组、遗传绘图与物理绘图

人类基因组含有近 30 亿核苷酸碱基对，携带着 30 000～100 000 个基因的遗传密码。二倍体基因组 DNA 构成 22 对常染色体和两条性染色体。每条染色体含有一条线性 DNA 分子，该分子的特点是含有 3 个功能元件，这些元件是细胞分裂时染色体成功复制所必需的：①自我复制序列；②着丝粒，即有丝分裂或减数分裂时纺锤体的连接点；③端粒，它能确保末端染色单体的完整复制。基因组图谱的定义是 DNA 分子中不同基因座（基因、调节序列、多态性标记序列等）的相对位置。有两种不同的绘制基因组基因座图谱的方法，即遗传绘图法（genetic mapping）与物理绘图法（physical mapping）。

遗传图谱的构建是采用减数分裂重组频率测定两基因座间的距离。DNA 分子中两基因座的距离越近，它们就越有可能一起遗传。正顺同线性（synteny）这一术语是指处于同一染色体上的基因座，而这些基因座不一定连锁在一起。位于不同染色体或在同一染色体上相距远的基因座，以 50%重组频率（recombination frequency）独立分开。

遗传距离用重组百分率（percentage recombination）或厘摩（centiMorgan，cM）表示。1cM 等于 1%重组率，即相当于约 $0.8\times10^{6}$ 个碱基对（bp）。人类基因组的遗传图谱跨距约为 3700cM。

遗传图谱是采用连锁分析方法（linkage analysis）来构建的，即对谱系中多样性标记的离散性进行分析。首先使用的遗传标记是表型特征（如色盲）和蛋白质多态性。然而这些多态性标记很少见。通过基因组中高度多态性序列（主要是微卫星序列）的发现使详细绘制遗传图谱成为可能，微卫星序列包括大量的由 2、3 或 4 个核苷酸组成的重复性短序列（Litt and Luty，1989；Stallings *et al.*，1991）。连锁分析是一种统计学方法，其分辨率的高低也取决于信息谱系的数量。

与遗传绘图法不同，物理绘图法是用核苷酸碱基对直接测定线性 DNA 分子上两基因座间的距离。因此，基因组的全部序列就被称为最终的物理图谱。遗传与物理绘图法二者都产生完全相同的基因顺序，但基因间的相对距离因重组频率的局部变异（local variation）而变化很大。分子生物学为物理图谱的构建提供了工具和手段，如能更详细地测定克隆基因的核苷酸序列，再如在分辨率较低的情况下，限制位点间的距离可用碱基对单位进行量化。限制位点是能被限制酶Ⅱ特异性识别并切割的短序列（主要是 4～8bp）。对限制位点的分析主要由 10～100kb（kilobase）组成的限制图谱完成，这些基因组片段能以质粒、噬菌体、黏粒、细菌人工染色体（bacterial artificial chromosome，BAC）（Shizuya *et al.*，1992）、噬菌体人工染色体（P1 artificial chromosome，PAC）（Ioannou *et al.*，1994）或酵母人工染色体（yeast artificial chromosome，YAC）（Burke *et al.*，1987）为载体被克隆和制备。

## 物理绘图法中使用的工具和试剂

### 采用人-啮齿动物的杂合体模型绘图

体细胞杂合体是用于在染色体上物理绘制人类基因图谱的工具。利用人与啮齿动物的细胞的融合及随后的培养筛选可获得稳定的杂合体，该杂合体通常含有一整套啮齿动物的染色体和一些人的染色体。该杂合体模型可用于特定人染色体上定位人类基因或基因产物（Zhang *et al.*，1990）。该方法可用来绘制同线性图谱。然而，在亚染色体区的基因定位则需要仅含部分人染色体的特殊杂合体。这已通过各种缺失绘图法而获得，包括微细胞介导的亚染色体的转移或融合（Zhang *et al.*，1989a），采用具有特定易位或间质缺失供体细胞。广泛使用的辐射杂合体方法也是基于这种缺失绘图的概念（Walter *et al.*，1994）。在此，一条人染色体与啮齿动物的染色体进行杂合，并用 X 射线将其切割成小片段。这种被照射的细胞群与啮齿动物细胞的融合及随后的对人染色体材料的筛选将产生含有不同人染色体片段的缺失杂合体模型。从源于 24 条人染色体特异性的啮齿动物杂合体的每一个中采集辐射诱导的杂合体，可得到一完整的人类辐射杂合体绘图谱模型。这使得采用聚合酶链反应（polymerase chain reaction，PCR）（Deloakas *et al.*，1998）法快速将基因、表达序列标签（expressed sequence tag，EST）、多态性 DNA 标记和序列标签位点（sequence tagged site，STS）（Olson *et al.*，1989；Weber and May，1989）绘制到亚染色体结构上成为可能。

### 采用染色体原位杂交绘图

采用原位杂交可直接确定基因在染色体上的位置，该法是指将含目标序列的DNA探针标记、变性，并与载玻片上失活的处于扩展中期的互补染色体DNA杂交。以往，探针通常用[$^{3}$H]核苷酸进行标记，杂交后用自显影法进行检测，用液体感光乳剂对染色体扩展进行覆盖，然后曝光和显影。银染（silver signal）类似于放射性探针，因此可确定目标DNA序列在染色体上的位置（Marynen *et al.*，1989；Zhang *et al.*，1989b）。这种放射性原位杂交的主要缺点是试验时间长（一般是2～3周）。此外，该检测通常会产生较强的背景色，且需依据统计结果方能确定染色体基因座。

荧光原位杂交（fluorescence *in situ* hybridization，FISH）可克服该缺点。探针用抗原（如生物素）标记的核苷酸来进行标记，然后采用荧光结合的抗体（如抗生物素蛋白）进行检测，背景干扰低，且多数扩展中期的染色体显示4个特异性的荧光信号（每种荧光信号代表两个同源染色体上的每个染色单体）。该方法的敏感性通常较低，因此需要使用大基因组探针（最好是大于10kb）。基因组噬菌体（15kb）、黏粒（40kb）、PAC（80～135kb）、BAC(130kb）甚至是YAC(200kb～2Mb）都可用作FISH绘图法的探针（Hardas *et al.*，1994）。

## 基因组绘图与疾病基因分离

遗传绘图法的最突出应用之一就是反向遗传学，即在不了解生化基础情况下绘制遗传疾病的图谱。根据遗传与物理绘图法，可绘出任何孟德尔表型特征的图谱。若信息谱系可以得到的话，那么疾病基因座与绘图的遗传标记之间的连锁关系可用于疾病诊断和载体鉴定。一旦疾病基因座被遗传定位，那么物理绘图法就可用来鉴别疾病基因及其主要缺陷（Collins，1995)。主要缺陷的识别对于设计疾病的特异性治疗方法也是必需的，且有可能促进未来体细胞基因疗法的发展。类似的方法也可用来研究多基因疾病，这将进一步扩展基因绘图法的应用。

比较基因绘图法（comparative gene mapping）是对不同物种基因组的同源基因进行绘图，也将有助于疾病基因的鉴别。在进化过程中，连锁基因组在一定程度上被保留下来，且含不同基因的同源染色体区域也能在如人、小鼠和大鼠这样的物种间保留下来。许多患有已确定的遗传图谱疾病或综合病症的大鼠或小鼠可用来绘制类似的人遗传基因座图谱（Zhang *et al.*，1989b)。

另外，疾病与细胞遗传损伤（如小染色体片段缺失或易位）之间的关系可用于在物理图谱上直接定位疾病基因座，并能用相同技术对位点进行鉴定。

## 癌症的恶性基因组绘图

癌症以细胞致癌性转化、组织侵染及最终发生转移为特征，是由已部分了解的特异基因的变异所致的一类“基因疾病”。当致癌基因因突变或基因扩增而被激活时，就会促进细胞致癌性演变。肿瘤抑制因子基因控制细胞的生长，这些基因的缺失或失活也会导致肿瘤发生。细胞学上，特定癌症常常表现为染色体异常，包括移位（translocation)、缺失（deletion)、倒置（inversion）和DNA扩增。因此，应用分子手段分析这些周期性发生的染色体异常的研究，已使人们能对大量与肿瘤发生和发展相关的基因进行鉴别。

常规染色体条带技术已为肿瘤恶性细胞的核型分析（karyotypical analysis）奠定了主要基础。由于染色体条带分析是一种完全依赖于经验的方法，故很难避免人为的误差和不确切的数据，尤其是在分析复杂核型（karyotype）而监测细微的结构变化时更是如此。FISH 在观测癌症的染色体异常特征上做出了显著贡献。关于恶性细胞中特定染色体或其衍生物的重排现象，能通过将染色体特异标记探针、染色体特异重复序列探针或染色体区域特异 DNA 探针与癌细胞杂交而进行直接观测。然而，采用这种方法来检测那些经常观测到的未知标记染色体或不明来源基因组片段的特征，需要用所有 24 条人染色体的探针。光谱核型法（spectrum karyotyping，SKY）（Schrock *et al.*，1996）的发展就是试图采用不同任意染料分别对 24 条人染色体进行显色观测而克服这种缺陷，但这种分析方法的分辨性仅局限于染色体水平。此外，小片段的标记染色体也可能检测不到。

比较基因组杂交（comparative genome hybridization，CGH）是一种用于不平衡遗传变异的全面检测方法，是以差异标记的癌变 DNA 和正常 DNA 与正常人的中期扩展 DNA 之间进行竞争性原位杂交技术为基础的（Kallioniemi *et al.*，1992）。DNA 序列的扩增区或缺失区被认为是用于检测标记 DNA 的两种荧光染料色率的增或减。然而，由该法获得的基因组信息仅限于在恶性基因组中发生基因扩增或缺失所在的区域。另外，这种方法也无法根据被测染色体区域的信息来合成 DNA。

染色体的显微解剖法（chromosome microdissection）可以直接从细胞学上可识别的区域分离 DNA。分离出的 DNA 可用于：①与区域特异性标记探针一起，用于特定染色体疾病的检测（Zhang *et al.*，1993a）；②用于基因扩增研究（Zhang *et al.*，1993b）；③与区域特异性 DNA 标记物一起，用于定位克隆（Zhang *et al.*，1995）。从技术手段上看，上述的分子细胞遗传技术是相辅相成的，有助于恶性基因组的快速扫描和染色体异常的定位，也有助于识别与肿瘤发生有关的基因。

## 基因组绘图的整合

基因组绘图已进入了最后的整合阶段，即把遗传和物理信息整合成更完善、更全面的图谱。该领域的里程碑之一是构建了含有 5264 个遗传标记的人类连锁图谱（Dib *et al.*，1996）。高分辨的遗传图谱可达到遗传绘图的 1cM 分辨限，标记间距小于 $1\times10^6$ 物理距离。这是用于遗传疾病研究中的最全面的工具之一。之后不久，又构建了含大于 30 000基因标记、平均间距约 100kb 的高密度物理图谱（Deloukas *et al.*，1998）。有了这些综合图谱，使人们将从 PAG 和 BAC 载体中分离的完整基因组片段组装成克隆图谱或综合毗连序列群成为可能。之后，典型 PAG 或 BAC 克隆被用作鸟枪质粒文库的 DNA 源，而文库中平均插入的片段大小约 1kb。最后，这些鸟枪克隆片段被用作高通量 DNA 测序的 DNA 模板，此时采用双脱氧终端生物化学和具有激光荧光监测的自动凝胶电泳进行检测（Venter *et al.*，1998）。人类基因组绘图还不完善，是由于染色体中大量分散和毗连的重复序列存在的缘故。尤其是依靠现有的技术，还无法对着丝粒和同臂内异染色质区中的这些重复序列进行组装。这些重复序列小到只含几个碱基对，或大到含 200bp。这些区域可在某一单一染色体区毗连重复成百上千次。尽管完成人类基因组测序还需要做很多工作，但现有的大量结构信息已促进了人类细胞分子机制的精确测定（Deloukas *et al.*，1998；Lander *et al.*，2001；Venter *et al.*，2001）。

# 功能基因组学

## 功能基因组学中微阵列技术的应用

功能基因组学标志着基因组研究已迈入一个新阶段：在基因组这样大的规模上评价基因的功能。这是以 DNA 微阵列技术的出现为标志的（Schena *et al.*，1995；DeRisi *et al.*，1997）。计算机基因（*in silico*）微阵列方法学是指将上万个 cDNA 克隆（即探针）

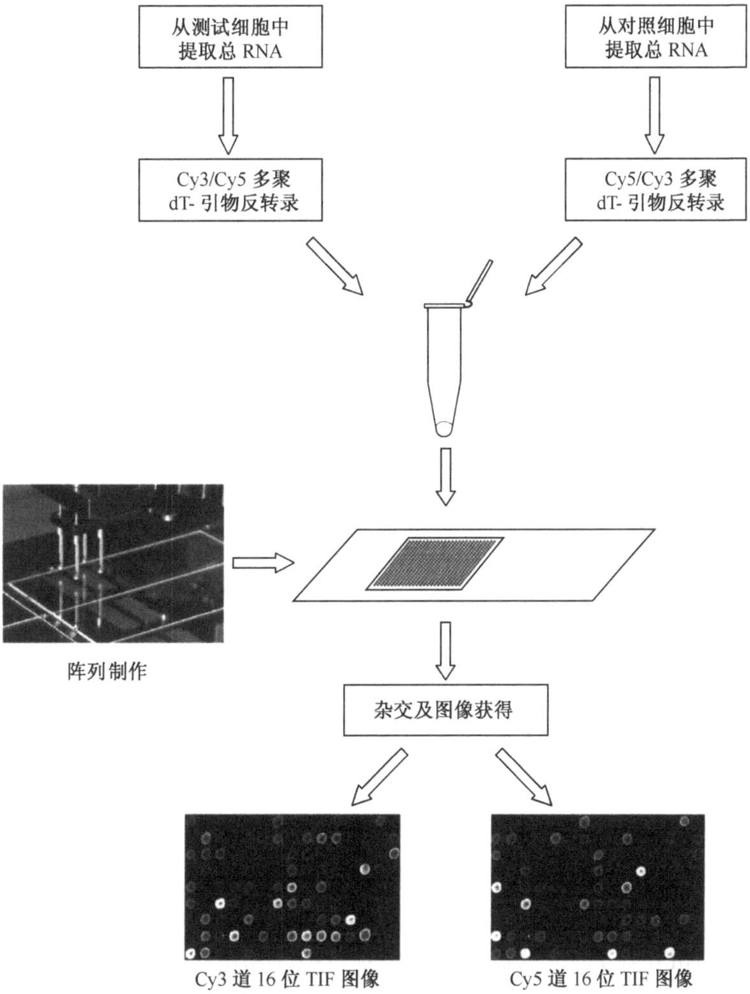


图 1.1　描述计算机基因（*in silico*）DNA 微阵列一般程序的图解
左图说明微阵列制作的印染过程，即采用 PCR 技术制备单一克隆的 cDNA 插入片段，并通过 GMS 417 阵列（Affymetrix）印染到覆盖有聚赖氨酸的载玻片上。与差异标记的测试探针和对照探针一起进行过夜杂交后，采用 GenePix 4000 扫描仪（Axon Instruments）进行扫描，以得到两种 TIF 图像：绿道和红道。每点的红与绿之间的像素比用作进一步数据分析的数值（因原书为黑白印刷，故无法显示彩色）

的插入片段于载玻片上（on to a glass slide）自动排列成整齐阵列，随后用两种差异标记的 RNA 工具（即目标基因）进行检测。RNA 样品通常用结合有荧光染料的核苷酸（如 Cy3-dUTP 或 Cy5-dUTP）来进行标记。比较至少两个处理组的 RNA（即目标基因），以鉴别如正常细胞与病变细胞间、野生动物与转基因动物间或一般对照与一系列研究样本间 mRNA 水平的差异。杂交后，采用适当波长的激光束来激活玻片，以产生两个 16 位 TIF 图。每个波长道中每个点的像素与荧光分子的数量成正比，因此可以用来定量分析与 cDNA 克隆（探针）杂交的目标分子的数量。每个波长信号强度的差异与杂交到同一 cDNA 探针上的两种不同目标的分子数量一致。DNA 微阵列的一般过程见图 1.1。用于该项技术的试验操作步骤已经很好地建立，因此，本章着重介绍数据分析。

## 微阵列数据分析

图 1.2 是典型的 DNA 微阵列图。由每一微阵列试验产生的数据量都是庞大的，相当于采用传统分子生物学方法（如 Northern 印迹）进行上万个单一核苷酸杂交试验所获得的数据量。如何将这样庞大的数据转变为有意义的生物学信息网是极富挑战性的。因此，生命科学家了解用于本领域的数据处理工具的工作原理是非常重要的。

### 数据前处理

目前，各种激光数据获取扫描仪均可在市场上买到。对于数据分析，通常必须先建立一个类似电子数据表的矩阵，其中行表示基因，列表示 RNA 样品，每格含表示特定样品中特定基因转录水平特点的比值（如 Cy5 像素/Cy3 像素）。用来研究该矩阵的方法有两种：矩阵行比较法与矩阵列比较法。通过寻找行中基因表达类型的相似性，可鉴别共同被调节的功能相关基因。通过比较样品的表达分布，能测定生物学上相关的样品或差异表达的基因。通常，矩阵需要选择性地删除一些丢失或含错误数值的基因。然后，将矩阵中数据用以 2 为底数的对数表示，以使数据处于正态分布，并减少可能存在的由极值导致的数据不均衡。当把一系列被测样品（如临床样品）与非配对对照（参比）样品进行比较时，考虑到测试样品的数据分析不依赖于非配对对照样品的基因表达水平，故需要通过平均或中值定心的方法来对以对数表示的比值做进一步处理。

### 相似性度量

现在对微阵列数据的研究主要集中在聚类分析（clustering）和图解分析（visualization）上。聚类分析是根据它们之间的相似性，把基因或 RNA 样品归类为有意义的组；图解分析是以容易理解的格式来描述聚类分析的结果。对于相似性的比较，欧几里得距离（Euclidean distance）的概念和相关系数的计算通常用于建立相似性度量。欧几里得距离是指两个 $n$ 维点（如 $X$ 和 $Y$）之间的距离。$X$ 对应值是 $X_1$，$X_2$，…，$X_N$；$Y$ 对应值是 $Y_1$，$Y_2$，…，$Y_N$。$X$ 与 $Y$ 之间的欧几里得距离是

$$\mathrm{d}(x,y)=\sqrt{\sum_{i}^{n}(x_i-y_i)^2}$$

式中，$n$ 是基因比较的 RNA 样品数或是样品比较的基因数。例如，对任意两个基因

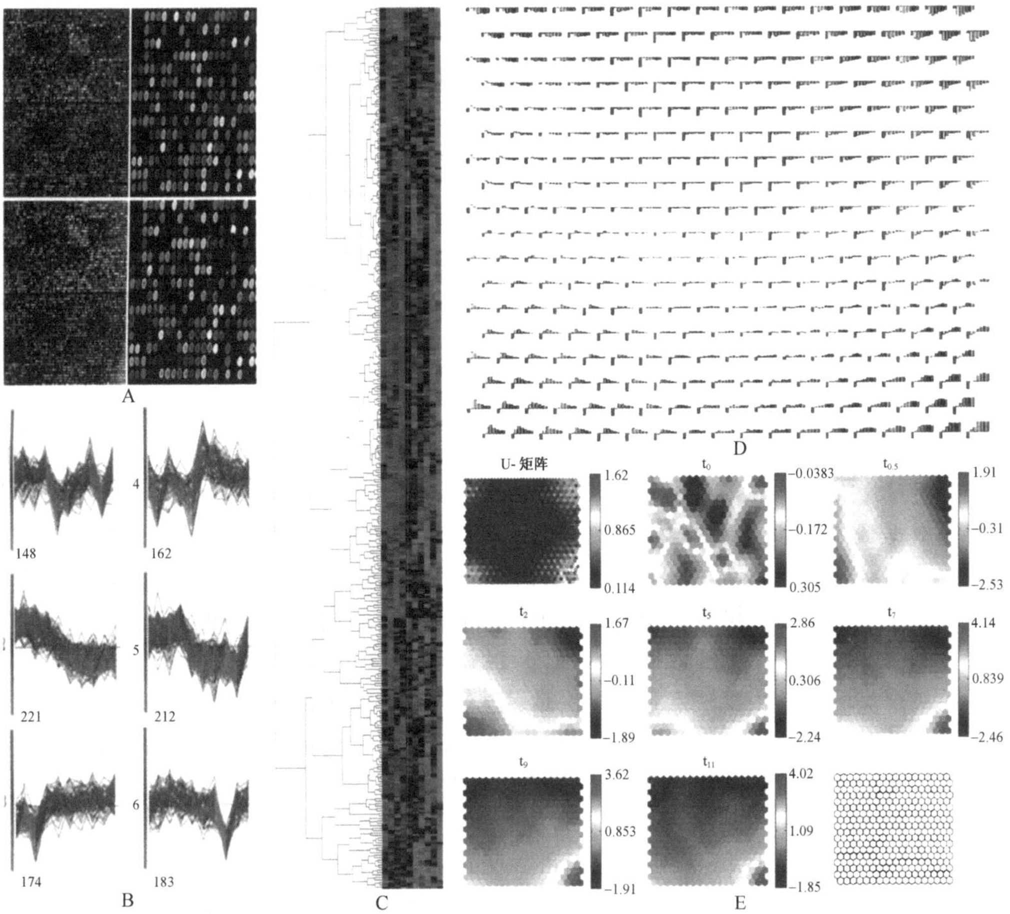


图 1.2　采用不同算法的微阵列数据图（见彩图）

A. 说明两个脑发育阶段之间的相似性和不相似性的彩图。左上图是 Cy3 标记交配后 11.5d 小鼠脑 cDNA 池与 Cy5 标记胚胎肝 cDNA 池的对比图，左下图是 Cy3 标记交配后 12.5d 小鼠脑 cDNA 池与对照组的对比图。右图详细说明部分左图。B. $K$值聚类分析得出的代表性聚类的图形分布。水平刻度表示从小鼠胚胎脑发育期间的 10 个不同时间点获得的 RNA 样品。垂直刻度表示基因表达从高（红色）到低（绿色），用对数比为单位，用中值减去得到的数值的变化。C. 通过对 10 个胚胎发育样品的 4608 个小鼠基因进行遗传（系统）聚类分析而获得的部分树图。红色表示上调，绿色表示下调。D. 表示在酵母孢子形成过程中被调控基因的基因聚类分析及表达类型的 SOM 条形图。全部基因均被组织进入 324（18×18）六角形的图单元。特定单元中的每个条形表示绘图到那个单元的基因的平均表达情况。E. U 矩阵和组分平面图。U 矩阵中的颜色编码代表欧几里得距离，颜色越深，距离越小。占据大部分图的大深蓝色区表示不受调控的基因，它们产生一些干扰的聚类。组分平面图（$t_0$～$t_{11}$）表示在酵母孢子形成过程中整个基因组中受调节基因的差异显示图。颜色编码指数表示基因的表达值，颜色越亮，数值越高。所有这些差异显示图在位置上相互关联：每个显示图中，某一位置的六角形与相同的图单元相对应，因而可直接比较不同图的相同位置的表达类型。最后一个标记图表示整个图上每个单元的位置

（如 $X$ 和 $Y$）进行三维空间（即 3 个样品）上的比较，$X$ 与 $Y$ 之间的欧几里得距离是

$$d(\vec{x},\vec{y}) = \sqrt{(x_1 - y_1)^2 + (x_2 - y_2)^2 + (x_3 - y_3)^2}$$

两点间的距离越近，相似性越大。

任意两个 $n$ 维点之间的相关系数被定义为

$$r = \frac{1}{n}\sum_{i=1}^{n}\left(\frac{X_i - \overline{X}}{\delta_X}\right)\left(\frac{Y_i - \overline{Y}}{\delta_Y}\right)$$

式中，$n$ 是基因比较的 RNA 样品数，或是样品比较的基因数；$\overline{X}$ 是 $X$ 点值的平均值；$\delta_X$ 是 $X$ 点值的标准差。

例如，若以所有样品或基因的 $X$ 值和 $Y$ 值绘制 $X$ 点与 $Y$ 点的曲线图，则 $\gamma$ 表示两曲线状形之间的相似性。相关系数总是在－1～1 之间。当 $\gamma$ 等于 1 时，表示两曲线形状完全相同；$\gamma$ 等于 0，表示两曲线形状完全独立；$\gamma$ 等于－1，表示两曲线形状呈负相关。欧几里得距离和相关系数二者均用于检测聚类分析的相似性，目前尚无清楚理由说明欧几里得距离和相关系数哪个更具有优势。

## 聚类分析算法

通常应用的基因聚类分析的算法包括等级聚类分析（hierarchical clustering）、*K* 值聚类分析（*K*-mean clustering）和自组图分析（self-organizing map，SOM）。等级聚类主要是根据个体（基因或样本）间的平均距离，通常用相关系数表示的相似程度而进行的聚类（Eisen *et al*.，1998；White *et al*.，1999）。该算法通过配对比较，最后把所有个体聚类成一个树图。树枝的长度表示个体之间的关系，即树枝越短，个体间的相似性越大（图 1.2C）。该算法已常用于微列阵数据分析，并已证明是一种有价值的分析工具。等级聚类分析的主要缺点是该算法是多基因树结构，该结构最适用于真正的遗传世代，如物种的进化（Tamayo *et al*.，1999），而不适用于活细胞中多种不同的途径。该法可能导致基因的不正确聚类分析，尤其在针对大量而复杂的数据集时更是如此。

*K* 值聚类分析是根据所有个体在每个类别中的平均向量计算的重复循环数。把所有个体分成一定数量（*K*）的独立但组内个体相同的组别，并将个体重排于中心最接近该个体的类别（图 1.2B）。欧几里得距离通常用来表示相似性。*K* 值聚类分析的局限性是随意确定的基因聚类数量不能反映活细胞中的真实情况，另外，聚类类别之间的关系也不明确。

SOM，一种以无指导学习为基础的人工智能算法（Kohonen，1995；Kohonen *et al*.，1996）似乎特别适用于微阵列数据分析。下面将进一步详细介绍该方法的应用。

## 自组图算法

该算法具有量化向量与作图向量的特点，随后把输出原型向量设定为原始多维输入数字数据的拓扑表示法。SOM 由通常在二维格上的一定数量的单元（neuron）组成。每个单元由一个多维原型向量表示。原型向量的维数等于输入向量的维数（即样品数）。输入向量的数目等于输入数目，即矩阵中的基因数。单元通过邻居关系与邻近单元相连决定拓扑结构，即图谱结构。原型向量从任意数值开始，并迭代排列。每个实际的输入向量与绘制方格中每个原型向量进行比较，比较的依据是：$\| \vec{x} - \vec{m}_c \| = \min_i \{ \| \vec{x} - \vec{m}_i \| \}$。其中，$\vec{x}$ 表示输入向量；$\vec{m}_c$ 表示输出向量。最佳匹配单位（best-matching unit，BMU）的定义是单元的原型向量与输入向量之间的最小欧几里得距离。同时，BMU 周围的拓扑接邻单元向排列输入向量延伸，结果被更新表示为：$\vec{m}_i(t+1) = \vec{m}_i(t) + \alpha(t)[\vec{x}(t) - \vec{m}_i(t)]$。SOM 排列处理通常需要两个阶段，即首先的粗

排与随后的细调两个过程。经迭代排列后，SOM 最后形成这样的格式，即相似特征的个体被绘制到相同的图谱单元或邻近图谱单元，以在整个图谱实现相关个体的平滑转换（Kohonen *et al.*，1996）。更重要的是，这种有序图谱为数字数据集的各种检验提供了一种常用平台。尽管 SOM 法已应用于几个以微阵列为基础的研究中（Tamayo *et al.*，1999；Toronen *et al.*，1999；Chen *et al.*，2001），但其全部潜力（特别是视图检验）还尚未被充分应用于微阵列数据分析中。最近，为了描述基因的转录变化，我们已引入组分平面图，即一种 SOM 的更深入的可视工具来表示微阵列数据。通过把这种组分平面图的特点与 SOM 进行整合，微阵列分析就不仅是基因的聚类分析，例如，还包括基因组上被调节基因的差异显示。

## 采用组分平面图整合的 SOM，同时进行基因聚类和基因组差异显示分析

为了说明这种方法相对于其他分析方法的优点，我们选择前面已分析了的含 6400 个基因和 7 个时间点的 7 个 RNA 样品的酵母孢子数据集（Chu *et al.*，1998）为例。酵母的孢子形成是二倍体细胞经减数分裂而产生单倍体生殖细胞的过程，包括两个相互重叠的步骤：减数分裂和孢子形成。该过程被划分为减数分裂Ⅰ、减数分裂Ⅱ和孢子形成三个过程。孢子形成的过程能用氮缺乏的培养基进行诱导。

对于 SOM 法及其图解，我们已采用 Vesanto 等（2000）编程的一种 SOM 工具箱。这种建立在 Matlab 5 计算程序环境上的工具箱能进行数据的前处理，采用各种不同拓扑结构来排列 SOM，并能以不同方式对 SOM 进行图示。为了从拓扑结构上使相邻连接的数量最大化，在进行 SOM 排列时，我们采用了六角形的原型向量来代替矩形的向量，然后，于二维网格（18×18 格）上采用 324 个原型向量进行该运算。对于图解分析，我们首先采用与前面出版图相似的条形图（图 1.2D），以对基因聚类分析和表达基因的表达类型进行全面认识。绘制到每个图单元的基因数为 7～62，在每个六角形单元中显示的条形图表示该单元中基因的平均表达类型。可看出，该图谱是以一定方式进行组织的，即把相关类型放置在邻近的图谱单元中，从而实现整个图谱表达类型的平滑转换。因此，基因聚类分析也可通过其核心单元外包围相邻图谱单元来进行认识。

为了表明孢子形成过程中调节基因聚类分析以外的特点，我们把 SOM 分析与组分平面图的强力视图工具整合起来。组分平面图为 SOM 变量图示提供了一种深入的方法。每个组分平面图被认为是 SOM 的一个平切图本，说明了在所有图谱单元中每个单一向量组分的数值。例如，图 1.2E 中第一个组分平面图（$t_0$）表示在时间点 0h 时的 SOM 切面图，而最后一个组分平面图（$t_{11}$）表示在孢子形成期间的时间点 11h 时的 SOM 切面图（Chu *et al.*，1998）。图谱单元的颜色是经过选择的，表示颜色越亮，则绘图到相应单元的基因的平均表达值就越大。每个 SOM 切面也可被认作是受调节基因在基因组范围内的差异显示图，其中上调的单元（红色六角形）、下调的单元（蓝色六角形）和中度转录的单元（绿色和黄色的六角形）均得到好的描绘。通过对这些基因组范围内的差异显示图进行比较，我们能获取细胞中受调节基因的许多其他特点。例如，这些差异显示图按一定顺序彼此相关，表明在转录水平上孢子的形成过程。绘图到两个上角的基因的相继失活提示，右边基因代表的功能组对氮缺乏培养基的诱导作用比左边

基因所代表的功能组更为敏感，尽管它们均被抑制向孢子形成的末期发展。绘图到两个底角的基因的相继激活，为我们提供了一张更加生动的孢子形成过程的画面。在孢子形成的早期，底左角和左边的基因被激活，表明这些基因特异性地与减数分裂Ⅰ有关。相比之下，右角基因的表达不断提高，则表明这些基因与减数分裂Ⅱ和孢子形成有关。这与参与减数分裂Ⅱ和孢子形成的已知基因已作图到这些角落单元中的结果一致。

SOM 法具有很大潜力，尤其是在数据图示方面更是如此。迄今，大多数用于微阵列数据作图的程序均限于基因聚类分析，图 1.2D 中的条形图示就是一个典型例子。相比之下，图 1.2E 中所显示的 U-矩阵（统一距离矩阵）是一种距离矩阵方法，可对邻近图谱单元中原型向量之间的配对距离进行图示，并有助于确定 SOM 的聚类结构。我们已采用这种作图法成功地确定了大脑发育期间表达的与发育相关基因的一些核心聚类。但若干扰高，此法对数据的解释可能困难。孢子形成数据集中存在大量不受调节的基因也支持了上述观点，这些基因以随机方式形成聚类，在 SOM 中心形成一个可见的聚类区（图 1.2D）。

组分平面图提供了一种对 SOM 的组分变量进行图示的深入研究方法。因此，SOM 能被切割成多个样品特异的、基因组范围的差异显示图。每个显示图均详细描述了特定样品在基因组范围的转录变化。这些基因组范围的差异显示图大大有助于认识微阵列数据的生物学意义。正如本节所述，我们能直接测定孢子形成过程中差异表达的基因在基因组范围的功能意义，而采用其他方法得到类似的结论则要做大得多的努力（Chu *et al.*，1998）。组分平面图也可应用于其他生物的微阵列数据。例如，我们已应用该方法分析早期大脑发育阶段的 10 个时间点的小鼠脑样的微阵列数据。通过这些研究，我们已鉴别出大量与大脑发育有关的基因。这些基因组范围的差异显示图可用于鉴别被调节基因的功能意义。而且，根据在这些图上相同位置的类型，它们还能把不同样品的数据相互关联起来，这对于来自临床研究的样品来说，前景是非常光明的。这对微阵列数据分析的潜在影响可能是巨大的。

## 小　　结

随着在基因组研究中取得的进展，基因组学的概念延伸超出了基因组结构分析的范畴，还包括基因组的功能分析。结构基因组学着重于采用各种分子生物学工具来进行基因组的遗传绘图和物理绘图。遗传图谱是以谱系中多态性标记的分离的连锁关系分析为基础。物理图谱是测定核苷酸碱基对中基因座间的距离。一个基因组的最终物理图谱是对其完整 DNA 序列的测定。基因组绘图最突出的应用之一，就是对疾病基因的研究，这方面的典型例子是反向遗传学。癌症遗传学也是疾病基因研究领域中的一个重要方面。尽管全部的人类基因组测序正接近完成，但认识基因组绘图中所用的工具和试剂可能仍有助于我们目前的研究。本章以 DNA 微阵列分析技术为代表，重点介绍了功能基因组学。这种技术可平行检测成千上万个基因，为揭示活细胞中的分子机制提供了最全面的方法。DNA 微阵列分析最具挑战性的部分，是要把大量的数据转化为具有生物学意义的信息网。与其他的数据开发处理工具相比，我们相信，SOM 当与组合平面图整合应用时，是这方面最强有力的工具。这种整合方法不仅可对基因进行聚类分析，而且也可使被调节基因在基因组范围规模进行差异显示分析。该方法的应用特别适合于临床病例研究，因为临床病例研究中常常需要对每个患者的转录组分模式间进行详细比较。随着基因组信息的大大丰富和分析技术的快速发展，研究活的人类细胞中的分子作用机制的时代已经到来。

## 致　　谢

作者感谢 Li Xiao 和 Yue Teng 在数据计算和作图方面给予的极大帮助。

### 参 考 文 献


Burke, D.T., Carle, G.F. and Olson, M.V. (1987) Cloning of large segments of exogenous DNA into yeast by means of artificial chromosome vectors. *Science* 236, 806–812.

Chen, J.J., Peck, K., Hong, T.M., Yang, S.C., Sher, Y.P., Shih, J.Y., Wu, R., Cheng, J.L., Roffler, S.R., Wu, C.W. and Yang, P.C. (2001) Global analysis of gene expression in invasion by a lung cancer model. *Cancer Research* 61, 5223–5230.

Chu, S., DeRisi, J., Eisen, M., Mulholland, J., Botstein, D., Brown, P.O. and Herskowitz, I. (1998) The transcriptional program of sporulation in budding yeast. *Science* 282, 699–705.

Collins, F.S. (1995) Positional cloning moves from perditional to traditional. *Nature Genetics* 9, 347–350.

Deloukas, P., Schuler, G.D., Gyapay, G., Beasley, E.M., Soderlund, C., Rodriguez-Tome, P., Hui, L., Matise, T.C., McKusick, K.B., Beckmann, J.S. *et al.* (1998) A physical map of 30,000 human genes. *Science* 282, 744–746.

DeRisi, J.L., Iyer, V.R. and Brown, P.O. (1997) Exploring the metabolic and genetic control of gene expression on a genomic scale. *Science* 278, 680–686.

Dib, C., Faure, S., Fizames, C., Samson, D., Drouot, N., Vignal, A., Millasseau, P., Marc, S., Hazan, J., Seboun, E., Lathrop, M., Gyapay, G., Morissette, J. and Weissenbach, J.A. (1996) Comprehensive genetic map of the human genome based on 5,264 microsatellites. *Nature* 380, 152–154.

Eisen, M.B., Spellman, P.T., Brown, P.O. and Botstein, D. (1998) Cluster analysis and display of genome-wide expression patterns. *Proceedings of the National Academy of Sciences USA* 95, 14863–14868.

Hardas, B.D., Zhang, J., Trent, J.M. and Elder, J. (1994). Direct evidence for homologous sequences on the paracentric regions of human chromosome 1. *Genomics* 21, 359–363.

Ioannou, P.A., Amemiya, C.T., Garnes, J., Kroisel, P.M., Shizuya, H., Chen, C., Batzer, M.A. and de Jong, P.J. (1994) A new bacteriophage P1-derived vector for the propagation of large human DNA fragments. *Nature Genetics* 6, 84–89.

Kallioniemi, A., Kallioniemi, O.P., Sudar, D., Rutovitz, D., Gray, J.W., Waldman, F. and Pinkel, D. (1992) Comparative genomic hybridization for molecular cytogenetic analysis of solid tumors. *Science* 258, 818–821.

Kohonen, T. (1995) *Self-organizing Maps. Springer Series in Information Sciences*, Vol. 30, Springer, Berlin.

Kohonen, T., Oja, E., Simula, O., Visa, A. and Kangas, J. (1996) Engineering applications of the self-organizing map. *Proceedings of the IEFE* 84, 1358–1384.

Lander, E.S., Linton, L.M., Birren, B., Nusbaum, C., Zody, M.C., Baldwin, J., Devon, K., Dewar, K., Doyle, M., FitzHugh, W. *et al.* (2001) Initial sequencing and analysis of the human genome. *Nature* 409, 860–921.

Litt, M. and Luty, J.A. (1989) A hypervariable microsatellite revealed by *in vitro* amplification of a dinucleotide repeat within the cardiac muscle actin gene. *American Journal of Human Genetics* 44, 397–401.

Marynen, P., Zhang, J., Cassiman, J.J., Van den Berghe, H. and David, G. (1989) Partial primary structure of the 48- and 90-kilodalton core proteins of cell surface-associated heparan sulfate proteoglycans of lung fibroblasts. *Journal of Biological Chemistry* 264, 7017–7024.

Olson, M., Hood, L., Cantor, C. and Botstein, D. (1989) A common language for physical mapping of the human genome. *Science* 245, 1434–1435.

Schena, M., Shalon, D., Davis, R.W. and Brown, P.O. (1995) Quantitative monitoring of gene expression patterns with a complementary DNA microarray. *Science* 270, 467–470.

Schrock, E., du Manoir, S., Veldman, T., Schoell, B., Wienberg, J., Ferguson-Smith, M.A., Ning, Y., Ledbetter, D.H., Bar-Am, I., Soenksen, D., Garini, Y. and Ried, T. (1996) Multicolor spectral karyotyping of human chromosomes. *Science* 273, 494–497.

Shizuya, H., Birren, B., Kim, U.J., Mancino, V., Slepak, T., Tachiiri, Y. and Simon, M. (1992)

Cloning and stable maintenance of 300-kilobase-pair fragments of human DNA in *Escherichia coli* using an F-factor-based vector. *Proceedings of the National Academy of Sciences USA* 89, 8794–8797.

Stallings, R.L., Ford, A.F., Nelson, D., Torney, D.C., Hildebrand, C.E. and Moyzis, R.K. (1991) Evolution and distribution of (GT)n repetitive sequences in mammalian genomes. *Genomics* 10, 807–815.

Tamayo, P., Slonim, D., Mesiror, J., Zhu, Q., Kitareewan, S., Dmitrovsky, E., Lander, E.S. and Gowb, T.R. (1999) Interpreting patterns of gene expression with self-organizing maps: methods and application to hematopoietic differentiation. *Proceedings of the National Academy of Sciences USA* 96, 2907–2912.

Toronen, P., Kolehmainen, M., Wong, G. and Castren, E. (1999) Analysis of gene expression data using self-organizing maps. *FEBS Letters* 451, 142–146.

Venter, J.C., Adams, M.D., Sutton, G.G., Kerlavage, A.R., Smith, H.O. and Hunkapiller, M. (1998) Shotgun sequencing of the human genome. *Science* 280, 1540–1542.

Venter, J.C., Adams, M.D., Myers, E.W., Li, P.W., Mural, R.J., Sutton, G.G., Smith, H.O., Yandell, M., Evans, C.A., Holt, R.A. *et al.* (2001) The sequence of the human genome. *Science* 291, 1304–1351.

Vesanto, J. (2000) Neural network tool for data mining: SOM toolbox. In: *Proceedings of Symposium on Tool Environments and Development Methods for Intelligent Systems, TOOLMET2000.* Oulun yliopistopaino, Oulu, Finland, pp. 184–196.

Walter, M.A., Spillett, D.J., Thomas, P., Weissenbach, J. and Goodfellow, P.N. (1994) A method for constructing radiation hybrid maps of whole genomes. *Nature Genetics* 7, 22–28.

Weber, J.L. and May, P.E. (1989) Abundant class of human DNA polymorphisms which can be typed using the polymerase chain reaction. *American Journal of Human Genetics* 44, 388–396.

White, K.P., Rifkin, S.A., Hurban, P. and Hogness, D.S. (1999) Microarray analysis of *Drosophila* development during metamorphosis. *Science* 286, 2179–2184.

Zhang, J., Marynen, P., Devriendt, K., Fryns, J.P., Van den Berghe, H. and Cassiman, J.J. (1989a) Molecular analysis of the isochromosome 12P in the Pallister–Killian syndrome. Construction of a mouse–human hybrid cell line containing an i(12p) as the sole human chromosome. *Human Genetics* 83, 359–363.

Zhang, J., Hemschoote, K., Peeters, B., De Clercq, N., Rombauts, W. and Cassiman, J.J. (1989b) Localization of the PRR1 gene coding for rat prostatic proline-rich polypeptides to chromosome 10 by *in situ* hybridization. *Cytogenetics and Cell Genetics* 52, 197–198.

Zhang, J., Devriendt, K., Marynen, P., Van den Berghe, H. and Cassiman, J.J. (1990) Chromosome mapping using polymerase chain reaction on somatic cell hybrids. *Cancer Genetics and Cytogenetics* 45, 217–221.

Zhang, J., Meltzer, P., Jenkins, R., Guan, X.Y. and Trent, J. (1993a) Application of chromosome microdissection probes for elucidation of BCR-ABL fusion and variant Philadelphia chromosome translocations in chronic myelogenous leukemia. *Blood* 81, 3365–3371.

Zhang, J., Trent, J.M. and Meltzer, P.S. (1993b) Rapid isolation and characterization of amplified DNA by chromosome microdissection: identification of IGF1R amplification in malignant melanoma. *Oncogene* 8, 2827–2831.

Zhang, J., Cui, P., Glatfelter, A.A., Cummings, L.M., Meltzer, P.S. and Trent, J.M. (1995) Microdissection based cloning of a translocation breakpoint in a human malignant melanoma. *Cancer Research* 55, 4640–4645.

# 2 后基因组营养研究的前景

Hannelore Daniel
（德国慕尼黑技术大学食品和营养系分子营养组）

## 引　　言

每个营养过程都依赖于特定细胞中表述的由 mRNA 编码的大量蛋白质之间的相互作用。mRNA 水平的变化及其对应的蛋白质水平的变化（尽管两者的变化不一定平行）是生化途径控制养分或代谢物质流动的关键指标。食品的营养成分和非营养成分、膳食及生活方式，都会影响遗传信息传递中的每个步骤，包括从基因表达到蛋白质合成再到蛋白质降解，通过最复杂的方式改变着代谢功能。毫无疑问，随着遗传信息的日益积累，我们将发现一些令人激动的手段，可以洞察正常及病理条件下人体代谢的分子基础。而且，相对稳定的哺乳动物基因组与其快速变化着的营养环境之间的相互作用，也无疑是后基因组研究中最引人关注和最有趣的领域之一。

## 从基因到功能和从基因组学到表型组学

虽然我们已在哺乳动物基因的数量、单个基因的染色体定位、基因组结构和部分编码蛋白质的功能方面积累了大量信息，但距认识这些单个因素协同调节代谢的机制仍有很大差距。

### 代谢的分子描述

基因组数据仅含有限的关于整个细胞过程中动态变化的信息。然而，现代技术的进展，已能通过测定单个 RNA 分子或大量 mRNA 分子的表达水平而分析出细胞或生物体遗传反应的变异性和动态变化。基因组学描述大规模的 DNA 排序，可提供基础的遗传信息，并可了解基因编码区和调节元件（即启动子）的序列多样性（即单核苷酸多态性，single nucleotide polymorphism，SNP）。转录组学，也叫表达谱（expression profiling）——可通过 DNA 杂交和（或）定量聚合酶链反应（PCR）技术同时检测生物样品中几个甚至几千个可读框的mRNA水平（Celis，2000；Lockhart and Winzeler，2000）。蛋白质组学可使蛋白质组作为在细胞或器官中表达的基因组的蛋白质补体来被鉴别，并可测定蛋白质表示方式和水平的变化。而且，对于单个蛋白质，对功能有关键作用的翻译后修饰，甚至是氨基酸的替换（多态性）也能被监测（Dutt and Lee，2000；Pandey and Mann，2000）。基因组、转录组和蛋白质组分析这些新技术的应用对于未来 10 年营养学的发展及进入快速发展的功能基因组学时代都是非常关键的。

功能基因组学是以基因驱动技术或表型驱动技术为基础的（图 2.1）。基因驱动技术是利用基因组信息，在分子水平上对基因进行鉴别、克隆、表达和鉴定。表型驱动技术是在不了解潜在分子机制的情况下，研究随机突变片段或自然变种的表型特征，以鉴别和克隆引起特殊表型的基因。当然，这两种技术实际在所有的分析水平上均具有高度的互补性，从而共同决定基因型与表型之间的相关性。在自然界没有出现可表明单个基因/蛋白质功能失调的表型表现的先天性代谢缺陷之处，单个基因或基因组在代谢组成中的作用可通过试验动物如果蝇（*Drosophila melanogaster*）、线虫（*Caenorhabditis elegans*）、小鼠、大鼠和人类细胞系基因失活（“敲除”）或选择性表达（“敲入”）与过度表达模型来进行分析，这些转基因技术已产生大量缺失一个或几个基因或过度表达其他基因的动物模型。这些技术作为了解代谢的新的遗传工具虽然非常好，但并不能常获得明显的表型。这告诉研究者，基因功能的缺失可能被其他机制所补偿。更先进的动物转基因技术如 Cre/lox 调控的细胞、器官和（或）时间依赖型的基因失活或表达诱导可以极其完善的方式来分析表型结果，这些技术似乎特别适合于因简单基因损坏而对胚胎或新生儿的发育具有致命作用的情况（Sauer，1998）。

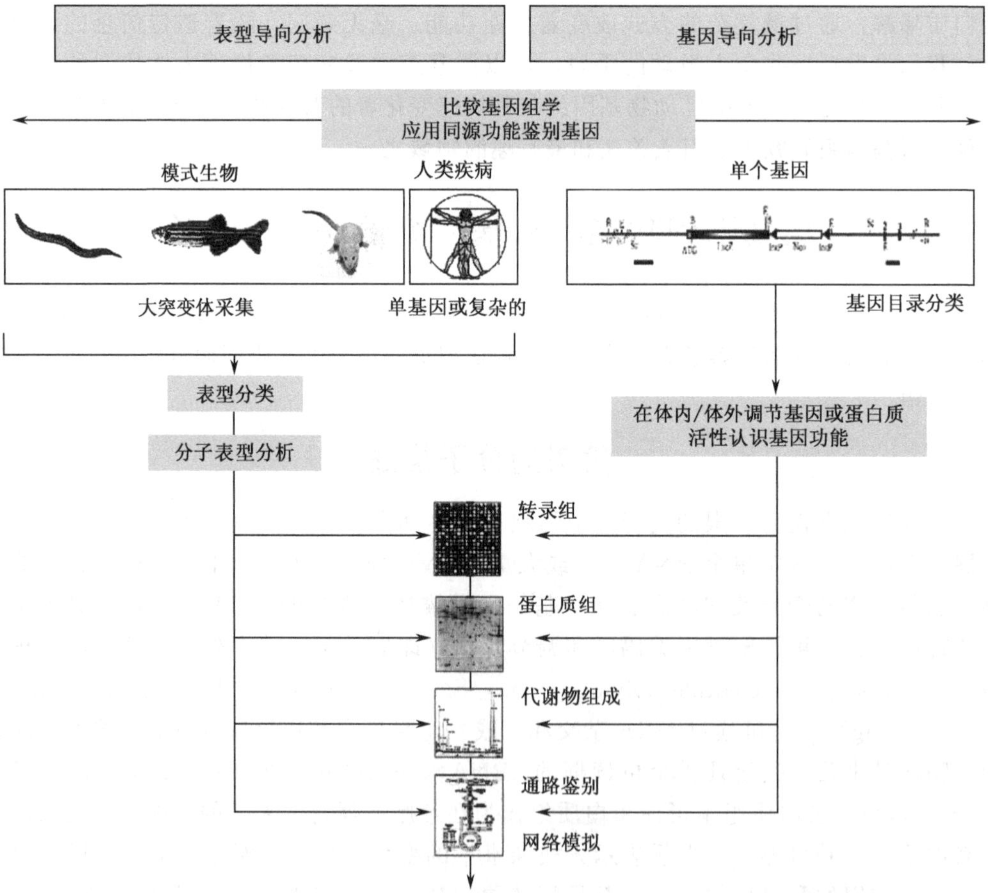


图 2.1　关于功能基因组学两种不同方法的概述图

方法一：通过对模式生物或人类单基因疾病的表型分析而导向，以鉴别隐伏基因及其功能。方法二：以分析单一基因的功能而导向。主要的分析工具和研究方法用不同水平信息之间的联系来表示尽管功能基因组学是随着传统的单基因技术开始，但通过采用高通量技术（DNA微阵列和蛋白质组学）结合目标基因缺失或选择性过度表达以探究遗传系统和细胞系统的工作变化，功能基因组学将迅速向更系统化的整体水平转变。这就意味着，我们将面临繁多庞大的基因和蛋白质表达数据库。为了使这些数据集具有意义和价值，我们需要有系统的技术和计算策略及算法来帮助描述代谢（Paton *et al.*，2000；Tomita，2001）。

## 闭合循环：从代谢到代谢组学

以蛋白质的生物合成和降解速率（蛋白质周转）来测定人类代谢的稳态水平，这些蛋白质可作为酶、受体、转运蛋白、通道、激素和其他信号分子而发挥代谢功能，或为细胞、器官或骨骼提供结构成分。在蛋白质之间，有许多代谢中间产物或作为蛋白质、糖类、脂肪或异寡聚体（heterooligomer）的组成成分，或用于ATP合成。过去我们主要通过测定某些单一代谢中间产物浓度来研究代谢的表型表达，而现代新的分子技术使我们能对从DNA到mRNA再到蛋白质及其功能这一生物信息流程中的每一步骤进行测定。但是，生物调节超出了对基因表达和蛋白质合成的调控，即代谢的主要决定因素包括蛋白质之间的互作及其代谢中间产物在动力学效应和（或）功能的变构调节上对蛋白质活性的影响。所以，在mRNA和蛋白质表达水平上获取的信息并不足以用来预测代谢的结果（ter Kuile and Weshoff，2001）。因此，从基因到mRNA，再到蛋白质及其功能这一流程测定的最后阶段就是对在蛋白质、细胞器、细胞和器官之间流动的代谢产物的类型和浓度进行分析。于是，我们不得不止步于开始研究代谢产物之处。然而，代谢组学可分析由所有可检测到的低、中分子质量化合物组成的代谢池，而不是分析单个代谢中间产物。不同物理方法可用来对生物样品中的代谢物进行全面分析。在大多数情况下，这些方法还与传统层析分离技术和含傅里叶转换的红外线光谱法、电喷雾电离质谱分析（electrospray ionization mass spectrometry，ESI-MS）及核磁共振光谱法（nuclear magnetic resonance，NMR）（Fiehn *et al.*，2000；Glassbrook *et al.*，2000）结合一起应用。当以野生型和转基因生物考察在细菌系统、酵母和植物上取得的研究进展时，代谢的全面分析与差别表型分析中聚类分析的统计方法联合应用的发展潜力就变得很明显（Delneri *et al.*，2001；Kose *et al.*，2001）。表型学（Schilling *et al.*，1999）最终以系统观点来揭示基因功能。它利用所有基因组信息、所有表达信息（在mRNA和蛋白质水平）和所有代谢物，依据生物调节网以最全面的方式描述代谢。

# 方法与技术

基因组学采用最经典的DNA序列技术，而转录组分析主要采用荧光检测系统来测定生物样品中的信使（mRNA）表达水平。表达模式已在第1章中进行了描述，故在此不再叙述。

## 蛋白质组学

术语“蛋白质组”作为基因组的补充而引入，包括在某一特定的基因组中所有被转录和翻译的可读框（蛋白质编码区域）。蛋白质组分析是以通过二维聚丙烯酰胺凝胶电泳（two-dimensional polyacrylamide gel electrophoresis，2D-PAGE）分离蛋白质为基础的。尽管2D-PAGE似乎已是一个“古老”的方法，但仍是目前可使用的分辨率最高的分离方法。然而，要获得标准化和可重复的分离条件仍是不容易的。类似地，凝胶中蛋白质的染色情况对于分析质量很关键。考马斯亮蓝（Coomassie blue）染色法和银染法（silver staining）均可采用，但银染法的灵敏度比考马斯亮蓝染色法至少高10倍。最近，敏感的荧光染料的发展又进一步提高了检测的灵敏度（Patton，2000）。2D-PAGE通过等电聚焦（isoelectric focusing，IEF）法根据其电荷等电点（isoelectric point，pI）对蛋白质进行第一向分离，SDS-PAGE法根据其分子大小（分子质量）对蛋白质进行第二向分离。因此，该法具有无可比拟的分辨复杂蛋白质混合物的能力，并能同时分析成百上千个基因产物（Görg *et al.*，2000）。然而，并不是所有的蛋白质都能用2D-PAGE法进行同等理想地分辨和分离，对碱性、疏水性很强的膜内在蛋白质和高分子质量的蛋白质进行分析仍是一个问题。在某些情况下，可能有必要根据细胞区室（细胞膜/微粒体、胞液、线立体）或根据蛋白质的溶解度对蛋白质进行预分离（Blackstock and Weir，1999；Cordwell *et al.*，2000）。此外，在大量持家蛋白存在的情况下，细胞低浓度的蛋白质仍很难分辨，而这些蛋白质可能具有极其重要的细胞功能（如在信号通路中）（Corthals *et al.*，2000）。但是，新技术正在不断形成，包括标签技术（Adam *et al.*，2001）和2D凝胶分离前微量蛋白质的浓缩技术。

以下两个发展对于蛋白质组学中2D-PAGE蛋白质分离技术的复兴至关重要：①高分辨率质谱仪的改进，可根据分子质量电离并分析肽和蛋白质；②多物种（包括人类）的基因组测序，以提供表达蛋白质的编码区的信息。最后，将未翻译（或部分翻译）的质谱与数据库中表达基因的核苷酸序列的翻译相匹配的计算机运算法的发展，对于在蛋白质组学研究领域中所取得的进展也是很重要的。

虽然目前鉴别凝胶中蛋白质斑点最常用的方法是肽绘图法或“指纹”分析，但许多其他技术和方法也可应用（Chalmers and Gaskell，2000；Gevaert and Vandekerckhove，2000）。例如，可把蛋白质从凝胶中转移到膜上，让它们与特异的抗体结合，然后通过免疫反应来检测蛋白质。也可把蛋白质从凝胶上切割下来，并用N端测序法进行经典的Edman降解或电喷雾电离质谱法（ESI-MS）分析，以获得某氨基酸序列；然后可将该序列与存在于数据库中的序列进行比较来鉴别该蛋白质。

为了获取序列信息，更快速、更先进的分析技术是肽质量分析法。此时，含蛋白质的斑点从凝胶中被切割下来，对该凝胶片段进行化学处理，以使该蛋白质易被蛋白酶如胰蛋白酶水解。蛋白质消化需要高纯度的胰蛋白酶，该酶不能被可切割含阳离子氨基酸残基的肽键的酶污染。根据胰蛋白酶的位点特异性水解，肽片段的特征类型可用作鉴别原蛋白质的肽质量指纹。

用蛋白酶消化产生的肽混合物通常被用于进行基质辅助激光解吸电离时间差质谱仪（matrix assisted laser desorption/ionization time-of-flight mass spectrometry，MALDI-TOFMS）分析，以测定某一特定蛋白质所特有的相应肽质量。所获质谱根据与某一特定基因组中已鉴别可读框（编码氨基酸序列）的“实际消化”所预测的质量进行比较，并用以不同算法来解析类型并预测蛋白质的软件进行分析。然而，质谱需要进一步仔细分析，以确保观测的准确性。例如，如果被测的质量超过了某一特定肽片段的预测质量，那么，人们即可预测到可能发生了原蛋白质的翻译后修饰，如磷酸基团的添加、赖氨酸或脯氨酸残基的羟基化、糖基化或脂肪酸的添加。由被预测质量测得的其他偏差可能是因编码序列的多态性所致，包括微细的氨基酸替换，甚至更明显的氨基酸缺失或插入。这样的原序列的变化能被分辨，但成功率取决于替代的类型，且分析可能需要内肽的 Edman 测序或更先进的 ESI-MS 质量分析（Gaskell，1997）。

如果对同质群的培养细胞或细胞系进行分析，则蛋白质组分析就变得很简单。相比之下，组织样品包含具有不同表达模式的各种细胞群，而这些不同表达模式可能以不同方式来影响获得的蛋白质组类型。生物样品的变异（即每种细胞的百分含量不同）进一步使分析复杂化，使分析变得特别困难。这样的变异可能使得通过细胞特异表面标记和免疫亲和技术或通过激光微切割技术来分离不同细胞群变得很有必要（Banks *et al.*，1999；Simone *et al.*，2000）。

将来会出现利用含所有已表达蛋白质特异性抗体的抗体文库进行简化的蛋白质组学分析方法。这将使蛋白质组分析变成对几乎每个已知蛋白质都能进行鉴别和定量的高通量微板分析测定（Brody and Gold，2000）。

## 代 谢 组 学

主要在以细菌、酵母和植物为研究对象的研究中模拟基因缺失或新基因的目标表达对生物代谢物组成模式影响的结果。代谢组分析作为比较显示基因功能的一种新工具，不仅能更深入地研究复杂的调节过程，还能直接测定其表型。自动气相色谱/质谱（gas chromatography/mass spectrometry，GC/MS）技术能对大量代谢物进行定量分析，而代谢物的具体成分仍需用化学法进行测定。针对这些庞大的数据集，采用各种用于组分分析的数据处理工具，以把某一特定的基因型与其特征性的代谢表型相配位（Glassbrook *et al.*，2000）。

在传统的营养研究中，大多数已用于代谢过程特点研究的分析方法都是以经典定量分析化学为基础的。重点研究特异的化合物或代谢途径需要特异的定量分析方法。发展的这些方法已被最优化，以产生能描述目标化合物的高质量数据。然而，这类分析方法不大适用于同时收集表征生物体营养过程的大量代谢物的信息。

代谢模式中的概念是不同的。为了同时监控数百甚至数千个代谢物，就需要用高通量分析方法来检测化合物的相对变化而不是绝对浓度。对小分子组成模式，最常用的分析方法包括与质谱仪联用的高效液相色谱（high performance liquid chromatograph，HPLC）或气相色谱（gas chromatograph，GC）。与其他种类的检测仪相比，质谱仪一般更灵敏并更具选择性，若配以恰当的样品导入和电离方法，质谱仪则能选择性地分析有机和无机化合物。但是，在检测前，代谢物必须经过与在线质量检测器联用的层析技术进行分离。GC 可根据相对气压及对层析柱中填料的亲和力而分离化合物，但只局限于挥发性和热稳定性的化合物。大多数生物化合物，如糖、氨基酸和有机酸，在其自然状态下都不足以挥发到可用 GC 进行分离，因此，必须进行衍生化处理后方能进行 GC 分离。HPLC 分离更适合于不稳定的高分子质量的化合物的分析和自然形态的非挥发性的极性化合物的分析。虽然以 GC 和 HPLC 为基础的模式分析方法不是真正的定量分析，但通过运用合适的标准，所检测的化合物及其相对数量即可在不同研究报道之间进行比较。用 GC-MS 和 HPLC-MC 技术的高通量筛选也会产生大量的分析数据，这些数据需要用先进的信息学技术来获取大量的信息。

代谢物模式技术（metabolite profiling technology，MPT）将使得收集关于代谢物经生物途径流通及对这些途径调控方面的信息成为可能，尤其是高分辨率的$^{1}$H-NMR光谱（其优点是能检测任何含质子的代谢物）似乎在代谢物模式研究中变得越来越重要。过去，NMR 技术主要用于分析哺乳动物体液和组织中代谢物的变化情况，现在还能扩展用于检测其他核素（如$^{31}$P)或天然同位素（如$^{13}$C)。

当把代谢组学应用于使用富含同位素（如$^{13}$C)的底物的研究中时，代谢物分析就可通过流通量的定量分析而达到动态水平（Brenna，2001)。这种自动的生化模式技术将成为代谢和功能基因组学研究中多学科的综合方法的一个重要组成部分。

## 以上技术应用于人类研究的现时局限性

在基础研究方面，基因组学、转录组学、蛋白质组学和代谢组学的以上所述的技术在营养研究中的应用似乎不受限制。在应用研究领域，这些技术也具有鉴别特异性标记物（生物标记物）的很大潜力，而这些标记物是对某一特定营养物质、非营养化合物、处理或饮食的状况的反应。营养状况的生物标记物可能包括单个 mRNA 或蛋白质的水平的变化，但也可能包括一大组 mRNA 或蛋白质的类型的变化。迄今为止，细胞功能的生物标记物已主要根据代谢知识而采取合理方法进行鉴别。当同时分析几千个可能受影响的指示分子（mRNA、蛋白质）时，这些新的筛选技术基本上是非逻辑的。

当培养的人类细胞或模型生物及试验动物用于营养研究时，这些新技术平台的使用几乎不受限制。然而，在人类研究方面存在局限性。尽管单个细胞的 mRNA 可通过 PCR 技术扩增和定量分析单个基因或几组基因，但 mRNA 水平上表达模式的研究还是受到能否获得含足够分析用高质量 RNA 细胞的限制。在蛋白质组学方面，只有通过侵入技术（invasive technique）才能获得评价蛋白质类型所需的足够数量的人类细胞。

虽然某些细胞（如毛发滤泡细胞、皮肤细胞，甚至是脱落的肠细胞）能通过非侵入技术（non-invasive technique）获得足够数量，但不同类型的血细胞是试验材料［RNA 和（或）蛋白质］的理想来源，并能用作“报道细胞”（reporter cell)。这些细胞能反映饲粮的变化，也具有不同的生命期、不同的基因表达谱和调探系统，且以不同体区室为作用目标。特别是外周血淋巴细胞，在采用 DNA 阵列技术进行的对环境因子反映的人类研究中，已用于鉴别潜在 mRNA 生物标记物（Amundson *et al.*，2000；Glynne *et al.*，2000；Marrack *et al.*，2000)。不同来源的 DNA 可用于 SNP 分析，相关 SNP 鉴别的大规模应用是生物医学所有领域中分子流行病学的快速发展的基础（Haugen，1999；Tan *et al.*，1999；Beeley *et al.*，2000；Perera and Weinstein，2000；Schork，2000)。表达谱、SNP 分析和蛋白质组学已成为药物发现研究中的已建立好的技术平台，且对于药物基因组学（pharmacogenomics）很关键，而药物基因组学是将遗传多态性（SNP）和药物代谢酶的选择性表达与药物代谢的动力学及动态学中的表型差异相联系的一个词组（Beeley *et al.*，2000；Rininger *et al.*，2000；Norton，2001)。目前，一些营养研究已应用了基因组学、蛋白质组学和代谢组学的新技术。这些研究包括应用基因阵列来研究对微量养分变化状况（见第 1 章）或热量限制的反应（Lee *et al.*，1999）与应用蛋白质组学来识别肥胖小鼠过氧化物酶体增殖受体的配体效应（Edvardsson *et al.*，1999）以及小鼠血浆成分和饲粮诱导变化的代谢表征（Vigneau-Callahan *et al.*，2001）。

## 小　结

基因-营养互作是基因组与环境之间相互影响的范例。我们已步入了后基因组研究的时代，毫无疑问，分子营养学将成为人们关注的中心，因为营养和其他食物成分是影响基因和蛋白质活性的关键因素。大量的遗传信息和高通量的新技术为营养研究提供了令人振奋的工具。表达阵列、蛋白质组分析和高通量代谢物模式能使我们在 mRNA 或蛋白质水平上认识哺乳生物对其营养环境变化的反应。由于这些工具能产生大量的数据集，因此，营养学对生物信息学的需求正日益明显，对在细胞生物学和代谢生物化学方面具有良好知识基础的营养研究者的需求也正变得日益突出。所有这些技术应用的最终目标，是扩展我们对代谢和营养的认识并测定其与健康和疾病之间是怎样相关的。

## 参　考　文　献


Adam, G.C., Cravatt, B.F. and Sorensen, E.J. (2001) Profiling the specific reactivity of the proteome with non-directed activity-based probes. *Chemistry and Biology* 8, 81–95.

Amundson, S.A., Do, K.T., Shahab, S., Bittner, M., Meltzer, P., Trent, J. and Fornace, A.J. Jr (2000) Identification of potential mRNA biomarkers in peripheral blood lymphocytes for human exposure to ionizing radiation. *Radiation Research* 154, 342–346.

Banks, R.E., Dunn, M.J., Forbes, M.A., Stanley, A., Pappin, D., Naven, T., Gough, M., Harnden, P. and Selby, P.J. (1999) The potential use of laser capture microdissection to selectively obtain distinct populations of cells for proteomic analysis – preliminary findings. *Electrophoresis* 20, 689–700.

Beeley, L.J., Duckworth, D.M. and Southan, C. (2000) The impact of genomics on drug discovery. *Progress in Medicinal Chemistry* 37, 1–43.

Blackstock, W.P. and Weir, M.P. (1999) Proteomics: quantitative and physical mapping of cellular proteins. *Trends in Biotechnology* 17, 121–127.

Brenna, J.T. (2001) Natural intramolecular isotope measurements in physiology: elements of the case for an effort toward high-precision position-specific isotope analysis. *Rapid Communications in Mass Spectrometry* 5, 1252–1562.

Brody, E.N. and Gold, L. (2000) Aptamers as therapeutic and diagnostic agents. *Journal of Biotechnology* 74, 5–13.

Celis, J.E., Kruhoffer, M., Gromova, I., Frederiksen, C., Ostergaard, M., Thykjaer, T., Gromov, P., Yu, J., Palsdottir, H., Magnusson, N. and Orntoft, T.F. (2000) Gene expression profiling: monitoring transcription and translation products using DNA microarrays and proteomics. *FEBS Letters* 480, 2–16.

Chalmers, M.J. and Gaskell, S.J. (2000) Advances in mass spectrometry for proteome analysis. *Current Opinion in Biotechnology* 11, 384–390.

Cordwell, S.J., Nouwens, A.S., Verrills, N.M., Basseal, D.J. and Walsh, B.J. (2000) Subproteomics based upon protein cellular location and relative solubilities in conjunction with composite two-dimensional electrophoresis gels. *Electrophoresis* 21, 1094–1103.

Corthals, G.L., Wasinger, V.C., Hochstrasser, D.F. and Sanchez, J.C. (2000) The dynamic range of protein expression: a challenge for proteomic research. *Electrophoresis* 21, 1104–1115.

Delneri, D., Brancia, F.L. and Oliver, S.G. (2001) Towards a truly integrative biology through the functional genomics of yeast. *Current Opinion in Biotechnology* 12, 87–91.

Dutt, M.J. and Lee, K.H. (2000) Proteomic analysis. *Current Opinion in Biotechnology* 11, 176–179.

Edvardsson, U., Alexandersson, M., Brockenhuus von Lowenhielm, H., Nystrom, A.C., Ljung, B., Nilsson, F. and Dahllof, B. (1999) A proteome analysis of livers from obese (ob/ob) mice treated with the peroxisome proliferator WY14,643. *Electrophoresis* 20, 935–942.

Fiehn, O., Kopka, J., Dormann, P., Altmann, T., Trethewey, R.N. and Willmitzer, L. (2000) Metabolite profiling for plant functional genomics. *Nature Biotechnology* 18, 1157–1161.

Gaskell, S.J. (1997) Electrospray: principles and practice.

*Journal of Mass Spectrometry* 32, 677–688.
Gevaert, K. and Vandekerckhove, J. (2000) Protein identification methods in proteomics. *Electrophoresis* 21, 1145–1154.
Glassbrook, N., Beecher, C. and Ryals, J. (2000) Metabolic profiling on the right path. *Nature Biotechnology* 18, 1142–1143.
Glynne, R., Ghandour, G., Rayner, J., Mack, D.H. and Goodnow, C.C. (2000) B-lymphocyte quiescence, tolerance and activation as viewed by global gene expression profiling on microarrays. *Immunological Reviews* 176, 216–246.
Gorg, A., Obermaier, C., Boguth, G., Harder, A., Scheibe, B., Wildgruber, R. and Weiss, W. (2000) The current state of two-dimensional electrophoresis with immobilized pH gradients. *Electrophoresis* 21, 1037–1053.
Haugen, A. (1999) Progress and potential of genetic susceptibility to environmental toxicants. *Scandinavian Journal of Work Environment and Health* 25, 537–540.
Kose, F., Weckwerth, W., Linke, T. and Fiehn, O. (2001) Visualizing plant metabolomic correlation networks using clique-metabolite matrices. *Bioinformatics* 17, 1198–1208.
Lee, C.K., Klopp, R.G., Weindruch, R. and Prolla, T.A. (1999) Gene expression profile of aging and its retardation by caloric restriction. *Science* 285, 1390–1393.
Lockhart, D.J. and Winzeler, E.A. (2000) Genomics, gene expression and DNA arrays. *Nature* 405, 827–836.
Marrack, P., Mitchell, T., Hildeman, D., Kedl, R., Teague, T.K., Bender, J., Rees, W., Schaefer, B.C. and Kappler, J. (2000) Genomic-scale analysis of gene expression in resting and activated T cells. *Current Opinion in Immunology* 12, 206–209.
Norton, R.M. (2001) Clinical pharmacogenomics: applications in pharmaceutical R&D. *Drug Discovery Today* 6, 180–185.
Pandey, A. and Mann, M. (2000) Proteomics to study genes and genomes. *Nature* 405, 837–846.
Paton, N.W., Khan, S.A., Hayes, A., Moussouni, F., Brass, A., Eilbeck, K., Goble, C.A, Hubbard, S.J. and Oliver, S.G. (2000) Conceptual modelling of genomic information. *Bioinformatics* 16, 548–557.
Patton, W.F. (2000) Making blind robots see: the synergy between fluorescent dyes and imaging devices in automated proteomics. *Biotechniques* 28, 944–948.
Perera, F.P. and Weinstein, I.B. (2000) Molecular epidemiology: recent advances and future directions. *Carcinogenesis* 21, 517–524.
Rininger, J.A., DiPippo, V.A. and Gould-Rothberg, B.E. (2000) Differential gene expression technologies for identifying surrogate markers of drug efficacy and toxicity. *Drug Discovery Today* 5, 560–568.
Sauer, B. (1998) Inducible gene targeting in mice using the Cre/lox system. *Methods* 14, 381–392.
Schilling, C.H, Edwards, J.S. and Palsson, B.O. (1999) Toward metabolic phenomics: analysis of genomic data using flux balances. *Biotechnology Progress* 15, 288–295.
Schork, N.J., Fallin, D. and Lanchbury, J.S. (2000) Single nucleotide polymorphisms and the future of genetic epidemiology. *Clinical Genetics* 58, 250–264.
Simone, N.L., Paweletz, C.P., Charboneau, L., Petricoin, E.F. and Liotta, L.A. (2000) Laser capture microdissection: beyond functional genomics to proteomics. *Molecular Diagnosis* 5, 301–307.
Tan, K.T., Dempsey, A. and Liew, C.C. (1999) Cardiac genes and gene databases for cardiovascular disease genetics. *Current Hypertension Report* 1, 51–58.
ter Kuile, B.H. and Westerhoff, H.V. (2001) Transcriptome meets metabolome: hierarchical and metabolic regulation of the glycolytic pathway. *FEBS Letters* 500, 169–171.
Tomita, M. (2001) Whole-cell simulation: a grand challenge of the 21st century. *Trends in Biotechnology* 19, 205–210.
Vigneau-Callahan, K.E., Shestopalov, A.I., Milbury, P.E., Matson, W.R. and Kristal, B.S. (2001) Characterization of diet-dependent metabolic serotypes: analytical and biological variability issues in rats. *Journal of Nutrition* 131, 924S–932S.

# 第 2 部分
# 细胞营养稳衡、细胞增殖与凋亡

# 3 有机养分浆膜转运蛋白的分子生理学

Hannelore Daniel
（德国慕尼黑技术大学食品和营养系分子营养组）

## 引 言

养分跨细胞浆膜的转运是代谢和营养稳衡的一个关键步骤，因为细胞膜是渗透的一道屏障，能把代谢过程区室化。细胞膜的磷脂双层本身对亲水性的低分子质量化合物和大分子具有很低的渗透性。但是，即使直到现在仍然被认为是以被动扩散方式穿过细胞膜的亲脂性养分，如胆固醇或脂肪酸，似乎需要特殊的膜转运蛋白来控制其渗透和代谢流程。维持与细胞外环境完全不同的内环境的稳定是生命所必需的，因此，进化过程中已出现的大量具有高度特异性功能的膜蛋白不仅控制着对离子、水、常量养分、微量养分和代谢中间产物的膜渗透性，而且还控制着对异生素的膜渗透性。

我们对膜蛋白的分子结构及功能的认识非常匮乏，因为在大多数情况下膜蛋白不能通过传统的蛋白质纯化技术进行分离。鉴于细胞外蛋白和细胞质蛋白能被足量纯化，使其结晶和测定其三维结构及与功能相关的结构改变成为可能，那么膜的转运功能就不可能与分离的结构蛋白成分有关。

随着分子生物学和新克隆技术的出现，对膜转运蛋白的研究发生了巨大变化。在20世纪80年代，第一个编码哺乳动物细胞膜蛋白的cDNA被分离，使对蛋白质结构及其在细胞膜区的整合的预测成为可能。获得的这些序列也能通过同源筛选技术（homology screening）用于鉴别相关序列和蛋白质。此外，还可采用表达克隆技术获取编码养分转运蛋白质的新序列（Daniel，2000）。这里，当把一个cDNA池引入某一模型细胞后，可通过检测其编码蛋白质的功能而筛选编码蛋白质的cDNA文库，而该cDNA文库是由含从目标组织分离的数千个单一mRNA的mRNA池中产生的。这主要是在南非蛙（*Xenopus laevis*）的卵母细胞中完成，因为这种蛙的卵母细胞大（约1mm），且cDNA池易通过微量加液器被引入细胞中。一旦卵母细胞转运测定发现其具有预期养分的转运功能，则该cDNA池被分离，再次注入不同的池，并再次筛选检测功能表达。最后，编码已知功能的蛋白质的单一cDNA被鉴别出来，该cDNA序列可提供编码区（可读框，open reading frame，ORF），且能翻译成相应的氨基酸序列。然后，根据序列中氨基酸残基的不同极性以及形成插入细胞膜磷脂双层的α螺旋区需要10～15个或更多连续的疏水性氨基酸残基的特性，采用不同的法则来对蛋白质的二级结构进行预测。因此，预测是更亲水性的区域把这些跨膜区域连接起来，从而形成细胞内和细胞外的连接区域。在很多情况下，将根据单个膜转运蛋白的亲水性分析结果预测出的拓扑结构用试验证实时，证明是正确的。分析膜蛋白拓扑结构主要通过导入能与细胞外或膜内反应渗透出细胞的特异性抗体或反应物结合的标记（flag）或专门的抗原决定序列（小片段的氨基酸序列）来完成。另外，还可采用一种半胱氨酸扫描方法，这种方法把蛋白质的转运功能用作报道系统。编码半胱氨酸残基的密码子通过任一目标位置的定点诱变整合入 cDNA，采用专门的最终会损害蛋白质功能的不透膜性或透膜性的反应物对半胱氨酸残基进行化学修饰，通过异种表达系统中的功能分析而测定接近细胞内或细胞外的半胱氨酸残基的位置。

## 哺乳动物转运分子生物学的开端：GLUT 转运蛋白

克隆人类营养转运蛋白的一个里程碑就是鉴别出了 HepG2 细胞的第一个 GLUT 家族成员（Mucckler *et al.*，1985）。在蛋白质水平上，已克隆的 cDNA 显示出与红细胞葡萄糖转运蛋白高度的相似性，甚至氨基酸组成也完全相同，而采用传统 Edman 测序法已获得红细胞葡萄糖转运蛋白的部分序列。根据人类基因组测序组织（human genome sequencing orgnization，HUGO）资料所推荐的术语，GLUT 转运蛋白现在被指定是具有连续编号的 SLC2A 组。目前，GLUT 家族（表 3.1）中的 GLUT-1～GLUT-5 蛋白是调节运送单糖进出哺乳动物细胞的主要转运蛋白（Seatter and Gould，1999）。GLUT-6-12 蛋白的功能还不清楚，而 GLUT-13 是主要存在于大脑中的质子依赖性肌醇转运蛋白（Uldry *et al.*，2001）。一般来说，GLUT 由最有可能形成 12 个跨膜（transmembrane，TM）片段的近 500 个氨基酸组成，而这些 TM 片段经排列形成孔道。GLUT 蛋白质的 N 端或 C 端都朝向胞内区室，靠近转运蛋白 N 端的大的细胞外环在成熟蛋白质中被糖基化，而连接 TM 片段 6 和片段 7 的大的细胞内环含有磷酸化的共同位点（Hruz and Muckler，2001）。最近已报道 GLUT-1 及其与浆细胞膜整合的三维

**表 3.1　参与己糖和相关溶质跨膜转运的哺乳动物 GLUT 转运蛋白家族及已鉴别的与其功能异常相关的人类疾病**

| SLC2a 家族 | 主要表达位点 | HUGO 代码 | 人类疾病 |
|---|---|---|---|
| GLUT-1 | 红细胞、脑（血管的） | SLC2A1 | →GLUT-1 缺乏综合征 |
| GLUT-2 | 肝脏、胰腺（小岛） | SLC2A2 | →Fanconi-Bickel 综合征，NIDDM |
| GLUT-3 | 脑（神经细胞的） | SLC2A3 | →胎儿葡萄糖中毒 |
| GLUT-4 | 肌肉、脂肪组织、心脏 | SLC2A4 | →HIV 的中毒性 latrogenic 糖尿病，NIDDM |
| GLUT-5 | 小肠、肾脏、睾丸 | SLC2A5 | |
| GLUT-6 | 新的异构体，功能还不清楚 | SLC2A6 | |
| GLUT-7 | | SLC2A7 | |
| GLUT-8 | | SLC2A8 | |
| GLUT-9 | | SLC2A9 | |
| GLUT-Ⅹ | | | |
| GLUT-13（HMIT）（质子依赖性肌醇转运蛋白） | 脑 | SLC2A13 | |

NIDDM：非胰岛素依赖性糖尿病。

模型，该模型显示出横穿整个蛋白质并可形成底物转运孔道的通道样中心结构（Zuniga *et al.*，2001）。由于尚未获得任何养分转运蛋白的结晶结构，故这些模型在认识转运蛋白的结构和功能方面具有重要作用。

从养分和葡萄糖稳衡的角度来看，GLUT-2 和 GLUT-4 转运蛋白具有核心作用。GLUT-2 控制着葡萄糖在肝细胞中的流入和流出，因此，不仅可通过清除门静脉血中的葡萄糖来控制采食后的血浆葡萄糖水平，而且还可通过影响糖异生或糖原分解产生的葡萄糖的流出维持饥饿期间葡萄糖水平的稳定。在上皮细胞（如肠上皮细胞）的基底外侧膜的 GLUT-2 表达使葡萄糖、半乳糖和果糖转运出上皮细胞而进入循环成为可能，而这些单糖主要通过钠依赖性葡萄糖转运蛋白 SGLT-1（葡萄糖、半乳糖）或经过 GLUT-5 转运蛋白（果糖）从肠腔进入上皮细胞。由于 GLUT 转运蛋白（除 GLUT-13 外）是依浓度梯度来调节单糖转运的，因此，代谢作为跨膜浓度梯度的调节器而在调节整体转运能力方面发挥关键作用。GLUT-2 作为向细胞提供葡萄糖，而后葡萄糖促进胰岛素分泌的体系而在胰岛的 β 细胞中具有极其重要的功能。GLUT-2 与葡糖激酶位于 β 细胞中血糖感应通路的中心。另一个最重要的 GLUT 家族成员是 GLUT-4——目前治疗非胰岛素依赖性的Ⅱ型糖尿病（non-insulin-dependent typeⅡ diabete，NIDDM）的一个关键目标分子。GLUT-4 转运蛋白通过细胞内囊泡转移而与脂肪细胞和骨骼肌细胞的浆膜进行胰岛素依赖性的融合，可用作说明信号通路复杂性及膜运输和膜融合所必需的细胞系统复杂性的例子。GLUT-4 的合成及在浆膜与储存囊泡之间的快速循环受到包括多种激素和代谢物在内的许多因素的调节，并受锻炼情况的影响（Watson and Pessin，2001）。GLUT-4 在以敏感促进性能系统为目标的 NIDDM 治疗中正发挥着关键作用。最近，在用葡萄糖代谢发生巨大变化的 *glut-4*（cre/loxP 基因定位）肌肉特异性基因失活的小鼠进行的一个很精确的研究中，表明了 GLUT-4 对葡萄糖稳衡的关键作用（Kim *et al.*，2001）。GLUT 家族的葡萄糖转运蛋白的各类功能失调均已报道（表 3.1），且最近人类免疫缺陷病毒（human immunodeficiency virus，HIV）的蛋白酶抑制剂已被发现是 GLUT-4 的抑制剂，这可能解释艾滋病（AIDS）患者能量和葡萄糖代谢的一些代谢性障碍。GLUT-1 缺乏导致抽搐，并伴随着脑脊髓液中葡萄糖浓度的降低；而 GLUT-2 缺乏是导致类似于Ⅰ型糖原积贮病的范可尼别克综合征（Fanconi-Bickel syndrome）的基本原因。

## 生电转运蛋白的 SGLT 家族

20 世纪 60 年代初期小肠葡萄糖转运依赖钠的发现，促进了葡萄糖进入肠细胞与钠离子转运相偶联的分子基础和这种主动转运的热力学理论的研究。每天，大量葡萄糖作为肠道碳水化合物被消化及作用二糖水解的终产物而被吸收，提示相关葡萄糖转运蛋白具有非常高的转运能力。葡萄糖运入肠细胞的热力学使葡萄糖能在细胞中积累，从而形成能使葡萄糖通过基底外侧膜上的 GLUT-2 而向下转运的浓度梯度，即只要血液循环中的葡萄糖浓度低于细胞内的浓度。由于根皮苷是一种高亲和力的摄入抑制剂，因而它被证明是一种对分析葡萄糖转运的非常有帮助的工具。

在最初提出肠葡萄糖转运的钠梯度假说后约 25 年，人们才鉴别出这种转运蛋白。

1987 年，在采用爪蟾（*Xenopus*）卵母细胞的表达克隆和选择在卵母细胞诱导钠依赖性和根皮苷敏感性葡萄糖转运的 mRNA 中，人们发现了第一个被编码为 SGLT-1 的生电葡萄糖转运蛋白的 cDNA（Hediger *et al.*，1987）。目前，HUGO 资料表明，许多转运蛋白属于同一家族，单个成员不仅能转运己糖，还能转运其他物质，例如，钠可依赖性地转运碘化物、肌醇和如生物素及泛酸这样的水溶性维生素。在原核细胞和无脊椎动物中也发现有类似于 SGLT-1 的蛋白质，该蛋白具有转运尿素、脯氨酸、糖或维生素的功能。

SGLT 家族的不同成员在异种系统中表达时，均明显地表现出功能上相似性。三个种类（人、兔和鼠）的高亲和力 $Na^+$-葡萄糖协同转运蛋白 SGLT-1、低亲和力 $Na^+$-葡萄糖协同转运蛋白 SGLT-2、$Na^+$-依赖性肌醇协同转运蛋白 SMIT1 和 $Na^+$-依赖性碘化物协同转运蛋白 NIS 均已在爪蟾卵母细胞中表达，且均已采用放射示踪技术和电生理学技术对它们进行了分析（Turk and Wright，1997）。在底物不存在的情况下，它们均表现出对 $Na^+$ 的渗透性降低，但在底物存在的情况下，则能记录下表明 $Na^+$-底物协同转运的电压依赖性的正向内电流。转运循环是个有序的过程，首先 $Na^+$ 与蛋白质结合，后与底物结合，并转运该复合物，然后释放出底物和协同转运的离子进入细胞质。该过程的限速步骤可能是没有底物但带有电荷的转运蛋白重新复位。到细胞膜外表面的过程。最近，已有人提出关于 SGLT-1 本身具有对水的渗透性的建议，即大量水分子在每个循环中与钠离子和葡萄糖一起被协同转运（Loo *et al.*，1996）。根据 SGLT-1 每天转运葡萄糖的数量，其水转运能力相当于由小肠上段吸收约 5L 液体。

在结构水平上，采用与突变、抗原决定簇标记、免疫技术以及其他细胞生物技术和生物物理技术相结合的功能分析，人们已对 SGLT-1 蛋白质进行了广泛的研究。根据试验数据和计算分析，提出了人 SGLT-1 含 14 个跨膜区域的模型（图 3.1）。所有 TM 区域都有可能是 α 螺旋，N 端位于细胞外，而多电荷的 C 端大区域位于细胞质中（Turk *et al.*，1996）。人 SGLT-1 有 664 个氨基酸残基，而 SGLT 家族的其他成员有 530～735 个氨基酸。采用斜截修饰的 SGLT-1 蛋白技术和生物物理技术分析，有两个区域已被确

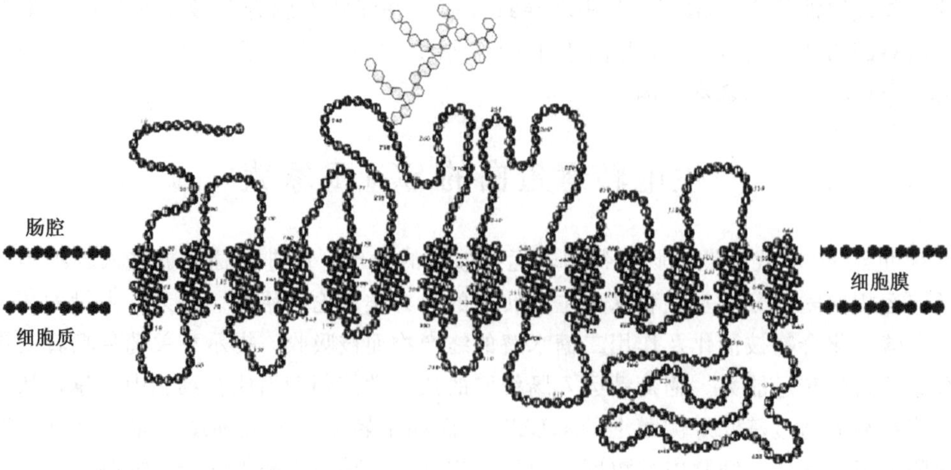


图 3.1　生电 $Na^+$-依赖性葡萄糖转运蛋白 SGLT-1 的预测的膜拓扑结构
该 SGLT-1 主要分布于肠细胞和肾上皮细胞的顶膜中

定为该蛋白质的功能区，其中一个是钠离子的结合部位，另一个是与底物结合，调节钠和葡萄糖转运入细胞内的区域。钠离子和底物通过该蛋白质转运所必需的构象改变，似乎只需要两个 TM 螺旋在膜平面上发生 10～15Å 倾斜，然后形成类似孔道的渗透结构（Eskandari *et al*.，1998）。

葡萄糖-半乳糖吸收障碍综合征是一种由 SGLT-1 功能失调引起的遗传疾病（Martin *et al*.，1996）。研究发现人类的基因突变已使鉴别 SGLT-1 蛋白的关键性氨基酸残基成为可能，而其基因突变可导致肠道葡萄糖和半乳糖吸收障碍，进而引起致命性腹泻。令人吃惊的是，在患者中发现，几乎所有基因突变都可引起 SGLT-1 蛋白生物合成途径的缺损及其与细胞膜融合的缺乏。这是膜转运蛋白生物学的重要领域之一。蛋白质是怎样从内质网到高尔基体再到特定细胞膜区室（顶部与基底外侧）的呢？

肠道 SGLT-1 的表达受多种激素和肠内因素的调节。肠葡萄糖的转运也表现出明显的昼夜规律。动力学研究表明，SGLT-1 的最大变化速度（$V_{max}$）与基因表达的变化有关。SGLT-1 可能与一些类黄酮类栎精苷（可能还有其他苷）相当高的吸收速率有关（Walgren *et al*.，2000；Ader *et al*.，2001）。这些次生植物代谢产物是非常好的抗氧化剂，并能作为各种信号途径中蛋白质的修饰剂。

人的肾脏每天从血浆中过滤掉约 180g D-葡萄糖，而葡萄糖在近端小管高效地被重吸收。$Na^+$-葡萄糖协同转运通过肾脏刷状缘膜和易化扩散通过基底外侧膜，这就可解释葡萄糖从原尿进入血液循环的再循环过程。在近曲小管，大量葡萄糖通过低亲和力、高容量的转运蛋白（可能是 SGLT-2）而被重吸收。剩余的葡萄糖则在近直小管通过高亲和力、低容量的转运蛋白（最有可能是 SGLT-1）而被摄取。葡萄糖重吸收中导致严重葡萄糖尿的先天性肾缺陷似乎主要与 SGLT-2 蛋白的功能失调有关，而患 SGLT-1 功能失调的患者仅有轻微的肾葡萄糖尿（Sankarasubbaiyan *et al*.，2001）。

## 核苷的转运过程

核苷为合成 DNA 和 RNA 所必需，也可作为代谢中间产物。各种组织自身不能合成核苷，因而需依靠外源核苷的供应。组织之间核苷的交换需要特定的核苷转运蛋白。此外，核苷也可作为 DNA、RNA 和核苷酸在胃肠道消化的终产物而由饲粮供给。肠上皮细胞表达的两类 $Na^+$-依赖性核苷转运蛋白能将核苷向上运输与钠离子顺电化学梯度向下运动偶联起来。这两种生电系统均为集中性的核苷转运蛋白（concentrative nucleoside transporter，CNT），CNT1 转运嘧啶碱基的核苷，CNT2 转运那些含嘌呤碱基的核苷（Pastor-Anglada *et al*.，2001）。核苷从上皮细胞流出而进入血液循环是通过平衡核苷转运蛋白（equilibrative nucleoside transporter，ENT）亚类来介导的。ENT 这个大家族的蛋白质成员广泛分布于人体各处，都是单向转运蛋白，顺着跨膜梯度双向转运核苷（Hyde *et al*.，2001）。ENT 家族的成员还对嘧啶碱基或嘌呤碱基具有一定的特异性（Ritzel *et al*.，2001）。集中性的 $Na^+$-依赖性核苷转运蛋白含约 650 个氨基酸残基，最有可能含 13 个跨膜区域。ENT 蛋白和 CNT 蛋白不仅对核苷代谢，而且对结构上类似于核苷的抗病毒和细胞抑制药物的摄取和运输也都是很重要的（Wang *et al*.，1997）。

## 单羧基化合物转运蛋白

组织器官之间单羧基化合物的交换为各种代谢过程所必需，这些代谢过程包括乳酸从肌肉组织和其他细胞中被释放出来，随后又被重摄取进入肝脏进行糖异生（克力循环，Coricycle），以及饥饿和酮酸中毒期间的酮体转运。其他单羧基化合物如丙酮酸、丁酸、丙酸和乙酸或者是支链氨基酸的酮衍生物（keto-derivative）在组织间的流通量也较大。图 3.2 对 MCT 家族各种单羧基化合物的已知转运蛋白的一些细胞功能进行了总结。人的单羧基化合物转运蛋白 MCT-1～MCT-9（HUGO：SLC16A1～SLC16A9）与其他原核生物和真核生物的大量相关基因已被一起鉴别（Halestrap and Price，1999）。MCT-1 普遍在哺乳动物上表达。心脏和红肌 MCT-1 的表达很强，且工作强度增加可上调心脏和红肌的 MCT-1，表明 MCT-1 在转运乳酸进行氧化方面具有特殊的作用（Bonen *et al.*，1997）。有趣的是，根据免疫染色的结果，横纹肌的 MCT-1 也在线粒体膜中被发现，提示 MCT-1 可定位于浆膜和线粒体膜上（Brooks *et al.*，1999）。MCT-2 是一种表达于肾小管近端、精子尾部和神经细胞中的高亲和力类型的转运蛋白。MCT-3 只在肾色素上皮中表达（Halestrap and Price，1999）。

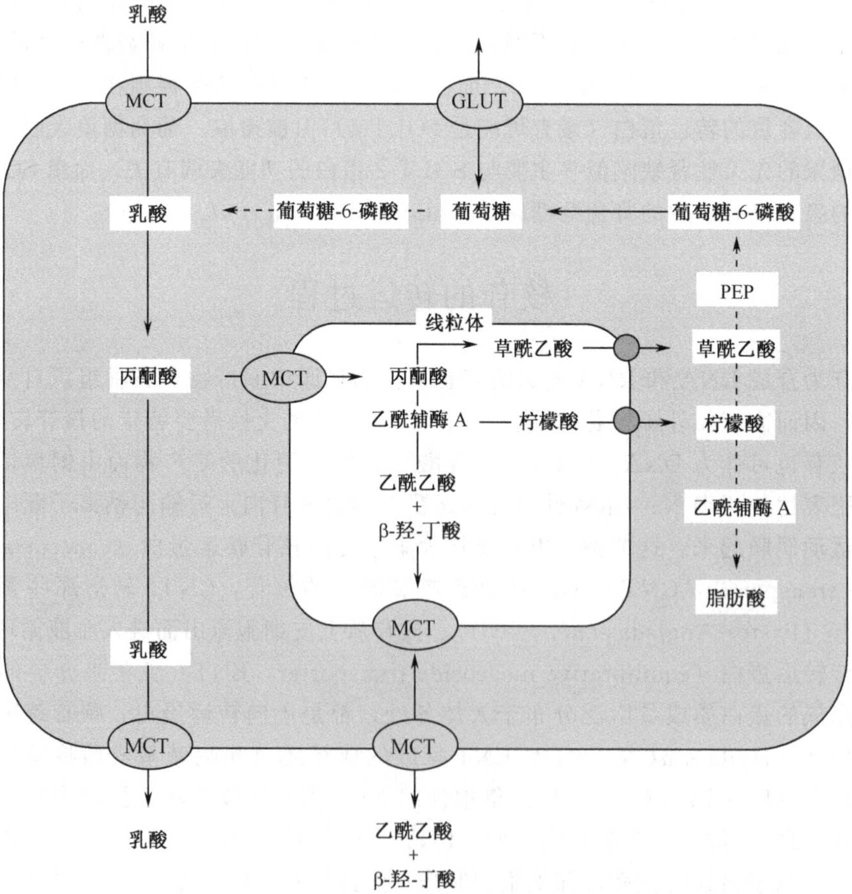


图 3.2 已鉴别的 MCT 载体的细胞功能及其在连接各种代谢途径的单羧基化合物的细胞流通中的作用

MCT-4 转运蛋白主要存在于白肌和其他糖酵解率高的细胞中，包括肿瘤细胞和白血细胞，这些细胞对乳酸的有效流出具有特别高的要求（Bonen，2000）。对异种表达的 MCT-4 作为白骨骼肌中主要转运蛋白同种型的功能分析表明，L-乳酸、D-乳酸和丙酮酸的 $K_m$ 值分别为 28mmol/L、519mmol/L 和 153mmol/L，因而把 MCT-4 鉴别为一种低亲和力类型的转运系统。MCT-1 的亲和力变化范围是从丙酮酸的 2.1mmol/L 到 L-乳酸的 4.4mmol/L 和 D-乳酸的大于 60mmol/L（Juel and Halestrap，1999）。在绝大多数情况下，单羧基化合物的转运是电中性的，阴离子与质子共同被转运。人们已全面了解了丙酮酸和乳酸转入红细胞的动力学特点，且发现 MCT-1 是血红细胞浆膜上存在的唯一的 MCT 蛋白形式（Poole and Halestrap，1993）。其转运过程是质子与转运蛋白结合，然后再与乳酸结合形成一复合物，移位后将溶质释放进入细胞质。转运蛋白能否进行双向转运，取决于底物及协同转运质子的浓度，当 $[乳酸]_{进}/[乳酸]_{出}=[H^+]_{出}/[H^+]_{进}$ 时，转运达到平衡（Halestrap and Price，1999）。

对 MCT 基因表达的调节是代谢适应各种环境条件的重要部分。增加机体的活动量能通过促进 MCT-1 表达而提高肌肉中乳酸/$H^+$ 转运的能力。相比之下，对于神经切除的大鼠肌肉，乳酸的转运速度降低，且 MCT-1 和 MCT-4 的表达减少。已知低氧条件（hypoxia）能提高乳酸脱氢酶 M 和其他糖分解酶以及 MCT-1 的表达。这种适应调节过程涉及各种转录因子和反应元件，包括低氧诱导因子 1、cAMP 反应元件和红细胞生成素低氧增强子。由于 MCT-4 在肌纤维中的分布情况与乳酸脱氢酶-M 类似，因此，低氧也能通过类似的分子机制来提高 MCT-4 的表达。在有充血性心脏病的大鼠模型中，心脏单羧基化合物转运蛋白 MCT-1 的表达增强提示，乳酸可用作心脏代谢的一个重要呼吸底物，这在发病条件下更依赖于碳水化合物（Halestrap and Price，1999）。

另一个由 MCT 家族成员介导的转运有机弱酸如丁酸、乳酸、丙酸和乙酸的重要组织为结肠上皮。结肠微生物能代谢可溶性纤维（如菊粉和果寡糖）及未被消化的淀粉，从而产生大量的单羧基化合物。肠腔短链脂肪酸的浓度可达 100～150mmol/L，且跨刷状缘膜摄入后，可为结肠细胞提供能量物质（丁酸），并将其他的物质（如乳酸、丙酸和乙酸）运到血液进行循环。除作为能量底物外，丁酸还能促进正常结肠细胞的生长，但能抑制已建立的结肠腺瘤细胞和结肠癌细胞的增殖。丁酸转运入结肠细胞的动力学特征与 MCT-1 的特征一致，因此，MCT-1 这种转运蛋白在组织稳衡和肠适应性调控方面似乎非常重要（Stein *et al.*，2000）。

有关特异性的短链脂肪酸转运蛋白的重要性，人们根据细胞膜对这些不带电荷的低分子质量化合物具有很高的渗透性曾进行过争论。膜两侧间 pH 依赖性分配与对非极性物质的无限制扩散曾被认为是转运的唯一机制。当然，该过程目前仍是单羧基化合物转运的基本路线，因为这是由弱酸的物理化学特性决定的。转运蛋白如 MCT-1 及其表达水平的调节可对各种有机弱酸进出细胞的流量进行调控，从而确保代谢的稳衡。

## 脂肪酸转运蛋白

过去人们认为，长链脂肪酸（long chain fatty acid，LCFA）不需要依靠特异性转运蛋白而穿过细胞膜，因为它们具有很高的疏水性，且能快速进入磷脂双层。然而，稳衡机制要求对膜转运进行调节，以与代谢相协调。越来越多的证据表明，不仅是质膜和胞质脂肪酸结合蛋白（早已被认识），膜通渗步骤的特异性转运蛋白也参与细胞脂肪酸的摄取过程。LCFA不仅可作为能量储存底物及代谢能量物质，而且还具有调节细胞生长和其他各种细胞功能。多种膜连接的脂肪酸结合/转运蛋白，如43kDa浆膜脂肪酸结合蛋白（fatty acid binding protein，FABPpm）、88kDa脂肪酸转移酶（fatty acid translocase，FAT）和各种脂肪酸转运蛋白（FATP类）已被鉴别（Bonen *et al.*，1998）。FATP/SLC27A家族中有6种已知的蛋白质能在组织中表达，并能使LCFA在哺乳动物中进行跨膜转运（Hirsch *et al.*，1998）。鉴于饱和脂肪酸是很好的底物，故多不饱和脂肪酸难以转运。除视黄酸之外，脂溶性维生素不能作为底物。脂肪酸的辅酶A（CoA）衍生物不能被FATP转运，而细胞中大多数FATP的过度表达使CoA合成酶活性提高了2～5倍，这提示FATP本身具有该酶活性或相关酶活性（Herrmann *et al.*，2001）。SLC27A家族具有一个保守的信号基序（motif）和一个保守的AMP结合区域，这是发挥转运功能所必需的。FATP的作用机制及其确切的膜拓扑结构目前尚不清楚。但是可以推测，LCFA是与质子一起被协同转运，以使其能克服内在的负膜电位而电中性地转运入细胞。

FATP的表达因组织种类而变化（Hirsch *et al.*，1998）。肝脏表达各种同工物（被称为FATP2-5），主要为FATP-5，而肌肉组织主要表达FATP-1。FATP-2存在于肾脏，FATP-3主要存在于肺部。心脏表达FATP-6同工物。FATP-4在肠上皮细胞刷状缘膜上的表达表明，该蛋白质参与LCFA的肠吸收。FAT/CD36表明与FAB表达的密切协同调节关系。*CD36*基因敲除小鼠脂肪酸的吸收与利用降低，胆固醇浓度升高，体重降低，且与正常的小鼠相比，血浆葡萄糖和胰岛素水平降低，代谢速率减慢（Coburn *et al.*，2000），此外耐力也降低。不同人群缺乏*CD36*的发生率为0.3%～11%，通常与心脏脂肪酸摄入不足和高血压以及葡萄糖耐受力降低有关。这里应提到的是，胰岛素敏化药物已表明对CD36有调节作用（Yamamoto *et al.*，1990）。

## 氨基酸和短链肽的跨膜转运

20种蛋白原性氨基酸及其衍生物是一类具有不同极性、净电荷和分子质量的异种化合物。它们不仅被用作合成蛋白质的结构单位、器官间代谢中氮和碳元素的载体和能量底质，而且也可被用作生物活性化合物（如神经递质）的前体物质。一些氨基酸及其衍生物本身就是神经递质，其他的用于化合物的缀合和分泌。氨基酸转运蛋白介导和调节这些营养物质跨浆膜的流入和流出，因而在器官间的代谢中起关键作用。对各类氨基酸具有不同特异性的大量膜转运蛋白，因氨基酸的不同理化特性而介导其转运过程。

早在20世纪60年代，Halvor N Christensen的研究组就在其开创性的工作中，主要根据对红细胞、肝细胞和纤维原细胞中转运情况的分析而测定了不同转运途径的底物特异性（Christensen，1990）。这一早期工作还揭示了一些普遍规律，如L-异构体对几乎所有转运系统都有较快转运的立体选择性及一些载体蛋白相当宽的底物特异性。一些转运蛋白对酸性氨基酸、碱性氨基酸或芳香族氨基酸具有相对特异性。此外，在鉴别平衡转运系统和离子依赖性的向上转运系统时，也观测到转运步骤热力学特性的差异。

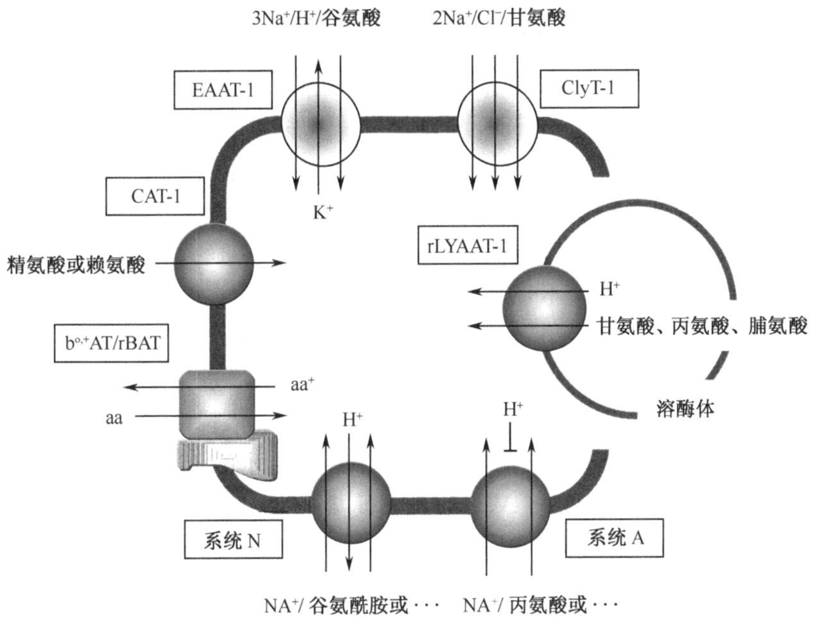


图 3.3　不同载体家族中有代表性的成员的氨基酸转运活性及其可能的作用方式、参与的离子和流通偶联的比率

图 3.3总结了氨基酸转运系统及代表性载体类型的不同作用机制和作用方式。

1991 年，第一个编码哺乳动物氨基酸转运蛋白的 cDNA 被鉴别为嗜亲性的鼠反转录病毒受体，该受体证实能转运阳离子氨基酸（Kim *et al.*，1991；Wang *et al.*，1991）。在此发现之前，神经递质 γ 氨基丁酸（γ-aminobutyric acid，GABA）的第一个转运蛋白被克隆（Guastella *et al.*，1990）。在过去的几年中，采用表达克隆和同源性筛选技术分离出了越来越多的转运蛋白。目前，约有 30 个编码不同氨基酸转运活性的蛋白质的 cDNA 已被鉴别（不包括剪接变体）。表 3.2 对哺乳动物细胞中氨基酸的主要转运途径（可细分为 $Na^+$ 依赖性及 $Na^+$ 非依赖性途径）和在靶细胞中表达时具有转运活性的已被克隆的 cDNA 进行了总结。

**表 3.2　在哺乳动物细胞浆膜上发现的不同氨基酸转运系统的分类转运活性及编码介导这种活性的蛋白质的 cDNA**

| 编码转运活性蛋白的已分离的 cDNA | | | |
|---|---|---|---|
| $Na^+$ 依赖性的 | | $Na^+$ 非依赖性的 | |
| A | ATA1-3 | L | 4F2hc/LAT(X) |
| N | SN1-3 | $y^+$ | CAT1-3 |
| GLY | GlyT1-3 | $b^{o,+}$ | rBAT/ $b^{o,+}$AT |
| ASC | ASC1-2 | PAT | LYAAT-1 |
| BETA | GAT1-3,BGT-1 | | |
| IMINO | 还未被鉴别 | | |
| $B^o$ | $ATB^o$ | | |
| $B^{o,+}$ | $ATB^{o,+}$ | | |
| $X_{AG}^-$ | EAAT1-5 | | |

# 两性氨基酸的转运途径

两性氨基酸转运系统 A、ASC、N 和 L 几乎存在于所有的细胞类型中。系统 A、ASC 和 N 通过与钠离子协同转运而介导小侧链氨基酸转运到细胞内。系统 L 转运蛋白则介导含大量侧链（即支链和芳香团）的氨基酸的转运，但不需要钠离子的参与。系统 L 已被认为在许多组织中用作氨基酸转出系统而不是转入系统。系统 A 和 N 优先转运谷氨酸、丙氨酸和丝氨酸，而系统 ASC 优先转运丙氨酸、丝氨酸和半胱氨酸。相比之下，小而无分支侧链的氨基酸不是系统 L 的理想底物，而类似物 BCH（2-氨基内侧环-2,2,1-庚烷基-2-羧酸）则是系统 L 进行钠非依赖性转运的模式底物。具有丰富侧链的两性氨基酸在多数情况下及在多数的非上皮细胞中，均通过钠非依赖性交换过程转运（Palacin *et al.*，1998）。在许多上皮细胞中，丙氨酸的钠依赖性转运能被 *N*-甲基氨基异丁酸（MeAIB）抑制，且 MeAIB 的钠依赖性转运可被用来测定系统 A 特异性的转运活性（Reimer *et al.*，2000）。系统 A 是高度 pH 敏感性和生电性的，而钠依赖性系统 ASC 对 pH 相对不敏感，且为电中性转运，表明系统 ASC 可能是与钠在两个方向运动有关的氨基酸的逆向转运蛋白。系统 A 的一个重要特点，是它在许多细胞类型中的转运活性受高度调节。这种调节包括许多细胞和组织中细胞循环进展和细胞生长的上调作用，以及胰岛素、胰高血糖素、儿茶酚胺、糖皮质激素、各种生长因子和促细胞分裂素通过完全不同的信号途径进行的激素调控（McGivan and Pastor-Anglada，1994）。胰高血糖素和表皮生长因子可立即提高肝细胞中系统 A 的活性，而胰岛素能以基因转录依赖性的方式上调肝细胞中系统 A 的活性，也可通过补充浆膜上已有转运蛋白的快速途径来实现这一上调作用（骨骼肌中也一样）。然而，胰岛素缺乏或胰岛素抵抗也与肝脏和骨骼肌中系统 A 活性的适应性上调有关，提示系统 A 转运蛋白是短期、中期和慢性适应的大量复杂调节因子的作用靶点。

至少从其分子结构和功能失调的病理生理学的角度看，最令人感兴趣的氨基酸转运系统是具有 $y^+L$ 和 $b^{0,+}$ 转运活性（表 3.2）的异二聚氨基酸转运蛋白，这类转运蛋白存在于各种不同的细胞中（Chillaron *et al.*，2001；Wagner *et al.*，2001）。其结构方面的新颖性是，它们由两条独立的蛋白质链组成，一条重链（已被鉴别的两条重链之一），一条轻链（7 条之一）。这两条单体通过细胞外的二硫键桥而寡聚合在一起，所形成复合物的转运活性及在膜上的位置取决于其两条亚链的性质。一般来说，形成有活性的氨基酸转运蛋白的两条亚链是由一条被称为 4F2hc 或 rBAT 且能与 7 条已鉴别轻链（LAT1、LAT2、$y^+$LAT1、$y^+$ LAT2、ascAT1、xCT 或 $b^{0,+}$ AT）之一结合的糖蛋白重链组成的（Chillaron *et al.*，2001）。重链 rBAT 主要表达于上皮细胞，与轻链结合后，形成一复合物，再移位到细胞顶膜，从而形成了钠非依赖性的氨基酸转运过程（Palacin *et al.*，2001a）。另一条重链 4F2hc 与各种轻链结合，而这些轻链能介导上皮细胞和非上皮细胞的基底侧膜上的氨基酸转运机制。这些重链都有一个附着于单个 TM 域上的类似糖苷酶的细胞外域（Verrey *et al.*，2000）。不同轻链的大小不一，但都含有广泛分布于膜蛋白的 12 个跨膜区域，且其 N 端和 C 端朝向细胞质。图 3.4 描述了不同轻链如何与重链 4F2hc 或 rBAT 结合以形成具有选择性转运不同氨基酸的异二聚体复合物的模式。重链的主要功能很可能是作为内部载体，使该复合物能易位并整合到靶细胞

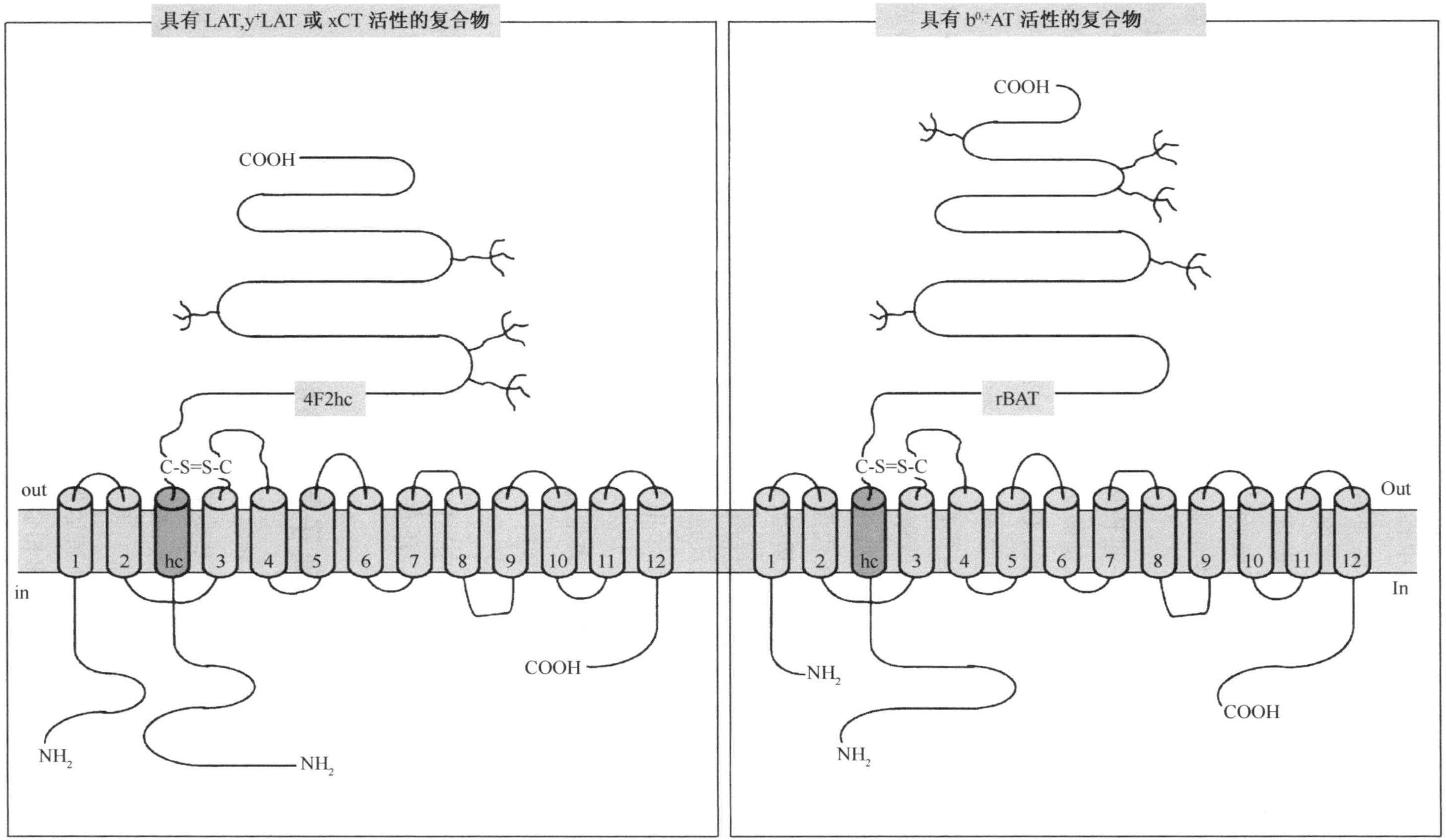


图 3.4 哺乳动物浆细胞膜上发现的异二聚体氨基酸转运复合物的可能的结构

它们由重链 4F2hc 或 rBAT 组成，这些重链通过细胞外二硫键连桥而接到各种轻链之一上。然后，这些不同的复合物构成具有底物特异性和膜定位不同特点的转运途径膜上。异二聚体转运蛋白能以强制交换的方式转运各种中性氨基酸，即它们在调节往细胞内转运某些氨基酸的同时迫使细胞内的氨基酸流出。此外，rBAT 相连复合物的 $b^{0,+}$ 活性能转运阳离子氨基酸来交换中性氨基酸，这样就形成转运电流，而 $y^+$ LAT1-rBAT 复合物则利用钠离子协同转运细胞内阳离子氨基酸来交换中性氨基酸。现已鉴别转运蛋白复合物中的各种突变，这些突变导致功能失调和严重代谢紊乱。$b^{0,+}$ AT 缺损引起具有肾临床症的非Ⅰ型半胱氨酸尿症（cysteinuria）（Palacin *et al.*，2001b），从而说明肾小管的顶膜对半胱氨酸的重吸收缺损。

$y^+$ LAT1 的功能缺失引起上皮细胞基底侧膜上的转运缺陷，从而导致溶尿素蛋白不耐症（lysinuric protein intolerance），而在正常情况下上皮细胞基底侧膜上的 $y^+$ LAT1 能把中性氨基酸与钠离子一起转运入细胞，以把阳离子氨基酸从细胞中交换出来。最近发表的两篇综述对这类氨基酸交换体的结构和功能进行了系统、深入的阐述，可进一步阅读（Verrey *et al.*，2000；Chillaron *et al.*，2001）。

最近，系统 N 氨基酸转运蛋白已被克隆（Chaudry *et al.*，1999；Nakanishi *et al.*，2001）。谷氨酰胺是这类氨基酸转运蛋白最重要的底物。作为碳和氮的载体，谷氨酰胺在器官间代谢中的核心作用及其与酸碱代谢的关系使谷氨酰胺转运蛋白及其调节成为各种代谢过程的焦点。加之谷氨酰胺是神经元谷氨酸的前体物质并能有效地被神经元摄取，因此，系统 N 转运蛋白在递质代谢方面也是令人感兴趣的。新颖性是谷氨酰胺流入与阳离子流通的离子偶联。由转运蛋白 SN1 和 SN2 介导的谷氨酰胺摄取是钠依赖性但电中性的。这是由一钠离子（或锂离子）与谷氨酰胺一起协同转运，同时伴随着质子流出细胞所致，而质子流出细胞又反过来使细胞内发生碱化。然而，谷氨酰胺的流通转运也能生电，因为 SN1 本身具备质子电导性质，这使质子流出既可与谷氨酰胺摄取偶联也可不与其偶联（Chaudry *et al.*，2001）。谷氨酰胺转运与细胞酸碱平衡的密切功能偶联，是认识代谢性酸中毒或碱中毒情况下谷氨酰胺代谢中代谢适应调节的关键。除谷氨酰胺外，转运蛋白 SN1 和 SN2 能转运几乎所有的两性氨基酸（甘氨酸、丝氨酸、丙氨酸、天冬氨酸和组氨酸，但不能转运甲氨基丁酸）。转运蛋白 SN1 主要存在于门静脉周围（periportal）和静脉周围（perivenous）区域的肝细胞及脑部的星形胶质细胞中。SN2 mRNA 不仅存在于肝脏和脑，还存在于肾脏、脾脏和睾丸中，且很可能有剪切变体（Nakanishi *et al.*，2001）。

## 阳离子氨基酸的转运

尽管阳离子氨基酸（精氨酸、赖氨酸和鸟氨酸）的转运过程已被认为是由异二聚交换体介导的交换过程，但是这些氨基酸最重要的转运蛋白是最初被认为是嗜亲性反转录病毒受体的 CAT 家族成员（Kim *et al.*，1991；Wang *et al.*，1991）。在 CAT 家族中，至少有三个基因编码四种不同的同工物（isoform）[CAT-1、CAT-2（A）、CAT-2（B）和 CAT-3]，但也可能编码 CAT-4 转运蛋白。它们都是含 12～14 个 TM 片段、分子质量约为 70kDa 的蛋白质。尽管结构相似，但它们在组织分布、动力学和调节特性上各不相同。系统 $y^+$（CAT 转运蛋白）及细胞对阳离子氨基酸的摄取过程是由细胞内的负膜电位单向驱动的。CAT 蛋白转运的阳离子氨基酸不仅可用于生长与发育，而且还可用于特定的细胞功能，如肌酸、肉毒碱和多胺的合成或为不同一氧化氮合成酶提供精氨酸（Closs，2002）。

## 阴离子氨基酸的转运

阴离子氨基酸（谷氨酸和天冬氨酸）是 XaG 和 Xc 转运系统的底物（Gadea and Lopez-Colome，2001a）。脑中 XAG 系统的四种同工物均属于 $Na^+$ 依赖性氨基酸转运蛋白家族，其转运方式有两种：一是生电性的转运方式，在转运过程中伴随一个以上钠离子运动；或是电中性的转运方式，具体依同工物而定。一些转运蛋白在进行 $Na^+$ 依赖性的谷氨酸盐转入的同时伴随着钾离子的流出。这些兴奋性氨基酸转运蛋白（excitatory amino acid transporter，EAAT）在通过去除突触间隙中的谷氨酸而使大脑中谷氨酸的神经传递失活方面具有重要的作用，但它们对谷胱甘肽的合成似乎也具有重要作用。所有的 EAAT 对阴离子氨基酸都具有非常高的特异性，且在有效和完全地去除突触间隙中谷氨酸所必需的微摩尔范围内具有非常高的亲和力。虽然脑部是主要的表达部位，但在其他许多组织中也发现有 EAAT3 和 EAAT4，而 EAAT5 很可能是普遍表达的。ASC 家族中的转运蛋白即 ASCT1 和 ASCT2 主要转运两性氨基酸（丙氨酸、半胱氨酸、丝氨酸、谷氨酸、亮氨酸、异亮氨酸和缬氨酸），但也能转运阴离子氨基酸。这些转运蛋白均以 $Na^+$ 依赖性的方式进行转运，且是生电性的或电中性的（Utsunomiya-Tate *et al.*，1996）。

其他氨基酸转运蛋白包括 GLYT 载体蛋白，它们优先转运具有各种异构体和剪切变体的甘氨酸。它们在介导甘氨酸的细胞摄入的同时协同转运 2 个 $Na^+$ 和 1 个 $Cl^-$，含有 12 个推定的 TM 区域（Gadea and Lopez-Colome，2001b）。最近，哺乳动物的第一类质子依赖性的生电氨基酸转运蛋白已被发现（Sage *et al.*，2001），它们可能是通过氨基酸与质子转运相偶联将短侧链氨基酸（甘氨酸、丙氨酸、丝氨酸和脯氨酸）运出溶酶体的最初载体。在小肠和肾脏上皮细胞的顶膜中也发现了这种转运小的中性氨基酸的质子依赖性途径。根据它们的作用方式，人们可推测，这些转运蛋白与在原核细胞、酵母和植物上发现的一样，与哺乳动物来源的古老质子偶联的氨基酸转运途径类似。

## 甲状腺激素的转运

尽管甲状腺激素的两种形式 T4 和 T3 都不是氨基酸而是酪氨酸的衍生物，但是它们的跨细胞膜转运也需要氨基酸转运蛋白的参与，并在甲状腺激素代谢中发挥非常重要的作用。根据其高亲脂性，人们最初认为，甲状腺激素是以被动扩散方式进入细胞的。但根据其转运的动力学，特别是立体选择性，显然特异性结合和载体介导过程肯定参与其中。现已鉴别发现在纳摩尔级范围对 T4 和 T3 具有亲和力的低转运力的机制，及另一个高结合、高转运但在微摩尔级范围具有低亲和力的转运系统。T4 和 T3 通过高亲和力位点的摄入过程是能量、温度且常常是钠离子依赖性的，而且最有可能表现为通过钠离子依赖性和非依赖性机制进行甲状腺素的跨浆膜转运（Hennemann *et al.*，2001）。

在各种组织和细胞（红细胞、垂体细胞、星形胶质细胞、血脑屏障和脉络膜结节）中，T4 和 T3 的摄取很可能是由系统 L 或 T（芳香族氨基酸）转运蛋白来介导的。T3 而非 T4 从某些细胞种类的流出也是饱和转运（Hennemann *et al.*，2001）。由于肝脏在甲状腺激素的清除和转化中发挥重要作用，故肝脏的转运过程为甲状腺激素代谢所必需。人肝细胞对激素的摄入是ATP依赖性的，且对后面的碘化甲状腺氨酸代谢很可能是限速的。在饥饿条件下，肝中对T4的摄取很可能由于非酯化的脂肪酸、胆红素或ATP耗竭的抑制作用而减少，且因T4向T3的转化量降低而导致血浆T3水平降低。最近通过鉴别发现，几个有机阴离子转运蛋白也是甲状腺激素的转运载体蛋白（Friesema *et al.*，1999）。

## 短链肽的转运

以肽结合形式摄取氨基酸是自然界普遍存在的生物现象。细菌、酵母和真菌及植物、无脊椎动物和脊椎动物的特定细胞均表达摄取二肽和三肽的膜蛋白。另外，一些物种（但不是哺乳动物）还具有能接收更大寡肽（≥4个氨基酸残基）的转运蛋白。根据其分子和功能特点，肽的膜转运蛋白已被归类为质子依赖性寡肽转运蛋白的PTR家族（Steiner *et al.*，1995）。在哺乳动物，两个基因已被鉴别出具有转运二肽和三肽的活性：即肽转运蛋白1（peptide transporter 1，PEPT1）和肽转运蛋白2（peptide transporter 2，PEPT2），相当于SLC15A1和SLC15A2（Fei *et al.*，1994；Ramamoorthy *et al.*，1995；Boll *et al.*，1996）。第三个蛋白质可能是转运二肽的组氨酸转运蛋白（PHT-1）（Ya结合mashita *et al.*，1997）。人的PEPT1和PEPT2蛋白分别由708和729个氨基酸残基组成，均含12个TM结构域，且N端和C端朝向细胞质。哺乳动物的肽载体能将肽的跨膜运动与质子（$H_3O^+$）沿着向内电化学质子梯度的运动偶联起来，这就使得肽能逆底物物梯度进行转运。通过与质子流的偶联，不论底物所带的净电荷如何，肽转运都是生电性的，且能导致细胞内酸化（Amasheh *et al.*，1997）。转入肽所必需的质子梯度主要但不专门由电中性质子/阳离子交换体如$Na^+$/$H^+$逆向转运蛋白提供，而$Na^+$/$H^+$逆向转运蛋白能再次输出质子以交换进入细胞的$Na^+$离子。然而，肽转运的主要驱动力是细胞内的负膜电位。摄入细胞的正常二肽迅速地被胞质中大量的肽酶水解，这些肽酶对二肽和三肽具有高亲和力和高水解力。游离氨基酸随后被转入血液循环，或被用于细胞内的蛋白质合成或其他目的。

由于进化过程中出现了大量氨基酸转运蛋白，人们当然很有兴趣推测肽转运蛋白进化保守性的优点。很明显，与单个氨基酸的单独转运过程所需的能量相比，几个氨基酸一起一步转运需要消耗的细胞能量更低。而且，肽载体具有转运所有可能存在的二肽和三肽的功能表明，必需氨基酸与非必需氨基酸或底物的理化特性方面均无差异，其底物的分子大小可从96.2Da（甘二肽）到552.6Da（色三肽）。肽转运蛋白的主要功能是提供氨基酸以满足生长、发育和代谢的营养需要。但有证据表明，它们在转运蛋白仅在特异细胞中表达的更复杂的生物中也具有特殊功能。

PEPT1在肠上皮细胞顶膜的表达量最高，很可能与二肽和三肽的大量摄入有关，而这些二肽和三肽是饲粮蛋白在肠腔和膜上的消化终产物。PEPT1还存在于肾管状细胞。然而，肾脏中表达的主要肽转运蛋白是PEPT2（Daniel and Herget，1997）。肾PEPT2与肠PEPT1同工物相比，约50%氨基酸序列完全相同，70%氨基酸序列具有同源性，其中在TM结构域中的相同性最高。当采用PEPT2特异性抗敏探针在兔肾脏进行原位杂交试验时，发现PEPT2 mRNA在近端管状细胞中呈高表达，且在S1和S2片段中的表达水平高于S3片段（Smith *et al.*，1998）。采用反转录聚合酶链反应（reverse transcription-polymerase chain reaction，RT-PCR）、Northern 印迹分析和免疫检测技术的研究结果表明，许多非肾组织也能表达 PEPT2，其中在脑、脉络膜结节、肺上皮和乳腺的上皮细胞中的表达量特别高（Doring *et al.*，1998；Berger and Hediger，1999）。但尚未对这些组织中肽转运蛋白的生物学作用进行测定。

肾管状细胞中 PEPT2 的主要功能是氨基酸氮的重吸收和保存作用。摄入蛋白质食物后，循环中肽结合氨基酸的数量增加，且二肽和三肽不断由循环中内源蛋白质和寡肽的降解来产生，和（或）从各种组织中被释放。尽管能鉴别血浆中单个二肽或三肽，但是外周血液中的所有循环短链肽的总浓度仍不清楚。对各种动物进行的研究表明，约50%的循环血浆氨基酸是肽结合形式的，且绝大多数是二肽和三肽（Seal and Parker，1991）。除了过滤后进入管状系统的短链肽外，各种高催化活性的肾刷状缘膜结合的肽水解酶存在，且能通过水解肾小球中过滤的较大寡肽而提供大量的二肽和三肽。由于 PEPT2 是高亲和力的转运系统（$K_m$ 值取决于肽片段，$5 < K_m < 250 \mu mol/L$），因此，它能有效地从管液中清除二肽和三肽。

通过 PEPT1 和 PEPT2 的转运具有立体选择性，含氨基酸 L-异构体的寡肽对转运的亲和力比含 D-异构体的寡肽高。仅由 D-氨基酸组成的寡肽对转运不具有任何相关亲和力。哺乳动物的肽转运蛋白能转运各种肽类似药物，如氨基头孢菌素和氨基青霉素类的抗生素或选择性的肽酶抑制剂，如苯丁抑制素（bestatin）或卡普托普利（captopril），这使得肽转运蛋白对药物运送和药物动力学分析具有重要性。肠上皮中的 PEPT1 使这些药物有极好的口服生物学利用率（Adibi，1997）。肾转运蛋白 PEPT2 通过对已在肾小球中过滤的药物的重吸收来增加这些药物在血浆中的半衰期。

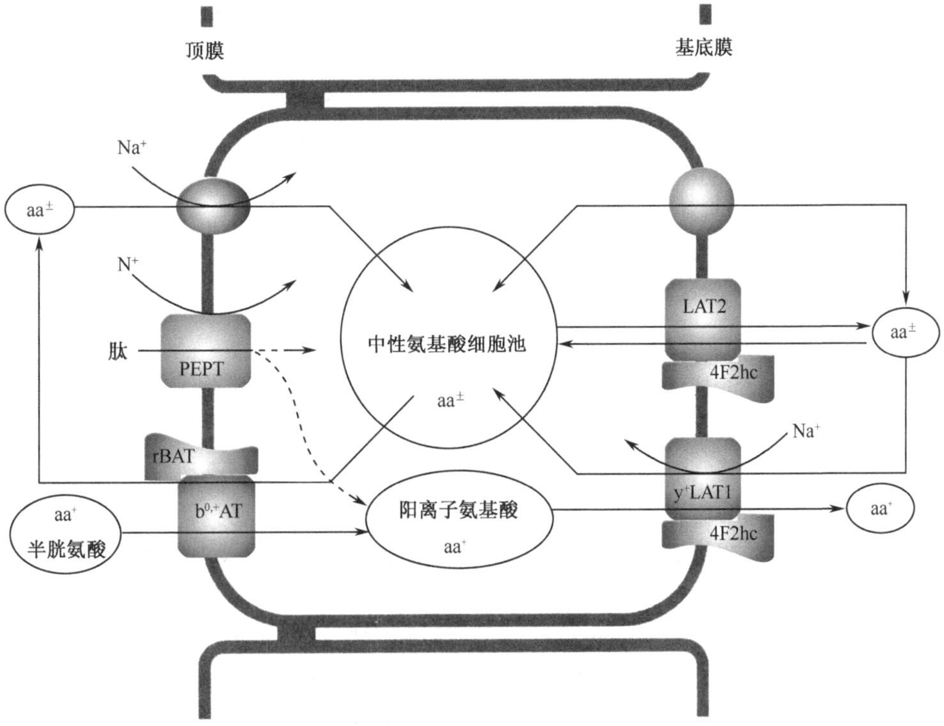


图 3.5　目前关于介导肠和肾上皮细胞中氨基酸氮跨细胞转运化的氨基酸和肽转运蛋白的协同作用的观点

关于 PEPT1 的临床重要性已在包括人的研究在内的各种研究中进行了阐述。与游离氨基酸相比，二肽已表明能更迅速地被小肠吸收，且由于它们具有较低的营养液渗透性而对肠腔营养更有用处。另外，在各种胃肠道疾病中，肽转运蛋白已被发现受病理生理学的影响比氨基酸转运蛋白小，甚至在黏膜功能受损的情况下，还能进行氨基酸氮的转运（Adibi，1997）。图 3.5 总结了上皮细胞中肽和游离氨基酸的不同转运途径，它们协同介导氨基酸氮的跨细胞吸收。

## 水的转运途径

机体中的一些细胞需要对水有非常高的膜渗透性。如在肾中，其上皮细胞的顶膜需要能维持水稳衡且受激素（主要是抗利尿激素）调控的水渗透性。肾脏系统的功能失调说明水的膜转运途径的定量重要性。如尿崩症是一种遗传疾病，能降低肾脏对水的重吸收能力，进而导致大量原尿的损失（Kwon *et al.*，2001）。根据尿崩症的严重程度，肾脏的水损失量每天可高达 50L。该病潜在损害了能调控特异膜水通道蛋白（aquaporin）的融合和适当功能的信号途径（Sansom and Law，2001）。当水通道蛋白表达并插入管状细胞的刷状缘膜上时，就形成了膜的水通道，它允许大量的水从内腔流到血液一侧，因此使尿液得以浓缩，并使水分得以保存。

水通道蛋白（AQP）是完整的膜蛋白，可在人的许多组织或特异细胞中表达，也能在细菌、植物和动物中表达。AQP 蛋白被细分为 AQP 固有家族（人类基因组至少含 10 个相关家族成员）与水甘油孔蛋白（aquaglyceroporin）（具有非常相似的结构，但还具有甘油通透性）（Borgnia *et al.*，1999）。这些蛋白质均含有可形成孔道的通道样核心结构。细菌 AQP 的高分辨 X 射线结构和人 AQP 的中等分辨电子显微镜方法使人们能详细研究这些膜蛋白的结构和功能。它们由 6 个跨膜螺旋和仅部分插入膜双层中的 2 个半 TM 螺旋组成，中心孔道长 20Å、内宽 2Å，这可解释它们对水和甘油有通透性的原因（Nollert *et al.*，2001）。对于 AQP 固有蛋白（AQP1），一个单独的蛋白质分子每秒可转运 $10^9$ 个水分子。

## 维　生　素

近来，人们在分子基础上仅对哺乳动物的介导维生素（除维生素 $B_{12}$ 外）跨细胞浆膜摄入或流出的转运蛋白进行了鉴别。尽管人们已通过流量检测分析了不同细胞系统中维生素的不同转运及其调节过程，但对其转运蛋白的分子特性仍不清楚。由于本节无法述及维生素转运情况的所有方面，故只能对一些重要的最新发现和新颖过程进行简要概述。图 3.6 总结了本节将要叙述的参与维生素转运的各种系统及其如何介导 TM 移位的建议方式。

根据不同水溶性维生素的理化特性，需要不同类型的转运蛋白把中性、阴性或阳性（硫胺素）的化合物转运入内带负膜电位的细胞区室。对于阴性维生素（如维生素 C、叶酸、泛酸和生物素），细胞摄取需要钠离子或质子的协同转运来进行电中性转运，或需要不同的配比（过量的钠离子或质子）来诱导生电转运。

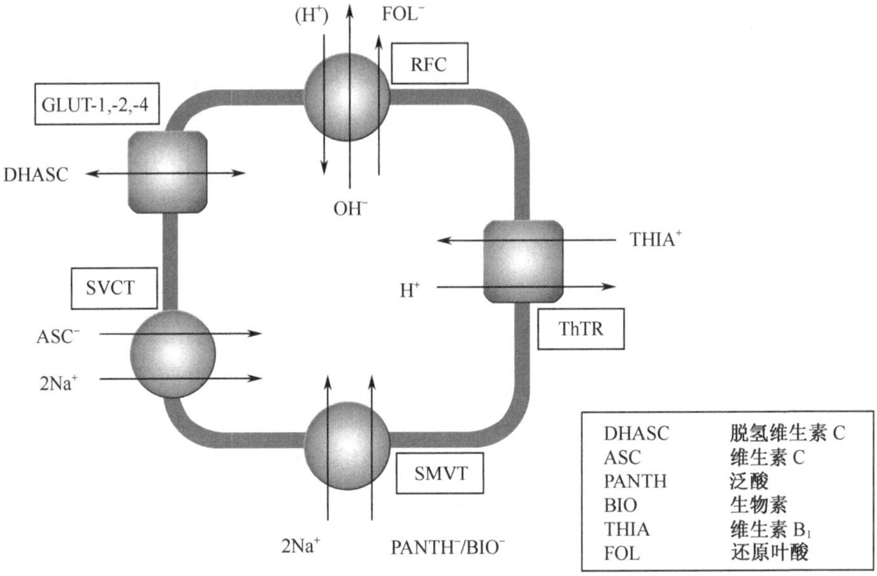


图 3.6 哺乳动物细胞浆膜上的维生素转运蛋白及其可能的作用机制

1999 年，采用表达克隆技术鉴别了维生素 C 转运蛋白 SVCT1 和 SVCT2（SLC23A2 和 SLC23A1）（Tsukaguchi *et al.*，1999）。当时人们已经知道，细胞的维生素 C 摄取是饱和的且是钠离子依赖性的，而还原性维生素 C 的转运是通过 GLUT 转运蛋白（GLUT-1～GLUT-4）以钠离子非依赖性的方式介导的，然后在细胞内快速被还原成维生素 C。人的钠依赖生电性维生素 C 转运蛋白 SVCT1 由 598 个氨基酸组成，对维生素 C 的亲和力约 250μmol/L，而 SVCT2 由 650 个氨基酸组成，是高亲和力（$K_m$ 约 20μmol/L）类的转运蛋白。这两种转运蛋白都是转运钠依赖性的协同转运蛋白，介导细胞维生素 C 的积累，但其表达方式很不同（Liang *et al.*，2001）。

介导泛酸、生物素和脂肪酸盐（lipoate）摄取的独特转运蛋白已被克隆，并被命名为 SMVT，即钠依赖性的多种维生素转运蛋白（SLC5A6）（Prasad *et al.*，1998）。人的 SMVT 由 635 个氨基酸组成，含 12 个已假设 TM 结构域，编码于 2p23 号染色体上。SMVT 按钠离子与底物分子的偶联以为 2∶1 的方式来转运底物，故为生电转运载体。这种蛋白质存在于被极化的上皮细胞的顶膜上，在小肠介导饲粮维生素的摄取及肾管状上皮细胞的重吸收。Northern 印迹分析表明，SMVT 转录物还存在于其他各种细胞和组织中。蛋白质数据库对比分析的结果表明，SMVT 与已知的 $Na^+$ 依赖性葡萄糖转运蛋白家族成员之间具有明显的序列相似性（Prasad and Ganapathy，2000）。

最近，许多还原型叶酸的转运蛋白被克隆和鉴别（Matherly，2001）。虽然人们认识叶酸结合蛋白 α 和 β 已有一些时间，但第一个被鉴定的哺乳动物膜上叶酸转运载体蛋白是 RFC-1（SLC19A1）。在进行克隆前，人们已对该蛋白质在完整组织中的功能进行了研究，表明它能介导钠离子非依赖性但 pH 强依赖性的叶酸摄取过程。随着细胞外 pH 降低，细胞对还原型叶酸、甲基-四氢叶酸和甲氨蝶呤的摄取量增加，提示这个转运过程是个叶酸-质子的协同转运过程，或是叶酸阴离子与细胞内的羟基离子进行交换的阴离子交换过程。比较人 RFC-1 的氨基酸序列与其他蛋白质序列，发现人 ThTr1，即硫胺素转运蛋白（SLC19A2）的序列与 RFC-1 的序列具有最高的相关性，两者约有 40%氨基酸序列完全相同，55%氨基酸序列相似（Dutta *et al.*，1999）。最近几年，该蛋白家族的第三成员（SLC19A3）已从人类和鼠中被克隆，并被确定为第二个硫胺素转运蛋白（Rajgopal *et al.*，2001）。作为水溶性维生素，硫胺素独有的特点是其阳离子特性。依 pH 而定，它能带一个或两个正电荷，这提示硫胺素可作为大量已被鉴别的有机阳离子转运蛋白（organic cation transporter，OCT）的底物。然而，OCT 家族中大多数的阳离子转运蛋白似乎都不能转运硫胺素。相反，由人胎盘克隆的 ThTr1 能诱导特异的硫胺素流进被转入细胞，这些被转入细胞具有进行硫胺素-质子交换的特性。因此，硫胺素流入细胞偶联于质子流出，硫胺素的转运具有明显的 pH 依赖性，即酸性的细胞外 pH 降低硫胺素的流入速度。最近发现，高亲和力的硫胺素转运蛋白 ThTr1（287delG）的碱基缺失是导致硫胺素效应的巨红细胞性贫血（thiamine-responsive megaloblastic anaemia，TRMA）综合征的原因，而该综合征与糖尿病和耳聋有关（Diaz *et al.*，1999）。两个患 TRMA 的患者均在与翻译起始部位相关的 cDNA 的核苷酸 287 处发生突变。遗传变异能导致移码和过早的终止密码子。患者的纤维原细胞几乎完全失去了硫胺素的摄入功能（Neufeld *et al.*，2001）。

尽管并不是所有介导不同水溶性维生素流入和流出的转运蛋白均已在分子水平上得以被鉴别，但最近几年，人们在这方面已取得了显著进展。迄今已知的基因和蛋白质已成为人们系统与先天性维生素代谢失调之间关系及根据功能差异筛选遗传异质性的有兴趣的研究靶标。

## 参 考 文 献


Ader, P., Block, M., Pietzsch, S. and Wolffram, S. (2001) Interaction of quercetin glucosides with the intestinal sodium/glucose co-transporter (SGLT-1). *Cancer Letters* 162, 175–180.

Adibi, S.A. (1997) The oligopeptide transporter (Pept-1) in human intestine: biology and function. *Gastroenterology* 113, 332–340.

Amasheh, S., Wenzel, U., Boll, M., Dorn, D., Weber, W., Clauss, W. and Daniel, H. (1997) Transport of charged dipeptides by the intestinal $H^+$/peptide symporter PepT1 expressed in *Xenopus laevis* oocytes. *Journal of Membrane Biology* 155, 247–256.

Berger, U.V. and Hediger, M.A. (1999) Distribution of peptide transporter PEPT2 mRNA in the rat nervous system. *Anatomy and Embryology (Berlin)* 199, 439–449.

Boll, M., Markovich, D., Weber, W.M., Korte, H., Daniel, H. and Murer, H. (1994) Expression cloning of a cDNA from rabbit small intestine related to proton-coupled transport of peptides, beta-lactam antibiotics and ACE-inhibitors. *Pflügers Archiv* 429, 146–149.

Boll, M., Herget, M., Wagener, M., Weber, W.M., Markovich, D., Biber, J., Clauss, W., Murer, H. and Daniel, H. (1996) Expression cloning and functional characterization of the kidney cortex high-affinity proton-coupled peptide transporter. *Proceedings of the National Academy of Sciences USA* 93, 284–289.

Bonen, A. (2000) Lactate transporters (MCT proteins) in heart and skeletal muscles. *Medicine in Science, Sports and Exercise* 32, 778–789.

Bonen, A., Baker, S.K. and Hatta, H. (1997) Lactate transport and lactate transporters in skeletal muscle. *Canadian Journal of Applied Physiology* 22, 531–532.

Bonen, A., Dyck, D.J. and Luiken, J.J. (1998) Skeletal muscle fatty acid transport and transporters. *Advances in Experimental Medicine and Biology* 441, 193–205.

Borgnia, M., Nielsen, S., Engel, A. and Agre, P. (1999) Cellular and molecular biology of the aquaporin water channels. *Annual Review of Biochemistry* 68, 425–458.

Brooks, G.A, Brownk, M.A., Butz, C.E., Sicurello, J.P. and Dubouchaud, H. (1999) Cardiac and skeletal muscle mitochondria have a monocarboxylate transporter MCT1. *Journal of Applied Physiology* 87, 1713–1718.

Chaudhry, F.A., Reimer, R.J., Krizaj, D., Barber, D., Storm-Mathisen, J., Copenhagen, D.R. and

Edwards, R.H. (1999) Molecular analysis of system N suggests novel physiological roles in nitrogen metabolism and synaptic transmission. *Cell* 99, 769–780.

Chaudhry, F.A, Krizaj, D., Larsson, P., Reimer, R.J., Wreden, C., Storm-Mathisen, J., Copenhagen, D., Kavanaugh, M. and Edwards, R.H. (2001) Coupled and uncoupled proton movement by amino acid transport system N. *EMBO Journal* 20, 7041–7051.

Chillaron, J., Roca, R., Valencia, A., Zorzano, A. and Palacin, M. (2001) Heteromeric amino acid transporters: biochemistry, genetics, and physiology. *American Journal of Physiology* 281, F995–F1018.

Christensen, H.N. (1990) Role of amino acid transport and countertransport in nutrition and metabolism. *Physiological Reviews* 70, 43–77.

Closs, E.I. (2002) Expression, regulation and function of carrier proteins for cationic amino acids. *Current Opinion in Nephrology and Hypertension* 11, 99–107.

Coburn, C.T., Knapp, F.F. Jr, Febbraio, M., Beets, A.L., Silverstein, R.L. and Abumrad, N.A. (2000) Defective uptake and utilization of long chain fatty acids in muscle and adipose tissues of CD36 knockout mice. *Journal of Biological Chemistry* 275, 32523–32529.

Daniel, H. (2000) Nutrient transporter function studied in heterologous expression systems. *Annals of the New York Academy of Sciences* 915, 184–192.

Daniel, H. and Herget, M. (1997) Cellular and molecular mechanisms of renal peptide transport. *American Journal of Physiology* 273, F1–F88.

Diaz, G.A., Banikazemi, M., Oishi, K., Desnick, R.J. and Gelb, B.D. (1999) Mutations in a new gene encoding a thiamine transporter cause thiamine-responsive megaloblastic anaemia syndrome. *Nature Genetics* 22, 309–312.

Doring, F., Walter, J., Will, J., Focking, M., Boll, M., Amasheh, S., Clauss, W. and Daniel, H. (1998) Delta-aminolevulinic acid transport by intestinal and renal peptide transporters and its physiological and clinical implications. *Journal of Clinical Investigation* 101, 2761–2767.

Dutta, B., Huang, W., Molero, M., Kekuda, R., Leibach, F.H., Devoe, L.D., Ganapathy, V. and Prasad, P.D. (1999) Cloning of the human thiamine transporter, a member of the folate transporter family. *Journal of Biological Chemistry* 274, 31925–31929.

Eskandari, S., Wright, E.M., Kreman, M., Starace, D.M. and Zampighi, G.A. (1998) Structural analysis of cloned plasma membrane proteins by freeze-fracture electron microscopy. *Proceedings of the National Academy of Sciences USA* 95, 11235–11240.

Fei, Y.J., Kanai, Y., Nussberger, S., Ganapathy, V., Leibach, F.H., Romero, M.F., Singh, S.K., Boron, W.F. and Hediger, M.A. (1994) Expression cloning of a mammalian proton-coupled oligopeptide transporter. *Nature* 368, 563–566.

Friesema, E.C., Docter, R., Moerings, E.P., Stieger, B., Hagenbuch, B., Meier, P.J., Krenning, E.P., Hennemann, G. and Visser, T.J. (1999) Identification of thyroid hormone transporters. *Biochemical and Biophysical Research Communications* 254, 497–501.

Gadea, A. and Lopez-Colome, A.M. (2001a) Glial transporters for glutamate, glycine and GABA I. Glutamate transporters. *Journal of Neuroscience Research* 63, 453–460.

Gadea, A. and Lopez-Colome, A.M. (2001b) Glial transporters for glutamate, glycine, and GABA III. Glycine transporters. *Journal of Neuroscience Research* 64, 218–222.

Guastella, J., Nelson, N., Nelson, N., Czyzyk, L., Keynan, S., Miedel, M.C., Davidson, N., Lester, H.A. and Kanner, B.I. (1990) Cloning and expression of a rat brain GABA transporter. *Science* 249, 1303–1306.

Halestrap, A.P. and Price, N.T. (1999) The proton-linked monocarboxylate transporter (MCT) family: structure, function and regulation. *Biochemical Journal* 343, 281–299.

Hediger, M.A., Coady, M.J., Ikeda, T.S. and Wright, E.M. (1987) Expression cloning and cDNA sequencing of the $Na^+$/glucose co-transporter. *Nature* 330, 379–381.

Hennemann, G., Docter, R., Friesema, E.C., deJong, M., Krenning, E.P. and Visser, T.J. (2001) Plasma membrane transport of thyroid hormones and its role in thyroid hormone metabolism and bioavailability. *Endocrine Reviews* 22, 451–476.

Herrmann, T., Buchkremer, F., Gosch, I., Hall, A.M., Bernlohr, D.A. and Stremmel, W. (2001) Mouse fatty acid transport protein 4 (FATP4): characterization of the gene and functional assessment as a very long chain acyl-CoA synthetase. *Gene* 270, 31–40.

Hirsch, D., Stahl, A. and Lodish, H.F. (1998) A family of fatty acid transporters conserved from mycobacterium to man. *Proceedings of the National Academy of Sciences USA* 95, 8625–8629.

Hruz, P.W. and Mueckler, M.M. (2001) Structural analysis of the GLUT1 facilitative glucose transporter. *Molecular Membrane Biology* 18, 183–193.

Hyde, R.J., Cass, C.E., Young, J.D. and Baldwin, S.A. (2001) The ENT family of eukaryote nucleoside and nucleobase transporters: recent advances in the investigation of structure/function relationships and the identification of novel isoforms. *Molecular Membrane Biology* 18, 53–63.

Juel, C. and Halestrap, A.P. (1999) Lactate transport in skeletal muscle – role and regulation of the monocarboxylate transporter. *Journal of Physiology* 517, 633–642.

Kim, J.K., Zisman, A., Fillmore, J.J., Peroni, O.D., Kotani, K., Perret, P., Zong, H., Dong, J., Kahn, C.R., Kahn, B.B. and Shulman, G.I. (2001) Glucose toxicity and the development of diabetes in

mice with muscle-specific inactivation of GLUT4. *Journal of Clinical Investigation* 108, 153–160.
Kim, J.W., Closs, E.I., Albittron, L.M. and Cunnigham, J.M. (1991) Transport of cationic amino acids by the mouse ecotropic retrovirus receptor. *Nature* 352, 725–728.
Kwon, T.H., Hager, H., Nejsum, L.N., Andersen, M.L., Frokiaer, J. and Nielsen, S. (2001) Physiology and pathophysiology of renal aquaporins. *Seminars in Nephrology* 21, 231.
Liang, W.J., Johnson, D. and Jarvis, S.M. (2001) Vitamin C transport systems of mammalian cells. *Molecular Membrane Biology* 18, 87–95.
Loo, D.D., Zeuthen, T., Chandy, G. and Wright, E.M. (1996) Cotransport of water by the $Na^+$/glucose cotransporter. *Proceedings of the National Academy of Sciences USA* 93, 13367–13370.
Martin, M.G., Turk, E., Lostao, M.P., Kerner, C. and Wright, E.M. (1996) Defects in $Na^+$/glucose cotransporter (SGLT1) trafficking and function cause glucose–galactose malabsorption. *Nature Genetics* 12, 216–220.
McGivan, J.D. and Pastor-Anglada, M. (1994) Regulatory and molecular aspects of mammalian amino acid transport. *Biochemical Journal* 299, 321–334.
Mueckler, M., Caruso, C., Baldwin, S.A., Panico, M., Blench, I., Morris, H.R., Allard, W.J., Lienhard, G.E. and Lodish, H.F. (1985) Sequence and structure of a human glucose transporter. *Science* 229, 941–945.
Nakanishi, T., Kekuda, R., Fei, Y.J., Hatanaka, T., Sugawara, M., Martindale, R.G., Leibach, F.H., Prasad, P.D. and Ganapathy, V. (2001) Cloning and functional characterization of a new subtype of the amino acid transport system N. *American Journal of Physiology* 281, C1757–C1768.
Neufeld, E.J., Fleming, J.C., Tartaglini, E. and Steinkamp, M.P. (2001) Thiamine-responsive megaloblastic anemia syndrome: a disorder of high-affinity thiamine transport. *Blood Cells, Molecules and Diseases* 27, 135–138.
Nollert, P., Harries, W.E., Fu, D., Miercke, L.J. and Stroud, R.M. (2001) Atomic structure of a glycerol channel and implications for substrate permeation in aqua(glycero)porins. *FEBS Letters* 504, 112–117.
Palacin, M., Estevez, R., Bertran, J. and Zorzano, A. (1998) Molecular biology of mammalian plasma membrane amino acid transporters. *Physiological Reviews* 78, 969–1054.
Palacin, M., Fernandez, E., Chillaron, J. and Zorzano, A. (2001a) The amino acid transport system b(o,+) and cystinuria. *Molecular Membrane Biology* 18, 21–26.
Palacin, M., Borsani, G. and Sebastio, G. (2001b) The molecular bases of cystinuria and lysinuric protein intolerance. *Current Opinion in Genetics and Development* 11, 328–335.
Pastor-Anglada, M., Casado, F.J., Valdes, R., Mata, J., Garcia-Manteiga, J. and Molina, M. (2001) Complex regulation of nucleoside transporter expression in epithelial and immune system cells. *Molecular Membrane Biology* 18, 81–85.
Poole, R.C. and Halestrap, A.P. (1993) Transport of lactate and other monocarboxylates across mammalian plasma membranes. *American Journal of Physiology* 264, C761–C782.
Prasad, P.D. and Ganapathy, V. (2000) Structure and function of mammalian sodium-dependent multivitamin transporter. *Current Opinion in Clinical Nutrition and Metabolic Care* 3, 263–266.
Prasad, P.D., Wang, H., Kekuda, R., Fujita, T., Fei, Y.J., Devoe, L.D., Leibach, F.H. and Ganapathy, V. (1998) Cloning and functional expression of a cDNA encoding a mammalian sodium-dependent vitamin transporter mediating the uptake of pantothenate, biotin, and lipoate. *Journal of Biological Chemistry* 273, 7501–7506.
Rajgopal, A., Edmondnson, A., Goldman, I.D. and Zhao, R. (2001) SLC19A3 encodes a second thiamine transporter ThTr2. *Biochimica et Biophysica Acta* 1537, 175–178.
Ramamoorthy, S., Han, H., Yang-Feng, T.L., Hediger, M.A., Ganapathy, V. and Leibach, F.H. (1995) Human intestinal $H^+$/peptide cotransporter. Cloning, functional expression, and chromosomal localization. *Journal of Biological Chemistry* 270, 6456–6463.
Reimer, R.J., Chaudhry, F.A., Gray, A.T. and Edwards, R.H. (2000) Amino acid transport system A resembles system N in sequence but differs in mechanism. *Proceedings of the National Academy of Sciences USA* 97, 7715–7720.
Ritzel, M.W., Ng, A.M., Yao, S.Y., Graham, K., Loewen, S.K., Smith, K.M., Hyde, R.J., Karpinski, E., Cass, C.E, Baldwin, S.A. and Young, J.D. (2001) Recent molecular advances in studies of the concentrative $Na^+$-dependent nucleoside transporter (CNT) family: identification and characterization of novel human and mouse proteins (hCNT3 and mCNT3) broadly selective for purine and pyrimidine nucleosides (system cib). *Molecular Membrane Biology* 18, 65–72.
Sagne, C., Agulhon, C., Ravassard, P., Darmon, M., Hamon, M., El Mestikawy, S., Gasnier, B. and Giros, B. (2001) Identification and characterization of a lysosomal transporter for small neutral amino acids. *Proceedings of the National Academy of Sciences USA* 98, 7206–7211.
Sankarasubbaiyan, S., Cooper, C. and Heilig, C.W. (2001) Identification of a novel form of renal glucosuria with overexcretion of arginine, carnosine, and taurine. *Americal Journal of Kidney Diseases* 37, 1039–1043.
Sansom, M.S. and Law, R.J. (2001) Membrane proteins: aquaporins – channels without ions. *Current Biology*

11, R71–R73.

Seal, C.J. and Parker, D.S. (1991) Isolation and characterization of circulating low molecular weight peptides in steer, sheep and rat portal and peripheral blood. *Comparative Biochemistry and Physiology B* 99, 679–685.

Seatter, M.J. and Gould, G.W. (1999) The mammalian facilitative glucose transporter (GLUT) family. *Pharmaceutical Biotechnology* 12, 201–228.

Smith, D.E., Pavlova, A., Berger, U.V., Hediger, M.A., Yang, T., Huang, Y.G. and Schnermann, J.B. (1998) Tubular localization and tissue distribution of peptide transporters in rat kidney. *Pharmaceutical Research* 15, 1244–1249.

Stein, J., Zores, M. and Schroder, O. (2000) Short-chain fatty acid (SCFA) uptake into Caco-2 cells by a pH-dependent and carrier mediated transport mechanism. *European Journal of Nutrition* 39, 121–125.

Steiner, H.Y., Naider, F. and Becker, J.M. (1995) The PTR family: a new group of peptide transporters. *Molecular Microbiology* 16, 825–834.

Tsukaguchi, H., Tokui, T., Mackenzie, B., Berger, U.V., Chen, X.Z., Wang, Y., Brubaker, R.F. and Hediger, M.A. (1999) A family of mammalian $Na^+$-dependent L-ascorbic acid transporters. *Nature* 399, 70–75.

Turk, E. and Wright, E.M. (1997) Membrane topology motifs in the SGLT cotransporter family. *Journal of Membrane Biology* 159, 1–20.

Turk, E., Kerner, C.J., Lostao, M.P. and Wright, E.M. (1996) Membrane topology of the human $Na^+$/glucose cotransporter SGLT1. *Journal of Biological Chemistry* 271, 1925–1934.

Uldry, M., Ibberson, M., Horisberger, J.D., Chatton, J.Y., Riederer, B.M. and Thorens, B. (2001) Identification of a mammalian H(+)-myo-inositol symporter expressed predominantly in the brain. *EMBO Journal* 20, 4467–4477.

Utsunomiya-Tate, N., Endou, H. and Kanai, Y. (1996) Cloning and functional characterization of a system ASC-like $Na^+$-dependent neutral amino acid transporter. *Journal of Biological Chemistry* 271, 14883–14890.

Verrey, F., Meier, C., Rossier, G. and Kuhn, L.C. (2000) Glycoprotein-associated amino acid exchangers: broadening the range of transport specificity. *Pflügers Archiv* 440, 503–512.

Wagner, C.A., Lang, F. and Broer, S. (2001) Function and structure of heterodimeric amino acid transporters. *American Journal of Physiology* 281, C1077–C1093.

Walgren, R.A., Lin, J.T., Kinne, R.K. and Walle, T. (2000) Cellular uptake of dietary flavonoid quercetin 4′-beta-glucoside by sodium-dependent glucose transporter SGLT1. *Journal of Pharmacology and Experimental Therapeutics* 294, 837–843.

Wang, H., Kavanaugh, M.P., North, R.A. and Kabat D. (1991) Cell-surface receptor for ecotropic murine retroviruses is a basic amino-acid transporter. *Nature* 352, 729–731.

Wang, J., Schaner, M.E., Thomassen, S., Su, S.F., Piquette-Miller, M. and Giacomini, K.M. (1997) Functional and molecular characteristics of Na(+)-dependent nucleoside transporters. *Pharmaceutical Research* 14, 1524–1532.

Watson, R.T. and Pessin, J.E. (2001) Subcellular compartmentalization and trafficking of the insulin-responsive glucose transporter, GLUT4. *Experimental Cell Research* 271, 75–83.

Wright, E.M. (2001) Renal Na(+)-glucose co-transporters. *American Journal of Physiology* 280, F10–F18.

Yamamoto, N., Ikeda, H., Tandon, N.N., Herman, J., Tomiyama, Y., Mitani, T., Sekiguchi, S., Lipsky, R., Kralisz, U. and Jamieson, G.A. (1990) A platelet membrane glycoprotein (GP) deficiency in healthy blood donors: Naka-platelets lack detectable GPIV (CD36). *Blood* 76, 1698–1703.

Yamashita, T., Shimada, S., Guo, W., Sato, K., Kohmura, E., Hayakawa, T., Takagi, T. and Tohyama, M. (1997) Cloning and functional expression of a brain peptide/histidine transporter. *Journal of Biological Chemistry* 272, 10205–10211.

Zuniga, F.A., Shi, G., Haller, J.F., Rubashkin, A., Flynn, D.R., Iserovich, P. and Fischbarg, J. (2001) A three-dimensional model of the human facilitative glucose transporter Glut1. *Journal of Biological Chemistry* 276, 44970–44975.

# 4 维生素及其生理活性形式的细胞内运输和区室化

Donald B. McCormick

（艾莫瑞大学医学院生物化学系，亚特兰大，佐治亚，美国）

## 概述和重点

各种各样的细胞构成了许多高级生物体，尤其是像人这样的复杂哺乳动物，大量的养分需供给于这些细胞。养分供给的细胞内运输和区室化的有关知识不可能在单一章节中做详细阐述，常量养分和微量养分在细胞内转运、利用的分子过程及途径将继续在有关生物化学和营养学的教科书中进行总结。目前适合于个人方面的书籍有：Devlin 主编的《生物化学与临床相关的教科书》（Wiley-Liss，纽约）和 Shils 等主编的《健康与疾病中的现代营养学》（Williams and Wilkias，巴尔迪摩）。更新的相关知识可参阅《生物化学年鉴》、《营养年鉴》和《营养的最新知识》简编以及研究期刊中的综述性文章。

本章的目的，至少将限于系统介绍维生素的主要研究进展。细胞内的运输和区室化显然一直没能引起人们特别的关注。在书和论文中仅有一些有关维生素的零散报道，如《维生素手册》第三版（Marcel Dekker，纽约），就对这些必需微量养分进行了一些介绍。

由于传统观点认为维生素通常经代谢转化为具有生理活性的形式如辅酶或激素后才能发挥其功能，因此本章对维生素的这些转化和随后的区室化方面的内容进行叙述是有必要的。为了连贯性，本章有理由对特定维生素及其功能形式细分下的每一方面进行叙述，其内容包括维生素经真核细胞浆膜上的摄入、随后在细胞质和细胞器（线粒体、内质网、溶酶体、细胞核等）中的分布及相关的代谢变化。

## 维 生 素 A

食物中维生素 A 前体（即带 $\beta$-白芷酮环的类胡萝卜素 A 前体）一旦在肠黏膜细胞上摄入，就在加氧酶催化作用下 15,15′-双键断裂形成视黄醛，或偏裂成 $\beta$-脱类胡萝卜醛，然后再进一步氧化成视黄醛（Olson，2001）。烟酰胺腺嘌呤二核苷酸（NAD）-依赖性脱氢酶能把视黄醛还原成视黄醇，与食物中的视黄醇一起大部分被转化成视黄酯。视黄酯再形成乳糜微滴，释放进入淋巴循环。肝对视黄酯的摄取导致酯催化的水解，然后释放出的视黄醇与细胞质中的视黄醇结合蛋白结合形成复合物。在肝实质性细胞中，一些全反式视黄醇与特异性蛋白（retinoid binding protein，RBP）结合，该特异蛋白与运甲状腺素蛋白一起被分泌到血浆中，以运往其他组织的细胞中。在细胞质和细胞核中，视黄醇及其在细胞质中被氧化成视黄醛和视黄酸的部分可与各种不同的细胞质类维生素 A 结合蛋白结合。这些类维生素 A 结合蛋白列于表 4.1。

**表 4.1 细胞类维生素 A 结合蛋白**

| 名称（缩写） | 主要配体 |
|---|---|
| 细胞视黄醇结合蛋白，Ⅰ型（CRBPⅠ） | 全反式视黄醇 |
| 细胞视黄醇结合蛋白，Ⅱ型（CRBPⅡ） | 全反式视黄醇 |
| 细胞视黄酸结合蛋白，Ⅰ型（CRABPⅠ） | 全反式视黄酸 |
| 细胞视黄酸结合蛋白，Ⅱ型（CRABPⅡ） | 全反式视黄酸 |
| 附睾视黄酸结合蛋白（ERABP） | 全反式和 9-顺式视黄酸 |
| 细胞视黄醛结合蛋白（CRALBP） | 11-顺式视黄醇和视黄醛 |
| 光感受器间视黄醇结合蛋白（IRBP） | 11-顺式视黄醛和全反式视黄醇 |

人们对这些结合蛋白的特性和功能已进行了综述（Ong *et al.*，1994；Saari，1994；Newcomer，1995；Li and Norris，1996）。细胞视黄醇结合蛋白Ⅰ（cellular retinol binding protein Ⅰ，CRBPⅠ）和细胞视黄醇结合蛋白Ⅱ及细胞视黄酸结合蛋白Ⅰ（cellular retinoic acid binding protein Ⅰ，CRABPⅠ）和细胞视黄酸结合蛋白Ⅱ是承载全反式异构体的相当小的蛋白质（15.0～15.7kDa）。在小肠细胞中，CRBPⅠ和 CRBPⅡ选择性地使视黄醇在卵磷脂：视黄醇-酰基转移酶（lecithin：retinol acyltransferase，LRAT）与乙酰基-CoA：视黄醇-酰基转移酶（Acyl-CoA：retinol acyltransferase，ARAT）催化作用下发生酯化。CRBPⅠ分布于许多组织，但 CRBPⅡ仅位于成年动物的皮肤上。CRBPⅠ和 CRBPⅡ都受类维生素 A 营养状况的影响（Kato，1985）。在类维生素 A 缺乏的小鼠小肠中，CRBPⅠ mRNA 降低，而 CRBP Ⅱ mRNA 提高。细胞视黄醛结合蛋白（cellular retinaldehyde-binding protein，CRALBP）（36kDa）和光感受器间视黄醇结合蛋白（interphotoreceptor retinol-binding protein，IRBP）（135kDa）作为视网膜细胞中 11-顺式异构体的转运蛋白而发挥功能。附睾视黄酸结合蛋白（epididymal retinoic acid-binding protein，ERABP）（18.5kDa）是睾丸细胞中全反式和 9-顺式视黄酸的载体。CRBP 通过调节细胞内类维生素 A 的代谢和影响核内受体的配体占有来影响类维生素 A 信号通路。

视黄酸（全反式和 9-顺式）和其他类维生素 A 一进入细胞核内，就与 α、β、γ 视黄酸受体（retinoic acid receptor，RAR）之中的一种或多种或类维生素 A X 受体（retinoid X receptor，RXR）紧密结合。与细胞核中激素结合的其他受体一样，这些受体含具有特异功能的 6 个蛋白质结构域（从 N 端到 C 端依次被命名为 A～F）。E 结构区与配体结合。对于 RAR 和 RXR 而言，DNA 上的激素反应元件（hormone response element，HRE）是相同的 AGGTCA 序列。许多基因都含有类维生素 A 受体的反应元件，其显著作用包括：通过视黄酸经视黄酸反应元件刺激某些胞液和核内类维生素 A 结合蛋白；刺激胚胎发育过程中的 *Hox a-1*（*Hox 1.6*）和 *Hox b-1*（*Hox 2.9*）起始基因；促进第Ⅰ类醇脱氢酶 3 的活性，从而诱导视黄醇向更多视黄酸转化。

## 维生素 D（钙化醇）

维生素 D（$V_D$）是一种激素原，需要在不同的组织中进行特异的、连续不断的转化（Collins and Norman，2001）。对于维生素 $D_3$（胆钙化醇）而言，主要位于皮肤表皮生发层的7-脱氢胆固醇在自然光照条件下，B环的9,10-键断裂，从而转化形成维生素$D_3$原。在维生素$D_2$（麦角钙化醇）形成过程中，植物性来源的麦角固醇含有一个22，23-Δ-24-甲基侧链，能人为地发生光化学转化，该过程类似于7-脱氢胆固醇转化为维生素$D_3$的过程。人工合成的$D_2$可用于丰富乳和乳产品。天然的和人工合成的维生素D发生羟基化后，能转化为具有激素活性的衍生物。维生素D与血浆中特异蛋白结合，主要转运至肝脏，在此通过微粒体和线粒体中的类P450羟化酶催化而发生25-羟基化作用（Saarem *et al.*，1984）。从肝脏释放出来的25-羟基-D与血浆维生素D结合蛋白形成复合物，再被转运至肾脏，并在近端管状细胞的线粒体中发生1α-羟基化。25-羟基-D 1α-羟化酶是一种可利用分子氧的混合功能氧化酶（Henry and Norman，1974）。该酶由均是线粒体膜结构组分的3个蛋白质分子（肾铁氧化还原蛋白、肾铁氧化还原蛋白还原酶和细胞色素P450）组成。调控肾1α-羟化酶是维生素D内分泌系统中最重要的调节点（Henry，1992）。影响活性的主要因素包括产物1α,25-二羟基-D、甲状旁腺激素（parathyroid hormone，PTH）以及血清$Ca^{2+}$和磷酸根的浓度（Henry *et al.*，1992）。除了天然的具有激素活性的二羟基-D外，还有近40种已知的$D_3$代谢产物，它们大部分是无活性的，并随粪排出体外。然而，24R,25-二羟基-$D_3$（也在肾脏中产生）很可能与1α,25-二羟基-$D_3$一起，为维生素D的某些生物反应（如骨骼和蛋壳的钙化）所必需（Collins and Norman，2001）。对于维生素$D_3$的储存情况，不同品种之间存在一些差异，但对人类而言，脂肪组织是$D_3$的主要储存部位，而肌肉储存着明显数量的25-羟基-$D_3$（Mawer *et al.*，1972）。1α,25-二羟基-$D_3$可经过许多途径发生代谢分解，从而使其从生物体内被快速清除（Kumar，1986）。一旦1α,25-二羟基-D被载体蛋白从血浆转运到靶组织时，该激素便与特异的、高亲和力的细胞内受体首先在胞质，随后在胞核中发生相互作用。据报道，超过24种靶组织和细胞含有高亲和力的1α,25-二羟基-$D_3$受体（Collins and Norman，2001）。该激素的核内受体（VDR）首先在维生素D缺乏的鸡的小肠中发现，表明是50kDa大小的DNA结合蛋白，属于同源性核受体的一个超家族。对维生素A核受体的情况而言，朝向VDR C端的E区与二羟基-D结合。作为第Ⅱ类受体，VDR（同RAR和RXR）能与其他受体结合形成异二聚体，从而促进生理效应的多样性。

核受体-激素复合物被激活后，与DNA上的HRE结合，以调节激素敏感基因的表达。基因转录调节可诱导或抑制特异mRNA，从而使最终反应生物学效应的蛋白质表达发生变化。已知有50多种基因受1α,25-二羟基-$D_3$的调节（Hannah and Norman，1994）。其中肠或肾组织中在mRNA水平上变化的一些基因列于表4.2中。

钙结合蛋白是受维生素D正调节的蛋白质之一。对钙结合蛋白的基因组诱导是1α,25-二羟基-$D_3$的主要作用之一。此外，在慢速的激素反应之前，还存在二羟基-D的非基因组作用。由二羟基-D介导的$Ca^{2+}$快速转运被称之为“转钙作用”（transcaltachia）（Nemere and Norman，1987）。在肠中，该过程很可能涉及刷状缘膜上内吞泡中$Ca^{2+}$的内在化，然后与溶酶体融合，并沿着微管移到基底侧膜进行胞外分泌。二羟基-D的其他非基因组作用可能包括磷酸肌醇的降解。现在清楚的是，维生素D主要通过1α,25-二羟基代谢物的作用而发挥广泛作用，且其核心作用在于促进肠对$Ca^{2+}$的吸收和骨的钙化。

**表 4.2　肠和肾脏中受 1α，25-二羟基-$D_3$ 调节的一些基因**

| 基因 | 组织 | 调控 |
|---|---|---|
| α-微管蛋白 | 肠 | 下调 |
| 醛缩酶亚基 B | 肾脏 | 上调 |
| 碱性磷酸酶 | 肠 | 上调 |
| ATP 合成酶 | 肠、肾脏 | 上、下调 |
| 钙结合蛋白 9K 和 28K | 肠、肾脏 | 上调 |
| 细胞色素氧化酶亚基Ⅰ、Ⅱ和Ⅲ | 肠、肾脏 | 上、下调 |
| 细胞色素 b | 肾脏 | 下调 |
| 铁氧还蛋白 | 肾脏 | 下调 |
| 1-羟基-D24-羟化酶 | 肾脏 | 上调 |
| 金属硫蛋白 | 肾脏 | 上调 |
| NADH 脱氢酶亚基Ⅰ、Ⅲ和Ⅳ | 肾脏、肠 | 下、上调 |
| 浆膜 $Ca^{2+}$ 泵 | 肠 | 上调 |
| VDR | 肠 | 上调 |

# 维 生 素 E

维生素 E（$V_E$）组的 8 个成员包括 4 种生育酚（α、β、γ 和 δ）和 4 种生育三烯醇（α、β、γ 和 δ），均仅由植物生物合成，且在食用植物油中含量特别高（Sheppard *et al.*，1993）。生物学最活性的形式是 RRR-（过去被称为 d-）α-生育酚，且多数研究均已揭示该形式 $V_E$ 的特性。$V_E$ 在小肠肠腔被吸收是一个被动扩散、非载体介导的、效率相对低的转运过程（Traber and Sies，1996；Chow，2001）。在肠细胞内，$V_E$ 包含在乳糜微粒中，且被分泌入细胞内的空间和淋巴系统。一旦乳糜微粒转变为残余微粒，$V_E$ 便释放进入循环的脂蛋白，并最终被送到各种组织中。肝的实质细胞在乳糜微粒由脂蛋白脂酶部分脂解而产生 apoE 后摄取残余微粒。肝脏把生育酚与极低密度脂蛋白（very low-density lipoprotein，VLDL）一起进行调控和包装。胞液 α-生育酚转运蛋白首先在小鼠肝脏中被发现（Catignani and Bieri，1977），最近又从人肝脏的 cNDA 序列中对其进行了鉴定（Arita *et al.*，1995），结果显示出对转运到 VLDL 上的 RRR-α-生育酚的选择性。纯化的小鼠肝脏蛋白有两种异构体，分子质量为 30～36kDa（Sato *et al.*，1991），均能促进 α-生育酚在膜之间的转运。具有类似作用的人肝脏蛋白分子质量为 36.6kDa（Kuhlenkamp *et al.*，1993）。肝脏和心脏的一种更小的 α-生育酚结合蛋白（14.2kDa）可能也参与了 α-生育酚的细胞内转运和代谢（Gordon *et al.*，1995）。

承载生育酚的 VLDL 从肝脏分泌进入血浆，以被转运到外周组织。尽管具体机制尚不完全清楚，但推测血浆中与 $V_E$ 结合的脂蛋白至少部分通过受体而把生育酚交换到细胞内（Traber and Sies，1996；Chow，2001）。多数组织具有聚积 α-生育酚的能力，但都不作为储存器官。体内许多 $V_E$ 都存在于脂肪组织中，而生育酚主要存在于周转慢的大量脂肪小滴中。$V_E$ 在肌肉、睾丸、脑部和脊髓中的代谢周转也慢。肾上腺中 α-生育酚浓度最高，而肺和脾也含有相对高浓度的 α-生育酚。α-生育酚主要被摄取并储存于实质细胞中（Bjorneboe *et al.*，1991），也有部分 α-生育酚被转运到非实质细胞。实质细胞能储存过量的生育酚，与非实质细胞相比不易被耗竭。在细胞内，光线粒体中 α-生育酚浓度最高，而胞液中的浓度较低。大部分生育酚位于膜上，包括周转快的红细胞膜。约 3/4 的线粒体 α-生育酚位于外膜，1/4 位于内膜。

生育酚一旦产生抗氧化作用，便转化为生育酚色氧自由基（tocopheryl chromanoxy radical，TCR）（Chow，2001）。TCR 可被抗坏血酸和谷胱甘肽这样的生理还原剂再还原为生育酚。部分自由基进一步氧化为 α-生育醌，而该醌能被肝细胞线粒体和微粒体中的 NADPH 依赖性还原酶转化为氢醌（Hayashi *et al.*，1992）。NADPH-细胞色素 P450 还原酶也能催化该生育醌还原为氢醌。该氢醌的侧链在细胞器内发生氧化形成共轭的 α-生育酸后，随尿排出体外。

# 维 生 素 K

天然的维生素 K（$V_K$）组是由植物叶绿醌（$K_1$）和甲萘醌类-7（$K_2$）及侧链类异戊二烯数目不同的其他细菌和动物甲萘醌类组成。$V_{K3}$ 是人工合成的维生素化合物，具有所有 $V_K$ 都共同含有的 2-甲基-1,4-萘醌的基本结构，$V_K$ 的大多数商品形式都是这种 $V_{K3}$ 的水溶性衍生物（Suttie，2001）。肠中，天然 $V_K$ 的吸收一般与饲粮中形成混合微团的其他脂肪和脂溶性物质的吸收类似。在肠细胞中，叶绿醌被包裹于乳糜微粒中，再排出进入淋巴循环。叶绿醌和甲萘醌在组织中的分布随时间而发生显著变化。甲萘醌在全身的分布速度虽比叶绿醌快，但在组织中的存留量却低。摄食植物性饲料的动物，其肝脏含有叶绿醌。大多数种类的动物肝脏中，都发现有烷基链含 6～13 异戊二烯基单位的甲萘醌类的存在。人肝约含 10% 叶绿醌以及各种甲萘醌类混合物（Matschiner，1971）。新生儿的肝脏缺乏甲萘醌（Shearer *et al.*，1988），但随着婴儿年龄的增长，甲萘醌水平增加（Kayata *et al.*，1989）。除肝脏外，$V_K$ 还集中于肾上腺、肺、骨髓、肾和淋巴结中。

在细胞内，叶绿醌集中于代表滑面内质网的高尔基体和微粒体中（Nyquist *et al.*，1971）。甲萘醌类（尤其是 MK-9）主要存在于线粒体中（Reedstrom and Suttie，1995）。初步试验证据表明，$V_K$ 结合的胞液蛋白能促进该维生素在细胞器内的运动（Kight *et al.*，1995）。当机体缺乏 $V_K$ 时，胞液中 $V_K$ 的消耗比膜上的更快（Knauer *et al.*，1976）。

$V_K$ 的代谢变化与对其的利用发生在细胞器膜尤其是内质网中。动物组织不能合成甲基氮菲环，但当 $V_{K3}$ 与肝匀浆物一起培养时，则在其 3 号位上加侧链以形成甲萘醌-4（Martius，1961）。这种活性存在于微粒体中（Dialameh *et al.*，1970）。除 $V_{K3}$ 发生烷基化形成甲萘醌外，叶绿醌的 3-叶绿基（phytyl）链也能被脱烷基然后再烷基化形成甲萘醌-4（Thijssen and Drittig-Reijnders，1994）。在 $V_K$ 的肝脏代谢物中，过去对2,3-环氧化物的认识（Matshiner *et al.*，1970）是理解 $V_K$ 循环的一个重要步骤，而 $V_K$ 循环涉及 $V_K$-依赖的羧化酶（Suttie，2001）。微粒体上 $V_K$ 相关代谢见图 4.1，这里强调 $V_K$ 在一些蛋白质的谷氨酰（Glu）残基形成 γ-羧基谷氨酰（Gla）残基中所发挥的作用。

$V_K$ 的一般醌式物的还原反应由嘧啶核苷酸依赖性的微粒体 K 醌还原酶与对抗凝血

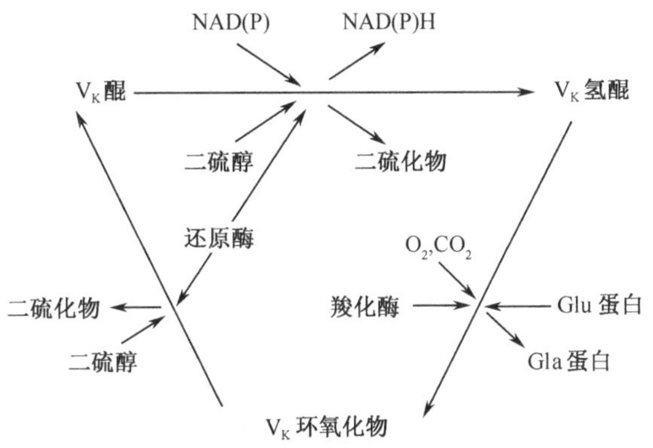


图 4.1　参与 γ-羧基谷氨酰（Gla）残基形成的 $V_K$ 循环

剂和杀鼠灵敏感的二硫醇（dithiol）依赖性还原酶催化。后者还原酶似乎与 $V_K$ 环氧化物还原酶的活性相同（Gardill and Suttie，1990）。研究表明，硫氧还蛋白/硫氧还蛋白还原酶系统就是参与其中的生理相关的二硫醇/二硫化物系统（van Haarlem *et al.*，1987；Silverman and Nandi，1988）。首次表明形成凝血因子合成中的 Gla 残基（Esmon *et al.*，1975）的羧化酶过程位于粗面内质网的腔侧面中（Carlisle and Suttie，1980），并可利用 $CO_2$ 和 $O_2$。这种人的羧化酶已被提纯（Wu *et al.*，1991a）并进行了 cDNA 的克隆和表达（Wu *et al.*，1991b）。Dowd 等（1995）已提出了 $V_K$ 依赖性羧化酶作用的化学机制。该过程中 $O_2$ 添加到 K 氢醌阴离子上，以形成能产生醇盐的二氧环烷（dioxetane）。醇盐被认为是从 Glu 残基的 γ-亚甲基团中夺取氢所必需的强碱基。$V_K$ 依赖的羧化作用不仅发生在凝血因子（如凝血素）形成的肝脏，而且也发生在因在骨钙素（osteocalcin）这样的蛋白质中 Gla 残基的钙结合特性而导致矿化的骨骼组织。在其他组织如肾脏（Griep and Friedman，1980）和精子（Soute *et al.*，1985）中也有报道其他含 Gla 的蛋白质。

## 硫胺素（$B_1$）

硫胺素与大多数水溶性维生素一样，以两种机制进入上皮细胞，即主动或载体介导摄取与被动扩散（Bowman *et al.*，1989）。低于 2μmol/L 时，硫胺素主要在小肠的空肠和回肠段，以不依赖 ATPase 的载体介导的过程而被吸收进入肠细胞（Tanpaichitr，2001）。该载体的性质目前仍不清楚，但已有关于肝脏及鸡蛋蛋白和蛋黄中含硫胺素结合蛋白的报道（Muniyappa *et al.*，1978）。硫胺素进入大多数哺乳动物细胞是一个也与焦磷酸化作用相关的串联过程（Rindi and Laforenza，1997），即一个需要含 ATP 和 $Mg^{2+}$ 的胞液硫胺素焦磷酸激酶的过程。上皮细胞中，大量硫胺素是被（焦）磷酸化的，而到达黏膜的浆膜侧面（serosal side）上的硫胺素则主要是游离的。硫胺素从浆膜侧面的向外转运取决于 $Na^+$ 和细胞浆膜极处（serosal pole）ATPase 的正常功能。血液中的硫胺素许多以硫胺素焦磷酸（thiamine pyrophosphate，TPP）的形式存在于红细胞中，一旦进入血液运输，约一半的硫胺素分布于骨骼肌，其余大部分分布于心脏、肝脏、肾脏、脑和脊髓，而脑和脊髓中硫胺素的含量约为外周神经的 2 倍。白细胞中硫胺素的浓度比红细胞中高 10 倍。在机体的总硫胺素（约 30mg）中，约 80%是 TPP，10%是三磷酸盐（triphosphate，TTP），而其余的是一磷酸盐（monophosphate，TMP）和硫胺素。已知参与磷酸酯形成的三种组织酶分别是合成 TPP 的硫胺素焦磷酸激酶、合成 TTP 的 TPP-ATP 磷酸转移酶和水解 TPP 形成 TMP 的硫胺素焦磷酸酶（Tanphaichitr，2001）。

在哺乳动物细胞中，硫胺素的活性辅酶形式是 TPP，而 TPP 为两类酶系统的辅基。一类是胞液转酮醇酶，在戊糖磷酸途径中具有重要作用，即催化 D-木酮糖 5-磷酸与 D-核糖 5-磷酸形成 D-景天庚酮糖 7-磷酸和 D-甘油醛 3-磷酸，或催化 D-木酮糖 5-磷酸与 D-赤藓糖 4-磷酸形成 D-果糖 6-磷酸和 D-甘油醛 3-磷酸的可逆转酮反应。第二类酶是在线粒体多酶脱氢酶系的 $\alpha$-酮酸脱羧酶亚基中，而该脱氢酶系可把丙酮酸转化为乙酰 CoA，把 $\alpha$-酮戊二酸转化为琥珀酰 CoA，以及把支链氨基酸转化为其他几种代谢产物。在这个多酶系的催化过程中，底物 $\alpha$-酮酸脱羧后，可通过 $\alpha$-羟烷基-TPP 转移到酰基转移酶中心的二氢脂基上，然后释放酰基 CoA，而该被氧化的脂基被与线粒体中 $NAD^+$ 结合的 FAD（flavin adenine dinucleotide）依赖性脱氢酶所还原。

尿中硫胺素的大量分解产物可反映其在细胞内的许多降解情况，包括硫胺素酶催化的该维生素嘧啶和噻唑的裂解、醇脱氢酶催化的噻唑上的 $\beta$-羟乙基的氧化等（McCormick，1988）。硫胺素及其酸代谢产物的损耗相对迅速，其周转速率高约 2 周，在任何组织的任何段时间都很少储存。

## 核 黄 素（$B_2$）

核黄素从多数膳食黄素蛋白在小肠腔中被碱性磷酸酶和焦磷酸酶催化发生裂解而被释放，然后进入肠细胞，在此经核黄素激酶催化的磷酸化作用而发生代谢并被利用（McCormick，1999；Rivlin and Pinto，2001）。许多这样形成的核黄素-5′-磷酸（FMN）进一步转化为 FAD，且这两种黄素辅酶都可发挥代谢功能。核黄素一旦穿出肠腔的浆膜侧面，便与高亲和力或低亲和力的蛋白结合（Whitehouse *et al.*，1991），以转运至机体的各种细胞。在多数细胞中，胞液核黄素激酶和 FAD 合成酶以序分别催化形成黄素辅酶 FMN 和 FAD。这些 ATP 利用酶首先从肝脏中被纯化，激酶最佳活性的发挥需要 $Zn^{2+}$（Merrill and McCormick，1980），合成酶最佳活性的发挥需要 $Mg^{2+}$（Oka and McCormick，1987）。磷酸酶作用于 FMN 和 FAD 焦磷酸酶，但这些水解酶是膜分离的，且参与该维生素的周转代谢及从细胞的释放（McCormick，1975）。黄素辅酶的生物合成受甲状腺素的影响（Rivlin，1970；Lee and McCormick，1985）。

与细胞器膜内的脱辅基蛋白共价结合的 FAD 仅有很小的一部分，却具有非常重要的作用（Yagi *et al.*，1976；Addison and McCormick，1978）。人和其他哺乳动物的线粒体内膜中含琥珀酸脱氢酶，线粒体基质中含肌氨酸和二甲基甘氨酸脱氢酶。在这些酶中，FAD 通过其 8$\alpha$-位点与组氨酰残基的咪唑 N 结合。线粒体外膜中的单胺氧化酶含 8$\alpha$-（*S*-半胱氨酰）FAD。对于能生物合成维生素 C 的哺乳动物，L-古洛糖酸内酯氧化酶含 8$\alpha$-（*N*-组氨酰）FAD，存在于微粒体，特别是肝脏和肾脏的微粒体中。机体还含有大量非共价的黄素辅酶依赖酶，其中一些存在于胞液中，另一些存在于细胞器中。例如，胞液中的FMN依赖性吡哆醇（吡哆胺）5′-磷酸氧化酶、线粒体基质中的β氧化系统的FAD依赖性脂酰CoA脱氢酶及微粒体中含FMN和FAD的NADPH-细胞色素P450还原酶。细胞内主要的黄素辅酶是FAD，约占总黄素的90%。只有约5%的黄素与形成前的脱辅基酶共价结合形成8α-结合FAD。

尿中存在的大量黄素反映出侧链和环甲基功能团的广泛分解代谢以及真皮光化学和小肠微生物对D-核糖醇基链的裂解现象（McCormick，1999）。7-羟甲基和8-羟甲基（7α-羟基和8α-羟基）核黄素、10-甲酸基和10-β-羟乙基黄素、光黄素和光色素均属于核黄素细胞因氧化作用而产生的分解产物（Chastain and McCormick，1991）。

# 烟　　酸

烟酸的两种形式（烟酸和烟酰胺）均可从胃中吸收，但更快而广泛的吸收还是在小肠（Kirkl and Rawling，2001）。这两种形式均能被摄取吸收，但也包括被动扩散。在肠细胞（及红细胞和其他细胞）中，烟酸和烟酰胺均能转化为烟酰胺腺嘌呤二核苷酸（nicotinamide ademine dinucleotide，NAD），前者通过Preiss-Handler途径转化（Preiss and Handler，1958），后者则通过烟酰胺焦磷酸化酶的Dietrich途径（Dietrich *et al.*，1966）而转化，具体步骤见图4.2。这些步骤需要的酶类包括：对烟酸和烟酰胺的磷酸核糖转移酶、对烟酸单核苷酸（nicotinic acid mononucleotide，NaMN）和烟酰胺单核苷（nicotinamide mononucleotide，NMN）的核腺苷酰转移酶、胞液NAD合成酶和仅在高浓度底物时发挥作用的烟酰胺脱氨酶。合成NADP需要胞液NAD激酶。部分烟酸进入血液，烟酰胺也如此，这是在调控细胞内NAD水平中起重要作用的NAD水解酶（glycohydrolase）释放的主要循环形式。肝脏是烟酸的核心加工器官，肝脏和肾脏能把色氨酸转化为喹啉酸，而喹啉酸在这些组织中的喹啉酸磷酸核糖转移酶的催化作用下与磷酸核糖焦磷酸（phosphoribosyl pyrophosphate，PRPP）发生反应，形成NaMN（Moat and Foster，1982）。

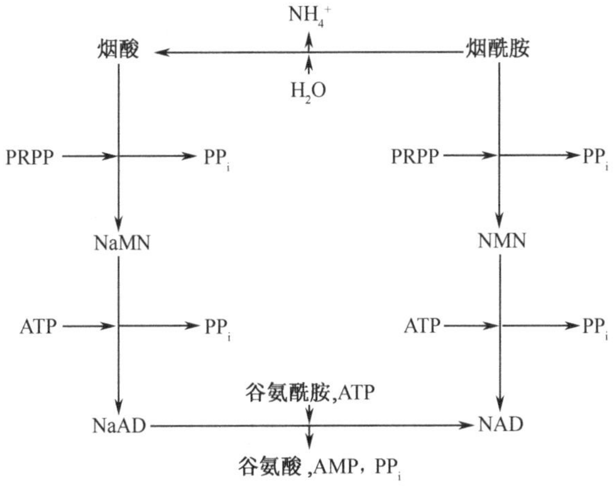


图4.2　烟酸转化为NAD的生物合成步骤

NAD起两个主要作用。其一是吡啶核苷酸和NADP可一起作为氧化还原作用的辅酶，以氢离子运输载体的形式参与许多氧化还原反应。细胞内的NAD大部分是被氧化的，而大部分NADP因磷酸戊糖途径而维持于还原状态，利用NAD或NADP的各种酶位于胞液和细胞器中。NAD的另一重要作用是作为酶系统的底物，尤其是那些与膜结合的、释放烟酰胺和作为单、多聚或环状产物的ADP-核糖的酚类。ADP-核糖化发生在各种系统中，包括受体蛋白的氨基酸残基。细胞核、胞液和浆膜中均发现有ADP-核糖转移酶蛋白。NAD：精氨酸ADP-核糖转移酶通过糖基磷脂酰肌醇被固定在浆膜的外表面（Zolkiewska *et al.*，1994）。多聚（ADP-核糖）由影响DNA修复的多聚酶合成，环状ADP-核糖影响钙离子的转运。而且，NADP（NAADP）的脱酰氨基的代谢产物是一种不同于环状ADP-核糖的有效$Ca^{2+}$-释放因子（Lee，2000）。

肝脏具有一些储存NAD的能力（Kirkl and Rawling，2001），同时也是形成大部分随尿排出的主要甲基化和羟基化分解产物的器官。在人上，烟酰胺被甲基化后，主要产生*N*-甲基烟酰胺，虽然有一部分被氧化产生*N*-甲基-2-吡啶酮-甲酰胺。烟酰胺主要与甘氨酸共轭结合形成烟酰尿酸（nicotinuric acid）。

## 维生素$B_6$

维生素$B_6$（$V_{B_6}$）组由三种维生素构体，即吡哆醇（pyridoxine，PN）、吡哆胺（pyridoxamine，PM）和吡哆醛（pyridoxal，PL）组成。此外，这些构体还有三种5′-磷酸盐形式，即PNP、PMP和PLP。PLP是主要的辅酶，也是多数天然食物中的主要存在形式。5′-磷酸盐可被小肠腔内的碱性磷酸酶水解（Henderson，1985）。与多数水溶性维生素一样，代谢诱捕（metabolic trapping）促进生理浓度下维生素$B_6$的吸收。一旦进入肠细胞和其他细胞，所有三种维生素构体均被磷酸化（McCormick，2001）。肝脏是承担维生素$B_6$代谢及其以后分布的主要器官，其主要以白蛋白结合的PLP形式分布。胞液中各种形式的维生素$B_6$之间的相互转化需要真核细胞中$Zn^{2+}$-ATP存在时活性最高的吡哆醛磷酸激酶（McCormick *et al.*，1961）和维生素$B_6$（吡哆胺）5′-磷酸氧化酶（Kazarinoff and McCormick，1975）催化的维生素$B_6$构体的磷酸化作用。而维生素$B_6$（吡哆胺）5′-磷酸氧化酶因其具FMN依赖性而对核黄素状态敏感（McCormick，1989）。已在人肝脏中测得这些酶活性（Merril *et al.*，1984），形成的PLP被捕获在肝细胞内，其原因不仅是磷酸基团带的负电荷，也是其醛基与内在蛋白质发生反应所致，并能产生可逆的西佛碱（Schiff base）（Li *et al.*，1974）。PLP经浆膜结合的磷酸酯酶水解后才能从细胞中被释放出来（Merrill and Henderson，1990）。维生素$B_6$及其代谢产物广泛分布在各组织中（McCormick，2001），但许多以PLP的形式分布于骨骼肌中（Coburn *et al.*，1991），且相当一部分是与磷酸酯酶结合的（Black *et al.*，1978）。值得注意的是，富含吡哆醛激酶的脑（McCormick and Snell，1959）在谷氨酰胺代谢和神经系统活性胺的形成中均需要PLP。

需要PLP的酶分布在胞液和细胞器中。辅酶因其与脱辅基酶蛋白结合而在胞液中形成。一部分脱辅基酶蛋白留在胞液中，而另一部分则转移到特异的细胞器中。PLP依赖性酶的细胞内定位一般涉及一序列片段，它能“决定”全酶最后坐落的位点。前线粒体形式的天门冬氨酸转氨酶 N 端序列与热激蛋白 70（Hsp70）伴侣分子相互作用，把该酶引入线粒体内（Artigue *et al.*，2000），而该酶的胞液存在形式则缺少这个序列。随着人的鸟氨酸脱羧酶被移动到浆膜，Ser167 被包围在受磷酸化作用调控的、p47phox 相关的、膜定位的基团内（Heiskala *et al.*，2000）。有关 PLP 的非酶功能，人们已对可能参与甾类激素调节和基因表达的报道进行了综述（Leklem，2001）。

人的维生素 $B_6$ 的分解代谢主要是由醛氧化酶催化的吡多醛向 4-吡哆酸的氧化。4-吡哆酸是随尿排出的一种代谢终产物。

# 泛 酸

据报道，细胞对泛酸的摄取依靠于 $Na^+$ 的协同转运（Fenstermacher and Rose，1986）。泛酸激酶能催化泛酸发生磷酸化，是 CoA 生物合成中的主要调节位点（Robishaw *et al.*，1982）。人和其他动物中具有辅酶功能的 4′-磷酸泛酸（phosphopantheine）生物合成的主要过程包括泛酸 4′-磷酸与半胱氨酸发生 ATP 依赖性缩合反应形成 4′-磷酸泛酰半胱氨酸，然后被 PLP 酶催化脱羧产生 4′-磷酸泛酸（Browm，1959）。4′-磷酸泛酸的用途之一是与线粒体的脂酰载体蛋白和脂肪酸合成酶结合；另一用途是通过添加 ATP 的 AMP 部分而转化成为 CoA，然后核糖-3′-羟基发生磷酸化。合成 CoA 所必需的所有酶均已从胞液中被分离，然而，由于 95％的 CoA 都在线粒体中，所以线粒体一定也含有类似的酶系统。但是，CoA 本身却不能穿过线粒体膜（Plesofsky，2001）。

泛酸的生化功能包括参与线粒体外膜中脂肪酸的合成及以酰基 CoA 形式在胞液、网状内皮组织系统、线粒体和细胞核中发挥多种功能。许多代谢过程都需要利用酰基 CoA 底物来变成以硫酯形式仍能携带 CoA 的产物，如线粒体内脂酰 CoA 的 $\beta$ 氧化过程。从 CoA 酯中转移酰基都需要蛋白质 N 端进行乙酰基、十四烷基或棕榈酰基修饰，这种类型的修饰影响上述蛋白质，包括具有信号转导作用的蛋白质的定位和活性。据估计，一半以上的可溶性真核蛋白被乙酰化（Driessen *et al.*，1985），核内组蛋白的乙酰化显著影响其作用。对于由十四烷基转移酶催化的蛋白质的十四烷基化作用（Towler *et al.*，1987），最早处于第二位置的 N 端甘氨酰残基是绝对必需的，且具有优先选择小的中性的近端氨基酸残基和第六位点的丝氨酰残基的特性。十四烷基化作用与蛋白质翻译同步进行，是不可逆的，并导致其与其他蛋白质的膜之间的结合力或相互作用减弱。棕榈酰化作用特异性低，发生在翻译之后，与半胱氨酰或丝氨酰残基结合形成一个可逆的酯键，这主要发生于与膜紧密结合的蛋白质。许多膜受体，如视紫质，都能被棕榈酰化。

CoA 经多步骤水解产生泛酸，最后一步是将泛酰巯基乙胺转化为巯基乙胺和泛酸，泛酸随尿排出体外。

# 生 物 素

生物素转运蛋白存在于肠道刷状缘膜（Mock，2001）上。多数研究指出，生物素转运蛋白对生物素的戊酸链和酰脲基具有特异性，且在电中性转运过程中与 $Na^+$ 偶联比为 1∶1。生物素穿过基底侧膜从肠细胞流出的过程也是由载体介导的，但不依赖于 $Na^+$，并且是生电的（Said *et al.*，1988）。尽管生物素在大鼠小肠近端的转运最为活跃，但结肠近端对生物素的吸收也很明显（Bowman and Rosenberg，1987）。人对生物素的吸收情况很可能也如此（Sorrell *et al.*，1971）。肝细胞（Bowers-Komro and McCormick，1985）和外周组织中生物素的摄取特点可能与小肠类似。

细胞内的生物素分解明显，并用于羧化酶，也可能用于其他蛋白质的特异修饰。在羧化全酶合成酶的作用下，生物素参与四种哺乳动物生物素依赖性羧化酶的组成，而细胞质和线粒体内均含有羧化全酶合成酶。该合成酶催化 ATP-依赖性的生物素酰 5′-腺苷酸的形成，然后生物素酰 5′-腺苷酸再与脱羧酶中特异识别的赖氨酰残基上的 ε 氨基缩合，以形成羧化全酶的氨基连接的生物胞素酰辅基。线粒体含三种生物素依赖的羧化酶，分别催化丙酮酸、甲基丁烯酰 CoA 和丙酰 CoA 而形成草酰乙酸、甲基戊烯二酰 CoA 和 D-甲基丙二酰 CoA。乙酰 CoA 羧化酶存在于线粒体和胞液中，有两种形式的线粒体乙酰 CoA 羧化酶的催化活性比胞液形式的小。胞液乙酰 CoA 羧化酶在脂肪酸生物合成的步骤中催化丙二酰 CoA 的形成。在所有情况下，这些羧化酶都需要 $HCO_3^-$ 和 $Mg^{2+}$ · ATP 来形成过渡的碳酰磷酸，以使附着于酶上的生物素的酰脲 N 发生羧化，然后再使激活的底物羧化。

大部分生物素进行的分解代谢形式包括戊酸侧链的 β 氧化和主要产生亚砜的 S 氧化（McCormick and Wright，1970；McCormick，1976）。双降生物素及其亚砜和砜、双降生物素甲基酮和四降生物素亚砜等分解代谢产物均已在人尿中被鉴别（Zempleni *et al.*，1996）。人奶中相当部分生物素相关的化合物也是生物素分解代谢产物（Stratton *et al.*，1996）。

## 叶　　酸

叶酸是能被还原和多聚谷氨酰化，且能在 $N^5$ 和 $N^{10}$ 位点进行一碳单位替换的天然叶酸的化学总称。游离叶酸只有在氧化和去双键后才能被吸收，且主要是在空肠中通过饱和的载体介导的方式被吸收（Brody and Shane，2001）。由还原的叶酸载体基因编码的小肠转运蛋白是一种跨膜蛋白，分布于大多数组织中（Moscow *et al.*，1995；Said *et al.*，1996），但对不同叶酸的特异性在不同组织及其顶膜和基底侧膜间存在差异。叶酸结合蛋白（有时也称为叶酸受体）也存在于小肠中。已发现介导肾脏和其他组织的细胞中叶酸的内吞作用（Said *et al.*，1996）。叶酸结合蛋白和膜转运蛋白位于一些极化细胞的膜两侧（Chancy *et al.*，2000）。在穿过肠黏膜的过程中，大部分叶酸被代谢成 5-甲基-四氢叶酸，通过还原型叶酸载体而在肝中摄取的叶酸代谢为还原型多聚谷氨酸（polyglutamate）并被保存下来。胞液利用 NADPH 的二氢叶酸还原酶使叶酸经二氢水平还原为四氢叶酸形式，然后利用 $Mg^{2+}$-ATP 的叶酸多聚谷氨酸（folypolyglutamate）合成酶将四氢叶酸转化成多聚谷氨酸。对于大多数叶酸依赖性的酶而言，多聚谷氨酸是比单谷氨酸更有效的底物，通过延长为三谷氨酸链而获得主要动力学特点（Shane，1989），对于参与蛋氨酸再合成循环的酶类则需要更长的链。在人类细胞中发现有六聚和七聚谷氨酸及更长链的衍生物（Foo *et al.*，1982）。在胞液和线粒体中还含有叶酸多聚谷氨酸（folypolyglutamate）合成酶的同工酶。这两种形式都是必不可少的，可提供表 4.3 中所列酶的辅酶还原型叶酸寡聚-γ-谷氨酸。

**表 4.3　参与叶酸依赖性代谢和相互转化的酶类的区室化**

| 酶 | 酶 |
|---|---|
| 细胞液 | 10-甲酰-四氢叶酸脱氢酶 |
| 二氢叶酸还原酶 | 细胞质 |
| 丝氨酸羟甲基转移酶 | 叶酸-多聚谷氨酸合成酶 |
| 5,10-亚甲基-氢叶酸脱氢酶、环化水解酶、10-甲酰-四氢叶酸合成酶[1] | γ-谷氨酰水解酶 |
| 甘氨基核苷酸转甲酰酶 | 线粒体 |
| 氨基咪唑羧基酰胺核苷酸转甲酰酶 | 甲酰-四氢叶酸合成酶 |
| 胸苷酸合成酶 | 甲酰-四氢叶酸脱氢酶 |
| 5,10-亚甲基四氢叶酸还原酶 | 次甲基-四氢叶酸脱氢酶 |
| 蛋氨酸合成酶 | 次甲基-四氢叶酸合成酶 |
| 四氢叶酸亚胺甲基转移酶，亚胺甲基四氢叶酸环化脱氨酶[2] | 丝氨酸羟甲基转移酶 |
| | 甘氨酸裂解酶复合物 |
| | 二甲基甘氨酸脱氢酶 |
| 谷氨酸转甲酰酶 | 肌氨酸脱氢酶 |
| 5-甲酰-四氢叶酸异构酶 | 甲硫氨酰-tRNA 转甲酰酶 |

1 一种具有三种功能的酶；2 一种具有两种功能的酶。

合成蛋氨酸、胸苷酸和嘌呤的主要途径位于胞液中，但线粒体中的叶酸代谢在甘氨酸代谢及为胞液一碳代谢提供一碳单位中起重要作用（Brody and Shane，2001）。参与叶酸代谢的许多酶都是多功能的，或是多种蛋白质复合物的组成部分，这就使得多聚谷氨酸中间代谢产物在活性部位间形成通道化，而不从复合物中释放中间代谢产物（Paquin *et al.*，1985）。

叶酸主要以 *N*-乙酰-ρ-氨基苯甲酰谷氨酸（Murphy *et al.*，1976）和少量 ρ-氨基苯甲酰谷氨酸的形式由尿中排出。5-或 10-甲酰四氢叶酸多聚谷氨酸积累时，叶酸分解代谢加强，反映叶酸被分解为蝶呤衍生物的重链铁蛋白已被分离（Stover，2001，个人交流）。

## 维 生 素 $B_{12}$（钴胺素）

维生素 $B_{12}$（$V_{B12}$）是天然的羟钴胺素或商品的氰钴胺素，其中氰基在细胞加工过程中被替换。维生素 $B_{12}$与内因子结合形成复合物后维生素 $B_{12}$迅速附着于表面受体如 cubilin（Kozyraki *et al.*，1998），然后以缓慢的、耗能的形式进入小肠细胞（Hines *et al.*，1968）。一旦进入，钴胺素即与免疫上类似于内因子的大分子结合（Rothenberg *et al.*，1972）。经门脉运输后，与 R 型结合蛋白结合的钴胺素被肝细胞加工，然后再进入血浆或随胆汁排出（Burger *et al.*，1975）。在哺乳动物组织中，仅有两种已知的维生素 $B_{12}$辅酶形式。一种是 5′-脱氧腺苷钴胺素，其合成过程是：含黄素蛋白的合成酶使维生

素 $B_{12a}$ 经维生素 $B_{12r}$ 还原为维生素 $B_{12s}$，然后与 ATP 反应形成钴胺素辅酶并释放三磷酸。5′-脱氧腺苷钴胺素的功能是与线粒体的 L-甲基丙二酰 CoA 突变酶一起把其底物转化成琥珀酰 CoA。维生素 $B_{12}$ 的另一种辅酶样功能是作为甲基钴胺素参与胞液蛋氨酸合成酶的构成，而蛋氨酸合成酶是一种甲基四氢叶酸-高胱氨酸甲基转移酶。维生素 $B_{12}$ 与叶酸之间的这种重要关系在临床上表现为维生素 $B_{12}$ 不足时的“甲基陷阱假说”（methyl trap hypothesis）（Beck，2001）。蛋氨酸合成酶的阻抑导致叶酸在 5-甲基-THF 水平上的蓄积。由于在没有蛋氨酸合成酶反应周转的情况下 5-甲基-THF 不能释放其甲基，因而叶酸被“截留”，其他叶酸依赖性反应的水平也减少。

## 维生素 C（L-抗坏血酸）

小肠细胞对 L-抗坏血酸的吸收也与其他水溶性维生素一样（Bowman *et al.*，1989；McCormick and Zhang，1992）。维生素 C（$V_C$）经过快速的门脉转运，广泛分布于身体的各种组织中。$V_C$ 一般作为氧化剂，也特异性作为含 $Fe^{2+}$ 和 $Cu^{2+}$ 氧化酶的氧化还原辅助因子。L-抗坏血酸与酶之间的关系很重要，其功能见表 4.4。

**表 4.4　与抗坏血酸功能相关的酶活性**

| 酶 | 功能 |
|---|---|
| 含 $Fe^{2+}$ 的加双氧酶 | |
| 脯氨酰-3-羟化酶 | 细胞外基质成熟 |
| 脯氨酰-4-羟化酶 | 细胞外基质成熟 |
| 赖氨酰羟化酶 | 细胞外基质成熟 |
| 6-*N*-三甲基-L-赖氨酸羟化酶 | 肉毒碱的生物合成 |
| γ-丁内铵羟化酶 | 肉毒碱的生物合成 |
| 酪氨酸-4-羟基苯丙酮酸羟化酶 | 酪氨酸代谢 |
| 含 $Cu^{2+}$ 的单氧酶 | |
| 多巴胺-β-单氧酶或羟化酶 | 去甲肾上腺素的生物合成 |
| 肽基甘氨酸-α-酰胺单氧酶 | 激素和激素释放因子的活化 |

L-抗坏血酸的代谢包括向 2-*O*-硫酸盐的转化和甲基化以及水解和氧化，以形成酸性物质作为分解代谢产物随尿排出体外。

## 辅酶的区室化

由于多数维生素都转化为极性更大、通常为阴离子的功能形式的辅酶，且这些带电的物质不易渗透过细胞器膜，因此，辅酶往往通过特殊的途径进入细胞器。正如前面对维生素的叙述一样，结合蛋白（有时是脱辅基酶蛋白）能与细胞质中活化的维生素（辅酶或激素）结合，并转移到细胞器膜，再在细胞膜上的受体或伴侣分子的协助下进入细胞器。例如，活化的维生素 A 和维生素 D 进入胞核内的基因组物质，或维生素 $B_6$ 作为 PLP-依赖性转氨酶全酶（holotransaminase）的辅酶进入线粒体。在其他情况下，如

NAD（H）、NADP（H）、FAD（$H_2$）和乙酰 CoA，通过线粒体内膜进入则需要如“穿梭”机制这样的装置。这些操作所必要的条件是：适当的酶位于膜的正确一侧，且适当的转运蛋白或转移酶位于膜上（中）以穿梭运输各种中间产物。如参与线粒体内外 NAD/NADH 氧化还原周转的苹果酸-天冬氨酸穿梭，就是通过分别定位于膜两侧的苹果酸脱氢酶和天冬氨酸氨基转移酶，以及膜转运蛋白使苹果酸和谷氨酸跨膜穿梭而实现的。α-甘油磷酸的穿梭与之类似，其中 FAD-脱氢酶位于膜内。乙酰 CoA 主要由线粒体内的丙酮酸脱氢酶系产生，还需要旁路机制把乙酰 CoA 等价转移到细胞质中。这可由柠檬酸提供，而柠檬酸是可渗透的，且一旦穿出线粒体，就能由 ATP-和 CoA-利用的柠檬素裂解酶裂解形成乙酰 CoA。最后，长链脂肪酸的酰基进入线粒体进行 β 氧化，这取决于肉碱酰基转移酶及酰基由 CoA 酯向有载体肉碱的 CoA 酯的转移。

## 参 考 文 献


Addison, R. and McCormick, D.B. (1978) Biogenesis of flavoprotein and cytochrome components in hepatic mitochondria from riboflavin-deficient rats. *Biochemical and Biophysical Research Communications* 81, 133–138.

Arita, M., Sato, Y., Miyata, A., Tanabe, T., Takahashi, E., Kayden, H.J., Arai, H. and Inoue, K. (1995) Human alpha-tocopherol transfer protein: cDNA cloning, expression and chromosomal localization. *Biochemical Journal* 306, 437–443.

Artigues, A., Bengoechea-Alonso, M.T., Crawford, D.L., Iriarte, A. and Martinez-Carrion, M. (2000) Biological implications of the different Hsp70 binding properties of mitochondrial and cytosolic aspartate aminotransferase. In: Iriarte, A., Kagan, H. and Martinez-Carrion, M. (eds) *Biochemistry and Molecular Biology of Vitamin $B_6$ and PQQ-dependent Proteins.* Birkhauser Verlag, Basel, pp. 111–116.

Beck, W.S. (2001) Cobalamin (vitamin $B_{12}$). In: Rucker, R.B., Suttie, J.W., McCormick, D.B. and Machlin, L.J. (eds) *Handbook of Vitamins*, 3rd edn. Marcel Dekker, New York, pp. 463–512.

Bjorneboe, A., Nenseter, M.S., Hagen, B.F., Bjorneboe, G.-E.A., Prydz, K. and Drevon, C.A. (1991) Effect of dietary deficiency and supplementation with all-rac-α-tocopherol in hepatic content in rat. *Journal of Nutrition* 121, 1208–1213.

Black, A.L., Guirard, B.M. and Snell, E.E. (1978) The behavior of muscle phosphorylase as a reservoir for vitamin $B_6$ in the rat. *Journal of Nutrition* 108, 670–677.

Bowers-Komro, D.M. and McCormick, D.B. (1985) Biotin uptake by isolated rat liver hepatocytes. *Annals of the New York Academy of Sciences* 447, 350–358.

Bowman, B.B. and Rosenberg, I. (1987) Biotin absorption by distal rat intestine. *Journal of Nutrition* 117, 2121–2126.

Bowman, B.B., McCormick, D.B. and Rosenberg, I.H. (1989) Epithelial transport of water-soluble vitamins. *Annual Review of Nutrition* 9, 187–199.

Brody, T. and Shane, B. (2001) Folic acid. In: Rucker, R.B., Suttie, J.W., McCormick, D.B. and Machlin, L.J. (eds) *Handbook of Vitamins*, 3rd edn. Marcel Dekker, New York, pp. 427–462.

Brown, G.E. (1959) The metabolism of pantothenic acid. *Journal of Biological Chemistry* 234, 370–378.

Burger, R.L., Schneider, C.S., Mehlman, C.S. and Allen, R.H. (1975) Human plasma R-type vitamin $B_{12}$-binding protein in the plasma transport of vitamin $B_{12}$. *Journal of Biological Chemistry* 250, 7707–7713.

Carlisle, T.L. and Suttie, J.W. (1980) Vitamin K dependent carboxylase: subcellular location of the carboxylase and enzymes involved in vitamin K metabolism in rat liver. *Biochemistry* 19, 1161–1167.

Catignani, G.L. and Bieri, J.G. (1977) Rat liver α-tocopherol binding protein. *Biochimica et Biophysica Acta* 497, 349–357.

Chancy, C.D., Kekuda, R. Huang, W., Prasad, P.D., Kuhnel, J., Sirotnak, F.M., Roon, P., Ganapathy, V. and Smith, S.B. (2000) Expression and differential polarization of the reduced-folate transporter I and the folate receptor alpha in mammalian retinal pigment epithelium. *Journal of Biological Chemistry* 275, 20676–20684.

Chastain, J.L. and McCormick, D.B. (1991) Flavin metabolites. In: Muller, F. (ed.) *Chemistry and Biochemistry of Flavins.* CRC Press, Boca Raton, Florida, pp. 195–200.

Chow, C.K. (2001) Vitamin E. In: Rucker, R.B., Suttie, J.W., McCormick, D.B. and Machlin, L.J. (eds) *Handbook of Vitamins*, 3rd edn. Marcel Dekker, New York, pp. 165–197.

Coburn, S.P., Ziegler, P.J., Costill, D.L., Mahuren, J.D., Fink, W.J., Schaltenbrand, W.E., Pauly, T.A., Pearson, D.R., Conn, P.J. and Guilarte, T.R. (1991) Response of vitamin $B_6$ content of muscle to changes in vitamin $B_6$ intake in men. *American Journal of Clinical Nutrition* 53, 1436–1442.

Collins, E.D. and Norman, A.W. (2001) Vitamin D. In: Rucker, R.B., Suttie, J.W., McCormick, D.B. and Machlin, L.J. (eds) *Handbook of Vitamins*, 3rd edn. Marcel Dekker, New York, pp. 51–113.

Dialameh, G.H., Yekundi, K.G. and Olson, R.E. (1970) Enzymatic akylation of menaquinone-O to menaquinones by microsomes from chick liver. *Biochimica et Biophysica Acta* 223, 332–338.

Dietrich, L.S., Fuller, L., Yero, I.L. and Martinez, L. (1966) Nicotinamide mononucleotide pyrophosphorylase activity in animal tissues. *Journal of Biological Chemistry* 241, 188–191.

Dowd, P., Ham, S.W., Naganathan, S. and Hershline, R. (1995) The mechanism of action of vitamin K. *Annual Review of Nutrition* 15, 419–440.

Driessen, H.P.C., de Jong, W.W., Tesser, G.I. and Bloemendal, H. (1985) The mechanism of N-terminal acetylation of proteins. *CRC Critical Reviews in Biochemistry* 18, 281–325.

Esmon, C.T., Sadowski, J.A. and Suttie, J.W. (1975) A new carboxylation reaction. The vitamin K-dependent incorporation of $H^{14}CO_3^-$ into prothrombin. *Journal of Biological Chemistry* 250, 4744–4748.

Fenstermacher, D.K. and Rose, R.C. (1986) Absorption of pantothenic in rat and chick intestine. *American Journal of Physiology* 250, G155.

Foo, S.K., McSloy, R.M., Rousseau, C. and Shane, B. (1982) Folate derivatives in human cells: studies on normal and 5,10-methylenetetrahydrofolate reductase-deficient fibroblasts. *Journal of Nutrition* 112, 1600–1608.

Gardill, S.L. and Suttie, J.W. (1990) Vitamin K epoxide and quinone reductase: evidence for reduction by a common enzyme. *Biochemical Pharmacology* 40, 1055–1061.

Gordon, M.J., Campbell, F.M., Duthie, G.G. and Dutta-Roy, A.K. (1995) Characterization of a novel alpha-tocopherol-binding protein from bovine heart cytosol. *Archives of Biochemistry and Biophysics* 318, 140–146.

Griep, A.E. and Friedman, P.A. (1980) Purification of a protein containing γ-carboxyglutamic acid from bovine kidney. In: Suttie, J.W. (ed.) *Vitamin K Metabolism and Vitamin K-dependent Proteins*. University Park Press, Baltimore, Maryland, pp. 307–310.

Hannah, S.S. and Norman, A.W. (1994) 1α,25-Dihydroxyvitamin $D_3$ regulated expression of the eukaryotic genome. *Nutrition Reviews* 52, 376–382.

Hayashi, T., Kanetoshi, A., Nakamura, M., Tamura, M. and Shirahama, H. (1992) Reduction of alpha-tocopherolquinone to alpha-tocopherolhydroquinone in rat hepatocytes. *Biochemical Pharmacology* 44, 489–493.

Heiskala, M., Zhang, J., Hayashi, S.-I., Holtta, E. and Anderson, L.C. (2000) Activation and transformation of cells induce translocation of ornithine decarboxylase (ODC) to the surface membrane. In: Iriarte, A., Kagan, H. and Martinez-Carrion, M. (eds) *Biochemistry and Molecular Biology of Vitamin $B_6$ and PQQ-dependent Proteins*. Birkhauser, Basel, pp. 227–232.

Henderson, L.M. (1985) Intestinal absorption of $B_6$ vitamers. In: Reynolds, R.D. and Leklem, J.E. (eds) *Vitamin $B_6$: Its Role in Health and Disease*. Alan R. Liss, New York, pp. 22–23.

Henry, H.L. (1992) Vitamin D hydroxylases. *Journal of Cellular Biochemistry* 49, 4–9.

Henry, H.L. and Norman, A.W. (1974) Studies on calciferol metabolism. IX. Renal 25-hydroxy-vitamin $D_3$-1-hydroxylase. Involvement of cytochrome P-450 and other properties. *Journal of Biological Chemistry* 249, 7529–7535.

Henry, H.L., Dutta, C., Cunningham, N., Blanchard, R., Penny, R., Tang, G., Marchetto, G. and Chou, S.-Y. (1992) The cellular and molecular recognition of $1,25(OH)_2D_3$ production. *Journal of Steroid Biochemistry and Molecular Biology* 41, 401–407.

Hines, J.D., Rosenberg, A. and Harris, J.W. (1968) Intrinsic factor-mediated radio-$B_{12}$ uptake in sequential incubator studies using everted sacs of guinea pig small intestine: evidence that IF is not absorbed into the intestinal cell. *Proceedings of the Society for Experimental Biology and Medicine* 129, 653–658.

Kato, M., Blaner, W.S., Mertz, J.R., Das, K., Kato, K. and Goodman, D.S. (1985) Influence of retinoid nutritional status on cellular retinol- and cellular retinoic acid-binding protein concentrations in various rat tissues. *Journal of Biological Chemistry* 260, 4832–4838.

Kayata, S., Kindberg, Greer, F.R. and Suttie, J.W. (1989) Vitamin $K_1$ and $K_1$ in infant human liver. *Journal of Pediatrics, Gastroenterology and Nutrition* 8, 304–307.

Kazarinoff, M.N. and McCormick, D.B. (1975) Rabbit liver pyridoxamine (pyridoxine) 5′-phosphate oxidase: purification and properties. *Journal of Biological Chemistry* 250, 3436–3442.

Kight, C.E., Reedstrom, C.K. and Suttie, J.W. (1995) Identification, isolation, and partial purification of a cytosolic binding protein for vitamin K from rat liver. *FASEB Journal* 9, A725.

Kirkland, J.B. and Rawling, J.M. (2000) Niacin. In: Rucker, R.B., Suttie, J.W., McCormick, D.B. and Machlin, L.J. (eds) *Handbook of Vitamins*, 3rd edn. Marcel Dekker, New York, pp. 213–254.

Knauer, T.E., Siegfried, C.M. and Matschiner, J.T. (1976) Vitamin K requirement and the concentration of vitamin K in rat liver. *Journal of Nutrition* 106, 1747–1756.

Kozyraki, R., Kristiansen, M., Silahtaroglu, A., Hansen, C., Jacobsen, C., Tommerup, N., Verroust, P.J. and Moestrup, S.K. (1998) The human intrinsic factor-vitamin $B_{12}$ receptor, cubilin: molecular chacterization and chromosomal mapping of the gene to 10p within the autosomal recessive megaloblastic anemia (MGA1) region. *Blood* 91, 3593–3600.

Kuhlenkamp, J., Ronk, M., Yusin, M., Stolz, A. and Kaplowitz, N. (1993) Identification and purification of a human liver cytosolic tocopherol binding protein. *Protein Expression and Purification* 4, 382–389.

Kumar, R. (1986) The metabolism and mechanism of action of 1,25-dihydroxyvitamin $D_3$. *Kidney International* 30, 793–803.

Lee, H.C. (2000) NAADP: an emerging calcium signaling molecule. *Journal of Membrane Biology* 173, 1–8.

Lee, S.-S. and McCormick, D.B. (1985) Thyroid hormone regulation of flavocoenzyme biosynthesis. *Archives of Biochemistry and Biophysics* 237, 197–201.

Leklem, J.E. (2001) Vitamin $B_6$. In: Rucker, R.B., Suttie, J.W., McCormick, D.B. and Machlin, L.J. (eds) *Handbook of Vitamins*, 3rd edn. Marcel Dekker, New York, pp. 339–396.

Li, E. and Norris, A.W. (1996) Structure/function of cytoplasmic vitamin A-binding proteins. *Annual Review of Nutrition* 16, 205–234.

Li, T.-K., Lumeng, L. and Veitch, R.L. (1974) Regulation of pyridoxal 5′-phosphate metabolism in liver. *Biochemical and Biophysical Research Communications* 61, 627–634.

Martius, C. (1961) The metabolic relationships between the different K vitamins and the synthesis of the ubiquinones. *American Journal of Clinical Nutrition* 9, 97–103.

Matschiner, J.T. (1971) Isolation and identification of vitamin K from animal tissue. In: *Symposium Proceedings on the Biochemistry, Assay, and Nutritional Value of Vitamin K and Related Compounds.* Association of Vitamin Chemists, Chicago, pp. 21–37.

Matschiner, J.T., Bell, R.G., Amelotti, J.M. and Knauer, T.E. (1970) Isolation and chacterization of a new metabolite of phylloquinone in the rat. *Biochimica et Biophysica Acta* 201, 309–315.

Mawer, E.B., Backhouse, J., Holman, C.A., Lumb, G.A. and Stanbury, S.W. (1972) The distribution and storage of vitamin D and its metabolites in human tissues. *Clinical Science* 43, 413–431.

McCormick, D.B. (1975) Metabolism of riboflavin. In: Rivlin, R.S. (ed.) *Riboflavin.* Plenum Press, New York, pp. 153–198.

McCormick, D.B. (1976) Biotin. In: Hegsted, D.M. (ed.) *Present Knowledge in Nutrition.* The Nutrition Foundation, New York, pp. 217–225.

McCormick, D.B. (1988) Thiamin. In: Shils, M.E. and Young, V.R. (eds) *Modern Nutrition in Health and Disease.* Lea and Febiger, Philadelphia, pp. 355–361.

McCormick, D.B. (1989) Two interconnected B vitamins: riboflavin and pyridoxine. *Physiology Reviews* 69, 1170–1198.

McCormick, D.B. (1999) Riboflavin. In: Shils, M.E., Olson, J.A., Shike, M. and Ross, A.C. (eds) *Modern Nutrition in Health and Disease.* Lea and Febiger, Malvern, Philadelphia, pp. 391–399.

McCormick, D.B. (2001) Vitamin $B_6$. In: Bowman, B.A. and Russell, R.M. (eds) *Present Knowledge in Nutrition.* ILSI-Nutrition Foundation, Washington, DC, pp. 207–213.

McCormick, D.B. and Snell, E.E. (1959) Pyridoxal kinase of human brain and its inhibition by hydrazine derivatives. *Proceedings of the National Academy of Sciences USA* 45, 1371–1379.

McCormick, D.B. and Wright, L.D. (1970) The metabolism of biotin and its analogues. *Comprehensive Biochemistry* 21, 81–110.

McCormick, D.B. and Zhang, Z. (1992) Cellular assimilation of water-soluble vitamins in the mammal: riboflavin, $B_6$, biotin, and C. *Proceedings of the Society for Experimental Biology and Medicine* 202, 265–270.

McCormick, D.B., Gregory, M.E. and Snell, E.E. (1961) Pyridoxal phosphokinases I. Assay, distribution, purification, and properties. *Journal of Biological Chemistry* 236, 2076–2084.

Merrill, A.H. and Henderson, J.M. (1990) Vitamin $B_6$ metabolism by human liver. *Annals of the New York Academy of Sciences* 585, 110–117.

Merrill, A.H. Jr and McCormick, D.B. (1980) Affinity chromatographic purification and properties of flavokinase (ATP: riboflavin 5′phosphotransferase) from rat liver. *Journal of Biological Chemistry* 255, 1335–1338.

Merrill, A.H., Henderson, J.M., Wang, E., McDonald, B.W. and Millikin, W.J. (1984) Metabolism of vitamin $B_6$ by human liver. *Journal of Nutrition* 114, 1664–1674.

Moat, A.G. and Foster, J.W. (1982) Biosynthesis and salvage pathways of pyridine nucleotides. In: Dolphin, D., Powanda, M. and Poulson, R. (eds) *Pyridine Nucleotide Coenzymes: Chemical, Biochemical and Medical Aspects.* John Wiley & Sons, New York, Part B, pp. 1–24.

Mock, D.M. (2001) Biotin. In: Rucker, R.B., Suttie, J.W., McCormick, D.B. and Machlin, L.J. (eds) *Handbook of Vitamins*, 3rd edn. Marcel Dekker, New York, pp. 427–462.

Moscow, J.A., Gong, M., He, R., Sgagias, M.K., Dixon, K.H., Anzick, S.L., Meltzer, P.S. and Cowan, K.H. (1995) Isolation of a gene encoding a human reduced folate carrier (RFC1) and analysis of its expression in transport-deficient methotrexate-resistant human breast cancer cells. *Cancer Research* 55, 3790–3794.

Muniyappa, K., Murphy, U.S. and Adiga, P.R. (1978) Estrogen induction of thiamin carrier protein in chicken liver. *Journal of Steroid Biochemistry* 9, 888.

Murphy, M., Keating, M., Boyle, P., Weir, D.G. and Scott, J.M. (1976) The elucidation of the mechanism of folate catabolism in the rat. *Biochemical and*

*Biophysical Research Communications* 71, 1017–1024.
Nemere, I. and Norman, A.W. (1987) Studies on the mode of action of calciferol. LII. Rapid action of 1,25-dihydroxyvitamin $D_3$ on calcium transport in perfused chick duodenum: effect of inhibitors. *Journal of Bone and Mineral Research* 2, 99–107.
Newcomer, M.E. (1995) Retinoid-binding proteins: structural determinants important for function. *FASEB Journal* 9, 229–239.
Nyquist, S.E., Matschiner, J.T. and James Morre, D.J. (1971) Distribution of vitamin K among rat liver cell fractions. *Biochimica et Biophysica Acta* 244, 645–649.
Oka, M. and McCormick, D.B. (1987) Complete purification and general characterization of FAD synthetase from rat liver. *Journal of Biological Chemistry* 262, 7418–7422.
Olson, J.A. (2001) Vitamin A. In: Rucker, R.B., Suttie, J.W., McCormick, D.B. and Mzchlin, L.J. (eds) *Handbook of Vitamins*, 3rd edn. Marcel Dekker, New York, pp. 1–50.
Ong, D.E., Newcomer, M.E. and Chytil, F. (1994) Cellular retinoid-binding proteins. In: Sporn, M.B., Roberts, A.B. and Goodman, D.S. (eds) *The Retinoids: Biology, Chemistry, and Medicine*, 2nd edn. Raven Press, New York, pp. 283–317.
Paquin, J., Baugh, C.M. and MacKenzie, R.E. (1985) Channeling between the active sites of formiminotransferase-cyclodeaminase. Binding and kinetic studies. *Journal of Biological Chemistry* 260, 14925–14931.
Plesofsky, N.S. (2001) Pantothenic acid. In: Rucker, R.B., Suttie, J.W., McCormick, D.B. and Machlin, L.J. (eds) *Handbook of Vitamins*, 3rd edn. Marcel Dekker, New York, pp. 317–337.
Preiss, J. and Handler, P. (1958) Biosynthesis of diphosphopyridine nucleotide I. Identification of intermediates. *Journal of Biological Chemistry* 233, 488–500.
Reedstrom, C.K. and Suttie, J.W. (1995) Comparative distribution, metabolism, and utilization of phylloquinone and menaquinone-9 in rat liver. *Proceedings of the Society for Experimental Biology and Medicine* 209, 403–409.
Rindi, G. and Laforenza, U. (1997) *In vitro* systems for studying thiamin transport in mammals. *Methods in Enzymology* 279, pp. 118–131.
Rivlin, R. (1970) Medical progress: riboflavin metabolism. *New England Journal of Medicine* 283, 463.
Rivlin, R. and Pinto, J.T. (2001) Riboflavin (Vitamin $B_2$). In: Rucker, R.B., Suttie, J.W., McCormick, D.B. and Machlin, L.J. (eds) *Handbook of Vitamins*, 3rd edn. Marcel Dekker, New York, pp. 255–273.
Robishaw, J.D., Berkich, D. and Neely, J.R. (1982) Rate-limiting step and control of coenzyme A synthesis in cardiac muscle. *Journal of Biological Chemistry* 257, 10967–10972.
Rothenberg, S.P., Weisberg, H. and Ficarra, A. (1972) Evidence for the absorption of immunoreactive intrinsic factor into the intestinal epithelial cell during vitamin $B_{12}$ absorption. *Journal of Laboratory and Clinical Medicine* 79, 587–597.
Saarem, K., Bergseth, S., Oftebro, H. and Pedersen, J.I. (1984) Subcellular localization of vitamin $D_3$ 25-hydroxylase in human liver. *Journal of Biological Chemistry* 259, 10936–10940.
Saari, J.C. (1994) Retinoids in photosensitive systems. In: Sporn, M.B., Roberts, A.B. and Goodman, D.E. (eds) *The Retinoids: Biology, Chemistry, and Medicine*, 2nd edn. Raven Press, New York, pp. 351–385.
Said, H.M., Redha, R. and Nylander, W. (1988) Biotin transport in basal lateral membrane vesicles of human intestine. *Gastroenterology* 94, 1157–1163.
Said, H.M., Nguyen, T.T., Dyer, D.L., Cowan, K.H. and Rubin, S.A. (1996) Intestinal folate transport: identifiction of a cDNA involved in folate transport and functional expression and distribution of its mRNA. *Biochimica et Biophysica Acta* 1281, 164–172.
Sato, Y., Hagiwara, K., Arai, H. and Inoue, K. (1991) Purification and characterization of the α-tocopherol transfer protein from rat liver. *FEBS Letters* 288, 41–45.
Shane, B. (1989) Folylpolyglutamate synthesis and role in the regulation of one-carbon metabolism. In: Aurbach, G.D. and McCormick, D.B. (eds) *Vitamins and Hormones.* Academic Press, Vol. 45, pp. 263–335.
Shearer, M.J., McCarthy, P.T., Crampton, O.E. and Mattock, M.B. (1988) The assessment of human vitamin K status from tissue measurements. In: Suttie, J.W. (ed.) *Current Advances in Vitamin K Research.* Elsevier, New York, pp. 437–452.
Sheppard, A.J., Pennington, J.A.T. and Weihrauch, J.L. (1993) Analysis and distribution of vitamin E in vegetable oils and foods. In: Packer, L. (ed.) *Vitamin E in Health and Disease.* Marcel Dekker, New York, pp. 9–31.
Silverman, R.B. and Nandi, D.L. (1988) Reduced thioredoxin: a possible physiological cofactor for vitamin K epoxide reductase. Further support for an active site disulfide. *Biochemical and Biophysical Research Communications* 155, 1248–1254.
Sorrell, M.F., Frank, O., Thomson, A.D., Aquino, H. and Baker, H. (1971) Absorption of vitamins from the large intestine *in vivo*. *Nutrition Reports International* 3, 143–148.
Soute, B.A.M., Muller-Ester, W., de Boer-van den Berg, M.A.G., Ulrich, M. and Vermeer, C. (1985) Discovery of a γ-carboxyglutamic acid-containing protein in human spermatozoa. *FEBS Letters* 190, 137–141.
Stratton, S., Mock, N. and Mock, D. (1996) Biotin and biotin metabolites in human milk: the metabolites are not negligible. *Journal of Investigative Medicine*
Suttie, J.W. (2001) Vitamin K. In: Rucker, R.B., Suttie,

J.W., McCormick, D.B. and Machlin, L.J. (eds) *Handbook of Vitamins*, 3rd edn. Marcel Dekker, New York, pp. 115–164.

Tanphaichitr, V. (2001) Thiamine. In: Rucker, R.B., Suttie, J.W., McCormick, D.B. and Machlin, L.J. (eds) *Handbook of Vitamins*, 3rd edn. Marcel Dekker, New York, pp. 275–316.

Thijssen, H.H.W. and Drittij-Reijnders, M.J. (1994) Vitamin K distribution in rat tissues: dietary phylloquinone is a source of tissue menaquinone-4. *British Journal of Nutrition* 72, 415–425.

Towler, D.A., Adams. S.P., Eubanks, S.R., Towery, D.S., Jackson-Machelski, E., Glaser, L. and Gordon, J.I. (1987) Purification and characterization of yeast myristoyl CoA: protein *N*-myristoyltransferase. *Proceedings of the National Academy of Sciences USA* 84, 2708–2712.

Traber, M.G. and Sies, H. (1996) Vitamin E in humans: demand and delivery. *Annual Review of Nutrition* 16, 321–347.

Van Haarlem, L.J.M., Soute, B.A.M. and Vermeer, C. (1987) Vitamin K-dependent carboxylase. Possible role for thioredoxin in the reduction of vitamin K metabolites in liver. *FEBS Letters* 222, 353.

Whitehouse, W.S.A., Merrill, A.H. Jr and McCormick, D.B. (1991) Riboflavin-binding protein. In: Muller, F. (ed.) *Chemistry and Biochemistry of Flavins*. CRC Press, Boca Raton, Florida, pp. 287–292.

Wu, S.-M., Morris, D.P. and Stafford, D. (1991a) Identification and purification to near homogeneity of the vitamin K-dependent carboxylase. *Proceedings of the National Academy of Sciences USA* 88, 2236–2240.

Wu, S.-M., Cheung, W.-F., Frazier, D. and Stafford, D. (1991b) Cloning and expression of the cDNA for human γ-glutamyl carboxylase. *Science* 254, 1634–1636.

Yagi, K., Nakaga, Y., Suzuki, O. and Ohishi, N. (1976) Incorporation of riboflavin into covalently-bound flavins in rat liver. *Journal of Biochemistry* 79, 841–843.

Zempleni, J., McCormick, D.B. and Mock, D.M. (1996) The identification of biotin sulfone, bisnorbiotin methyl ketone, and tetranorbiotin-*l*-sulfoxide in human urine. *American Journal of Clinical Nutrition* 65, 508–511.

Zolkiewska, A., Okazaki, I.J. and Moss, J. (1994) Vertebrate mono-ADP-ribosyltransferases. *Molecular and Cellular Biochemistry* 138, 107–112.

# 5 增殖细胞中的养分稳衡

Janos Zempleni
（内布拉斯加州-林肯大学营养科学与饮食系，林肯，内布拉斯加州，美国）

## 哺乳动物的细胞周期

细胞周期是细胞生长并分化成两个子细胞的一套有序的过程，即细胞增殖（Murray and Hunt，1993）。哺乳动物有两种基本不同的细胞周期：①细胞有丝分裂，结果产生两个二倍体子细胞；②细胞减数分裂，产生 4 个单倍体子细胞（精子和卵的形成）。本章将专门介绍细胞的有丝分裂。

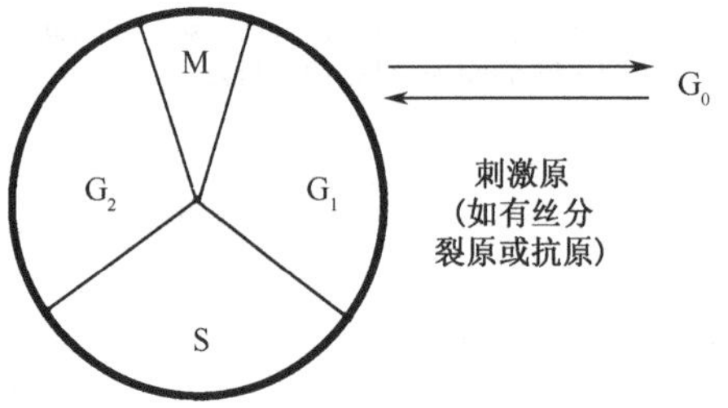


图 5.1 细胞周期

圆截各部分基本与细胞周期的各时相所占的百分率成比例。$G_0$＝静止状态；$G_1$＝开始第 1 阶段；S＝DNA 合成期；$G_2$＝开始第 2 阶段；M ＝有丝分裂期

养分或生长因子的缺乏将导致细胞进入特殊的休眠状态，称之为 $G_0$，在此阶段细胞不进行分化（图 5.1）（Murray and Hunt，1993）。如果提供的养分充足，细胞即可在化合物如激素、抗原和有丝分裂原的刺激下重新进入细胞周期。细胞有丝分裂可分为两大部分：分裂间期（包括 $G_1$、S、$G_2$）和有丝分裂期。细胞周期随细胞的分裂而完结。

（1）分裂间期　分裂间期包括 3 个阶段（Murray and Hunt，1993）：在 $G_1$ 期，膜合成，线粒体和大部分细胞蛋白出现。在 S 期，合成 DNA 以进行染色体复制。一旦 S 期结束，细胞即进入 $G_2$ 期，该时期的细胞生长是为细胞有丝分裂（“M”）做准备。

（2）有丝分裂期　有丝分裂期可再划分为 4 个时期（Murray and Hunt，1993）：在第一时期（前期），染色体被压缩（图 5.2）；在前中期，核膜断裂，染色体与微管连接，然后微管再与中心体相连；在中期，微管连接的染色体在两个中心体的中间排成列；两个姐妹染色体的连接溶解表明细胞分裂后期的开始，姐妹染色体分离并分别朝纺锤体的相反极相互移开；最后，细胞质分裂（细胞分裂的物理过程）开始。

细胞增殖与一些基因表达的显著增加有关。例如，有丝分裂原诱导的人淋巴细胞增殖导致 *c-myc*、4F1、JE-3、KC-1（Kaczmarek *et al.*，1985）、HsRAD15（Flygare *et al.*，1996）和 P120（Wilson and Freeman，1996）这些基因的表达增加。同样，细胞增殖与许多代谢途径的代谢率的提高有关，例如，细胞增殖需要细胞的生长（如细胞膜的合成）和 DNA 的复制（Murray and Hunt，1993）。养分是必不可少的辅酶（如维生素 B 族）、能量供应的底物（如葡萄糖），也是对细胞增殖所必需的许多代谢途径中的前体

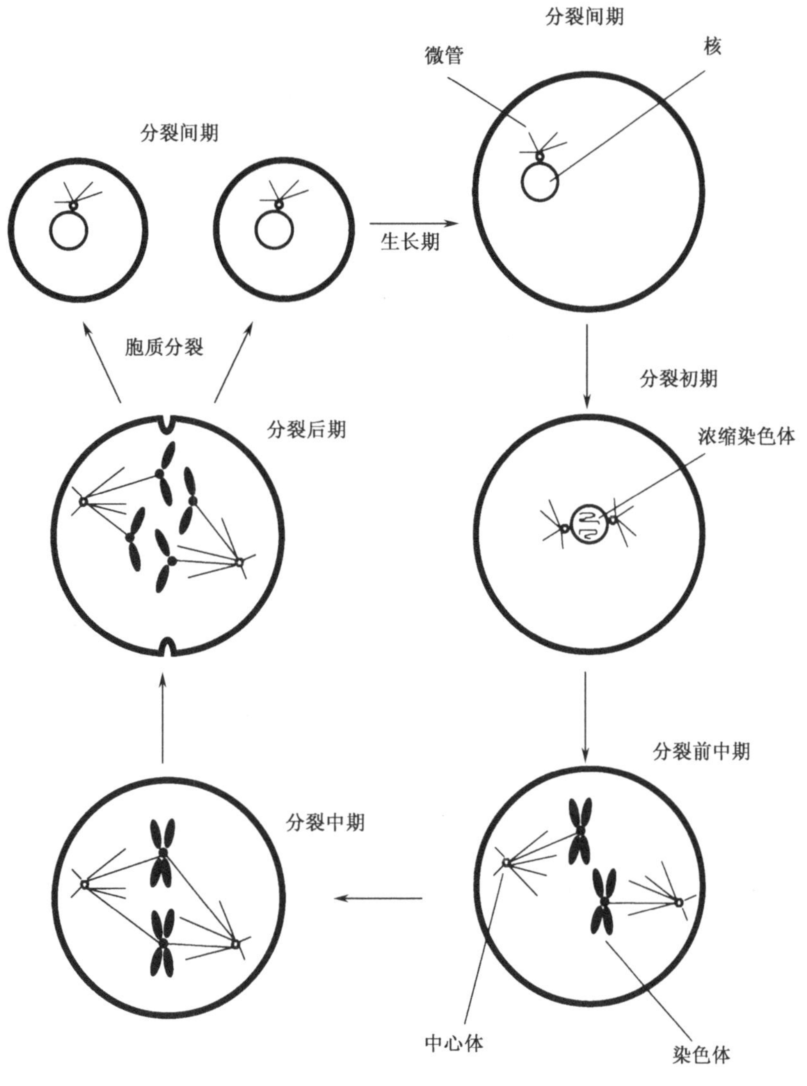


图 5.2　细胞的有丝分裂过程

物质（如必需脂肪酸）。在本章，我们将述证据说明细胞增殖与参与养分摄取和代谢的基因表达的增加有关。而且我们还将证明，细胞增殖可能将提高对养分的需求量。若养分供应不足，这种增加的需求可能会减慢细胞增殖速率并将损害细胞的功能。

## 生物素作为研究增殖细胞中养分稳衡的模型

生物素这种维生素将用作阐明细胞增殖对细胞养分摄取和利用影响的模型。我们选择生物素作为模型，是因为人们已对哺乳动物细胞中生物素转运和利用的特点做了充分的研究，这为分析细胞增殖对生物素稳衡的影响奠定了基础。本部分将对生物素在静止、非增殖细胞中的细胞转运和利用进行综述，以为读者提供这方面的基础知识。

## 细胞的生物素转运

对小鼠肝细胞中生物素转运的早期研究表明，生物素吸收是一个能量、钠离子和Na-K-ATPase依赖性的由转运蛋白介导的过程（Bowers-Komro and McCormick，1985）。在人源性 HepG2 细胞（Said *et al.*，1994）和胎盘刷状缘膜（Karl and Fisher，1992；Schenker *et al.*，1993；Hu *et al.*，1994）也发现了类似的生物素吸收特征。最近，一种跨小鼠胎盘（Prasad *et al.*，1998）和人胎盘（Wang *et al.*，1999）转运生物素、泛酸和硫辛酸的蛋白质已被克隆并进行功能表达，该转运蛋白被称为钠离子依赖性的多种维生素转运蛋白（sodium-dependent multivitamin transporter，SMVT）。随后，在小肠、肾脏、肝脏、脑、肺、心脏、胰腺和骨骼肌等各种组织中均已检测到了编码 SMVT 的 mRNA（Prasad *et al.*，1998；Wang *et al.*，1999）。

尽管 SMVT 在细胞对生物素的吸收方面发挥了重要作用，但是人组织中还有其他转运蛋白也可参与了（一些）生物素的吸收。例如，人单核细胞可通过一种可能是生物素特异性的转运蛋白来积累生物素，泛酸和硫辛酸不会与生物素竞争结合这种转运蛋白（Zempleni and Mock，1998，1999b）。人单核细胞中的这种生物素转运蛋白目前还未被克隆和测序。这样，SMVT 与单核细胞中的生物素转运蛋白之间的同源性尚不清楚。初步的研究结果表明，单羧基化合物转运蛋白可解释人单核细胞中部分生物素的转运（见下文）。介导生物素转入单核细胞的相同转运蛋白也能把生物素转出这些细胞，这已被逆向转运试验所证实（Zempleni and Mock，1998，1999a）。

## 生物素依赖性的羧化酶

在哺乳动物中，生物素作为下面 4 种羧化酶的共价结合辅酶：乙酰-CoA 羧化酶（EC 6.4.1.2）、丙酮酸羧化酶（EC 6.4.1.1）、丙酰-CoA 羧化酶（EC 6.4.1.3）和3-甲基丁烯酰-CoA 羧化酶（EC 6.4.1.4）（Wood and Barden，1977；Knowles，1989）。

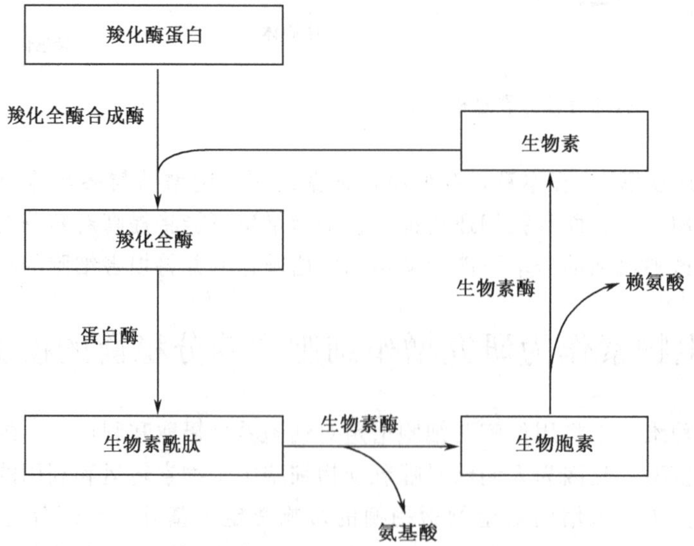


图 5.3 生物素的利用和循环

生物素与每一种羧化酶蛋白中赖氨酸的ε-氨基之间形成的共价结合是由羧化全酶合成酶（EC 6.3.4.10；图 5.3）来催化的（Dakshinamurti and Chauhan，1994）。这 4 种哺乳动物羧化酶催化重要代谢途径如脂肪酸合成、氨基酸代谢、奇链脂肪酸和葡萄糖异生中的基本步骤（Zempleni，2001）。

羧化全酶被蛋白酶降解形成生物素酰肽，然后再被生物素酶（EC 3.5.1.12）降解释放出生物胞素（生物素酰-ε-赖氨酸）。最后，生物素酶催化生物胞素释放出生物素，用于新的羧化全酶的再循环（Wolf *et al.*，1985）。

## 组蛋白的生物素化

组蛋白是介导 DNA 折叠成染色质的主要蛋白质（Wolffe，1998）。DNA 与组蛋白的结合具有静电学性质，即 DNA 中带负电荷的磷酸基团与组蛋白中带正电荷的ε-氨基（赖氨酸残基）和胍基（精氨酸残基）结合。

组蛋白可在体内进行翻译后修饰，包括乙酰化（Ausio and Holde，1986；Hebbes *et al.*，1988；Lee *et al.*，1993）、甲基化（Wolffe，1998）、磷酸化（Wolffe，1998）、遍在蛋白化（Wolffe，1998）和多聚（ADP-核糖基化）（Chambon *et al.*，1966；Boulikas，1988；Boulikas *et al.*，1990）。这些基团与组蛋白的氨基酸残基如赖氨酸残基的ε-氨基和精氨酸残基的胍基共价结合。部分修饰作用［如乙酰化和多聚（ADP-核糖基化）］使组蛋白失去一个正电荷，导致 DNA 与组蛋白之间的结合减弱。试验表明，一些组蛋白的修饰与 DNA 的转录增强（如组蛋白的乙酰化）和 DNA 修复机制［组蛋白的多聚（ADP-核糖基化）］有关（Wolffe，1998）。

Hymes 等（1995）提出了生物素酶催化生物素与组蛋白共价结合的反应机制，该反应导致组蛋白中带正电荷的氨基数量减少。组蛋白的生物素化表明生物素在 DNA 的转录、复制或修复中有一定的作用，组蛋白的乙酰化、甲基化、磷酸化、遍在蛋白化和 ADP-核糖化也与之类似。组蛋白生物素化的酶催化机制及其潜在的生理作用将在第 17 章进一步讨论。

# 外周血单核细胞作为细胞模型

本章综述了近期的试验研究，直接说明了有关细胞增殖与生物素代谢之间的相互关系。绝大多数试验采用外周血单核细胞（peripheral blood mononuclear cell，PBMC）作为细胞模型，PBMC 是由骨髓中多功能的造血干细胞产生的免疫细胞（B 细胞、T 细胞和各种粒细胞）的异质群体（Janeway *et al.*，1999）。

PBMC 为什么是研究细胞增殖对生物素内稳衡影响的理想模型？原因是 PBMC 很容易从人外周血液中获取，而且采用此类人细胞得到的研究结果可直接用于人的营养。刚分离出的 PBMC 一般处于静止状态，但在采用抗原或有丝分裂原进行刺激后，一些 PBMC（B 细胞和 T 细胞）将快速增殖。有关静止的人 PBMC（吸收、代谢和排出）的生物素内稳衡，人们已进行了很成功的描述（Velazquez *et al.*，1990，1995；Zempleni and Mock，1998，1999a，b），这为增殖 PBMC 的研究提供了坚实的基础。

# 细胞增殖对生物素吸收的影响

## 生物素吸收的速度

在有丝分裂原刺激下增殖的 PBMC 对生物素的吸收速度比非刺激对照组快 5 倍（Zempleni and Mock，1998，1999c）。例如，若采用美洲商陆有丝分裂原（从美洲商陆中提取的凝集素）诱导细胞增殖，3 天后，吸收进入 PBMC 的生物素以剂量依赖的方式增加，是对照组静止细胞的 481%～722%（图 5.4）。当分别用伴刀豆凝集素 A 或植物凝集素诱导细胞增殖时，生物素吸收与上面的结果类似。

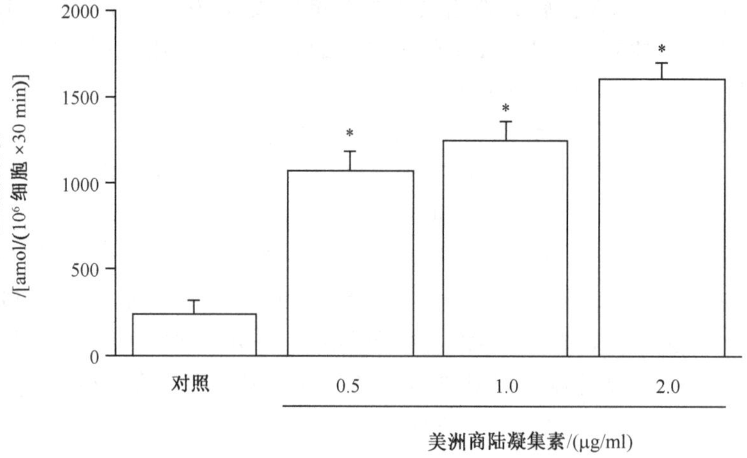


图 5.4　增殖的人 PBMC 和静止对照细胞中对［$^3$H］生物素的吸收情况。采用美洲商陆凝集素（0.5、1.0 或 2.0μg/ml）来诱导细胞增殖，在 37℃下培养 3 天，对照组不添加美洲商陆凝集素。数值用平均数±SD 表示，$n=6$，* 表示与对照组比较 $P<0.01$，数据采用 Dunnett 的程序进行 ANOVA 分析（Zempleni and Mock，1999c）

之后，PBMC 与有丝分裂原一起培养至多 4 天，测定在不同时间点的生物素转运速度。当用 PBMC 对细胞增殖标记胸苷的吸收量来判断时，生物素吸收速度的增加与细胞增殖速度一致（Zempleni and Mock，1999c）。在培养基中添加美洲商陆有丝分裂原 48～72h 后，生物素和胸苷的吸收速度达到最大（Zempleni and Mock，1999c）。这些数据证实了由于增殖导致的生物素需要量的增加将增强生物素转运进入刺激有丝分裂原的 PBMC 中。另外，有丝分裂原对 PBMC 进行的短期（小于 15min）刺激不影响生物素的转运速度。这表明，有丝分裂原不能通过与生物素或细胞表面的转运蛋白之间化学上的相互作用来提高生物素的转运。但有丝分裂原的长期（1～4 天）刺激由于诱导了细胞的增殖，从而提高了生物素的转运。

## 细胞周期中特定时相的生物素吸收

前面叙述的研究是采用 PBMC 非共同培养基，即细胞周期的不同时相的细胞群体进行的试验。因此，这些研究无法确定提高生物素吸收具体是发生在细胞周期的哪个时相，如 $G_1$、S、$G_2$ 或 M 时期。针对这种缺陷，有人另外进行了一系列的研究（Stanley *et al.*，2002)。增殖的 PBMC 与下面的某一种化学物质一起培养而被固定在细胞周期的某个指定时相：①渥曼青霉素（100nmol/L）使细胞固定在 $G_1$ 期；②蚜栖菌素（118μmol/L）使细胞固定在 S 期；③亚德里亚霉素（1μmol/L）使细胞固定在 $G_2$ 期；④秋水仙碱（5 μmol/L)使细胞固定在 M 期。以静止的 PBMC（$G_0$ 期）作为对照。生物素进入这些细胞中的吸收速率采用[$^3$H]生物素来测定。

在细胞周期的 $G_1$ 期，生物素吸收速率提高（是对照组在 $G_0$ 期的 4.5 倍）；在后面的时相（S、$G_2$ 和 M 期；图 5.5），生物素的吸收速率也提高了。但 $G_1$、S、$G_2$ 或 M 期之间的差异不显著。这些结果提示：①细胞表面的生物素转运蛋白数目在细胞周期的早期就增加了（图 5.5)；②在整个细胞周期，转运蛋白的数目都有增加。

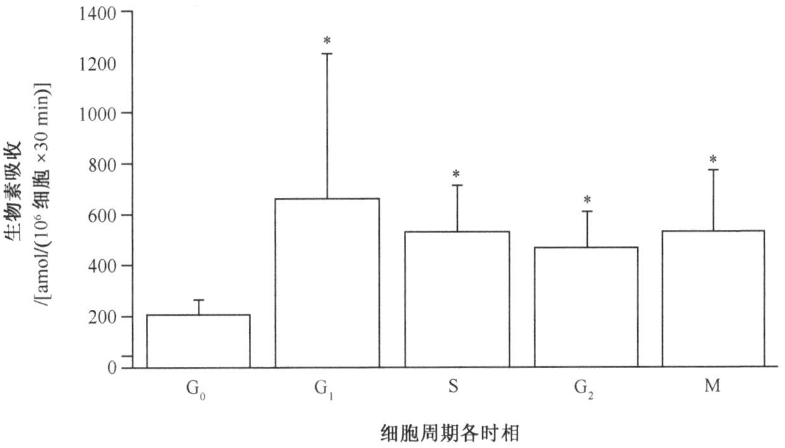


图 5.5　细胞周期固定的人 PBMC 中［$^3$H］生物素的吸收情况。数值用平均数±SD 表示，$n=6$. * 表示与对照组比较差异极显著（$P<0.01$）(Stanley *et al.*，2002)

## 转运蛋白的合成

增殖 PBMC 生物素吸收的增加是由细胞表面的生物素转运蛋白数量的增加所致还是由转运蛋白对生物素的亲和力增加所致？动力学资料表明，生物素吸收的提高是由转运蛋白数量的增加所致。生物素的最大转运速度（$V_{max}$）可作为 PBMC 表面的生物素转运蛋白数目的估测指标。理论上讲，增殖 PBMC 中生物素转运蛋白的合成增加将提高生物素的 $V_{max}$。增殖 PBMC 的生物素吸收 $V_{max}$ 比非增殖 PBMC 大约高 4 倍，分别为 9.1±6.6 和 2.3±1.6fmol/（$10^6$ 细胞×30min）(Zempleni and Mock，1999c)。而转运蛋白对生物素的亲和力（采用米氏常数来判断）在增殖与静止 PBMC 之间差异不显著。总之，这些结果表明，增殖 PBMC 生物素吸收的提高是通过增加转运蛋白的合成而不是提高底物的亲和力来实现的。

以 mRNA 的浓度进行判断，SMVT 的基因表达与增殖 PBMC 中 SMVT 的合成增加是一致的。SMVT 基因表达采用从静止和增殖 PBMC 分离得到的总 RNA 进行的反转录聚合酶链反应（reverse transcriptase-polymerase chain reaction，RT-PCR）来确定。对某些持家基因进行了量化比较（Zempleni *et al.*，2001），发现增殖PBMC的编码 SMVT 的 RNA 浓度是静止细胞的（10.6±1.6）倍（图 5.6)。在增殖 PBMC 中，SMVT

表达的提高幅度超过了编码转铁蛋白受体的持家基因表达的提高幅度［是对照组的(5.6±6.4）倍］，而甘油醛-3-磷酸脱氢酶是对照组的（1.0±0.1）倍，组蛋白 H1.3 是对照组的（1.8±0.1）倍，组蛋白 H4 是对照组的（1.1±0.1）倍。典型例子见图 5.6。

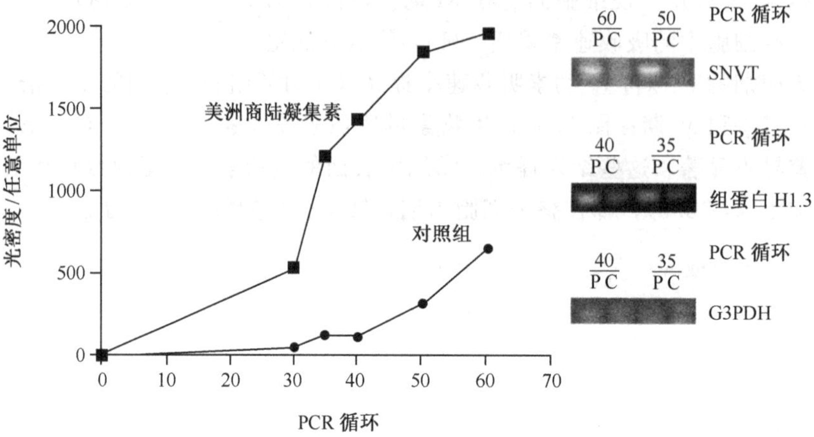


图 5.6 人 PBMC 中钠离子依赖性的多种维生素转运蛋白（SMVT）基因表达情况

细胞采用 2.0μg/ml 美洲商陆凝集素（P）培养 48h；对照组（C）不添加任何有丝分裂原。提取总 RNA，利用 SMVT 特异性的引物，采用 RT-PCR 方法进行基因序列扩增；组蛋白 H1.3 和甘油醛-3-磷酸脱氢酶（G3PDH）作为对照采用 RT-PCR 进行量化分析（Zempleni *et al.*，2001）

从理论上讲，增殖 PBMC 的生物素吸收的提高也可能是由其他转运蛋白而不是由 SMVT 的合成的提高所致。然而，随后的一系列试验表明，增殖和静止的 PBMC 表达的生物素转运蛋白完全相同：①在增殖和静止的 PBMC 中，转运蛋白对生物素的亲和力（采用米氏常数来判断）完全相同（Zempleni and Mock，1999c）；②在增殖和静止的 PBMC 中，生物素转运蛋白对底物的特异性完全相同（Zempleni and Mock，1999c）；③在增殖和静止的 PBMC 中，编码 SMVT 的 mRNA 已经被鉴定出来（Zempleni *et al.*，2001）。

尽管 SMVT 在生物素转运方面具有重要作用，但初步的研究结果表明，其他转运蛋白也能将部分生物素转入 PBMC。例如，某些有机酸如己酸、丙酮酸和乙酸可与生物素竞争进入细胞（Zempleni，资料未发表）。这些有机酸是哺乳动物细胞的单羧基化合物转运蛋白的底物（Halestrap and Price，1999）。而且，人血细胞的两种单羧基化合物转运蛋白（monocarboxylate transporter，MCT4 和 MCT7）均已被克隆并测序（Halestrap and Price，1999）。总之，在 PBMC 中单羧基化合物转运蛋白可能转运部分生物素的假说与这些初步的研究结果一致。

## 细胞增殖对编码生物素依赖性的羧化酶和羧化全酶合成酶的基因表达的影响

### 羧化酶活性

PBMC 在被有丝分裂原诱导增殖的同时，生物素依赖性羧化酶的活性提高（Stanley *et al.*，2002)。把美洲商陆凝集素添加到培养基中 3 天后，3-甲基丁烯酰-CoA 羧化酶的活性是静止对照组的（2.8±0.8）倍（图 5.7）。当用伴刀豆凝集素 A 诱导细胞增殖时，羧化酶活性的提高也出现类似情况［对照组的（3.1±0.1）倍］。3-甲基丁烯酰-CoA 羧化酶的最大活性与生物素进入增殖 PBMC 的最大速度一致。

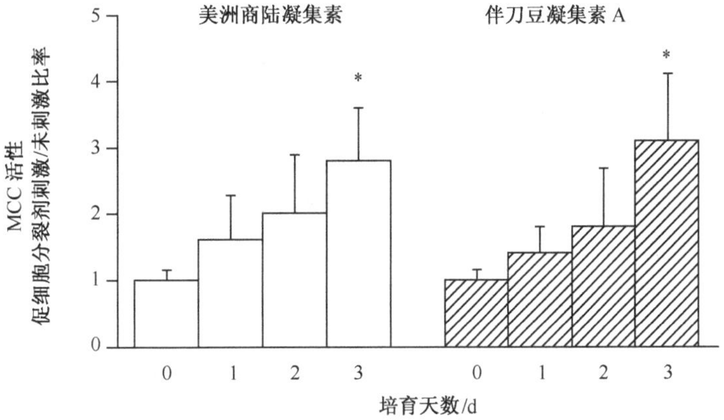


图 5.7 有丝分裂原诱导的人 PBMC 3-甲基丁烯酰-CoA 羧化酶（methylcrotonyl-coA carboxylase，MCC）活性

PBMC 采用美洲商陆凝集素（2.0μg /ml）或伴刀豆凝集素 A（20μg/ml）培养 3 天，在每个时间间隙点，检测 MCC 的活性。数值用平均数±SD 表示，$n=6$. ＊表示与对照组比较差异极显著（$P<0.01$）（Stanley *et al.*，2002）

当用有丝分裂原诱导细胞增殖时，PBMC 中丙酰-CoA 羧化酶活性也提高（Stanley *et al.*，2002）。然而，丙酰-CoA 羧化酶活性的提高幅度低于 3-甲基丁烯酰-CoA 羧化酶。在美洲商陆凝集素添加到培养基后的第 3 天，丙酰-CoA 羧化酶活性是静止对照组的（1.5±0.5）倍（图 5.8）。采用不同的有丝分裂原来刺激细胞，其结果类似：在培养基中添加伴刀豆凝集素 A 后的第 3 天，丙酰-CoA 羧化酶活性是静止对照组的(1.2±

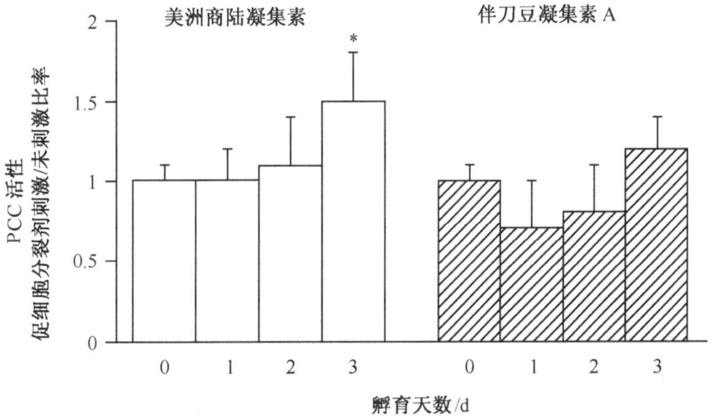


图 5.8 有丝分裂原诱导的人 PBMC 的丙酰-CoA 羧化酶（propionyl-CoA carboxylase，PCC）活性

PBMC 采用美洲商陆凝集素（2.0μg/ml）或伴刀豆凝集素 A（20μg/ml）培养 3 天，在每个时间点，MCC 活性被检测。数值用平均数±SD 表示，$n=6$. ＊表示与对照组差异显著（$P<0.05$）(Stanley *et al.*，2002)

0.2）倍。其他两种羧化酶（乙酰-CoA 羧化酶和丙酮酸羧化酶）的活性太低，未能测出。

人们推测，为了增加羧化酶的活性，增殖 PBMC 将增加生物素的需要量。上述试验结果与此推测一致，表明细胞通过增加细胞对生物素的吸收以满足生物素需要量的增加。

### 编码生物素依赖性的羧化酶和羧化全酶合成酶的 mRNA 水平

也许细胞可以通过提高羧化酶蛋白的生物素化来增加羧化全酶的合成，从而提高羧化酶的活性。理论上讲，羧化酶蛋白有以下两种来源：①来自细胞羧化酶池以前合成的羧化酶蛋白；②应对细胞增殖，特异性地合成新的羧化酶蛋白。后者很可能与编码生物素依赖性的羧化酶的 mRNA 水平升高有关。为了确定 PBMC 增殖是否伴随有新羧化酶蛋白合成的增加，这里总结了对编码生物素依赖性羧化酶 mRNA 水平进行定量分析的研究结果。

编码生物素依赖性羧化酶的 mRNA 水平，PBMC 增殖组比静止对照组要高（Stanley *et al.*，2002）。例如，采用定量 PCR 分析，用美洲商陆凝集素刺激后的 PBMC 组中编码丙酰-CoA 羧化酶 mRNA 水平大约是静止对照组的 4.2 倍（图 5.9）。同样，刺激后的 PBMC 组的编码 3-甲基丁烯酰-CoA 羧化酶 mRNA 水平是静止对照组的 2.3 倍（数据未列出）。这些编码酶的 mRNA 水平升高的原因目前仍不清楚，可能是由于转录的增强，也可能是由于 mRNA 降解的降低，或是两者皆有。

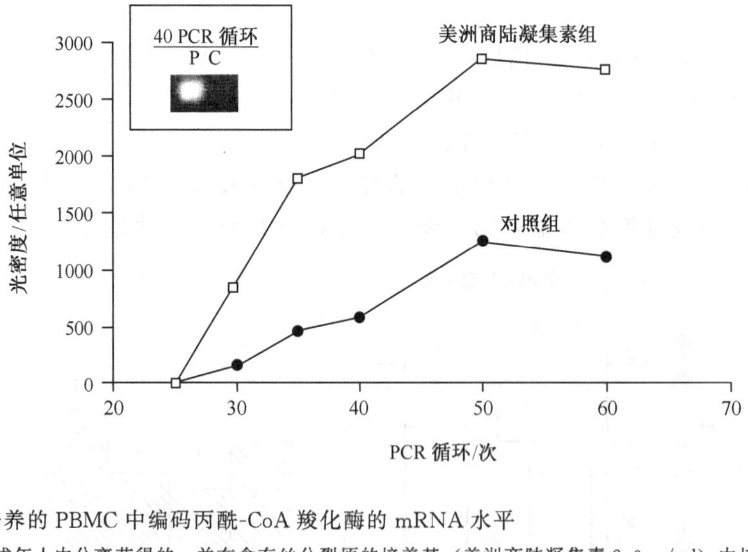


图 5.9　在培养的 PBMC 中编码丙酰-CoA 羧化酶的 mRNA 水平

细胞是从健康成年人中分离获得的，并在含有丝分裂原的培养基（美洲商陆凝集素 2.0μg/ml）中培养 3 天，对照组的培养基不含美洲商陆凝集素。提取总 RNA，编码丙酰-CoA 羧化酶的 mRNA 采用 RT-PCR 进行定量分析。P＝美洲商陆凝集素组，C＝对照组（Stanley *et al.*，2002）

羧化全酶合成酶催化生物素与羧化酶蛋白共价结合形成羧化全酶。凝集素刺激 PBMC增殖后的编码羧化全酶合成酶的 mRNA 约是静止对照组的 2.4 倍（图 5.10）。这一发现与 PBMC 增殖时羧化全酶合成酶基因表达增强，从而使生物素与羧化酶结合增多的假设是一致的。编码持家基因（甘油醛-3-磷酸脱氢酶和组蛋白 H4）的 mRNA 水平在增殖和静止期 PBMC 中是相同的，提示生物素依赖性羧化酶和羧化全酶合成酶的效应不是由在凝胶上的上样量不等的人为因素造成的。

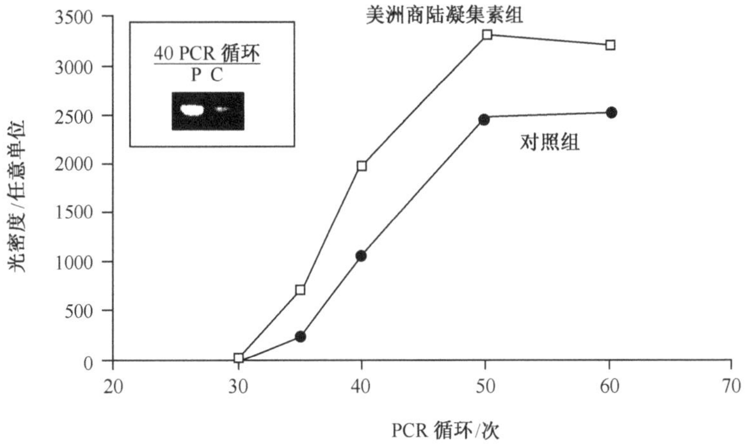


图 5.10 在培养的 PBMC 中编码羧化全酶合成酶的 mRNA 水平。细胞是从健康成年人中分离获得的，并在含有丝分裂原的培养基（美洲商陆凝集素 2.0μg/ml）中培养 3 天，对照组的培养基不含美洲商陆凝集素。提取总 RNA，编码羧化全酶合成酶的 mRNA 采用 RT-PCR 进行定量分析。P＝美洲商陆凝集素组，C＝对照组（Stanley *et al.*，2002）

总之，这些研究结果证实，PBMC 随着 3-甲基丁烯酰-CoA 羧化酶、丙酰-CoA 羧化酶和羧化全酶合成酶合成的增加而增殖。随后，羧化全酶合成酶介导新合成的 3-甲基丁烯酰-CoA 羧化酶和丙酰-CoA 羧化酶的生物素化。

## 细胞增殖对组蛋白生物素化的影响

Wolf 及其同事提出了生物素酶（biotinidase）催化组蛋白进行生物素化的反应机制（Hymes *et al.*，1995）。随后的研究表明，组蛋白可在体内发生生物素化，并且细胞增殖提高了组蛋白的生物素化程度（Stanley *et al.*，2001）。与 $G_0$ 期相比，$G_1$、S、$G_2$ 和 M 期的组蛋白生物素化提高了约 4 倍，该发现与 PBMC 增殖时组蛋白生物素化的增加导致对生物素需求量增加的推测一致。静止和增殖的 PBMC 中组蛋白的生物素化将在第 17 章进行专门的叙述。

# 营养状况对细胞增殖的影响

本章已证实，由于细胞增殖增强了羧化酶和组蛋白的生物素化，细胞增殖将提高生物素的需求量。增殖细胞通过增加细胞内生物素的吸收来满足其需求量的提高。同样，细胞增殖也将提高其他养分如烟酰胺腺嘌呤二核苷酸（NAD）（Williams *et al.*，1985）、核黄素（Zempleni and Mock，2000）、泛酸（Zempleni *et al.*，2001）、维生素 $B_{12}$（Hall，1984）和各种氨基酸（Carter and Halvorson，1973；van den Berg and Betel，

1973）的吸收。这些研究结果与细胞增殖对此类养分的需要量增加一致。如果养分的供应不足，细胞增殖速度将降低。

一些细胞培养试验所提供的重要信息是：在养分缺乏的培养基中，细胞周期受阻。例如，若在不含丝氨酸的培养基中进行培养，海拉细胞（子宫颈癌组织的细胞株，Hela）在 $G_0$ 期受阻（Dakshinamurti *et al.*，1985）。对于有丝分裂原刺激后的淋巴细胞，14 种氨基酸是维持蛋白质合成和正常增殖速率所必需的（Waithe *et al.*，1975）。同样，生物素和亚油酸也是细胞增殖所必需的营养素（Dakshinamurti *et al.*，1985；Lernhardt，1990）。生物素可刺激细胞产生未知生长因子（Moskowitz and Cheng，1985）。虽然营养素是细胞增殖所必需的，但在组织培养基中一些营养素在超生理剂量时能降低细胞生长和增殖水平（Tsai and Gardner，1994）。总之，营养素缺乏或有毒均会严重影响细胞的快速增殖。这些细胞包括小肠组织细胞、胎儿细胞和抗原刺激后的免疫细胞。

## 致　　谢

感谢国家卫生研究所（DK 60447）和 USDA/CSREES 项目基金 2001-35200-10187 的支持。

## 参 考 文 献


Ausio, J. and van Holde, K.E. (1986) Histone hyperacetylation: its effect on nucleosome conformation and stability. *Biochemistry* 25, 1421–1428.

Boulikas, T. (1988) At least 60 ADP-ribosylated variant histones are present in nuclei from dimethylsulfate-treated and untreated cells. *EMBO Journal* 7, 57–67.

Boulikas, T., Bastin, B., Boulikas, P. and Dupuis, G. (1990) Increase in histone poly(ADP-ribosylation) in mitogen-activated lymphoid cells. *Experimental Cell Research* 187, 77–84.

Bowers-Komro, D.M. and McCormick, D.B. (1985) Biotin uptake by isolated rat liver hepatocytes. In: Dakshinamurti, K. and Bhagavan, H.N. (eds) *Biotin*. New York Academy of Sciences, New York, pp. 350–358.

Carter, B.L.A. and Halvorson, H.O. (1973) Periodic changes in rate of amino acid uptake during yeast cell cycle. *Journal of Cell Biology* 58, 401–409.

Chambon, P., Weill, J.D., Doly, J., Strosser, M.T. and Mandel, P. (1966) On the formation of a novel adenylic compound by enzymatic extracts of liver nuclei. *Biochemical and Biophysical Research Communications* 25, 638–643.

Dakshinamurti, K. and Chauhan, J. (1994) Biotin-binding proteins. In: Dakshinamurti, K. (ed.) *Vitamin Receptors: Vitamins as Ligands in Cell Communication*. Cambridge University Press, Cambridge, pp. 200–249.

Dakshinamurti, K., Chalifour, L.E. and Bhullar, R.J. (1985) Requirement for biotin and the function of biotin in cells in culture. In: Dakshinamurti, K. and Bhagavan, H.N. (eds) *Biotin*. New York Academy of Sciences, New York, pp. 38–54.

Flygare, J., Benson, F. and Hellgren, D. (1996) Expression of the human RAD51 gene during the cell cycle in primary human peripheral blood lymphocytes. *Biochimica et Biophysica Acta* 1312, 231–236.

Halestrap, A.P. and Price, N.T. (1999) The proton-linked monocarboxylate transporter (MCT) family: structure, function and regulation. *Biochemical Journal* 343, 281–299.

Hall, C.E. (1984) The uptake of vitamin B12 by human lymphocytes and the relationships to the cell cycle. *Journal of Laboratory and Clinical Medicine* 103, 70–81.

Hebbes, T.R., Thorne, A.W. and Crane-Robinson, C. (1988) A direct link between core histone acetylation and transcriptionally active chromatin. *EMBO Journal* 7, 1395–1402.

Hu, Z.-Q., Henderson, G.I., Mock, D.M. and Schenker, S. (1994) Biotin uptake by basolateral membrane of human placenta: normal characteristics and role of ethanol. *Proceedings of the Society for Biology and Experimental Medicine* 206, 404–408.

Hymes, J., Fleischhauer, K. and Wolf, B. (1995) Biotinylation of histones by human serum biotinidase: assessment of biotinyl-transferase activity in sera from normal individuals and children with biotinidase deficiency. *Biochemical and Molecular Medicine* 56, 76–83.

Janeway, C.A., Travers, P., Walport, M. and Capra, J.D. (1999) *Immuno Biology*, Garland Publishing/Elsevier, London.

Kaczmarek, L., Calabretta, B. and Baserga, R. (1985) Expression of cell-cycle-dependent genes in phytohemagglutinin-stimulated human lymphocytes. *Proceedings of the National Academy of Sciences USA* 82, 5375–5379.

Karl, P. and Fisher, S.E. (1992) Biotin transport in microvillous membrane vesicles, cultured trophoblasts and the isolated perfused cotyledon of the human placenta. *American Journal of Physiology* 262, C302–C308.

Knowles, J.R. (1989) The mechanism of biotin-dependent enzymes. *Annual Review of Biochemistry* 58, 195–221.

Lee, D.Y., Hayes, J.J., Pruss, D. and Wolffe, A.P. (1993) A positive role for histone acetylation in transcription factor access to nucleosomal DNA. *Cell* 72, 73–84.

Lernhardt, W. (1990) Fatty acid requirement of B lymphocytes activated *in vitro*. *Biochemical and Biophysical Research Communications* 166, 879–885.

Moskowitz, M. and Cheng, D.K.S. (1985) Stimulation of growth factor production in cultured cells by biotin. In: Dakshinamurti, K. and Bhagavan, H.N. (eds) *Biotin*. New York Academy of Sciences, New York, pp. 212–221.

Murray, A. and Hunt, T. (1993) *The Cell Cycle*. Oxford University Press, New York.

Prasad, P.D., Wang, H., Kekuda, R., Fujita, T., Fei, Y.-J., Devoe, L.D., Leibach, F.H. and Ganapathy, V. (1998) Cloning and functional expression of a cDNA encoding a mammalian sodium-dependent vitamin transporter mediating the uptake of pantothenate, biotin, and lipoate. *Journal of Biological Chemistry* 273, 7501–7506.

Said, H.M., Ma, T.Y. and Kamanna, V.S. (1994) Uptake of biotin by human hepatoma cell line, Hep G(2): a carrier-mediated process similar to that of normal liver. *Journal of Cellular Physiology* 161, 483–489.

Schenker, S., Hu, Z., Johnson, R.F., Yang, Y., Frosto, T., Elliott, B.D., Henderson, G.I. and Mock, D.M. (1993) Human placental biotin transport: normal characteristics and effect of ethanol. *Alcohol Clinical and Experimental Research* 17, 566–575.

Stanley, J.S., Griffin, J.B. and Zempleni, J. (2001) Biotinylation of histones in human cells: effects of cell proliferation. *European Journal of Biochemistry* 268, 5424–5429.

Stanley, J.S., Mock, D.M., Griffin, J.B. and Zempleni, J. (2002) Biotin uptake into human peripheral blood mononuclear cells increases early in the cell cycle, increasing carboxylase activities. *Journal of Nutrition* 132, 1854–1859.

Tsai, F.C.H. and Gardner, D.K. (1994) Nicotinamide, a component of complex culture media, inhibits mouse embryo development *in vitro* and reduces subsequent developmental potential after transfer. *Fertility and Sterility* 61, 376–382.

van den Berg, K.J. and Betel, I. (1973) Selective early activation of a sodium dependent amino acid transport system in stimulated rat lymphocytes. *FEBS Letters* 29, 149–152.

Velazquez, A., Zamudio, S., Baez, A., Murguia-Corral, R., Rangel-Peniche, B. and Carrasco, A. (1990) Indicators of biotin status: a study of patients on prolonged total parenteral nutrition. *European Journal of Clinical Nutrition* 44, 11–16.

Velazquez, A., Teran, M., Baez, A., Gutierrez, J. and Rodriguez, R. (1995) Biotin supplementation affects lymphocyte carboxylases and plasma biotin in severe protein-energy malnutrition. *American Journal of Clinical Nutrition* 61, 385–391.

Waithe, W.I., Dauphinais, C., Hathaway, P. and Hirschhorn, K. (1975) Protein synthesis in stimulated lymphocytes. II. Amino acid requirements. *Cellular Immunology* 17, 323–334.

Wang, H., Huang, W., Fei, Y.-J., Xia, H., Fang-Yeng, T.L., Leibach, F.H., Devoe, L.D., Ganapathy, V. and Prasad, P.D. (1999) Human placental $Na^+$-dependent multivitamin transporter. *Journal of Biological Chemistry* 274, 14875–14883.

Williams, G.T., Lau, K.M.K., Coote, J.M. and Johnstone, A.P. (1985) NAD metabolism and mitogen stimulation of human lymphocytes. *Experimental Cell Research* 160, 419–426.

Wilson, A. and Freeman, J.W. (1996) Regulation of P120 mRNA levels during lymphocyte stimulation: evidence that the P120 gene shares properties with early and late genes. *Journal of Cellular Biochemistry* 60, 458–468.

Wolf, B., Heard, G.S., McVoy, J.R.S. and Grier, R.E. (1985) Biotinidase deficiency. *Annals of the New York Academy of Sciences* 447, 252–262.

Wolffe, A. (1998) *Chromatin*. Academic Press, San Diego, California.

Wood, H.G. and Barden, R.E. (1977) Biotin enzymes. *Annual Review of Biochemistry* 46, 385–413.

Zempleni, J. (2001) Biotin. In: Bowman, B.A. and Russell, R.M. (eds) *Present Knowledge in Nutrition*, 8th edn. International Life Sciences Institute, Washington, DC, pp. 241–252.

Zempleni, J. and Mock, D.M. (1998) Uptake and metabolism of biotin by human peripheral blood mononuclear cells. *American Journal of Physiology* 275, C382–C388.

Zempleni, J. and Mock, D.M. (1999a) The efflux of biotin from human peripheral blood mononuclear cells. *Journal of Nutritional Biochemistry* 10, 105–109.

Zempleni, J. and Mock, D.M. (1999b) Human peripheral blood mononuclear cells: inhibition of biotin transport by reversible competition with pantothenic

acid is quantitatively minor. *Journal of Nutritional Biochemistry* 10, 427–432.
Zempleni, J. and Mock, D.M. (1999c) Mitogen-induced proliferation increases biotin uptake into human peripheral blood mononuclear cells. *American Journal of Physiology* 276, C1079–C1084.
Zempleni, J. and Mock, D.M. (2000) Proliferation of peripheral blood mononuclear cells increases riboflavin influx. *Proceedings of the Society for Experimental Biology and Medicine* 225, 72–79.
Zempleni, J., Stanley, J.S. and Mock, D.M. (2001) Proliferation of peripheral blood mononuclear cells causes increased expression of the sodium-dependent multivitamin transporter gene and increased uptake of pantothenic acid. *Journal of Nutritional Biochemistry* 12, 465–473.

# 6　营养和细胞凋亡

John C. Mathers
（英国纽卡斯尔大学临床医学院人类营养研究中心，泰恩河的纽卡斯尔）

## 生命和细胞死亡

地球上最普通的多细胞生物体是线虫（*Caenorhabditis elegans*），它可在混合肥料中大量繁殖，在人肠道中定殖并可传播疾病，如河盲症和象皮病。由于体型小（长1mm）、易生长和生命周期短（约 2 周），线虫广泛被应用于发育生物学和衰老过程的研究。发育初期线虫有 1090 个细胞，但成熟的雌雄同体只含 959 个体细胞，其余的 131 个细胞因凋亡而被清除。大部分凋亡细胞在胚胎发生的早期就死亡（受精后的 250～450min），并且其中很多（105/131）是神经细胞（Xue *et al.*，2002）。凋亡是一种程序性细胞死亡，根据组织学和生化特性鉴定，包括细胞膜起泡、细胞收缩、染色质浓缩和 DNA 碎裂。与细胞坏死（necrosis）不同，凋亡细胞的内容物没有被释放到细胞间隙，因此细胞凋亡不会引发免疫反应。在凋亡末期，被称为凋亡小体（apoptotic body）的细胞残留体迅速被周围的细胞吞噬。在正常情况下，氨基磷酸酯（aminophospholipid）如磷脂酰丝氨酸或磷脂酰乙醇胺位于浆膜内侧，凋亡时位于细胞膜表面。凋亡细胞上的氨基磷酸酯与糖蛋白乳脂球表皮生长因子（epidemal growth factor，EGF）-8 结合，然后被噬菌细胞吞噬（Hanayama *et al.*，2002）。

自 Kerr 等（1972）首次公布后，凋亡在约 30 年里一直被认为是细胞死亡的一种方式。但最近，它在正常生长发育及许多疾病的病原学和治疗中的重要性非常明显地显现出来。另一种程序性细胞死亡被称为类凋亡（paraptosis），其形态学和生化特性与细胞凋亡完全不同，不受凋亡抑制剂的影响（Sperandio *et al.*，2000）。类凋亡在神经系统的发育和疾病的恶化方面具有重要作用。

## 细胞凋亡在健康和疾病中的作用

从受精卵发育成婴儿包括 4 个关键过程：细胞分裂、分化、形态形成和细胞凋亡。正如在线虫（*C. elegans*）的发育过程中一样，由凋亡导致的细胞死亡是人类胚胎发育中不可缺少的组成部分，并且这种固有的死亡程序能对一系列细胞内外信号做出反应。胚胎中发生细胞凋亡的明显迹象包括器官的成形及手指间和足趾间网膜的去除。神经系统的发育过程先是产生过量的细胞，然后是无法建立突触连接的细胞发生凋亡（Renehan *et al.*，2001）。凋亡是内分泌腺胰腺发育的重要机制，更是新生儿 β 细胞群重新塑造的重要机制（Scaglia *et al.*，1997）。

胸腺退化一直被认为是营养不良（Prentice，1999）和免疫功能受损的敏感标志。

将母鼠哺乳的小鼠数量增加一倍来诱导小鼠严重营养不良，试验发现，营养不良组的小鼠胸腺细胞发生自然凋亡的速率比正常饲喂组的高 7 倍（Ortiz *et al.*，2001）。凋亡速率的提高导致营养不良时的胸腺萎缩（Ortiz *et al.*，2001），同时严重影响短期或长期的健康状况（Moore *et al.*，1997）。T 淋巴细胞的内稳衡反映了细胞凋亡与细胞增殖之间的平衡。细胞凋亡是由局部的白介素-2（IL-2）浓度所致，细胞增殖由不同抗原包括食物来源的抗原刺激（Lenardo *et al.*，1999）。Nur77 是一对孤儿核受体（orphan nuclear receptor）中的成员之一（另一种是 Nurr1），可与核受体类维生素 A X 受体（retinoid X receptor，RXR）形成异源二聚体，然后再与视黄酸（具有转录激活作用的维生素 A 衍生物）结合，激活基因转录（McCaffre *et al.*，2001）。Nur77 可能在免疫系统的凋亡和 T 细胞的阴性选择中发挥着重要作用（McCaffrey *et al.*，2001）。

除了极少数特例（如神经系统），在整个生命活动期，所有组织的细胞都在不断进行分裂，因此，凋亡是成人期必需的一个生命过程，以维持新生细胞和死亡细胞之间的平衡，保证细胞的正常数量和组织器官的完整性。事实上，人们发现，细胞增殖和凋亡途径是相互联系的（Evan and Littlewood，1998），尽管它们的生化机制完全不同。凋亡在伤口愈合期的组织修复、分娩后的子宫衰退和乳腺组织的泌乳后变化中也都具有重要作用。

## DNA 受损对细胞凋亡的影响

在电离辐射（使单链或双链断裂）、活性氧类物质（reactive oxygen species，ROS）（由于缺乏抗氧化剂攻击嘌呤和嘧啶环）和烷化剂（如食物中可形成 DNA 加合物的杂环胺）等因素不断作用下，细胞 DNA 受损，而且产生的错误也能同 DNA 一起进行复制。对于单细胞生物，若细胞要存活并发挥正常功能，DNA 修复是唯一的办法。相反，若修复的代价或风险太大，多细胞动物能通过凋亡来杀死受损细胞。凋亡可能是一种谨慎的作用方式，可清除受损的干细胞或具有巨大增殖能力的其他细胞（除有丝分裂后的细胞外）或生命周期短的细胞（肠上皮或表皮细胞）；而修复可能是一种“安全”的选择（Evan and Littlewood，1998）。凋亡的进化起源还不清楚，但有人提出，单细胞生物应该存在一种细胞死亡程序，作为应急策略来抑制相关个体的感染（Vaux，2002）。

作为一种防御方式，当 *ATM* 基因的蛋白质产物检测到损伤时，DNA 受损伤的细胞就阻滞在细胞周期 $G_1$-S 检查点。*ATM* 基因的遗传突变将导致罕见的隐性疾病共济失调-毛细血管扩张（ataxia-telangiectasia），其特征是小脑共济失调、眼睛血管扩张、免疫缺陷和生长受阻，极易发生癌变（Strachan and Read，1999）。ATM 蛋白把 DNA 损伤的信号传递给 *TP53*，导致 p53 肿瘤抑制蛋白的浓度提高。p53 被称为“基因组的监护者”，但具体途径目前还不清楚。四聚体形式的 p53 是转录因子，通过与 MDM-2 蛋白的相互作用，p53 被降解，所以正常情况下 p53 含量低（Evan and Littlewood，1998）。当 p53 浓度增加时，细胞周期被终止，为启动 DNA 修复或凋亡提供时间。DNA 损伤后，一些细胞会停止生长并进行修复，而其他细胞则发生凋亡，其原因目前仍不清楚。有人猜测，随着细胞修复能力的降低，细胞主要发生凋亡（Liu and Kulesz-Martin，2001）。最原始形式的 p53 在果蝇上被鉴定，它介导细胞凋亡，但细胞生长不停止，这表明传递凋亡信号是 p53 最初的作用（Liu and Kulesz-Martin，2001）。p53 非依赖性的凋亡可能通过促进 DNA 修复或导致胎儿死亡在防止畸形发生中发挥重要作用（Norimura *et al.*，1996）。*TP53* 是肿瘤中突变率或丢失率最高的基因，这表明它在基因组保护中起着关键作用。50%以上的人类肿瘤均含突变的 *TP53*，其中大部分（大于90%）是进化上较保守的 DNA 结合区域上的错义突变（Hollstein *et al.*，1994）。

p53 与 DNA 之间的结合可以是序列特异性的，也可以是序列非特异性的，这在很大程度上取决于 p53 分子的还原状态，因为这种结合需要一个锌指蛋白和超过 7 个半胱氨酸残基的参与（Liu and Kulesz-Martin，2001）。越来越多的试验证实，氧化应激途径和 DNA 链断裂产生的信号由 p53 整合（Liu and Kulesz-Martin，2001），这也进一步证实了 p53 在保护基因组免受饲粮诱导或其他形式的损伤中发挥关键作用。家族性乳房癌基因 *BRCA1* 和 *BRCA2* 也可能是 $G_1$-S 期的损伤检查点（Strachan and Read，1999）。越来越多的试验证实，一些化学物质能使 DNA 发生烷基化，形成$O^6$-烷基鸟嘌呤加合物。它们还能利用 DNA 错配修复（MMR）途径的 MutSα 支路产生信号，启动凋亡，该作用不受细胞 p53 状态的影响（Hickman and Samson，1999）。Fishel（2001）提出，这些 MMR 蛋白作为特殊的“直接感受器”，可以连接 DNA 修复途径或细胞凋亡途径。

## 凋亡的过程

尽管人们认识到凋亡对平衡有丝分裂的作用已有一段时间了，知道凋亡过程的主要特征也已近 20 年（Raff，1992）。但以前的研究集中在细胞增殖上，仅在过去的十多年，人们才对凋亡进行了广泛、专门的研究。人们现在已经知道，大多数但不是所有的人类细胞都能表达细胞自杀性凋亡的必需蛋白，而且当周围细胞不发出存活信号时，自杀性凋亡就成了一种默认的死亡途径（Raff，1992；Thompson，1995）。凋亡过程非常迅速，从噬菌作用的启动到整个过程的结束仅需要不到一个小时（McCarthy，2002）。细胞凋亡也是个随机过程，也就是说，某一组织中不是所有的细胞都同时死亡，所以凋亡的细胞数量/速率的组织学评定非常困难（McCarthy，2002）。从参与凋亡过程的分子角度来说，它包括 4 个不同的阶段：①存活信号；②死亡信号；③调节基因和蛋白质；④效应分子。Jacobson 和 McCarthy（2002）对凋亡的分子生物学进行了全面的阐述，凋亡过程见图 6.1。

### 存 活 信 号

Raff（1992）首先提出了细胞存活的群体调节观点，他引用的例子是靶细胞分泌的神经营养因子对脊椎动物神经细胞的存活的作用。在没有睾酮的情况下，腹侧前列腺的上皮细胞发生凋亡；在没有促肾上腺皮质激素的情况下，肾上腺皮质细胞同样也会发生凋亡。T 淋巴细胞需要 IL-2，而内皮细胞需要纤维原细胞生长因子（属于其他生长因子）来抑制凋亡。作为只允许组织中正常细胞存活的机制，存活信号和凋亡之间的对立关系普遍存在（图 6.2）。若失去与周围正常细胞的联系，细胞很可能死亡（Ashkenazi and Dixit，1998）。癌细胞只有经过突变获得了在没有外部存活信号情况下也能存活的能力，才有可能转移（Raff，1992）。

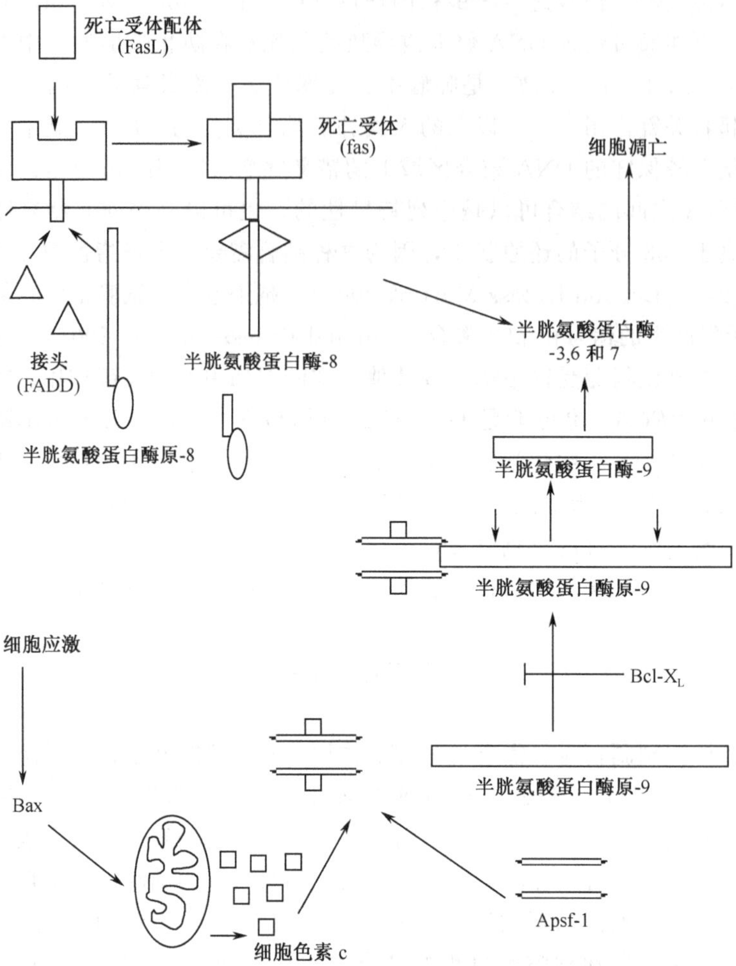


图 6.1　细胞凋亡的信号转导和发生过程中的主要步骤

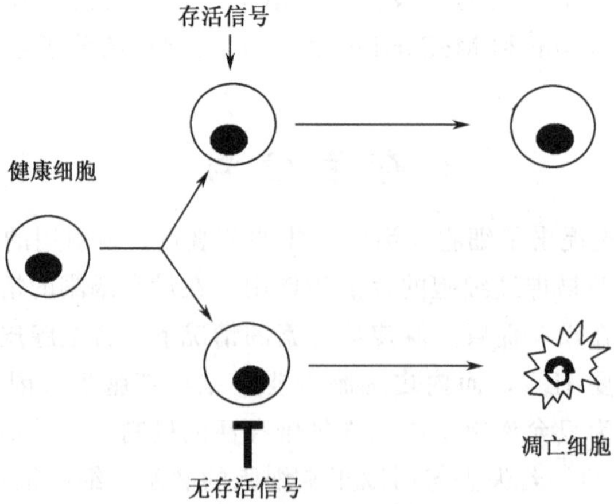


图 6.2　存活信号的缺失能启动细胞凋亡

# 死亡信号

很多刺激物质都能促进凋亡，但它们主要通过两个凋亡信号途径来发挥作用，即死亡受体途径和死亡受体非依赖性或线粒体途径（Gupta，2001）。

## 死亡受体途径

哺乳动物存在一种不正常细胞凋亡的直接死亡机制（Ashkenazi and Dixit，1998）。这种利他型自杀是由配体如肿瘤坏死因子（tumour necrosis factor，TNF）和 Fas 配体（FasL 或 CD95L）来介导的。这些死亡信号受体存在于细胞浆膜上，其编码基因属于 TNF 受体基因超家族，其中人们认识比较透彻的是 TNF 受体 1（TNF recepter，TN-FR1）和 Fas 受体（也称为 CD95 或 Apo1）（Ashkenazi and Dixit，1998）。这些受体均含细胞外结构域和细胞质结构域，细胞质结构域作为死亡结构域（death domain，DD）。TNF 和 FasL 与各自的受体结合后所发生的一系列反应有些不同（Ashkenazi and Dix-it，1998；Gupta，2001），但整个过程比较相似。FasL 是同源三聚体，在与受体（3 个 Fas 分子）结合后，使受体的 DD 聚集。然后再与接头蛋白 FADD（Fas 相关死亡结构域，Fas-associated death domain）结合，FADD 反过来可聚集半胱氨酸蛋白酶原-8（半胱氨酸蛋白酶-8 的酶原形式，下面将讨论）（Ashkenazi and Dixit，1998）。Fas 的 DD、FADD 和半胱氨酸蛋白酶原-8 结合形成一种死亡诱导信号复合物（death inducing sig-nalling complex，DISC），通过半胱氨酸蛋白酶原-8 的自催化裂解激活半胱氨酸蛋白酶-8，从而介导细胞凋亡（Gupta，2001）。

细胞凋亡的 Fas- FasL 途径在确保细胞免疫功能的正常发挥方面具有重要的作用：①细胞毒性 T 细胞和自然杀伤性细胞（natural killer，NK）对靶细胞的杀死；②T细胞库的选择和自我识别性 T 细胞的清除；③免疫应答结束后活化淋巴细胞的清除；④“免疫特权”组织如睾丸和眼睛中炎症/免疫细胞的清除（Gupta，2001）。*Fas* 基因或 *FasL* 基因突变与自身免疫性紊乱有关（Lenado *et al.*，1999）。

机体感染后，被激活的巨噬细胞和 T 细胞分泌 TNF，TNF 再与 TNFR1 结合，激活转录因子 NF-κB 和 AP-1，从而启动前炎症基因和免疫调节基因的转录（Ashkenazi and Dixit，1998；Gupta，2001）。然而，若蛋白质合成受到抑制，TNF 将促进细胞凋亡。当三聚体 TNF 与 3 个 TNFR1 分子结合后，TNFR1 的 DDS 与各种信号蛋白（包括 TRADD、FADD、TRAF-2 和 RIP）结合，并把它们聚集在一起。NF-κB 的激活可能需要 TRAF-2 和 RIP，而凋亡的信号转导需要 FADD。FADD 聚集并激活效应分子，然后调节半胱氨酸蛋白酶（Ashkenazi and Dixit，1998；Gupta，2001 ）。

## 线粒体途径

应激分子（ROS 和各种活性氮类物质）、化疗药物和紫外线照射均能导致线粒体损伤，促进细胞通过死亡受体非依赖性途径凋亡（Gupta，2001 ）。在这种情况下，线粒体释放的细胞色素 c（也称为 Apaf-2）可转导凋亡信号，它与凋亡蛋白酶激活因子-1（Apaf-1）（Zou *et al.*，1997）和 Apaf-3（也称为半胱氨酸蛋白酶-9）结合形成“凋亡小体”，从而催化激活半胱氨酸蛋白酶-3，导致凋亡（Li *et al.*，1997）。在凋亡敏感性

B 细胞小鼠淋巴瘤细胞的凋亡早期，UCP2（线粒体解偶联蛋白的小家族成员之一）的表达增强；而对于凋亡非敏感性 B 细胞小鼠淋巴瘤细胞，UCP2 的表达没有增强（Voehringer *et al.*，2000）。然而，人们目前还不能确定 UCP 是否直接诱导凋亡（Collins *et al.*，2001）。

## 调节基因和蛋白质

Bcl-2 家族蛋白在决定细胞是存活还是凋亡方面可能具有关键作用（Tsujimoto，2002）。这个家族包括促凋亡蛋白和抗凋亡蛋白，在很多情况下，有 4 个保守的 Bcl-2 同源（BH）结构域，分别被命名为 BH1～BH4（Reed，1998；Gross *et al.*，1999）。在正常细胞中，抗凋亡蛋白是线粒体、细胞核和内质网的内在膜蛋白，而促凋亡蛋白位于细胞质和细胞骨架的不同位置（Gross *et al.*，1999）。Bcl-2 家族蛋白对细胞死亡的调节机制还不十分清楚，但促凋亡蛋白和抗凋亡蛋白的相对丰度及翻译后修饰（如磷酸化或裂解）可能对激活这些蛋白质起决定性作用。接到细胞死亡信号或细胞损伤后，促凋亡蛋白可能发生构象变化，使 BH3 结构域暴露，移位并停留在线粒体外膜（Gross *et al.*，1999）。在此，它们可能与同类分子或家族中的抗凋亡蛋白结合形成二聚体或多聚体，使膜的通透性发生变化，并释放细胞色素 c（Gupta，2001）。Gross 等（1999）认为，Bcl-2 家族的抗凋亡蛋白可能"守卫线粒体门户"，而促凋亡蛋白只有在获取死亡信号后"才能进入"。这些促凋亡分子可能占据了整个细胞的战略位置，作为细胞损伤的哨兵。因此，在受到各种损伤后，普通的凋亡途径可能被激活（Gross *et al.*，1999）。

## 效应分子

对于线虫（*C. elegans*），诱导凋亡需要 *ced-3* 和 *ced-4* 两种基因，这两种基因常被 *bcl-2* 类似基因 *ced-9* 所抑制（White，1996）。*ced-3* 的哺乳动物同源蛋白是 IL-1β 转移酶（interleukin-1β-converting enzyme，ICE 或半胱氨酸蛋白酶-1），它能把 IL-1β 加工成为参与炎症反应的成熟形式（Thornberry *et al.*，1992；Yuan *et al.*，1993）。在凋亡过程中，因特定的细胞蛋白质被半胱氨酸蛋白酶裂解，从而导致细胞死亡。半胱氨酸蛋白酶是一系列具有相似的氨基酸序列、结构和底物特异性的蛋白酶（Thornberry and Lazebnik，1998）。这些蛋白酶的活性部位含有半胱氨酸残基，能在靶蛋白的特异天冬氨酸残基部位进行切割（Cohen，1997）。半胱氨酸蛋白酶基因的表达物是半胱氨酸蛋白酶酶原，蛋白质降解后，产生一个大亚基（约 20kDa）和一个小亚基（约 10kDa），这两个亚基再结合，形成一个异源二聚体，两个异源二聚体再结合成具有催化活性的四聚体（Thornberry and Lazebnik，1998）。至少有 14 种哺乳动物半胱氨酸蛋白酶已被分离鉴定，并根据最适作用底物的断裂位点将它们进行了分类（Roy and Cardone，2002）。

半胱氨酸蛋白酶杀死细胞的途径为：①使凋亡抑制因子失活，如 Bcl-2 蛋白；②破坏细胞结构，如核纤层蛋白（组成核纤层）；③把调节部位和催化部位分开，从而解除蛋白质调节，如 DNA-$PK_{CS}$（参与 DNA 修复）（Thornberry and Lazebnik，1998）。目前认识最透彻的半胱氨酸蛋白酶作用底物是 DNA 修复酶：多聚（ADP-核糖）聚合酶（PARP），其他的作用底物包括细胞周期调节蛋白 Rb、固醇调节元件结合蛋白（SREBP-1 和 SREBP-2）（Cohen，1997）和 APC（Wnt 信号传递途径的关键成分）

(Browne *et al.*, 1994)。

级联活化过程始于半胱氨酸蛋白酶原-8转化成半胱氨酸蛋白酶-8（死亡受体途径）或半胱氨酸蛋白酶原-9转化成半胱氨酸蛋白酶-9（线粒体途径），导致效应分子（半胱氨酸蛋白酶-3和-7）的激活（Cohen，1997）。如果（可能）所有细胞（包括生命周期长的神经细胞）都存在半胱氨酸蛋白酶的构成性表达，则所有细胞都有可能发生凋亡（默认途径），所以半胱氨酸蛋白酶的激活受到严格的控制（Thornberry and Lazebnik，1998）。似乎对半胱氨酸蛋白酶的调节发生在细胞凋亡的早期阶段，由Bcl-2及其家族成员发挥主要作用（White，1996；Gross *et al.*，1999）。

## 疾病预防和治疗中细胞凋亡的调节

组织内稳衡是指有丝分裂产生的新细胞数目与通过凋亡或其他方式死亡的细胞数目相同。越来越多的研究表明，许多疾病的发生都涉及凋亡过度或凋亡抑制（表6.1）。例如，对谷蛋白有遗传性过敏的人，在食入谷蛋白后，肠上皮细胞的凋亡速率提高了4倍，超过小肠上皮腺窝中细胞的增殖速率，导致小肠绒毛萎缩（Moss *et al.*，1996）。而且，由于食物中大分子物质没有被降解成结构单体，不能被吸收，会产生一系列主动的腹内疾病症状（乳糜泄的典型症状，译者注）。在摄入不含谷蛋白（面筋）的食物后，小肠凋亡指数恢复到正常水平（Moss *et al.*，1996）。细胞凋亡可能被其他以“平坦性病变”为特征的肠道疾病抑制，也可能被如化疗、叶酸或维生素$B_{12}$的缺乏等因素诱发（Lewin and Weinstein，1996）。肠黏膜细胞异常高的凋亡速率可能是溃疡性结肠炎的主要原因（Strater *et al.*，1997）。获得性免疫缺陷综合征（acquired immunodeficiency syndrome，AIDS）的发病与$CD4^+$ T细胞的缺失密切相关，当可溶性病毒产物gp120与CD4受体结合后，可介导$CD4^+$ T细胞发生凋亡（Thompson，1995）。细胞凋亡的增加可能还与其他几种疾病有关，包括阿尔茨海默氏病、骨关节炎、骨质疏松症（Thompson，1995）及年龄相关性视网膜退化（被称为色素性视网膜炎）（Davidson and Steller，1998）。癌症是由凋亡抑制导致的组织不正常生长的最好例证，下面将详细叙述。

**表6.1　与细胞增殖和凋亡失衡有关的疾病**

| 与凋亡抑制有关的疾病 | 与凋亡增强有关的疾病 |
|---|---|
| 癌症 | 艾滋病 |
| 自身免疫性疾病 | 神经变性性疾病 |
| 病毒感染性疾病 | 脊髓发育不良综合征 |
| | 缺血性损伤 |
| | 毒素诱导的肝病 |

摘自Thompson（1995）。

异常凋亡在许多疾病发生过程中及程序性细胞死亡途径的关键步骤中起基本作用，对这些知识的认识使人们对凋亡调节及其治疗措施开展了更广泛的研究。主要研究对象包括：①死亡信号分子；②*bcl-2*基因家族（White，1996）；③半胱氨酸蛋白酶（Thornberry and Lazebnik，1998）。有些物质可能通过选择性地阻遏死亡信号来抑制细胞凋亡，如与 Fas 或 TNFR1 结合的物质，而有些肿瘤表达假受体来阻止 FasL 结合的 NK 细胞对肿瘤细胞的致死作用（Pitti *et al.*，1998）。Bcl-2 的过度表达使细胞免受凋亡（White，1996），因此，对不同 Bcl-2 家族成员表达的调节为促进凋亡或抑制凋亡提供了思路。血管紧张素转换酶抑制剂和人类免疫缺陷病毒（human immunodeficiency virus，HIV）蛋白酶抑制剂是两类抑制蛋白酶活性的药物（Thornberry and Lazebnik，1998）。然而，目前面临的主要难题是组织细胞特异性的问题，即开发出的有效药物要能选择性地促进或抑制疾病组织的细胞凋亡。

# 肿　　瘤

从本质上说，肿瘤是一种由突变引起的遗传疾病，是由约 6 个特异基因中的两个拷贝发生缺失或遗传沉默所致。这些突变包括所有 DNA 损伤形式，它们使细胞获得选择性生长的权利。同时，大多数肿瘤发生的试验模型表明，肿瘤的生长是由细胞分裂的增强所致。但 Tomlinson 和 Bodmer（1995）提出了包括细胞分化和凋亡变化的数学模型，该模型认为细胞分化和凋亡对肿瘤的生长有极大的影响，并能更好地解释良性肿瘤、恶化前的生长及肿瘤生长速率（包括衰退）的变化，这也提示肿瘤的生长是通过细胞增殖的加强和（或）细胞凋亡的抑制来实现的。细胞增殖的增强是指在“增殖衰老”（proliferative senescence）（可能的传递信号是端粒缩短）之前更多细胞发生分裂。随着肿瘤块不断增大，肿瘤细胞常处于缺氧状态，从而激活 p53 并促进凋亡（Graeber *et al.*，1996）。p53 是发生由端粒功能失调导致的凋亡所不可缺少的（Karleseder *et al.*，1999）。Lowe 和 Lin（2000）推测，直到肿瘤发展到缺氧状态或端粒缩短达到一定程度，p53 突变的细胞才具备选择优势，这也解释了 p53 突变通常发生在肿瘤发育的晚期。

# 结肠直肠癌

肠上皮由于其独特的结构特点如快速的细胞动力学且比较容易获取，已成为研究细胞增殖、细胞移行、细胞分化的理想组织，最近也常被用于细胞死亡的研究。结肠直肠癌（colorectal cancer，CRC）是一种由基因损伤所致的结肠上皮病，这些基因的正常表达物参与调节结肠细胞的基本代谢过程。事实上，我们对肿瘤发生的遗传突变的认识大部分来自于对 CRC 的研究。研究发现，CRC 的发生发展是包含多个步骤的过程。首先是形成利用显微镜可观测到的异常隐窝病灶，再形成良性的腺瘤性息肉（腺瘤），发生癌变，并迅速转移。这些组织学上划分的阶段是由至少五、六个基因发生一系列遗传变化（由于突变或启动子甲基化引起的基因沉默）所致（Fearon and Vogelstein，1990）。根据 Kinzler 和 Vogelstein（1996）持家基因假说，增殖细胞群含有一种基因，它具有整体调节的作用，以确保组织中细胞数目保持稳定。肠上皮的持家基因是 *APC*，它编码一种具有多功能的大蛋白质，主要调节 Wnt 信号传递途径（Peifer and Polakis，2000）。顾名思义，持家基因的突变（肠上皮的 *APC*）促进细胞增殖，从而使细胞在生成和死亡之间失去平衡，这使得突变细胞具有选择优势。研究表明，CRC 形成过程中凋亡受到抑制（Bedi *et al.*，1995），其原因可能是 *APC* 的缺失。因为在几乎所有的

CRC 肿瘤的早期，均能观察到 *APC* 的缺失（Fearon and Vogelstein，1990）。该观点已被 HT29 细胞（缺失功能性 APC 蛋白）的全长 *APC* 的转染和表达试验证实，试验发现，全长 *APC* 的转染和表达促进了细胞凋亡，降低了细胞生长（Morin *et al.*，1996）。在凋亡早期，APC 蛋白在 Asp777 后被半胱氨酸蛋白酶-3 降解形成一个特殊、稳定的 90kDa 片段（Browne *et al.*，1994，1998；Webb *et al.*，1999）。对于结肠直肠癌，*TP53* 基因发生突变的频率非常高，而对于腺瘤，*TP53* 基因通常不发生突变（Fearon and Vogelstein，1990）。*TP53* 基因的突变使肿瘤晚期的凋亡调节系统失去一个重要成员。

## 凋亡作为结肠直肠癌化学预防的调节目标

流行病学研究表明，饮食差异是世界各地 CRC 发生率产生差异的原因（Doll and Peto，1981）。然而，个体内和个体间的饮食习惯和食物化学复杂性有相当大的变化，传统的流行病学是个粗略的研究手段，无法分辨出食物中发挥作用的具体成分，也无法确定多大的剂量具有预防保护作用，而又是多大的剂量才具有潜在的毒性（Mathers and Burn，1999）。通过体外和动物研究及对自愿者的干预研究，CRC 的化学预防已取得巨大进展（Burn *et al.*，1998；Mathers，2000）。在 CRC 肿瘤发生过程中，细胞凋亡受到抑制，如果杀死 DNA 受损细胞是避免肠细胞发生 *APC* 基因或其他关键基因突变的有效方法，那么，具有增强细胞凋亡作用的养分或其他食物成分的鉴定则能促进高效化学预防药物的开发。

### 长链脂肪酸

食物中脂肪的含量影响 CRC 的发病率，但脂肪的类型影响程度更大（英国卫生部，1998；Whelan and McEntee，2001）。以啮齿动物为模型的研究结果表明，高脂肪饲粮提高了肿瘤的发生率。同时，当脂肪的摄入量固定时，用 *n*-3 系列的脂肪酸（特别是二十碳五烯酸，eicosapentaenoic acid，EPA）替代 *n*-6 系列的脂肪酸（特别是花生四烯酸）能降低肠癌的发生率（Petrik *et al.*，2000；Whelan and McEntee，2001）。饲喂鱼油［富含 *n*-3 多不饱和脂肪酸（polyunsaturated fatty acid，PUFA）包括 EPA 和二十二碳六烯酸（docosahexaenoic acid，DHA）］能抑制二甲肼（dimethylhydrazine，DMH）对小鼠结肠癌的诱导作用（Latham *et al.*，1999）。令人奇怪的是，给 $Apc^{\Delta716}$ 小鼠饲喂 DHA 使雌性小鼠的肠息肉减少，但对雄性小鼠无作用（Oshima *et al.*，1995）。多种 PUFA 对上皮细胞有细胞毒性，而凋亡可能是细胞毒性的一个要素（Johnson，2001）。EPA 促进 CRC 细胞系的凋亡，而且鱼油中的脂肪酸对 CRC 的预防作用可能是由 DNA 损伤的腺窝细胞更多地发生凋亡所引起的（Latham *et al.*，1999）。*n*-3 PUFA 对细胞凋亡的诱导作用可能是通过脂肪过氧化产物来介导的，尤其对于 P450 过度表达的肿瘤细胞更是如此（Stoll，2002）。试验表明，细胞谷胱甘肽的缺失促进了凋亡，而抗氧化剂则阻遏了凋亡（Clarke *et al.*，1999；Latham *et al.*，2001），因此有人猜测 *n*-3 PUFA 对细胞凋亡的促进作用可能是由氧化应激的提高所引起的，但也可能存在其他机制（Johnson，2001）。越来越多的试验表明，过氧物酶体增殖激活受体（peroxisome proliferator-activated receptor，PPAR）γ 在这些作用机制中起着关键性作用。PPARγ 是核激素受体超家族的成员之一，在结肠直肠腺瘤中的表达水平比较高（Sarraf *et al.*，1998）。长链 PUFA 是激活 PPARγ 的配体之一（Kliewer *et al.*，1997），可诱导细胞的末期分化和凋亡。活化的 PPARγ 抑制环氧化酶（cyclooxygenase，COX）-2 的活性，促进人结肠癌细胞系 HT29 的凋亡（Yang and Frucht，2001）。

共轭亚油酸（conjugated linoleic acid，CLA）是一类 C18 脂肪酸的各种几何异构体和位置异构体的总称。由于瘤胃细菌能使 PUFA 发生部分氢化而合成 CLA，所以天然存在于反刍动物的奶和组织中。在过去几年，CLA 已成为具有巨大发展潜力的抗癌物质（Parodi，1997），但对其作用方式的认识却较少。Park 等（2001）推测 CLA 可能是 COX 的作用底物。同亚油酸、亚麻酸和花生四烯酸一样，CLA 也能进行链的延长和去饱和。摄入的 CLA 与脂肪酸之间存在竞争，且能产生各种形式的类花生酸终产物。他们指出，在 DMH 处理的大鼠饲粮中添加 1%CLA 可降低结肠癌的发生率，这可能与结肠黏膜的前列腺素 $E_2$ 和凝血噁烷 $B_2$ 浓度的降低有关（Park *et al.*，2001）。饲喂 CLA 大鼠与对照组相比，结肠末端扁平黏膜（flat mucosa）的凋亡指数高 2.5 倍，但 Park 等（2001）无法确定是 CLA 提高细胞凋亡的信号转导途径。形态正常的结肠黏膜细胞凋亡率提高从而去除 DMH 损伤细胞可能是 CLA 抗癌的原因。或者这种损伤可能是 CLA 直接的作用结果，细胞凋亡速率的提高是机体正常自我平衡过程，而与肿瘤抑制无关。饲喂含 3% CLA 的饲粮对 $Apc^{Min/+}$ 小鼠的肠肿瘤增殖无影响（Petrick *et al.*，2000），但值得注意的是，以 $Apc^{Min/+}$ 小鼠模型进行的试验，绝大多数肿瘤发生在小肠，而 CLA 对肿瘤发生的作用可能具有位点特异性。

## 脂溶性维生素

流行病学研究表明，多吃水果和蔬菜能有效地预防 CRC（和其他癌症），这已无可非议（英国卫生部，1998），但这些食物中具有预防功能的具体成分还不清楚。蔬菜和水果都富含多种抗氧化物质，其中研究最广泛的是 β-胡萝卜素。然而，研究发现，高剂量的 β-胡萝卜素对中年吸烟男子或用石棉进行过前处理的 β-胡萝卜素对人体不但不具有预防 CRC 的作用，还提高了肺癌的发生率（α-生育酚、β-胡萝卜素癌症预防研究组，1994）。在某一特定剂量下，β-胡萝卜素在氧化应激高的肺组织中可能以 β-胡萝卜素前体物质的形式发挥作用，而不是作为抗氧化剂发挥作用。最近研究表明，β-胡萝卜素能诱导细胞周期滞留在 $G_2$-M 期，促进细胞凋亡，并以剂量依赖方式抑制 CRC 细胞的生长（Palozza *et al.*，2002）。细胞周期停滞与细胞周期蛋白 A（$G_2$-M 过程中的关键调节蛋白）的表达下调有关，同时，随着凋亡的增强，抗凋亡蛋白 Bcl-2 和 Bcl-xL（但促凋亡蛋白 Bax 不发生变化）的表达受到抑制（Palozza *et al.*，2002）。各种 CRC 细胞系把 β-胡萝卜素不同程度地融合到细胞膜上，β-胡萝卜素融合量越多，细胞生长受抑制的程度越强，细胞凋亡的提高程度越大，然而发生这种现象的原因尚不清楚（Palozza *et al.*，2002）。

用高剂量 β-胡萝卜素培养的结肠直肠腺癌细胞的凋亡可能被 ROS 诱导（Palozza *et al.*，2001），正如前面对长链 PUFA 的讨论。β-胡萝卜素对凋亡的诱导作用可能是通过 p53 非依赖性的机制发生的，因为 Palozza 等（2002）研究的几种细胞系均含有 p53 的突变形式。这些细胞的氧化损伤可能是通过 MMR-依赖性的机制来察觉的，然后引起死亡信号并通过线粒体途径来传导。试验所采用的 $\beta$-胡萝卜素浓度可在人血浆中被观察到，从而使这些研究具有特殊的实际意义（Palozza *et al.*，2002）。维生素 A 的衍生物 9-顺式-视黄酸是 RXR 的配体，与特异性反应元件结合后，诱导基因表达。而 RXR 可与 PPAR 结合形成异源二聚体。HT29 结肠癌细胞与特异 PPARγ 配体和 9-顺式-视黄酸一起培养，增强了 PPARγ 配体对凋亡的促进作用，但是 9-顺式-视黄酸单独作用的效果甚微（Yang and Frucht，2001）。

研究也发现，维生素 E 能抑制 CRC 细胞的生长，促进 CRC 细胞的凋亡（Chinery *et al.*，1997）。与 $\beta$-胡萝卜素的作用方式类似，维生素 E 的这些作用并不依赖于 p53，可能是由于 $p21^{WAF1/CIP1}$ 和 c/EBPB 的激活引起细胞周期停滞所致（Chinery *et al.*，1997）。维生素 E 的作用机制还存在一些矛盾，因为 ROS 可能通过激活前列腺凋亡反应蛋白-4（Par-4）来发挥其对凋亡的诱导作用（Meydani *et al.*，2001）。再则，若用维生素 E 对细胞培养进行前处理，则可抑制 Par-4 的诱导作用（Chan *et al.*，1999）。

## 植物次级代谢产物

### 异硫氰酸酯

十字花科植物（如椰菜和花椰菜）含有相当高浓度的次级代谢产物——硫代葡糖苷和黑芥子酶（硫葡糖甙裂解酶，EC3.2.3.1）。植物组织的物理损伤（如砍伤或由咀嚼造成的损伤）使硫代葡糖苷充分接触黑芥子酶，且被分解成异硫氰酸酯和其他产物。异硫氰酸酯被认为是芸苔发挥抑癌作用的重要中介物质（Hecht，1999），能促进体内的细胞凋亡（Smith *et al.*，1998）。根据其抗癌作用的强弱，发现最有效的异硫氰酸酯之一是从椰菜获取的莱菔子硫（sulphoraphane）。莱菔子硫把 HT29 癌细胞的生长阻抑在 $G_2$-M 期（与细胞周期蛋白 A 和 B1 表达的提高有关的效果），同时提高细胞凋亡，阻遏细胞生长（Gamet-Payrastre *et al.*，2000）。人们在随后的研究中发现，在 HT29 癌细胞中未检测到抗凋亡 Bcl-2 蛋白，但莱菔子硫提高了促凋亡蛋白 Bax 的表达。Fimognari 等（2002）报道了在人 T 细胞白血病细胞中也观察到类似结果，因此，判断莱菔子硫对细胞周期的阻滞作用及对凋亡的诱导效应可能是一种普遍存在的现象。

### 姜黄色素

姜黄色素[1,7-二(4-羟基-3-甲氧苯基)1,6-庚二烯-3,5-二酮]是姜黄香科（spice tumeric）的主要色素，是从姜黄（*Curcuma longa*）中提取的，能抑制结肠腺瘤细胞系 HT29 和 HCT15 的生长（Hanif *et al.*，1997），降低氧化偶氮甲烷诱导的 F344 大鼠（Kawamori *et al.*，1999）和 $Apc^{min}$ 小鼠（Mahmoud *et al.*，2000）的肠肿瘤发生率。这些抗肿瘤效应与转录因子 NF-κB 活性降低有关（Singh and Aggarwal，1995），还与体外（Kuo *et al.*，1996）和体内（Samaha *et al.*，1997）细胞凋亡的增强有关。用 PhIP [2-氨基-1-甲基-6-苯基咪唑（4,5-b）吡啶——煮熟后的肉中含量最丰富的杂环胺] 饲喂 $Apc^{min}$ 小鼠，小肠近端的肿瘤增殖加大了 1 倍，同时使黏膜细胞的凋亡速率降低了 1

半。对野生型C57B1/6动物进行试验也出现类似结果（图6.3和图6.4）（Collett *et al*.，2001）。姜黄色素部分削弱了 $Apc^{min}$ 小鼠对PhIP诱导的凋亡的抵抗力，抑制了小肠近端PhIP诱导的肿瘤发生（图6.3和图6.4）（Collett *et al*.，2001）。研究已表明，姜黄色素能增强细胞的凋亡抵抗力，因为它抑制了TNF对NF-κB的激活作用（Plummer *et al*.，1999），提高了p53的表达（Jee *et al*.，1998）。姜黄色素对细胞凋亡抵抗力的增强作用可能与转录因子p53和NF-κB的平衡发生改变引起的 *Apc* 突变有关（Collett *et al*.，2001）。但是，姜黄色素抗肿瘤作用的具体机制仍有待进一步研究。

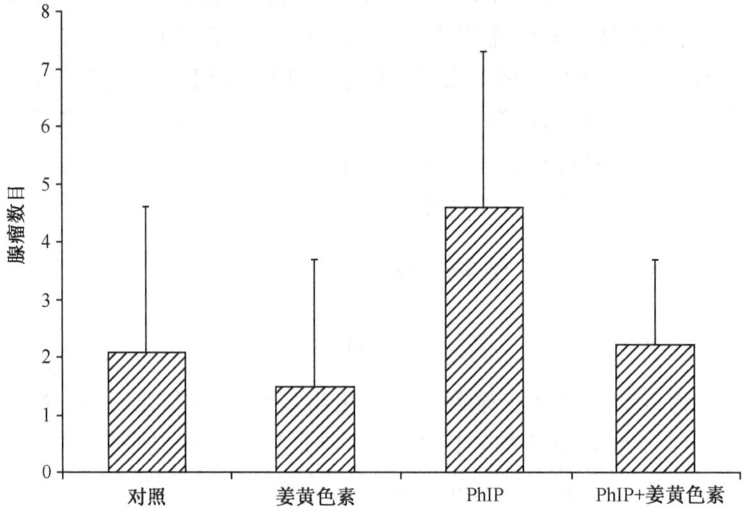


图6.3 断奶 $Apc^{min}$ 小鼠用姜黄色素和(或)PhIP饲喂10周后近端小肠的腺瘤增殖（每只小鼠的肿瘤数）情况（Collett *et al*.，2001）

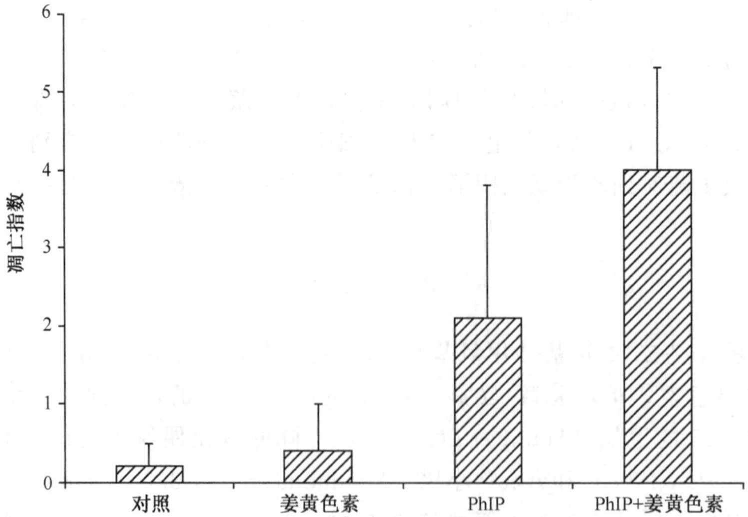


图6.4 断奶 $Apc^{min}$ 小鼠用姜黄色素和(或)PhIP饲喂10周后近端小肠黏膜的凋亡指数（Collett *et al*.，2001）

## 水 杨 酸

流行病学试验已有力地证明，有规律地服用非类固醇一类消炎药物（non-steroidal anti-imflammatory drug，NSAID）——阿司匹林（乙酰水杨酸）与 CRC 发病率的降低有关（Giovannucci *et al.*，1994）。这种效应对男女都一样，可能受剂量和时间的影响。由于阿司匹林是众所周知的 COX 抑制剂，而 COX-2 的增加又出现在肿瘤中，所以，人们过去认为阿司匹林的预防作用是由 COX 依赖性介导的（DuBois and Smalley，1996）。事实上是 COX-2 的特异性抑制剂的干扰作用抑制了小鼠（Oshima *et al.*，1996）和患家族性腺瘤息肉病者（Steinback *et al.*，2000）的肠息肉形成。然而，该保护作用不一定受到 COX-2 相关性途径的影响。阿司匹林的代谢产物（水杨酸）通过增强凋亡来抑制 CRC 细胞系的生长（Elder *et al.*，1996），而且 NSAID 对 CRC 细胞系的生长抑制可能不受 COX-1 和 COX-2 的活性的影响，而是增殖抑制与凋亡促进的共同作用结果（Richter *et al.*，2001）。PPARδ 是 NSAID 的受 APC 调节的目标分子（He *et al.*，1999），被 PPARδ 诱导的基因可能调节细胞末期分化和凋亡。水杨酸广泛分布于各种食物中，作为重要的植物次级代谢产物，在防御系统中占有重要地位。在食物加工过程中，也可以产生这种物质。由于食物成分数据库没有食物中水杨酸盐含量的资料，所以有关水杨酸盐摄入量的资料非常匮乏。人每天的摄入量可能只有几毫克（Janssen *et al.*，1996），这远远低于降低 CRC 发病率所需要的阿司匹林的量（Giovannuci *et al.*，1994），因此对于典型的西方消费者来说，食物来源的水杨酸不可能产生有效的预防 CRC 作用。

## 丁 酸

短链羧酸——丁酸是最有效的抗肿瘤药之一（D'Argenio，1996），由结肠中碳水化合物发酵产生，浓度可达 25mmol/L（Cummings *et al.*，1987）。1～5mmol/L 丁酸能抑制 CRC 细胞系的细胞生长，使细胞积聚在 $G_1$ 期，并通过 p53 非依赖性途径诱导细胞凋亡（Hague *et al.*，1993，1995；Heerdt *et al.*，1994）。尽管其具体途径目前还不清楚，但是研究已发现，丁酸是有效的组蛋白脱乙酰酶抑制剂（Wu *et al.*，2001），乙酰化作用在使基因组的构象发生变化后，使得其在正常情况下不能表达的促凋亡基因可能发生表达（Johnson，2001）。cDNA 微阵列技术的运用正在为研究丁酸处理后的 CRC 细胞的分化和凋亡途径提供新的思路（Mariadason *et al.*，2000；Della Ragione *et al.*，2001）。在大鼠饲粮中添加小麦麸提高了盲肠丁酸的浓度，同时提高了结肠上皮细胞的组蛋白乙酰化作用（Boffa *et al.*，1992）。这间接证明了丁酸对大鼠的抗癌作用（D'Argenio，1996）可能是通过组蛋白乙酰化来介导的。然而，COX-2 特异性物质增强了丁酸对表达 COX-2 的 CRC 细胞的生长抑制作用（Crew *et al.*，2000）。

一些碳水化合物（如抗性淀粉）能抵抗小肠酶的降解，而在大肠中被发酵成丁酸，因此这些抗性碳水化合物的摄入是提高丁酸摄入量的最简便的途径，但抗性碳水化合物的摄入量与结肠丁酸之间的数量关系还没有完全确定（Mathers *et al.*，1997）。丁酸及其常用的盐——丁酸钠均具有刺激性气味，不适合直接作为化学预防药物，因此开发有效、实用的丁酸盐衍生物已引起人们极大的兴趣。苯基丁酸和三丁酸甘油酯可阻遏细胞

周期由 $G_1$ 时段向 S 时段的过渡（与 CDK2 表达的降低有关），并抑制 Rb 蛋白的轻度磷酸化（hypophosphorylation），从而阻止 HT29 细胞的生长（Clarke *et al.*，2001）。苯基丁酸和三丁酸甘油酯也能通过激活半胱氨酸蛋白酶-3 来诱导 HT29 细胞的凋亡（Clarke *et al.*，2001）。同丁酸一样，苯基丁酸能抑制组蛋白脱乙酰酶的活性，因此三丁酸甘油酯也极有可能具有该作用，因为内源脂肪酶可降解三丁酸甘油酯产生丁酸。

## 总　结

人们现在普遍认为，细胞凋亡是正常生长发育所必需的生命活动，细胞凋亡的失调是许多疾病的重要发病原因。阐明凋亡发生及其调节的分子机制一直以来都是细胞生物学和分子生物学基础研究的重要内容。越来越多的证据表明，整个生命周期的营养供给在健康和疾病中发挥基础性的作用，并且许多作用是通过对凋亡的调节来实现的。已有试验（尤其是 CRC 领域中的研究）表明，几种食物来源的物质诱导了癌细胞的凋亡，因此它们作为化学预防药物具有巨大的发展潜力。然而，在这一领域中大部分研究都还停留在表象阶段，而对于这些物质的作用机制也只有粗略的认识。另外，在体外和动物试验中采用的剂量通常比人们从常规食物中摄食的量要高出许多倍。后基因组技术为全世界基因表达的检测做了充分的准备（虽然其成本仍较高），如微阵列技术和蛋白质组学，为营养研究工作者开始研究营养及食物中其他生物活性物质对细胞凋亡及最终对健康的影响提供了极大的机遇。

## 致　谢

感谢道恩·马兰德（Dawn Marland）对图 6.1 和莉斯·威廉斯（Liz Williams）对图 6.3 和图 6.4 提供的帮助。本实验室对营养和凋亡的研究受到食品标准局（合同号：NO3002）及生物技术和生物科学研究委员会的资助（授权号 13/DO9671）

### 参 考 文 献

Alpha-Tocopherol, Beta Carotene Cancer Prevention Study Group (1994) The effect of vitamin E and beta-carotene on lung cancer incidence and other cancers in male smokers. *New England Journal of Medicine* 330, 1029–1035.

Ashkenazi, A. and Dixit, V.M. (1998) Death receptors: signalling and modulation. *Science* 281, 1305–1308.

Bedi, A., Pasricha, P.J., Akhtar, A.J., Barber, J.B., Bedi, G.C., Giardello, F.M., Zehnbauer, B.A., Hamilton, S.R. and Jones, R.J. (1995) Inhibition of apoptosis during development of colorectal cancer. *Cancer Research* 55, 1811–1816.

Boffa, L.C., Lupton, J.R., Mariani, M.R., Ceppi, M., Newmark, H.L., Scalmati, A. and Lipkin, M. (1992) Modulation of epithelial cell proliferation, histone acetylation, and luminal short chain fatty acids by variation of dietary fiber (wheat bran) in rats. *Cancer Research* 52, 5906–5912.

Browne, S.J., Williams, A.C., Hague, A., Butt, A.J. and Paraskeva, C. (1994) Loss of APC protein expressed by human colonic epithelial cells and the appearance of a specific low-molecular-weight form is associated with apoptosis *in vitro*. *International Journal of Cancer* 59, 56–64.

Browne, S.J., Macfarlane, M., Cohen, G.M. and Paraskeva, C. (1998) The adenomatous polyposis coli protein and retinoblastoma protein are cleaved early in apoptosis and are potential substrates for caspases. *Cell Death and Differentiation* 5, 206–213.

Burn, J., Chapman, P.D., Bishop, D.T. and Mathers, J. (1998) Diet and cancer prevention: the Concerted Action Polyp Prevention (CAPP) Studies. *Proceedings of the Nutrition Society* 57, 183–186.

Chan, S.L., Tammariello, S.P., Estus, S. and Mattson,

M.P. (1999) Prostate apoptosis response-4 mediates trophic factor withdrawal-induced apoptosis of hippocampal neurones: actions prior to mitochondrial dysfunction and caspase activation. *Journal of Neurochemistry* 73, 502–512.

Chinery, R., Brockman, J.A., Peeler, M.O., Shyr, Y., Beauchamp, R.D. and Coffey, R.J. (1997) Antioxidants enhance the cytotoxicity of chemotherapeutic agents in colorectal cancer: a p53-independent function of p21WAF1/CIP1 via c/EBP beta. *Nature Medicine* 3, 1233–1241.

Clarke, R.G., Lund, E.K., Latham, P., Pinder, A.C. and Johnson, I.T. (1999) Effect of eicosapentaenoic acid on the proliferation and incidence of apoptosis in the colorectal cancer cell line HT29. *Lipids* 34, 1287–1295.

Cohen, G.M. (1997) Caspases: the executioners of apoptosis. *Biochemical Journal* 326, 1–16.

Collett, G.P., Robson, C.N., Mathers, J.C. and Campbell, F.C. (2001) Curcumin modifies $Apc^{min}$ apoptosis resistance and inhibits 2-amino 1-methyl-6-phenylimidazo[4,5-b] pyridine (PhIP) induced tumour formation in $Apc^{min}$ mice. *Carcinogenesis* 22, 821–825.

Collins, S., Cao, W., Dixon, T.M., Daniel, K.W., Onuma, H. and Medvedev, A.V. (2001) Body weight regulation, uncoupling proteins, and energy metabolism. In: Moustaïd-Moussa, N. and Berdanier, C.D. (eds) *Nutrient–Gene Interactions in Health and Disease*. CRC Press, Boca Raton, Florida, pp. 262–281.

Crew, T.E., Elder, D.J.E. and Paraskeva, C. (2000) A cyclooxygenase-2 (COX-2) selective non-steroidal anti-inflammatory drug enhances the growth inhibitory effect of butyrate in colorectal carcinoma cells expressing COX-2 protein: regulation of COX-2 by butyrate. *Carcinogenesis* 21, 69–77.

Cummings, J.H., Pomare, E.W., Branch, W.J., Naylor, C.P.E. and Macfarlane, G.T. (1987) Short chain fatty acids in the human large intestine, portal, hepatic and venous blood. *Gut* 28, 1221–1227.

D'Argenio, G., Cosenza, V., Delle Cave, M., Iovino, P., Della Valle, N., Lombardi, G. and Mazacca, G. (1996) Butyrate enemas in experimental colitis and protection against large bowel cancer in a rat model. *Gastroenterology* 110, 1727–1734.

Davidson, F.F. and Steller, H. (1998) Blocking apoptosis prevents blindness in *Drosophila* retinal degeneration mutants. *Nature* 391, 587–591.

Della Ragione, F., Criniti, V., Della Pietra, V., Borriello, A., Oliva, A., Indaco, S., Yamamato, T. and Zappia, V. (2001) Genes modulated by histone acetylation as new effectors of butyrate activity. *FEBS Letters* 499, 199–204.

Department of Health (1998) *Nutritional Aspects of the Development of Cancer*. Report on Health and Social Subjects 48. The Stationery Office, London.

Doll, R. and Peto, R. (1981) The causes of cancer: quantitative estimates of avoidable risks of cancer in the United States today. *Journal of the National Cancer Institute* 66, 1191–1308.

DuBois, R.N. and Smalley, W.E. (1996) Cyclooxygenase, NSAIDs, and colorectal cancer. *Journal of Gastroenterology* 31, 898–906.

Elder, D.J., Hague, A., Hicks, D.J. and Paraskeva, C. (1996) Differential growth inhibition by the aspirin metabolite salicylate in human colorectal tumor cell lines: enhanced apoptosis and *in vitro*-transformed adenoma relative to adenoma cell lines. *Cancer Research* 56, 2273–2276.

Evan, G. and Littlewood, T. (1998) A matter of life and cell death. *Science* 281, 1317–1322.

Fearon, E.R. and Vogelstein, B. (1990) A genetic model for colorectal tumorigenesis. *Cell* 61, 759–767.

Fimognari, C., Nüsse, M., Cesari, R., Iori, R., Cantelli-Forti, G. and Hrelia, P. (2002) Growth inhibition, cell-cycle arrest and apoptosis in human T-cell leukemia by the isothiocyanate sulforaphane. *Carcinogenesis* 23, 581–586.

Fishel, R. (2001) The selection for mismatch repair defects in hereditary nonpolyposis colorectal cancer: revising the mutator hypothesis. *Cancer Research* 61, 7369–7374.

Gamet-Payrastre, L., Li, P., Lumeau, S., Cassar, G., Dupont, M.A., Chevolleau, S., Gasc, N., Tulliez, J. and Tercé, F. (2000) Sulforaphane, a naturally occurring isothiocyanate, induces cell cycle arrest and apoptosis in HT29 human colon cancer cells. *Cancer Research* 60, 1426–1433.

Giovannicci, E., Rimm, E.B., Stampfer, M.J., Colditz, G.A., Ascherio, A. and Willett, W.C. (1994) Aspirin use and risk for colorectal cancer and adenoma in male health professionals. *Annals of Internal Medicine* 121, 241–246.

Graeber, T.G., Osmanian, C., Jacks, T., Housman, D.E., Koch, C.J., Lowe, S.W. and Giaccia, A.J. (1996) Hypoxia-mediated selection of cells with diminished apoptotic potential in solid tumours. *Nature* 379, 88–91.

Gross, A., McDonnell, J.M. and Korsmeyer, S.J. (1999) BCL-2 family members and the mitochondria in apoptosis. *Genes and Development* 13, 1899–1911.

Gupta, S. (2001) Molecular steps of death receptor and mitochondrial pathways of apoptosis. *Life Sciences* 69, 2957–2964.

Hague, A., Manning, A.M., Hanlon, K.A., Huschtscha, L.I., Hart, D. and Paraskeva, C. (1993) Sodium butyrate induces apoptosis in human colonic tumour cell lines in a p53-independent pathway: implications for the possible role of dietary fibre in the prevention of large-bowel cancer. *International Journal of Cancer* 55, 498–505.

Hague, A., Elder, D.J., Hicks, D.J. and Paraskeva, C. (1995) Apoptosis in colorectal tumour cells: induction by the short chain fatty acids butyrate, propionate and acetate and by the bile salt deoxycholate. *International Journal of Cancer* 60, 400–406.

Hanayama, R., Tanaka, M., Miwa, K., Shinohara, A., Iwamatsu, A. and Nagata, S. (2002) Identification of a factor that links apoptotic cells to phagocytes. *Nature* 417, 182–187.

Hanif, R., Qaio, L., Shiff, S.J. and Rigas, B. (1997) Curcumin, a natural plant phenolic food additive, inhibits cell proliferation and induces cell cycle changes in colon adenocarcinoma cell lines by a prostaglandin-independent pathway. *Journal of Laboratory and Clinical Medicine* 130, 576–584.

He, T.-C., Chan, T.A., Vogelstein, B. and Kinzler, K.W. (1999) PPARδ is a APC-regulated target of non-steroidal anti-inflammatory drugs. *Cell* 99, 335–345.

Hecht, S.S. (1999) Chemoprevention of cancer by isothiocyanates, modifiers of carcinogen metabolism. *Journal of Nutrition* 129, 768S–774S.

Heerdt, B.G., Houston, M.A. and Augenlicht, L.H. (1994) Potentiation by specific short-chain fatty acids of differentiation and apoptosis in human colonic carcinoma cell lines. *Cancer Research* 54, 3288–3293.

Hickman, M.J. and Samson, L.D. (1999) Role of DNA mismatch repair and p53 in signalling induction of apoptosis by alkylating agents. *Proceedings of the National Academy of Sciences USA* 96, 10764–10769.

Hollstein, M., Rice, K., Greenblatt, M.S., Soussi, T., Fuchs, R., Sorlie, T., Horvig, E., Smith-Sorensen, B., Montesano, R. and Harris, C.C. (1994) Database of p53 gene somatic mutations in human tumours and cell lines. *Nucleic Acids Research* 22, 3551–3555.

Janssen. P.L., Hollman, P.C., Reichman, E., Venema, D.P., van Staveren, W.A. and Katan, M.B. (1996) Urinary salicylate excretion in subjects eating a variety of diets shows that amounts of bioavailable salicylates are low. *American Journal of Clinical Nutrition* 64, 743–747.

Jee, S.H., Shen, S.C., Tseng, C.R., Chiu, H.C. and Kuo, M.L. (1998) Curcumin induces a p53-dependent apoptosis in human basal cell carcinoma cells. *Journal of Investigative Dermatology* 111, 656–666.

Jacobson, M.D. and McCarthy, N. (eds) (2002) *Apoptosis.* Frontiers in Molecular Biology 40, Oxford University Press, Oxford.

Johnson, I.T. (2001) Mechanisms and anticarcinogenic effects of diet-related apoptosis in the intestinal mucosa. *Nutrition Research Reviews* 14, 229–256.

Karlseder, J., Broccoli, D., Dai, Y., Hardy, S. and de Lange, T. (1999) p53- and ATM-dependent apoptosis induced by telomeres lacking TRF2. *Science* 283, 1321–1325.

Kawamori, T., Lubert, R., Steele, V.E., Kelloff, G.K., Kaskey, R.B., Rao, C.V. and Reddy, B.S. (1999) Chemopreventive effect of curcumin, a naturally-occurring anti-inflammatory agent, during promotion/progression stages of colon cancer. *Cancer Research* 59, 597–601.

Kerr, J.F., Wyllie, A.H. and Currie, A.R. (1972) Apoptosis: a basic biological phenomenon with wide-ranging implications in tissue kinetics. *British Journal of Cancer* 26, 239–257.

Kinzler, K.W. and Vogelstein, B. (1996) Lessons from hereditary colorectal cancer. *Cell* 87, 159–170.

Kliewer, S.A., Sundseth, S.S., Jones, S.A., Bron, P.J., Wisely, G.B., Koble, C.S., Devchand, P., Wahli, W., Wilson, T.M., Lenhard, J.M. and Lehmann, J.M. (1997) Fatty acids and eicosanoids regulate gene expression through direct interactions with peroxisome proliferator-activated receptors α and γ. *Proceedings of the National Academy of Sciences USA* 94, 4318–4323.

Kuo, M.L., Huang, T.S. and Lin, J.K. (1996) Curcumin, an antioxidant and anti-tumour promoter, induces apoptosis in human leukemia cells. *Biochimica et Biophysica Acta* 1317, 95–100.

Latham, P., Lund, E.K. and Johnson, I.T. (1999) Dietary *n*-3 PUFA increases the apoptotic response to 1,2-dimethylhydrazine, reduces mitosis and suppresses the induction of carcinogenesis in the rat colon. *Carcinogenesis* 20, 645–650.

Latham, P., Lund, E.K., Brown, J.C. and Johnson, I.T. (2001) Effects of cellular redox balance on induction of apoptosis by eicosapentaenoic acid in HT29 colorectal adenocarcinoma cells and rat colon *in vivo*. *Gut* 49, 97–105.

Lenardo, M., Chan, K.M., Hornung, F., McFarland, H., Siegel, R., Wang, J. and Zheng, L. (1999) Mature lymphocyte apoptosis – immune regulation in a dynamic and unpredictable antigenic environment. *Annual Reviews in Immunology* 17, 221–253.

Lewin, D. and Weinstein, W.M. (1996) Cell death – where is thy sting? *Gut* 39, 883–884.

Li, P., Nijhawan, D., Budihardjo, I., Srinivasula, S.M., Ahmad, M., Alnemri, E.S. and Wang, X. (1997) Cytochrome c and dATP-dependent formation of Apaf-1/Caspase-9 complex initiates and apoptotic protease cascade. *Cell* 91, 479–489.

Liu, Y. and Kulesz-Martin, M. (2001) p53 protein at the hub of cellular DNA damage response pathways through sequence-specific and non-sequence-specific DNA binding. *Carcinogenesis* 22, 851–860.

Lowe, S.W. and Lin, A.W. (2000) Apoptosis in cancer. *Carcinogenesis* 21, 485–495.

Mahmoud, N.N., Carothers, A.M., Grunberger, D., Bilinski, R.T., Churchill, M.R., Martucci, C., Newmark, H. and Bertagnolli, M.M. (2000) Plant phenolics decrease intestinal tumours in a murine model of familial adenomatous polyposis. *Carcinogenesis* 21, 921–927.

Mariadason, J.M., Corner, G.A. and Augenlicht, L.H.

(2000) Genetic reprogramming in pathways of colonic cell maturation induced by short chain fatty acids: comparison with trichostatin A, sulindac, and curcumin and implications for chemoprevention of colon cancer. *Cancer Research* 60, 4561–4572.

Mathers, J.C. (2000) Food and cancer prevention: human intervention studies. In: Johnson, I.T. and Fenwick G.R. (eds) *Dietary Anticarcinogens and Antimutagens: Chemical and Biological Aspects*. Royal Society of Chemistry, Cambridge, pp. 395–403.

Mathers, J.C. and Burn, J. (1999) Nutrition in cancer prevention. *Current Opinion in Oncology* 11, 402–407.

Mathers, J.C., Smith, H. and Carter, S. (1997) Dose–response effects of raw potato starch on small intestinal escape, large-bowel fermentation and gut transit time in the rat. *British Journal of Nutrition* 78, 1015–1029.

McCaffrey, P., Andreola, F., Giandomenico, V. and De Luca, L.M. (2001) Vitamin A and gene expression. In: Moustad-Moussa, N. and Berdanier, C.D. (eds) *Nutrient–Gene Interactions in Health and Disease*. CRC Press, Boca Raton, Florida, pp. 283–319.

McCarthy N.J. (2002) Why be interested in death? In: Jacobson, M.D and McCarthy, N.J. (eds) *Apoptosis*. Oxford University Press, Oxford, pp. 1–22.

Meydani, S.N., Claycombe, K.J. and Sacristán, C. (2001) Vitamin E and gene expression. In: Moustaïd-Moussa, N. and Berdanier, C.D. (eds) *Nutrient–Gene Interactions in Health and Disease*. CRC Press, Boca Raton, Florida, pp. 393–424.

Moore, S.E., Cole, T.J., Poskitt, E.M., Sonko, B.J., Whitehead, R.G., McGregor, I.A. and Prentice, A.M. (1997) Season of birth predicts mortality in rural Gambia. *Nature* 388, 434.

Morin, P.J., Vogelstein, B. and Kinzler, K.Z. (1996) Apoptosis and *APC* in colorectal tumorigenesis. *Proceedings of the National Academy of Sciences USA* 93, 7950–7954.

Moss, S.F., Attia, L., Scholes, J.V., Walters, J.R.F. and Holt, P.R. (1996) Increased small intestinal apoptosis in coeliac disease. *Gut* 39, 811–817.

Norimura, T., Nomoto, S., Katsuki, M., Gondo, Y. and Kondo, S. (1996) *p53*-independent apoptosis suppresses radiation-induced teratogenesis. *Nature Medicine* 2, 577–580.

Ortiz, R., Cortéz, L., González-Márquez, H., Gómez, J.L., González, C. and Cortéz, E. (2001) Flow cytometric analysis of spontaneous and dexamethazone-induced apoptosis in thymocytes from severely malnourished rats. *British Journal of Nutrition* 86, 545–548.

Oshima, M., Takahashi, M., Oshima, H., Tsutsumi, M., Yazawa, K., Sugimura, T., Nishimura, S., Wakabayashi, K. and Taketo, M.M. (1995) Effects of docosahexaenoic acid (DHA) on intestinal polyp development in Apc delta 716 knockout mice. *Carcinogenesis* 16, 2605–2607.

Oshima, M., Dinchuk, J.E., Kargman, S.L., Oshima, H., Hancock, B., Kwong, E., Trzaskos, J.M., Evans, J.F. and Taketo, M.M. (1996) Suppression of intestinal polyps in Apc$\Delta$716 knockout mice by inhibition of cyclooxygenase-2 (COX-2). *Cell* 87, 803–809.

Palozza, P., Calviello, G., Serini, S., Maggiano, N., Lanza, P., Raneletti, F.O. and Bartoli, G.M. (2001) β-carotene at high concentrations induces apoptosis by enhancing oxy-radical production in human adenocarcinoma cells. *Free Radicals in Biology and Medicine* 30, 1000–1007.

Palozza, P., Serini, S., Maggiano, N., Angeline, M., Boninsegna, A., Di Nicuolo, F., Ranelletti, F.O. and Calviello, G. (2002) Induction of cell cycle arrest and apoptosis in human colon adenocarcinoma cell lines by β-carotene through down-regulation of cyclin A and Bcl-2 family proteins. *Carcinogenesis* 23, 11–18.

Park, H.S., Ryu, J.H., Ha, Y.L. and Park, J.H.Y. (2001) Dietary conjugated linoleic acid (CLA) induces apoptosis of colonic mucosa in 1,2-dimethlyhydrazine-treated rats: mechanism of the anticarcinogenic effect by CLA. *British Journal of Nutrition* 86, 549–555.

Parodi, P.W. (1997) Cows' milk fat components as potential anticarcinogenic agents. *Journal of Nutrition* 127, 1055–1060.

Peifer, M. and Polakis, P. (2000) Wnt signalling in oncogenesis and embryogenesis – a look outside the nucleus. *Science* 287, 1606–1609.

Petrik, M.B., McEntee, M.F., Johnson, B.T., Obukowicz, M.G. and Whelan, J. (2000) Highly-unsaturated (*n*-3) fatty acids, but not α-linolenic, conjugated linoleic and gamma-linolenic acids, reduce tumorigenesis in $Apc^{Min/-}$ mice. *Journal of Nutrition* 130, 2434–2443.

Plummer, S.M., Holloway, K.A., Manson, M.M., Munks, R.J., Kaptein, A., Farrow, S. and Howells, L. (1999) Inhibition of cyclooxygenase 2 expression in colon cells by the chemopreventive agent curcumin involves inhibition of NFκB activation via the NIK/IKK signalling complex. *Oncogene* 18, 6013–6020.

Pitti, R.M., Marsters, S.A., Lawrence, D.A., Roy, M., Kischkel, F.C., Dowd, P., Huang, A., Donahue, C.J., Sherwood, S.W., Baldwin, D.T., Godowski, P.J., Wood, W.I., Gurney, A.L., Hillan, K.J., Cohen, R.L., Goddard, A.D., Botstein, D. and Ashkenazi, A. (1998) Genomic amplification of a decoy receptor for Fas ligand in lung and colon cancer. *Nature* 396, 699–703.

Prentice, A.M. (1999) The thymus: a barometer of malnutrition. *British Journal of Nutrition* 81, 345–347.

Reed, J.C. (1998) Bcl-2 family proteins. *Oncogene* 17, 3225–3236.

Renehan, A.G., Booth, C. and Potten, C.S. (2001) What is

apoptosis, and why is it important? *British Medical Journal* 322, 1536–1538.

Roy, N. and Cardone, M.H. (2002) The caspases: consequential cleavage. In: Jacobson, M.D and McCarthy, N.J. (eds) *Apoptosis*. Oxford University Press, Oxford, pp. 93–135.

Samaha, H.S., Kelloff, G.J., Stelle, V., Rao, C.V. and Reddy, B.S. (1997) Modulation of apoptosis by sulindac, curcumin, phenylethyl-3-methyl cafeate and 6-phenylhexyl isothiocyanate: apoptotic index as a biomarker in colon cancer chemoprevention and promotion. *Cancer Research* 59, 1301–1305.

Sarraf, P., Mueller, E., Jones, D., King, F.J., DeAngelo, D.J., Partridge, J.B., Holden, S.A., Chen, L.B., Singer, S., Fletcher, C. and Spiegelman, B.M. (1998) Differentiation and reversal of malignant changes in colon cancer through PPARγ. *Nature Medicine* 4, 1046–1052.

Scaglia, L., Cahill, C.J., Finegood, D.T. and Bonner-Weir, S. (1997) Apoptosis participates in the remodelling of the endocrine pancreas in the neonatal rat. *Endocrinology* 138, 1736–1741.

Singh, S. and Aggarwal, B.B. (1995) Activation of transcription factor NF-κB is suppressed by curcumin (diferuloylmethane). *Journal of Biological Chemistry* 270, 24995–25000.

Smith, T.K., Lund, E.K., Musk, S.S.R. and Johnson, I.T. (1998) Inhibition of DMH-induced aberrant crypt foci, and induction of apoptosis in rat colon, following oral administration of a naturally occurring glucosinolate. *Carcinogenesis* 19, 267–273.

Sperandio, S., de Belle, I. and Bredesen, D.E. (2000) An alternative, nonapoptotic form of programmed cell death. *Proceedings of the National Academy of Sciences USA* 97, 14376–14381.

Steinback, G., Lynch, P.M., Phillips, R.K., Wallace, M.H., Hawk, E., Gordon, G.B., Wakabayashi, N., Shen, Y., Fujimura, T., Su, L.K. and Levin, B. (2000) The effect of celecoxib, a cyclooxygenase-2 inhibitor, in familial adenomatous polyposis. *New England Journal of Nutrition* 342, 1946–1952.

Stoll, B.A. (2002) *n*-3 Fatty acids and lipid peroxidation in breast cancer inhibition. *British Journal of Nutrition* 87, 193–198.

Strachan. T. and Read, A.P. (1999) *Human Molecular Genetics 2*. BIOS Scientific Publishers, Oxford.

Sträter, J., Wellish, I., Riedl, S., Walczak, H., Koretz, K., Tandara, A., Krammer, P.H. and Möller, P. (1997) CD95 (APO-1/Fas)-mediated apoptosis in colon epithelial cells: a possible role in ulcerative colitis. *Gastroenterology* 113, 160–167.

Thompson, C.B. (1995) Apoptosis in the pathogenesis and treatment of disease. *Science* 267, 1456–1462.

Thornberry, N.A. and Lazebnik, Y. (1998) Caspases: enemies within. *Science* 281, 1312–1316.

Thornberry, N.A., Bull, H.G., Calaycay, J.R., Chapman, K.T., Howard, A.D., Kostura, M.J., Miller, D.K., Molineau, S.M., Weidner, J.R. and Auins, J. (1992) A novel heterodimeric cysteine protease is required for interleukin-1β processing in monocytes. *Nature* 356, 768–774.

Tomlinson, I.P.M. and Bodmer, W.F. (1995) Failure of programmed cell death and differentiation as causes of tumors: some simple mathematical models. *Proceedings of the National Academy of Sciences USA* 92, 11130–11134.

Tsujimoto, Y. (2002) Regulation of apoptosis by the Bcl-2 family of proteins. In: Jacobson, M.D and McCarthy, N.J. (eds) *Apoptosis*. Oxford University Press, Oxford, pp. 136–160.

Vaux, D.L. (2002) Viruses and apoptosis. In: Jacobson, M.D and McCarthy, N.J. (eds) *Apoptosis*. Oxford University Press, Oxford, pp. 262–277.

Voehringer, D.W., Hirschberg, D.L., Xiao, J., Lu, Q., Roederer, M., Lock C.B., Herzenberrg, L.A., Steinman, L. and Herzenberg, L.A. (2000) Gene microarray identification of redox and mitochondrial elements that control resistance or sensitivity to apoptosis. *Proceedings of the National Academy of Sciences USA* 97, 2680–2685.

Webb, S.J., Nicholson, D., Bubb, V.J. and Wyllie, A.H. (1999) Caspase-mediated cleavage of APC results in an amino-terminal fragment with an intact armadillo repeat domain. *FASEB Journal* 13, 339–346.

Whelan, J. and McEntee, M.F. (2001) Dietary fats and APC-driven intestinal tumorigenesis. In: Moustaïd-Moussa, N. and Berdanier, C.D. (eds) *Nutrient–Gene Interactions in Health and Disease*. CRC Press, Boca Raton, Florida, pp. 231–260.

White, E. (1996) Life, death, and the pursuit of apoptosis. *Genes and Development* 10, 1–15.

Wu, J.T., Archer, S.Y., Hinnebusch, B., Meng, S. and Hodin, R.A. (2001) Transient vs. prolonged histone hyperacetylation: effects on colon cancer cell growth, differentiation, and apoptosis. *American Journal of Physiology* 280, G482–G490.

Xue, D., Wu, C.-I. and Shah, M.S. (2002) Programmed cell death in *C. elegans*: the genetic framework. In: Jacobson, M.D and McCarthy, N.J. (eds) *Apoptosis*. Oxford University Press, Oxford, pp. 23–55.

Yang, W.-L. and Frucht, H. (2001) Activation of the PPAR pathway induces apoptosis and COX-2 inhibition in HT-29 human colon cancer cells. *Carcinogenesis* 22, 1379–1383.

Yuan, J., Shaham, S., Ledoux, S., Ellis, H.M. and Horvitz, H.R. (1993) The *C. elegans* death gene *ced-3* encodes a protein similar to mammalian interleukin-1β-converting enzyme. *Cell* 75, 641–652.

Zou, H., Henzel, W.J., Liu, X., Lutschg, A. and Wang, X. (1997) Apaf-1, a human protein homologous to *C. elegans* CED-4, participates in cytochrome c-dependent activation of caspase-3. *Cell* 90, 405–413.

# 第 3 部分

# 营养物质在信号转导、基因表达和蛋白水解中的作用

# 7　哺乳动物中葡萄糖对基因表达的调节

Fabienne Foufelle 和 Pascal Ferré
（因瑟姆 465 单元，巴黎第六大学，瑞查池生物医药中心，巴黎，法国）

## 引　　言

对从细菌到人类的所有生物而言，通过对基因表达的调节来适应营养环境是共同的需要。在单细胞生物中，营养物质本身就是诱导其适应营养环境的信号。大肠杆菌中的乳糖操纵子就是营养素调节基因的第一个例子。哺乳动物中，营养素对营养性环境的潜在影响由于激素的作用而变得模糊不清。直到现在，营养素的直接作用才被认可。

在本文中，我们将重点讨论葡萄糖对基因表达的调节，特别是对肝脏的基因表达的调节，因为近来对肝细胞的细胞和分子机制的研究取得了重要的进展。

## 我们对于营养中葡萄糖反应基因的观点

在营养范畴中，将一个基因归类为特殊的葡萄糖反应基因所遵循的原则是什么呢？首先，在体内试验中，此基因的表达能够受日粮中可利用葡萄糖的调节。第二，在体外试验中，葡萄糖对此基因的表达必须能够同激素（如胰岛素）的影响区分开。第三，葡萄糖对基因表达产生影响所需要的浓度范围必须是合理的。尽管利用极端浓度去考察一个特定的现象是可以接受的，但此现象必须能够适用于葡萄糖更大的生理浓度变化（通常的范围是 4～15mmol/L）。在体外，在将低浓度（或零浓度）葡萄糖同高浓度葡萄糖进行比较时必须小心，因为在这种情况下，与葡萄糖调节蛋白如 GRP78/Bip 所引起的效应一样，与能量耗竭或糖基化不全所引起的应激相比，基因调节或许是第二位的。葡萄糖对基因表达的影响需要同其他细胞应激相区别。

## 葡萄糖调节肝脏基因表达的生理基础

在哺乳动物中，葡萄糖不断地被大脑（人类为 120g/d）、红细胞和肾髓质等器官利用的比率很高。当含有碳水化合物的食物被吸收时，可诱导几个代谢过程以降低肝（糖原合成与分解）内源葡萄糖的产生，并增加肝脏和肌肉中葡萄糖的吸收和以糖原形式的葡萄糖的储存。被吸收的葡萄糖通过门静脉被运送到肝脏，一旦肝糖原的储存达到极限，肝细胞就将葡萄糖转化成脂类（脂肪生成），它们以极低密度脂蛋白（VLDL）的形式运出，最后以甘油三酯的形式储存在脂肪组织中。相反，如果食物中葡萄糖的利用率降低，葡萄糖利用途径被抑制，则糖生成途径就被激活。代谢途径的调节包括由特定蛋白质（酶和转运蛋白）的活性所引起的快速调节和由这些蛋白质数量变化所引起的慢速调节。这些调节主要通过调节转录速率来实现。

## 糖分解/脂肪生成和糖异生的基因表达由食物中碳水化合物利用率调节

在肝脏中，高碳水化合物日粮可以诱导糖酵解和脂肪生成过程中的几个关键酶的表达，这几个关键酶分别为用于糖酵解的葡糖激酶（Iynedjian *et al.*，1987）、6-磷酸果糖-1-激酶（Rongnoparut *et al.*，1991）、6-磷酸果糖-2-激酶/果糖-2-6-二磷酸酶（Colosia *et al.*，1988）、醛缩酶 B（Weber *et al.*，1984）和 L-丙酮酸激酶（liver-pyruvate kinase，L-PK）(Vaulont *et al.*，1986)，用于脂肪生成的 ATP-柠檬酸裂解酶（Elshourbagy *et al.*，1990）、乙酰辅酶 A 羧化酶（acetyl-COA carboxylase，ACC）（Pape *et al.*，1988）、脂肪酸合成酶（fatty acid synthase，FAS）（Paulauskis and Sul，1989；Katsurada *et al.*，1990）、硬脂酰辅酶 A 脱饱和酶（Ntambi，1992），以及用于磷酸戊糖途径的葡萄糖-6-磷酸脱氢酶（Kletzien *et al.*，1985；Katsurade *et al.*，1989）和 6-磷酸葡萄糖脱氢酶（Miksicek and Towle，1983）等。这里应该提到的是，看似与脂肪生成有关的酸性小多肽就是因被诱导产生的 *S14* 基因编码。*S14* 基因在脂肪生成组织（白色和褐色脂肪组织、肝脏和哺乳期乳腺）中表达，其表达的调节途径与脂肪生成过程中基因的调节类似（Mariash *et al.*，1986；Kinlaw *et al.*，1987；Clarke *et al.*，1990）。尽管有些研究表明 S14 蛋白（基因）可能参与脂肪生成过程中酶基因表达的调节（Kinlaw *et al.*，1995；Zhu *et al.*，2001），但其确切的生理功能目前还不清楚。富含碳水化合物的日粮对葡萄糖利用过程中多数基因 mRNA 合成的影响是大量（4～25倍）、快速的（激活后 1～2h），并包括一转录机制。

相反，高碳水化合物日粮抑制了糖异生酶如磷酸烯醇式丙酮酸羧激酶（PEPCK）(Granner *et al.*，1983）和葡萄糖-6-磷酸酶（Argaud *et al.*，1997）的基因表达。这种抑制作用速度很快，一旦碳水化合物利用率降低，则很容易逆转。

碳水化合物的吸收导致了一些底物如葡萄糖和乳糖浓度的增加，从而使胰腺激素胰岛素和胰高血糖素的浓度提高。在体内试验中，将激素与其底物在基因表达调节过程中的各自作用区分开是很困难的。在体外研究中，利用初步成熟的肝细胞或细胞系来揭示激素和底物调节基因表达的机制取得了较大的进展。从这些研究中，发现了不同类型的基因调节：一些基因（如葡糖激酶）在不依赖葡萄糖的情况下能够被胰岛素诱导表达(Iynedjian *et al.*，1989)；其他一些基因如 *L-PK*、*FAS*、*ACC*、*S14* 以及硬脂酰辅酶 A 脱饱和酶基因的诱导表达则需要胰岛素和葡萄糖浓度的共同提高（Decaux *et al.*，1989；Waters and Ntambi，1994；Prip-Buus *et al.*，1995；Koo *et al.*，2001；O'Callaghan *et al.*，2001）。胰岛素（Sasaki *et al.*，1984）和葡萄糖（Scott *et al.*，1998；Cournarie *et al.*，1999）可以独立对 PEPCK 的表达进行负调节。最后，胰岛素虽然抑制了葡萄糖-6-磷酸酶的表达，但与之相矛盾的是，高浓度的葡萄糖又可以提高其表达（Argaud *et al.*，1997；VanSchaftingen and Gerin，2002）。在这里，我们将集中探讨葡萄糖单独调节肝脏中基因表达的机制。胰岛素对基因表达的影响已在别处进行了总结（Foufelle and Ferré，2002）。

## 葡萄糖对糖分解酶和脂肪生成酶基因表达的影响

如上所述，肝脏中一些糖分解和脂肪合成基因要得到最大限度的表达，需要增加胰岛素和葡萄糖的浓度（Girard *et al.*，1997；Towle *et al.*，1997；Vaulont *et al.*，2000）。这种调节包括对其转录速率的刺激。在葡萄糖不存在的情况下，胰岛素不能诱导它们的表达。在胰岛素不存在的情况下，葡萄糖在脂肪组织中的作用大大减弱，在成熟肝细胞中的作用几乎消失。这些基因表达的同时需要高浓度胰岛素和葡萄糖的一个潜在的原因是：饭后，代谢优先补充肝糖原的储存，只有在葡萄糖特别丰富的情况下，其中的碳原子才会在脂肪合成中被利用。

## 葡萄糖的作用：代谢假说

葡萄糖通过其有较高米-曼氏常数（$K_m$）的葡萄糖转运蛋白（GLUT2）的作用进入肝细胞。与存在于脂肪和肌肉组织中的 GLUT4 转运蛋白相反，GLUT2 的活性不依赖于胰岛素。葡萄糖一旦进入细胞，就被葡糖激酶（一种高 $K_m$ 己糖激酶）磷酸化产生葡萄糖-6-磷酸。GLUT2 和葡糖激酶介导的磷酸化速率与血浆中葡萄糖的浓度相平行。这时，葡萄糖-6-磷酸可以进入糖原合成、糖分解或磷酸戊糖途径。

有证据表明，葡萄糖只有在进行代谢后才能影响转录（Foufelle *et al.*，1996）。例如，在肝脏中，葡萄糖起作用需要有葡糖激酶的存在，此酶负责将葡萄糖磷酸化为葡萄糖-6-磷酸（Doiron *et al.*，1994；Prip-Buus *et al.*，1995）。胰岛素可以显著提高葡糖激酶的表达，因此，基因的胰岛素依赖性最初可以由胰岛素介导的葡糖激酶的诱导表达进行解释。不过，也有证据表明胰岛素可以直接通过胰岛素应答转录因子（insulin-responsive transcription factor）SREBP-1c 发挥作用（Foufelle and Ferré，2002）。

目前引起关注的问题是：①影响转录的葡萄糖代谢物的性质（是哪种代谢物?）；②葡萄糖代谢物影响转录机制的传导机制；③基因启动子中的葡萄糖应答元件；④参与代谢作用的转录因子的性质。

因为葡萄糖对基因表达的影响是浓度依赖性的，作为信号的代谢物的浓度变化必须与细胞外葡萄糖浓度相平行。再则，代谢物浓度的变化必须发生在 mRNA 浓度变化之前。在以前的研究中，我们认为葡萄糖-6-磷酸或许是影响 FAS（Prip-Buus *et al.*，1995）及属于这一类范畴的基因如 *L-PK* 和 *S14* 的转录的代谢物（Foufelle *et al.*，1992；Prip-Buus *et al.*，1995）。这种想法是建立在以下观察的基础上的：①在脂肪组织和一 β 细胞系（INS1）中，葡萄糖类似物 2-脱氧葡萄糖能模仿葡萄糖的作用，而且这种葡萄糖类似物必须被磷酸化成在细胞中累积的 2-脱氧葡萄糖-6-磷酸后才能进行代谢（Foufelle *et al.*，1992；Brun *et al.*，1993；Marie *et al.*，1993）；②细胞内葡萄糖-

6-磷酸浓度与基因的表达相平行（Foufelle *et al.*，1992；Prip-Buus *et al.*，1995；Mourrieras *et al.*，1997）；③在体内，已糖磷酸浓度变化的时间过程符合基因诱导的相关时间模式（Munnich *et al.*，1987）。与此假说相一致，葡萄糖-6-磷酸因为是糖原合成酶活性的激动剂，并可作为肝脏糖原代谢中葡萄糖可利用性的信号（Bollen *et al.*，1998）。有趣的是，*GLUT2* 敲除的禁食小鼠 *L-PK* 基因的表达和糖原含量均增加，并伴有较高浓度的葡萄糖-6-磷酸（Burcelin *et al.*，2000）。注射葡萄糖-6-磷酸系统组成成分之一的葡萄糖-6-磷酸转移酶蛋白活性抑制剂的大鼠也出现了上述现象，肝脏葡萄糖-6-磷酸、糖原和甘油三酯浓度大大增加，并激活了 *FAS* 和 *ACC* 基因的表达（Bandsma *et al.*，2001）。

葡萄糖代谢的非氧化途径中的一种中间产物木酮糖-5-磷酸可能也在信号方面起作用（Doiron *et al.*，1996），这一假设是基于以下观察：①在一些细胞中，2-脱氧葡萄糖-6-磷酸可进一步代谢为磷酸戊糖途径中的中间代谢产物；②较低浓度（0.5mmol/L）的木酮糖-5-磷酸前体物木糖醇在葡萄糖-6-磷酸浓度没发生变化的情况下，可通过作用肝细胞系（AT3F）中 *L-PK* 基因的启动子刺激报道基因（reporter gene）的转录；③在原培养肝细胞中，木糖醇（5～10mmol/L）能够诱导 L-PK mRNA 的浓度增加 6 倍（Doiron *et al.*，1996）；④木酮糖-5-磷酸能够激活由磷酸酶 2A 介导的脱磷酸作用，此磷酸酶也参与了转录因子的脱磷酸作用（Nishimura and Uyeda，1995）。葡萄糖、木糖醇和二羟基丙酮以不同的浓度进入糖分解/磷酸戊糖途径。与葡萄糖-6-磷酸不同，在不同浓度葡萄糖、木糖醇和二羟基丙酮培养的肝细胞中，*FAS* 和 *S14* 基因的表达不与木酮糖-5-磷酸的浓度相平行（Mourrieras *et al.*，1997）。有趣的是，木糖醇引起了葡萄糖-6-磷酸浓度的剂量依赖性增加。这个结论并不是不可预料的，因为磷酸戊糖途径中的代谢产物可以在糖分解过程中被重复利用。因此，在这些试验中，木酮糖-5-磷酸并不适合作为诱导 *FAS* 和 *S14* 基因的信号代谢物。只有将全部的传递机制确定后，葡萄糖代谢物的作用才能得到明确阐明。

## 葡萄糖的作用：非代谢假说

Johnston（1999）描述了酵母中的葡萄糖的膜传感器。尽管不能转运葡萄糖，这些被称为 SnF3 和 Rgt2 的蛋白质的结构与葡萄糖转运蛋白相类似。它们有一个大的胞质内环，此环可能参与了葡萄糖信号途径。目前认为 GLUT2 可能是肝细胞中葡萄糖的传感器。有证据表明，在对依赖葡萄糖的 *L-PK* 基因表达的控制中，GLUT2 的大胞质内环可能发挥了一定的作用（Guillemain *et al.*，2000，2002）。葡萄糖与 GLUT2 的相互作用可能是通过此环产生了一个信号，使信号的转导发生改变。近来，利用 GLUT2 环的双杂交系统证明了 karyopherin α2（包含在核输入中的一个受体）是一潜在结合伴侣（Guillemain *et al.*，2002）。肝癌细胞中无活性 karyopherin α2 的过量表达导致由葡萄糖诱导的 *L-PK* 基因表达的减少。因此，这种作用机制与包含在肝脏中葡萄糖代谢的信号系统的作用有关。

# 何谓从葡萄糖到转录机制的细胞级联?

为了解释葡萄糖的组织反应性，提出以下假设：①葡萄糖的存在可能改变转录因子的细胞核数量（包括细胞核定位的改变）；②在葡萄糖存在的情况下，转录因子经过翻译后修饰改变，致使与基础转录机制或辅激活蛋白因子间的相互反应也发生改变。转录因子的翻译后修饰可能包括通过与葡萄糖相关信号代谢物结合而产生的蛋白质异构化，或核蛋白激酶/磷酸酶活性调节而引起的磷酸化/去磷酸化作用改变，或两者都发生变化。从本文可看出蛋白质磷酸酶抑制因子抑制了葡萄糖对 *FAS*、*ACC* 和 *S14* 基因表达的影响（Sudo and Mariash，1994；Daniel *et al.*，1996；Foretz *et al.*，1998)，这说明对葡萄糖的刺激作用可能还有去磷酸化机制。

AMP 激活蛋白激酶（AMP-activated protein kinase，AMPK）是葡萄糖激活基因调节中的蛋白激酶的潜在候选者。AMPK 为一丝氨酸/苏氨酸激酶，通过对胆固醇和脂肪酸代谢过程中关键酶的磷酸化起代谢“主开关”的作用。实际上，AMPK 的磷酸化、ACC 和 3-羟基-3-甲基戊二酰辅酶 A 还原酶的失活抑制了脂肪生成和胆固醇合成（Carling *et al.*，1987）。AMPK 可被类似于低氧和肌肉收缩等导致的细胞内 ATP 耗竭和 AMP 增加的应激所激活。AMP 激活 AMPK 的作用机制有以下两种：①AMPK 的别构激活；②AMPK 激酶的刺激，导致 AMPK 磷酸化（Hardie *et al.*，1998）。

AMPK 在结构和功能上与酵母蛋白激酶复合物 Snf1（sucrose non-fermenting，非发酵蔗糖）相关，这为 AMPK 在基因转录调节中的可能作用机制提供了重要线索（Woods *et al.*，1994）。在酵母中，许多基因的转录受到高浓度葡萄糖的抑制（Gancedo，1998）。在限制葡萄糖的情况下生长的酵母中，Snf1 激酶活性对这些基因的去阻遏是非常必要的。AMPK 和 Snf1 都来自于由一个催化亚基和两个调节亚基组成的异源三聚体（Hardie and Carling，1997；Hardie *et al.*，1998）。哺乳动物中 AMPK 亚基与 Snf1 复合物中的相应部分的氨基酸序列具有高度的同源性，这两种激酶在功能上也有相似性（Woods *et al.*，1994）。

这些发现使几个研究团队的人们想到 AMPK 或许参与了哺乳动物体内葡萄糖对基因转录的调节，下面的研究为这一想法提供了证据。研究发现，肝细胞在 AMPK 细胞渗透激活剂 5-氨基-咪唑氨甲酰（AICA）核苷中培养时，AMPK 被激活，从而导致葡萄糖诱导 *FAS*、*L-PK* 和 *S14* 基因表达的抑制。（Foretz *et al.*，1998；Leclerc *et al.*，1998）。在培养的肝细胞中，组成型 AMPK 活性形式的过度表达导致 AMPK 活性的增加，也可引起 *FAS*、*L-PK* 和 *S14* 基因表达的抑制（Woods *et al.*，2000）。这些结果意味着 AMPK 抑制了肝细胞中葡萄糖诱导的基因的表达。

由于 AMPK 活性的增加抑制了葡萄糖激活基因的转录，因此 AMPK 活性的降低可能是葡萄糖刺激基因转录机制中的一部分。然而，利用激酶的显性失活形式抑制 AMPK 的活性后没有观察到任一葡萄糖诱导基因表达的变化（Woods *et al.*，2000）。而且，肝细胞培养基中葡萄糖浓度 5～25mmol/L 的变化对 AMPK 的活性没有抑制作用（Foretz *et al.*，1998）。以上结果表明，肝脏中葡萄糖没有通过直接抑制 AMPK 的作用影响基因的表达。

相反，当细胞外葡萄糖浓度从30mmol/L减到3mmol/L时，β-细胞系中AMPK活性增加（Salt *et al.*，1998；da Silva Xavier *et al.*，2000），这与转染的*L-PK*基因启动子活性的增加相一致。向在3mmol/L葡萄糖中培养的MIN6细胞中注入α2AMPK抗体，细胞质和细胞核中AMPK活性受到抑制，此过程与高浓度葡萄糖对L-PK启动子活性的影响效果相似（da Silva Xavier *et al.*，2000）。因此，在一些对低浓度葡萄糖特别敏感的细胞（如β-细胞）中，当葡萄糖浓度增加时，AMPK活性受到抑制可能是导致基因转录增加作用机制的一部分。

总之，AMPK抑制肝脏中编码脂肪生成和糖分解酶的基因表达。然而，AMPK的抑制作用并不影响葡萄糖对基因表达的诱导。因为另外一些不受葡萄糖控制的基因同样受到AMPK的抑制（Hubert *et al.*，2000；Lochhead *et al.*，2000；Zhou *et al.*，2000；Zheng *et al.*，2001；MacLean *et al.*，2002），这表明AMPK对编码消耗ATP途径中酶的基因有普遍的抑制作用。

## 葡萄糖反应中顺式作用DNA序列

被确认的第一个葡萄糖应答元件（glucose response element，G1RE）在*L-PK*基因上（Thompson and Towle，1991；Bergot *et al.*，1992），这个元件位于L-PK基因转录起始位点上游183bp内。在转基因动物中，这一区域介导了组织表达的特异性以及营养素和激素的反应（Cuif *et al.*，1992）。后来的研究将G1RE定位在转录起始位点－183bp和－96bp的区域内，这与在分离的肝细胞中进行转染实验所得出的结果一样（Thompson and Towle，1991；Bergot *et al.*，1992；Diaz-Guerra *et al.*，1993）。另外一个被称作ChoRE（carbohydrate response element）的葡萄糖应答元件随之在*S14*基因中被定性。在大鼠分离的肝细胞中，*S14*基因起始位点－1439bp和－1423bp之间的区域对葡萄糖的反应性是必要的（Shih and Towle，1992）。在小鼠*S14*启动子中，ChoRE定位于转录起始位点－1450bp和－1425bp区域内（Koo and Towle，2000）。*L-PK*和*S14*基因中与G1RE和ChoRE相结合的复合物分别与其相邻的辅助位点协同作用，构成葡萄糖的整个完全反应性（Bergot *et al.*，1992；Diaz-Guerra *et al.*，1993；Shih *et al.*，1995）。对于*L-PK*基因来说，葡萄糖的反应性是通过与L3（－126～144bp）以及L4（－145～168bp）元件密切的协同作用得到的（Bergot *et al.*，1992；Liu *et al.*，1993）。L4元件的一个多聚体能够将葡萄糖反应性赋予一个对葡萄糖无应答的异源启动子（Bergot *et al.*，1992）。只有在L3和L4元件同时存在的情况下，才能得到葡萄糖的完全反应性。*S14*基因转录起始位点－1467bp和－1448bp之间的元件对葡萄糖的完全反应性是必要的（Shih *et al.*，1995；Kaytor *et al.*，1997）。通过分别对*L-PK*和*S14*基因中G1RE和ChoRE的比较，发现了以下的相似性：葡萄糖应答元件包含两个相隔5bp的E-box或与E-box序列相似的CANNTG序列，这两个E-box基元的存在对葡萄糖反应是很关键的。最近，ChoRE/G1RE又在小鼠*FAS*基因启动子转录起始位点－7214bp和－7190bp之间被识别（Rufo *et al.*，2001）。这个区域包含了与*L-PK*和*S14*基因中ChoRE相似的一段回文序列［CATGTG$(n)_5$GGCGTG］。正如以前*L-PK*和*S14*基因中ChoRE所显示的一样，此ChoRE能将葡萄糖反应性赋予一异源启动子。

*ACC* 基因中的葡萄糖敏感区域首先在 PⅡ启动子转录起始位点上游－340bp 和－249bp区域间被识别。PⅡ启动子在所有组织中都可被表达，而且对食物变化无反应（Kim *et al.*，1996）。在细胞系 30A5 脂肪细胞中，PⅡ启动子的转录活性由葡萄糖诱导增加了 3 倍（Daniel and Kim，1996）。SREBP-1 和 Sp1 在此区域内的结合位点已被识别（Daniel and Kim，1996；Lopez *et al.*，1996）。与 PⅡ相比较，ACC PⅠ启动子主要在脂肪组织和肝脏中表达，它的活性因诱导脂肪生成的食物的变化而显著升高（Lopez-Casillas *et al.*，1991；Kim *et al.*，1996）。在培养的肝细胞中，PⅠ转录活性可在葡萄糖的诱导下增加 5 倍。PⅠ启动子在转录起始位点－126bp 和－102bp 之间存在 ChoRE（O'Callaghan *et al.*，2001），此元件与那些在 *FAS*、*S14* 和 *L-PK* 启动子中的元件是同源的。这些不同 ChoRE/G1RE 序列的比较在表 7.1 中列出。ChoRE 被认为是由两个相隔一定距离的 E-box 或与 E-box 序列相似的 CANNTG 构成。每一个 E-box 的前 4bp 及 E-box 间隔序列长度对葡萄糖反应来说都是很关键的。

**表 7.1　各种 ChoRE/GIRE 序列的比较**

| 基因 | 位置 | 葡萄糖反应序列 |
|---|---|---|
| 大鼠 *L-PK* | －166 | CACGGGGCACTCCCGTG |
| 大鼠 *S14* | －1439 | CACGTGGTGGCCCTGTG |
| 小鼠 *S14* | －1442 | CACGCTGGAGTCAGCCC |
| 大鼠 *FAS* | －7210 | CATGTGCCACAGGCGTG |
| 大鼠 *ACC P1* | －122 | CATGTGAAAACGTCGTG |

S14＝Spot14；L-PK＝肝脏丙酮酸激酶；ACC＝乙酰辅酶 A 羧化酶；FAS＝脂肪酸合成酶。来源于 Bergot 等(1992)；Liu 等(1993)；Shih 等(1995)；Koo 和 Towle(2000)；Rufo 等(2001)。

## 反式作用因子的特点

在*L-PK* 和 *S14* 葡萄糖应答元件中所述的 E-box 是碱性的螺旋-环-螺旋（basic helix-loop-helix，bHLH）类转录因子的结合位点。此类转录因子都有以下结构：①一个碱性的结构域，一段富含赖氨酸和精氨酸的氨基酸序列；②两个两亲性 α 螺旋通过环相连（Gregor *et al.*，1990）。这些结构参与了与 DNA 的结合以及蛋白质与蛋白质之间的相互作用。在属于这一类的蛋白质转录因子中，上游刺激因子（upstream stimulatory factor，USF）是与 ChoRE/G1RE 相结合的主要蛋白质，被认为是碳水化合物反应因子的潜在候选者。USF 最初是以与腺病毒晚期启动子相结合的蛋白质形式被发现的（Sawadago and Roeder，1985）。Sawadogo（1988）及 Sirito 等（1994）对 DNA 结合性质相同的两个 USF 蛋白（USF1 和 USF2）的特性进行了描述：它们只是 N 端结构域不同（USF1-USF2），二聚体形式（肝脏中 USF1-USF2）优先与 DNA 相结合（Viollet *et al.*，1996），组织分布是普遍的，并且好像不随营养或激素的水平发生改变。在 *L-PK*与 *S14* 基因中已证明了 USF 与 G1RE 的 E-box 的结合（Vaulont *et al.*，1989；Diaz-Guerra *et al.*，1993；Shih and Towle，1994；Lefrancois-Martinez *et al.*，1995）。

一系列的研究检验了这个假说，包括培养细胞中 USF 野生型和显性失活形式过量表达的研究（Kaytor *et al.*，1997，2000）、小鼠 *USF1* 和 *USF2* 基因敲除的研究（Vallet *et al.*，1997，1998）、USF 合成及其相应于食物的结合活性研究及在 G1RE/ChoRE 野生型和突变型中 USF 的结合活性的研究。从这些研究中得出了这样一个结论，仅仅只有这些因子还不能对葡萄糖通过 G1RE/ChoRE 发挥作用做出解释（Kaytor *et al.*，2000）。

后来证明，与 G1RE/ChoRE 结合的还有其他蛋白质，其中一些蛋白质虽然还没有进一步定性，但也被作为葡萄糖应答转录因子的新候选者（Hasegawa *et al.*，1999；Lou *et al.*，1999；Koo and Towle，2000）。只是在最近，其中一个与 G1RE/ChoRE 结合的蛋白质才被证实为一个潜在的葡萄糖应答转录因子。Uyeda 的课题组根据转录因子与 *L-PK* 启动子结合的能力，从小鼠的肝脏中纯化了一个转录因子，这个蛋白质被命名为碳水化合物应答结合元件蛋白质（carbohydrate response element-binding protein，ChREBP）（Yamashita *et al.*，2001）。这个因子满足了 ChoRE 结合蛋白的许多标准。ChREBP 是一个有二重核定位信号的 bHLH 亮氨酸拉链蛋白，其与突变的 ChoRE 结合的能力与这些 ChoRE 与肝细胞培养物中葡萄糖浓度反应的能力相对应。当转染入肝细胞后，ChREBP 能够刺激包含 ChoRE 的 L-PK 启动子，而且在有葡萄糖存在的情况下，此作用显著加强。尽管此因子的 mRNA 存在于小脑、小肠、肾脏和肝脏等组织中，但只在肝脏中发现此因子与 ChoRE 结合（Yamashita *et al.*，2001）。

葡萄糖对基因转录的影响可能与 ChREBP 因回应葡萄糖而从细胞质到细胞核中的异位有关。当葡萄糖的浓度低时，ChREBP 定位于细胞质中，当葡萄糖浓度高时即进入细胞核（Kawaguchi *et al.*，2001）。这相对于特定丝氨酸（Ser166）的去磷酸化作用是第二位的。Ser166 是在 cAMP 浓度高（如禁食过程中）的情况下，由蛋白激酶 A（protein kinase A，PKA）将其磷酸化。磷酸化后，靠近 DNA 结合域的另一个 PKA 磷酸化位点阻止了 ChREBP 与 DNA 的结合，这个位点在高浓度葡萄糖存在的情况下同样发生去磷酸化（Kawaguchi *et al.*，2001）。因此，葡萄糖的主要作用是激活磷酸酶，抵消 cAMP 的作用，诱导 ChREBP 进入细胞核并增加其 DNA 结合活性，这与胰高血糖素和胰岛素/葡萄糖在 *L-PK* 基因和脂肪生成基因中的拮抗作用相一致。

## 葡萄糖对糖异生酶基因表达的影响

葡萄糖抑制了肝脏中 PEPCK 的转录，这与众所周知的胰岛素的抑制作用是同时发生的（Kahn *et al.*，1987；Meyer *et al.*，1991），且这个作用不依赖于胰岛素的存在。正如上面对脂肪生成酶所描述的一样，葡萄糖发挥作用时需要将葡萄糖磷酸化为葡糖-6-磷酸。葡萄糖的抑制作用可在成年小鼠的肝细胞中观察到，但在没有葡糖激酶表达的乳鼠的肝细胞中观察不到（Cournarie *et al.*，1999）。而且，在 H4IIE 肝癌细胞中不存在葡萄糖对 PEPCK 表达的这种作用，在此细胞中表达的是己糖激酶Ⅰ而不是葡糖激酶（Scott *et al.*，1998）。腺病毒使葡糖激酶过量表达，可以恢复葡萄糖的作用（Scott *et al.*，1998）。目前对 PEPCK 启动子中葡萄糖敏感区域的位置以及相关的转录因子的同一性都还不清楚。

葡萄糖-6-磷酸酶是另外一个关键的糖异生酶。正如以前所阐明的一样，它的表达随日粮中葡萄糖的储存量增加而增加，但被高碳水化合物日粮和胰岛素抑制。胰岛素的抑制作用可能涉及一个叉头（forkhead）家族的转录因子（Nakae *et al*.，2001；Foufelle and Ferré，2002）。令人惊奇的是，在肝癌和原始培养的肝细胞培养基中高浓度葡萄糖能诱导葡萄糖-6-磷酸酶的表达。这种作用至少有一部分是由转录机制介导的

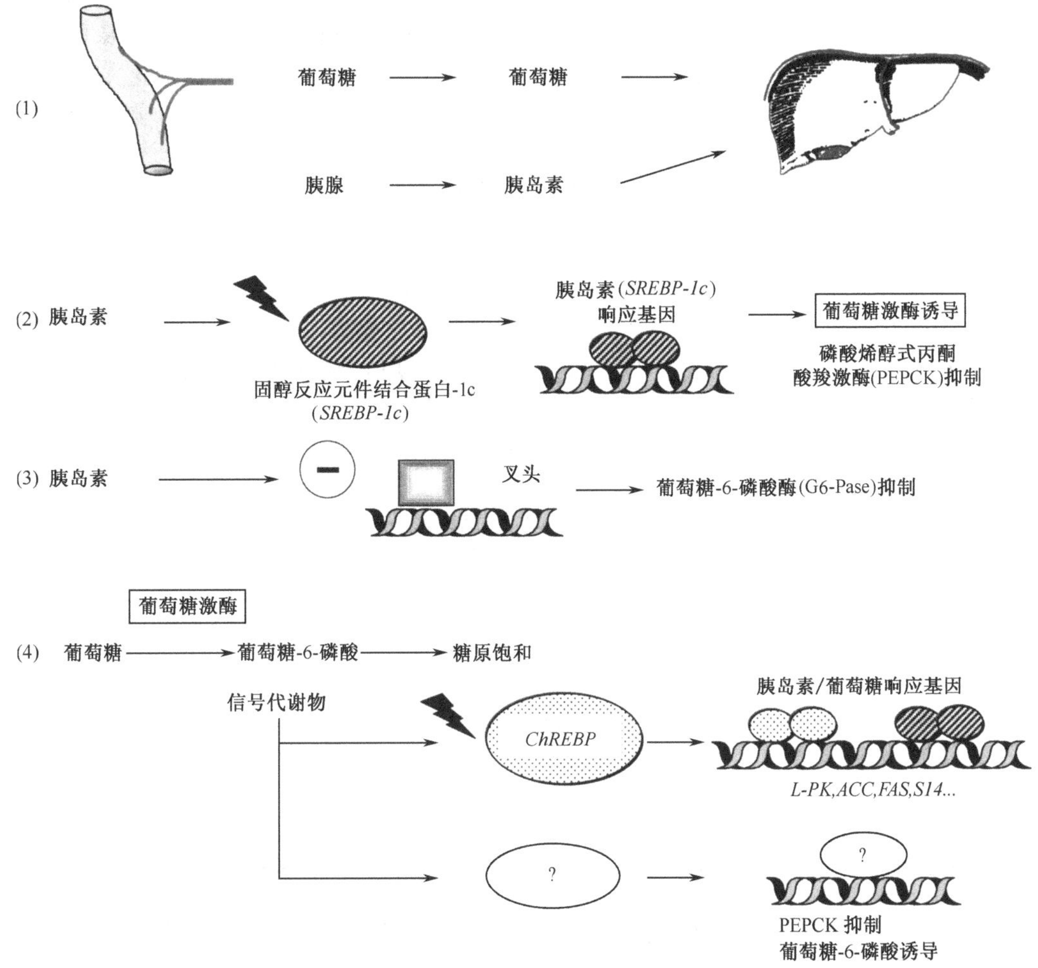


图 7.1　胰岛素和葡萄糖对肝脏基因表达的调节：(1) 碳水化合物丰富的膳食引起血浆葡萄糖浓度增加，导致胰腺 $\beta$ 细胞胰岛素的分泌，葡萄糖和胰岛素都被运送至肝脏；(2) 胰岛素刺激 SREBP-1c，导致葡糖激酶转录的激活和 PEPCK 转录的抑制（Foufelle and Ferré，2002）；(3) 胰岛素同样抑制了叉头家族的一转录因子，因此抑制了葡萄糖-6-磷酸酶基因的表达；(4) 葡糖激酶合成的增加使葡萄糖磷酸化为葡萄糖-6-磷酸。葡糖激酶步骤下游的信号代谢物（葡萄糖-6-磷酸或木酮糖-5-磷酸）可能通过去磷酸化作用介导机制将转录因子（ChREBP）激活。ChREBP 定位于核中，与 SREBP-1c 一起，ChREBP 激活糖酵解/脂肪生成途径中基因，如 *L-PK*，*ACC*，*FAS* 和 *S14*。一个到目前还没有弄清楚的葡萄糖信号系统抑制了 PEPCK 的表达（与胰岛素相连接），激活了葡萄糖-6-磷酸酶的表达（与胰岛素的作用相反）。这个机制解释了葡萄糖过量所引起的葡萄糖生成抑制、糖原储存和脂类合成增多现象

(Argaud *et al.*，1997)。因为糖酵解激活物果糖2,6-二磷酸以及在不同步骤中进入糖酵解过程的底物（木糖醇和果糖）浓度的增加模拟了葡萄糖的作用，这说明一个葡萄糖代谢物可能参与了这一刺激作用（Argaud *et al.*，1997；Massillon，2001）。木糖醇对葡萄糖-6-磷酸酶基因的体内诱导作用已被阐明（Massillon *et al.*，1998），这与木酮糖-5-磷酸浓度加倍而葡萄糖-6-磷酸浓度没有变化相对应。

至于对胰岛素和葡萄糖都敏感的所有基因而言，胰岛素和葡萄糖作为激活剂或抑制剂协同发挥作用。很明显这不符合葡萄糖-6-磷酸酶基因的情况。这种奇怪现象的生理相关性目前还不清楚。这也表明，葡萄糖-6-磷酸酶在葡萄糖的可利用性中尚有目前未知的功能。

## 总　结

肝脏中葡萄糖对基因转录的调节是葡萄糖稳衡调节作用机制的重要组成部分（图7.1）。研究这些信号的级联反应是否存在于其他葡萄糖和脂类代谢的关键组织如肌肉、脂肪组织和β细胞中是很有意义的。而且，这对葡萄糖信号途径中一些成分是否与一些疾病如Ⅱ型糖尿病或脂类紊乱相关的研究也是很重要的。最后，关于葡萄糖对基因的诱导和抑制作用机制是否相同有待于进一步的研究。

## 致　谢

我们感谢Jean Girard博士在这些研究过程中的主要贡献。我们实验室的工作由INSERM和欧洲共同体（CT Fair 97-3011）提供支持。

### 参 考 文 献

Argaud, D., Kirby, T.L., Newgard, C.B. and Lange, A.J. (1997) Stimulation of glucose 6 phosphatase gene expression by glucose and fructose-2,6-biphosphatase. *Journal of Biological Chemistry* 272, 12854–12861.

Bandsma, R.H., Wiegman, C.H., Herling, A.W., Burger, H.J., ter Harmsel, A., Meijer, A.J., Romijn, J.A., Reijngoud, D.J. and Kuipers, F. (2001) Acute inhibition of glucose-6-phosphate translocator activity leads to increased *de novo* lipogenesis and development of hepatic steatosis without affecting VLDL production in rats. *Diabetes* 50, 2591–2597.

Bergot, M.O., Diaz-Guerra, M.J.H., Puizenat, N., Raymondjean, M. and Kahn, A. (1992) *Cis*-regulation of the L-type pyruvate kinase gene promoter by glucose, insulin and cyclic AMP. *Nucleic Acids Research* 20, 1871–1878.

Bollen, M., Keppens, S. and Stalmans, W. (1998) Specific features of glycogen metabolism in the liver. *Biochemical Journal* 336, 19–31.

Brun, T., Roche, E., Kim, K.H. and Prentki, M. (1993) Glucose regulates acetyl-CoA carboxylase gene expression in a pancreatic beta-cell line (INS-1). *Journal of Biological Chemistry* 268, 18905–18911.

Burcelin, R., del Carmen Munoz, M., Guillam, M.T. and Thorens, B. (2000) Liver hyperplasia and paradoxical regulation of glycogen metabolism and glucose-sensitive gene expression in GLUT2-null hepatocytes. Further evidence for the existence of a membrane-based glucose release pathway. *Journal of Biological Chemistry* 275, 10930–10936.

Carling, D., Zammit, V.A. and Hardie, D.G. (1987) A common bicyclic protein kinase cascade inactivates the regulatory enzymes of fatty acid and cholesterol biosynthesis. *FEBS Letters* 223, 217–222.

Clarke, S.D., Armstrong, M.K. and Jump, D.B. (1990) Nutritional control of rat liver fatty acid synthase and S14 mRNA abundance. *Journal of Nutrition* 120, 218–224.

Colosia, A.D., Marker, A.J., Lange, A.J., el-Maghrabi, M.R., Granner, D.K., Tauler, A., Pilkis, J. and Pilkis, S.J. (1988) Induction of rat liver 6-phosphofructo-2-kinase/fructose-2,6-

bisphosphatase mRNA by refeeding and insulin. *Journal of Biological Chemistry* 263, 18669–18677.

Cournarie, F., Azzout-Marniche, D., Foretz, M., Guichard, C., Ferré, P. and Foufelle, F. (1999) The inhibitory effect of glucose on phosphoenolpyruvate carboxykinase gene expression in cultured hepatocytes is transcriptional and requires glucose metabolism. *FEBS Letters* 460, 527–532.

Cuif, M.H., Cognet, M., Boquet, D., Tremp, G., Kahn, A. and Vaulont, S. (1992) Elements responsible for hormonal control and tissue specificity of L-type pyruvate kinase gene expression in transgenic mice. *Molecular and Cellular Biology* 12, 4852–4861.

Daniel, S. and Kim, H.K. (1996) Sp1 mediates glucose activation of the acetyl-CoA carboxylase promoter. *Journal of Biological Chemistry* 271, 1385–1392.

Daniel, S., Zhang, S., DePaoli-Roach, A.A. and Kim, K.H. (1996) Dephosphorylation of sp1 by protein phosphatase 1 is involved in the glucose-mediated activation of the acetyl-CoA carboxylase gene. *Journal of Biological Chemistry* 271, 14692–14697.

da Silva Xavier, G., Leclerc, I., Salt, I.P., Doiron, B., Hardie, D.G., Kahn, A. and Rutter, G.A. (2000) Role of AMP-activated protein kinase in the regulation by glucose of islet beta cell gene expression. *Proceedings of the National Academy of Sciences USA* 97, 4023–4028.

Decaux, J.F., Antoine, B. and Kahn, A. (1989) Regulation of the expression of the L-type pyruvate kinase gene in adult rat hepatocytes in primary culture. *Journal of Biological Chemistry* 264, 11584–11590.

Diaz-Guerra, M.J.M., Bergot, M.O., Martinez, A., Cuif, M.H., Kahn, A. and Raymondjean, M. (1993) Functional characterisation of the L-type pyruvate kinase gene glucose response complex. *Molecular and Cellular Biology* 13, 7725–7733.

Doiron, B., Cuif, M.H., Kahn, A. and Diaz-Guerra, M.J. (1994) Respective roles of glucose, fructose, and insulin in the regulation of the liver-specific pyruvate kinase gene promoter. *Journal of Biological Chemistry* 269, 10213–10216.

Doiron, B., Cuif, M.H., Chen, R. and Kahn, A. (1996) Transcriptional glucose signaling through the glucose response element is mediated by the pentose phosphate pathway. *Journal of Biological Chemistry* 271, 5321–5324.

Elshourbagy, N.A., Near, J.C., Kmetz, P.J., Sathe, G.M., Southan, C., Strickler, J.E., Gross, M., Young, J.F., Wells, T.N. and Groot, P.H. (1990) Rat ATP citrate-lyase. Molecular cloning and sequence analysis of a full-length cDNA and mRNA abundance as a function of diet, organ, and age. *Journal of Biological Chemistry* 265, 1430–1435.

Foretz, M., Carling, D., Guichard, G., Ferré, P. and Foufelle, F. (1998) AMP-activated protein kinase inhibits the glucose-activated expression of fatty acid synthase gene in rat hepatocytes. *Journal of Biological Chemistry* 273, 14767–14771.

Foufelle, F. and Ferré, P. (2002) New perspectives in the regulation of hepatic glycolytic and lipogenic genes by insulin and glucose: a role for the transcription factor SREBP-1c. *Biochemical Journal* 366, 377–391.

Foufelle, F., Gouhot, B., Pégorier, J.P., Perdereau, D., Girard, J. and Ferré, P. (1992) Glucose stimulation of lipogenic enzyme gene expression in cultured white adipose tissue. *Journal of Biological Chemistry* 267, 20543–20546.

Foufelle, F., Girard, J. and Ferré, P. (1996) Regulation of lipogenic enzyme expression by glucose in liver and adipose tissue: a review of the potential cellular and molecular mechanisms. *Advances in Enzyme Regulation* 36, 199–226.

Gancedo, J.M. (1998) Yeast carbon catabolite repression. *Microbiology and Molecular Biology Reviews* 62, 334–361.

Girard, J., Ferré, P. and Foufelle, F. (1997) Mechanisms by which carbohydrates regulate expression of genes for glycolytic and lipogenic enzymes. *Annual Review of Nutrition* 17, 325–352.

Granner, D.K., Andreone, T.L., Sasaki, K. and Beale, E.G. (1983) Inhibition of transcription of PEPCK gene by insulin. *Nature* 305, 549–551.

Gregor, P.D., Sawadogo, M. and Roeder, R.G. (1990) The adenovirus major late transcription factor USF is a member of the helix–loop–helix group of regulatory proteins and binds to DNA as a dimer. *Genes and Development* 4, 1730–1740.

Guillemain, G., Loizeau, M., Pincon-Raymond, M., Girard, J. and Leturque, A. (2000) The large intracytoplasmic loop of the glucose transporter GLUT2 is involved in glucose signaling in hepatic cells. *Journal of Cell Science* 113, 841–847.

Guillemain, G., Munoz-Alonso, M.J., Cassany, A., Loizeau, M., Faussat, A.M., Burnol, A.F. and Leturque, A. (2002) Karyopherin alpha2: a control step of glucose-sensitive gene expression in hepatic cells. *Biochemical Journal* 364, 201–209.

Hardie, D.G. and Carling, D. (1997) The AMP-activated protein kinase – fuel gauge of the mammalian cell? *European Journal of Biochemistry* 246, 259–273.

Hardie, D.G., Carling, D. and Carlson, M. (1998) The AMP-activated/SNF1 protein kinase subfamily: metabolic sensors of the eukaryotic cell? *Annual Review of Biochemistry* 67, 821–855.

Hasegawa, J., Osatomi, K., Wu, R.F. and Uyeda, K. (1999) A novel factor binding to the glucose response elements of liver pyruvate kinase and fatty acid synthase genes. *Journal of Biological Chemistry* 274, 1100–1107.

Hubert, A., Husson, A., Chedeville, A. and Lavoinne, A. (2000) AMP-activated protein kinase counteracted the inhibitory effect of glucose on the phosphoenolpyruvate carboxykinase gene expression in rat hepatocytes. *FEBS Letters* 481, 209–212.

Iynedjian, P.B., Ucla, C. and Mach, B. (1987) Molecular cloning of glucokinase cDNA. Developmental and dietary regulation of glucokinase mRNA in rat liver. *Journal of Biological Chemistry* 262, 6032–6038.

Iynedjian, P.B., Jotterand, D., Nouspikel, T., Asfari, M. and Pilot, P.R. (1989) Transcriptional regulation of glucokinase gene by insulin in cultured liver cells and its repression by the glucagon–cAMP system. *Journal of Biological Chemistry* 264, 21824–21829.

Johnston, M. (1999) Feasting, fasting and fermenting. Glucose sensing in yeast and other cells. *Trends in Genetics* 15, 29–33.

Kahn, C.R., Lauris, V., Koch, S., Crettaz, M. and Granner, D.K. (1987) Acute and chronic regulation of phosphoenolpyruvate carboxykinase mRNA by insulin and glucose. *Molecular Endocrinology* 3, 840–845.

Katsurada, A., Iritani, N., Fukuda, H., Matsumura, Y., Noguchi, T. and Tanaka, T. (1989) Effects of nutrients and insulin on transcriptional and post-transcriptional regulation of glucose-6-phosphate dehydrogenase synthesis in rat liver. *Biochimica et Biophysica Acta* 1006, 104–110.

Katsurada, A., Iritani, N., Fukuda, H., Matsumara, Y., Nishimoto, N., Noguchi, T. and Tanaka, T. (1990) Effects of nutrients and hormones on transcriptional and post-transcriptional regulation of fatty acid synthase in rat liver. *European Journal of Biochemistry* 190, 427–435.

Kawaguchi, T., Takenoshita, M., Kabashima, T. and Uyeda, K. (2001) Glucose and cAMP regulate the L-type pyruvate kinase gene by phosphorylation/dephosphorylation of the carbohydrate response element binding protein. *Proceedings of the National Academy of Sciences USA* 98, 13710–13715.

Kaytor, E.N., Shih, H.M. and Towle, H.C. (1997) Carbohydrate regulation of hepatic gene expression: evidence against a role for the upstream stimulatory factor. *Journal of Biological Chemistry* 272, 7525–7531.

Kaytor, E.N., Qian, J., Towle, H.C. and Olson, L.K. (2000) An indirect role for upstream stimulatory factor in glucose-mediated induction of pyruvate kinase and S14 gene expression. *Molecular and Cellular Biochemistry* 210, 13–21.

Kim, T.S., Leahy, P. and Freake, H.C. (1996) Promoter usage determines tissue specific responsiveness of the rat acetyl-CoA carboxylase gene. *Biochemical and Biophysical Research Communications* 225, 647–653.

Kinlaw, W.B., Perez-Castillo, A.M., Fish, L.H., Mariash, C.N., Schwartz, H.L. and Oppenheimer, J.H. (1987) Interaction of dietary carbohydrate and glucagon in regulation of rat hepatic messenger ribonucleic acid S14 expression: role of circadian factors and 3′,5′-cyclic adenosine monophosphate. *Molecular Endocrinology* 1, 609–613.

Kinlaw, W.B., Church, J.L., Harmon, J. and Mariash, C.N. (1995) Direct evidence for a role of the spot 14 protein in the regulation of lipid synthesis. *Journal of Biological Chemistry* 270, 16615–16618.

Kletzien, R.F., Prostko, C.R., Stumpo, D.J., McClung, J.K. and Dreher, K.L. (1985) Molecular cloning of DNA sequences complementary to rat liver glucose-6-phosphate dehydrogenase mRNA. Nutritional regulation of mRNA levels. *Journal of Biological Chemistry* 260, 5621–5624.

Koo, S.H. and Towle, H.C. (2000) Glucose regulation of mouse S(14) gene expression in hepatocytes. Involvement of a novel transcription factor complex. *Journal of Biological Chemistry* 275, 5200–5207.

Koo, S.H., Dutcher, A.K. and Towle, H.C. (2001) Glucose and insulin function through two distinct transcription factors to stimulate expression of lipogenic enzyme genes in liver. *Journal of Biological Chemistry* 276, 9437–9445.

Leclerc, I., Kahn, A. and Doiron, B. (1998) The 5′-AMP-activated protein kinase inhibits the transcriptional stimulation by glucose in liver cells, acting through the glucose response complex. *FEBS Letters* 431, 180–184.

Lefrancois-Martinez, A.M., Martinez, A., Antoine, B., Raymondjean, M. and Kahn, A. (1995) Upstream stimulatory factor proteins are major components of the glucose response complex of the L-type pyruvate kinase gene promoter. *Journal of Biological Chemistry* 270, 2640–2643.

Liu, Z., Thompson, K.S. and Towle, H.C. (1993) Carbohydrate regulation of the rat L-type pyruvate kinase gene requires two nuclear factors: LF-A1 and a member of the c-myc family. *Journal of Biological Chemistry* 268, 12787–12795.

Lochhead, P.A., Salt, I.P., Walker, K.S., Hardie, D.G. and Sutherland, C. (2000) 5-Aminoimidazole-4-carboxamide riboside mimics the effects of insulin on the expression of the 2 key gluconeogenic genes PEPCK and glucose-6-phosphatase. *Diabetes* 49, 896–903.

Lopez, J., Bennett, M.K., Sanchez, H.B., Rosenfeld, J.M. and Osborne, T.F. (1996) Sterol regulation of acetyl CoA carboxylase: a mechanism for coordinate control of cellular lipid. *Proceedings of the National Academy of Sciences USA* 93, 1049–1053.

Lopez-Casillas, F., Ponce-Castaneda, M.V. and Kim, K.H. (1991) *In vivo* regulation of the activity of the two promoters of the rat acetyl coenzyme-A carboxylase gene. *Endocrinology* 129, 1049–1058.

Lou, D.Q., Tannour, M., Selig, L., Thomas, D., Kahn, A. and Vasseur-Cognet, M. (1999) Chicken ovalbumin upstream promoter-transcription factor II, a new partner of the glucose response element of the L-type pyruvate kinase gene, acts as an inhibitor of the glucose response. *Journal of Biological Chemistry* 274, 28385–28394.

MacLean, P.S., Zheng, D., Jones, J.P., Olson, A.L. and

Dohm, G.L. (2002) Exercise-induced transcription of the muscle glucose transporter (GLUT 4) gene. *Biochemical and Biophysical Research Communications* 292, 409–414.

Mariash, C.N., Seelig, S., Schwartz, H.L. and Oppenheimer, J.H. (1986) Rapid synergistic interaction between thyroid hormone and carbohydrate on mRNAS14 induction. *Journal of Biological Chemistry* 261, 9583–9586.

Marie, S., Diaz-Guerra, M.J., Miquerol, L., Kahn, A. and Iynedjian, P.B. (1993) The pyruvate kinase gene as a model for studies of glucose-dependent regulation of gene expression in the endocrine pancreatic beta-cell type. *Journal of Biological Chemistry* 268, 23881–23890.

Massillon, D. (2001) Regulation of the glucose-6-phosphatase gene by glucose occurs by transcriptional and post-transcriptional mechanisms. Differential effect of glucose and xylitol. *Journal of Biological Chemistry* 276, 4055–4062.

Massillon, D., Chen, W., Barzilai, N., Prus-Wertheimer, D., Hawkins, M., Liu, R., Taub, R. and Rossetti, L. (1998) Carbon flux via the pentose phosphate pathway regulates the hepatic expression of the glucose-6-phosphatase and phosphoenolpyruvate carboxykinase genes in conscious rats. *Journal of Biological Chemistry* 273, 228–234.

Meyer, S., Höppner, W. and Seitz, H.J. (1991) Transcriptional and post-transcriptional effects of glucose in liver phosphoenolpyruvate carboxykinase gene expression. *European Journal of Biochemistry* 202, 985–991.

Miksicek, R.J. and Towle, H.C. (1983) Use of a cloned cDNA sequence to measure changes in 6-phosphogluconate dehydrogenase mRNA levels caused by thyroid hormone and dietary carbohydrate. *Journal of Biological Chemistry* 258, 9575–9579.

Mourrieras, F., Foufelle, F., Foretz, M., Morin, J., Bouché, S. and Ferré, P. (1997) Induction of fatty acid synthase and S14 gene expression by glucose, xylitol and dihydroxyacetone in cultured rat hepatocytes is closely correlated with glucose 6-phosphate concentrations. *Biochemical Journal* 323, 345–349.

Munnich, A., Lyonnet, S., Chauvet, D., Van Schaftingen, E. and Kahn, A. (1987) Differential effects of glucose and fructose on liver L-type pyruvate kinase gene expression *in vivo*. *Journal of Biological Chemistry* 262, 17065–17071.

Nakae, J., Kitamura, T., Silver, D.L. and Accili, D. (2001) The forkhead transcription factor Foxo1 (Fkhr) confers insulin sensitivity onto glucose-6-phosphatase expression. *Journal of Clinical Investigation* 108, 1359–1367.

Nishimura, M. and Uyeda, K. (1995) Purification and characterization of a novel xylulose 5-phosphate-activated protein phosphatase catalysing dephosphorylation of fructose-6-phosphate 2-kinase: fructose-2,6-biphosphatase. *Journal of Biological Chemistry* 270, 26341–26346.

Ntambi, J.M. (1992) Dietary regulation of stearoyl-CoA desaturase 1 gene expression in mouse liver. *Journal of Biological Chemistry* 267, 10925–10930.

O'Callaghan, B.L., Koo, S.H., Wu, Y., Freake, H.C. and Towle, H.C. (2001) Glucose regulation of the acetyl-CoA carboxylase promoter PI in rat hepatocytes. *Journal of Biological Chemistry* 276, 16033–16039.

Pape, M.E., Lopez-Casillas, F. and Kim, K.H. (1988) Physiological regulation of acetyl-CoA carboxylase gene expression: effects of diet, diabetes, and lactation on acetyl-CoA carboxylase mRNA. *Archives of Biochemistry and Biophysics* 267, 104–109.

Paulauskis, J.D. and Sul, H.S. (1989) Hormonal regulation of mouse fatty acid synthase transcription in liver. *Journal of Biological Chemistry* 264, 574–577.

Prip-Buus, K., Perdereau, D., Foufelle, F., Maury, J., Ferré, P. and Girard, J. (1995) Induction of fatty acid synthase gene expression by glucose in primary culture of rat hepatocytes. *European Journal of Biochemistry* 230, 309–315.

Rongnoparut, P., Verdon, C.P., Gehnrich, S.C. and Sul, H.S. (1991) Isolation and characterization of the transcriptionally regulated mouse liver (B-type) phosphofructokinase gene and its promoter. *Journal of Biological Chemistry* 266, 8086–8091.

Rufo, C., Teran-Garcia, M., Nakamura, M.T., Koo, S.H., Towle, H.C. and Clarke, S.D. (2001) Involvement of a unique carbohydrate-responsive factor in the glucose regulation of rat liver fatty-acid synthase gene transcription. *Journal of Biological Chemistry* 276, 21969–21975.

Salt, I.P., Johnson, G., Ashcroft, S.J. and Hardie, D.G. (1998) AMP-activated protein kinase is activated by low glucose in cell lines derived from pancreatic beta cells, and may regulate insulin release. *Biochemical Journal* 335, 533–539.

Sasaki, K., Cripe, T.P., Koch, S.R., Andreone, T.L., Petersen, D.D., Beale, E.G. and Granner, D.K. (1984) Multihormonal regulation of PEPCK gene transcription. The dominant role of insulin. *Journal of Biological Chemistry* 259, 15242–15251.

Sawadogo, M. (1988) Multiple forms of the human gene-specific transcription factor USF. *Journal of Biological Chemistry* 263, 11994–12001.

Sawadago, M. and Roeder, R. (1985) Interaction of a gene-specific transcription factor with the adenovirus major late promoter upstream of the TATA box region. *Cell* 43, 165–175.

Scott, D.K., O'Doherty, R.M., Stafford, J.M., Newgard, C.B. and Granner, D.K. (1998) The repression of hormone-activated PEPCK gene expression by glucose is insulin-independent but requires glucose metabolism. *Journal of Biological Chemistry* 273, 24145–24151.

Shih, H.M. and Towle, H.C. (1992) Definition of the carbohydrate response element of the rat S14 gene. *Journal of Biological Chemistry* 267, 13222–13228.

Shih, H.M. and Towle, H.C. (1994) Definition of the carbohydrate response element of the rat S14 gene. Context of the CACGTG motif determines the specificity of carbohydrate regulation. *Journal of Biological Chemistry* 269, 9380–9387.

Shih, H.M., Liu, Z. and Towle, H.C. (1995) Two CACGTG motifs with proper spacing dictate the carbohydrate regulation of hepatic gene transcription. *Journal of Biological Chemistry* 270, 21991–21997.

Sirito, M., Walker, S., Lin Q., Kozlowski, M., Klein, W. and Sawadago, M. (1994) Ubiquitous expression of the 43- and 44-kDa forms of transcription factor USF in mammalian cells. *Nucleic Acids Research* 22, 427–433.

Sudo, Y. and Mariash, C.N. (1994) Two glucose-signaling pathways in S14 gene transcription in primary hepatocytes: a common role of protein phosphorylation. *Endocrinology* 134, 2532–2540.

Thompson, K.S. and Towle, H.C. (1991) Localization of the carbohydrate response element of the rat L-type pyruvate kinase gene. *Journal of Biological Chemistry* 266, 8679–8682.

Towle, H.C., Kaytor, E.N. and Shih, H.M. (1997) Regulation of the expression of lipogenic enzyme genes by carbohydrate. *Annual Review of Nutrition* 17, 405–433.

Vallet, V.S., Henrion, A.A., Bucchini, D., Casado, M., Raymondjean, M., Kahn, A. and Vaulont, S. (1997) Glucose-dependent liver gene expression in USF2 −/− mice. *Journal of Biological Chemistry* 272, 21944–21949.

Vallet, V.S., Casado, M., Henrion, A.A., Bucchini, D., Raymondjean, M., Kahn, A. and Vaulont, S. (1998) Differential roles of upstream stimulatory factors 1 and 2 in the transcriptional response of liver genes to glucose. *Journal of Biological Chemistry* 273, 20175–20179.

Van Schaftingen, E. and Gerin, I. (2002) The glucose-6-phosphatase system. *Biochemical Journal* 362, 513–532.

Vaulont, S., Munnich, A., Decaux, J.F. and Kahn, A. (1986) Transcriptional and post-transcriptional regulation of L-type pyruvate kinase gene expression in rat liver. *Journal of Biological Chemistry* 261, 7621–7625.

Vaulont, S., Puzenat, N., Levrat, F., Cognet, M., Kahn, A. and Raymonjean, M. (1989) Proteins binding to the liver-specific pyruvate kinase gene promoter. *Journal of Molecular Biology* 209, 205–219.

Vaulont, S., Vasseur-Cognet, M. and Kahn, A. (2000) Glucose regulation of gene transcription. *Journal of Biological Chemistry* 275, 31555–31558.

Viollet, B., Lefrancois-Martinez, A.M., Henrion, A., Kahn, A., Raymondjean, M. and Martinez, A. (1996) Immunochemical characterization and transacting properties of upstream stimulatory factor isoforms. *Journal of Biological Chemistry* 271, 1405–1415.

Waters, K.M. and Ntambi, J.M. (1994) Insulin and dietary fructose induce stearoyl-CoA desaturase 1 gene expression of diabetic mice. *Journal of Biological Chemistry* 269, 27773–27777.

Weber, A., Marie, J., Cottreau, D., Simon, M.P., Besmond, C., Dreyfus, J.C. and Kahn, A. (1984) Dietary control of aldolase B and L-type pyruvate kinase mRNAs in rat. Study of translational activity and hybridization with cloned cDNA probes. *Journal of Biological Chemistry* 259, 1798–1802.

Woods, A., Munday, M.R., Scott, J., Yang, X., Carlson, M. and Carling, D. (1994) Yeast SNF1 is functionally related to mammalian AMP-activated protein kinase and regulates acetyl CoA carboxylase *in vivo*. *Journal of Biological Chemistry* 269, 19509–19515.

Woods, A., Azzout-Marniche, D., Foretz, M., Stein, S.C., Lemarchand, P., Ferre, P., Foufelle, F. and Carling, D. (2000) Characterization of the role of AMP-activated protein kinase in the regulation of glucose-activated gene expression using constitutively active and dominant negative forms of the kinase. *Molecular and Cellular Biology* 20, 6704–6711.

Yamashita, H., Takenoshita, M., Sakurai, M., Bruick, R.K., Henzel, W.J., Shillinglaw, W., Arnot, D. and Uyeda, K. (2001) A glucose-responsive transcription factor that regulates carbohydrate metabolism in the liver. *Proceedings of the National Academy of Sciences USA* 98, 9116–9121.

Zheng, D., MacLean, P.S., Pohnert, S.C., Knight, J.B., Olson, A.L., Winder, W.W. and Dohm, G.L. (2001) Regulation of muscle GLUT-4 transcription by AMP-activated protein kinase. *Journal of Applied Physiology* 91, 1073–1083.

Zhou, M., Lin, B.Z., Coughlin, S., Vallega, G. and Pilch, P.F. (2000) UCP-3 expression in skeletal muscle: effects of exercise, hypoxia, and AMP-activated protein kinase. *American Journal of Physiology, Endocrinology and Metabolism* 279, E622–E629.

Zhu, Q., Mariash, A., Margosian, M.R., Gopinath, S., Fareed, M.T., Anderson, G.W. and Mariash, C.N. (2001) Spot 14 gene deletion increases hepatic *de novo* lipogenesis. *Endocrinology* 142, 4363–4370.

Regulation of muscle GLUT-4 transcription by AMP-activated protein kinase. *Journal of Applied Physiology* 91, 1073–1083.
Zhou, M., Lin, B.Z., Coughlin, S., Vallega, G. and Pilch, P.F. (2000) UCP-3 expression in skeletal muscle: effects of exercise, hypoxia, and AMP-activated protein kinase. *American Journal of Physiology, Endocrinology and Metabolism* 279, E622–E629.
Zhu, Q., Mariash, A., Margosian, M.R., Gopinath, S., Fareed, M.T., Anderson, G.W. and Mariash, C.N. (2001) Spot 14 gene deletion increases hepatic *de novo* lipogenesis. *Endocrinology* 142, 4363–4370.

# 8 哺乳动物细胞中依赖氨基酸的转录调节

Michael S. Kilberg，Van Leung-Pineda 和 Chin Chen
（生物化学与分子生物学系，佛罗里达大学医学院，
盖恩斯维尔，佛罗里达，美国）

## 氨基酸作为信号分子

就公共营养而言，蛋白质/氨基酸的利用率是一个重要因素，特别是在发育（Morgane *et al.*，1993）期间，在一系列的疾病如糖尿病（Hoffer，1993）、恶性营养不良（Roediger，1995）和肝性脑（Mizock，1999）的发展中以及在以下将要讨论的癌化学疗法的一些病例中特别重要。除了上述明确的关系，我们对哺乳动物基本细胞过程中氨基酸的调控了解很少。完整动物营养缺乏法是研究营养作用机制的一个必要但复杂的模型，因为由营养诱导产生的激素变化是一混杂效应，在研究过程中激素本身就可能改变活性表达。因此，许多实验室应用培养氨基酸缺乏细胞作为有用的试验模型，并应用此模型鉴定了由氨基酸缺乏诱导产生的特定 mRNA、特异蛋白质和特异活性（mRNA 例子见表 8.1）。在这些情况下，氨基酸并不是人们所熟知的代谢或蛋白质合成前体物，而是反映生物体营养状况的信号分子。这种氨基酸依赖信号所产生的结果之一就是改变特异基因的转录率。从机制的观点来看，研究的目的是为了了解这个被称为氨基酸反应（amino acid response，AAR）途径的信号传递过程中的每一个步骤。当了解了 AAR 途径中每一个步骤的重要性后，本文将重点讨论氨基酸作为基因转录调节者的作用。

**表 8.1 哺乳动物细胞中氨基酸限制引起蛋白质 mRNA 含量升高的例证**

| | |
|---|---|
| 氨基酸转运蛋白 | |
| CAT1 | Hyatt 等(1997) |
| ATA2 | Gazzola 等(2001) |
| 天冬酰胺合成酶 | Gong 等(1991) |
| C/EBPα | Marten 等(1994) |
| C/EBPβ | Marten 等(1994) |
| CHOP | Marten 等(1994) |
| IGFBP-1 | Straus 等(1993) |
| | Jousse 等(1998) |
| 核糖体蛋白 | |
| L17 | Laine 等(1991) |
| S25 | Laine 等(1991) |
| L35 | Hitomi 等(1993) |
| S13 | Hitomi 等(1993) |

# 酵母中氨基酸对基因表达的调节

酿酒酵母（*Saccaromyces cerevisiae*）使用的是普通调节机制（general control non-repressible，GCN，非阻遏一般调节），酵母细胞中一个氨基酸缺乏就可调节 1000 多个基因的表达（Natarajan *et al.*，2001）。因哺乳动物蛋白质介质与酵母 GCN 信号途径中的蛋白质介质相似，这表明我们可以从研究日益广泛的酵母中得到很多资料。GCN4 为转录激活物 bZIP 家族中的一员，氨基酸缺乏可导致 GCN4 表达的增加（Hinnebusch，1997）。后来在氨基酸缺乏的调节下，GCN4 与基因启动子区域中的 5′ATGA C/G TCAT-3′（Hope and Struhl，1985；Sellers *et al.*，1990）序列相结合。在哺乳动物细胞中，bZIP 蛋白二聚体与 DNA 上一些位点如 AP-1 位点、5′-ATGACTCAT-3′和 ATF/CREB 位点、5′-ATGACGTCAT-3′等相结合（Sellers *et al.*，1990）。如下所述，酵母中 GCN4 识别序列与哺乳动物细胞中负责氨基酸调节的序列并不相同，而且没有哺乳动物 GCN4 同系物的报道，但是至少有一些氨基酸信号的上游事件是相似的。GCN4p 合成的翻译调节依赖于翻译起始时对 GCN4 蛋白编码序列前端起始密码子 AUG 的识别。在 GCN4 mRNA 5′区有 4 个 AUG 密码子，在达到框内终止密码子前，每一个 AUG 密码子后都跟着一个或两个有义密码子（Hinnebusch，1997）。在氨基酸丰富的情况下，在这些可读框（open reading frame，ORF）内从起始到结束反复识别，最后导致很少或没有真正 GCN4 编码序列的开始。由于氨基酸的缺乏，重新起始的效率很低，以至于许多核糖体在最后短的可读框内不能再起始，而是从 GCN4 蛋白起始密码子起始。

GCN4 合成的这种翻译调节是由 GCN2 蛋白激酶来调控的，不带电的 tRNA 的增多可将 GCN2 蛋白激酶激活。一旦被激活，GCN2 蛋白激酶即在 Ser51 位点将酵母翻译起始因子 eIF-2 的“α”亚基磷酸化。当 eIF-2α 被磷酸化后，与鸟嘌呤交换因子 eIF-2B 的结合更紧密，减少了 eIF-2-GTP 的水平，结果造成了 40S 亚基在与 eIF-2-GTP 结合之前越过上游 ORF4（Uore4）的识别（Hinnebusch，1997）。这些变化引起了 GCN4 翻译起始的相应变化，因此代表了缺乏反应信号途径中的一个重要成分。作为与不带电 tRNA 检测机制的一个附加证据，在酵母和含有温度敏感 tRNA 合成酶的哺乳动物细胞中，尽管细胞内的氨基酸池是正常的，却仍然发生了氨基酸剥夺途径的激活。同样，将 Fao 肝癌用 $tRNA^{His}$合成酶抑制剂组氨醇进行处理后也导致了 AAR 途径的激活（Hutson and Kilberg，1994）。

哺乳动物中与酵母 GCN2 相对应的序列已由两个课题组分别进行了克隆和定性（Berlanga *et al.*，1999；Sood *et al.*，2000）。尽管哺乳动物 GCN2 及其酵母对应物还有待定性，但其结构是同源的，包括一个组氨酰-tRNA 合成酶结构域和一个 eIF-2α 激酶结构域。许多研究表明，哺乳动物 GCN2 作为 eIF-2α 激酶发挥作用（Berlanga *et al.*，1999；Sood *et al.*，2000），其作用是介导氨基酸依赖信号途径。关于后面一点，哺乳动物 GCN2 代替酵母内源 GCN2 蛋白在酵母中的表达证明氨基酸缺乏能够将哺乳动物 GCN2 激活。哺乳动物细胞中转录因子 ATF-4 的合成可由氨基酸的缺乏而提高，在 GCN2 敲除的小鼠中不存在对转录因子 ATF-4 的翻译调节（Harding *et al.*，2000）。

## 氨基酸有效性改变哺乳动物的活动

无论在体内或体外，酶活动、转运活动、蛋白质含量、mRAN 含量及特异基因的转录都受到氨基酸有效性的调节。本段的重点是氨基酸缺乏对转录的调节。表 8.1 所列为一系列蛋白质，在哺乳动物中，这些蛋白质的 mRNA 由于氨基酸缺乏而增加。多数情况下，mRNA 含量增加与转录相关，在有些情况下或许与 mRNA 的稳定性相关。尽管有转录调节的直接证据，但是由于氨基酸有效性对它们的调节与其生物学功能密切相关，C/EBPα、C/EBPβ、ATA2 转运蛋白及核糖体蛋白 L35 和 S13 也列在了表中。Marten 等（1994）证明大鼠肝癌细胞中氨基酸缺乏增加了 C/EBPα 和 C/EBPβ mRNA 量。C/EBP 家族成员表达的增加使人们对 C/EBP 同源蛋白（C/EBP homologous protein，CHOP）的氨基酸依赖的转录调节产生特别的兴趣，这将在下文中做更详细的讨论。同样引人注意的还有 C/EBPβ 作为转录因子之一参与人类天冬酰胺合成酶基因的诱导，此诱导作用由 AAR 和内质网应激反应（endoplasmic reticulum stress response，ERSR）感觉营养途径的共同激活引起（Siu *et al.*，2001）。对转录因子表达过程中依赖于氨基酸所发生的变化的进一步的定性研究将为洞察氨基酸限制和细胞内发生特定变化情况下基因表达的机制提供有价值的证据。

尽管在氨基酸受限时核糖体蛋白基因的转录增加似乎是矛盾的，但几种核糖体蛋白的 mRNA 含量却是增加的（Laine *et al.*，1994），而且这种增加对核糖体蛋白 L17 和 S25 来说实际上是在转录（Laine *et al.*，1994）。有趣的是这两种蛋白质增加的 mRNA 含量在氨基酸缺乏时仍滞留在核内，只是在氨基酸补充后才释放入细胞质中进行翻译（Laine *et al.*，1994）。Adilakshmi 和 Laine（2002）研究表明，p53 在核内与 S25 mRNA 的结合可能是核滞留发生的部分原因。将来还要进一步研究氨基酸有效性是如何调节这种核滞留的，这种研究会有利于从机制上了解有趣而新颖的细胞内氨基酸调节过程。

Bruhat 等（1999）总结了氨基酸依赖性调节对胰岛素样生长因子结合蛋白-1（insulin like growth factor binding protein1，IGFBP-1）的重要性。这种对特殊基因产物的调节是有意义的，因为营养依赖性基因表达的调控可能对许多组织和器官都有深远的影响。到目前为止，许多已知受调节的其他基因可能影响了特定细胞的代谢和生长状况或一个器官内的局部细胞群体。目前还没有关于氨基酸有效性对激素和细胞因子表达影响的综合性研究，但这样的研究会有助于我们了解器官间蛋白质营养对细胞生长和代谢的影响。例如，鼠胰腺细胞中组氨酸缺乏会抑制胰高血糖素的合成（Paul *et al.*，1998）。

人们对钠依赖兼性离子氨基酸转运系统 A 中的底物依赖调节研究已有 30 年（Gazzola *et al.*，1972）。在此期间，刊物上常有以该主题发表的综述（Kilberg *et al.*，1993；Palacín *et al.*，1998）。最近对识别编码此活性的多种基因进行的一些研究表明，相应于氨基酸有效性，*ATA2* 基因（又称 *SAT2*）是负责调节系统 A 转运活性的（Gazzola *et al.*，2001）。在过去的 5 年中，人们对阳离子氨基酸转运蛋白（cationic amino acid transporter，CAT1）的营养调节进行了描述（Hyatt *et al.*，1997；Aulak *et al.*，1999；Fernandez *et al.*，2001）。这种特定氨基酸转运蛋白营养调节的一个有趣的特点是其调节作用发生在转录和翻译水平。在氨基酸限制期间，为提高 CAT1 mRNA 的翻译水平而优先利用了核糖体内入口部位，这可能代表了氨基酸对特定蛋白质生物合成控制的早期典型模型（Fernandez *et al.*，2001）。

基因芯片或阵列为我们提供了新的筛选技术，可以预料将有更多在转录和翻译水平上受氨基酸有效性调节的哺乳动物基因被发现。为了举例说明氨基酸缺乏在整个细胞过程中的调控潜力，已证实酵母转录因子 GCN4p 调节了超过 1000 个基因的表达（Natarajan *et al.*，2001）。

# 转录调控模型

## CHOP

### 功　　能

营养缺乏时，肯定需要细胞对该特定营养素的代谢做出反应，目前同样知道任何一种营养素的缺乏常引起大范围的应激反应。这种应激反应将激活多个信号传递途径，这些信号传递途径通过转录和转录后机制改变细胞状态。许多研究证明，个别转录因子的激活是相对应于多种细胞内应激的刺激物而发生的，这其中包括 C/EBP 同源蛋白/生长停滞和 DNA 损伤蛋白 153（CHOP/GADD153）。*CHOP* 基因最初被认为是由 DNA 损伤诱导产生的基因，现在已认识到 *CHOP* 会被许多应激刺激物包括（葡萄糖缺乏引发的）ERSR（endoplasmic reticulum stress response，ERSR）（Yoshida *et al.*，2000）和氨基酸缺乏激活的 AAR 途径（Jousse *et al.*，1999；Fafournoux *et al.*，2000）所激活。而且，*CHOP* 基因敲除已证明了 *CHOP* 表达与细胞内应激引起的细胞程序性死亡之间的联系（Zinszner *et al.*，1998）。正如上面所提到的一样，*CHOP* 是转录因子家族 C/EBP 中的一员（Ron and Habener，1992），而其他两个成员（C/EBPα 和 C/EBPβ）也受氨基酸有效性的调节（Marten *et al.*，1994）。最初，人们认为 *CHOP* 是通过与其他 C/EBP 家族成员形成异源二聚体抑制它们的活性，从而发挥负调节子的作用（Ron and Habener，1992）。然而，最近较多的证据表明 *CHOP*-C/EBP 异源二聚体能够激活许多基因（Wang *et al.*，1998）。

### AAR 和 ERSR 途径

尽管氨基酸的缺乏会通过转录后机制而增加 *CHOP* 的表达（Bruhat *et al.*，1997；Abcouwer *et al.*，1999），但也有证据表明，*CHOP* 基因是由于氨基酸或葡萄糖缺乏而对转录进行调节（Price and Calderwood，1992；Jousse *et al.*，1999）。真核细胞中葡萄糖缺乏引起内质网（endoplasmic reticulum，ER）中糖蛋白的异常积累导致了 ERSR 的发生，与之相对应，在酵母中发生的反应则为非折叠蛋白反应（unfolded protein response，UPR）（Kaufman，1999；Pahl，1999；Patil and Walter，2001）。ERSR 信号传递途径最后导致许多基因转录的增加，其中的许多基因包含在蛋白质过程中并在内质网内运输。典型例子就是内质网固有伴侣蛋白 GRP78。值得注意的是氨基酸缺乏（如

AAR途径）并不能诱导ERSR激活基因（如GRP78）的表达，这或许是因为只有蛋白质合成速度的降低并不会引起内质网中错误折叠蛋白的积累。哺乳动物细胞中ERSR途径的靶基因中包含着一个或多个高度保守的顺式元件（ER应激元件，ERSE）的拷贝，其共同序列是5′-CCAAT-$N_9$-CCACG-3′（Yoshida *et al.*，1998；Roy and Lee，1999）。人类*CHOP*启动子包含两个定位在相反方向的ERSE序列（*CHOP*-ERSE1和CHOP-ERSE2）（Yoshida *et al.*，2000）。突变分析证明*CHOP*-ERSE2是无功能的，而CHOP-ERSE1（核苷酸－93～－75）介导了ER应激造成的基因激活，ER应激是葡萄糖缺乏对基因诱导的基础（Yoshida *et al.*，2000）。Jousse等（1999）对人类*CHOP*启动子进行了缺失分析，表明ERSR途径中转录激活所需的顺式元件与氨基酸缺乏经被AAR途径激活引起的转录增加所需的元件不同。他们的缺失分析与Yoshida等（2000）对ERSE序列的鉴定结果一致，而且正如下面讨论的一样，*CHOP*基因中氨基酸反应元件（amino acid response element，AARE）后来被证明位于核苷酸－302～－310（Bruhat *et al.*，2000；Fafournoux *et al.*，2000）。因此，*CHOP*基因经AAR途径激活所需的基因组元件与葡萄糖缺乏时经ERSR途径激活所需的基因组元件完全不同。下面将对比介绍氨基酸或葡萄糖缺乏时两组不同的基因组元件在人类天冬酰胺合成酶启动子区发挥转录调控的机制。

## CHOP AARE

在知道了人类*CHOP*启动子内AARE核心序列为5′-TGATGCAAT -3′（核苷酸－302～－310）后，Bruhat等（2000）应用体外电泳迁移率变动分析（electrophoretic mobility shift assay，EMSA）考察了该序列与C/EBP以及ATF/CREB结合位点的相似性。研究证实了C/EBPβ和ATF-2存在于用*CHOP* AARE序列作为探针形成的蛋白质-DNA复合物中，但在氨基酸缺乏的细胞提取物中，这些复合物的丰度并没有提高。为了考察这两个转录因子在体内是否都可激活*CHOP*基因，研究者对缺乏ATF-2或C/EBPβ的鼠胚胎成纤维细胞（mouse embryonic fibroblast，MEF）中CHOP mRNA的含量进行了分析。结果发现，在两种敲除类型的细胞中都可发生由ERSR途径激活引起的*CHOP*基因激活，而由氨基酸限制引起的*CHOP*基因的激活只在C/EBPβ敲除的细胞中发生，在缺乏ATF-2的细胞中并不发生。他们的后续研究表明，在缺乏ATF-2的细胞中转染入ATF-2表达质粒后能够恢复氨基酸控制，而且ATF-2的一个显性失活形式的表达能抑制其在野生型MEF中的诱导作用（Bruhat *et al.*，2000）。Bruhat等（2000）证明，用EMSA为基础使某种特异转录因子（在本例中为C/EBPβ）在体外结合，不能得出此因子在体内也能发挥作用的结论。

为了证明在缺乏某种特定氨基酸的细胞中，内源天冬酰胺合成酶（AS）和*CHOP*基因的调节不一样，Jousse等（2000）对AS和CHOP mRNA进行了Northern分析，并对启动子荧光素酶报道基因结构物（promoter-luciferase reporter constructs）进行了瞬时转染。虽然当HeLa细胞缺乏亮氨酸、半胱氨酸、天冬氨酸或组氨酸时，AS启动子的相对活性要大于*CHOP*启动子，但在细胞缺乏蛋氨酸时，*CHOP*启动子的活性几乎是原来的2倍（Jousse *et al.*，2000）。因每一氨基酸都有各自不同的调节作用，Jousse等（2000）提出了细胞内存在多种途径感知每一氨基酸的利用率，这些途径可使不同基因被激活的假设。他们还证明在不引起蛋白质合成显著抑制的低亮氨酸浓度（30μmol/L）下，培育 HeLa 细胞（子宫颈癌组织的细胞株）可导致 *CHOP* 和 AS mRNA 表达增加，增加的水平只是略低于亮氨酸完全缺乏时增加的水平。而且由放线酮诱导的蛋白质合成抑制（抑制的程度与所有氨基酸限制时所导致的蛋白质合成抑制的程度相同）没有导致任一基因的激活。Jousse 等（2000）根据这些结果得出蛋白质合成抑制不是 AAR 途径激活的前提条件的结论。虽然这些研究者利用 HeLa 细胞进行研究，但其结果与我们实验室应用低水平的放线酮处理大鼠 Fao 肝癌细胞所得的试验结果相似（Hutson and Kilberg，未出版的结果）。

有趣的是，缺乏某一氨基酸所导致的某一特定基因激活的程度可能存在组织特异性差异。Bruhat 等（2000）研究表明，在 HeLa 细胞、Caco-2 和 HepG2 肝癌细胞中，亮氨酸、蛋氨酸或赖氨酸缺乏通过 *CHOP* AARE 所引起的转录激活的程度是不同的。这些结果与体内 AAR 途径早期步骤的组织特异性之间的关系还有待于进一步证实。Entingh 等（2001）研究了小鼠成纤维细胞中氨基酸缺乏对 *CHOP* 的诱导，结果表明在这些细胞中 *CHOP* 基因的诱导需要血清，并且进一步证明了所需血清成分中需含有胰岛素样生长因子-Ⅰ（insulin-like growth factor-Ⅰ，IGF-Ⅰ）。通过对抑制子的研究，他们又得出了 *CHOP* 诱导途径包括磷酯酰肌醇-3-激酶（phosphatidylinositol-3 kinase，PI3K）和哺乳动物的雷帕霉素靶（mammalian target of rapamycin，mTOR）的结论。氨基酸限制引起的基因诱导对血清的要求可能依赖于靶基因或所研究的细胞的类型。例如，当氨基酸缺乏时，人类成纤维细胞中系统 A 氨基酸转运蛋白的诱导依赖于血清（Gazzola *et al.*，1981；Kilberg *et al.*，1985），而在大鼠肝癌细胞中的诱导不依赖血清（Kilberg *et al.*，1985）。同样，氨基酸缺乏对 AS 的诱导不需要血清（M. S. Kilberg *et al.*，未出版结果）。Jousse 等（1998）研究了氨基酸限制时肝癌细胞和培养的原肝细胞中的 IGF-Ⅰ、IGF-Ⅱ以及 IGFBP-Ⅰ的表达，结果表明，随着氨基酸的限制，IGFBP-Ⅰ mRNA 及蛋白质含量显著增加，而胰岛素样生长因子Ⅰ的表达则不受影响或只轻微的减少。与 Jousse 等（1998）体外研究数据一致，蛋白质营养不良已被证明减少了 IGF-Ⅰ循环水平并提高了 IGFBP-Ⅰ。虽然这些结果看起来与 *CHOP* 基因激活需要 IGF-Ⅰ的结论相矛盾，但 Entingh 等（2001）证明即使是 IGF-Ⅰ的水平降低，但仍可以满足 *CHOP* 基因诱导的需要。很明显，在调节 AAR 途径中，IGF-Ⅰ确切的作用及其中包含的明显的细胞特异性还需要作进一步的研究。

## 天冬酰胺合成酶

### 酶　　学

人类天冬酰胺合成酶（asparagine synthetase，AS）催化天冬酰胺的合成并催化天冬氨酸、ATP 和谷氨酰胺合成谷氨酸。按序列分析，AS 蛋白属于第二类谷氨酰胺转移酶（GAT）超家族。在结构上，AS 蛋白的 N 端区域包含了谷氨酰胺利用（GAT）结构域，C 端区域包含了可催化天冬氨酸激活的酶的激活部位。

许多物种的 AS cDNA 都已被克隆，并且分析证明在这些物种间有着高度的同源性（Richards and Schuster，1998）。大鼠、仓鼠和人类细胞都表达约 2.0kb 的 AS mRNA。

在仓鼠细胞中还表达一个较大的约 2.5kb 的 mRNA，在大鼠细胞中也已观察到三种 AS mRNA，分别为 2.0kb、2.5kb 和 4.0kb。所有这三种 mRNA 都同等地受到氨基酸缺乏的诱导（Hutson and Kilberg，1994）。通过对大鼠 2.5kb cDNA 的 3′-非翻译区进行 Northern 分析发现，只有 2.5kb 和 4.0kb 的 AS mRNA 发生了杂交，这表明不同长度的 mRNA 间聚腺苷酸长度不等或者是发生了剪接。

## 氨基酸对 AS 表达的调节

Gong 等（1990）在检测 *ts11* 突变细胞中完成细胞周期阻断的基因时鉴定了 AS。同一实验室（Gong *et al.*，1991）证明在缺乏天冬氨酸、亮氨酸、异亮氨酸或谷氨酸的细胞中，AS mRNA 的水平将升高，同 Andurlis 等（1979）早期应用 tRNA 合成酶突变体检测酶活性所得的结果一致。这些研究表明 AS mRNA 的表达并不是只对天冬酰胺特异，也受其他氨基酸利用率的调节。Hutson 和 Kilberg（1994）证明在大鼠正常的肝组织中和培养的 Fao 肝癌细胞中，相应于所有氨基酸的缺乏，AS mRNA 的含量均提高。同样，培养基中单独一种必需氨基酸如组氨酸、苏氨酸或色氨酸的缺乏同样能使稳衡状态的 AS mRNA含量升高，苯丙氨酸、亮氨酸和异亮氨酸的缺乏也有同样的作用，但作用范围较小。这些结果也说明了哺乳动物细胞中 AAR 途径调节的广谱性。

有趣的是，在另外的无氨基酸培养介质中只添加一种氨基酸即可以抑制 Fao 肝癌细胞中 AS mRNA 表达的增加（Hutson and Kilberg，1994）。与重新在氨基酸完全介质中培养的细胞相比较，谷氨酸在很大程度上抑制了 AS mRNA 的诱导，但天冬氨酸、组氨酸和亮氨酸也同样有效。脯氨酸、丝氨酸和苏氨酸能适度地抑制 AS mRNA 的诱导作用，作用最弱的抑制子是天冬氨酸、甘氨酸和谷氨酸。在只缺乏组氨酸的介质中培养 Fao 细胞 12h，的确导致了细胞内几种必需氨基酸（天冬氨酸、丝氨酸、甘氨酸、丙氨酸和脯氨酸）含量的升高，但谷氨酸和组氨酸的含量分别下降了 50%和 35%（Hutson *et al.*，1996）。从含所有氨基酸的介质中单独剔除一种氨基酸能够诱导 AS 的表达，这说明其他 19 种氨基酸的存在不足以将基因维持在抑制状态。另一方面，在无氨基酸的 Krebs-Ringer 碳酸氢盐介质中单独添加一种氨基酸对基因产生了抑制作用，这说明要将基因维持在基础或抑制状态并不是所有的氨基酸都必须存在（Hutson and Kilberg，1994）。这种自相矛盾的结论还需要进一步的验证，也可能包括 AAR 途径中前期信号的传导。

升高的 AS mRNA 同多聚核糖体的结合（Hutson *et al.*，1996）以及 AS 蛋白合成脉冲追踪标记（Hutson *et al.*，1997）证明，随着氨基酸的缺乏，升高的 AS mRNA 将被翻译为蛋白质。与此相一致，Arfin 等（1977）报道氨基酸缺乏 24h 后，中国仓鼠卵巢（Chinese hamster overy，CHO）细胞中 AS 酶活性升高。组织中 AS 的含量变化很大，但在大鼠胰、睾丸、脑和脾中最高（Hongo *et al.*，1992）。早在 AS cDNA 分离之前，Arfin 等（1977）就证明将 CHO 细胞转移到缺乏天冬酰胺的介质中，$tRNA^{Asn}$的氨酰化程度会降低，而 AS 活性水平将升高。与之相似，当将一个包含温度敏感的天冬酰胺酰-tRNA 合成酶的突变细胞系转移至非许可温度中时，天冬酰胺酰-$tRNA^{Asn}$的含量下降而 AS 活性升高（Andrulis *et al.*，1979）。当生长在非许可温度中时，即使天冬酰胺酰-tRNA$^{Asn}$的水平未发生变化，含有温度敏感的亮氨酰-tRNA、蛋氨酰-tRNA 和赖氨酰-tRNA 合成酶的 CHO 细胞突变体的 AS 活性也将增加（Andrulis *et al.*，1979）。这些结果与 Berlanga 等（1999）和 Sood 等（2000）所提出的 GCN2 激酶通过结合广谱的无负载 tRNA 来感知氨基酸缺乏的建议一致。Hansen 等（1972）研究表明氨基组氨醇通过抑制相应的 tRNA 合成酶阻止了组酰胺酰-tRNA$^{His}$的形成，用 5mmol/L 氨基组氨醇处理 Fao 肝癌细胞后，AS mRNA 的水平升高到细胞中全部氨基酸缺乏时 AS mRNA 的水平甚至是更高（Hutson and Kilberg，1994）。组氨醇处理在没有减少细胞质中游离组氨酸含量的情况下增加了 AS mRNA 的含量，这个事实说明 AAR 信号途径不是由游离氨基酸的水平来启动的。总体来说，这些结果不仅证明了 tRNA 的水平调节在感知氨基酸缺乏时是很重要的，而且也表明了哺乳动物细胞中的感觉机制与酵母中相应的一般调节是相似的，在酵母中任何一种氨基酸的缺乏都是有作用的。

## AS 表达和细胞周期的调节

当生长在非许可温度下时，突变仓鼠的 BHK *ts11* 细胞在细胞周期的 $G_1$ 期受到特定的阻碍（Greco *et al.*，1989；Gong and Basilico，1990）。Basilico 及其同事鉴定了一种能使突变仓鼠的 BHK *ts11* 细胞完成细胞周期的人类克隆。通过序列同源性分析，该人类克隆被认定为 AS，由于其具有通过外源添加天冬酰胺避过 *ts11* 突变的能力而进一步被确认。当生长在非许可的温度下时，BHK *ts11* 细胞能产生一种无活性的天冬酰胺合成酶（Gong and Basilico，1990），导致细胞天冬酰胺的耗竭和 AS mRNA 的相应增加。在介质中加入天冬酰胺可导致 AS mRNA 水平的降低（Greco *et al.*，1989），这也显示出天冬酰胺含量与细胞周期之间的联系。在无血清的 Balb/c 3T3 细胞中重新加入血清可导致 AS mRNA 大量增加。在人类、鼠和仓鼠细胞中，AS mRNA 大概在 $G_1$ 中期被诱导，而在培养介质中加入天冬酰胺后阻止了血清激活细胞中 $G_1$ 诱导的发生。Hongo 等（1989）证明，AS 活性是在淋巴细胞激活的过程中由植物凝集素来诱导的，而且其活性的增加与 DNA 合成率一致。Colletta 和 Cirafici（1992）用促甲状腺素处理休眠大鼠甲状腺细胞后，细胞周期进入 S 期，同时引起 AS mRNA 含量的增加，这为 AS 基因可能受细胞周期依赖方式的调节提供了另外的证据。

## AS 表达与白血病天冬酰胺酶疗法的关系

癌细胞为了维持快速的生长和细胞分裂，对营养物质的需要有所增加，在相应的代谢途径中通常有高水平酶的表达。人们根据某一特定癌症的这种需要已设计出相应的临床措施，从而使关键酶无法足够地表达。例如，L-天冬酰胺酶（L-asparaginase，ASNase）是儿童急性原淋巴细胞白血病（acute lymphoblastic leukaemia，ALL）化学疗法中的一种辅助药物（Capizzi and Holcenberg，1993；Chabner and Loo，1996；Chakrabarti and Schuster，1996；Muller and Boos，1998）。这种疗法可以非常成功地减轻疾病，但由于产生耐药性而引起疾病复发仍是一个问题。ASNase 催化 AS 的逆反应，该治疗方法之所以有效是因为细胞中天冬酰胺被耗竭，使那些有足够的 AS 抵抗这种作用的细胞存活下来。然而，儿童急性原淋巴细胞白血病（ALL）细胞 AS 的表达都处在一个特别低的水平，因此，ASNase 处理在阻止这种形式白血病的生长方面是很有效的。而且，临床上应用的细菌 ASNase 酶显示了低水平的谷氨酰胺酶活性，所以谷氨酰胺的耗竭可能也起了一定的作用。

通过 ALL 细胞与 ASNase 直接接触可以对有耐药性的 ALL 细胞进行选择（Hutson *et al.*，1997）。Aslanian 和 Kilberg（2001）证明，在对 ASNase 有抵抗力的细胞中会发生许多适应性的代谢变化，总体来说有两方面的影响：①增加细胞内 AS 底物天冬氨酸和谷氨酸浓度，有可能提高天冬酰胺的转化率；②通过双向不依赖 $Na^+$ 转运蛋白将天冬酰胺的流出量降到最少，并通过 $Na^+$ 驱动转运蛋白增加天冬酰胺的流入量，从而改变天冬酰胺的质膜流量。另外，Aslanian 等（2001）证明，在没有药物选择的情况下，ASNase 敏感的亲本 ALL 细胞中 AS 的过量表达可导致其对 ASNase 的抵抗性，这说明单独的高 AS 就足以产生耐药性表型。而且，这些研究人员证明即使在培养介质中将药物移去 6 个星期或更长时间，AS 的表达仍会提高，表明 *AS* 基因表达率发生了永久的变化。与此结果相一致，人们已经证明移去药物后 ASNase 抵抗性也不会逆转（Aslanian *et al.*，2001）。

## *AS* 启动子分析

Guerrini 等（1993）证明氨基酸对 AS mRNA 的调节发生在转录水平。氯霉素乙酰转移酶（chloramphenicol acetyltransferase，CAT）报道基因处于含启动子区域和前两个外显子和内含子序列的 3.4kb 片段的人类基因组 DNA 调节之下。当细胞中转染入这种报道基因时，相应于天冬酰胺和亮氨酸的缺乏，CAT mRNA 的活性升高。相反，在同样的情况下，受猿猴病毒 40（SV40）早期启动子控制的 CAT mRNA 的水平和活性降低。缺失分析表明，*AS* 基因内－164～＋44 的核苷酸序列保持了氨基酸缺乏时的可诱导性（Guerrini *et al.*，1993）。通过诱变对 *AS* 启动子进行进一步的观测发现，在－70～－64 核苷酸（5′-CATGATG-3′）序列中存在着氨基酸对 *AS* 基因调节所必需的 AARE。EMSA 用含有这种 AARE 序列的双链寡核苷酸证明了体外特定蛋白质-DNA 复合物的形成，但是 AARE 序列却不精确的符合任一已知转录因子所共有的共同序列。正如下面将要描述的一样，后来的研究说明了 *AS* 基因组序列的作用比营养调节更重要。

## *AS* 基因对复合营养素的感知

Barbosa-Tessmann 等（1999a）证明人类 *AS* 基因也可由葡萄糖缺乏诱导，这种诱导作用是通过 ERSR（UPR）途径介导的（Barbosa-Tessmann *et al.*，1999b）。葡萄糖缺乏 8h 后可观察到人类肝癌 HepG2 细胞中 AS mRNA 含量的增加，在 12h 后达到高峰。同样也有文献报道随葡萄糖缺乏会引起 AS 蛋白表达的增加（Barbosa-Tessmann *et al.*，1999a）。为了证明葡萄糖缺乏所引起的 AS 转录的诱导是 ERSR 途径的结果，此途径中的其他被识别的激活剂如蛋白糖基化抑制剂衣霉素和脯氨酸结构类似物铃兰氨（Aze）都被认为是激活剂（Barbosa-Tessmann *et al.*，1999b）。这些结果表明人类 AS 是 ERSR 途径的一个靶基因，而且在氨基酸代谢和 ERSR 途径之间起到了联系的作用。在 ER 应激过程中天冬酰胺生物合成增加的确切代谢功能目前还不清楚。

ERSR 途径通过转录机制诱导了 *AS* 基因。Barbosa-Tessmann 等（1999b）报道显示当一个报道基因由人类 *AS* 启动子启动时，其表达水平会由于葡萄糖的缺乏而显著的提高。缺失分析表明，*AS* 基因中 ERSR 控制有关的顺式元件位于 *AS* 启动子－111～－34 核苷酸，但是已知的存在于其他所有以前被识别的 ERSR 可诱导基因中的哺乳动物 ERSE 共同序列（5′-CCAAT-$N_9$-CCACG-3′）并不存在。Barbosa-Tessmann 等（2000）相继证明，氨基酸（AAR 途径）或葡萄糖（ERSR 途径）缺乏所引起的 *AS* 基因转录激活是由 *AS* 近端启动子中一套共同的独特基因组元件介导的。

## 养分感测反应元件

正如图 8.1 所示，硫酸二甲酯体内印迹法证明了人类 *AS* 启动子区主要转录起始部位的直接上游包含着 6 个分开的蛋白质结合位点（Barbosa-Tessmann *et al.*，2000）。其中的 5 个包含在人类 *AS* 基因的营养调控中，为 3 个 GC 框（GC-Ⅰ、GC-Ⅱ和 GC-Ⅲ）和 2 个养分感测应答元件（nutrient sensing response element，NSRE-1 和-2）。所有的 3 个 GC 框都为维持基础转录和氨基酸限制时 *AS* 基因达到最大激活所必需（Leung-Pineda and Kilberg，2002）。然而，当将其分别进行功能性分析时，这三种 GC 序列间不存在完全重复，在转录中它们的重要性程度不同（GC-Ⅲ＞GC-Ⅱ＞ GC-Ⅰ）。在体外，其中的两种 GC 序列（GC-Ⅱ和 GC-Ⅲ）与 Sp1 或 Sp3 一起形成蛋白质 DNA 复合物，但这些复合物的绝对总含量和 Sp1 或 Sp3 蛋白质总量并没有随氨基酸的缺乏而升高。在缺乏 SP 蛋白的果蝇 SL2 细胞中，Sp1 和 Sp3 的体内表达增加了 *AS* 启动子的活性，但这两个因子间在功能上存在着一定的差异（Leung-Pineda and Kilberg，2002）。Sp1 的表达通过 *AS* 启动子增加了基础转录，但在 SL2 细胞缺乏氨基酸时却不能引起转录的进一步增加。与之相反，Sp3 的表达不仅提高了基础转录也提高了氨基酸缺乏诱导的 AS 驱动的转录。

体内印迹试验所识别的两个蛋白质结合部位显示了相应于氨基酸缺乏时蛋白质保护的变化（Barbosa-Tessmann *et al.*，2000）。这两个最初称为位点Ⅴ和位点Ⅵ的位点现在被重新命名为 NSRE-1 和 NSRE-2（图 8.1）。在这两个位点的整个区域内的单个核苷酸突变将这两个元件的边界限定得更加清楚（图 8.2）。NSRE-1 序列（5′-TGATGAAAC-3′）位于 *AS* 近端启动子－68～－60 核苷酸，与 Guerrini 等（1993）第一次鉴定的有 AARE 活性的序列相重叠。Barbosa-Tessmann 等（2000）证明了 NSRE-1 序列对 AAR 途径和 ERSE 途径激活所引起的 *AS* 基因的诱导是必需的，很明显这个元件的功能要比单纯的一个 AARE 的作用要重要得多。为了反映这种更广泛的营养-检测能力，人们创造了 NSRE-1 这个术语。当在用 EMSA 对从氨基酸缺乏的 HepG2 细胞中提取的核提取物进行测试时发现，蛋白质 NSRE-1 复合物数量上升（Barbosa-Tessmann *et al.*，2000）。而且，位于 NSRE-1 下游 11 个核苷酸位置的第二个元件 5′-GTTACA-3′（核苷酸－48～－43）同样被证明为氨基酸和葡萄糖缺乏对 *AS* 基因诱导所完全必需（Barbosa-Tessmann *et al.*，2000），并再一次通过单核苷酸突变限定了这个位点的边界（图 8.2），但是这个序列与任何已知转录因子的共同序列都不完全符合。总体来说，启动子分析表明，为了使 *AS* 基因对氨基酸或葡萄糖缺乏有最适的转录反应，至少需要 3 个分离的顺式调节元件，一个或多个 GC 框、NSRE-1 和 NSRE-2。为了描述这些元件

```
                                      GC-I                       GC-II
-173  CAAAAGAGCT CCTCCTTGCG CCCTTCCGCC GCCCCACTTA GTCCTGCTCC GCCCCGGACA
      GTTTTCTCGA GGAGGAACGC GGGAAGGCGG CGGGGTGAAT CAGGACGAGG CGGGGCCTGT

               GC-III                                    NSRE-1
-113  CCCCGCGGCC CCGCCCCTGT GCGCGCTGGT TGGTCCTCGC AGGCATGATG AAACTTCCCG
      GGGGCGCCGG GGCGGGGACA CGCGCGACCA ACCAGGAGCG TCCGTACTAC TTTGAAGGGC

        NSRE-2
-53   CACGCGTTAC AGGAGCCAGG TCGGTATAAG CGCCAGCGGC CTCGCCGCCC GTCaagctgt
      GTGCGCAATG TCCTCGGTCC AGCCATATTC GCGGTCGCCG GAGCGGCGGG CAGttcgaca

+8    CCACATCCCT GGCCTCAGCC CGCCACATCA CCCTGACCTG CTTA
      GGTGTAGGGA CCGGAGTCGG GCGGTGTAGT GGGACTGGAC GAAT
```

图 8.1 人类天冬酰胺合成酶基因的近端启动子序列。6 个潜在的蛋白质结合位点通过体内印迹法被认可（Barbosa-Tessmann *et al.*，2000）。其中 5 个位点对转录的营养调控有作用：GC 框Ⅰ～Ⅲ，NSRE-1 和 NSRE-2。正如体内印迹或体外 EMSA 所检测到的一样，结合在 3 个 GC 框上的蛋白质在对照和营养缺乏细胞中是相同的，而结合在 NSRE-1 和 NSRE-2 上的蛋白质则通过 AAR 或 ERSR 途径的激活而被提高。带阴影的框是建立在如图 8.2 所显示的单核苷酸突变的基础上，显示出了 NSRE-1 和 NSRE-2 的边界

的集体作用，出现了养分感测响应装置（nutrient-sensing response unit，NSRU）这个术语。

如上所述，人类 *CHOP* 基因是由 AAR 和 ERSR 途径通过 *CHOP* 启动子区域内两套完全独立的顺式作用元件激活的，其中 ERSE-1 在－93～－75 之间（Yoshida *et al.*，2000），AARE 在－302～－310 之间（Bruhat *et al.*，2000）。正如刚才所描述的一样，人类 *AS* 基因的转录也可由 AAR 和 ERSR 途径激活，但与 *CHOP* 不同，这种激活作用是通过一套相同的顺式作用元件一起作用组成 NSRU 完成的（Barbosa-Tessmann *et al.*，2000）。有趣的是，在 *CHOP* 启动子内 NSRE-1 和 AARE 序列有高度的一致性，只是两个核苷酸不同（图 8.3）。尽管 *CHOP* 启动子中的 AARE 序列与 *AS* 启动子中的 AARE 序列出现在相反的链中，但不论其方向如何，这些序列作为增强子元件在基因激活中的作用已被观察到（Barbosa-Tessmann *et al.*，2000）。除了 *CHOP* AARE 和 *AS* NSRE-1 间这两个核苷酸的差异，两个启动子间的主要区别好像是第二个顺式作用元件 NSRE-2 的存在（Barbosa-Tessmann *et al.*，2000），而 *CHOP* 启动子无此元件，从而引起对 ERSR 途径的反应机制不同。如果 *CHOP* 启动子中 *AS* NSRE-1 序列与 AARE 有相似性，人们或许会推测删除 *AS* NSRE-2 序列可能会阻断 *AS* 基因激活的 ERSR 途径，但会保留 AAR 途径对基因的激活。然而，仅只 NSRE-2 序列的突变就会导致两条营养调节途径的完全消失（Barbosa-Tessmann *et al.*，2000）。由于一些未知的原因，*CHOP* 启动子中 NSRE-1 样 AARE 序列的存在足以引起通过 AAR 途径的转录诱导，而与之紧密联系的 NSRE-1 序列却不能。有趣的是，在 *CHOP* 启动子中插入

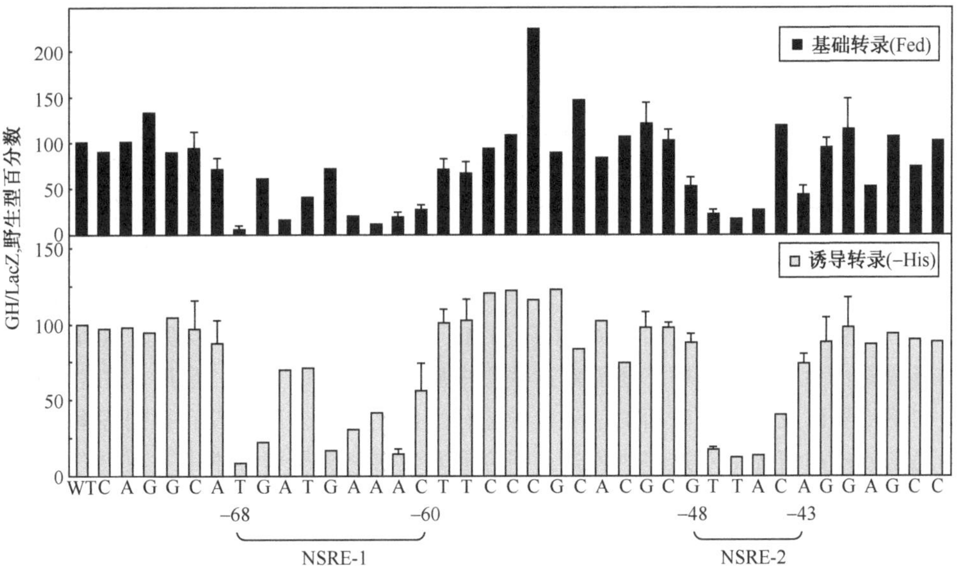


图 8.2 应用单核苷酸突变界定 NSRE-1 和 NSRE-2 调节位点的边界。按照 Barbosa-Tessmann 等(2000)描述的方法，进行了突变和瞬时表达研究。将人类 *AS* 基因中－173～＋51 核苷酸序列连接到作为报道基因的人类生长激素基因（GH）上，通过分析其 GH mRNA 含量相对于共转染对照物（为细胞肥大病毒启动子启动的 *lacZ* 基因）的含量来确定转录程度。细胞被放在完全介质（MEM）（图中数据标为“Fed”）或缺乏组氨酸的 MEM（标为“-His”）中培养 18h。没有标准差符号的柱状图的数据代表的是单一分析，并且只对一个核苷酸替代物进行了测试，嘌呤对嘌呤，嘧啶对嘧啶。那些带有标准差的柱状图的部位的结果代表 3 个独立试验，并且野生型的核苷酸最少改变成 3 个其他核苷酸中的 2 个。为了更好地界定 NSRE-1 和 NSRE-2 结合部位，后面的位点考察的更详细

AS NSRE-2 序列（*CHOP* AARE 下游 11 个核苷酸），可将反应性传送到 ERSR 途径（Bruhat *et al.*，2002）。

| | | | |
|---|---|---|---|
| AS | nt–68 | 5′ -TGATGAAAC- 3′<br>3′ -ACTACTTTG- 5′ | nt–60 |
| CHOP | nt–310 | 5′ -ATTGCATCA- 3′<br>3′ -TAACGTAGT- 5′ | nt–302 |

图 8.3 *AS* NSRE-1 和 *CHOP* AARE 调节位点序列的比较。图中显示了在人类 *AS* 和 *CHOP* 启动子中介导 AAR 的顺式作用元件，该图也阐明了与转录起始部位相关的双链序列和相应的核苷酸号码。注意这两个元件位于相反的链中

*CHOP* AARE 和 *AS* NSRE-1 好像也能与不同的转录因子结合。如上所述，在体外 CHOP 与 ATF-2 和 C/EBPβ 两者都可以结合，但在体内只有 ATF-2 可能与此序列具有功能上的联系（Bruhat *et al.*，2000）。与之相反，*AS* NSRE-1 序列在体外不与 ATF-2 结合，但与 C/EBPβ 结合（Siu *et al.*，2000）。而且，C/EBPβ 过量表达，通过 NSRE-1 序列对基础和诱导转录都有正调节作用（Siu *et al.*，2000）。结合在 *CHOP* AARE 和 *AS* NSRE-1 上的转录因子的显著区别目前还不清楚，但有人认为在单一或多个 AAR 途径的上游信号步骤中存在着异质性。将来对这两个基因及其氨基酸转录控制的比较研究将是很有趣的，并可进一步增进我们的认识。

# 小结和未来研究机遇

哺乳动物细胞中营养素调控，特别是氨基酸调节研究还处在刚刚起步的阶段，对应于蛋白质/氨基酸缺乏的全部靶基因目前还不清楚，但在不久的将来应该可以通过DNA微芯片和阵列技术对其进行描述。当然，正如目前已经清楚的一样，通过转录而激活的基因名单和通过翻译而调控的mRNA种类名单是不同的，并且或许是显著的不同。对这些转录和翻译活动的作用机制还需要做更详细的阐述。现在已采取了一些措施并鉴定了一些必需蛋白质，但从上面对*CHOP*和*AS*对比的陈述中可以很明显地看出，有更多的蛋白质也参与到了其中，还需对这些蛋白质做进一步的鉴定。最后，在氨基酸缺失的最初感受器以及后来被激活的信号转导途径方面还有很多问题没有得到答案。在正常体内稳态或疾病情况下对这些过程进行研究和定性将是分子营养学中一项很有趣并很重要的内容。

# 致　　谢

本工作为国家健康研究院M. S. K.资助项目（DK-52064，DK-59315）。

## 参 考 文 献

Abcouwer, S.F., Schwarz, C. and Meguid, R.A. (1999) Glutamine deprivation induces the expression of *GADD45* and *GADD153* primarily by mRNA stabilization. *Journal of Biological Chemistry* 274, 28645–28651.

Adilakshmi, T. and Laine, R.O. (2002) Ribosomal protein S25 mRNA partners with MTF-1 and La to provide a p53-mediated mechanism for survival or death. *Journal of Biological Chemistry* 277, 4147–4151.

Andrulis, I.L., Hatfield, G.W. and Arfin, S.M. (1979) Asparaginyl-tRNA aminoacylation levels and asparagine synthetase expression in cultured Chinese hamster ovary cells. *Journal of Biological Chemistry* 254, 10629–10633.

Arfin, S.M., Simpson, D.R., Chiang, C.S., Andrulis, I.L. and Hatfield, G.W. (1977) A role for asparaginyl-tRNA in the regulation of asparagine synthetase in a mammalian cell line. *Proceedings of the National Academy of Sciences USA* 74, 2367–2369.

Aslanian, A.M. and Kilberg, M.S. (2001) Multiple adaptive mechanisms affect asparagine synthetase substrate availability in asparaginase resistant MOLT-4 human leukemia cells. *Biochemical Journal* 358, 59–67.

Aslanian, A.M., Fletcher, B.S. and Kilberg, M.S. (2001) Asparagine synthetase expression alone is sufficient to induce L-asparaginase resistance in MOLT-4 human leukaemia cells. *Biochemical Journal* 357, 321–328.

Aulak, K.S., Mishra, R., Zhou, L., Hyatt, S.L., de Jonge, W., Lamers, W., Snider, M. and Hatzoglou, M. (1999) Post-transcriptional regulation of the arginine transporter Cat-1 by amino acid availability. *Journal of Biological Chemistry* 274, 30424–30432.

Barbosa-Tessmann, I.P., Pineda, V.L., Nick, H.S., Schuster, S.M. and Kilberg, M.S. (1999a) Transcriptional regulation of the human asparagine synthetase gene by carbohydrate availability. *Biochemical Journal* 339, 151–158.

Barbosa-Tessmann, I.P., Chen, C., Zhong, C., Schuster, S.M., Nick, H.S. and Kilberg, M.S. (1999b) Activation of the unfolded protein response pathway induces human asparagine synthetase gene expression. *Journal of Biological Chemistry* 274, 31139–31144.

Barbosa-Tessmann, I.P., Chen, C., Zhong, C., Siu, F., Schuster, S.M., Nick, H.S. and Kilberg, M.S. (2000) Activation of the human asparagine synthetase gene by the amino acid response and the endoplasmic reticulum stress response pathways occurs by common genomic elements. *Journal of Biological Chemistry* 275, 26976–26985.

Berlanga, J.J., Santoyo, J. and De Haro, C. (1999) Characterization of a mammalian homolog of the GCN2 eukaryotic initiation factor-2 alpha kinase. *European Journal of Biochemistry* 265, 754–762.

Bruhat, A., Jousse, C., Wang, X.-Z., Ron, D., Ferrara, M.

and Fafournoux, P. (1997) Amino acid limitation induces expression of CHOP, a CCAAT/enhancer binding protein-related gene, at both transcriptional and post-transcriptional levels. *Journal of Biological Chemistry* 272, 17588–17593.

Bruhat, A., Jousse, C. and Fafournoux, P. (1999) Amino acid limitation regulates gene expression. *Proceedings of the Nutrition Society* 58, 625–632.

Bruhat, A., Jousse, C., Carraro, V., Reimold, A.M., Ferrara, M. and Fafournoux, P. (2000) Amino acids control mammalian gene transcription: activating transcription factor 2 is essential for the amino acid responsiveness of the CHOP promoter. *Molecular and Cellular Biology* 20, 7192–7204.

Bruhat, A., Averous, J., Carraro, V., Zhong, C., Reimold, A.M., Kilbery, M.A. and Fafournoux, P. (2002) Differences in the molecular mechanisms involved in the transcriptional activation of the CHOP and asparagine synthetase genes in response to amino acid deprivation or activation of the unfolded protein response. *Journal of Biological Chemistry* 277, 48107–48114.

Capizzi, R.L. and Holcenberg, J.S. (1993) Asparaginase. In: Holland, J.F., Frei, E. III, Bast, R.C. Jr, Kufe, D.W., Morton, D.L. and Weichselbaum, R.R. (eds) *Cancer Medicine*, 3rd edn., R.R. Lea & Febiger, Philadelphia, pp. 796–805.

Chabner, B.A. and Loo, T.L. (1996) Enzyme therapy: L-asparaginase. In: Chabner, B.A. and Longo, D.L. (eds) *Cancer Chemotherapy and Biotherapy*, 2nd edn. Lippincott-Raven Publishers, Philadelphia, pp. 485–492.

Chakrabarti, R. and Schuster, S.M. (1996) L-Asparaginase: perspectives on the mechanisms of action and resistance. *International Journal of Pediatric Hematology/Oncology* 4, 597–611.

Colletta, G. and Cirafici, A.M. (1992) TSH is able to induce cell cycle-related gene expression in rat thyroid cell. *Biochemical and Biophysical Research Communications* 183, 265–272.

Entingh, A.J., Law, B.K. and Moses, H.L. (2001) Induction of the C/EBP homologous protein (CHOP) by amino acid deprivation requires insulin-like growth factor I, phosphatidylinositol 3-kinase, and mammalian target of rapamycin signaling. *Endocrinology* 142, 221–228.

Fafournoux, P., Bruhat, A. and Jousse, C. (2000) Amino acid regulation of gene expression. *Journal of Biochemistry* 351, 1–12.

Fernandez, J., Yaman, I., Mishra, R., Merrick, W.C., Snider, M.D., Lamers, W.H. and Hatzoglou, M. (2001) Internal ribosome entry site-mediated translation of a mammalian mRNA is regulated by amino acid availability. *Journal of Biological Chemistry* 276, 12285–12291.

Gazzola, G.C., Franchi-Gazzola, R., Saibene, V., Ronchi, P. and Guidotti, G.G. (1972) Regulation of amino acid transport in chick embryo heart cells I. Adaptive system of mediation for neutral amino acids. *Biochimica et Biophysica Acta* 266, 407–421.

Gazzola, G.C., Dall'Asta, V. and Guidotti, G.G. (1981) Adaptive regulation of amino acid transport in cultured human fibroblasts. *Journal of Biological Chemistry* 256, 3191–3198.

Gazzola, R.F., Sala, R., Bussolati, O., Visigalli, R., Dall'Asta, V., Ganapathy, V. and Gazzola, G.C. (2001) The adaptive regulation of amino acid transport system A is associated to changes in ATA2 expression. *FEBS Letters* 490, 11–14.

Gong, S.S. and Basilico, C. (1990) A mammalian temperature-sensitive mutation affecting $G_1$ progression results from a single amino acid substitution in asparagine synthetase. *Nucleic Acids Research* 18, 3509–3513.

Gong, S.S., Guerrini, L. and Basilico, C. (1991) Regulation of asparagine synthetase gene expression by amino acid starvation. *Molecular and Cellular Biology* 11, 6059–6066.

Greco, A., Ittmann, M. and Basilico, C. (1987) Molecular cloning of a gene that is necessary for $G_1$ progression in mammalian cells. *Proceedings of the National Academy of Sciences USA* 84, 1565–1569.

Greco, A., Gong, S.S., Ittmann, M. and Basilico, C. (1989) Organization and expression of the cell cycle gene, *ts 11*, that encodes asparagine synthetase. *Molecular and Cellular Biology* 9, 2350–2359.

Guerrini, L., Gong, S.S., Mangasarian, K. and Basilico, C. (1993) *Cis-* and *trans*-acting elements involved in amino acid regulation of asparagine synthetase gene expression. *Molecular and Cellular Biology* 13, 3202–3212.

Hansen, B.S., Vaughan, M.H. and Wang, L.-J. (1972) Reversible inhibition by histidinol of protein synthesis in human cells at the activation of histidine. *Journal of Biological Chemistry* 247, 3854–3857.

Harding, H.P., Novoa, I.I, Zhang, Y., Zeng, H., Wek, R., Schapira, M. and Ron, D. (2000) Regulated translation initiation controls stress-induced gene expression in mammalian cells. *Molecular Cell* 6, 1099–1108.

Hinnebusch, A.G. (1997) Translational regulation of yeast GCN4. *Journal of Biological Chemistry* 272, 21661–21664.

Hitomi, Y., Ito, A., Naito, Y. and Yoshida, A. (1993) Liver-specific induction of ribosomal protein gene expression by amino acid starvation in rats. *Bioscience, Biotechnology and Biochemistry* 57, 1216–1217.

Hoffer, L.J. (1993) Are dietary-protein requirements altered in diabetes-mellitus? *Canadian Journal of Physiology and Pharmacology* 71, 633–638.

Hongo, S., Takeda, M. and Sato, T. (1989) Induction of asparagine synthetase during lymphocyte activation by phytohemagglutinin. *Biochemistry International*

18, 661–666.
Hongo, S., Fujimori, M., Shioda, S., Nakai, Y., Takeda, M. and Sato, T. (1992) Immunochemical characterization of rat testicular asparagine synthetase. *Archives of Biochemistry and Biophysics* 295, 120–125.
Hope, I.A. and Struhl, K. (1985) GCN4 protein, synthesized *in vitro*, binds HIS3 regulatory sequences: implications for general control of amino acid biosynthetic genes in yeast. *Cell* 43, 177–188.
Hutson, R.G. and Kilberg, M.S. (1994) Cloning of rat asparagine synthetase and specificity of the amino acid-dependent control of its mRNA content. *Biochemical Journal* 303, 745–750.
Hutson, R.G., Warskulat, U. and Kilberg, M.S. (1996) An example of nutrient control of gene expression: amino acid-dependent regulation of asparagine synthetase. *Clinical Nutrition* 51, 327–331.
Hutson, R.G., Kitoh, T., Amador, D.A.M., Cosic, S., Schuster, S.M. and Kilberg, M.S. (1997) Amino acid control of asparagine synthetase: relation to asparaginase resistance in human leukemia cells. *American Journal of Physiology* 272, C1691–C1699.
Hyatt, S.L., Aulak, K.S., Malandro, M., Kilberg, M.S. and Hatzoglou, M. (1997) Adaptive regulation of the cationic amino acid transporter-1 (Cat-1) in Fao cells. *Journal of Biological Chemistry* 272, 19951–19957.
Jousse, C., Bruhat, A., Ferrara, M. and Fafournoux, P. (1998) Physiological concentration of amino acids regulates insulin-like-growth-factor-binding protein 1 expression. *Biochemical Journal* 334, 147–153.
Jousse, C., Bruhat, A., Harding, H.P., Ferrara, M., Ron, D. and Fafournoux, P. (1999) Amino acid limitation regulates CHOP expression through a specific pathway independent of the unfolded protein response. *FEBS Letters* 448, 211–216.
Jousse, C., Bruhat, A., Ferrara, M. and Fafournoux, P. (2000) Evidence for multiple signaling pathways in the regulation of gene expression by amino acids in human cell lines. *Journal of Nutrition* 130, 1555–1560.
Kaufman, R.J. (1999) Stress signaling from the lumen of the endoplasmic reticulum: coordination of gene transcriptional and translational controls. *Genes and Development* 13, 1211–1233.
Kilberg, M.S., Han, H.P., Barber, E.F. and Chiles, T.C. (1985) Adaptive regulation of neutral amino acid transport system A in rat H4 hepatoma cells. *American Journal of Physiology* 122, 290–298.
Kilberg, M.S., Stevens, B.R. and Novak, D. (1993) Recent advances in mammalian amino acid transport. *Annual Review of Nutrition* 13, 137–165.
Laine, R.O., Laipis, P.J., Shay, N.F. and Kilberg, M.S. (1991) Identification of an amino acid-regulated mRNA from rat liver as the mammalian equivalent of bacterial ribosomal protein L22. *Journal of Biological Chemistry* 266, 16969–16972.
Laine, R.O., Shay, N.F. and Kilberg, M.S. (1994) Nuclear retention of the induced mRNA following amino acid-dependent transcriptional regulation of mammalian ribosomal proteins L17 and S25. *Journal of Biological Chemistry* 269, 9693–9697.
Leung-Pineda, V. and Kilberg, M.S. (2002) Role of Sp1 and Sp3 in the nutrient-regulated expression of the human asparagine synthetase gene. *Journal of Biological Chemistry* 277, 16585–16591.
Marten, N.W., Burke, E.J., Hayden, J.M. and Straus, D.S. (1994) Effect of amino acid limitation on the expression of 19 genes in rat hepatoma cells. *FASEB Journal* 8, 538–544.
Mizock, B.A. (1999) Nutritional support in hepatic encephalopathy. *Nutrition* 15, 220–228.
Morgane, P.J., Austinlafrance, R., Bronzino, J., Tonkiss, J., Diazcintra, S., Cintra, L., Kemper, T. and Galler, J.R. (1993) Prenatal malnutrition and development of the brain. *Neuroscience and Behavioral Reviews* 17, 91–128.
Muller, H.J. and Boos, J. (1998) Use of L-asparaginase in childhood ALL. *CRC Critical Reviews in Oncology/Hematology* 28, 97–113.
Natarajan, K., Meyer, M.R., Jackson, B.M., Slade, D., Roberts, C., Hinnebusch, A.G. and Marton, M.J. (2001) Transcriptional profiling shows that Gcn4p is a master regulator of gene expression during amino acid starvation in yeast. *Molecular and Cell Biology* 21, 4347–4368.
Pahl, H.L. (1999) Signal transduction from the endoplasmic reticulum to the cell nucleus. *Physiological Reviews* 79, 683–701.
Palacin, M., Estevez, R., Bertran, J. and Zorzano, A. (1998) Molecular biology of mammalian plasma membrane amino acid transporters. *Physiological Reviews* 78, 969–1054.
Patil, C. and Walter, P. (2001) Intracellular signaling from the endoplasmic reticulum to the nucleus: the unfolded protein response in yeast and mammals. *Current Opinion in Cell Biology* 13, 349–355.
Paul, G.L., Waegner, A., Gaskins, H.R. and Shay, N.F. (1998) Histidine availability alters glucagon gene expression in murine alphaTC6 cells. *Journal of Nutrition* 128, 973–976.
Price, B.D. and Calderwood, S.K. (1992) Gadd45 and Gadd153 messenger RNA levels are increased during hypoxia and after exposure of cells to agents which elevate the levels of the glucose-regulated proteins. *Cancer Research* 52, 3814–3817.
Richards, N.G.J. and Schuster, S.M. (1998) Mechanistic issues in asparagine synthetase catalysis. *Advances in Enzymology* 72, 145–198.
Roediger, W.E.W. (1995) New views on the pathogenesis of Kwashiorkor – methionine and other amino-acids. *Journal of Pediatric Gastroenterology and Nutrition* 21, 130–136.
Ron, D. and Habener, J.F. (1992) CHOP, a novel

developmentally regulated nuclear protein that dimerizes with transcription factors C/EBP and LAP and functions as a dominant-negative inhibitor of gene transcription. *Genes and Development* 6, 439–453.

Roy, B. and Lee, A.S. (1999) The mammalian endoplasmic reticulum stress response element consists of an evolutionarily conserved tripartite structure and interacts with a novel stress-inducible complex. *Nucleic Acids Research* 27, 1437–1443.

Sellers, J.W., Vincent, A.C. and Struhl, K. (1990) Mutations that define the optimal half-site for binding yeast GCN4 activator protein and identify an ATF/CREB-like repressor that recognizes similar DNA sites. *Molecular and Cellular Biology* 10, 5077–5086.

Siu, F.Y., Chen, C., Zhong, C. and Kilberg, M.S. (2001) CCAAT/enhancer-binding protein beta (C/EBPβ) is a mediator of the nutrient sensing response pathway that activates the human asparagine synthetase gene. *Journal of Biological Chemistry* 276, 48100–48107.

Sood, R., Porter, A.C., Olsen, D., Cavener, D.R. and Wek, R.C. (2000) A mammalian homologue of GCN2 protein kinase important for translational control by phosphorylation of eukaryotic initiation factor-2α. *Genetics* 154, 787–801.

Straus, D.S., Burke, E.J. and Marten, N.W. (1993) Induction of insulin-like growth factor binding protein-1 gene expression in liver of protein-restricted rats and in rat hepatoma cells limited for a single amino acid. *Endocrinology* 132, 1090–1100.

Wang, X.-Z., Kuroda, M., Sok, J., Batchvarova, N., Kimmel, R., Chung, P., Zinszner, H. and Ron, D. (1998) Identification of novel stress-induced genes downstream of *chop*. *EMBO Journal* 17, 3619–3630.

Wek, S.A., Zhu, S. and Wek, R.C. (1995) The histidyl-tRNA synthetase-related sequence in the eIF-2 alpha protein kinase GCN2 interacts with tRNA and is required for activation in response to starvation for different amino acids. *Molecular and Cellular Biology* 15, 4497–4506.

Yoshida, H., Haze, K., Yanagi, H., Yura, T. and Mori, K. (1998) Identification of the *cis*-acting endoplasmic reticulum stress response element responsible for transcriptional induction of mammalian glucose-regulated proteins. *Journal of Biological Chemistry* 273, 33741–33749.

Yoshida, H., Okada, T., Haze, K., Yanagi, H., Yura, T., Negishi, M. and Mori, K. (2000) ATF6 activated by proteolysis binds in the presence of NF-Y (CBF) directly to the *cis*-acting element responsible for the mammalian unfolded protein response. *Molecular and Cellular Biology* 20, 6755–6767.

Zinszner, H., Kuroda, M., Wang, X.Z., Batchvarova, N., Lightfoot, R.T., Remotti, H., Stevens, J.L. and Ron, D. (1998) CHOP is implicated in programmed cell death in response to impaired function of the endoplasmic reticulum. *Genes and Development* 12, 982–995.

# 9 脂肪酸与基因表达

Ulrike Beisiegel，Joerg Heeren 和 Frank Schnieders
（细胞分子生物学学院，汉堡-Eppendorf 大学医院
汉堡，德国）

## 引 言

脂肪酸（FA）是所有动物能量供应中最重要的分子（营养素），同时也是类二十烷酸（前列腺素、凝血噁烷和白三烯）合成的前体物。但是近年来，越来越多的证据表明脂肪酸参与了基因表达的转录调节过程（Duplus *et al.*，2000；Jump and Clarke，1999；Jump，2001；Hihi *et al.*，2002）。本章主要阐述脂肪作为转录因子的配体在调节脂肪和葡萄糖代谢相关基因表达以及细胞分化、生长及炎症反应中所起的作用。

## 脂肪酸及其活性衍生物

脂肪酸的基本结构为疏水的多碳链，其结构因链长度和饱和度的不同而不同。小于14个碳原子的中链和短链脂肪酸通常为饱和脂肪酸，主要用于能量供应。长链脂肪酸（LCFA）中有饱和脂肪酸，也有单不饱和脂肪酸和多不饱和脂肪酸（PUFA）。长链脂肪酸除了供能外还具有其他的功能。含量最丰富的单不饱和脂肪酸是油酸（C18∶1，ω-9）[①]。多不饱和脂肪酸主要分为两类，即ω-3和ω-6脂肪酸。ω后的数字表示从疏水碳链的甲基端起第一个双键的位置。因为动物细胞不能在C-9以后的位置[②]形成双键，所以这些脂肪酸必须由饲粮提供。亚油酸（C18∶2，ω-6）是一种必需的多不饱和脂肪酸，为花生四烯酸（C20∶4，ω-6）的前体物，而α-亚麻酸（C18∶3，ω-3）为二十碳五烯酸（C20∶5，ω-3；eicosapentaenoic acid，EPA）和二十二碳六烯酸（C22∶6，ω-3；docosahexaenoic acid，DHA）的前体物（Goodridge，1991）。花生四烯酸和EPA都为类二十烷酸（白三烯、前列腺素和凝血噁烷）以及一些长链脂肪酸（二十二碳四烯酸和二十二碳五烯酸）的前体物。许多由这两种不同的前体物衍生而来的类二十烷酸都是血凝固和炎症反应的相关因子，它们相对于ω-3和ω-6类二十烷酸来说具有独立的功能（Serhan and Oliw，2001）。

多不饱和脂肪酸是引起脂质过氧化的自由基反应的底物，其与连续双键相邻的甲基

---

① 脂肪酸名称后圆括号中的数字意义：如油酸（C18∶1，ω-9）表示一个含有18个碳原子的脂肪酸，并在ω碳原子后第9位碳原子上有一双键。ω碳原子是脂肪酸烃链的第一个甲基。

② 欧米加编号方式系统从甲基端开始，而德耳塔编号方式系统从羧基（-COOH）端开始：

```
                  3    2    1        德耳塔编号方式
H3C-(CH2)n-CH2-CH2-COOH
   ω          β    α                 欧米加编号方式
```

上的碳原子很容易被氧化（Halliwell and Gutteridge，1989）。这些被氧化的脂肪酸不仅能分解代谢物而且能影响基因表达（Nagy *et al.*，1998；Jump and Clarke，1999；Delerive *et al.*，2000）。

共轭脂肪酸是由18个碳的多不饱和脂肪酸衍生而来的，大多数共轭脂肪酸为亚油酸的衍生物，因此被命名为“共轭亚油酸”（conjugated linoleic acid，CLA）。CLA包括不同的共轭双键模式，这决定了它们在代谢中的特定功能。植烷酸是一类由植物叶绿醇衍生而来的具有支链结构的脂肪酸，这种稀有脂肪酸因其在基因表达中的作用而变得越来越重要。

## 日粮中脂肪的代谢

摄入日粮中的脂肪后，小肠细胞将其合成为含甘油三酯丰富的乳糜微粒，通过胸导管进入血液循环。乳糜微粒以载脂蛋白B-48为结构蛋白，主要包含长链脂肪酸。中等长度和短链的脂肪酸在小肠中与白蛋白相结合，直接被转运到血液中。乳糜微粒中的甘油三酯在脂蛋白脂酶（lipo-protein lipase，LPL）的催化下水解。LPL是一种在肌肉和脂肪组织中大量表达的酶（Eckel，1989）。这种脂类分解主要为上述两组织同时也为其他一些消耗LCFA的组织提供长链脂肪酸（Goldberg，1996）。在胰岛素促进合成代谢的作用下，LCFA重新酯化并以脂肪滴的形式沉积在脂肪组织中。禁食时，脂肪组织中的激素敏感脂酶在儿茶酚胺的刺激下被激活，生成长链脂肪酸，在白蛋白的作用下被转运到其他组织而被利用（Holm *et al.*，2000）。

生理条件下细胞内长链脂肪酸摄取的主要机制是一个饱和过程，可能是由目前尚未明确的某种白蛋白受体介导的（Berk and Stump，1999）。长链脂肪酸及其衍生物也可通过脂肪酸转运蛋白（fatty acid transporter，FAT）的作用而被摄取（Motojima *et al.*，1998）。细胞内不同部位长链脂肪酸的代谢途径不同，包含了许多不同的脂肪酸结合蛋白（fatty acid binding protein，FABP）（Abumrad *et al.*，1998；McArthur *et al.*，1999；Norris and Spector，2002）。在细胞胞质中，脂肪酸结合蛋白将脂肪酸释放，使其与它们相应的可溶性受体（转录因子）发生相互作用，从而影响细胞核中基因的表达。在微粒体中，不饱和的长链脂肪酸（单不饱和脂肪酸和多不饱和脂肪酸）通过环加氧酶、脂肪酸氧化酶和单加氧酶途径被氧化，形成类二十烷酸和环氧或羟基脂肪酸（Sprecher，2000），这些产物能够通过与不同的转录因子相结合而影响基因的表达（Jump and Clarke，1999）。在过氧化物酶体和线粒体中，被激活的多不饱和脂肪酸是β-氧化的底物。另外，被激活的脂肪酸能够作为磷脂和甘油三酯合成的底物，这些物质后来被用于形成细胞膜和脂蛋白。所有这些途径，包括相关的代谢途径如糖类代谢都由日粮中脂肪直接调节，也受到激素作用的调节（如胰岛素和三碘甲腺原氨酸）（Kaytor *et al.*，1997；Towle *et al.*，1997；Jump *et al.*，2001；Vaulont *et al.*，2000）。

## 转录因子和基因表达

健康生物中，特定细胞的能量代谢和健康水平会受到环境的严格调节，而该环境是由生物的发育状态、营养水平、微环境（pH、离子梯度）以及生长因子和细胞因子的存在状态而决定的。为了在特定的条件下能够提供全部所需的酶和蛋白质，因而在细胞内存在着一个复杂的转录因子系统调节基因的表达。人类基因组中，目前已知被编码的转录因子超过 2000 个（Brivanlou and Dranell，2002）。近来，Brivanlou 和 Dranell（2002）提出了新的真核细胞转录因子分类方法。这种分类将正调节的真核细胞转录因子划分为两组：①恒定活性核因子，在任何时间都存在于所有细胞的细胞核中，体外分析研究表明其具有激活转录的潜在能力。②调节转录因子，又分成两个主要的类型，一为发育或特定细胞型转录因子，二为信号依赖型转录因子。信号依赖型转录因子又可根据其不同的信号机制划分为三个功能组分，即细胞核受体超家族因子、胞内信号反应因子和细胞表面受体（配体相互作用和信号级联）激活因子。

脂肪酸激活转录因子是细胞核受体超家族因子中的一员，包括内分泌受体如雌激素受体、糖皮质激素受体（glucocorticoid receptor，GR）和维生素 D 受体（VDR），同时还包括未知配体的孤儿受体和最近才确定配体的“接受孤儿受体”（Mangelsdorf and Evans，1995；Schoonjans *et al.*，1997）。脂肪酸可能是在与“接受孤儿受体”中的成员如视黄酸 X 受体（retinoid X receptor，RXR）和过氧化物酶体增殖子活化受体（peroxisome proliferator-activated receptor，PPAR）结合后激活了基因的表达。

## 脂肪酸作为 PPAR 系统的配体

1990 年配体诱导转录因子引起过氧化物酶体增殖的发现开辟了基因表达调控研究的新领域（Issemann and Green，1990）。接着，有证据表明 PPAR 系统可能在脂质内稳衡调节中起着重要的作用。通过对过氧化物酶体的增殖作用而被明显识别的 PPAR 主要对啮齿动物发挥这种作用，而在人类肝脏中对其他基因的调节作用也在后来的研究中被证实（Palmer *et al.*，1998）。

关于 PPAR 激动剂作用的早期研究是用 PPARα 的潜在激活剂合成配体降血脂药物（Schoonjans *et al.*，1996a）和 PPARγ 的高亲和配体抗糖尿病药物噻唑烷（Henry，1997；Berger and Moller，2002；O′Moore-Sullivan and Prins，2002）进行的。1992 年，Auwerx 证实脂肪酸是 PPAR 的生理相关性配体，它们参与了一系列基因的调节。所有这些基因在它们的启动子部位都存在一个 PPAR 反应元件（PPAR-response element，PPRE）。这种受体至少有三种相关的类型：PPARα、PPARγ 和 PPARβ（或 δ）（Michalik and Wahli，1999；Berger and Moller，2002），这些受体的组织分布和主要作用如图 9.1。

PPAR 的主要靶基因组是与脂类和葡萄糖代谢相关的基因（表 9.1）。因此，它们能够调节能量代谢的关键功能并建立一个详细的系统，以满足生物不同的代谢需要。随着对 PPAR 系统的逐步深入了解，我们认识到此系统具有较高的复杂性，这正好与其负担的特殊任务相符合（Berger and Moller，2002）。这种复杂性是由三种不同的同型受体 α、β、γ 引起的，这三种受体以不同的亲和力与许多生理配体相结合，通过与 RXR 形成异二聚体而成为有活性的转录因子（Miyata *et al.*，1994）。另外，此异二聚体的最适活性依赖于辅因子，辅因子在配体与 PPAR 受体结合后被激活并得到补充（Xu *et al.*，1999）。依赖于受体-配体复合物的特性和共受体及辅因子的可利用性，靶基因的表达得到了特定的调节（图 9.2）（Berger and Moller，2002）。

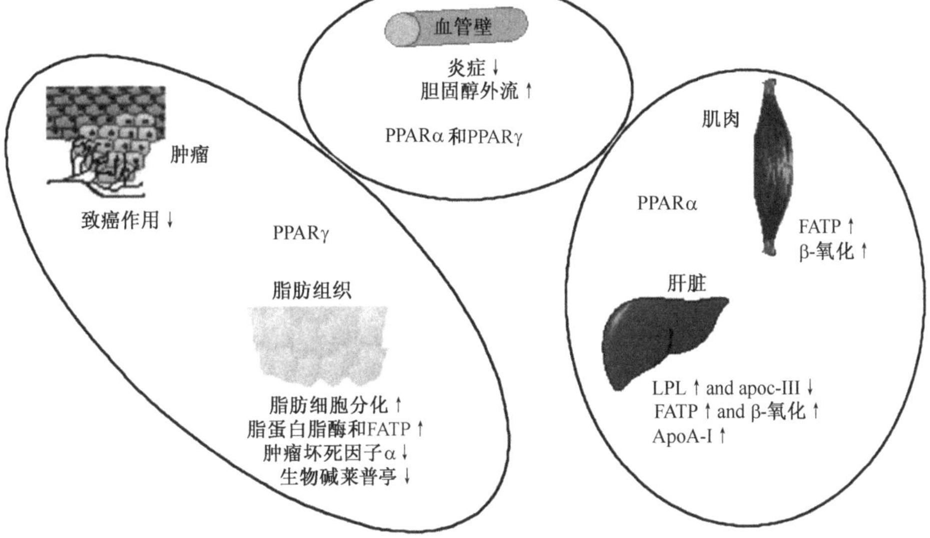


图 9.1 不同组织中 PPAR 的作用。PPARα 主要在肝脏和肌肉起作用，而 PPARγ 主要影响脂肪组织和肿瘤细胞中基因的表达。动脉内壁中内皮细胞和巨噬细胞对 PPARα 和 PPARγ 都有反应

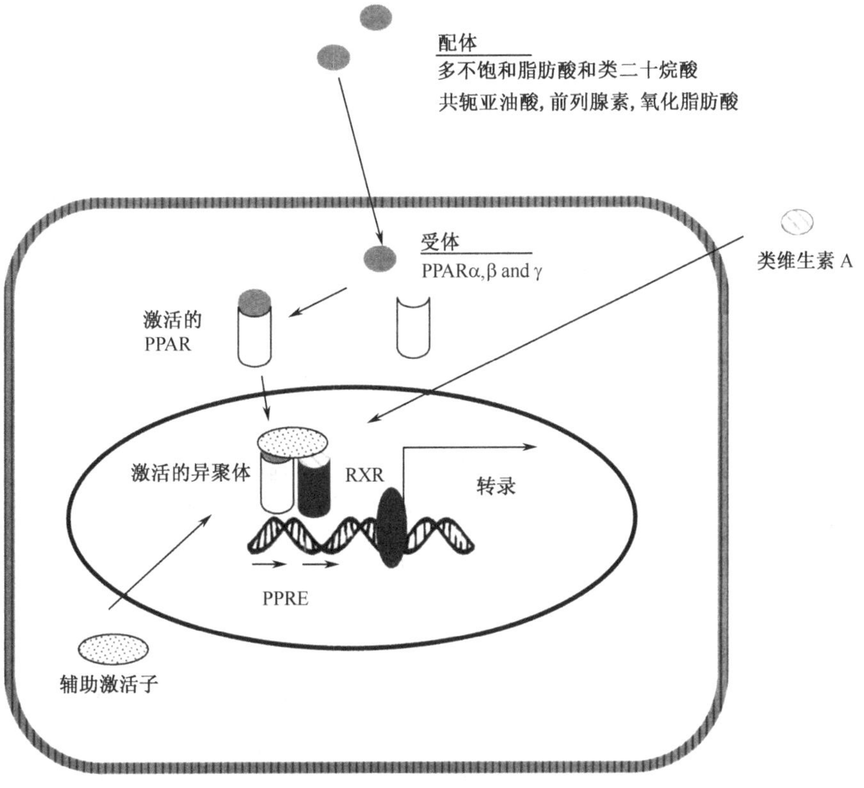


图 9.2 脂肪酸对 PPAR 的激活

脂肪酸能够通过脂肪酸转运蛋白质的作用被摄入细胞。摄入的脂肪酸在细胞质中可以激活 PPAR，而与之相对应，视黄酸受体（RXR）可被类维生素 A 或脂肪酸激活。这两种激活的核受体形成一异源二聚体，结合在靶基因启动子的反应元件上。一个额外的激活作用能够通过对辅助激活子的补充而得到，此辅激活子可以结合到异源二聚体上

在最近几年，脂肪酸作为核受体尤其是作为 PPAR 的生理配体（Desvergne *et al.*，1998）得到了人们越来越多的关注（Hihi *et al.*，2002）。表 9.1 列出了三种同型 PPAR 生理配体及其主要的生理功能。

**表 9.1 脂类激活转录因子的主要生理配体及其生理功能**

| 核受体 | 主要脂类配体 | 激活模式 | 调节基因 | 基因功能 |
|---|---|---|---|---|
| PPARα | ω-3 多不饱和脂肪酸<br>如 γ-亚麻酸<br>ω-6 多不饱和脂肪酸<br>如花生四烯酸<br>类二十烷酸<br>如 8S-HETE、LTB4 | 直接与脂类作用，与 RXR 形成异二聚体 | apoA-Ⅰ，apoA-Ⅱ，apoC-Ⅲ<br>FABP<br>CPTI<br>乙酰基辅酶 A 氧化酶<br>乙酰基辅酶 A 脱氢酶<br>Cyp4A1/6 P450 家族 | 脂蛋白的运输<br>细胞内脂肪酸运输<br>脂肪酸进入线粒体<br>过氧化物酶体 β-氧化<br>线粒体 β-氧化<br>过氧化物酶体 ω-氧化 |
| PPARγ | ω-3 多不饱和脂肪酸<br>如 γ-亚麻酸<br>ω-6 多不饱和脂肪酸<br>如花生四烯酸<br>类二十烷酸<br>如前列腺素 J2 | 直接与脂类作用，与 RXR 形成异二聚体 | FABP，FATP，CD36<br>LPL<br>乙酰基辅酶 A 合成酶<br>UCP<br>TNF-α<br>瘦素 | 脂肪酸运输<br>脂蛋白水解<br>脂肪生成<br>产热<br>促炎细胞因子<br>食物摄入调控者 |
| PPARδ | ω-3 多不饱和脂肪酸<br>如 γ-亚麻酸<br>ω-6 多不饱和脂肪酸<br>如花生四烯酸<br>类二十烷酸<br>如前列腺素 J2 | 直接与脂类作用，与 RXR 形成异二聚体 | ABCA1<br>FABP<br>环加氧酶<br>胚胎发育过程中复杂模式基因 | 胆固醇逆向运输<br>脂肪酸运输<br>前列腺素合成<br>胎儿发育 |
| HNF4α | 激动性的：<br>饱和脂肪酸 C14/C16<br>拮抗性的：<br>饱和脂肪酸 C18<br>ω-3/6 多不饱和脂肪酸 | 直接与激活的脂类作用<br>（脂酰辅酶 A）<br>同源二聚体 | ApoA-Ⅰ，apoB，apoC-Ⅲ<br>MTP<br>FABP<br>乙酰基辅酶 A 脱氢酶<br>HNF1α/PXR<br>Cyp3A4-6 P450 家族 | 脂蛋白运输<br>脂蛋白装配<br>脂肪酸运输<br>线粒体 β-氧化<br>转录因子<br>过氧化物酶体羟基化 |
| LXRs | 激动性的氧化型低密度脂蛋白<br>拮抗性的 ω-3/6 多不饱和脂肪酸 | 直接与氧化型低密度脂蛋白作用与多不饱和脂肪酸竞争，与 RXR 形成异二聚体 | Cyp7A<br>CETP<br>ABCA1<br>LXR 对 SREBP1c 的影响：<br>脂肪酸，硬脂酰辅酶 A 脱饱和酶 ACL<br>苹果酸酶<br>6-磷酸葡萄糖脱氢酶 | 胆酸代谢<br>胆固醇交换<br>胆固醇逆向转运<br>转录因子<br>脂肪酸合成<br><br>脂肪生成途径 |

Jump 和 Clarke（1999）总结了日粮脂肪对基因表达调节的研究进展，并指出不同脂肪酸与合成配体相比激活 PPAR 的相对潜力。在生理配体中，亚油酸具有最高的潜力，其次是花生四烯酸和油酸，饱和脂肪酸对这些受体只有低亲和性。Wolfrum 等（2001）研究了不同脂肪酸对肝脏脂肪酸结合蛋白（FABP）的特异影响，并肯定了亚油酸和 γ-亚麻油酸在其反式激活系统中的高活性。最近的研究进一步阐明了脂肪酸及其衍生物在基因表达调节过程中的特定作用（Berger and Moller，2002；Hibi *et al.*，

2002）。

许多研究报道了由于ω-3多不饱和脂肪酸（PUFA）在脂类代谢中的特殊作用，在健康和疾病方面具有特殊功能。这些作用可能也受基因表达水平的调控，ω-3多不饱和脂肪酸作为PPARα的配体比ω-6多不饱和脂肪酸更有效（Price *et al.*，2000），能够引起脂类氧化和产热的增多（Baillie *et al.*，1999）。因此ω-3多不饱和脂肪酸可以使脂肪酸由储藏能量转为被氧化，这个过程对能量平衡和胰岛素敏感性都是必要的（Mori *et al.*，1999）。CLA已被证明是PPARα的潜在自然发生配体和激动剂（Moya-Camarena *et al.*，1999）。正如最近的报道所总结的那样，这些数据产生的前提条件是把CLA看作人类健康功能性食品的营养性产物（Whigham *et al.*，2000）。有一点必须强调，对CLA的大多数实验是以啮齿动物为研究对象进行的，由于PPAR系统在人类和啮齿动物间存在着种间差异，在啮齿动物中的发现不能直接应用于人类。有数据表明，不同的CLA对PPARα有不同的影响（Moya-Camarena *et al.*，1999）。在剔除PPARα小鼠上的研究表明，CLA不仅能通过这种转录因子系统起作用，而且也能够应用其他的调节方式起作用（Peters *et al.*，2001）。

植烷酸和降植烷酸在结构上与植物叶绿醇密切相关。最近它们被证明为RXR、PPARα和PPARδ的配体（Lampen *et al.*，2001）。它们在褐色脂肪分化中作用的研究表明，植烷酸为一促分化剂而降植烷酸不是（Schluter *et al.*，2002）。

花生四烯酸的代谢产物类二十烷酸已被描述为PPAR的配体。这对前列腺素如PGA1和PGA2、PGD1和PGD2以及PGJ2来说是正确的，但不能断定其他的类二十烷酸的产物也是这样（Yu *et al.*，1995）。Yu等描述了羟基花生四烯酸（8S-hydroxyeicosatetraenoic acid，8S-HETE）对PPARα的影响。1995年，两个独立的研究组分别通过研究发现了PGJ2通过PPARγ对脂肪细胞分化的影响（Forman *et al.*，1995；Kliwewe *et al.*，1995）。

在下面的章节中，我们将焦点集中在脂肪酸的影响上，对辅因子和共激动剂不做进一步的描述。

## 脂肪酸调节基因表达

### 脂类代谢和动脉粥样硬化过程中基因的调节

Chawla等（2001）最近已经对核受体在脂类代谢中的作用做了综述。PPARα的主要靶基因对禁食和饲喂过程中能量的储存起着重要的作用，而PPARγ是脂肪生成过程中一个关键的调节者，并在细胞分化、胰岛素致敏、动脉粥样硬化以及致癌过程中发挥重要作用。在这里有一点必须强调，关于这些PPAR功能的数据大多是通过对在高脂血症和糖尿病药理学研究中使用的合成PPAR激动剂（如非诺贝特类和噻唑烷二酮类药）的研究得到的（Staels *et al.*，1997）。正如上面合成配体所揭示的那样，由于PUFA可与PPAR以低亲和力结合，因此它们被认为在基因调节中有相同的影响，虽然它们的水平要低得多。

PPARα激动剂刺激肝脏脂蛋白脂酶基因的表达（Schoonjans *et al.*，1996b），导致乳糜微粒中甘油三酯的水解和营养性脂肪酸向脂肪组织、肌肉和其他组织的转运。乳糜微粒残余物被肝脏中残余物受体摄取，并通过极低密度脂蛋白（very low density lipoprotein，VLDL）的作用向其他组织分布。ApoCⅢ是残余物摄取的重要抑制因子，而PPARα可以抑制 *apoCⅢ* 的表达（Staels *et al.*，1995），因此导致残余物的清除。PPARα还能够影响 *apoAⅠ* 基因表达（Vu-Dac *et al.*，1994），并能增加肝脏来源的高密度脂蛋白（high density lipoprotein，HDL）前体物的数量，这种前体物在低密度脂蛋白（LDL）介导的甘油三酯含量丰富的脂蛋白的脂解作用下形成，导致 $HDL_3$ 增加。PPARα激动剂对肝脏脂蛋白代谢中这三种基因的影响，使得脂蛋白的分布向着人们想要得到的低甘油三酯和较高的高密度脂蛋白含量方向发展，从而产生了抗动脉粥样硬化模式（图 9.1）。

在细胞水平，PPARα的活化对FABP（Wolfrum *et al.*，2001）以及促进PUFA向过氧化物酶体转运的脂肪酸转运蛋白A、B、C、D2和D3（Fourcade *et al.*，2001）的转录起着正调节作用。作为PPAR配体完成由PPAR系统调节的代谢级联反应的最后分解代谢步骤，脂肪酸氧化过程中包含的酶被PPARα激动剂和细胞色素P450酶CYP4A活化（Muerhoff *et al.*，1992）。

PPARγ在脂类代谢中的作用主要是通过脂肪细胞中基因的功能来介导的（Walczak and Tontonoz，2002），这将在下面的章节中加以描述。PPARγ激动剂已被证明在动脉粥样硬化和免疫系统中发挥着重要的作用。

PPAR对包含在内皮细胞生物学和炎症反应中的基因的调节作用是另外一个重要的领域，它有助于我们进一步的了解日粮脂肪如何在抵抗动脉硬化中起作用。Plutzky（2001）的研究使我们更深入的了解了在内皮细胞中表达的PPAR及其对血管生物学的作用（图 9.1）。细胞因子诱导的基因表达的抑制在动脉粥样硬化的病理机制中起着重要的作用，但对这个领域的讨论超出了这篇报道的范围。

PPAR和炎症反应的数据很让人迷惑（Clark，2002），这可能不仅为动脉硬化的日粮治疗，同时也可能为炎症反应过程引起的其他变性疾病，如老年性痴呆病（Landreth and Heneka，2001）和关节炎开辟了一个全新的领域（Fahmi *et al.*，2002）。

## 脂肪组织中基因的调节

许多研究者都认为PPARγ在脂肪生成过程中是必需的而且也是充足的（Brun *et al.*，1997；Spiegelman，1997；Fajas *et al.*，1998；Rangwala and Lazar，2000），再则，它还影响脂肪细胞来源的信号分子的分泌（Walczak and Tontonoz，2002）（图 9.1 和图 9.3）。

Berger和Moller（2002）在最近的一篇综述中总结了PPARγ对白色脂肪组织中基因的调节作用。他们列举了其对包括调节细胞内能量稳衡的基因和控制脂类代谢的基因脂蛋白脂肪酶（LPL）、脂肪酸转运蛋白（FAT）和解偶联蛋白（uncoupling protein，UCP）基因的调节作用。通过刺激脂肪细胞中乙酰辅酶A合成酶基因的表达，PPARγ起到促进脂肪生成的作用。PPARγ还能影响编码分泌型激素如生物碱莱普亭和肿瘤坏死因子α（TNF-α）的基因，而TNF-α又介导胰岛素抗性。在脂肪组织中受PPARγ调节的另外一组重要的基因为葡萄糖代谢中的基因，如胰岛素受体底物2（insulin receptor substrate-2，IRS-2）和丙酮酸脱氢激酶4（pyruvate dehydrogenase kinase 4，

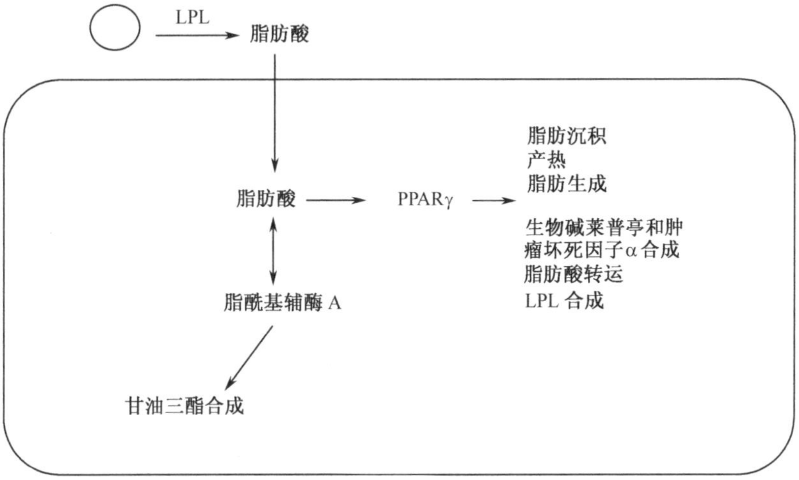


图 9.3 脂肪酸介导的脂肪基因表达中的核受体和主要的代谢途径

通过 LPL 介导的脂解作用释放的脂肪酸被摄入脂肪细胞后，可以被激活用于合成甘油三酯或被用作 PPARγ 的激活子。脂肪细胞中与脂类代谢有关的几个重要基因都由 PPARγ 调节

PDK4）。

Rangwala 和 Lazar（2000）总结了包含在脂肪生成过程中的其他转录因子，这些转录因子中有一部分也是由脂类激活的。Holst 和 Grimaldi（2002）讨论了由多不饱和脂肪酸作为转录因子 PPARδ 的激动剂在脂肪细胞分化过程中的作用。

多不饱和脂肪酸作为 PPARγ 的激动剂所影响的基因谱很复杂，不能在此做讨论，但也已清楚地表明，营养性脂肪酸及其衍生物对 PPARγ 的刺激作用对高脂血症和糖尿病的发展有重要且长期的影响，而且能够阻止动脉粥样硬化和冠心病的发生（Hsueh and Law，2001）。

## 肝脏中基因的调节

尽管 PUFA 及激活的 PUFA（FA-CoA）在细胞内的浓度很低（Knudsen *et al.*，1999；Elholm *et al.*，2000），但 PUFA 及其代谢物通过影响核转录因子的活性对肝脏基因表达产生影响的作用是很清楚的（图 9.4）。这些转录因子除了 PPARα 外，还包括固醇元件结合蛋白 1（sterol element-binding protein 1，SREBP1）、肝脏 X 受体（liver X receptor，LXR）和肝核因子 4α（hepatic nuclear factor 4α，HNF 4α）（图 9.1）（Kersten *et al.*，2000；Osborne，2000；Schultz *et al.*，2000；Hayhurst *et al.*，2001；Zannis *et al.*，2001；Berger and Moller，2002）。PPARα 是第一个被证实的在肝实质细胞中大量表达的 PUFA 受体的转录因子。不同的肝脏靶基因为一组具有相对同源性的基因，它们参与脂类分解代谢，如 PUFA 的跨膜转运（如脂肪酸转运蛋白；Motojima *et al.*，1998）、PUFA 胞内运输（如肝脏脂肪酸结合蛋白；Issemann *et al.*，1992）、PUFA 在微粒体和线粒体内的氧化［如乙酰辅酶 A 合成酶（Schoonjans *et al.*，1995）和乙酰辅酶 A 氧化酶（Dreyer *et al.*，1992）］以及脂蛋白代谢［如脂蛋白脂酶（Schoonjans *et al.*，1996b）、apoAⅠ（Vu-Dac *et al.*，1994）和 apoCⅢ（Staels *et al.*，1995）］等（图 9.1）。

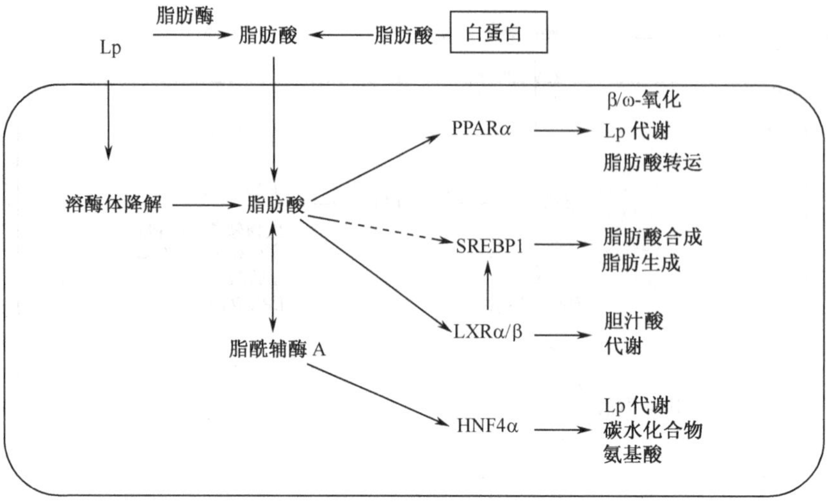


图 9.4　脂肪酸介导的肝基因表达中的核受体及其主要的代谢途径

肝脏中脂肪酸对脂类代谢的调节不仅受 PPARα 的调节，HNF4α 也包含在多种基因的调节中，这些基因参与了血浆脂蛋白的产生和清除。而且，固醇与 LXR 的结合能调节胆固醇的内稳衡和胆酸代谢。LXR 活性受到 PUFA 的抑制，因此减少了 SREBP1c 的表达。低浓度的 SREBP1c 导致调节肝脂肪生成途径中关键酶表达的减少

除了在禁食适应过程中的作用，油酸和多不饱和脂肪酸同样抑制脂类重新合成和碳水化合物代谢过程中的基因，这点能够通过肝脏中基因表达受 LXRα 和 LXRβ 介导的依赖于 FA 的抑制得到解释（Repa *et al.*，2000）。LXR 调节固醇元素结合蛋白（SREBP1c）的转录（Peet *et al.*，1998；Repa and Mangelsdorf，2000）。PUFA 能够拮抗氧化固醇（LXR 的主要生理配体）的结合，结果造成 SREBP1c 的 mRNA 水平下降（Ou *et al.*，2001）。SREBP1c 为存在于内质网膜上的转录因子，通过蛋白质两步水解作用而被释放，此蛋白质水解作用是由细胞内低胆固醇浓度诱导产生的。成熟的转录因子进入胞核（Brown and Goldstein，1997），这对参与脂肪酸和甘油三酯合成代谢途径的基因是一个很强的刺激（如脂肪酸合成酶、硬脂酰辅酶 A 去饱和酶、微粒体转移蛋白）（Shimano *et al.*，1997；Sato *et al.*，1999）。因此，PUFA 和 LXR 配体的竞争导致了活性 SREBP1c 水平的降低（Mater *et al.*，1999；Hannah *et al.*，2001；Ou *et al.*，2001），这能够为 PUFA 对肝脂肪合成的抑制作用提供解释（Jump and Clarke，1999）。

除了 PUFA，不同的 CoA 硫脂能够与核受体 HNF4α 结合（Hertz *et al.*，1998）。在由配体诱导产生的受体构象改变的情况下（Petrescu *et al.*，2002），HNF4α 被激活，并能以同源二聚体的形式与 DNA 序列相结合（Jiang *et al.*，1995）。效应基因对转录因子（如 HNF1α）和参与脂肪酸、脂蛋白和脂类代谢的蛋白质（如 apoAⅠ、apoCⅢ和微粒体转移蛋白）及其他蛋白质（Hayhurst *et al.*，2001；Zannis *et al.*，2001）进行编码。由于 PPARα 在人类肝脏中的表达量相当低（Palmer *et al.*，1998），所以对人类而言，HNF4α 可能对一些由 PUFA 介导的转录作用是很重要的（Jump and Clarke，1999）（图 9.4）。

# 肿瘤细胞中基因的调节

生活方式和营养影响了许多病理生理过程。对免疫系统而言，许多报道强调了营养对癌的发育和生长的影响。Helmlinger 等（2000）最近的研究成果支持了肿瘤代谢主要是由底物可利用性决定而不是由肿瘤细胞新陈代谢需要而决定的结论。这一结论及其他的结论暗示了日粮组成在肿瘤治疗中发挥着一定的作用。

PPAR 已被证明是脂肪酸影响细胞增殖的主要介导物。令人感兴趣的是，不同的异构体有相反的作用。从已发表的数据中可得出这样一条基本原则：PPARα 和 PPARγ 有抗增殖的效应，而 PPARβ（也称为 PPARδ）可能具有一定的致肿瘤作用。正如下面所述，这条规则也有例外的情况。

脂肪酸能影响不同组织中癌的发育，对患者的预后效果有正面的也有负面的影响。ω-3 脂肪酸对乳腺癌的抑生长作用已得到证实（图 9.1）。Maillard 等（2002）的研究阐明了患者乳房脂肪组织中 ω-3 与 ω-6 脂肪酸含量的比值与乳腺癌的关系。这些研究人员假设此含量反映了这些脂肪酸的营养摄入量，并得出 ω-3 脂肪酸可以抵抗乳腺癌的生长的结论。Ge 等（2002）通过对线虫（*Caenorhabditis elegans*）MCF-7 细胞中脱氢酶的过量表达证实了 ω-3 脂肪酸的抗肿瘤作用、Escrich 等（2001）报道，内源合成的 ω-3 脂肪酸加速了癌细胞的死亡。总之，上述所有的研究都表明，ω-3 脂肪酸对肿瘤有抗增殖作用。有一点应指出，高脂肪日粮促进了富脂器官（如乳房和脂肪组织）中肿瘤的生长。

作为 PPARγ 的激动剂，脂肪酸能够影响肿瘤细胞生长和发育过程中细胞增殖的调节（Rosen and Spiegelman，2001）。PPARγ 激活的结果造成了癌细胞的再分化并诱导了细胞凋亡（Eibl *et al.*，2001；Takashima *et al.*，2001）。研究已证实 PPARγ 对人类的肾癌细胞有正调节作用，可以提高它们对由 PPAR 介导的生长抑制作用的易感性（Inoue *et al.*，2001）。另外，在脂肪肉瘤（Tontonoz *et al.*，1997）、乳腺癌（Mueller *et al.*，1998）、结肠癌以及前列腺癌（Mueller *et al.*，2000）中也有相似的发现。最近的一项研究证实了 PPARγ 介导的抗增殖效应的明确途径。PPARγ 对 PTEN 肿瘤抑制基因起正调节作用。剔除 PPARγ 后小鼠的这种作用也消失（Patel *et al.*，2001）。

很特别的是，PPARγ 在结肠上皮中的表达与它在脂肪组织中的表达水平相近。这种因子在结肠肿瘤中也显示出较高的水平。这些细胞对 PPAR 激动剂的反应为表达分化标识，从而生成恶性较弱的表型（Sarraf *et al.*，1998）。结肠癌细胞中 PPARγ 基因的突变可导致其功能的丧失。因此，PPARγ 在结肠中有抑制肿瘤活性的作用（Sarraf *et al.*，1999）。

各种 PPAR 在前列腺组织中都有表达。Segawa 等（2002）应用反转录聚合酶链反应（reverse transcription-polymerase chain reaction，RT-PCR）和免疫组织学技术分析正常、良性和恶性肿瘤前列腺组织中 PPARα、PPARβ 和 PPARγ 的表达。同正常前列腺组织相比较，PPARα 和 PPARβ 在所有测试组织中都有表达，而 PPARγ 在良性以及恶性肿瘤前列腺组织中被诱导表达。Shappell 及其同事的研究表明，前列腺瘤细胞中 PPARγ 能够被 15S 羟基花生四烯酸（15S-hydroxyeicosatetraenoicin，15S-HETE）激活，从而抑制了肿瘤的生长（Masamune *et al.*，2001）。15S-HETE 是在 15-脂肪氧化酶的作用下由日粮中的花生四烯酸转化而来的。15-脂肪氧化酶在前列腺瘤细胞系中被负调控。该发现符合肿瘤负调控外部增殖调节途径的范例，同时也表明只要 PPARγ 存在，这种负调控就是可逆的。在血管平滑肌细胞中，PPARα 通过激活 p38 蛋白激酶诱导细胞凋亡（Collett *et al.*，2000）。在另一个研究中，Diep（2001）发现 PPARα 在未分化的前列腺癌上皮中被正调控。目前对 PPARα 的这种诱导功能还不清楚。然而，PPARα 的依赖性调节，如脂肪酸转运蛋白的正调节在肿瘤的日粮能量供应中确有一定的作用。

总之，日粮脂肪酸可对肿瘤细胞增殖产生不同的影响。在许多情况下，根据肿瘤病原学和代谢，PPAR 在引发增殖的同时也能引起抗增殖效果。

## 小　结

许多流行病学研究和人群调查结果表明了 PUFA 在人类健康方面的积极作用，在这些观察的生化基础上发表了大量的资料。脂类，特别是 PUFA 已被证明是调节基因表达的重要因子，这些基因包含在脂类代谢、葡萄糖水平调节、细胞分化、肿瘤生长和炎症反应中。今天，我们可能只是了解到饲粮对健康生命重要性的一小部分内容，就像冰山的一角。为此，我们还要进一步深入研究特定的遗传倾向与饲粮作用间互作的危险性。

## 参 考 文 献


Abumrad, N., Harmon, C. and Ibrahimi, A. (1998) Membrane transport of long-chain fatty acids: evidence for a facilitated process. *Journal of Lipid Research* 39, 2309–2318.

Auwerx, J. (1992) Regulation of gene expression by fatty acids and fibric acid derivatives: an integrative role for peroxisome proliferator activated receptors. The Belgian Endocrine Society Lecture 1992. *Hormone Research* 38, 269–277.

Baillie, R.A., Takada, R., Nakamura, M. and Clarke, S.D. (1999) Coordinate induction of peroxisomal acyl-CoA oxidase and UCP-3 by dietary fish oil: a mechanism for decreased body fat deposition. *Prostaglandins, Leukotrienes and Essential Fatty Acids* 60, 351–356.

Berger, J. and Moller, D.E. (2002) The mechanisms of action of PPARs. *Annual Review of Medicine* 53, 409–435.

Berk, P.D. and Stump, D.D. (1999) Mechanisms of cellular uptake of long chain free fatty acids. *Molecular and Cellular Biochemistry* 192, 17–31.

Brivanlou, A.H. and Darnell, J.E. Jr (2002) Signal transduction and the control of gene expression. *Science* 295, 813–818.

Brown, M.S. and Goldstein, J.L. (1997) The SREBP pathway: regulation of cholesterol metabolism by proteolysis of a membrane-bound transcription factor. *Cell* 89, 331–340.

Brun, R.P., Kim, J.B., Hu, E. and Spiegelman, B.M. (1997) Peroxisome proliferator-activated receptor γ and the control of adipogenesis. *Current Opinion in Lipidology* 8, 212–218.

Chawla, A., Repa, J.J., Evans, R.M. and Mangelsdorf, D.J. (2001) Nuclear receptors and lipid physiology: opening the X-files. *Science* 294, 1866–1870.

Clark, R.B. (2002) The role of PPARs in inflammation and immunity. *Journal of Leukocyte Biology* 71, 388–400.

Collett, G.P., Betts, A.M., Johnson, M.I., Pulimood, A.B., Cook, S., Neal, D.E. and Robson, C.N. (2000) Peroxisome proliferator-activated receptor α is an androgen-responsive gene in human prostate and is highly expressed in prostatic adenocarcinoma. *Clinical Cancer Research* 6, 3241–3248.

Delerive, P., Furman, C., Teissier, E., Fruchart, J., Duriez, P. and Staels, B. (2000) Oxidized phospholipids activate PPARα in a phospholipase A2-dependent manner. *FEBS Letters* 471, 34–38.

Desvergne, B., Ijpenberg, A., Devchand, P.R. and Wahli, W. (1998) The peroxisome proliferator-activated receptors at the cross-road of diet and hormonal signalling. *Journal of Steroid Biochemistry and Molecular Biology* 65, 65–74.

Diep, Q.N., Touyz, R.M. and Schiffrin, E.L. (2000)

Docosahexaenoic acid, a peroxisome proliferator-activated receptor-α ligand, induces apoptosis in vascular smooth muscle cells by stimulation of p38 mitogen-activated protein kinase. *Hypertension* 36, 851–855.

Dreyer, C., Krey, G., Keller, H., Givel, F., Helftenbein, G. and Wahli, W. (1992) Control of the peroxisomal β-oxidation pathway by a novel family of nuclear hormone receptors. *Cell* 68, 879–887.

Duplus, E., Glorian, M. and Forest, C. (2000) Fatty acid regulation of gene transcription. *Journal of Biological Chemistry* 275, 30749–30752.

Eckel, R.H. (1989) Lipoprotein lipase. A multifunctional enzyme relevant to common metabolic diseases. *New England Journal of Medicine* 320, 1060–1068.

Eibl, G., Wente, M.N., Reber, H.A. and Hines, O.J. (2001) Peroxisome proliferator-activated receptor gamma induces pancreatic cancer cell apoptosis. *Biochemical and Biophysical Research Communications* 287, 522–529.

Elholm, M., Garras, A., Neve, S., Tornehave, D., Lund, T.B., Skorve, J., Flatmark, T., Kristiansen, K. and Berge, R.K. (2000) Long-chain acyl-CoA esters and acyl-CoA binding protein are present in the nucleus of rat liver cells. *Journal of Lipid Research* 41, 538–545.

Escrich, E., Solanas, M., Soler, M., Ruiz de Villa, M.C., Sanchez, J.A. and Segura, R. (2001) Dietary polyunsaturated *n*-6 lipids effects on the growth and fatty acid composition of rat mammary tumors. *Journal of Nutritional Biochemistry* 12, 536–549.

Fahmi, H., Pelletier, J.P. and Martel-Pelletier, J. (2002) PPARγ ligands as modulators of inflammatory and catabolic responses in arthritis. An overview. *Journal of Rheumatology* 29, 3–14.

Fajas, L., Fruchart, J.C. and Auwerx, J. (1998) Transcriptional control of adipogenesis. *Current Opinion in Cell Biology* 10, 165–173.

Forman, B.M., Tontonoz, P., Chen, J., Brun, R.P., Spiegelman, B.M. and Evans, R.M. (1995) 15-Deoxy-delta 12, 14-prostaglandin J2 is a ligand for the adipocyte determination factor PPAR gamma. *Cell* 83, 803–812.

Fourcade, S., Savary, S., Albet, S., Gauthe, D., Gondcaille, C., Pineau, T., Bellenger, J., Bentejac, M., Holzinger, A., Berger, J. and Bugaut, M. (2001) Fibrate induction of the adrenoleukodystrophy-related gene (ABCD2): promoter analysis and role of the peroxisome proliferator-activated receptor PPARα. *European Journal of Biochemistry* 268, 3490–3500.

Ge, Y., Chen, Z., Kang, Z.B., Cluette-Brown, J., Laposata, M. and Kang, J.X. (2002) Effects of adenoviral gene transfer of *C. elegans n*-3 fatty acid desaturase on the lipid profile and growth of human breast cancer cells. *Anticancer Research* 22, 537–543.

Goldberg, I.J. (1996) Lipoprotein lipase and lipolysis: central roles in lipoprotein metabolism and atherogenesis. *Journal of Lipid Research* 37, 693–707.

Goodridge, A.G. (1991) Fatty acid desaturation and chain elongation in eucaryotes. In: Vance, D.E. and Vance, J. (eds) *Biochemistry of Lipids, Lipoproteins and Membranes.* Elsevier Science, Amsterdam, pp. 141–168.

Halliwell, B. and Gutteridge, J.M.C. (1989) Lipid peroxidation: a radical chain reaction. In: *Free Radicals in Biology and Medicine.* Oxford University Press, Oxford, pp. 188–276.

Hannah, V.C., Ou, J., Luong, A., Goldstein, J.L. and Brown, M.S. (2001) Unsaturated fatty acids down-regulate srebp isoforms 1a and 1c by two mechanisms in HEK-293 cells. *Journal of Biological Chemistry* 276, 4365–4372.

Hayhurst, G.P., Lee, Y.H., Lambert, G., Ward, J.M. and Gonzalez, F.J. (2001) Hepatocyte nuclear factor 4α (nuclear receptor 2A1) is essential for maintenance of hepatic gene expression and lipid homeostasis. *Molecular and Cellular Biology* 21, 1393–1403.

Helmlinger, G., Sckell, A., Dellian, M., Forbes, N.S. and Jain, R.K. (2002) Acid production in glycolysis-impaired tumors provides new insights into tumor metabolism. *Clinical Cancer Research* 8, 1284–1291.

Henry, R.R. (1997) Thiazolidinediones. *Endocrinology and Metabolism Clinics of North America* 26, 553–573.

Hertz, R., Magenheim, J., Berman, I. and Bar-Tana, J. (1998) Fatty acyl-CoA thioesters are ligands of hepatic nuclear factor-4α. *Nature* 392, 512–516.

Hihi, A.K., Michalik, L. and Wahli, W. (2002) PPARs: transcriptional effectors of fatty acids and their derivatives. *Cellular and Molecular Life Sciences* 59, 790–798.

Holm, C., Osterlund, T., Laurell, H. and Contreras, J.A. (2000) Molecular mechanisms regulating hormone-sensitive lipase and lipolysis. *Annual Review of Nutrition* 20, 365–393.

Holst, D. and Grimaldi, P.A. (2002) New factors in the regulation of adipose differentiation and metabolism. *Current Opinion in Lipidology* 13, 241–245.

Hsueh, W.A. and Law, R.E. (2001) PPARγ and atherosclerosis: effects on cell growth and movement. *Arteriosclerosis, Thrombosis and Vascular Biology* 21, 1891–1895.

Inoue, K., Kawahito, Y., Tsubouchi, Y., Kohno, M., Yoshimura, R., Yoshikawa, T. and Sano, H. (2001) Expression of peroxisome proliferator-activated receptor gamma in renal cell carcinoma and growth inhibition by its agonists. *Biochemical and Biophysical Research Communications* 287, 727–732.

Issemann, I. and Green, S. (1990) Activation of a member of the steroid hormone receptor superfamily by peroxisome proliferators. *Nature* 347, 645–650.

Issemann, I., Prince, R., Tugwood, J. and Green, S. (1992) A role for fatty acids and liver fatty acid binding protein in peroxisome proliferation? *Biochemical*

*Society Transactions* 20, 824–827.
Jiang, G., Nepomuceno, L., Hopkins, K. and Sladek, F.M. (1995) Exclusive homodimerization of the orphan receptor hepatocyte nuclear factor 4 defines a new subclass of nuclear receptors. *Molecular and Cellular Biology* 15, 5131–5143.
Jump, D.B, (2002) The biochemistry of N3-polyunsaturated fatty acids. *Journal of Biological Chemistry* 277, 8755–8758.
Jump, D.B. and Clarke, S.D. (1999) Regulation of gene expression by dietary fat. *Annual Review of Nutrition* 19, 63–90.
Jump, D.B., Thelen, A.P. and Mater, M.K. (2001) Functional interaction between sterol regulatory element-binding protein-1c, nuclear factor Y, and 3,5,3′-triiodothyronine nuclear receptors. *Journal of Biological Chemistry* 276, 34419–34427.
Kaytor, E.N., Shih, H. and Towle, H.C. (1997) Carbohydrate regulation of hepatic gene expression. Evidence against a role for the upstream stimulatory factor. *Journal of Biological Chemistry* 272, 7525–7531.
Kersten, S., Desvergne, B. and Wahli, W. (2000) Roles of PPARs in health and disease. *Nature* 405, 421–424.
Kliewer, S.A., Lenhard, J.M., Willson, T.M., Patel, I., Morris, D.C. and Lehmann, J.M. (1995) A prostaglandin J2 metabolite binds peroxisome proliferator-activated receptor gamma and promotes adipocyte differentiation. *Cell* 83, 813–819.
Knudsen, J., Jensen, M.V., Hansen, J.K., Faergeman, N.J., Neergaard, T.B. and Gaigg, B. (1999) Role of acylCoA binding protein in acylCoA transport, metabolism and cell signaling. *Molecular and Cellular Biochemistry* 192, 95–103.
Lampen, A., Meyer, S. and Nau, H. (2001) Phytanic acid and docosahexaenoic acid increase the metabolism of all-*trans*-retinoic acid and CYP26 gene expression in intestinal cells. *Biochimica et Biophysica Acta* 1521, 97–106.
Landreth, G.E. and Heneka, M.T. (2001) Anti-inflammatory actions of peroxisome proliferator-activated receptor gamma agonists in Alzheimer's disease. *Neurobiology of Aging* 22, 937–944.
Maillard, V., Bougnoux, P., Ferrari, P., Jourdan, M.L., Pinault, M., Lavillonniere, F., Body, G., Le Floch, O. and Chajes, V. (2002) N-3 and N-6 fatty acids in breast adipose tissue and relative risk of breast cancer in a case–control study in Tours, France. *International Journal of Cancer* 98, 78–83.
Mangelsdorf, D.J. and Evans, R.M. (1995) The RXR heterodimers and orphan receptors. *Cell* 83, 841–850.
Masamune, A., Kikuta, K., Satoh, M., Sakai, Y., Satoh, A. and Shimosegawa, T. (2001) Ligands of peroxisome proliferator-activated receptor-γ block activation of pancreatic stellate cells. *Journal of Biological Chemistry* 277, 141–147.
Mater, M.K., Thelen, A.P., Pan, D.A. and Jump, D.B. (1999) Sterol response element-binding protein 1c (SREBP1c) is involved in the polyunsaturated fatty acid suppression of hepatic S14 gene transcription. *Journal of Biological Chemistry* 274, 32725–32732.
McArthur, M.J., Atshaves, B.P., Frolov, A., Foxworth, W.D., Kier, A.B. and Schroeder, F. (1999) Cellular uptake and intracellular trafficking of long chain fatty acids. *Journal of Lipid Research* 40, 1371–1383.
Michalik, L. and Wahli, W. (1999) Peroxisome proliferator-activated receptors: three isotypes for a multitude of functions. *Current Opinion in Biotechnology* 10, 564–570.
Miyata, K.S., McCaw, S.E., Marcus, S.L., Rachubinski, R.A. and Capone, J.P. (1994) The peroxisome proliferator-activated receptor interacts with the retinoid X receptor *in vivo*. *Gene* 148, 327–330.
Mori, T.A., Bao, D.Q., Burke, V., Puddey, I.B., Watts, G.F. and Beilin, L.J. (1999) Dietary fish as a major component of a weight-loss diet: effect on serum lipids, glucose, and insulin metabolism in overweight hypertensive subjects. *American Journal of Clinical Nutrition* 70, 817–825.
Motojima, K., Passilly, P., Peters, J.M., Gonzalez, F.J. and Latruffe, N. (1998) Expression of putative fatty acid transporter genes are regulated by peroxisome proliferator-activated receptor alpha and gamma activators in a tissue and inducer-specific manner. *Journal of Biological Chemistry* 273, 16710–16714.
Moya-Camarena, S.Y., Vanden Heuvel, J.P., Blanchard, S.G., Leesnitzer, L.A. and Belury, M.A. (1999) Conjugated linoleic acid is a potent naturally occurring ligand and activator of PPARα. *Journal of Lipid Research* 40, 1426–1433.
Mueller, E., Sarraf, P., Tontonoz, P., Evans, R.M., Martin, K.J., Zhang, M., Fletcher, C., Singer, S. and Spiegelman, B.M. (1998) Terminal differentiation of human breast cancer through PPAR gamma. *Molecular Cell* 1, 465–470.
Mueller, E., Smith, M., Sarraf, P., Kroll, T., Aiyer, A., Kaufman, D.S., Oh, W., Demetri, G., Figg, W.D., Zhou, X.P., Eng, C., Spiegelman, B.M. and Kantoff, P.W. (2000) Effects of ligand activation of peroxisome proliferator-activated receptor gamma in human prostate cancer. *Proceedings of the National Academy of Sciences USA* 97, 10990–10995.
Muerhoff, A.S., Griffin, K.J. and Johnson, E.F. (1992) The peroxisome proliferator-activated receptor mediates the induction of CYP4A6, a cytochrome P450 fatty acid omega-hydroxylase, by clofibric acid. *Journal of Biological Chemistry* 267, 19051–19053.
Nagy, L., Tontonoz, P., Alvarez, J.G., Chen, H. and Evans, R.M. (1998) Oxidized LDL regulates macrophage gene expression through ligand activation of PPARγ. *Cell* 93, 229–240.
Norris, A.W. and Spector, A.A. (2002) Very long chain *n*-3 and *n*-6 polyunsaturated fatty acids bind

strongly to liver fatty acid-binding protein. *Journal of Lipid Research* 43, 646–653.

O'Moore-Sullivan, T.M. and Prins, J.B. (2002) Thiazolidinediones and type 2 diabetes: new drugs for an old disease. *Medical Journal of Australia* 176, 381–386.

Osborne, T.F. (2000) Sterol regulatory element-binding proteins (SREBPs): key regulators of nutritional homeostasis and insulin action. *Journal of Biological Chemistry* 275, 32379–32382.

Ou, J., Tu, H., Shan, B., Luk, A., DeBose-Boyd, R.A., Bashmakov, Y., Goldstein, J.L. and Brown, M.S. (2001) Unsaturated fatty acids inhibit transcription of the sterol regulatory element-binding protein-1c (SREBP-1c) gene by antagonizing ligand-dependent activation of the LXR. *Proceedings of the National Academy of Sciences USA* 98, 6027–6032.

Palmer, C.N., Hsu, M.H., Griffin, K.J., Raucy, J.L. and Johnson, E.F. (1998) Peroxisome proliferator activated receptor-α expression in human liver. *Molecular Pharmacology* 53, 14–22.

Patel, L., Pass, I., Coxon, P., Downes, C.P., Smith, S.A. and Macphee, C.H. (2001) Tumor suppressor and anti-inflammatory actions of PPARγ agonists are mediated via upregulation of PTEN. *Current Biology* 11, 764–768.

Peet, D.J., Turley, S.D., Ma, W., Janowski, B.A., Lobaccaro, J.M., Hammer, R.E. and Mangelsdorf, D.J. (1998) Cholesterol and bile acid metabolism are impaired in mice lacking the nuclear oxysterol receptor LXR alpha. *Cell* 93, 693–704.

Peters, J.M., Park, Y., Gonzalez, F.J. and Pariza, M.W. (2001) Influence of conjugated linoleic acid on body composition and target gene expression in peroxisome proliferator-activated receptor alpha-null mice. *Biochimica et Biophysica Acta* 1533, 233–242.

Petrescu, A.D., Hertz, R., Bar-Tana, J., Schroeder, F. and Kier, A.B. (2002) Ligand specificity and conformational dependence of the hepatic nuclear factor-4α (HNF-4α). *Journal of Biological Chemistry* 277, 23988–23999.

Plutzky, J. (2001) Peroxisome proliferator-activated receptors in endothelial cell biology. *Current Opinion in Lipidology* 12, 511–518.

Price, P.T., Nelson, C.M. and Clarke, S.D. (2000) Omega-3 polyunsaturated fatty acid regulation of gene expression. *Current Opinion in Lipidology* 11, 3–7.

Rangwala, S.M. and Lazar, M.A. (2000) Transcriptional control of adipogenesis. *Annual Review of Nutrition* 20, 535–559.

Repa, J.J. and Mangelsdorf, D.J. (2000) The role of orphan nuclear receptors in the regulation of cholesterol homeostasis. *Annual Review in Cell and Developmental Biology* 16, 459–481.

Repa, J.J., Liang, G., Ou, J., Bashmakov, Y., Lobaccaro, J.M., Shimomura, I., Shan, B., Brown, M.S., Goldstein, J.L. and Mangelsdorf, D.J. (2000) Regulation of mouse sterol regulatory element-binding protein-1c gene (SREBP-1c) by oxysterol receptors, LXRα and LXRβ. *Genes and Development* 14, 2819–2830.

Rosen, E.D. and Spiegelman, B.M. (2001) PPARγ: a nuclear regulator of metabolism, differentiation, and cell growth. *Journal of Biological Chemistry* 276, 37731–37734.

Sarraf, P., Mueller, E., Jones, D., King, F.J., DeAngelo, D.J., Partridge, J.B., Holden, S.A., Chen, L.B., Singer, S., Fletcher, C. and Spiegelman, B.M. (1998) Differentiation and reversal of malignant changes in colon cancer through PPARγ. *Nature Medicine* 4, 1046–1052.

Sarraf, P., Mueller, E., Smith, W.M., Wright, H.M., Kum, J.B., Aaltonen, L.A., de la Chapelle, A., Spiegelman, B.M. and Eng, C. (1999) Loss-of-function mutations in PPAR gamma associated with human colon cancer. *Molecular Cell* 3, 799–804.

Sato, R., Miyamoto, W., Inoue, J., Terada, T., Imanaka, T. and Maeda, M. (1999) Sterol regulatory element-binding protein negatively regulates microsomal triglyceride transfer protein gene transcription. *Journal of Biological Chemistry* 274, 24714–24720.

Schluter, A., Giralt, M., Iglesias, R. and Villarroya, F. (2002) Phytanic acid, but not pristanic acid, mediates the positive effects of phytol derivatives on brown adipocyte differentiation. *FEBS Letters* 517, 83–86.

Schoonjans, K., Watanabe, M., Suzuki, H., Mahfoudi, A., Krey, G., Wahli, W., Grimaldi, P., Staels, B., Yamamoto, T. and Auwerx, J. (1995) Induction of the acyl-coenzyme A synthetase gene by fibrates and fatty acids is mediated by a peroxisome proliferator response element in the C promoter. *Journal of Biological Chemistry* 270, 19269–19276.

Schoonjans, K., Staels, B. and Auwerx, J. (1996a) Role of the peroxisome proliferator-activated receptor (PPAR) in mediating the effects of fibrates and fatty acids on gene expression. *Journal of Lipid Research* 37, 907–925.

Schoonjans, K., Peinado-Onsurbe, J., Lefebvre, A.M., Heyman, R.A., Briggs, M., Deeb, S., Staels, B. and Auwerx, J. (1996b) PPARα and PPARγ activators direct a distinct tissue-specific transcriptional response via a PPRE in the lipoprotein lipase gene. *EMBO Journal* 15, 5336–5348.

Schoonjans, K., Martin, G., Staels, B. and Auwerx, J. (1997) Peroxisome proliferator-activated receptors, orphans with ligands and functions. *Current Opinion in Lipidology* 8, 159–166.

Schultz, J.R., Tu, H., Luk, A., Repa, J.J., Medina, J.C., Li, L., Schwendner, S., Wang, S., Thoolen, M., Mangelsdorf, D.J., Lustig, K.D. and Shan, B. (2000) Role of LXRs in control of lipogenesis. *Genes and*

*Development* 14, 2831–2838.
Segawa, Y., Yoshimura, R., Hase, T., Nakatani, T., Wada, S., Kawahito, Y., Kishimoto, T. and Sano, H. (2002) Expression of peroxisome proliferator-activated receptor (PPAR) in human prostate cancer. *Prostate* 51, 108–116.
Serhan, C.N. and Oliw, E. (2001) Unorthodox routes to prostanoid formation: new twists in cyclooxygenase-initiated pathways. *Journal of Clinical Investigation* 107, 1481–1489.
Shimano, H., Horton, J.D., Shimomura, I., Hammer, R.E., Brown, M.S. and Goldstein, J.L. (1997) Isoform 1c of sterol regulatory element binding protein is less active than isoform 1a in livers of transgenic mice and in cultured cells. *Journal of Clinical Investigation* 99, 846–854.
Spiegelman, B.M. (1997) Peroxisome proliferator-activated receptor gamma: a key regulator of adipogenesis and systemic insulin sensitivity. *European Journal of Medical Research* 2, 457–464.
Sprecher, H. (2000) Metabolism of highly unsaturated *n*-3 and *n*-6 fatty acids. *Biochimica et Biophysica Acta* 1486, 219–231.
Staels, B., Vu-Dac, N., Kosykh, V.A., Saladin, R., Fruchart, J.C., Dallongeville, J. and Auwerx, J. (1995) Fibrates downregulate apolipoprotein C-III expression independent of induction of peroxisomal acyl coenzyme A oxidase. A potential mechanism for the hypolipidemic action of fibrates. *Journal of Clinical Investigations* 95, 705–712.
Staels, B., Schoonjans, K., Fruchart, J.C. and Auwerx, J. (1997) The effects of fibrates and thiazolidinediones on plasma triglyceride metabolism are mediated by distinct peroxisome proliferator activated receptors (PPARs). *Biochimie* 79, 95–99.
Takashima, T., Fujiwara, Y., Higuchi, K., Arakawa, T., Yano, Y., Hasuma, T. and Otani, S. (2001) PPAR-γ ligands inhibit growth of human esophageal adeno-carcinoma cells through induction of apoptosis, cell cycle arrest and reduction of ornithine decarboxylase activity. *International Journal of Oncology* 19, 465–471.
Tontonoz, P., Singer, S., Forman, B.M., Sarraf, P., Fletcher, J.A., Fletcher, C.D., Brun, R.P., Mueller, E., Altiok, S., Oppenheim, H., Evans, R.M. and Spiegelman, B.M. (1997) Terminal differentiation of human liposarcoma cells induced by ligands for peroxisome proliferator-activated receptor gamma and the retinoid X receptor. *Proceedings of the National Academy of Sciences USA* 94, 237–241.
Towle, H.C., Kaytor, E.N. and Shih, H.M. (1997) Regulation of the expression of lipogenic enzyme genes by carbohydrate. *Annual Review of Nutrition* 17, 405–433.
Vaulont, S., Vasseur-Cognet, M. and Kahn, A. (2000) Glucose regulation of gene transcription. *Journal of Biological Chemistry* 275, 31555–31558.
Vu-Dac, N., Schoonjans, K., Laine, B., Fruchart, J.C., Auwerx, J. and Staels, B. (1994) Negative regulation of the human apolipoprotein A-I promoter by fibrates can be attenuated by the interaction of the peroxisome proliferator-activated receptor with its response element. *Journal of Biological Chemistry* 269, 31012–31018.
Walczak, R. and Tontonoz, P. (2002) PPARadigms and PPARadoxes: expanding roles for PPARγ in the control of lipid metabolism. *Journal of Lipid Research* 43, 177–186.
Whigham, L.D., Cook, M.E. and Atkinson, R.L. (2000) Conjugated linoleic acid: implications for human health. *Pharmacology Research* 42, 503–510.
Wolfrum, C., Borrmann, C.M., Borchers, T. and Spener, F. (2001) Fatty acids and hypolipidemic drugs regulate peroxisome proliferator-activated receptors α- and γ-mediated gene expression via liver fatty acid binding protein: a signaling path to the nucleus. *Proceedings of the National Academy of Sciences USA* 98, 2323–2328.
Xu, L., Glass, C.K. and Rosenfeld, M.G. (1999) Coactivator and corepressor complexes in nuclear receptor function. *Current Opinion in Genetics and Development* 9, 140–147.
Yu, K., Bayona, W., Kallen, C.B., Harding, H.P., Ravera, C.P., McMahon, G., Brown, M. and Lazar, M.A. (1995) Differential activation of peroxisome proliferator-activated receptors by eicosanoids. *Journal of Biological Chemistry* 270, 23975–23983.
Zannis, V.I., Kan, H.Y., Kritis, A., Zanni, E. and Kardassis, D. (2001) Transcriptional regulation of the human apolipoprotein genes. *Frontiers in Biosciences* 6, D456–D504.

# 10　视黄酸受体和视黄酸 X 受体在调节维生素 A 作用分子机制中的作用

Dianne R. Soprano[1,2]和 Kenneth J. Soprano[2,3]

[[1] 生物化学系；[2] 癌症研究与分子生物学费尔斯（Fels）研究所及[3] 微生物与免疫学系，神殿大学医学院，费城，宾夕法尼亚州，美国]

## 引　　言

1925 年，Wolbach 和 Howe 首次证明维生素 A 是一种必需养分。从那时起，已证明维生素 A（视黄醇）（图 10.1）及其生物活性衍生物（统称为类维生素 A）是许多生物功能包括生长、视力、生殖、胚胎发育、上皮组织分化和免疫功能所必需的（Sporn *et al.*，1994）。除了夜间视力外，所有这些功能都是受视黄酸（retinoic acid，RA）调节的（图 10.1）。

如图 10.2 所示，血浆中多数视黄醇是结合到视黄醇结合蛋白（retinol-binding protein，RBP）上而被运输到需要维生素 A 的靶组织细胞中。在被这些细胞吸收后，视黄醇可以视黄醇脂类形式被储存，或被氧化为 RA。在这些细胞的细胞质内，视黄醇和 RA 能分别结合到被称为细胞视黄醇结合蛋白（cellular retinol-binding protein，CRBP）和细胞视黄酸结合蛋白（cellular retinoic acid binding protein，CRABP）的细胞结合蛋白上。RA 被转运到细胞核，而其大多数生物学效应是在细胞核内通过调节基因表达的水平发挥的。为调节基因表达，RA 必须结合到核蛋白上（依赖于配体的转录因子），这些核蛋白被称为视黄酸受体（retinoic acid receptor，RAR）和类维生素 A 的 X 受体（retinoid X receptor，RXR）。既然 RAR 和 RXR 是理解 RA 作用的分子机制的关键，那么本章将集中在 RAR 和 RXR 及其在调节 RA 多效中的作用上。近年来有许多关于 RBP 和视黄醇转运（Gottesman *et al.*，2001）、维生素 A 代谢（Napoli，1999a，b；Duester，2000）以及细胞类维生素 A 结合蛋白（Noy，2000）的优秀综述性文章。

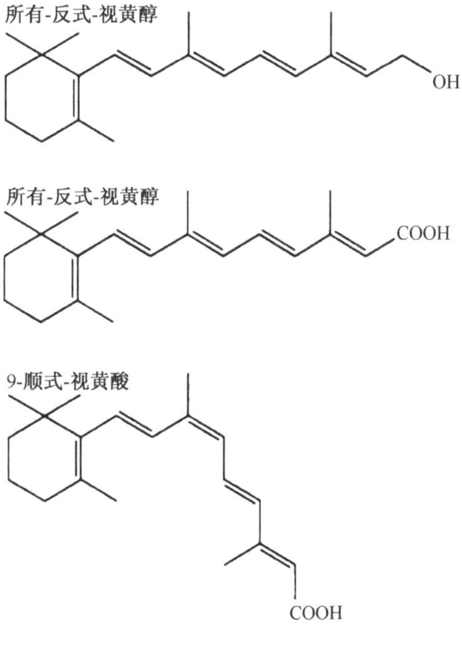


图 10.1　视黄醇和视黄酸的结构

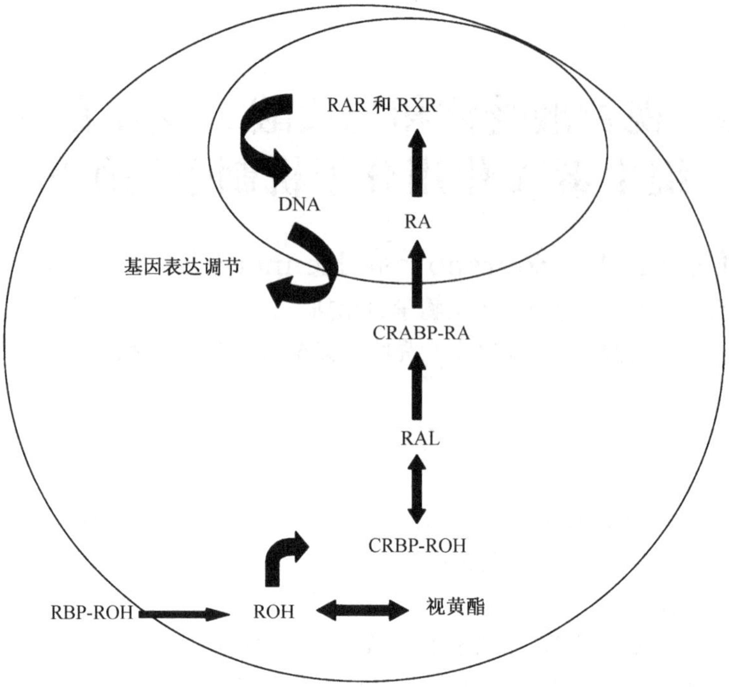


图 10.2 维生素 A 信号转导途径的概述
缩写：CRABP，细胞视黄酸 A 结合蛋白；CRBP，细胞视黄醇结合蛋白；RA，视黄酸 A；RAL，视黄醛；RAR，视黄酸受体；RBP，视黄醇结合蛋白；ROH，视黄醇；RXR，类维生素 A 的 X 受体

## RAR 和 RXR 的识别

1987 年，两个小组在研究类固醇激素受体的超家族时分别独立报道了第一个 RAR（Giguere *et al.*，1987；Petkovich *et al.*，1987）。在这个 RAR 被发现之后不久，另外两个 RAR 也被识别发现，并将它们命名为 RARβ（Benbrook *et al.*，1988；Brand *et al.*，1988）和 RARγ（Krust *et al.*，1989；Zelent *et al.*，1989），而第一个被发现的 RAR 就被命名为 RARα。这 3 个 RAR 亚型的每一个都被位于基因组不同染色体上的不同基因所编码。

在第一个 RAR 被发现后几年，Mangelsdorf 等（1990）描述了另一个孤儿受体的性质，这个受体导致新型的称为类维生素 A X 受体（RXR）的 RA 应答受体被识别。短时间以后，两个其他不同基因编码的且高度同源的 RXR 亚型又被发现（Mangelsdorf *et al.*，1992）。因此，RXR 家族包括 3 个成员，即 RXRα、RXRβ 和 RXRγ。令人吃惊的是，比较 RAR 家族和 RXR 家族受体的氨基酸序列表明它们的整个蛋白质长度仅具有较弱的同源性，且发现 DNA 结合域相似度最高（为 60%），而 RA 结合域仅有约 25%的氨基酸相同性。有趣的是，在称为超气孔的果蝇中发现有一个 RAR 同类物，该同类物与蜕皮激素形成二聚体，但在一般的果蝇中未发现有 RAR 同类物（Mangelsdorf *et al.*，1992）。这些结果表明，RAR 和 RXR 是在进化期间单独出现的，且受体的二聚化（见下面）在进化期间被很好地保存了下来。

# RAR 和 RXR 的剖析

## 域 结 构

核受体超家族包括 RAR 和 RXR，它们具有由 5 个或 6 个功能不同的域即 A～F 域组成的相同结构（Chambon，1996）（图 10.3）。N 端 A/B 域具有可变长度，且受体间的同源性最小。A/B 域的主要功能是配体独立的转录激活，被称为激活功能-1（AF-1）（Nagpal *et al.*，1993）。此 AF-1 具有能与 E 域的 AF-2 功能产生协同作用的独立功能（见下面）（Nagpal *et al.*，1993；Folkers *et al.*，1995）。C 域结构是最保守的核受体（95%在 3 个 RAR 亚型之间，92%～95%在 3 个 RXR 亚型间），且负责 DNA 结合。在 C 域的 66 个氨基酸中心区域有一对被称为锌指的结合锌结构，负责 DNA 主要凹槽内部的连接和 DNA 反应元件半点间隔的区分。D 域是铰链区域，与受体反应元件的半位点间隔需要量（Stunnenberg，1993）和辅阻遏物结合位点（Chen and Evans，1995；Kurokawa *et al.*，1995）的确定有关。E 域或配体结合域（ligand binding domain，LBD）是一个功能复杂的区域，由 220 个含有配体结合域、受体二聚形（Nagpal *et al.*，1993）和被称为激活功能-2（activation function 2，AF-2）的配体依赖型转录激活活性的氨基酸残基组成（Nagpal *et al.*，1993）。E 域的氨基酸组成在 3 个 RAR 亚型间（82%～85%）和 3 个 RXR 亚型间（86%～89%），也是高度保守的。F 域位于 RAR 的 C 端，但其功能还没被确定。另外，RXR 不含 F 域。

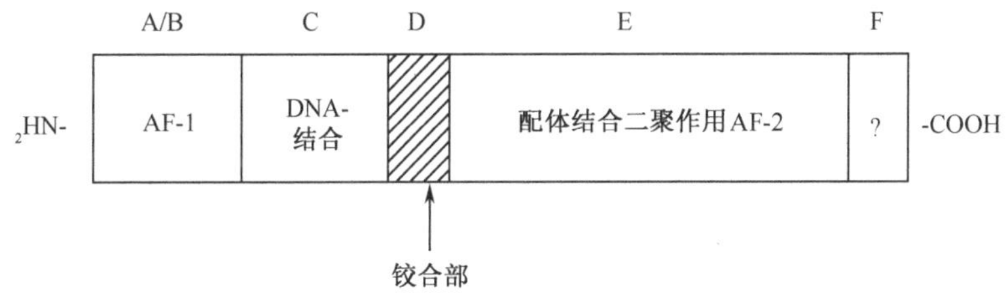


| | | |
|---|---|---|
| RAR$\alpha$ | 100% | 100% |
| RAR$\beta$ | 95% | 82% |
| RAR$\gamma$ | 95% | 85% |
| RAR$\alpha$ | 61% | 27% |
| RAR$\beta$ | 56% | 25% |
| RAR$\gamma$ | 58% | 24% |

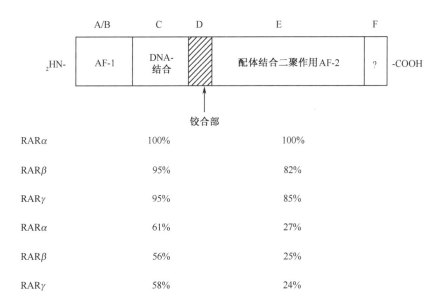

图 10.3 视黄酸受体和类维生素 A 的 X 受体的功能域图示

DNA 和配体结合域的氨基酸相同性百分数与 RARα 相关

## RAR 和 RXR 的配体特异性

RXR 能特异地结合 9-顺式-视黄酸（9-*cis*-RA）（$K_d$＝1～10nmol/L），而 RAR 以同

等的亲和力（$K_d$=1～5nmol/L）结合全-反式-视黄酸（*trans*-RA）和9-顺式-RA（Heyman *et al.*，1992；Levin *et al.*，1992）。有趣的是，9-顺式-RA对RAR和RXR都具有高的亲和力，然而，在体内却很难检测到大量9-顺式-RA。这提示可能有一个未知的RXR-特异性配体存在（deUrquiza *et al.*，2000）。

随着6种不同类维生素A受体亚型被发现，而每一个亚型都对结合RA具有相同的亲和力，人们对设计和合成能定位个体RAR或RXR亚型或能区别RAR和RXR亚型的选择性类维生素A已进行了大量的研究。到目前为止，已研发出大量RAR亚型选择性类维生素A竞争物和拮抗物以及几个RXR-选择性类维生素A竞争物和拮抗物（表10.1）（Chandraratna，1998；Johnson and Chandrartna，1999；Nagpal and Chandraratna，2000）。这些选择性类维生素A对研究RAR和RXR的结构和功能非常有用。更重要的是，这些类维生素A原型与另外更多尚待研发的受体选择性类维生素A的增效剂、反增效剂和拮抗剂对于治疗各种皮肤病、肿瘤、二型糖尿病、动脉硬化和肥胖症都具有很大临床益处的潜力。这些类维生素A很可能有较好的理疗价值和较少的毒性问题。

**表10.1 类维生素A核受体与代表性受体亚型选择性配体**

| 受体 | 异构体 | 配体 | | |
|---|---|---|---|---|
| RARα | α1、α2 | 全-反式-RA<br>Am80<br>AGN190521<br>AGN193109(拮抗物) | 9-顺式-RA<br>Am580/CD336<br>CD2366 | AGN193312<br>TD550 |
| RARβ | β1、β2、β3、β4 | 全-反式-RA<br>CD2314<br>AGN193273<br>CD2665<br>AGN190521<br>AGN193109(拮抗物) | 9-顺式-RA<br>Ch55<br>CD270<br>SR3985<br>CD2366 | AGN193174<br>AGN193078<br>CD271<br>TTNN |
| RARγ | γ1、γ2 | 全-反式-RA<br>AHPN/CD437<br>SR11254/MM11254<br>MM11389<br>CD270<br>SR3985<br>CD2366<br>AGN193109(拮抗物) | 9-顺式-RA<br>CD666<br>SR11363<br>AGN193078<br>CD271<br>TTNN<br>MM11253(拮抗物) | 4-HPR<br>CD2325<br>SR11364<br>AGN193273<br>CD2665<br>AGN190521 |
| RXRα | α1、α2 | 9-顺式-RA<br>SR11217<br>SR11246 | LG1069<br>SR11234 | SR11203<br>SR11236 |
| RXRβ | β1、β2 | 9-顺式-RA<br>SR11217<br>SR11246 | LG1069<br>SR11234 | SR11203<br>SR11236 |
| RXRγ | γ1、γ2 | 9-顺式-RA<br>SR11217<br>SR11246 | LG1069<br>SR11234 | SR11203<br>SR11236 |

## RAR 和 RXR 的异构体

RAR 和 RXR 除了有三种亚型外，每种亚型都有几种异构体。每种 RAR 和 RXR 亚型的不同异构体仅在 N 端 A 域的氨基酸序列和（或）其 mRNA 5′未翻译区的核酸序列上存在差异，而 B～F 域的氨基酸序列是完全相同的。这些异构体的出现是由于替代性增强子的使用、特异拼接和为启动转录而使用内部 CUG 密码子所致。RARβ 的 4 个主要异构体（RARβ1、β2、β3 和 β4）与 RARα 的两个主要异构体（α1 和 α2）及 RAR 的两个主要异构体（γ1 和 γ2）都已被很好地阐述（Giguere *et al.*，1990；Kastner *et al.*，1990；Leroy *et al.*，1991；Zelent *et al.*，1991；Nagpal *et al.*，1992）。此外，RARα（α3～α7）和 RARγ（γ3～γ7）的另外 5 个次要异构体已被描述。最后，三种 RXR 亚型的每一种亚型的 2 个异构体（α1、α2、β1、β2、γ1 和 γ2）都已被报道（Liu and Linney，1993；Brocard *et al.*，1994；Nagata *et al.*，1994）。

## RAR 和 RXR 配体结合域的结构

近几年，apo-RXRα 的 LBD（Bourguet *et al.*，1995）、holo-RARγ 的 LBD（Renaud *et al.*，1995）、结合到选择性拮抗物上的 RXRα 的 LBDs 与基本有活性的 RXRαF318A 突变体之间的异二聚体（Bourguet *et al.*，2000）和与 9-顺-RA 结合的 holo-RARα LBD（Egea *et al.*，2000）的高分辨率 X 射线晶体结构已被报道。对 RAR 和 RXR 的 LBD 结构的阐述将导致新的亚型选择性类维生素 A 增效剂和拮抗物的设计以及其潜在的药理学应用。

对这些晶体结构与几个其他类固醇/甲状腺激素核受体家族成员的分析已表明，每个受体共有一个新型蛋白折叠，即反平行 α 螺旋三明治，它由 12 个 α 螺旋（数字标号为 H1～H12）和位于螺旋 H5 和 H6 之间的 β 转角组成（Wurtz *et al.*，1996）。在配体结合袋中的 RA 与大量位于 α 螺旋 H1、H3、H5、H11 和 H12、环 6-7 和 11-12 及 β 片层 sl 上的氨基酸残基连接。比较 apo- RXRαLBD 和 holo-RARγLBD 的 X 射线晶体结构发现，主要结构的改变发生在配体结合区，被称为“开放”apo-结构和“闭合”holo-结构。近来人们已能进一步直接比较结合到 9-顺式-RA 上的 apo- RXRαLBD 和 holo-RARγLBD 的 X 射线晶体结构。9-顺式-RA 连接到 LBD 上最初导致螺旋 H3 重排，然后通过排出螺旋 H11 而重排螺旋 H12。最大的构形变化发生在螺旋 H12 上，H12 以 apo-形式突出于蛋白核心，暴露于溶液中，而 holo-形式的 H12 螺旋发生旋转并向配体结合袋状物折回。这诱导 LBD 的压缩和新表面的形成，使它能结合到共同激活子上参与调节转录活性（见下面）。而且，增效剂与抗增效剂结合的 LBD 结构的比较表明了螺旋 H12 有两个不同构型存在，同时表明螺旋 H12 能适应 holo-增效剂结合形式和 holo-抗增效剂结合形式。这种在 H12 上的构型差异至少部分导致调节增效剂和抗增效剂结合受体的不同转录活性。因此，螺旋 H12 大的结构改变和配体结合于 RAR 和 RXR 区域的螺旋 H3、H6 和 H11 的相关结构改变，对通过产生以下结合辅助激活子及移走辅助抑制子的表面而测定这些受体的转录活性至关重要。

此外，大量点突变试验研究了完整 RARγ 晶体结构中对 RARγ 配体结合区与 RA

羧基互作至关重要的几个保守氨基酸残基在 RARα、RARβ 和 RARγ 中的作用（Tairis *et al.*，1994，1995；Scafonas *et al.*，1997；Wolfgang *et al.*，1997；Zhang *et al.*，1998，2000）。在所有三种 RAR 异构型中，同源精氨酸残基（αArg276、βArg269 和 γArg278）向丙氨酸或谷氨酸的突变表明，这种氨基酸残基在通过与 RA 羧基互作而结合 RA 中起关键作用。这与 LBD holo-RARγ 的晶体结构是一致的，其中 γArg278 与 RA 的羧基盐 O-22 形成盐桥（Renaud *et al.*，1995）。在 RARβ 中，这个精氨酸残基（Arg269）的功能是在结合 RA 中连接赖氨酸 220 和丝氨酸 278，而在 RARα 和 RARγ 两者中，同源精氨酸残基（分别为 Arg276 和 Arg278）在与结合的 RA 分子羧基的协同上是相对独立的。这表明每个 RAR 亚型中这种氨基酸残基在配体结合位点中的定位与带电环境是不同的。

RAR LBD 的 X 射线晶体结构分析和定位基因突变研究表明，在 RAR 的配体结合区内 3 个非保守氨基酸残基，它们对 RAR 亚型配体的特异性非常重要（Ostrowski *et al.*，1995，1998；Gehin *et al.*，1999）。在 RARβ 中 Ala225 突变为丝氨酸，能将 RARβ 转化为 RARα 配体选择性受体（Ostrowski *et al.*，1998；Gehin *et al.*，1999）。另外，RARβ 中的 Ile263 突变为蛋氨酸（Ostrowski *et al.*，1998），或 RARβ 中的 Ile263 和 Val388 各自突变为蛋氨酸和丙氨酸（Gehin *et al.*，1999），能将 RARβ 转化为 RARγ 配体选择受体。因此，这 3 个氨基酸残基在区分 RAR 亚型-选择性配体上而不是在区分 RA 中起关键作用。然而，RARβ 的羧基协调氨基酸（Arg269）与 RARγ 配体特异性氨基酸（Ile263 和 Vel388）或 RARβ 的 RARα 配体特异性氨基酸（Ale225）的同时突变对 RA 结合和 RA 依赖的受体转录活性仅有小的影响（Zhang *et al.*，2003）。这表明与类维生素 A 羧基基团互作的配体结合袋区域与参与亚型配体特异性的区域的功能是独立的。

## RAR 和 RXR 转录调节的机制

### 视黄酸反应元件

RAR-依赖性靶基因和 RXR-依赖性靶基因的转录激活需要受体与位于靶基因启动子区的特定 DNA 序列互作。这些区域被称为视黄酸反应元件（retinoic acid response element，RARE）和类维生素 A 的 X 受体反应元件（retinoid X receptor response element，RXRE）。对大量 RA 反应基因启动子的分析导致了 RARE 和 RXRE 特性的确定。图 10.4 表示了 RAR-RXR 异质二聚体结合到 RARE 以及 RXR 同质二聚体结合到 RXRE 上的模式。

RARE 和 RXRE 由具有一致的 AGGTGA 序列的半位点组成，对核苷酸的替换有相当的柔性。这个半位点序列能形成一个直接重复即一个回文结构，不需任何一致结构也可被发现。但是，它们最常作为一个直接重复被排列（Umesono *et al.*，1988，1991；Umesono and Evans，1989；Naar *et al.*，1991）。RAR-RXR 异质二聚体（见下面）紧密结合到被 2 个或 5 个核酸（分别称为 DR-2 和 DR-5）分离的半位点序列上。DR-5RARE 像在 RARβ 基因的启动子中发现的一样（de The *et al.*，1990；Sucov *et al.*，1990），是靶基因的启动子中最常被观测到的。另一方面，RXRE 是典型的相同一致序列 AGGTGA 与一个核酸分离半位点序列（DR-1）的直接重复。

## 同源二聚化与异源二聚化

RAR 形成具有混杂 RXR 的异源二聚体，并与靶基因启动子上的 DR-5 或 DR-2 RARE 结合（图 10.4）。RAR 和 RXR 的异源二聚化为紧密结合和通过 RARE 而转录激活基因表达所必需。在溶液中，RAR 和 RXR 的二聚化作用弱，但一旦结合 DNA 上，这种互作便加强。7 个一组的重复位于 LBD 中，为 RAR-RXR 和 RXR-RXR 的二聚化所必需（Forman *et al.*，1992)。而且，位于 DNA-结合域（域 C）中的锌指 Dbox 是二聚化和通过二聚物识别 RARE 所必需的（Umesono and Evans，1988)。RAR-RXR 异源二聚体以特定的位置和极性结合到 RARE DNA 序列上。RXR 占据 5′上游半位点，而 RAR 占据 3′下游半位点（Predki *et al.*，1994；Perlmann *et al.*，1993；Kurokawa *et al.*，1993)。

除了与 RAR 二聚化外，RXR 还能同源二聚化并能与 DR-1 RXRE 结合，或能与其他核酸受体，包括维生素 D 受体、甲状腺素受体和过氧化物激活增生受体（peroxesome proliferator activated receptor，PPAR）及许多其他孤儿受体形成异源二聚体。

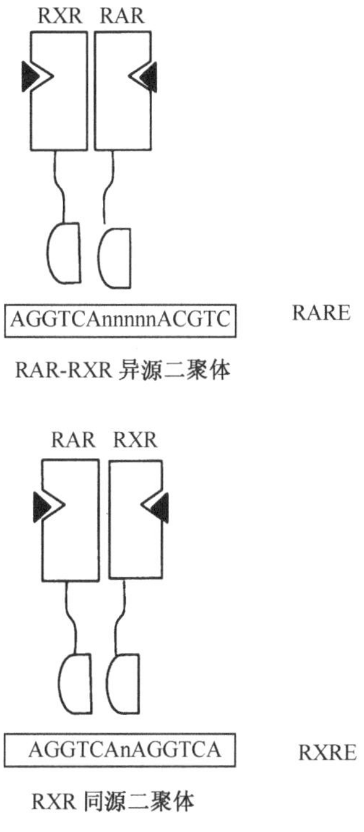


图 10.4　RAR-RXR 异源二聚体与 RXR 同源二聚体模式

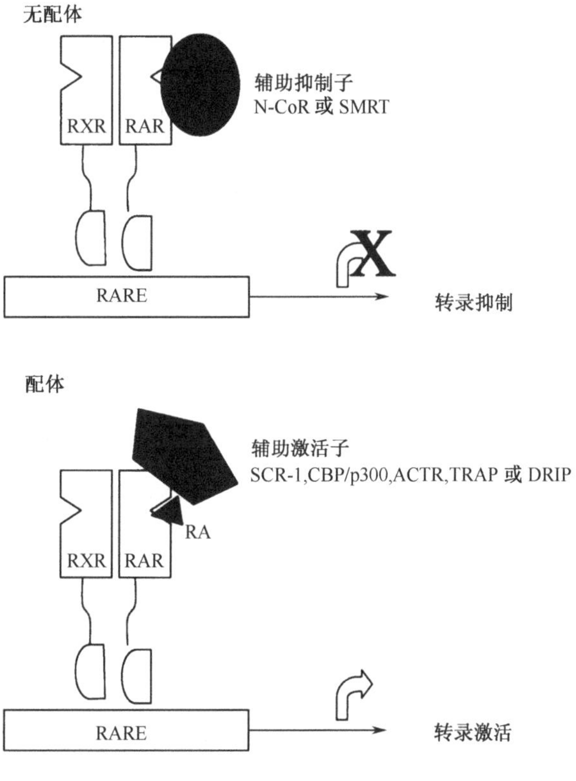


图 10.5　通过 RAR-RXR 异源二聚体的转录抑制和激活机制的模式

缺少 RA 时，RAR-RXR 异源二聚体与抑制基因的转录的辅助抑制子复合物互作。一旦发生 RA 结合，辅助抑制子复合物便分离，并被辅助激活子复合物所取代，引起转录活化

## 辅助抑制子与辅助激活子

与其他核受体的情况一样，配体与受体结合将导致抑制转录的辅助抑制子分子释放，也使转录活性物聚集的辅助激活子得以补充（Tong *et al.*，1995）（图 10.5）。与非配位 RAR 结合的两个辅助抑制子已被识别，即核受体辅助抑制子（nuclear receptor co-repressor，N-CoR）与 RAR 和 TR 的沉默调节子（silencing mediator for RAT and TR，SMRT）（Chen and Evans，1995；Horlein *et al.*，1995；Kurokawa *et al.*，1995）。这两种辅助抑制子仅在没有配体时与受体的 D 域结合。这些辅助抑制子通过补充组蛋白脱乙酰基酶而调节它们的负转录作用（Alland *et al.*，1997；Heinzel *et al.*，1997）。组蛋白脱乙酰基酶诱导染色质结构的变化，致使 DNA 无法接近转录结构。一旦发生配体结合和 LBD 结构的构型变化（以前已讨论），辅助抑制子便被释放，辅助激活子与 RAR-RXR 异源二聚体结合。以配体依赖方式与 RAR 和 RXR 互作的几个辅助激活子已被识别。在这些辅助激活子中，有 SRC-1、CBP/p300、ACTR、TRAP 和 DRIP（Bannister and Kouzarides，1996；Chen *et al.*，1997；Spencer *et al.*，1997）。这些辅助激活子与配位受体的 AF-2 区直接互作。一些辅助激活子具有内在固有的组蛋白乙酰基转移活性，使基本转录元件能较大地结合到 DNA 上。

# 不同 RAR 和 RXR 亚型的功能作用

最有趣的难题之一是为什么会有 6 个性质不同的类维生素 A 受体亚型和每个亚型都有几个的异构体存在。几方面的证据表明，单个 RAR 和 RXR 亚型以及它们各自的异构体至少具有一些特定的不交叉的功能，因此在调节 RA 影响的广谱生物过程中具有独一无二的作用。这些作用包括在胚胎发育和成熟过程中的单个 RAR 和 RXR 亚型的独特表达模式与转基因鼠及 F9 畸形癌细胞中的特定 RAR-RXR 亚型或异构体的灭活作用。

## RAR 和 RXR 的表达模式

从各种成人的组织中分离的 RNA 经 Northern 印迹分析表明，RARα mRNA 的表达是多种多样的，而最高水平是在大脑的特定区域，包括海马和小脑（Giguere *et al.*，1987；de The *et al.*，1989；Krezel *et al.*，1999）。RARα1 mRNA 在脑、皮肤、肌肉、心脏和肾脏中更为丰富，而 RARα2 mRNA 则在肺、肝脏和肠道中略高（Leroy *et al.*，1991）。RARβ 的 mRNA 分布模式表现出可变性。高水平的 RARβ 表达被发现于肾、前列腺、脊髓束、小脑皮质、垂体和肾上腺中；平均水平在肝脏、脾、子宫、卵巢和睾丸中；而低表达水平则在乳腺和眼睛中被检测到（Benbrook *et al.*，1988；de The *et al.*，1989）。利用上游 P1 启动子转录的 RARβ1mRNA 和 RARβ3 mRNA，在脑、皮肤和肺中表达丰富，而由第二下游 P2 启动子产生的 RARβ2 mRNA 和 RARβ mRNA4 则在肾脏、肝脏、肺、皮肤、心脏、肌肉、肠道和大脑中被观测到（Zelent *et al.*，1991；Nagpal *et al.*，1992）。RARγ 转录在其表达中是最有限的，RARγ1 和 RARγ2 两者主要被发现于皮肤和肺中（Noji *et al.*，1989；Zelent *et al.*，1989；Kastner *et al.*，1990）。三个 RXR mRNA 广泛存在于成人组织中，至少有一个 RXR 亚型存在于每个受测组织中（Mangelsdorf *et al.*，1990，1992）。这与 RXR 作为许多核受体超家族成员的混杂二聚化参与者的作用相一致。

在不同发育阶段的鼠胚胎中通过原位杂交测验各种 RAR 和 RXR 亚型的表达模式也表明，每个亚型有其各自独特的可能或不可能与其他亚型交叠的表达模式。RARα 的表达在突变胚胎发育中是普遍存在的，而 RARβ 和 RARγ 约表达则有限（Dolle *et al.*，1990a，b；Ruberte *et al.*，1990，1991，1993；Mendelsohn *et al.*，1994c）。类似地，RARβ 以通常的模式表达，而 RARα 和 RARγ 的表达则更有限（Mangelsdorf *et al.*，1992；Dolle *et al.*，1994）。而且，每个受体亚型的异构体在胚胎发育期间也显示出其各自独特的表达模式。

## 类维生素 A 受体敲除突变小鼠

另一类试验方法是利用基因打靶技术破坏特定目的基因，从而使小鼠的特定 RAR-RXR 亚型或其异构体的功能失活。既然 RA 为胚胎发育所必需，且 RAR 和 RXR 被认为有调节 RA 的作用，那么受体中的一种没有突变的鼠会表现出类似于维生素 A 缺乏产生的异常似乎是有道理的。发现某些特定失常与敲除小鼠中缺乏个别核受体亚型有关。然而，总体上讲，在单个 RAR 和 RXR 之间似乎存在一定程度的被迫功能累赘。例如，纯合的 RARα 突变小鼠表现初生致死和睾丸退化（Lufkin *et al.*，1993）；纯合 RARα 突变小鼠表现出与维生素 A 缺乏有关的症状，包括生长缺陷、早期死亡和雄性不育（Lohnes *et al.*，1993）；纯合 RXRα 突变小鼠被发现死于 GD13.5 与 16.5 之间（Sucov *et al.*，1994）；而纯合的 RXRβ 小鼠除雄性不育外都是正常的（Kastner *et al.*，1996）。在 RARα 缺失小鼠上观察到胚胎死亡，这是由心室发育不全，导致心室壁很薄和心室分隔缺损所致，因此，这将 RARα 的功能与心脏形成相关的重要结果相关联。

不像单个 RAR 突变小鼠，两个 RAR 基因或 RAR 和 RXR 两个基因同时失活将导致鼠发生各种异常，其中许多症状都与以往报道的胎儿维生素 A 缺乏所致症状相似，包括眼睛、四肢、心脏、头部和神经系统的缺陷（Wilson *et al.*，1953）。Kastner 等（1994）报道，当杂合 RARα 突变小鼠与携带 RARα 或 RARγ 基因中的无用等位基因小鼠杂交时会有功能趋同现象。这种功能趋同现象表明，RARα-RAR 异质二聚体在体内对类维生素 A 信号传输有调节作用。有报道说，在子宫内或出生后不久就有更严重畸形或死亡出现是由于小鼠携带 2 个不同 RAR 亚型表达缺失的双纯合 RAR 突变，或携带 1 个 RAR 亚型和 1 个 RXR 亚型的纯合突变所致（Lohnes *et al.*，1994；Mendelsohn *et al.*，1994b；Luo *et al.*，1996；Chiang *et al.*，1998；Dupe *et al.*，1999）。有趣的是，对 RARα/RARγ 突变小鼠和 RARα/RARβ 突变小鼠的后脑模式分析表明，RARα 和（或）RARγ 在调节具体阐明预期菱脑原节 5/菱脑原节 6 领域方面的 RA 依赖过程中发挥重要作用，但 RARβ 在后来设定该领域的尾部界线中也是重要的（Wendling *et al.*，2001）。最后，突变小鼠因缺乏有功能的 RARβ 和 RARγ 基因且仅有一个有功能的 RARα 基因（$RAR\alpha^{+/-}$；$RXR\beta^{-/-}$；$RAR\gamma^{-/-}$）而能存活，且因缺少 RARβ 仅表现出生长缺陷和雄性不育，表明仅一个 RXRα 拷贝就足以发挥 RXR 的大多数功能（Krezel *et al.*，1996）。

有关 RAR 和 RXR 亚型的单个异构体的信息很少。发现通过基因打靶技术而使两个 RARα1 基因、两个 RARβ2 基因或两个 RARγ2 基因功能失活的小鼠仍能存活，且表现正常而没有明显的表型异常（Li *et al.*，1993；Lohnes *et al.*，1993；Mendelsohn *et al.*，1994a）。另一方面，在 RARβ2 和 RARγ2 复合突变小鼠中发现了视网膜发育异常和退化（Grondona *et al.*，1996）。近来，Mascrez 等（2001）通过研究专门删除大多数 RXRα1 A/B 区域的小鼠突变的影响而检测了 RXRα1 的 AF-1 区的作用，其中包括配体独立激活功能 AF-1。这些研究表明，RXRα1 的 AF-1 区在胚胎发育期通过 RAR-RXR 转导类维生素 A 信号中的功能是非常重要的（然而不如配体依赖激活功能 AF-2）。

近来，基因敲除技术已进步到能应用 Cre/lox 系统以位点特异和时空控制的方式将细胞突变有效地导入特定基因中（Metzger and Chambon，2001），这就可以在特定的细胞类型中和小鼠生命期的特定时间对目的基因进行功能灭活。初期的研究就是用这种方法对出生后各个时期的鼠肝细胞（Imai *et al.*，2001b；Wan *et al.*，2000a，b）、脂肪细胞（Imai *et al.*，2001a）和皮肤表皮细胞（Li *et al.*，2000，2001）中的 RXRα 进行功能灭活。总之，鼠可存活，但也已经获得了关于特定组织中 RXRα 功能的重要信息。在肝细胞中，通过 RXRα 异质二聚体化调节的所有途径似乎是相互折中的。RXRα 似乎在与成熟肝细胞和再生肝细胞的生命期相关的机制中发挥重要作用。在皮肤中，RXRα 对出生前的表皮发育并非是必需的，然而却在出生后的皮肤突变尤其在毛囊生长中发挥重要作用。最后，RXRα 似乎在脂肪合成中可能与 PPARγ 一起作为异质二聚体相伴者而发挥中心作用。该方法在阐述胚胎产生和成年鼠不同时间特定组织中每个 RAR 和 RXR 亚型及其各自异构体的特异功能方面很有前景。

## F9 畸胎癌细胞中类维生素 A 受体的失活

一种可选择的方法就是利用基因打靶技术在功能上对 F9 畸胎癌细胞中的特定 RAR 亚型进行灭活，并对这些细胞中的 RA-依赖性基因的表达模式进行检测。F9 畸胎癌细胞是一个已得到很好研究的模型系统，该方法系统对 RA 处理分化为原始内胚层、体壁内胚层或脏腑内胚层高度灵敏且取决于培养条件（Hogan *et al.*，1983）。而且，大量由 RA 诱导的基因组在这些细胞中已被充分阐述。对两个 RARγ 基因或两个 RARα 基因遭到破坏的 F9 细胞的测验已表明，RAR 中的每一个亚型在调节 RA-依赖的分化特异性基因表达方面都显现出特异性（Boylan *et al.*，1993，1995；Taneja *et al.*，1995；Chiba *et al.*，1997）。RARα 的缺失被证明与 RA 诱导的 CRABP-Ⅱ和 Hoxb-1 表达的减少有关，而 RARγ 的缺失则与 RA 诱导的 Hoxa-1 和 Hoxa-3、昆布氨酸 B1、胶原酶Ⅳ、GATA-4、CRBPI、Stra4、Stra6、Stra8、Cdx-1、GAP43 和 BMP-2 表达的丢失相关。但是，几个 RARγ 特异基因的表达能在 RARγ 缺失的细胞中通过再次表达 RARγ、过度表达 RARα 或在较低程度上通过过度表达 RARβ 而得以恢复（Taneja *et al.*，1995）。因此，在这些“敲除”的 F9 细胞中也可以观测到一定量的功能冗余，然而其生理意义尚不清楚。近来，Faria 等（1999）把 F9 细胞中 RARβ2 的两个等位基因予以破坏，表明 RARβ2 在调节对 RA 的后期反应中起重要作用。包括 CRABPⅡ和昆布氨酸 B1 在内的几个基因已表明需要 RARβ2 表达，以在 RA 诱导分化的过程中实现其广泛和最大的表达。

# 小 结

已积累的大量证据支持这样的概念，即至少一些基因的RA-依赖性转录调控是由特定RAR亚型（功能特异性）调节的。然而，在RAR敲除小鼠和F9细胞中已观测到功能冗余。几个报道（Chiba *et al.*，1997；Plassat *et al.*，2000；Rochette-Egly *et al.*，2000）已提供证据表明，F9细胞中的基因敲除会在细胞内产生人工条件，从而使在正常条件下不会出现的启动子依赖性功能冗余表现出来。因此，在基因敲除研究中观测到的功能冗余不能被作为野生型生理条件下缺乏每一RAR亚型功能特异性的证据。为了解决这一问题，应用不改变细胞中内源类维生素A受体亚型水平的不同试验方法进行另外的研究是必需的。

## 参 考 文 献


Alland, L., Muhle, R., Hou, H., Potes, J., Chin, L., Schreiber-Agus, N. and DePinho, R.A. (1997) Role for N-CoR and histone deacetylase in Sin-3 mediated transcriptional repression. *Nature* 387, 49–55.

Bannister, A.J. and Kouzarides, T. (1996) The CBP co-activator is a histone acetyltransferase. *Nature* 384, 641–643.

Benbrook, D., Lernhardt, E. and Pfahl, M. (1988) A new retinoic acid receptor identified from a hepatocellular carcinoma. *Nature* 333, 669–672.

Bourguet, W., Ruff, M., Chambon, P., Gronemeyer, H. and Moras, D. (1995) Crystal structure of the ligand-binding domain of the human nuclear receptor RXR-α. *Nature* 375, 377–382.

Bourguet, W., Vivat, V., Wurtz, J.-M., Chambon, P., Gronemeyer, H. and Moras, D. (2000) Crystal structure of a heterodimeric complex of RAR and RXR ligand-binding domains. *Molecular Cell* 5, 289–298.

Boylan, J.F., Lohnes, D., Taneja, R., Chambon, P. and Gudas, L.J. (1993) Loss of retinoic receptor γ function in F9 cells by gene disruption results in aberrant *Hoxa-1* expression and differentiation upon retinoic acid treatment. *Proceedings of the National Academy of Sciences USA* 90, 9601–9605.

Boylan J.F., Lufkin, T., Achkar, C.C., Taneja, R., Chambon, P. and Gudas, L.J. (1995) Targeted disruption of retinoic acid receptor α (RARα) and RARγ results in receptor-specific alterations in retinoic acid-mediated differentiation and retinoic acid metabolism. *Molecular and Cellular Biology* 15, 843–851.

Brand, N., Petkovich, M., Krust, A., Chambon, P., de The, H., Marchio, A., Tiollais, P. and Dejean, A. (1988) Identification of a second human retinoic acid receptor. *Nature* 332, 850–853.

Brocard J., Kastner, P and Chambon, P. (1994) Two novel RXR alpha isoforms from mouse testis. *Biochemical and Biophysical Research Communications* 229, 211–218.

Chambon, P. (1996) A decade of molecular biology of retinoic acid receptors. *FASEB Journal* 10, 940–954.

Chandraratna, R.A. (1998) Future trends: a new generation of retinoids. *Journal of the American Academy of Dermatology* 39, S149–S152.

Chen, J.D. and Evans, R.M. (1995) A transcriptional co-repressor that interacts with nuclear hormone receptors. *Nature* 377, 454–457.

Chen, H., Lin, R.J., Schiltz, R.L., Chakravarti, D., Nash, A., Nagy, L., Privalsky, M.L., Nakatani, Y. and Evans, R.M. (1997) Nuclear receptor coactivator ACTR is a novel histone acetyltransferase and forms a multimeric activation complex with P/CAF and CBP/p300. *Cell* 90, 569–580.

Chiang, M.-Y., Misner, D., Kempermann, G., Schikorski, T., Giguere, V., Sucov, H.M., Gage, F.H., Stevens, C.F and Evans, R.M. (1998) An essential role for retinoid receptors RARβ and RXRγ in long-term potentiation and depression. *Neuron* 21, 1353–1361.

Chiba, H., Clifford, J., Metzger, D. and Chambon, P. (1997) Distinct retinoid X receptor–retinoic acid receptor heterodimers are differentially involved in the control of expression of retinoid target genes in F9 embryonal carcinoma cells. *Molecular and Cellular Biology* 17, 3013–3020.

de The, H., Marchio, A., Tiollais, P. and Dejean, A. (1989) Differential expression and ligand regulation of retinoic acid receptor α and β genes. *EMBO Journal* 8, 429–433.

de The, H., Vivanco-Ruiz, M.D., Tiollais, P., Stunnenberg, H. and Dejean, A. (1990) Identification of a retinoic acid responsive element in the retinoic acid receptor β gene. *Nature* 343, 177–180.

deUrquiza, A.M., Liu, S., Sjoberg, M., Zetterstrom,

R.H., Griffiths, W., Sjovall, J. and Perlmann, T. (2000) Docosahexanoic acid, a ligand for the retinoid X receptor in mouse brain. *Science* 290, 2140–2144.

Dolle, P., Ruberte, E., Kastner, P., Petkovich, M., Stoner, C.M., Gudas, L. and Chambon, P. (1990a) Differential expression of genes encoding α, β, and γ retinoic acid receptors and CRABP on the developing limbs of the mouse. *Nature* 342, 702–705.

Dolle, P., Ruberte, E., Leroy, P., Morriss-Kay, G. and Chambon, P. (1990b) Retinoic acid receptors and cellular retinoid binding proteins I. A systematic study of their differential pattern of transcription during mouse organogenesis. *Development* 110, 1133–1151.

Dolle, P., Fraulob, V., Kastner, P. and Chambon, P. (1994) Developmental expression of the murine retinoid X receptor (RXR) genes. *Mechanics of Development* 45, 91–104.

Duester, G. (2000) Families of retinoid dehydrogenases regulating vitamin A function. Production of visual pigment and retinoic acid. *FEBS Letters* 267, 4315–4324.

Dupe V., Ghyselinck, N.B., Wendling, O., Chambon, P. and Mark, M. (1999) Key roles of retinoic acid receptors alpha and beta in the patterning of the caudal hindbrain, pharyngeal arches and otocytst in the mouse. *Development* 126, 5051–5119.

Egea, P., Mitschler, A., Rochel, N., Ruff, M., Chambon, P. and Moras, D. (2000) Crystal structure of the human RXRα ligand-binding domain bound to its natural ligand: 9-*cis* retinoic acid. *EMBO Journal* 19, 2592–2601.

Faria, T.N., Mendelsohn, C., Chambon, P. and Gudas, L.J. (1999) The targeted disruption of both allelles of RARβ2 in F9 cells results in the loss of retinoic acid-associated growth arrest. *Journal of Biological Chemistry* 274, 26783–26788.

Folkers, G.E., van Heerde, E.C. and van der Saag, P.T. (1995) Activation function 1 of retinoic acid receptor beta 2 is an acidic activator resembling VP16. *Journal of Biological Chemistry* 270, 23552–23559.

Forman, B.M., Casanova, J., Raaka, B.M., Ghysdael, J. and Samuels, H.H. (1992) Half-site spacing and orientation determines whether thyroid hormone and retinoic acid receptors and relative factor bind to DNA response elements as monomers, homodimers or heterodimers. *Molecular Endocrinology* 6, 429–442.

Gehin, M., Vivat, V., Wurtz, J.M., Losson, R., Chambon, P., Moras, D. and Gronemeyer, H. (1999) Structural basis for engineering of retinoic acid receptor isotype-selective agonists and antagonists. *Chemistry and Biology* 6, 519–529.

Giguere, V., Ong, E.S., Segui, P. and Evans, R.M. (1987) Identification of a receptor for the morphogen retinoic acid. *Nature* 330, 624–629.

Giguere, V., Shago, M., Zirngibl, R., Tate, P., Rossant, J. and Varmuza, S. (1990) Identification of a novel isoform of the retinoic acid receptor expressed in the mouse embryo. *Molecular and Cellular Biology* 10, 2335–2340.

Gottesman, M.E., Quadro, L. and Blaner, W.S. (2001) Studies of vitamin A metabolism in mouse model systems. *BioEssays* 23, 409–419.

Grondona, J.M., Kastner, P., Gansmuller, A., Decimo, D. and Chambon, P. (1996) Retinal dysplasia and degeneration in RARβ2/RARγ2 compound mutant mice. *Development* 122, 2173–2188.

Heinzel, T., Lavinski, R.M., Mullen, T.M., Soderstrom, M., Laherty, C.D., Torchia, J., Yang, W.M., Brard, G., Ngo, S.D., Davie, J.R., Seto, E., Eisenman, R.N., Rose, D.W., Glass, C.K. and Rosenfeld, M.G. (1997) A complex containing N-CoR, mSin3 and histone deacetylase mediates transcriptional repression. *Nature* 387, 43–48.

Heyman, R.A., Mangelsdorf, D.J., Dyck, J.A., Stein, R.B., Eichele, G., Evans, R.M. and Thaller, C. (1992) 9-*Cis* retinoic acid is a high affinity ligand for the retinoid X receptor. *Cell* 68, 397–406.

Hogan, B.L.M., Barlow, D. and Tilly, R. (1983) F9 teratocarcinoma cells as a model for the differentiation of parietal and visceral endoderm in the mouse embryo. *Cancer Surveys* 2, 115–140.

Horlein, A.J., Naar, A.M., Heinzel, T., Torchia, J., Gloss, B., Kurokawa, R., Ryan, A., Kamei, Y., Soderstrom, M. and Glass, C.K. (1995) Ligand-independent repression by thyroid hormone receptor mediated by a nuclear receptor co-repressor. *Nature* 377, 397–404.

Imai, T., Jiang, M., Chambon, P. and Metzger, D. (2001a) Impaired adipogenesis and lipolysis in the mouse upon selective ablation of the retinoid X receptorα mediated by a tomoxifen-inducible chimeric Cre recombinase (Cre-$ER^{T2}$) in adipocytes. *Proceedings of the National Academy of Sciences USA* 98, 224–228.

Imai, T., Jiang, M., Kastner, P., Chambon, P and Metzger, D. (2001b) Selective ablation of retinoid X receptor alpha in hepatocytes impairs their lifespan and regenerative capacity. *Proceedings of the National Academy of Sciences USA* 98, 4581–4586.

Johnson, A. and Chandraratna, R.A. (1999) Novel retinoids with receptor selectivity and functional selectivity. *British Journal of Dermatology* 140, 12–17.

Kastner, P., Krust, A., Mendelsohn, C., Garnier, J.M., Zelent, A., Leroy, P., Staub, A. and Chambon, P. (1990) Murine isoforms of retinoic acid receptor gamma with specific patterns of expression. *Proceedings of the National Academy of Sciences USA* 87, 270–274.

Kastner, P., Grondona, J.M., Mark, M., Gansmuller, A., Le Meur, M., Decimo, D., Vonesch, J.-L., Dolle, P. and Chambon, P. (1994) Genetic analysis of RXR-α development function: convergence of RXR and RAR signaling pathways in heart and eye morphogenesis. *Cell* 78, 987–1003.

Kastner, P., Mark, M., Leid, M., Gansmuller, A., Chin, W., Grondona, J.M., Decimo, D., Krezel, W., Dierich, A and Chambon, P. (1996) Abnormal spermatogenesis in RXR beta mutant mice. *Genes and Development* 10, 80–92.

Krezel, W., Dupe, V., Mark, M., Dierich, A., Kastner, P. and Chambon, P. (1996) RXR gamma null mice are apparently normal and compound RXR alpha +/−/RXR beta −/−/RXR gamma −/− mutant mice are viable. *Proceedings of the National Academy of Sciences USA* 93, 9010–9014.

Krezel, W., Kastner, P. and Chambon, P. (1999) Differential expression of retinoid receptors in the adult mouse central nervous system. *Neuroscience* 89, 1291–1300.

Krust, A., Kastner, P., Petkovich, M., Zelent, A. and Chambon, P. (1989) A third human retinoic acid receptor, hRAR-γ. *Proceedings of the National Academy of Sciences USA* 86, 5310–5314.

Kurokawa, R., Yu, V.C., Naar, A., Kyakumoto, S., Han, Z., Silverman, S., Rosenfeld, M.G. and Glass, C.K. (1993) Differential orientations of the DNA-binding domain and carboxy-terminal dimerization regulate binding site selection by nuclear receptor heterodimers. *Genes and Development* 7, 1423–1435.

Kurokawa, R., Soderstrom, M., Horlein, A., Halachmi, S., Brown, M., Rosenfeld, M.G. and Glass, C.K. (1995) Polarity-specific activities of retinoic acid receptors by a co-repressor. *Nature* 377, 451–454.

Levin, A.A., Sturzenbecher, L.J., Kazmer, S., Bosakowski, T., Huselton, C., Allenby, G., Speck, J., Kratzeisen, C., Rosenberger, M., Lovey, A. and Grippo, J.F. (1992) 9-*Cis* retinoic acid stereoisomer binds and activates the nuclear receptor RXRα. *Nature* 355, 359–361.

Leroy, P., Krust, A., Zelent, A., Mendelsohn, C., Garnier, J.-M., Kastner, P., Dierich, A. and Chambon, P. (1991) Multiple isoforms of the mouse retinoic acid receptor α are generated by alternative splicing and differential induction by retinoic acid. *EMBO Journal* 10, 59–69.

Li, E., Sucov, H.M., Lee, K.-F., Evans, R.M. and Jaenisch, R. (1993) Normal development and growth of mice carrying a targeted disruption of the α1 retinoic acid receptor gene. *Proceedings of the National Academy of Sciences USA* 90, 1590–1594.

Li, M., Indra, A.K., Warot, X., Brocard, J., Messaddeq, N., Kato, S., Metzger, D. and Chambon, P. (2000) Skin abnormalities generated by temporally controlled RXRα mutations in mouse epidermis. *Nature* 407, 633–636.

Li, M., Chiba, H., Warot, X., Messaddeq, N., Gerard, C., Chambon, P. and Metzger, D. (2001) RXR-alpha ablation in skin keratinocytes results in alopecia and epidermal alteration. *Development* 128, 675–688.

Liu, Q. and Linney, E. (1993) The mouse retinoid-X receptor-gamma gene: genomic organization and evidence for functional isoforms. *Molecular Endocrinology* 7, 651–658.

Lohnes, D., Kastner, P., Dierich, A., Mark, M., LeMeur, M. and Chambon, P. (1993) Function of retinoic acid receptor γ in the mouse. *Cell* 73, 643–658.

Lohnes, D., Mark, M., Mendelsohn, C., Dolle, P., Dierich, A., Gorry, P., Gansmuller, A. and Chambon, P. (1994) Function of the retinoic acid receptors (RARs) during development. (I). Craniofacial and skeletal abnormalities in RAR double mutants. *Development* 120, 2723–2748.

Lufkin, T., Lohnes, D., Mark, M., Dierich, A., Gorry, P., Gaub, M.P., LeMeur, M. and Chambon, P. (1993) High postnatal lethality and testis degeneration in retinoic acid receptor alpha mutant mice. *Proceedings of the National Academy of Sciences USA* 90, 7225–7229.

Luo, J., Sucov, H.M., Bader, J.A., Evans, R.M. and Giguere, V. (1996) Compound mutants for retinoic acid receptor (RAR) beta and RAR alpha1 reveal developmental functions for multiple RAR beta isoforms. *Mechanics of Development* 55, 33–44.

Mangelsdorf, D.J., Ong, E.S., Dyck, J.A. and Evans, R.M. (1990) Nuclear receptor that identified a novel retinoic acid response pathway. *Nature* 345, 224–229.

Mangelsdorf, D.J., Borgmeyer, U., Heyman, R.A., Zhou, J.Y., Ong, E.S., Oro, A.E., Kakizuka, A. and Evans, R.M. (1992) Characterization of three RXR genes that mediate the action of 9-*cis* retinoic acid. *Genes and Development* 6, 329–344.

Mascrez, B., Mark, M., Krezel, W., Dupé, V., LeMeur, M., Ghyselinck, N.B. and Chambon, P.A. (2001) Differential contributions of AF-1 and AF-2 activities to the developmental functions of RXRα. *Development* 128, 2049–2062.

Mendelsohn, C., Mark, M., Dolle, P., Dierich, A., Gaub, M.P., Krust, A., Lampron, C. and Chambon, P. (1994a) Retinoic acid receptor beta2 (RAR beta 2) null mutant mice appear normal. *Developmental Biology* 166, 246–258.

Mendelsohn, C., Lohnes, D., Decimo, D., Lufkin, T., LeMeur, M., Chambon, P. and Mark, M. (1994b) Function of the retinoic acid receptors (RARs) during development (II). Multiple abnormalites at various stages of organogenesis in RAR double mutants. *Development* 120, 2749–2771.

Mendelsohn, C., Larkin, S., Mark, M., Lemeur, M., Clifford, J., Zelent, A. and Chambon, P. (1994c) RARβ isoforms: distinct transcriptional control by retinoic acid and specific spatial patterns of promoter activity during mouse embryonic development. *Mechanics of Development* 45, 227–241.

Metzger, D. and Chambon, P. (2001) Site- and tissue-specific gene targeting in the mouse. *Methods* 24, 71–80.

Naar, A.M., Boutin, J.M., Lipkin, S.M., Yu, V.C., Holloway, J.M., Glass, C.K. and Rosenfeld, M.G. (1991) The orientation and spacing of core DNA-binding motifs dictate selective transcriptional responses to three nuclear receptors. *Cell* 65, 1267–1279.

Nagata, T., Kanno, Y., Ozato, K. and Taketo, M. (1994) The mouse RXRβ gene encoding RXR beta: genomic organization and two mRNA isoforms generated by alternative splicing of transcripts initiated by CpG island promoters. *Gene* 142, 183–189.

Nagpal, S. and Chandraratna, R.A. (2000) Recent developments in receptor-selective retinoids. *Current Pharmaceutical Design* 6, 919–931.

Nagpal, S., Zelent, A. and Chambon, P. (1992) RAR-β4, a retinoic acid receptor isoform is generated from RAR-β2 by alternative splicing and the usage of a CUG initiator codon. *Proceedings of the National Academy of Sciences USA* 89, 2718–2722.

Nagpal, S., Friant, S., Nakshatri, H. and Chambon, P. (1993) RARs and RXRs: evidence for two autonomous transactivation functions (AF-1 and AF-2) and heterodimerization *in vivo*. *EMBO Journal* 12, 2349–2360.

Napoli, J.L. (1999a) Retinoic acid: its biosynthesis and metabolism. *Progress in Nucleic Acid Research and Molecular Biology* 63, 139–188.

Napoli, J.L. (1999b) Interactions of retinoid binding proteins and enzymes in retinoid metabolism. *Biochimica et Biophysica Acta* 1440, 139–162.

Noji, S., Yamaii, T., Koyama, E., Nohno, T., Fujimoto, W., Arata, J. and Taniguichi, S. (1989) Expression of retinoic acid receptor genes in keratinizing front skin. *FEBS Letters* 259, 86–90.

Noy, N. (2000) Retinoid-binding proteins: mediators of retinoid action. *Biochemical Journal* 348, 481–495.

Ostrowski, J., Hammer, L., Roalsvig, T., Pokornowski, K. and Reczek, P.R. (1995) The N-terminal portion of the E domain of retinoic acid receptor alpha and beta is essential for the recognition of retinoic acid and various analogs. *Proceedings of the National Academy of Sciences USA* 92, 1812–1816.

Ostrowski, J., Roalsvig, T., Hammer, L., Starret, J., Yu, K.-L. and Reczek, P. (1998) Serine 232 and methionine 272 define the ligand binding pocket in retinoic acid receptor subtypes. *Journal of Biological Chemistry* 273, 3490–3495.

Perlmann, T., Rangarajan, P.N., Umesono, K. and Evans, R.M. (1993) Determinants for selective RAR and TR recognition of direct repeat HREs. *Genes and Development* 7, 1411–1422.

Petkovich, M., Brand, N.J., Krust, A. and Chambon, P. (1987) A human retinoic acid receptor which belongs to the family of nuclear receptors. *Nature* 330, 444–450.

Plassat, J.-L., Penna, L., Chambon, P. and Rochette-Egly, C. (2000) The conserved amphipatic α-helical core motif of RARγ and RARα activating domains is indispensable for RA-induced differentiation of F9 cells. *Journal of Cell Science* 113, 2887–2895.

Predki, P.F., Zamble, D., Sarkar, B. and Giguere, V. (1994) Ordered binding of retinoic acid and retinoid X receptors to asymmetric response elements involves determinants adjacent to the DNA-binding domain. *Molecular Endocrinology* 8, 31–39.

Renaud, J.P., Rochel, N., Ruff, M., Vivat, V., Chambon, P., Gronemeyer, H. and Moras, D. (1995) Crystal structure of the RAR-γ ligand-binding domain bound to all-*trans*-retinoic acid. *Nature* 378, 681–689.

Rochette-Egly, C., Plassat, J.-L., Taneja, R. and Chambon, P. (2000) The AF-1 and AF-2 activating domains of retinoic acid receptor-α (RARα) and their phosphorylation are differentially involved in parietal endodermal differentiation of F9 cells and retinoid-induced expression of target genes. *Molecular Endocrinology* 14, 1398–1410.

Ruberte, E., Dolle, P., Krust, A., Zelent, A., Morriss-Kay, G. and Chambon, P. (1990) Specific spatial and temporal distribution of retinoic acid receptor gamma transcripts during mouse embryogenesis. *Development* 108, 213–222.

Ruberte, E., Dolle, P., Chambon, P. and Morriss-Kay, G. (1991) Retinoic acid receptors and cellular retinoic acid binding proteins. II. Their differential pattern of transcription during early morphogenesis in mouse embryos. *Development* 111, 45–60.

Ruberte, E., Friederich, V., Chambon, P. and Morriss-Kay, G. (1993) Retinoic acid receptors and cellular retinoid binding proteins. III. Their differential transcript distribution during mouse nervous system development. *Development* 118, 267–282.

Scafonas, A., Wolfgang, C.L., Gabriel, J.L., Soprano, K.J. and Soprano, D.R. (1997) Differential role of homologous positively charged residues for ligand binding in retinoic acid receptor α compared with retinoic acid receptor β. *Journal of Biological Chemistry* 272, 11244–11249.

Spencer, T.E., Jenster, G., Burcin, M.M., Allis, C.D., Zhou, J., Mizzen, C.A., McKenna, N.J., Onate, S.A., Tsai, S.Y., Tsai, M.J. and O'Malley, B.W. (1997) Steroid receptor coactivator-1 is a histone acetyltransferase. *Nature* 389, 194–198.

Sporn, M.B., Roberts, A.B. and Goodman, D.S. (1994) *The Retinoids: Biology, Chemistry and Medicine*. Raven Press, New York.

Stunnenberg, H.G. (1993) Mechanism of transactivation by retinoic acid receptors. *BioEssays* 15, 309–315.

Sucov, H.M., Murakami, K.K. and Evans, R.M. (1990) Characterization of an autoregulated response element in the mouse retinoic acid receptor type b gene. *Proceedings of the National Academy of Sciences USA* 87, 5392–5396.

Sucov, H.M., Dyson, E., Gumeringer, C.L., Price, J., Chien, K.R. and Evans, R.M. (1994) RXR-α mutant mice establish a genetic basis for vitamin A signaling in heart morphogenesis. *Genes and Development* 8, 1007–1018.

Tairis, N., Gabriel, J.G., Gyda, M., Soprano, K.J. and Soprano, D.R. (1994) Arg269 and Lys220 of retinoic acid receptor-β are important for the binding of retinoic acid. *Journal of Biological Chemistry* 269, 19516–19522.

Tairis, N., Gabriel, J.L., Soprano, K.J. and Soprano, D.R. (1995) Alteration in the retinoid specificity of retinoic acid receptor-β by site-directed mutagenesis of Arg269 and Lys220. *Journal of Biological Chemistry* 270, 18380–18387.

Taneja, R., Bouillet, P., Boylan, J.F., Gaub, M.-P., Roy, B., Gudas, L.J. and Chambon, P. (1995) Reexpression of retinoic acid receptor (RAR) γ or overexpression of RARα or RARβ in RARγ-null F9 cells reveals a partial functional redundancy between the three RAR types. *Proceedings of the National Academy of Sciences USA* 92, 7854–7858.

Tong, G.X., Tanen, M.R. and Bagchi, M.K. (1995) Ligand modulates the interaction of thyroid hormone receptor beta with the basal transcriptional machinery. *Journal of Biological Chemistry* 270, 10601–10611.

Umesono, K. and Evans, R.M. (1989) Determinants of target gene specificity for steroid/thyroid hormone receptors. *Cell* 57, 1139–1146.

Umesono, K., Giguere, V., Glass, C.K., Rosenfeld, M.G. and Evans, R.M. (1988) Retinoic acid and thyroid hormone induce gene expression through a common responsive element. *Nature* 336, 262–265.

Umesono, K., Morakami, K.K., Thompson, C.C. and Evans, R.M. (1991) Direct repeats as selective response elements for the thyroid hormone, retinoic acid and vitamin D3 receptors. *Cell* 65, 1255–1266.

Wan, Y.J., Cai, Y., Lungo, W., Fu, P., Locker, J., French, S. and Sucov, H.M. (2000a) Peroxisome proliferator-activated receptor α-mediated pathways are altered in hepatocyte-specific retinoid X receptor α-deficient mice. *Journal of Biological Chemistry* 275, 28285–28290.

Wan, Y.J., An, D., Cai, Y., Repa, J.J., Hung-Po Chen, T., Flores, M., Postic, C., Magnuson, M.A., Chen, J., Chien, K.R., French, S., Mangelsdorf, D.T. and Sucov, H.M. (2000b) Hepatocyte-specific mutation establishes retinoid X receptor alpha as a heterodimeric integrator of multiple physiological processes in the liver. *Molecular and Cellular Biology* 20, 4436–4444.

Wendling, O., Ghyselinck, N.B., Chambon, P. and Mark, M. (2001) Roles of retinoic acid receptors in early embryonic morphogenesis and hindbrain patterning. *Development* 128, 2031–2038.

Wilson, J.G., Roth, C.B. and Warkany, J. (1953) An analysis of the syndrome of malformations induced by maternal vitamin A deficiency. Effects of restoration of vitamin A at various times during gestation. *American Journal of Anatomy* 92, 189–217.

Wolbach, S.B. and Howe, P.R. (1925) Tissue changes following deprivation of fat soluble A vitamin. *Journal of Experimental Medicine* 6, 753–757.

Wolfgang, C.L., Zhang, Z.-P., Gabriel, J.L., Pieringer, R.A., Soprano, K.J. and Soprano, D.R. (1997) Identification of sulfhydryl modified cysteine residues in the ligand binding pocket of retinoic acid receptor β. *Journal of Biological Chemistry* 272, 746–753.

Wurtz, J.-M., Bourquet, W., Renaud, J.-P., Vivat, V., Chambon, P., Moras, D. and Gronemeyer, H. (1996) A canonical structure for the ligand-binding domain of nuclear receptors. *Nature Structural Biology* 3, 87–94.

Zelent, A., Krust, A., Petkovich, M., Kastner, P. and Chambon, P. (1989) Cloning of a murine alpha and beta retinoic acid receptor and a novel receptor gamma predominantly expressed in the skin. *Nature* 339, 714–717.

Zelent, A., Mendelsohn, C., Kastner, P., Krust, A., Garnier, J.-M., Ruffenach, F., Leroy, P. and Chambon, P. (1991) Differentially expressed isoforms of the mouse retinoic acid receptor β are generated by usage of two promoters and alternative splicing. *EMBO Journal* 10, 71–81.

Zhang, Z.-P., Gambone, C.J., Gabriel, J.L., Wolfgang, C.L., Soprano, K.J. and Soprano, D.R. (1998) $Arg^{278}$, but not $Lys^{229}$ or $Lys^{236}$, plays an important role in the binding of retinoic acid by retinoic acid receptor γ. *Journal of Biological Chemistry* 273, 4016–4021.

Zhang, Z.-P., Shukri, M., Gambone, C.J., Gabriel, J.L., Soprano, K.J. and Soprano, D.R. (2000) Role of $Ser^{289}$ in RARγ and its homologous amino acid residue in RARα and RARβ in the binding of retinoic acid. *Archives of Biochemistry and Biophysics* 380, 339–346.

Zhang, Z.-P., Hutcheson, J.M., Poynton, H.C., Gabriel, J.L., Soprano, K.J. and Soprano, D.K. (2003) Arginine and retinoic acid receptor B which coordinates with the carboxyl group of retinoic acid functions independent of the amino acid residues responsible for retinoic acid receptor subtype ligand specificity. *Archives of Biochemistry and Biophysics* 409, 375–384

# 11　生物素、维生素 $B_6$ 和维生素 C 对基因表达的调节

Krishnamurti Dakshinamurti

（生物化学与医学遗传系，医学院，马尼托巴大学与神经病毒学及神经退化病研究室，圣波尼菲斯医院研究中心，温尼伯湖，加拿大）

## 引　　言

真核细胞中仅有小比例的基因组在任何时间都是转录活跃的。通常是位于受影响结构基因上游的反应元件调节单个基因转录的速率。细胞分化期间基因表达的调节以及通过细胞因子、生长因子和激素对基因表达的影响已被充分证明。多细胞生物中的大多数专门化细胞在对细胞外信号反应时能改变其基因表达的模式。不同细胞类型以不同方式对同一信号进行反应。基因表达能在从 DNA 到 RNA 到蛋白质的任何步骤上进行调节，通过基因调节蛋白对单个基因的转录进行打开或关闭。一些基因在恒定水平转录，并通过转录后调节过程实现打开或关闭。这些过程需要通过调节蛋白或调节 RNA 分子来识别特定的结构或序列。

在高等生物中，调节基因表达的细胞外信号包括影响激素、神经调节或免疫反应系统的因素。生物的营养状态也是其中之一，因为营养状态与不同的调节系统有密切关系。除了这些间接途径外，某些食物成分通过与基因组中的调节元件直接互作而影响基因表达，导致特定基因转录速率的改变。应对养分供给变化挑战的适应性机制已产生，因此，养分供给的变化能诱导或抑制特定基因或基因组的转录。在许多情况下，转录调节（上调或下调）是通过改变转录因子的有效性或通过改变因子与特定调节序列结合的能力来实现的。维生素的生理影响随其进入细胞、代谢转化及与特定脱辅基酶蛋白的关系引起代谢途径的加强。在这方面，维生素是代谢的启动子。在一些情况下，特定酶（脱辅基酶蛋白）数量的变化是对辅酶（辅因子）的供给产生的转录或翻译应答反应的结果。这样的反应需要一种能感觉辅因子水平变化并能将这种变化的信号转换成基因表达水平应答的机制。维生素功能除了能作为酶的辅酶或者辅基外，还有调节细胞蛋白合成的作用（Dakshinamurti，1994，1997）。本文主要集中讨论水溶性维生素如生物素、维生素 $B_6$ 和抗坏血酸对基因表达的调节作用，同时也将讨论细胞增生对这些维生素的营养需求。

## 生　物　素

在高等生物中，生物素最为人知的功能是作为 4 个含生物素羧化酶的辅基。其在乙

酰辅酶 A 羧化酶、丙酰辅酶 A 羧化酶、3-甲基丁烯酸-辅酶 A 羧化酶和丙酮酸羧化酶中的作用可解释它参与碳水化合物、脂类和一些氨基酸脱氨基代谢中的必需性。许多年前，我们曾假设，生物素除了作为 4 个含生物素羧化酶的辅基外，在细胞功能上也有作用（Daskshinamurti *et al.*，1985；Daskshinamurti and Chauhan，1989）。这个假设是建立在我们已证明细胞培养基中需要生物素的基础上的。这个观点多年来已得到我们和其他研究者研究工作的坚定支持（Zempleni and Mock，2001）。单个生长/分化因子或其受体的合成与分泌可能是由生物素启动的，且生物素的其他作用可能也源于此。根据动物缺乏生物素的不同症状，生物素的作用似乎不只有这些。生物素除了在细胞增生和分化中的作用外，其在胚胎发育、睾丸发育、免疫防御机能和特定蛋白的调节作用也已被发现（Dakshinamurti and Chauhan，1994）。生物素调节关键糖酵解和糖异生激酶。它诱导蛋黄中生物素结合蛋白的合成（White and Whitehead，1987）。促分裂原诱导的增殖通过增加外周血单核细胞（peripheral blood mononuclear cell，PBMC）膜上特定生物素转运蛋白的数量来增加 PBMC 对生物素的摄取（Zempleni and Mock，1999）。据报道，生物素在胰腺对胰岛素的释放（Sone *et al.*，1999）、一些生物素羧化酶和生物素羧化全酶合成酶的合成（Rodriguez-Melendez *et al.*，2001）以及 HepG2 细胞中的脱唾液酸糖蛋白受体的表达（Collins *et al.*，1988）中起作用。多种作用机制，如生物素通过 cGMP 介导的鸟苷酸激酶参与转录起始、基因表述的调控并通过组蛋白生物素化调控 DNA 复制都已被阐明。

## 核生物素结合蛋白

在大鼠和鸡的不同组织中，生物素的细胞内分馏表明有相当数量的生物素与细胞核部分相联系（Dakshinamurti and Mistry，1963a，b）。缺乏生物素的大鼠肝脏中生物素的含量大概是正常鼠肝含量的 1/10，且其中多达 20%存在于细胞核。生物素存在于由培养细胞的不同组织制备的细胞核中，即使这种细胞核不拥有任何固定 $CO_2$ 的能力，这表明在细胞核中的生物素不充当羧化酶的辅基。在生物素缺乏期间，细胞核部分的生物素似乎被保存下来，而其他细胞器的生物素则优先损失（Boeckx and Dakshinamurti，1974，1975）。核内生物素已被证明以非共价键形式结合到蛋白质上。Dakshinamurti 等（1985）从鼠肝脏细胞核中分离到一种生物素结合蛋白。这种蛋白质在体外能可逆地结合到生物素上，最大结合为 3.54pmol/μg 蛋白质。生物素的分离常数为 $2.2\times10^{-7}$mol/L。在十二烷基硫酸钠存在时，聚丙烯酰胺凝胶电泳表明，这个蛋白质的一个明显亚基的分子质量为 60kDa。Meisler 和 Thannasi（1990）报道了鼠肝细胞瘤来源的 HTC 细胞中的一种相似的磷酸吡哆醛结合蛋白。

已有关于光清晰细胞核内（optically clear nuclei，OCN）存在着高浓度生物素的其他研究报道。已有报道证明，结肠腺瘤和癌、子宫内膜组织以及类似胎儿肺脏的肺内胚层肿瘤中有类似于桑椹的固体细胞巢（Nakatani *et al.*，1994；Yang *et al.*，1995）。这几个器官中的桑椹细胞核被称作 OCN，并已发现含有显著量的生物素（Nakatani *et al.*，1994；Okamoto *et al.*，1995；Yang *et al.*，1995）。含有相当数量生物素与桑椹发育无关的 OCN 出现在怀孕期间的子宫内膜及卵巢内膜肿瘤中（Yokoyama *et al.*，1993；Sickel and Sant'Agnese，1994）。生物素化多肽或积累的生物素结合蛋白的类型导致积累的机制以及其临床病理学意义尚需进一步研究（Sasaki *et al.*，1999）。

## 培养细胞对生物素的需要

根据生物素是碳水化合物、脂类和某些氨基酸脱氨残基代谢所必需的预测培养细胞需要生物素。可是早期研究表明，培养细胞并不需要生物素。Keranan（1972）报道，生长在生物素缺乏培养基中的 HeLa 细胞比生长在补充生物素的培养基中的 HeLa 细胞含有更多的生物素，这可能是由于转化的细胞具有合成生物素的能力。应用生物素耗竭的胎牛血清（fetal bovine serum，FBS）和 Eagle 极限必需培养基，根据细胞活力、生物素含量及生物素依赖或不依赖酶的活性（Dakshinamurti and Chalifour，1981；Chalifour and Dakshinamurti，1982a，b），我们发现 HeLa 细胞、人成纤维细胞和被劳斯肉瘤变形病毒转化（rous sarcoma virus-transformed，RST）的幼仓鼠肾（baby hamster kidney，BHK）细胞需要生物素。我们也证明了与生长在补充生物素的培养基中的细胞相比，亮氨酸掺入生物素缺乏的 HeLa 细胞的匀浆或胞液蛋白中的量会显著下降。当把生物素添加到生物素缺乏的培养基中时，亮氨酸掺入蛋白质的量可增加两倍。应用嘌呤霉素和蛹虫草菌素的试验表明，当缺乏生物素的细胞补充生物素时，细胞质中会出现新的 RNA。当哺乳动物体外培养细胞不接受特定信号包括特定生长因子时，就停顿在 $G_1$ 期的静止不分裂状态即 $G_0$ 期中。此时，细胞周期控制系统的成分的合成被关闭。

有些正常细胞由于丝氨酸缺乏而被滞留在 $G_1$ 期中，一旦向培养基中补充丝氨酸，因丝氨酸缺乏而处于 $G_1$ 滞留期的正常细胞便开始将[$^3$H] 胸腺嘧啶结合到 DNA 中。但在相似条件下，缺乏生物素的 HeLa 细胞却没有将[$^3$H] 胸腺嘧啶结合到 DNA 中，即使也给培养基恢复补充生物素。但是在给生物素缺乏的培养物中补充生物素的 4h 之内，[$^3$H] 胸腺嘧啶与 DNA 的结合达到最大。此时存在刺激蛋白质合成的作用。这两个现象是相关的，且生物素的促生长作用可能通过某些蛋白质的合成来实现（Bhullar and Dakshinamurti，1995）。Moscowitz 和 Cheng（1985）认为，生长在含有生物素的培养基中的 RST BHK 细胞产生了一种刺激细胞增殖的非透析因子。

试验表明，当 Madin-Darby 犬肾细胞在用 Dulbecco 改进的 Eagle 培养基（dulbecco's modified eagle's medium，DMEM）与 F-12 培养基的混合物（1∶1）培养时，在补充转铁蛋白的 DMEM 中的生长需要高密度脂蛋白（ligh density lipoprotein，HDL）（Gospodarowicz and Cohen，1985）。在没有 HDL 时支持细胞生长的 F-12 培养基的成分包含生物素（DMEM 中缺乏）和胆碱（DMEM 中浓度不足）。重要的是，在血浆中有相当数量的生物素是与 HDL 部分结合的，这种生物素不能透析并以非共价键形式与蛋白质相结合。

## 生物素在细胞分化中的作用

各种转化细胞系产生能使其在补充转铁蛋白和胰岛素的无血清培养基中生长的转化生长因子。由 3T3 鼠纤维原细胞系来源的 L1 亚系具有在休眠状态分化成有脂肪细胞特征的细胞类型的能力。当它们达到融合并开始分化时，能大大增加其甘油三酯合成的速度。脂肪生成速度的增加与脂肪酸生物合成途径中关键酶活性的协同增加相一致。这种增加与分化期间这些 mRNA 的核决定性转录速度的明显升高相关联（Bernlohr *et al.*，

1985)。分化过程可以许多方式加速，如在培养物中增加血清用量或添加胰岛素或生物素。当培养物中的血清被大量透析时，在没有脂肪沉积的情况下，脂肪细胞分化的形态和酶变化特征被诱导（Kuri-Harcuch *et al.*，1978）。人们推测，透析从血清中除去了生物素，且生物素的作用与乙酰辅酶A羧化酶的增加有关。然而，将生物素添加到培养基中后需要24～28h才能沉积甘油三酯。这种延迟表明，在生物素的影响下可能形成诱导整套脂肪生成酶（除乙酰辅酶A羧化酶以外的非生物素酶）所需的一些因子。为了表达这种分化，培养基中融合的完成伴随着脂肪生成因子的存在（Kuri-Harcuch and Marsch-Morino，1983）。该因子存在于除家猫以外的大多数动物的血清中。猫血清中生物素含量不到牛血清中生物素含量的1/10，这一点具有重要意义。

## 生物素与胚胎发育

Watanabe（1990）报道，饲喂生物素缺乏饲粮的家禽的胚胎先天畸形。母体生物素缺乏在鼠上极易产生胚胎或胎儿畸形，即使母鼠没有表现出任何典型的生物素缺乏症状。在妊娠中期，生物素缺乏的胚胎比正常胚胎的重量轻，并有外观畸形如小颌畸形和短肢畸形。在妊娠的15.5d时，颚突起的大小明显下降，这可能是由于间质增生的改变所致。肢芽间质的增生情况可能与之相同。

Watanabe等（1995）对培养中颚突起的发育进行了研究。在器官培养72h后，来自正常鼠胚胎（生物素充足）的90%以上外植体处于第六发育阶段，而来源于生物素缺乏胚胎（在缺乏生物素的培养基中培养）的外植体中的相应数字为6.5%。如果将生物素缺乏外植体在含有生物素（$10^{-8}$mol/L）的培养基中培养，那么处于第六发育阶段的百分数则上升到30%以上。在胚胎去除前24h给缺乏生物素的母鼠口服生物素(20mg)，当在生物素缺乏培养基中培养时导致33%外植体处于第六发育阶段，当在含有生物素（$10^{-7}$mol/L）的培养基中培养时使处于第六阶段的外植体数量大于50%。

就生物素羧化酶在主要代谢途径的作用而言，生物素缺乏造成的畸形可能是由于中间代谢物或次生代谢物的积累所致。然而，当这些复合物以$10^{-4}$mol/L的浓度添加到器官培养基中时，任何有机酸中间代谢物或次生代谢物对外植体的颚闭合都没有产生有害影响。这些结果强调了间质增生过程中对生物素的连续需要，这可能是为了器官形成过程中生长因子的合成。

## 特定功能的生物素需要

在早期的文献中已报道了鼠因生物素缺乏导致精子发生延迟和精子数量降低的证据。哺乳动物精子的产生主要依赖于睾酮，而睾酮由间质细胞产生，作用于输精管的足细胞及生精小管细胞驱动精子的产生。Paulose等（1989）研究表明，在生物素缺乏的鼠中，睾丸和血清中的睾酮水平降低。在这些大鼠中生物素缺乏引起生精小管上皮脱落程度明显。用促性腺激素或生物素处理生物素缺乏大鼠，可增加其血清中的睾酮水平。然而，即使通过植入睾酮使生物素缺乏大鼠中的睾酮保持在高水平上，但血清中睾酮的增加仍不能导致正常的精子的产生。给生物素缺乏的大鼠单独服用生物素或联合服用生物素和睾酮，可导致正常精子产生，表明生物素可能参与了局部睾丸因子的形成，而这些因子除为睾酮和促卵泡激素所必需外还为间质细胞、足细胞与生精小管细胞间的正常互作所必需。尽管我们已表明睾丸蛋白的合成需要生物素，但还不清楚这些因子的特征。

脑来源的神经营养因子（brain derived neurotrophic factor，BDNF）是细胞信号分子家族（神经营养蛋白）的一个成员，在神经元的发育和可塑性上具有重要作用。BDNF影响控制鸣禽声音学习行为的端脑区域内的发育和成年可塑性。BDNF 的表达似乎与鸣禽声音学习的特定阶段相关。Johnson 等（2000）已在幼年雄性动物的特定端脑核（RA 和 HVC）中识别发现了高水平的生物素。在幼年雄性动物的 HVA 和 RA 核内可能存在着一个生物素化蛋白的特定上调作用。它的发育表达与声音的学习相关。Wang 和 Pevsner（1999）也已报道了对学习和记忆很重要的脑区-海马中有高水平的生物素，强调了生物素调节机制可能在神经可塑性上发挥的重要作用。

## 生物素与生物素相关蛋白质的合成

### 蛋黄中的生物素结合蛋白

蛋黄中的生物素与两个分开的蛋白质 BBPⅠ和 BBPⅡ结合，而这两个蛋白质与抗生物素蛋白不同，其浓度直接与食物中的生物素含量有关（White and Whitehead, 1987）。在食物生物素浓度较低时，BBPⅠ是主要的转运蛋白，而当食物中生物素浓度较高时，BBPⅡ则占优势。这两种蛋白质都受性激素的诱导，而食物生物素对它们的调控似乎要超过激素的调节。

### 生物素羧化酶

Rodriguez-Melandez 等（1999）研究了生物素对鼠肝脏线粒体羧化酶基因表达的影响。生物素缺乏鼠的肝中丙酰 CoA 羧化酶（propionyl-CoA carboxylase，PCC）活性比对照组低 20%。当将生物素添加到生物素缺乏鼠肝细胞的原代培养物中后，至少需要 24h PCC 才能达到对照组活性和生物素化水平，而丙酮酸羧化酶（pyruvate carboxylase，PC）在生物素添加培养基 1h 后就有活性并被完全生物素化。生物素缺乏对 PC 酶总量的影响最小，而 PCC 的 α 亚单位和 3-甲基丁烯酸-辅酶 A 羧化酶（3-methylcrotonyl-CoA carboxylase，MCC）的总量在缺乏条件下显著下降。在生物素缺乏的肝脏中，这两种蛋白质的 mRNA 水平没有显著变化，表明生物素在转录后水平调节 PCC 和 MCC 的表达。在生物素缺乏时，PC 羧化酶蛋白的量不受影响。

### 生物素羧化全酶合成酶

生物素作为生物素羧化酶的辅基，通过羧化全酶合成酶（holocarboxylase synthetase，HCS）共价结合到羧化酶蛋白上。HCS mRNA 在生物素缺乏鼠中显著降低，且在给生物素缺乏鼠服用生物素后增加，并在其服用生物素后 24h 达到对照组水平（Rodriguez-Melendez *et al.*，2001）。这些结果表明了生物素在转录水平上调节的可能性。然而，在 HCS mRNA 水平恢复中，长时间的延迟可能表明存在一个需要代谢处理的间接作用，而该代谢处理影响蛋白质合成、内分泌或其他信号途径。这些报道为生物素作为参与其转运或作为辅基的蛋白质的遗传表达的调节子提供了证据。

## 生物素与免疫系统

生物素羧化全酶合成酶或生物素酶（一个将生物素从生物素蛋白上分离的酶）活性的缺失导致婴儿多种羧化酶缺乏综合征的出现。这是一个代谢性遗传疾病，且与包括对服用生物素有反应的免疫功能异常的临床表现有关（Roth *et al.*，1982）。生物素缺乏鼠对白喉毒素有弱的抗体反应。经绵羊红细胞免疫的生物素缺乏鼠，其脾中抗体形成细胞数量下降，且胸腺的大小和细胞结构也减小。Baez-Saldana 等（1998）在研究生物素缺乏对鼠免疫功能指标的影响时发现，脾细胞的绝对数量以及携带不同表型标记的脾细胞比例都有显著变化，表达表面免疫球蛋白（$sIg^{+}$）的细胞（B 细胞）的百分数显著下降。来自缺乏鼠脾细胞的有丝分裂诱导原增生比对照组鼠细胞的低。T 细胞对有丝分裂原刺激的增生反应的抑制和脾中 B 细胞数量的下降影响了免疫系统对抗原刺激的充分的反应能力。

淋巴细胞是免疫系统的部分。它们对抗原刺激的反应是引起增生。Zempleni 和 Mock（1999，2000，2001）研究了生物素在人淋巴细胞增生中的作用。与非增生细胞的生物素摄入相比较，增生淋巴细胞生物素的摄入量增加了 3～7 倍（见第 5 章）。

## 组蛋白的生物素化作用

生物素蛋白的蛋白水解产生生物素肽，其中最小的生物素肽为生物胞素（ε-N-生物素酰-L-赖氨酸）。生物胞素和合成生物素肽可被生物素酶水解，来自人血清的这种酶已被纯化至同质程度（Chauhan and Dakshinamurti，1986）。尽管生物素酶作为水解酶从小生物素肽回收生物素的作用已普遍被接受，但我们的研究表明，这可能不是生物素酶的主要功能，因为生物素酶水解生物胞素的最适 pH 为 4.5～5，而血清的 pH 是 7.4。生物胞素在血清中的浓度是在纳摩尔范围，而生物素酶对生物胞素的亲和常数在微摩尔范围。Chauhan 和 Dakshinamurti（1998）提出，生物素酶是作为生物素携带蛋白质而发挥作用的。Hymes 等（1995）随后的工作表明，生物素酶本身被生物胞素生物素化。他们进一步研究表明，组氨酸是从生物素化的生物胞素酶转换而来的生物素的特异受体，这增加了组氨酸作为生物素化转移的内源底物的可能性（Hymes and Wolf，1999）。Stanley 等（2001）进一步研究表明，体内组蛋白受乙酰化、甲基化、磷酸化、遍在蛋白化和聚合 ADP-核糖基化作用在体内人类细胞的组蛋白发生生物素化，并且在细胞增殖时生物素化程度增加修饰（见第 17 章）（Wolffe，1998）。组蛋白的生物素化和组蛋白的其他修饰一样，组蛋白的生物素化能引起 DNA 的转录增加（Sommerville *et al.*，1993；Pham and Sauer，2000）。

## 生物素对特异蛋白质合成的调节

### 脱唾液酸糖蛋白和胰岛素的受体

脱唾液酸糖蛋白受体（asialoglycoprotein receptor，ASGR）是完全分化的肝细胞所特有的。人的肝胚细胞瘤系 HepG2 的最大受体活性仅出现在融合培养中。在 10% FBS 的最少必需培养基（minimal essential medium，MEM）中生长到汇合的 HepG2

细胞表明 ASGR 具有配体特征，且其分子质量与从人肝脏中纯化的受体的分子质量相似。然而，在蛋白质合成与总细胞蛋白质含量与对照细胞相似的条件下，通过透析或超滤作用将 FBS 的低分子质量部分除去后，显著减少了 ASGR 的表达。在对数生长期将生物素或生物胞素添加到补充了透析过的 FBS 的培养基中，通过融合的 HepG2 细胞支持了 ASGR 的正常表达（Collins *et al.*，1988）。生物素在 $10^{-8}$mol/L 浓度时最为有效。从 HepG2 细胞中分离 mRNA 没有显示出 ASGR 转录的差异，而此时细胞生长在添加 20%FBS 的 MEM 中或补充 10%透析的 FBS 的 MEM 中，说明依赖于生物素的翻译后过程负责 HepG2 细胞 ASGR 的最终表达。

De la Vega 和 Stockert（2000）已报道，在补充透析的 FBS 的 MEM 中生长到汇合的人肝胚细胞瘤系 HepG2 或 HuH-7 中，$^{125}$I标记的胰岛素的结合下降了 75%以上。细胞表面胰岛素结合的损失是由胰岛素受体（insulin receptor，IR）的减少所致。与 ASGR 的结果相似，在添加透析过的 FBS 的培养基中补足生物素后可完全恢复 IR 的表达。同样，对 ASGR 来说，无论细胞生长在添补 FBS、透析的 FBS 的 MEM 中还是生长在添加透析的 FBS 加生物素的 MEM 中，IR mRNA 的丰度都没有差异。这些结果表明，由生物素介导的转录后过程是 IR 表达下降的原因。

第二信使 8-溴-cGMP 以非加性方式模仿了生物素的作用，表明生物素的作用是通过鸟苷酸环化酶激活所引起的 cGMP 水平的变化介导的（Vesely，1982；Singh and Dakshinamurti，1988）。这个结论被其他的研究发现所证实，即将心房促尿钠排泄因子或硝普钠（颗粒及可溶的鸟苷酸环化酶激活子）分别添加到培养基中，这引起了 ASGR 合成的恢复（Stockert and Ren，1997）。调节 ASGR 和 IR 表达的生物素介导的信号传递途径中共同的下游元件被认为是一个 cGMP 依赖的蛋白激酶（cGk）。De la Vega 和 Stockert（2000）假定，对 cGk 的 cGMP 诱导反应的外被体蛋白 α-COP 的磷酸化作用可阻止外被体复合物与 ASGR mRNA 5′非翻译区的高亲和度结合，并允许核糖体对翻译起始位点进行扫描。cGk 调节 IR 表达的分子水平目前还不清楚。

### 葡糖激酶的诱导

Dakshinamurti 和 Cheah-Tan（1968a，b）研究表明，肝葡糖激酶的活性随鼠的生物素状况而发生改变。生物素也在青年鼠的葡糖激酶早熟的发育中发挥作用（Dakshinamurti and Hong，1969）。葡糖激酶的合成是在发育、营养和激素的调控下完成的（Meglasson and Matschinsky，1984）。生物素对葡糖激酶 mRNA 的体内调节研究是在葡糖激酶活性低的饥饿鼠上进行的（Chauhan and Dakshinamurti，1991）。在给饥饿鼠注射生物素 1h 后，与正常饲喂鼠相比，生物素注射饥饿鼠的葡糖激酶 mRNA 浓度增加了约 4 倍，比没注射生物素的饥饿鼠高 19 倍。在给患糖尿病鼠进行胰岛素治疗后，葡糖激酶的活性和数量都增加，在 4h 中达到非糖尿病对照水平的 165%。尽管胰岛素和生物素的作用一致，但生物素对葡糖激酶的诱导作用似乎比胰岛素的作用更快。在给饥饿鼠服用生物素后的不同时间，对葡糖激酶基因转录的相对速率通过连续转录试验进行了跟踪。用作内部控制（内标）的肌动蛋白基因的转录不受生物素的影响。服用生物素将葡糖激酶基因转录增加了约 6.7 倍。

### 磷酸烯醇式丙酮酸羧激酶（PEPCK）的抑制

在肝脏和肾脏中，PEPCK参与葡萄糖生成，且通常被认为是限速酶。酶合成受食物和多种激素的调控（Granner and Pilkis，1990）。在禁食鼠和糖尿病鼠中，肝PEPCK的活性都明显增加。给饥饿鼠服用生物素使PEPCK的活性降低了2.6倍（Chauhan and Dakshinamurti，1991）。Dakshinamurti和Li（1994）研究了糖尿病鼠肝脏中PEPCK的生物素调节。服用生物素后3h，肝PEPCK mRNA下降至约为非注射生物素对照水平的15%。在给糖尿病鼠服用胰岛素的平行试验中，服用后3h肝PEPCK mRNA下降至约为非注射生物素对照水平的10%。在连续转录试验中，生物素在30min时使肝PEPCK基因转录速率抑制了55%。生物素对转录的抑制显著大于其他刺激。绝食或糖尿病情况下，血浆胰高血糖素含量升高，通过促进PEPCK基因的转录而诱导PEPCK的合成。胰岛素与生物素间在其对葡萄糖代谢酶的作用上有很多相似性。这两者都可诱导编码葡萄糖激酶（一种关键的糖酵解酶）的mRNA，并抑制编码葡萄糖生成关键酶PEPCK的mRNA。

Sone等（1999）应用离体的胰腺灌注研究了生物素对生物素缺乏鼠和对照鼠胰岛素分泌的影响。生物素缺乏鼠中胰岛素对20mmol/L葡萄糖的反应约为对照鼠水平的20%，而对20mmol/L葡萄糖加上1mmol/L生物素的反应分别增加到对20mmol/L葡萄糖灌注反应水平的165%和185%。生物素有加强葡萄糖诱导的胰岛素反应的作用。

Romero-Navarro等（1999）已报道，生物素可刺激培养中鼠胰岛的葡糖激酶活性。通过分支DNA分析，他们表明，用$10^{-6}$mol/L生物素处理的胰岛中葡糖激酶mRNA的水平在12h和24h分别相对增加了约40%和80%。生物素处理（$10^{-6}$mol/L）也增加了胰岛素的分泌。生物素缺乏鼠胰岛葡糖激酶活性和mRNA下降了50%。在从生物素缺乏鼠中分离的胰岛中，对葡萄糖反应的胰岛素分泌也受到损害，表明生物素影响胰岛葡糖激酶的活性和表达以及培养的胰岛中的胰岛素分泌。

葡萄糖是主要的胰岛素促分泌素，其胰岛素释放的刺激促进其代谢。Matchinsky（1996）表明，葡糖激酶是胰腺β细胞上的葡萄糖传感器和代谢信号发生器。生物素通过对葡糖激酶合成的转录影响似乎可增加葡糖激酶对胰腺β细胞胰岛素分泌的影响。

# 维生素$B_6$

维生素$B_6$代谢为其活性形式吡哆醛-5-磷酸（pyridoxal-5-phosphate，PLP），而PLP可作为60多种酶的辅酶。另外，糖原磷酸化酶活性对PLP的需要使其成为碳水化合物、脂类和氨基酸的代谢中的一种多用途因子。维生素$B_6$通过PLP在不同系统如内分泌、神经调节、免疫和心血管系统中发挥作用。PLP作为PLP依赖酶的辅因子外的功能也已被认识，其中具有特殊意义的功能是它在基因表达中的作用。

## 细胞核中的磷酸吡哆醛

Meisler和Thanassi（1990）表明，在鼠肝细胞瘤来源的HTC细胞核中存在着PLP和PLP结合蛋白。在饲喂足够维生素$B_6$饲粮的鼠中，细胞核PLP的数量占细胞内 PLP 总量的 21%，而在饲喂维生素 $B_6$ 缺乏饲粮的鼠中，此数值上升至 39%，表明该维生素能储存于细胞核中。这与细胞核内生物素的分布相类似（Dakshinamurti and Mistry，1963a）。PLP 通过改变类固醇受体复合物与 DNA、染色质和核的相互作用调节了类固醇激素活性的较早报道（Litwack，1998；Maksymowyck *et al.*，1990），Meisler and Thanassi（1990）强调了 PLP 在细胞核中存在的生理意义。

## 维生素 $B_6$ 与细胞增殖

PBMC 和鼠脾细胞已被用作模型来研究体外培养细胞生长的营养和激素需要。Mathews 等（1994）研制出一种无血清无蛋白培养基，可维持受促分裂原刺激的人外周血淋巴细胞的短期生长。一种相似的鼠脾细胞生长的培养基也已出现。研究者在体外将[$^3$H]胸腺嘧啶脱氧核苷掺入细胞作为细胞生长的反映，发现维生素 $B_6$ 从鼠饲粮中去除后将导致生长反应减少，而此生长反应只与培养基中维生素 $B_6$ 的去除有关。

据报道，维生素 $B_6$ 缺乏的啮齿类动物的初级和次级抗体的产生都受损，且抗体产生细胞的数量减少。Grimble（1997）也报道了此情况下胸导管淋巴细胞、体外淋巴细胞增生和 T 细胞介导的细胞毒素显著减少。

## PLP 与类固醇激素作用

在 5mmol/L 维生素 $B_6$ 存在的情况下，生长的细胞中对酶如酪氨酸氨基转换酶的糖皮质激素依赖性诱导作用降低（Disorbo and Litwack，1981）。Allgood 等（1990）表明，维生素 $B_6$ 通过 HeLa 细胞中人糖皮质激素受体调节转录激活。这种转录调节作用不仅限于糖皮质激素受体，而且还延伸到其他的类固醇激素超家族中（Allgood and Cidlowski，1992）。PLP 在细胞内的浓度对类固醇诱导的基因表达具有显著影响，且 PLP 水平的增加导致其对各种类固醇激素的转录反应降低，反之亦然（Tully *et al.*，1994）。

## PLP 与肝脏酶的基因表达

Natori 及其同事提供的证据表明，PLP 在鼠肝脏中直接与糖皮质激素受体互作，且调节胞质天冬氨酸转氨酶（cytosolic aspartate amino transferase，cAST）的基因表达（Oka *et al.*，1995）。他们的研究表明，尽管通过免疫化学法测定的缺乏鼠和对照鼠中的 cAST 蛋白含量相同，但维生素 $B_6$ 缺乏鼠肝脏中的 cAST mRNA 数量比对照鼠的高 7 倍。他们表明，这与脱辅酶蛋白优先螯合入溶酶体所引起 cAST 的加速降解相关。糖皮质激素受体与糖皮质激素反应元件上诱导的酶的基因相结合并能增加它们的转录。人们研究了糖皮质激素受体与糖皮质激素受体元件的结合。体外与 PLP 一起培养的肝脏提取物 DNA 结合活性的失活表明，cAST 基因表达的调节是由糖皮质激素与 PLP 的直接互作引起的。

Sato 等（1996）通过相似的研究表明，维生素 $B_6$ 缺乏鼠肝脏中胱硫醚酶 mRNA 增加了几倍，且其溶酶体中胱硫醚酶的数量也相应增加，说明脱辅酶蛋白被特异地螯合入溶酶体。这就解释了为什么在 cAST 基因的表达增加几倍的情况下，维生素 $B_6$ 缺乏鼠肝中胱硫醚酶蛋白质的浓度依然不发生变化的原因。

## 维生素 $B_6$ 与白蛋白基因的表达

在维生素 $B_6$ 缺乏鼠肝脏中能观测到包括如 $\beta$-肌动蛋白和 3-磷酸甘油醛基因表达的总体增加（Oka *et al.*，1993）。这要归因于维生素 $B_6$ 缺乏鼠肝脏中 RNA 聚合酶Ⅰ和聚合酶Ⅱ的激活。在不依赖维生素 $B_6$ 的蛋白质中，白蛋白基因的表达在维生素 $B_6$ 缺乏时增加。维生素 $B_6$ 缺乏鼠肝脏中白蛋白 mRNA 的含量比对照组高 7 倍，这种增加远大于 RNA 聚合酶活性的增加（Oka *et al.*，1995）。Natori 等（2000）认为，维生素 $B_6$ 通过与 PLP 直接互作而使组织特异性转录因子失活，由此来调节白蛋白基因的表达。

## 维生素 $B_6$ 与癌症

Natori 等（2000）报道，在培养基中添加吡哆醇或吡哆醛抑制了肝细胞瘤 HepG2 细胞的生长，且白蛋白的合成和分泌也同时受到抑制。他们也报道，在给鼠服用大量的吡哆醇后，移植入 C3H/He 鼠中的 MH-134 肝细胞瘤细胞的生长显著降低。其他报道表明，维生素 $B_6$ 抑制了含氮甲烷诱导的结肠癌的发生（Komatsu *et al.*，2001）。

# 维 生 素 C

抗坏血酸（维生素 C）的抗氧化活性是被认识的最好的功能。维生素 C 和维生素 E 通过氧化还原循环作用而消除细胞毒素自由基。抗坏血酸在细胞外基质（extracellular matrix，ECM）蛋白质合成中的作用已被很好地认识。维生素 C 在胶原蛋白、弹性蛋白、补充的 C1q 和乙酰胆碱酯酶中的脯氨酸和赖氨酸残基羟基化方面发挥作用。抗坏血酸在其他细胞过程如神经调节、激素和神经传递素合成和免疫系统方面也有作用。除了作为抗氧化剂和羟基化作用的辅助因子，近期的研究主要集中于其在基因表达方面的直接作用。

## 维生素 C 在细胞外基质产生中的作用

抗坏血酸开始被认为是抗坏血病维生素。人的维生素 C 缺乏导致坏血病，是一种以缺乏 ECM 的产生为特征的疾病。延迟伤口的愈合是维生素 C 缺乏的一个特征。大多数动物，除了灵长类、豚鼠和蝙蝠外均能合成维生素 C，因此不需要食物源的维生素 C。患坏血病豚鼠服用维生素 C 后促进了胶原中脯氨酸羟基化。在这些动物中，软骨中的蛋白多糖合成也受到损害。

胶原是在骨、软骨、腱和皮肤这样的组织中发现的一种结构蛋白，这些胶原以 3 倍螺旋结构为特征。对于形成稳定螺旋的三种肽而言，肽链中大量脯氨酸残基需被脯氨酸羟基化酶羟化，而该过程需要抗坏血酸。弹性蛋白是另外一种 ECM 蛋白，在弹性组织中被发现，且含有被羟化的脯氨酸残基。维生素 C 除了在羟基化反应中作用外，还可刺激大量细胞类型中的胶原合成。维生素 C 也影响弹性蛋白的合成。

## 细胞增殖和细胞分化对维生素 C 的需要量

对大多数物种来说，维生素 C 不是必需的微量养分，又因为大多数培养基都含有血清添加物，所以培养中细胞对维生素 C 的需要一般不加以考虑，直到鼠血浆细胞瘤细胞对添加到培养物中的抗坏血酸盐出现反应时为止。人们观测到了依赖于维生素 C 的动物的细胞对维生素 C 的需要量（Park and Kimler，1991）。已有关于培养基中加抗坏血酸引起的细胞毒性的报道。细胞毒性是由于在没有保护的培养基中抗坏血酸盐引起的超氧化物和氧自由基的产生所致。抗坏血酸的刺激细胞增生的作用即使在紧密相关的细胞中也并不是一致的。人的早幼粒肿瘤细胞 HL-60 对维生素 C 呈正反应（Alcain *et al.*，1990），而来自各种急性骨髓白血病患者的骨髓细胞的反应并不均衡（Park *et al.*，1992）。形成结缔组织的间质细胞对抗坏血酸的反应是增生，而胚胎纤维原细胞却表现出生长停滞。抗坏血酸被添加到被培养的类成骨细胞中后，会诱导该细胞的分化和特定的与该细胞表型相关的基因的表达，如 L/K/B 碱性磷酸酶的同工酶和骨钙素。这还伴随着对胶原分泌的刺激（Franceshi *et al.*，1994）。

## 维生素 C 与软骨生成

在骺和断裂愈合处内软骨纵向骨生长期间，软骨被成骨细胞所取代。在发育过程中，软骨组织经历有序的形态改变，包括增生、细胞肥大、ECM 的产生和矿物化（Brigelius-Flohe and Flohe，1996）。这也伴随着生物化学的变化（Castagnola *et al.*，1988）。未成熟细胞产生Ⅰ型和Ⅱ型胶原和纤连蛋白，增生的软骨细胞合成和分泌大量的Ⅰ型和Ⅳ型胶原以及软骨特性蛋白多糖，肥大的非分离细胞产生Ⅹ型胶原，矿物化组织表达碱性磷酸酶和Ⅹ型胶原。胶原类型的这些变化伴随着其 mRNA 水平的变化。前肥大细胞增生其向肥大细胞的分化需要维生素 C（Gerstenfeld and Landis，1991）。随着胶原Ⅹ型的表达，碱性磷酸酶活性增加。软骨细胞中的矿物化也是抗坏血酸依赖性的。尽管维生素 C 对一般细胞的生长不是必需的，但对 ECM 的产生却是必需的。

## 维生素 C 与胶原的基因表达

维生素 C 通过刺激原胶原基因的转录和稳定 mRNA 而增加原胶原 mRNA 的数量。用抗坏血酸治疗人的皮肤纤维素化能使原胶原 mRNA 增加 2 倍。在原代禽类腱（primary avian tendon，PAT）细胞中，原胶原 mRNA 增加约 6 倍。在用 α，α 联吡啶（一种羟化酶抑制剂）处理的 PAT 细胞中，原胶原合成速率显著降低，但对原胶原 mRNA 没有任何抑制影响。这些结果表明，抗坏血酸可通过与原胶原基因上的抗坏血酸特异性顺式调节元件互作而直接增加原胶原基因的转录（Kurata *et al.*，1993）。

## 维生素 C 与非胶原基因表达

与抗坏血酸在正常 ECM 产生中的作用有关的就是抗坏血酸对Ⅳ型胶原酶（基质金属蛋白酶-2，matrix metallo proteinase，MMP-2）作用的报道。在怀孕期间维生素 C 的低进食量与高风险的未成熟膜破裂（premature rupture of the membrane，PROM）有关。除了其有在胶原基因表达方面的作用外，Pfeffer 等（1998）还报道了抗坏血酸作为 MMP-2 基因表达在转录水平的调节子的一种新作用，表明胶原降解在 PROM 中可能是一个重要因子。

患有 Ehlers-Danlos 综合征Ⅵ型（Ehlers-Danlos syndrome type Ⅵ，EDSⅥ）（一种遗传

性的结缔组织失调症）的患者的皮肤纤维化降低了赖氨酸羟基化酶（lysyl hydroxylase，LH）的活性。LH 使赖氨酸残基以肽键方式羟基化，这是胶原翻译后修饰中关键的一步。LH 缺乏的 EDS 患者有极度的关节高度活动性、皮肤高度延展性、易受损伤性和延迟伤口愈合。使用抗坏血酸治疗 ECM 患者的皮肤纤维素化增加了 LH 水平，也同时增加 LH mRNA 水平。抗坏血酸盐使总胶原产生增加 2 倍（Yeowell *et al.*，1995）。

抗坏血酸缺乏曾用抗坏血酸合成有遗传缺陷的 ODS 鼠（*ad*/*od* 基因型）进行过研究（Ikeda *et al.*，1996，1997，1998）。抗坏血酸缺乏降低了血清载脂蛋白 A-1 的浓度，而血清载脂蛋白 A-1 也受其 mRNA 水平降低的调节。抗坏血酸缺乏也降低了肾脏脂肪酸结合蛋白的肾水平和 $\alpha_{2M}$球蛋白的血清水平。抗坏血酸缺乏鼠 $\alpha_{2M}$球蛋白 mRNA 的肝水平显著降低。肾脏脂肪酸结合蛋白是一种蛋白质水解的 $\alpha_{2M}$球蛋白部分，而这种球蛋白在肝脏中合成。人们用 ODS 鼠对抗坏血酸缺乏对阳性急性期蛋白质、结合珠蛋白和 $\alpha_1$-酸糖蛋白与阴性急性期蛋白质、载脂蛋白 A-1 和白蛋白的肝基因表达进行了进一步研究。抗坏血酸缺乏显著增加血清中结合珠蛋白的浓度和降低载脂蛋白 A-1 及白蛋白的浓度。结合珠蛋白和 $\alpha_1$-酸糖蛋白的增加与其 mRNA 水平的增加一致，相应的，抗坏血酸缺乏的 ODS 大鼠中载脂蛋白 A-1 和白蛋白的降低与肝中这些蛋白质的 mRNA 水平的降低一致。

抗坏血酸作为多巴胺-$\beta$-羟基化酶的辅助因子而在交感神经系统的儿茶酚胺合成中有重要作用。Seitz 等（1998）报道，神经细胞瘤细胞系 SK-N-SH 在用抗坏血酸培养 2h 后导致 3，4 二羟（基）苯丙氨酸（3,4-dihydroxyphenylalanine，DOPA）和多巴胺的合成增加。这些细胞用高水平的抗坏血酸培养 5d 后，色氨酸羟化酶的基因表达增加 3 倍，而多巴胺-$\beta$-羟化酶的基因表达并没改变。以上结果表明，抗坏血酸可能是治疗早期帕金森综合征的一个有用辅助因子。

Mori 等（1992，1997）报道，豚鼠抗坏血酸缺乏使其肝微粒体中各种形式细胞色素 P450 的量减少，也降低了细胞色素 P450 mRNA 的表达，尤其是对 1A1 和 1A2 亚型更是如此，表明豚鼠 P450 的转录受到抗坏血酸的调节。

Bowie 和 O'Neill（2000）的研究表明，维生素 C 通过肿瘤坏死因子（tumour necrosis factor，TNF）并经 p38 有丝分裂原激活的蛋白激酶（mitogen activated protein kinase，MAPK）抑制 NF-κB 的活化。NF-κB 是调节大量参与免疫和炎症反应的基因表达的一种真核转录因子。它被包括促炎症细胞浆中间白细胞素 1（IL-1）和 TNF 这样的致病信号所激活。在细胞胞液中，NF-κB 以一种由结合到抑制蛋白 IκB 上的转录激活二聚体（p50-p65）组成的潜在形式存在，一旦受到细胞浆诱导物刺激，IκB 就被 IκB 激酶（IKK）磷酸化，然后被 26S 蛋白体降解。被释放的 NF-κB 二聚体随后易位到核上，与启动子中的 κB 元件结合并激活靶基因表达。维生素 C 抑制 NF-κB 的胞液刺激，这种作用并不是由于维生素 C 的抗氧化作用所致，因为其他抗氧化剂没有模仿或增大维生素 C 的这种作用。维生素 C 的影响通过应激激活的蛋白激酶 p38MAPK 的活化来调节，目前还不清楚维生素 C 是怎样激活的 p38MAPK 的。激活的 p38MAPK 抑制 NF-κB 的 TNF 活化，也抑制 IκB 的磷酸化和 IKK 的活化。

Reddy 等（2001）提出了维生素 C 在增加子宫颈癌 HeLa 细胞化疗反应中的作用。人的乳头状淋瘤病毒（human papilloma virus，HPV）参与子宫颈癌的发病机制。高度危险的 HPV 类型致癌能力归因于其能改变感染细胞的 E6 和 E7 癌基因蛋白。E6 癌基因蛋白以降解的 p53 蛋白作为靶蛋白，影响细胞的周期调节。这些癌基因蛋白也激活端粒酶，从而有助于被改变细胞的存活。以毫摩尔浓度的维生素 C 培养 HPV 阳性 HeLa 细胞来研究其对 HVTE6/7 致癌基因表达的转录翻译调节的影响。维生素 C 的处理导致滤过性病毒癌基因蛋白 E6 的下调，同时引起激活蛋白（activator protein，AP-1）成员 c-jun 和 c-fos 以时间和剂量依赖方式降低。E6 的下调与预凋亡 p53 和 Bax 蛋白的上调以及细胞凋亡抑制因子 Bcl-2 的下调有关。

Catani 等（2001）表明，维生素 C 能调节细胞做出反应以对抗紫外线（UV）照射所致的细胞损伤和细胞死亡。抗坏血酸直接通过清除各种自由基来保护角质形成细胞免受 UV 介导的细胞毒素的影响。另外，抗坏血酸诱导 IL-1α mRNA 的转录下调，因 UV 辐射引起的氧化应激能激活氧化还原敏感的转录因子，包括 NF-κB 和 AP-1 复合物的成员如 c-jun 和 c-fos。抗坏血酸盐能抑制 AP-1 依赖性特异启动子的转移活化。抗坏血酸是 AP-1 和 AP-1 依赖性转录的一种负调节因子。

芳香胺致癌物经历由肝脏细胞色素 P450 催化的 *N*-氧化而形成反应代谢物。这些代谢物结合到组织或器官的大分子（如 DNA）上而形成加合物。这样的 DNA 致癌物加合物的形成是启动靶组织中芳香胺诱导致癌作用中的一个重要步骤。Hung 和 Lu（2001）表明，维生素 C 降低了 C6 神经胶质瘤细胞中芳香胺 DNA 加合物的形成。

## 参 考 文 献


Alcain, F.J., Buron M.I., Rodriguez-Aguilera, J.C., Villalba, J.M.Z. and Navas, P. (1990) Ascorbate free radical stimulates the growth of a human promyelocytic leukemia cell line. *Cancer Research* 50, 5887–5891.

Allgood, V.E. and Cidlowski, J.A. (1992) Vitamin B6 modulates transcriptional activation by multiple members of the steroid hormone receptor superfamily. *Journal of Biological Chemistry* 267, 3819–3824.

Allgood, V.E., Powell-Oliver, F.E. and Cidlowski, J.A. (1990) Vitamin B6 influences glucocorticoid receptor-dependent gene expression. *Journal of Biological Chemistry* 265, 12424–12433.

Baez-Saldana, A. Diaz, G., Espinoza, B. and Ortega, E. (1998) Biotin deficiency induces changes in subpopulations of spleen lymphocytes in mice. *American Journal of Clinical Nutrition* 67, 431–437.

Bernlohr, D.A., Bolanowski, M.S., Kelly, T.J. Jr and Lane, M.D. (1985) Evidence for an increase in transcription of specific mRNA during differentiation of 373-L1 preadiposites. *Journal of Biological Chemistry* 260, 5563–5567.

Bhullar, R.P. and Dakshinamurti, K. (1985) The effects of biotin on cellular functions of HeLa cells. *Journal of Cellular Physiology* 122, 425–430.

Boeckx, R.L. and Dakshinamurti, K. (1974) Biotin-mediated protein synthesis. *Biochemical Journal* 140, 549–556.

Boeckx, R.L. and Dakshinamurti, K. (1975) Effect of biotin on RNA synthesis. *Biochimica et Biophysica Acta* 383, 282–289.

Bowie, A.G. and O'Neill, L.A.J. (2000) Vitamin C inhibits NF-κB activation by TNF via the activation of p38 mitogen-activated protein kinase. *Journal of Immunology* 165, 7180–7188.

Brigelius-Flohe, R. and Flohe, L. (1996) Ascorbic acid, cell proliferation and cell differentiation in culture. *Subcellular Biochemistry* 25, 83–107.

Castagnola, P., Dozin, B., Moro, G. and Cancedda, R. (1988) Changes in the expression of collagen genes show two stages in chondrocyte differentiation *in vitro*. *Journal of Cell Biology* 106, 461–467.

Catani, M.V., Ross, A., Costanzo, A., Sabatini, S. and Levrero, M. (2001) Induction of gene expression via activator protein-1 in the ascorbate protection against 4v-induced damage. *Biochemical Journal* 356, 77–85.

Chalifour, L.E. and Dakshinamurti, K. (1982a) The requirement of human fibroblasts in culture. *Biochemical and Biophysical Research Communications* 104, 1047–1053.

Chalifour, L.E. and Dakshinamurti, K. (1982b) The characterization of the uptake of avidin–biotin complex by HeLa cells. *Biochimica et Biophysica Acta* 721, 64–69.

Chauhan, J. and Dakshinamurti, K. (1986) Purification and characterization of human serum biotinidase. *Journal of Biological Chemistry* 261, 4268–4275.

Chauhan, J. and Dakshinamurti, K. (1988) Role of human serum biotinidase as biotin-binding protein. *Biochemical Journal* 256, 365–270.

Chauhan, J. and Dakshinamurti, K. (1991) Transcriptional regulation of the glucokinase gene by biotin in starved rats. *Journal of Biological Chemistry* 266, 10035–10038.

Collins, J.C., Paietta, E., Green, R., Morell, A.G. and Stockert, R.J. (1988) Biotin-dependent expression of the asialoglycoprotein receptor in HepG2. *Journal of Biological Chemistry* 263, 11280–11283.

Dakshinamurti, K. (1997) Vitamin receptors. In: Myers, R.A. (ed.) *Encyclopedia of Molecular Biology and Molecular Medicine*, Vol. 6. VCH Verlagsgesellschaft mbh, Weinheim, pp. 235–244.

Dakshinamurti, K. and Chalifour, L.E. (1981) The biotin requirement of HeLa cells. *Journal of Cellular Physiology* 107, 427–438.

Dakshinamurti, K. and Chauhan, J. (1989) Biotin. *Vitamins and Hormones* 45, 337–384.

Dakshinamurti, K. and Chauhan, J. (1994) Biotin binding proteins. In: Dakshinamurti, K. (ed.) *Vitamin Receptors*. Cambridge University Press, Cambridge, pp. 200–249.

Dakshinamurti, K. and Cheah-Tan, C. (1968a) Liver glucokinase of the biotin-deficient rat. *Canadian Journal of Biochemistry* 46, 75–80.

Dakshinamurti, K. and Cheah-Tan, C. (1968b) Biotin-mediated synthesis of hepatic glucokinase in the rat. *Archives of Biochemistry and Biophysics* 127, 17–21.

Dakshinamurti, K. and Hong, H.C. (1969) Regulation of key glycolytic enzymes. *Enzymologia Biologica Clinica* 11, 422–428.

Dakshinamurti, K. and Li, W. (1994) Transcriptional regulation of liver phosphoenolpyruvate carboxykinase by biotin in diabetic rats. *Molecular and Cellular Biochemistry* 132, 127–132.

Dakshinamurti, K. and Mistry, S.P. (1963a) Tissue and intracellular distribution of biotin-$C^{14}OOH$ in rats and chickens. *Journal of Biological Chemistry* 238, 294–296.

Dakshinamurti, K. and Mistry, S.P. (1963b) Amino acid incorporation in biotin deficiency. *Journal of Biological Chemistry* 238, 297–301.

Dakshinamurti, K., Chalifour, L. and Bhullar, R.P. (1985) Requirement for biotin and the function of biotin in cells in culture. *Annals of the New York Academy of Sciences* 447, 38–55.

De La Vega, L.A. and Stockert, R.J. (2000) Regulation of the insulin and asialoglycoprotein receptors via cGMP-dependent protein kinase. *American Journal of Physiology* 279, C2037–C2042.

Disorbo, D.M. and Litwack, G. (1981) Changes in the intracellular levels of pyridoxal-5-phosphate affect the induction of tyrosine aminotransferase by glucocorticoids. *Biochemical and Biophysical Research Communications* 99, 1203–1208.

Franceschi, R.T., Iyer, B.S. and Cui, Y. (1994) Effects of ascorbic acid on collagen matrix formation and osteoblast differentiation in murine MC3T3-F1 cells. *Journal of Bone and Mineral Research* 9, 843–854.

Gerstenfeld, L.C. and Landis, W.J. (1991) Gene expression and extracellular matrix ultrastructure of a mineralizing chondrocyte cell culture system. *Journal of Cell Biology* 112, 501–513.

Gospodarowicz, D. and Cohen, D.C. (1985) Biotin and choline replace the growth requirement of Madin–Darby canine kidney cells for high-density lipoprotein. *Journal of Cellular Physiology* 124, 96–106.

Granner, D. and Pilkes, S. (1990) The genes of hepatic glucokinase metabolism. *Journal of Biological Chemistry* 265, 10173–10176.

Grimble, R.F. (1996) Interaction between nutrients, proinflammatory cytokines and inflammation. *Clinical Science (Lond)* 91, 121–130.

Hung, C.F. and Lu, K.H. (2001) Vitamin C inhibited DNA adduct formation and arylamine *N*-acetyltransferase activity and gene expression in rat glial tumor cells. *Neurochemical Research* 26, 1107–1112.

Hymes, J. and Wolf, B. (1999) Human biotinidase isn't just for recycling biotin. *Journal of Nutrition* 129, 485S–489S.

Hymes, J., Fleischhauer, K. and Wolf, B. (1995) Biotinylation of biotinidase following incubation with biocytin. *Clinica Chimica Acta* 233, 39–45.

Ikeda, S., Horio, F., Yoshida, A. and Kakinuma, A. (1996) Ascorbic acid deficiency reduces hepatic apolipoprotein A-1 mRNA in scurvy-prone ODS rats. *Journal of Nutrition* 126, 2505–2511.

Ikeda, S., Takasu, M., Satsuda, T. and Kakinyma A. (1997) Ascorbic acid deficiency decreases the renal level of kidney fatty acid-binding protein by lowering the $\alpha_{2U}$-globulin gene expression in liver in scurvy-prone ODS rats. *Journal of Nutrition* 127, 2173–2178.

Ikeda, S., Horio, F. and Kakinuma, A. (1998) Ascorbic acid deficiency changes hepatic gene expression of acute phase proteins in scurvy-prone ODS rats. *Journal of Nutrition* 128, 832–838.

Johnson, R., Norstrom, E. and Soderstrom, K. (2000) Increased expression of endogenous biotin, but not BDNF, in telencephalic song regions during zebra finch vocal learning. *Developmental Brain Research* 120, 113–123.

Keranan, A.J.A. (1972) The biotin synthesis of HeLa cells *in vivo*. *Cancer Research* 32, 119–124.

Komatsu, S.I., Watanabe, H., Oka, T., Tsuge, H., Nii, H. and Kato, N. (2001) Vitamin $B_6$-supplemented diets compared with a low vitamin $B_6$ diet suppresses azomethane-induced colon tumorigenesis in mice by reducing cell proliferation. *Journal of Nutrition* 131, 2204–2207.

Kurata, S., Senoo, H. and Hata, R. (1993) Transcriptional activation of type I collagen genes by ascorbic acid-2-phosphates in human skin fibroblasts and its failure in cells from a patient with alpha 2(I)-chain-defective Ehlers–Danlos syndrome. *Experimental Cell Research* 206, 63–71.

Kuri-Harcuch, W. and Marsch-Morino, M. (1983) DNA

synthesis and cell division related to adipose differentiation of 3T3 cells. *Journal of Cellular Physiology* 114, 39–44.

Kuri-Harcuch, W., Wise, L.S. and Green, H. (1978) Interruption of adipose conversion of 3T3 cells by biotin deficiency: differentiation without triglyceride accumulation. *Cell* 14, 53–59.

Litwack, G. (1988) The glucocorticoid receptor at the protein level. *Cancer Research* 48, 2636–2640.

Matchinsky, F.M. (1996) A lesson in metabolic regulation inspired by the glucokinase glucose sensor paradigm. *Diabetes* 45, 223–242.

Mathews, K.S., Mrowczynski, E. and Mathews, R. (1994) Dietary deprivation of B-vitamins reflected in murine splenocyte proliferation *in vitro*. *Biochemical and Biophysical Research Communications* 198, 451–458.

Meglasson, M.D. and Matchinsky, F.M. (1984) New perspective on pancreatic islet glucokinase. *American Journal of Physiology* 246, E1–E13.

Meisler, N.T. and Thanassi, J.W. (1990) Pyridoxine-derived vitamin $B_6$ vitamers and pyridoxal-5-phosphate-binding protein in cytosolic and nuclear fractions of HTC cells. *Journal of Biological Chemistry* 265, 1193–1198.

Mori, T., Kitamura, R., Imaoka, S., Funae, Y., Kitada, M. and Kamataki, T. (1992) Examination for lipid peroxidation in liver microsomes of guinea pigs as a causal factor in the decrease in the content of cytochrome P450 due to ascorbic acid deficiency. *Research Communications in Chemical Pathology and Pharmacology* 75, 209–219.

Mori, T., Itoh, S., Ohgiya, S., Ishizaki, K. and Kamataki, T. (1997) Regulation of CYPIA and CYP3A mRNAs by ascorbic acid in guinea pigs. *Archives of Biochemistry and Biophysics* 348, 268–277.

Moskowitz, M. and Cheng, D.K.S. (1985) Stimulation of growth factor production in cultured cells by biotin. *Annals of the New York Academy of Sciences* 447, 212–221.

Nakatani, Y., Kitamura, H., Inayama, Y. and Ogawa, N. (1994) Pulmonary endodermal tumor resembling fetal lung. The optically clear nucleus is rich in biotin. *American Journal of Surgical Pathology* 18, 637–642.

Natori, Y., Oka, T. and Kuwahata, M. (2000) Modulation of gene expression by vitamin $B_6$. In: Iriavte, A., Kagan, H.M. and Martinez-Carrion, M. (eds) *Biochemistry and Molecular Biology of Vitamin $B_6$ and PQQ-dependent Proteins*. Birkhauser Verlag, Basel, pp. 301–306.

Oka, T., Komori, N., Kuwahata, M., Sasa, T., Suzuki, I., Okada, M. and Natori, Y. (1993) Vitamin B6 deficiency causes activation of RNA polymerase and general enhancement of gene expression in rat liver. *FEBS Letters* 330, 409–413.

Oka, T., Komori, N., Kuwahata, M., Hiroi, Y., Shimoda, T., Okada, M. and Natori, Y. (1995) Pyridoxal 5-phosphate modulates expression of cytosolic aspartate amino transferase gene by inactivation of glucocorticoid receptor. *Journal of Nutritional Science and Vitaminology* 41, 363–375.

Okamoto, Y., Kashima, K., Daa, T., Yokyama, S., Nakayama, I. and Noguchi, S. (1995) Morule with biotin-containing intranuclear inclusions in thyroid carcinoma. *Pathology International* 45, 573–579.

Ozyhar, A. , Kiltz, H.H. and Pongs, O. (1990) Pyridoxal phosphate inhibits the DNA-binding activity of the ecdysteroid receptor. *European Journal of Biochemistry* 192, 167–174.

Park, C.H. and Kimler, B.F. (1991) Growth modulation of human leukemic, pre-leukemic and myeloma progenitor cells by L-ascorbic acid. *American Journal of Clinical Nutrition* 54, 1241S–1246S.

Park, C.H., Kimler, B.F., Bodensteiner, D., Lynch, S.R. and Hassanien, R.S. (1992) *In vitro* growth modulation by L-ascorbic acid of colony forming cells from bone marrow of patients with myelodysplastic syndromes. *Cancer Research* 52, 4458–4466.

Paulose, C.S., Thliveris, J., Viswanathan, M. and Dakshinamurti, K. (1989) Testicular function in biotin-deficient adult rats. *Hormone and Metabolic Research* 21, 661–665.

Pfeffer, F., Casanuera, E., Kamer, J., Guerra, A. and Perichart, O. (1998) Modulation of 72 kilodalton type IV collagenase (matrix metalloproteinase-2) by ascorbic acid in cultured human amnion-derived cells. *Biology of Reproduction* 59, 326–329.

Pham, A.D. and Sauer, F. (2000) Ubiquitin-activating/conjugating activity of $TAF_{II}250$, a mediator of activation of gene expression in *Drosophila*. *Science* 289, 2357–2360.

Reddy, V.G., Khanna, N. and Singh, N. (2001) Vitamin C augments chemotherapeutic response of cervical carcinoma HeLa cells by stabilizing p53. *Biochemical and Biophysical Research Communications* 282, 409–415.

Rodriguez-Melandez, R., Perez-Andrade, M.E., Diaz, A., Deolarte, A., Camacho-Arroyo, Ciceron, I., Ibarra, I. and Velazquez, A. (1999) Different effects of biotin deficiency and replenishment on rat liver pyruvate and propionyl-CoA carboxylases and on their mRNAs. *Molecular Genetics and Metabolism* 66, 16–23.

Rodriguez-Melendez, R., Cano, S., Mendez, S.T. and Velazquez, A. (2001) Biotin regulates the genetic expression of holocarboxylase synthetase and mitochondrial carboxylases in rats. *Journal of Nutrition* 131, 1909–1913.

Romero-Navarro, G., Cabrera-Valladares, G., German, M.S., Matchinsky, F.M., Velazquez, A., Wang, J. and Fernandez-Mejia, C. (1999) Biotin regulation of pancreatic glucokinase and insulin in primary cultured rat islets and in biotin-deficient rats. *Endocrinology* 140, 4595–4600.

Roth, K.S., Yang, W., Alan, L., Saunders, J., Gravel, R.A. and Dakshinamurti, K. (1982) Prenatal administration of biotin in biotin responsive multiple carboxylase deficiency. *Pediatric Research* 16, 126–129.

Sasaki, A., Yokoyama, S., Arita, R., Inomata, M., Kashima, K. and Nakayama, I. (1999) Morules with biotin-containing optically clear nuclei in colonic tubular adenoma. *American Journal of Surgical Pathology* 23, 336–341.

Sato, A., Nishioka, M., Awata, S., Nakayama K., Okada, M., Horiuchi, S., Okabe, N., Sassa, T., Oka, T. and Natori, Y. (1996) Vitamin B6 deficiency accelerates metabolic turnover of cystathionase in rat liver. *Archives of Biochemistry and Biophysics* 330, 409–413.

Seitz, G., Gerbhardt, S., Beck, J.F., Bohm, W., Lode, H.N., Niethammer, D. and Bruchelt, G. (1998) Ascorbic acid stimulates DOPA synthesis and tyrosine hydroxylase gene expression in the human neuroblastoma cell line SK-N-SH. *Neuroscience Letters* 244, 33–36.

Sickel, J.Z. and Sant'Agnese, P.A. (1994) Ananalous immunostaining of 'optically clear' nuclei in gestational endometrium: a potential pitfall in the diagnosis of pregnancy-related herpesvirus infection. *Archives of Pathology and Laboratory Medicine* 118, 831–833.

Singh, I.N. and Dakshinamurti, K. (1988) Stimulation of guanylate cyclase and RNA polymerase II activities in HeLa cells and fibroblasts by biotin. *Molecular and Cellular Biochemistry* 79, 47–55.

Sommerville, J., Baird, J. and Turner, B.M. (1993) Histone H4 acetylation and transcription in amphibian chromatin. *Journal of Cell Biology* 120, 277–290.

Sone, H., Ito, M., Suguyama, K., Ohneda, M., Maebashi, M. and Furukawa, Y. (1999) Biotin enhances glucose stimulated insulin secretion in the isolated perfused pancreas of the rat. *Journal of Nutritional Biochemistry* 10, 237–243.

Stanley, J.S., Griffin, J.B. and Zempleni, J. (2001) Biotinylation of histones in human cells. Effects of cell proliferation. *European Journal of Biochemistry* 268, 5424–5429.

Stockert, R.J. and Ren, Q. (1997) Cytoplasmic protein mRNA interaction mediate cGMP-modulated translational control of the asialoglycoprotein receptor. *Journal of Biological Chemistry* 272, 9161–9165.

Tully, D.B., Allgood, V.E. and Cidlowski, J.A. (1994) Modulation of steroid receptor-mediated gene expression by vitamin $B_6$. *FASEB Journal* 8, 343–349.

Vesely, D.L. (1982) Biotin enhances guanylate kinase. *Science* 216, 1329–1330.

Wang, H. and Pevsner, J. (1999) Detection of endogenous biotin in various tissues: novel functions in the hippocampus and implications for its use in avidin–biotin technology. *Cell and Tissue Research* 296, 511–516.

Watanabe, T. (1990) Micronutrients and congenital anomalies. *Congenital Anomalies* 30, 79–92.

Watanabe, T., Dakshinamurti, K. and Persaud, T.V.N. (1995) Effect of biotin on palatal development of mouse embryos in organ culture. *Journal of Nutrition* 125, 2114–2121.

White, H.B. III and Whitehead, C.C. (1987) The role of avidin and other biotin-binding proteins in the deposition and distribution of biotin in chicken eggs: discovery of a new biotin-binding protein. *Biochemical Journal* 241, 677–684.

Wolffe, A. (1998) *Chromatin*. Academic Press, San Diego, California.

Yang, P., Morizumi, H. and Sato, T. (1995) Pulmonary blastoma: an ultrastructural and immunohistochemical study with special reference to nuclear filament aggregation. *Ultrastructural Pathology* 19, 501–509.

Yeowell, H.N., Walker, L.C., Marshall, M.K., Murad, S. and Pinnell, S.R. (1995) The mRNA and activity of lysyl hydroxylase are up-regulated by the administration of ascorbate and hydralazine to human skin fibroblasts from a patient with Ehlers-Danlos syndrome type VI. *Archives of Biochemistry and Biophysics* 321, 510–516.

Yokayama, S., Kashima, K., Inoue, S., Daa, T., Nakayama, I. and Moriuchi, A. (1993) Biotin-containing intranuclear inclusions in endometrial glands during gestation and puerperium. *American Journal of Clinical Pathology* 99, 13–17.

Zempleni, J. and Mock, D.M. (1999) Mitogen-induced proliferation increases biotin uptake into human peripheral blood mononuclear cells. *American Journal of Physiology* 276, C1079–C1084.

Zempleni, J. and Mock, D.M. (2000) Utilization of biotin in proliferating human lymphocytes. *Journal of Nutrition* 130, 335S–337S.

Zempleni, J. and Mock, D.M. (2001) Biotin homeostasis during the cell cycle. *Nutrition Research Reviews* 14, 45–63.

# 12　硒与维生素 E

Alexandra Fischer[1]，Josef Pallauf[1]，Jonathan Majewicz[2]，
Anne Marie Minihane[2] 和 Gerald Rimbach[2]
（1 动物营养与营养生理学院，贾斯特斯-李比希-大学，吉森，德国；
2 食品生物科学学院，休辛克莱　人营养研究组，
瑞丁大学，英国）

## 维 生 素 E

维生素 E 这种最重要的脂溶性抗氧化剂是 Evans 和 Bishop 于 1922 年在加利福尼亚大学伯克利分校发现的（表 12.1）。自发现以来，人们对生育酚和生育三烯酚类的抗氧化进行了研究。在 1991 年，Angelo Azzi 的研究组首次对维生素 E 的非抗氧化剂、细胞信号功能进行了描述，表明维生素 E 可调节平滑肌细胞中蛋白激酶 C（protein kinase C，PKC）的活性（Boscoboinik *et al.*，1991a，b）。在转录水平上，维生素 E 可调节肝中 α-生育酚转运蛋白（tocopherol transfer protein，TTP）(Fechner *et al.*，1998）以及肝胶原α1（Chojkier *et al.*，1998）、胶原酶（Ricciarelli *et al.*，1999）和 α-原肌球蛋白基因（Aratri *et al.*，1999）的表达。近来，生育酚依赖性转录因子（生育酚相关蛋白，tocopherol associated

**表 12.1　维生素 E 试验研究中的重要发现**

| | |
|---|---|
| 早期历史 | |
| 1922 | Evans 和 Bishop 发现维生素 E 的存在，这种脂溶性因子（X 因子）可预防饲喂酸败猪油引起的胎儿死亡 |
| 1924 | Sure 给 X 因子命名为维生素 E，作为维生素的第 5 排序字母名称 |
| 1925 | Evans 由在希腊语中“tos”为孩子出生，“phero”意为出生，“ol”为乙醇分子的部分而提出单词 tocopherol（生育酚） |
| 早期研究（结构和功能特征的描述） | |
| 1930 | 不同动物维生素 E 缺乏症的特征（睾丸萎缩、胎儿重吸收、脑软化、肌肉营养不良的瘫痪） |
| 1938 | Fernholz 阐明维生素 E 的结构 |
| 1938 | Karrer 合成维生素 E |
| 1955 | Gordon 等揭示，成熟婴儿血液生育酚水平低，且红细胞在过氧化氢存在时进行培养出现异常溶解 |
| 1967 | Bunyan 等研究维生素 E 对多不饱和脂肪酸的抗氧化作用 |
| 1968 | 维生素 E 通过被列入美国食品和营养委员会的推荐膳食供给表中而被正式认作人的一种必需养分近期 |
| 维生素 E 研究（维生素 E 的非抗氧化功能及其对基因表达的影响） | |
| 1991 | Boscoboinik 等发现 α-生育酚通过对蛋白激酶 C 的调节抑制平滑肌细胞增生 |
| 1998 | Fechner 等发现 α-和 β-生育酚诱导肝中 α-生育酚转运蛋白的表达 |
| 1998 | Chojkier 等 α-生育酚调控肝胶原α1 基因转录 |
| 1999 | Aratri 等发现 α-生育酚增加 α-原肌球蛋白的转录水平 |
| 2001 | Yamauchi 等发现 α-生育酚通过与转录因子 TAP 结合而作为基因表达的转录调节子 |

protein，TAP）已被发现（Stocker *et al.*，1999；Yamauchi *et al.*，2001）。分子生物学和基因组技术的进展导致新的维生素 E 敏感基因和信号转导途径被发现。本文主要集中描述动脉粥样化形成中起重要作用的那些转录因子、信号分子、基因和蛋白质。

## 维生素 E 的化学和抗氧化特性

在植物中，可由同源酸合成生育酚和生育三烯酚类，它们全都是具有脂肪族类异戊二烯侧链的 6-色原烷醇衍生物。4 个生育酚的同源物（α、β、γ 和 δ）都有全饱和的 16 个碳植醇侧链，但生育三烯酚类含有相似的带有 3 个双键的类异戊二烯链。单个生育酚是根据甲基基团在苯酚环上的数量的位置来命名，即 α-、β-、γ-和 δ-维生素分别含有三、二、二和一个甲基基团（图 12.1）。这些结构的差异决定了它们的生物学活性，其中以 α-同源物的生物学活性最高。

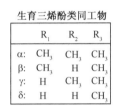

生育三烯酚类同工物

| | $R_1$ | $R_2$ | $R_3$ |
|---|---|---|---|
| α: | $CH_3$ | $CH_3$ | $CH_3$ |
| β: | $CH_3$ | H | $CH_3$ |
| γ: | H | $CH_3$ | $CH_3$ |
| δ: | H | H | $CH_3$ |

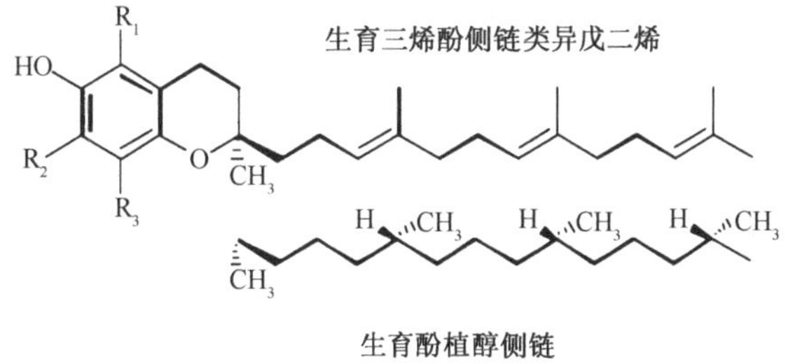


图 12.1　维生素 E 立体异构体的分子结构

维生素 E 的主要功能是其作为抗氧化剂的作用，保持体内几乎所有细胞的结构完整性。其抗氧化功能是通过自由基的减少来调节的，因此保护身体免于受到衰老病理生理异常和大量慢性疾病如动脉硬化症、癌症和风湿症性关节炎（Halliwell，1996；Parthasarathy *et al.*，1999；Malins *et al.*，2001）中产生的高活性氧类物质（reactive oxygen species，ROS）和高活性氮类物质（reactive nitrogen species，RNS）的有害影响。ROS 包括过氧化氢（$H_2O_2$）、超氧自由基（$O_2^-$）和高还原羟基自由基（OH·），都是在呼吸和吞噬过程中以及在微粒体细胞色素 P450 代谢期间形成的正常需氧代谢的副产物。

生物膜中的多不饱和脂肪酸（PUFA）由于其具高度不饱和性而对自由基的攻击特别敏感。简言之，此过程由自由基如 OH·，从 PUFA 上抽取氢形成 PUFA·自由基复合物开始，然后发生分子重排形成二轭二烯，对分子氧的攻击很敏感，因而产生过氧自由基（PUFAOO·），过氧自由基再从邻近的 PUFA 上抽取氢原子，从而形成一个链式反应。这种自动氧化的持续进行严重影响组织功能，除非清除自由基。维生素 E 由于其含量丰富、具脂溶性和有关消除自由基的效率高而被认为是在细胞膜中最重要的抗氧化剂（Ingold *et al.*，1987；Halliwell，1996；Brigelius-Flohé and Traber，1999）。

维生素 E 的抗氧化特性是通过酚的羟基实现的，而羟基很容易把其氢供给过氧化氢自由基，进而形成稳定的脂类。维生素 E 在提供氢原子中，变成相对惰性的自由基，因为不配对电子离开原位进入芳香环。这种保护的效率取决于两个因素：第一，由脂质尾测定的膜上分子的流动性；第二，色原烷醇环上的甲基数量，每个甲基基团都被赋予额外的抗氧化能力。另外，甲基对羟基的亲合性也是一个重要因素。因此，具有最多甲基基团且侧面与羟基基团相连的 α-同源物被认为比其他同源物更有效。

与 α-生育酚比较，α-生育三烯酚已被证明在防止脂肪过氧化方面更为有效（Serbinova *et al.*，1991；Suzuki *et al.*，1993）。对此提出的一个原因是脂质尾的性质。α-生育三烯酚的异戊二烯链与 α-生育酚相比，有更强的使膜功能失调的影响。这导致膜内更大的流动性和更均匀的分布。核磁共振研究也显示，α-生育三烯酚的色原烷醇环紧贴近膜表面。这些因子有助于 α-生育三烯酚有更大能力与自由基互作，且使分子更快速循环到其活性氧化的形式（Serbinova *et al.*，1991；Suzuki and Packer，1993）。

尽管维生素 E 在膜内起独特作用，但其在分离时却不发挥作用。保护细胞免于氧化应激的有害作用涉及一系列其他膜和水溶抗氧化剂及抗氧化酶，它们一起形成抗氧化防御系统（图 12.2）。在这种多因子系统中，胞质金属酶可阻止自由基的形成。超氧化物歧化酶可把 $O_2^-$ 转化为 $H_2O_2$，而谷胱甘肽超氧化物酶和过氧化氢酶进一步使 $H_2O_2$ 还原，因而阻止高活性 OH・的形成。水溶性抗氧化剂可作为抗氧化酶的辅助因子，也可作为独立的抗氧化剂或在维生素 E 的再循环方面发挥作用。维生素 E 以一个分子/2000～3000 个磷脂的浓度存在于膜中，因此会被快速耗竭，除非它再生为其活性形式。体外试验证据表明，抗坏血酸盐可使与膜结合的维生素 E 再生而发挥作用，转化生育酚自由基为其原来形式，且导致抗坏血酸自由基的形成（Kagan and Tyurina，1998；May *et al.* 1993）。然而，当前还缺乏体内试验证据。

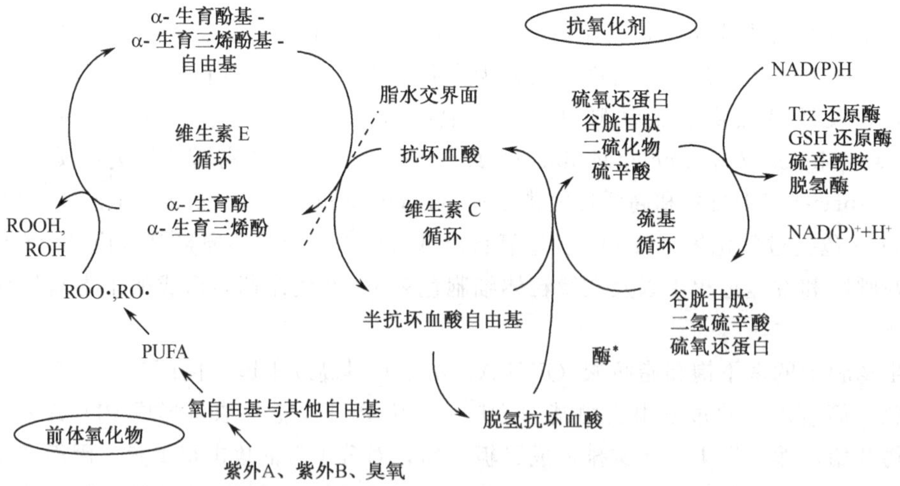


图 12.2　维生素 E、维生素 C 与硫醇氧化还原循环间互作的抗氧化网络图示

* 1）硫醇转移酶；2）GSH-依赖性脱氢抗坏血酸还原酶；
3）蛋白二硫化物异构酶；4）硫氧还蛋白（Trx）还原酶

## 吸收与转运

到目前为止，关于维生素 E 吸收和转运的大部分信息都是建立在生育酚基础上的

(Cohn *et al.*，1992；Kayden and Traber，1993；Herrera and Barbas，2001)。在小肠中，生育酚酯水解成游离的维生素 E，后在胆汁盐和胰液的作用下并入混合微团。如在发生胰腺炎、胆囊纤维化或胆固醇郁滞性肝脏疾病而引起这些小肠分泌物缺乏时，会导致维生素 E 的吸收异常和由此产生的营养缺乏综合征。微粒经被动扩散进入肠细胞，且生育酚与磷脂、胆固醇、甘油三酯和部分载脂蛋白一起形成乳糜微粒。一旦经淋巴系统进入循环系统，乳糜微粒则在附着于靶组织如肌肉和脂肪组织中的毛细血管内质上的脂蛋白脂酶的作用下被水解，一部分生育酚被释放并被内皮细胞摄取。残余的乳糜微粒通过受体调节的胞饮过程被肝脏摄取。与维生素 A 和维生素 D 相比，维生素 E 在循环系统中似乎没有专门的载体蛋白。然而，维生素 E 在被吸收后以一种非特异的方式进入脂蛋白微粒。在肝细胞中，来自乳糜微粒残余部分的生育酚结合到胞液 α-生育酚转运蛋白（α-TTP）上（Catignani and Bieri，1977），它能调节并将其转运至极低密度脂蛋白（VLDL）合成的位点上（粗面内质网和高尔基体）。这种 32kDa 的蛋白质几乎只在肝脏中表达。且近期的动物试验证据表明，食物中 α-生育酚调节肝 α-TTP mRNA 水平（Fechner *et al.*，1998）。并不像生育酚的吸收是异构体非选择性的，α-TTP 表现出立体异构体特异性，α-生育酚几乎只结合进新生的 VLDL 微粒。生育酚类似物对 α-TTP 的相对亲合性通过竞争研究计算如下：如 α-生育酚为 100%，则 β-生育酚为 38%，γ-生育酚为 9%，δ-生育酚为 2%（Hosomi *et al.*，1997）。非 α-异构体则大部分经胆汁排出（Traber and Kayden，1989）。α-TTP 现在被认为是血浆生育酚水平的主要决定因素。α-*TTP* 基因突变导致血浆和组织中 α-生育酚降低，最终导致被称作具有维生素 E 缺乏症的共济失调（ataxia vitamin E with Defficiency，AVED）的严重情况，且与神经和视网膜损伤有关（Traber *et al.*，1990；Ben Hamida *et al.*，1993）。在最近的研究中，α-TTP 敲除鼠（$Ttpa^{+/-}$ 和 $Ttpa^{-/-}$）被用作研究维生素 E 缺乏与动脉硬化症关系的一种模型（Terasawa *et al.*，2000）。血浆和组织中的 α-生育酚从对照到 $Ttpa^{+/-}$ 到 $Ttpa^{-/-}$ 逐步降低，同时 $Ttpa^{-/-}$ 的肝匀浆中缺乏 α-*TTP*，而在 $Ttpa^{+/-}$ 动物中，蛋白水平降低 50%。这种维生素 E 的缺乏症与近心端的大动脉损伤增加和脂肪过氧化速率的增加有关。这些发现进一步表明了该转运蛋白对生育酚代谢和最终的冠心病（coronary heart disease，CHD）的作用。

分泌的 VLDL 总量的近 50%～70%被水解成低密度脂蛋白（low-density-lipoprotein，LDL），同时生育酚转运进入 LDL 部分（Welty *et al.*，2000）。在循环中，LDL 和高密度脂蛋白（High-Density-Lipoprotein，LDL）部分尽管含有超过 90%的生育酚，但其在脂蛋白微粒间交换迅速（Behrens *et al.*，1982）。75kDa 血浆磷脂转运蛋白有助于生育酚在 HDL 与 VLDL 间的交换（Lagrost *et al.*，1998）。

尽管通过受体介导的脂蛋白胞饮或脂肪酸结合蛋白可同时摄取生育酚，人们对外周细胞对维生素 E 摄取的机制了解很少。最近的证据表明，特异膜生育酚结合蛋白（tocopherol-binding protein，TBPpms）也可调节生育酚的摄取（Dutta-Roy，1999）。

关于细胞内的维生素 E 转运的信息目前还很缺乏。由于其具强大的疏水性，转运到细胞位点需要特异的转运蛋白。但是，有多少其他 α-生育酚结合蛋白存在和它们通过哪些机制调节外周细胞中生育酚的转运，目前仍不清楚。近来，一种新的结合蛋白 TAP 已被识别（Stocker *et al.*，1999；Zimmer *et al.*，2000；Blatt *et al.*，2001；

Yamauchi *et al.*，2001），这个 46kDa 蛋白质表现出了与 α-TTP 序列显著的同源性，且普遍表达，尽管其在肝、脑和前列腺中被发现有最高的水平（Zimmer *et al.*，2000）。已提出该蛋白质在细胞内的生育酚代谢中发挥重要作用。TAP 的结构分析表明，它是广泛的类 SEC14 蛋白家族的一个成员，在细胞中的磷脂交换中起作用。近来的配体竞争研究表明，TAP 是结合到 α-生育酚上而不是其他生育酚异构体上（Blatt *et al.*，2001）。尽管研究处于早期阶段，但 TAP 很可能将被证明是与细胞生育酚过程有关的一个重要分子。

## 蛋白激酶 C 与蛋白磷酸酶 2A 活性

有些 α-生育酚的细胞反应是在转录和转录后水平被调节的。例如，α-生育酚抑制平滑肌肉细胞（smooth muscle cell，SMC）的增殖和 PKC 活性。SMC 增殖的抑制对 α-生育酚是特异性的，而对 trolox（一种类似于维生素 E 的物质）、植醇、β-生育酚和 α-生育酚脂类却没有作用（Boscoboinik *et al.*，1991a，b）。由于 α-生育酚和 β-生育酚有非常相似的自由基清除活性，故 α-生育酚作用于 PKC 的机制与其抗氧化特性无关。随后研究表明，PKC 在大量其他细胞类型包括单核细胞（Devaraj *et al.*，1996）、中性白细胞（Kanno *et al.*，1995）、纤维细胞（Hehenberger and Hansson，1997）和肾小球细胞（Tada *et al.*，1997）中受到抑制。最重要的是，在接近人的血浆浓度时，PKC 的这种抑制作用是通过 α-生育酚产生的（Azzi *et al.*，2001）。维生素 E 的抗增殖作用在 HeLa 细胞中却未见到，表明在维生素 E 发挥作用时有不同的细胞特异性细胞增殖途径（Fazzio *et al.*，1997）。另外，PKC 的抑制作用对 α-生育酚与该酶的直接互作无关，也与该酶表达的减少无关。然而，α-生育酚抑制 PKC 与蛋白磷酸酶 2A 的激活相关，反过来又可使 PKC-α 去磷酸化，从而抑制其活性（Clement *et al.*，1997；Ricciarelli *et al.*，1998）。α-生育酚对 PKC 的抑制作用仅在细胞水平上能观测到，对重组 PKC 不明显。

## 环 加 氧 酶

环加氧酶有两种异构体，即 COX-1 和 COX-2。COX-1 在大多数细胞中基本表达，而 COX-2 受到生长因子、肿瘤启动子、细胞因子、糖（肾上腺）皮质激素和脂多糖（LPS）的调节。环加氧酶把花生四烯酸（arachidonic acid，AA）转化为前列腺素 E2（PGF2），即凝血噁烷和类二十烷酸合成的前体物质。上皮细胞高水平的 COX-2 与细胞凋亡的抑制相关，且 COX-2 的过量表达与外科疾病的发病机制有关。在多数人结肠直肠癌中出现了 COX-2 的转录上调（Fosslien，2001）。有趣的是，AA 代谢的变化经 PKC 激活刺激细胞增生，表明 PKC 可能是某些肿瘤被启动或维持的主要信号传导途径之一。近年来，COX-2 在动脉粥样化形成中的作用已被发现。把抗体用于 COX-2 的免疫细胞化学研究，研究表明，COX-2 局限于带有冠状动脉疾病的动脉粥样化的患者的巨噬细胞（Baker *et al.*，1999）。在从老龄鼠中获得的单核细胞中，已表明维生素 E 诱导的 PGE2 产生的降低是通过 COX 活性降低而介导的（Wu *et al.*，2001）。然而，维生素 E 对 COX mRNA 和蛋白质水平没有影响，表明这是一个对 COX 酶的翻译后调节机制。其他非 α-生育酚同源物在抑制 COX 活性上也是有效的，但它们的抑制程度不同，表明抑制程度与其抗氧化能力相一致，提示可能有另外的机制参与。

在 LPS 刺激的 RAW264.7 巨噬细胞和白细胞介素（IL）-1β 处理的 A549 人上皮细胞中已表明，γ-生育酚因直接抑制 COX-2 而抑制 PGE2 的产生（Jiang *et al.*，2000）。而且，食物中 γ-生育酚的主要代谢物在这些细胞中也表现出抑制效应。α-生育酚在 50μmol/L 时轻微降低了（25%）巨噬细胞中 PGE2 的形成，但对上皮细胞中的没有影响。与以前提到的研究相似，γ-生育酚和羧乙基转羟基苯并氢（化）吡喃（carboxyethyl hydroxychromam，γ-CEHC）的抑制效应是因为它们抑制了 COX-2 活性，而不是因为影响了蛋白质的表达或底物利用率所致，且似乎与其抗氧化活性无关。

## 核因子-κB

核因子（nuclear factor，NF）-κB/Rel 家族的转录因子调控着参与炎症和增殖反应的各种基因的表达。典型的 NF-κB 二聚体由 p50 和 p65 亚基组成，且在胞液中以非活性形式结合到抑制蛋白 IκB 上。在被多种刺激因素如炎症或高度增生的细胞因子、ROS 和细菌细胞壁成分等激活后，IκB 便发生磷酸化，并从复合物中蛋白水解移去。激活的 NF-κB 随即进入核内，与启动子和增强子区域的调节 κB 元件互作，调控可诱导基因的转录（Baeuerle and Henkel，1994；Baeuerle and Baltimore，1996）。重要的是，激活的 NF-κB 在人的动脉粥样硬化的血小板中出现，而不是在没有动脉粥样硬化的正常血管的细胞中（Brand *et al.*，1996）以及在动脉损伤模型中（Lindner and Collins，1996）被原位识别。而且，NF-κB 可被导致动脉粥样化的食物（Liao *et al.*，1993）和氧化的 LDL（Brand *et al.*，1997）以及高级糖化作用的终端产物（Yan *et al.*，1994）激活。总结这些结果表明，NF-κB 对动脉粥样硬化形成起关键作用。

已知参与动脉硬化症发生的关键基因谱已表明受到 NF-κB 的调节，包括那些编码肿瘤坏死因子 α（TNF-α）、IL-1、巨噬细胞或粒细胞群体激活因子（macrophage orgrarulocyte colony stimulating factor，M/G-CSF）、单核细胞趋药性蛋白-1（monocyte chemotactic protein-1，MCP-1）、c-myc 和血管细胞黏附分子-1（vascular cell adhesion molecule-1，VCAM-1）和胞内黏附分子-1（intracellular adhesion molecule-1，ICAM-1）的基因（Rimbach *et al.*，2000；Collins and Cybulski，2001）。在动脉硬化损害的早期阶段，不同类型的细胞（巨噬细胞、SMC 和内皮细胞）相互作用，使动态平衡和自增生系统受到破坏，导致动脉壁发育受损和功能紊乱。图 12.3显示了 NF-κB 激活调节的略图，并列出了参与动脉粥样硬化形成的一些主要基因。

包括通过各种抗氧化剂抑制在内的多方证据表明，NF-κB 受氧化还原作用的调节。因为它在炎症反应中的关键作用，所以人们把精力集中在研发可调节 NF-κB 活性的治疗药物上。在此过程中，维生素 E 可通过直接影响 NF-κB 激活途径中的关键步骤，或通过调节细胞内的氧化还原状况起重要作用，而这又是 NF-κB 激活的主要决定因素之一。不断积累的恒定试验数据表明，维生素 E 的抗炎症特性部分是由于下调 NF-κB 所致。Suzuki 和 Packer（1993）研究了维生素 E 衍生物对 TNF-α 诱导的 NF-κB 激活的影响。用维生素 E 乙酸盐或 α-生育酚琥珀酸盐培养人的 Jurkat 细胞能产生 NF-κB 激活的浓度依赖性抑制。相似地，用 α-生育酚的琥珀酸盐前处理和随后用 LPS 激活的巨噬细胞系 THP-1 所进行的凝胶转换研究表明，NF-κB 的活性在 50μmol/L 的浓度中与在没有处理的对照组中相比被抑制了 43%（Nakamura *et al.*，1998）。然而，α-生育酚对

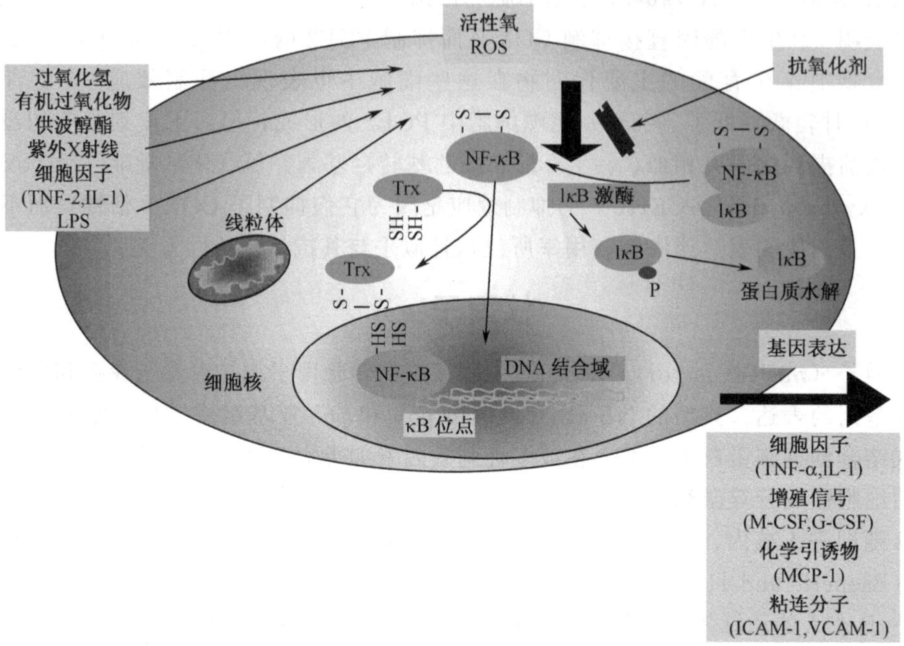


图 12.3　NF-κB 活性的调节

NF-κB 活性没有影响。在这个研究中，通过使用高效液相色谱同步测定维生素 E 和其衍生物来分析细胞维生素 E 的摄入。从培养物中回收的维生素 E 表明，α-生育酚和 α-生育酚琥珀酸盐是等效积累，且没有被代谢。这些观测结果表明，α-生育酚琥珀酸盐可能抑制 NF-κB 活化和（或）以其不变的形式移位到细胞核上。

## α-原肌球蛋白、细胞吸附蛋白、趋化因子及清除剂受体

已表明原肌球蛋白参与动脉粥样硬化形成的发生和恶化（Kocher *et al.*，1991）。在气囊损伤后的初期，血管中层的 SMC 和那些移进内膜的 SMC 含有较少的原肌球蛋白，而在气囊损伤后的后期，原肌球蛋白就恢复到正常值。在 1999 年，Aratri 等使用差异显示技术发现，大鼠血管 SMC 中的 α-原肌球蛋白的表达是通过 α-生育酚诱导的，当使用 β-生育酚时没有观测到 mRNA 水平有明显变化。因此作者把 α-生育酚诱导的原肌球蛋白的过度表达认为是在分子水平认识维生素 E 降低血压的一种相关发现，因为这种作用可能是含有较多原肌球蛋白的 SMC 较小收缩力的结果。

内皮细胞的激活导致血管细胞因子如 IL-1 和 TNF-α 的释放，这些细胞因子反过来诱导细胞表面黏附分子如 VCAM-1 和 ICAM-1 的表达，而这些分子又主要参与内皮嗜中性粒细胞的补充（Cybulski and Gimbrone，1991）。在食物和遗传引起的兔动脉粥样硬化模型的覆盖早期泡沫细胞损伤的动脉内质中已报道 ICAM-1 和 VACM-1 的病灶性表达（Thiery *et al.*，1996）。该表达与 MCP-1 的激活一起被引起单核细胞渗入血管壁、LDL 氧化和清除增加、充满脂肪的泡沫细胞的形成和动脉粥样硬化斑的发生或发展中（Rubanyi，1993）。

正如前面提到的，VCAM-1、ICAM-1 和 MCP-1 的转录至少部分依赖于 NF-κB 的

激活。细胞培养研究表明，用氧化的 LDL 处理内皮细胞显著增加了 VCAM-1 和 ICAM-1 的 mRNA 和蛋白质水平的表达（Yoshida *et al.*，2000）。然而，用α-生育酚进行前处理以剂量依赖方式减少了细胞黏附蛋白的表达。而且，多型核白细胞（polymorphonuclear leukocyte，PMN）或单核白细胞（mononuclear leukocyte，MNC）黏附到氧化的 LDL 激活的内皮细胞上的能力比未受刺激的内皮细胞的高很多。添加 α-生育酚的内皮细胞抑制了 PMN 或 MNC 的粘连。并且，维生素 E 能强化人内皮细胞，剂量依赖性地抑制了 IL-1β 诱导的 MCP-1 的产生（Zapolska-Downar *et al.*，2000）。本研究及其他研究表明，α-生育酚的抗动脉粥样硬化的作用可能部分是由细胞黏附蛋白和趋化因子的下调所致。尽管有证据表明维生素 E 在体外对细胞黏附蛋白有下调作用，但目前还缺乏体内证据。Ricciarelli 等（2000）近来表明，转运氧化的 LDL 进入细胞质的 CD36 清除剂受体在人的平滑肌细胞中表达。有趣的是，α-生育酚通过下调 CD36 mRNA和蛋白质表达的机制抑制氧化的 LDL 的摄取。据推测，α-生育酚对心血管有益的作用至少部分是通过降低 LDL 的摄取来调节的，而 LDL 随后导致泡沫细胞形成的减少。

## 一氧化氮与血小板聚集

一氧化氮（nitric oxide，NO）是调节心血管健康状况的一个关键分子，其通过内皮 NO 合成酶（endothelial NO synthase，eNOS）的产生抑制促炎细胞因子、粘连分子（De Caerina *et al.*，1995）和 MCP-1（Busse and Fleming，1995）的表达。NO 也抑制血小板粘连到内皮（de Graaf *et al.*，1992），能降低动脉管壁的渗透性（Cardonna-Sanclemente and Born，1995；Forster and Weinberg，1997），抑制血管 SMC 增生和迁移（Garg and Hassid，1989），还能作为抗氧化剂发挥作用（Patel *et al.*，2000）。在试验模型中，NO 合成的抑制会加重病情（Naruse *et al.*，1994），而增加 NO 的合成则减轻病情（Cooke *et al.*，1992）。动脉粥样硬化产生的主要危险因素，如年龄（Matz *et al.*，2000）、血胆固醇过多（Stroes *et al.*，1995）、糖尿病（Williams *et al.*，1996）、高血压（Panza *et al.*，1995）、吸烟（Celrmajer *et al.*，1993）和低出生重（Leeson *et al.*，1997）都与 NO 活性的降低有关。重要的是，NO 显著抑制 NF-κB（Matthews *et al.*，1996），这可解释其对粘连蛋白、MCP-1 和其他的基因转录的影响。NO 和 NF-κB 两者的关键作用不是相互排斥的。

Li 等（2001）研究了不同维生素 E 异构体对人的富含血小板的血浆中 NO 活性和血小板聚集的影响作用。维生素 E 的所有 3 个异构体（α-、β-和 δ-生育酚）显著降低 ADP 诱导的血小板聚集并依赖于剂量方式增加 NO 释放。这些异构体不影响 cNOS 蛋白的表达，但增加 cNOS 的磷酸化。进一步表明，α-生育酚的口服（400～1200IU/d）导致血小板生育酚浓度的增加，而这种浓度的增加与人上佛波醇-12-豆蔻酸盐 13-醋酸盐（phorbol 12-myristate 13-acetate，PMA）介导的血小板聚集的明显抑制相关（freedman *et al.*，1996）。来源于这些患者的血小板也表明了对 PKC 明显的完全抑制。这些发现可表示生育酚在预防心脏冠状动脉疾病发生中有一种潜在的机制。

## 硒

### 硒蛋白的生物合成

硒（Se）的最为人知的生物学功能是在细菌、原始细菌和真核细胞中发现的硒蛋白所发挥的。对于哺乳动物而言，硒蛋白为生命所必需，因为对硒蛋白合成非常重要的$tRNA^{sec}$基因的敲除对胚胎形成早期是致命的（Bösl *et al.*，1997）。硒以硒代半胱氨酸的形式被结合进硒蛋白。对真核生物中硒的基本作用机制了解很少，正在深入研究中，但从 Bock 和其合作者对埃希式大肠杆菌硒蛋白合成的遗传和生物化学研究中可获得一些了解（Ehrenreich *et al.*，1992）。参与该过程的 4 个基因（*selA*～*selD*）已被识别，其基因产物分别是硒代半胱氨酸合酶（selenocysteine synthase，SELA）、硒代半胱氨酸特异延长因子（selenocysteine-specific elongation factor，SELB）、硒代半胱氨酸特异tRNA（$tRNA^{sec}$ *selC* 基因产物）和硒磷酸合成酶（selenophosphate synthetase，SELD）。

真核细胞中硒代半胱氨酸结合形成的模型如图 12.4。通过丝氨酰-tRNA 合成酶首先使 $tRNA^{sec}$装满丝氨酸，硒代半胱氨酸合酶随之采用硒磷酸作为活性硒供体催化丝氨酰-$tRNA^{sec}$转化为硒代半胱氨酰-$tRNA^{sec}$（Low and Berry，1996），硒磷酸合成酶催化硒化物和 ATP 合成硒磷酸。硒代半胱氨酸的插入使其成为 3 个 UGA 编码，而此通常作为终止密码子发挥功能。UGA 作为硒代半胱氨酸密码子的再编码需要一个被称为硒代半胱氨酸的插入序列（selenocysteine insertion sequence，SECIS）的 mRNA 二级结构，该结构位于真核细胞 3′非翻译区，因而从较大距离上解码是必需的（Low and Berry，1996）。该结合还涉及两个后续蛋白质，即延长因子 eEFsec（SELB 同源物）和

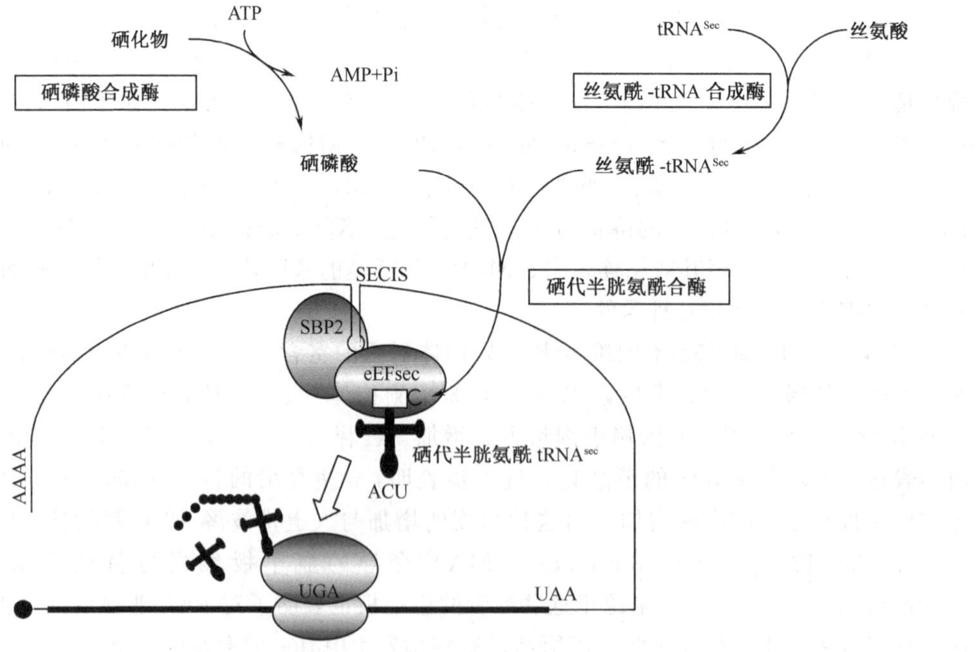


图 12.4　根据 Mansell 和 Berry（2001）改进的真核细胞中硒代半胱氨酸结合形成的模型

SBP2 (Copeland and Driscoll，1999)。据推测，SBP2 结合到 SECIS 后又结合到 tRNAsec-eEFsec 复合物上，该复合物随后能与核糖体互作并诱导硒代半胱氨酸插入 (Mansell and Berry，2001)。假设 UGA 是最新进化的终止密码子，且唯一作为硒代半胱氨酸的感觉密码子以前可能已发挥了作用 (Leinfelder et al.，1988)。既然它容易被氧化，那么把氧引入到空气能以硒代半胱氨酸为对照而反选。硒蛋白表达现在仅限于厌氧生长条件或很好保护的隔离氧的化学环境。到目前为止，识别的大多数硒酶能催化氧化还原反应，此反应中硒代半胱氨酸残基位于活性位点上。与结构上相关的氨基酸半胱氨酸取代硒代半胱氨酸产生了功能酶，但显著降低了 $K_{cat}$，表明硒代半胱氨酸的存在可赋予特定的生物化学特性。硒蛋白在缺硒的动物上能选择性被 $^{75}Se$ 标记，并经电泳分离后通过放射自显影进行观测 (Behne *et al.*，1996)。根据这类试验，在哺乳动物中硒蛋白的数量估计达到 30～50 (Köhrle *et al.*，2000)。到目前为止，通过序列分析而进行特征鉴别的带少于 20 个，其中具个有酶功能的带的数量大于 10。已被识别的和认识较清楚的硒蛋白有 4 个谷胱甘肽过氧化酶 (glutathione peroxidase，GPx)、3 个硫氧还蛋白还原酶 (thioredoxin reductase，TrxR)、3 个脱碘酶 (deiodinase，DI)、硒磷酸合成酶-2、硒蛋白 P (Selenoprotein P，SelP)、在肌肉中的硒蛋白 W 和其他一些未知功能的硒蛋白 (表 12.2)。

**表 12.2　哺乳动物中含硒代半胱氨酸的蛋白质**

| 硒蛋白 | 通常缩写 | 表达 | 功能 |
|---|---|---|---|
| 谷胱甘肽过氧化酶 | ROOH＋2RSH→ | ROH＋$H_2O$＋RSSR | |
| 胞液 GPx | cGPx，GPx1 | 普遍存在 | 抗氧化 |
| 磷脂过氧化氢 GPx | PHGPx，GPx4 | 胃肠道 | 氧化还原调节，精子成熟 |
| 血浆 GPx | PGPx，GPx3 | 普遍存在，尤其在肾 | 氧化还原缓冲？前列腺类物质代谢的调节？ |
| 胃肠 GPx | G1-GPx，GPx2 | 普遍存在 | 局部氧化还原保护？ |
| 碘甲腺原氨酸脱碘酶 | | | |
| 5′-脱碘酶，Ⅰ型 | 5′DI-Ⅰ | 甲状腺、肝、肾、中枢神经系统 | T3 合成：T4＋$2e^-$＋$H^+$→T3＋$1^-$<br>T3 和 T4 的降解 |
| 5′-脱碘酶，Ⅱ型 | 5′DI-Ⅱ | 垂体、甲状腺、胎盘、心脏和骨骼肌、中枢神经系统、棕色脂肪 | T3 合成：<br>T4＋$2e^-$＋$H^+$→T3＋$1^-$ |
| 5′-脱碘酶，Ⅲ型 | 5′DI-Ⅲ | 胎盘、中枢神经系统、皮肤 | T3 和 T4 的降解 |
| 硫氧还原蛋白还原酶 | | | |
| 硫氧还原蛋白还原酶 | TR，TrxR | 普遍存在 | 硫氧还原蛋白的还原、DNA 合成、硫醇二硫化物平衡 |
| 硫氧还原蛋白还原酶 2 | TR2 | 睾丸 | 未知 |
| 线粒体硫氧还原蛋白还原酶 | TR3 | 普遍存在 | 未知 |
| 硒磷酸合成酶-2 | SPS2 | 普遍存在 | 硒蛋白合成 |
| 15kDa 硒蛋白 | Sep15 | 普遍存在 | 未知，蛋白折叠？ |
| 硒蛋白 P | SelP | 血浆 | 未知，氧化还原保护？ |
| 硒蛋白 W | SelW | 普遍存在 | 未知，氧化还原保护？ |
| 硒蛋白 X、N、R、T | SelX、SelN、SelR、SelT | ？ | 未知 |

## 硒蛋白表达的调节

硒蛋白的生物合成主要取决于硒的生物学利用率和最终所形成的 $tRNA^{sec}$，而 $tRNA^{sec}$ 对所有的硒蛋白的水平都有调节作用。在硒缺乏时，UGA 被翻译作终止密码子，导致在 UGA 密码子上链的提前终止。硒蛋白合成的降低又进一步伴随各自 mRNA 水平的下降。这不是由较低的转录速率而是由稳定性的损失，即 mRNA 的降解增加所致（Bermano *et al.*，1996）。

对于硒蛋白的最优翻译而言，除了硒的供给外，它的 mRNA 似乎需要最优 UGA 区位、最优 UGA 环境和优化硒代半胱氨酸结合的 SECIS 元件。令人吃惊的是，单个硒蛋白在其对硒耗竭的特异性反应上存在差异，这种现象被称为“硒蛋白等级”现象（Flohé *et al.*，2000）。硒蛋白中有一些如 cGPx，在硒耗竭中很快消失（Knight 和 Sunde，1987；Li *et al.*，1990；Müller and Pallauf，2002），而当重新供给时又可缓慢再合成（Knight and Sunde，1988）。然而，其他的硒蛋白如 PHGPx，在有限硒供给时逐渐下降，而再补充后立即反弹（Weitzel *et al.*，1990；Bermano *et al.*，1996）。这种独一无二和特异性的硒蛋白表达的调节已表明至少其部分是通过 SECIS 元件调节的，因为在等级上具有相似位置的硒蛋白在 SECIS 区有相似的结构。在硒蛋白 mRNA SECIS 元件方面的差异能改变对硒代半胱氨酸翻译复合物的相对亲和性，因而赋予它们不同的翻译能力，当在硒有限时尤其如此。无论涉及什么机制，硒蛋白生物合成中的等级被认为可反应它们的生物重要性。Sunde（1997）认为有如下等级排序：

$$\text{GI-GPx} > \text{TrxR, SelP, 5'DI-I} > \text{cGPx}$$

除了硒依赖因子影响硒蛋白的表达，也存在不依赖硒的因子。例如，5′DI-Ⅰ的组织特异性表达主要见于甲状腺、肝脏、肾脏和脑垂体，5′DI-Ⅱ的组织特异表达主要在啮齿动物的棕色脂肪中，以及胎盘、甲状腺、脑垂体，和中枢神经系统，5′DI-Ⅲ的表达主要在胎盘、皮肤，也在胎儿和新生儿肝脏中，但不在成年人肝脏中（Köhrle *et al.*，2000）。在睾丸中可发现 PHGPx 的最高表达量（Maiorino *et al.*，1998）。GI-GPx 仅在胃肠道中表达（Wingler *et al.*，2000），而 pGPx 则优先在代谢表面（肾近曲小管、肠上皮、皮肤和肺）形成和分泌。在结构和功能尚未确定的硒蛋白中，硒蛋白 W 存在于心脏和骨骼肌上，其他较少阐述的硒蛋白见于睾丸、前列腺和胰腺中（Behne *et al.*，1997）。有时不寻常表达的模式具有专一的作用，可用来识别 PHGPx（精子发生）、GI-GPx（阻止过氧化氢吸收的屏障）和 DI（甲状腺激素合成）（Köhrle *et al.*，2000）。然而，到目前还没有一个具有组织特异表达的分子基础得到明确阐述。而且，有证据表明在硒耗竭的条件下含硒酶在某些组织中的减少比在其他组织中的减少更迅速。在两代硒耗竭的大鼠中，总 GPx 活性在肝脏、肾脏、肺、心脏、肾上腺、胰腺和肌肉中均小于正常组的 5%，在雄性胸腺和睾丸中小于正常组的 10%，但在脑中下降到仅约 50%，在卵巢、雌性胸腺中下降到 25%（Thompson *et al.*，1995）。从这些结果中可以得出结论，即肯定存在将硒导向特异组织中的特异酶的机制，这很可能是通过 mRNA 稳定性的组织特异性调节而实现的。cGPx 和 PHGPx 在硒有限的情况下，优先在脑、生殖器官和内分泌组织中存留。Brigelius-Flohé（1999）提出了 cGPx 的组织特异稳定性的如下顺序（不与绝对活性相混）：

脑≫胸腺>甲状腺>心脏>肝脏、肾脏和肺

对某些硒蛋白而言，能观测到激素依赖性调节。在单核细胞中，人的 TrxR 通过 1，25-二羟维生素 D3（Schütze *et al.*，1999）和在胎儿类造骨细胞中通过某些细胞因子和生长因子（Schütze *et al.*，1998）而被快速诱导。3 个 DI 同工酶被甲状腺激素、类维生素 A、性激素、糖皮质类固醇和一系列生长因子及细胞因子所调节。已表明 PHGPx 是依赖于性激素表达的。已经提出了关于 GPx 基因通过氧化应激的诱导的假设，但还未获得有说服力的体内证明。CGPx 通过氧化反应元件被诱导仅在体外被阐明（Cowan *et al.*，1993）。

## 为什么硒蛋白是重要的?

GPx 存在于所有有氧化过程发生的哺乳动物组织中。通过还原氢过氧化物为相应的醇类，这些酶能阻止活性氧自由基的产生，因此有助于保护生物体的大分子和生物膜免被氧化。GPx1 催化谷胱甘肽依赖性的过氧化氢的还原和各种有机过氧化氢物的还原。尽管人们把 GPx1 与过氧化氢酶和超氧化物歧化酶（一种主要的抗氧化酶）放在一起考虑，但其作为必需抗氧化剂的作用已受到质疑。在硒缺乏的动物上，GPx1 的活性和表达水平出现巨大降低，但对细胞的代谢却没有明显影响。另外，甚至在过氧化物的条件下，在 GPx1 敲除的鼠上并未观测到表型变化（Ho *et al.*，1997）。然而，当接触无致病性的库克萨基病毒（coxsackievirus）株时，cGPx 缺乏鼠会患心肌炎。令人吃惊的是，以前的无毒株在通过 $cGPx^{-/-}$ 鼠后也证实在对照鼠上是致病性的（Beck *et al.*，1998）。因缺乏 cGPx 而致的过氧化物水平的增加，导致了不能被任何其他硒蛋白或维生素 E 的作用阻止的突变率的增加。这些发现表明，cGPx 在环境应激拯救细胞中可能是非常重要的，而其在正常条件下的功能却能由其他细胞成分补偿。

虽然 PHGPx、pGPx、GI-GP 和 TrxR 通过减少硫氧还蛋白而催化氧化还原反应，但它们在过氧化氢物的解毒中不竞争支持 cGPx。因此，它们的可能作用在于对一系列氧化还原敏感酶进行调节。5-脂氧酶和 GPx 与还原型谷胱甘肽一起在体外能通过环加氧酶、15-脂氧酶阻止对花生四烯酸盐的利用，因为这些前列腺素和白细胞三烯合成的关键酶需要一定的过氧化物变为有活性的酶（Flohé *et al.*，2000）。因此，$H_2O_2$ 和烷基氢过氧化物能再次激活这些酶。已表明，PHGPx 在 RBL-2H3 细胞中的过度表达显著减少白细胞三烯 C4 和白细胞三烯 B4 的合成（Imai *et al.*，1998）。而且，在细胞外空间通过去除 $H_2O_2$ 和脂类氢过氧化物，PGPx 可成为炎症反应的调节因子。硒蛋白 P 结合到内皮细胞上，且推测可保护它们免受促炎症的氢过氧化物（包括过氧化亚硝酸盐）的损害（Sies *et al.*，1998）。

此外，过氧化物配合的降低能导致氧化还原敏感性转录因子如 NF-κB 的激活的减少（Viita *et al.*，1999）。IL-1 和依赖于 TNF-α 的 NF-κB 的激活可因 cGPx（Kretz-Remy *et al.*，1996）和在 ECV304 细胞中 PHGPx（Brigelius-Flohé *et al.*，1997）的过度表达而受到抑制。硫氧还蛋白（Trx）是 TrxR 的主要底物，它一方面似乎抑制了 NF-κB 的活化，因为 Trx 在细胞中的过度表达显著降低 NF-κB 活性，而另一方面，它似乎又为 NF-κB 中的半胱氨酸残基的还原所必需，而这又是 NF-κB 的 DNA 结合所必需（Hayashi *et al.*，1993）。

细胞凋亡过程是又一个受氧化还原调节的信号传导的例子，在体外能通过硒的添加（Kayanoki *et al.*，1996）以及 cGPx 和 PHGPx 过度表达（Packham *et al.*，1996）而被受刺激的 GPx 活性所抑制。由这些结果引出一个观念，即细胞凋亡信号传导级联反应的上部分可被氢过氧化物促进。用紫外光线刺激细胞通过细胞凋亡导致 80%的细胞死亡，而硒的添加显著改善了细胞的存活。硒的这种抗细胞凋亡的功能是通过对 caspases-3 的抑制，也可能是通过调节半胱氨酸残基而介导的（Park *et al.*，2000）。用人的免疫缺失病毒（human immunodeficiency virus，HIV）感染 T 淋巴细胞后，导致 GPx 活性降低，这与对脂类氢过氧化物和 $H_2O_2$ 诱导的细胞凋亡的较高敏感性有关（Sandstrom *et al.*，1994）。

硒及其在防止细胞凋亡方面的功能已关联到参与细胞周期以致癌症的发生过程。早期的流行病学研究表明，食物中硒的摄入量与某些恶性肿瘤的发生呈负相关（Clark，1985）。在肿瘤发生的早期阶段，ROS 对 DNA 的氧化损伤有重要作用。供给足量硒和维生素 E 的细胞似乎不易产生癌变（Halliwell，2000），因此紫外线诱导的皮肤癌与添补硒的无毛鼠的 GPx 活性呈负相关（Pence *et al.*，1994）。在曝露紫外线照射前或添加致癌物苯并芘或色氨酸热解物前，将细胞用亚硒酸盐或 $\alpha$-生育酚培养 24h，在所有情况下均可抑制细胞的变形，而同时添加这两种营养物质则显示出最大的效应。加硒培养会伴以 GPx 活性的增加和过氧化物的降低（Borek *et al.*，1986）。然而，就防止氧化 DNA 损伤而言，硒的保护作用还不易被解释清楚，因为硒在抗肿瘤发生时的摄入量显著大于与已知硒蛋白的最大表达有关的摄入量。因此当前讨论认为硒的药理浓度是通过促进细胞凋亡而产生直接的抗增生作用。过量的硒供给可能导致细胞内还原型谷胱甘肽的耗竭（Combs，1999），而且某些硒的代谢物可抑制细胞增殖。谷胱甘肽硒可能诱导肿瘤抑制基因 p53（Lanfear *et al.*，1994），且硒甲基硒代半胱氨酸在 S 时段能抑制乳腺细胞生长（Sinha and Medina，1997）。

此外，据报道，CHD 的发生与硒的供给呈反相关。动脉粥样硬化的产生属于病理过程，很可能与氧化应激相关。一旦摄入氧化的 LDL（oxLDL）时，巨噬细胞就转变为泡沫细胞，且 SMC 也增生，这就是启动动脉粥样硬化形成的过程。因此，任何能阻止或对抗 LDL 氧化的抗氧化剂系统均应抑制包括 GPx 在内的该过程。而且，需要一定过氧化物环境才具活性的脂肪氧酶，能在 LDL 颗粒内产生过氧羟基脂类，此后可进一步产生脂类过氧化。作为细胞外酶的 pGPx 和有效降低氧化的 LDL 中氢过氧化物的 PHGPx，是防止动脉粥样硬化的可能候选物。然而，pGPx 并无最优的专一性，因为 pGPx 作用于 $H_2O_2$ 和在某种程度上作用于过氧化的复合脂类不能降低胆固醇酯氢过氧化物（Yamamoto and Takahashi，1993）。PHGPx 确有最优的专一性，但不存在于细胞外，而 LDL 的氧化即在细胞外进行（Köhrle *et al.*，2000）。因此，一些矛盾的发现仍有待解释，且需要进一步研究来充分阐明硒在动脉粥样硬化产生中的作用。

## 硒和（或）维生素 E 的缺乏对基因表达的影响

为了测检与鼠缺乏硒和维生素 E 有关的分子结果，cDNA 阵列技术已应用于确定饲喂硒和维生素 E 缺乏的饲粮 7 周后的鼠肝脏中的转录反应（Fischer *et al.*，2000）。使用来自克隆技术的 Atlas$^{TM}$ 鼠的 cDNA 毒理学阵列Ⅱ同时监控 465 个基因的表达，但只有 2 倍或更多倍的变化才被认为是显著的（表 12.3）。

**表 12.3 大鼠肝基因表达中硒和维生素 E 缺乏症相关变化的选择**

| 基因库登记 | △-硒-E(倍数) | 基因 | 功能 |
| --- | --- | --- | --- |
| 凋亡/细胞周期 | | | |
| Y13336 | ↓2.0 | 防止细胞死亡 1 蛋白的防卫因子(DAD1) | 防细胞凋亡 |
| AF081503 | ↓2.6 | 凋亡蛋白 1 的抑制因子 | 防细胞凋亡 |
| U72350 | ↓3.2 | Bc12-L1 | 防细胞凋亡 |
| M22413 | ↓2.0 | 碳酸酐酶(CAⅢ) | 抗氧化剂,防细胞凋亡 |
| D90345 | ↓2.2 | T-复合蛋白 1(CCT)α-亚基 | 伴侣,蛋白质折叠 |
| X82021 | ↓2.2 | HSC70-互作蛋白(HIP) | 伴侣 HSC70 的稳定 |
| J03969 | ↓2.9 | 核仁磷酸蛋白(NPM) | 刺激正常细胞生长 |
| D14014 | ↓3.1 | $G_1$/S-特异性细胞周期蛋白 D1(CCND1) | 细胞周期的起始和致癌基因 |
| J04154 | ↑2.1 | 早期生长反应蛋白 1 | 抑制生长和诱导凋亡 |
| U77129 | ↑2.0[1] | SPS1/Ste20 同等物 KHS1 | 在 MAP 激酶通路中的信号传感器 |
| 抗氧化剂防/应激/炎症反应 | | | |
| X12367 | ↓18.8 | 细胞谷胱甘肽过氧化酶 I | 过氧化物脱毒 |
| J05181 | ↓3.4 | γ-谷氨酰半胱氨酸合成酶(γ-GCS) | 谷胱甘肽合成 |
| U22424 | ↓2.2 | 11-β-羟基类固醇脱氢酶 2 | 转化皮质脂酮为 11-脱氢皮脂酮 |
| L49379 | ↓2.3 | 多特异有机阴离子运出因子(cMOAT) | 脱毒,白三烯素 C4 的运出 |
| Jo2608 | ↑15.3 | DT-硫辛酰胺脱氢酶 | 异型生物质的代谢 |
| D00753 | ↑2.1 | SPl-3 丝氨酸蛋白酶抑制因子 | 急性期反应蛋白 |
| J00696 | ↑2.3 | α-1 酸糖蛋白 | 急性期反应蛋白 |
| J00734 | ↑2.3 | 纤维蛋白原 γ 链 | 急性期反应蛋白 |
| S65838 | ↑3.6 | 金属硫因蛋白 1 | 急性期反应蛋白,抗氧化剂 |

1 一个阵列中背景水平的基因信号。

仅只缺乏维生素 E 没有显著影响任何受监控的基因，很可能是不存于 cDNA 膜上的其他基因可能被维生素 E 以不同方式调节。另外，肝脏以外的组织都可能对维生素 E 诱导的不同基因表达的变化更加敏感。

除了 *cGPx* 基因的 13.9 倍下调外，仅只缺乏硒还伴随着 UDP-葡糖醛酸基转移酶 1 和胆红素-UDP 葡糖醛酸基转移酶同工酶 2 表达的增加。已知这两种酶对肝脏中异生素的解毒有重要作用。同样，也参与异生素代谢并通过糖皮质激素诱导的鼠肝脏细胞色素 P450 4B1 被诱导了 2.3 倍。与对照比较，缺硒动物的花生四烯酸 12-脂肪氧合酶 ALOX12）的 mRNA 水平高 2.4 倍。已表明，ALOX12 和 PHGPx 是作用相反的平衡细胞内过氧羟基脂类浓度的酶（Schnurr *et al.*，1999），但 PHGPx 活性的抑制增加 ALOX12 的酶催化作用（Chen *et al.*，2000）。硒和维生素 E 的缺乏影响到全部被检基因的 5%，缺乏的特征是抑制程序性细胞死亡基因的显著下负调，包括防止细胞死亡蛋白 1 防卫因子、细胞凋亡蛋白 1 的抑制因子和 Bc12-L1 的抑制因子。而且，被认为是生长和变形的抑制因子和细胞凋亡的诱导物的早期生长反应蛋白 1 的表达水平增加了 2 倍。因此，近期报道的作为抗氧化剂而发挥防止 $H_2O_2$ 诱导的细胞凋亡作用（Raisanen *et al.*，1999）的碳酸酐酶Ⅲ（carbonic anhydrase Ⅲ，CAⅢ）被下调了 2 倍。在二者都缺乏的鼠肝中更强的细胞负增长趋势是由核仁磷酸蛋白和 G/S 特异细胞周期蛋白 D1

的下调而进一步提出的，而该蛋白 D1 的特征是作为抗细胞凋亡机制中的一个重要信号。

而且，硒和维生素 E 的同时缺乏导致急性期反应蛋白（金属硫因蛋白、DT-硫辛酰胺脱氢酶、α-1 酸糖蛋白）和 SPI-3 丝氨酸蛋白酶抑制子（serine proteinase inhibitor-3，SPI-3）的诱导。因此，在饲喂缺硒和维生素 E 饲粮鼠的促炎症反应的进一步标志是鼠纤维蛋白原 γ-链的表达较高，而该表达已表明其在炎症期间的鼠肝中被显著上调（Simpson-Haidaris *et al.*，1995）。促炎症基因的诱导伴随着抗炎症酶 11-β-羟基类固醇脱氢酶 2 的一致抑制，而该酶在鼠的肝脏中转化糖（肾上腺）皮质激素皮质脂酮为其无活性的 11-脱氢形式，因而调控糖（肾上腺）皮质激素接近受体。这些结果表明，硒缺乏对硒蛋白表达有负影响，且很可能作为次级影响而对炎症和细胞周期依赖性基因有调节作用。

## 参 考 文 献


Aratri, E., Spycher, S.E., Breyer, I. and Azzi, A. (1999) Modulation of alpha-tropomyosin expression by alpha-tocopherol in rat vascular smooth muscle cells. *FEBS Letters* 447, 91–94.

Azzi, A., Breyer, I., Feher, M., Ricciarelli, R., Stocker, A., Zimmer, S. and Zingg, J. (2001) Nonantioxidant functions of alpha-tocopherol in smooth muscle cells. *Journal of Nutrition* 131, 378S–381S.

Baeuerle, P.A. and Baltimore, D. (1996) NF-kappa B: ten years after. *Cell* 87, 13–20.

Baeuerle, P.A. and Henkel, T. (1994) Function and activation of NF-kappa B in the immune system. *Annual Review of Immunology* 12, 141–179.

Baker, C.S., Hall, R.J., Evans, T.J., Pomerance, A., Maclouf, J., Creminon, C., Yacoub, M.H. and Polak, J.M. (1999) Cyclooxygenase-2 is widely expressed in atherosclerotic lesions affecting native and transplanted human coronary arteries and colocalizes with inducible nitric oxide synthase and nitrotyrosine particularly in macrophages. *Arteriosclerosis, Thrombosis, and Vascular Biology* 19, 646–655.

Beck, M.A., Esworthy, R.S., Ho, Y.S. and Chu, F.F. (1998) Glutathione peroxidase protects mice from viral-induced myocarditis. *FASEB Journal* 12, 1143–1149.

Behne, D., Kyriakopoeulos, A., Weiss-Nowak, C., Kalckloesch, M., Westphal, C. and Gessner, H. (1996) Newly found selenium-containing proteins in the tissues of the rat. *Biological Trace Element Research* 55, 99–110.

Behne, D., Kyriakopoulos, A., Kalcklosch, M., Weiss-Nowak, C., Pfeifer, H., Gessner, H. and Hammel, C. (1997) Two new selenoproteins found in the prostatic glandular epithelium and in the spermatid nuclei. *Biomedical and Environmental Science* 10, 340–345.

Behrens, W.A., Thompson, J.N. and Madere, R. (1982) Distribution of alpha-tocopherol in human plasma lipoproteins. *American Journal of Clinical Nutrition* 35, 691–696.

Ben Hamida, C., Doerflinger, N., Belal, S., Linder, C., Reutenauer, L., Dib, C., Gyapay, G., Vignal, A., Le Paslier, D. and Cohen, D. (1993) Localization of Friedreich ataxia phenotype with selective vitamin E deficiency to chromosome 8q by homozygosity mapping. *Nature Genetics* 5, 195–200.

Bermano, G., Arthur, J.R. and Hesketh, J.E. (1996) Selective control of cytosolic glutathione peroxidase and phospholipid hydroperoxide glutathione peroxidase mRNA stability by selenium supply. *FEBS Letters* 387, 157–160.

Blatt, D.H., Leonard, S.W. and Traber, M.G. (2001) Vitamin E kinetics and the function of tocopherol regulatory proteins. *Nutrition* 17, 799–805.

Borek, C., Ong, A., Mason, H., Donahue, L. and Biaglow, J.E. (1986) Selenium and vitamin E inhibit radiogenic and chemically induced transformation *in vitro* via different mechanisms. *Proceedings of the National Academy of Sciences USA* 83, 1490–1494.

Boscoboinik, D., Szewczyk, A. and Azzi, A. (1991a) Alpha-tocopherol (vitamin E) regulates vascular smooth muscle cell proliferation and protein kinase C activity. *Archives of Biochemistry and Biophysics* 286, 264–269.

Boscoboinik, D., Szewczyk, A., Hensey, C. and Azzi, A. (1991b) Inhibition of cell proliferation by alpha-tocopherol. Role of protein kinase C. *Journal of Biological Chemistry* 266, 6188–6194.

Bösl, M.R., Takaku, K., Oshima, M., Nishimura, S. and Taketo, M.M. (1997) Early embryonic lethality caused by targeted disruption of the mouse selenocysteine tRNA gene. *Proceedings of the National Academy of Sciences USA* 94, 5531–5534.

Brand, K., Page, S., Rogler, G., Bartsch, A., Brandl, R., Knuechel, R., Page, M., Kaltschmidt, C., Baeuerle, P.A. and Neumeier, D. (1996) Activated transcription factor nuclear factor-kappa B is present in the atherosclerotic lesion. *Journal of Clinical Investigation* 97, 1715–1722.

Brand, K., Page, S., Walli, A.K., Neumeier, D. and Baeuerle, P.A. (1997) Role of nuclear factor-kappa B in atherogenesis. *Experimental Physiology* 82, 297–304.

Brigelius-Flohé, R. (1999) Tissue-specific functions of individual glutathione peroxidases. *Free Radical Biology and Medicine* 27, 951–965.

Brigelius-Flohé, R. and Traber, M.G. (1999) Vitamin E: function and metabolism. *FASEB Journal* 13, 1145–1155.

Brigelius-Flohé, R., Friedrichs, B., Maurer, S., Schultz, M. and Streicher, R. (1997) Interleukin-1-induced nuclear factor kappa B activation is inhibited by overexpression of phospholipid hydroperoxide glutathione peroxidase in a human endothelial cell line. *Biochemical Journal* 328, 199–203.

Busse, R. and Fleming, I. (1995) Regulation and functional consequences of endothelial nitric oxide formation. *Annals of Medicine* 27, 331–340.

Cardona-Sanclemente, L.E. and Born, G.V. (1995) Effect of inhibition of nitric oxide synthesis on the uptake of LDL and fibrinogen by arterial walls and other organs of the rat. *British Journal of Pharmacology* 114, 1490–1494.

Catignani, G.L. and Bieri, J.G. (1977) Rat liver alpha-tocopherol binding protein. *Biochimica et Biophysica Acta* 497, 349–357.

Celermajer, D.S., Sorensen, K.E., Georgakopoulos, D., Bull, C., Thomas, O., Robinson, J. and Deanfield, J.E. (1993) Cigarette smoking is associated with dose-related and potentially reversible impairment of endothelium-dependent dilation in healthy young adults. *Circulation* 88, 2149–2155.

Chen, C.J., Huang, H.S., Lin, S.B. and Chang, W.C. (2000) Regulation of cyclooxygenase and 12-lipoxygenase catalysis by phospholipid hydroperoxide glutathione peroxidase in A431 cells. *Prostaglandins, Leukotrienes and Essential Fatty Acids* 62, 261–268.

Chojkier, M., Houglum, K., Lee, K.S. and Buck, M. (1998) Long- and short-term D-alpha-tocopherol supplementation inhibits liver collagen alpha1(I) gene expression. *American Journal of Physiology* 275, G1480–G1485.

Clark, L.C. (1985) The epidemiology of selenium and cancer. *Federation Proceedings* 44, 2584–2589.

Clement, S., Tasinato, A., Boscoboinik, D. and Azzi, A. (1997) The effect of alpha-tocopherol on the synthesis, phosphorylation and activity of protein kinase C in smooth muscle cells after phorbol 12-myristate 13-acetate down-regulation. *European Journal of Biochemistry* 246, 745–749.

Cohn, W., Gross, P., Grun, H., Loechleiter, F., Muller, D.P. and Zulauf, M. (1992) Tocopherol transport and absorption. *Proceedings of the Nutrition Society* 51, 179–188.

Collins, T. and Cybulski, M.I. (2001) NF-κB: pivotal mediator or innocent bystander in atherogenesis. *Journal of Clinical Investigation* 107, 255–264.

Combs, G.F. Jr (1999) Chemopreventive mechanisms of selenium. *Medizinische Klinik* 94 Supplement 3, 18–24.

Cooke, J.P., Singer, A.H., Tsao, P., Zera, P., Rowan, R.A. and Bilingam, M.E. (1992) Antiatherogenic effects of L-arginine in the hypercholesterolemic rabbit. *Journal of Clinical Investigation* 90, 1168–1172.

Copeland, P.R. and Driscoll, D.M. (1999) Purification, redox sensitivity, and RNA binding properties of SECIS-binding protein 2, a protein involved in selenoprotein biosynthesis. *Journal of Biological Chemistry* 274, 25447–25454.

Cowan, D.B., Weisel, R.D., Williams, W.G. and Mickle, D.A. (1993) Identification of oxygen responsive elements in the 5′-flanking region of the human glutathione peroxidase gene. *Journal of Biological Chemistry* 268, 26904–26910.

Cybulsky, M.I. and Gimbrone, M.A. (1991) Endothelial expression of a mononuclear leukocyte adhesion molecule during atherogenesis. *Science* 251, 788–791.

De Caerina, R., Libby, P., Peng, H.B., Thannickal, V.J., Rajavashisth, T.B., Gimbrone, M.A.J., Shin, W.S. and Liao, J.K. (1995) Nitric oxide decreases cytokine-induced endothelial activation: nitric oxide selectively reduces endothelial expression of adhesion molecules and proinflammatorry cytokines. *Journal of Clinical Investigation* 96, 60–68.

de Graaf, J.C., Banga, J.D., Moncado, S., Palmer, R.M., de Groot, P.G. and Sixma, J.J. (1992) Nitric oxide functions as an inhibitor of platelet adhesion under flow conditions. *Circulation* 85, 2284–2290.

Devaraj, S., Li, D. and Jialal, I. (1996) The effects of alpha tocopherol supplementation on monocyte function. Decreased lipid oxidation, interleukin 1 beta secretion, and monocyte adhesion to endothelium. *Journal of Clinical Investigation* 98, 756–763.

Dutta-Roy, A.K. (1999) Molecular mechanism of cellular uptake and intracellular translocation of alpha-tocopherol: role of tocopherol-binding proteins. *Food Chemistry and Toxicology* 37, 967–971.

Ehrenreich, A., Forchhammer, K., Tormay, P., Veprek, B. and Böck, A. (1992) Selenoprotein synthesis in *E. coli*. Purification and characterisation of the enzyme catalysing selenium activation. *European Journal of Biochemistry* 206, 767–773.

Evans, H.M. and Bishop, K.S. (1922) On the existence of a hitherto unrecognized dietary factor essential for reproduction. *Science* 56, 650–651.

Fazzio, A., Marilley, D. and Azzi, A. (1997) The effect of alpha-tocopherol and beta-tocopherol on proliferation, protein kinase C activity and gene expression in different cell lines. *Biochemistry and Molecular Biology International* 41, 93–101.

Fechner, H., Schlame, M., Guthmann, F., Stevens, P.A. and Rustow, B. (1998) Alpha- and delta-tocopherol induce expression of hepatic alpha-tocopherol-transfer-protein mRNA. *Biochemical Journal* 331, 577–581.

Fischer, A., Pallauf, J., Gohil, K., Weber, S.U., Packer, L. and Rimbach, G. (2001) Effect of selenium and vitamin E deficiency on differential gene expression in rat liver. *Biochemical and Biophysical Research Communications* 285, 470–475.

Flohé, L., Andreesen, J.R., Brigelius-Flohé, R., Maiorino, M. and Ursini, F. (2000) Selenium, the element of the moon, in life on earth. *IUBMB Life* 49, 411–420.

Forster, B.A. and Weinberg, P.D. (1997) Changes with age in the influence of endogenous nitric oxide on transport properties of the rabbit aortic wall near branches. *Arteriosclerosis, Thrombosis, and Vascular Biology* 17, 1361–1368.

Fosslien, E. (2001) Review: molecular pathology of cyclooxygenase-2 in cancer-induced angiogenesis. *Annals of Clinical Laboratory Science* 31, 325–348.

Freedman, J.E., Farhat, J.H., Loscalzo, J. and Keaney, J.F. Jr (1996) Alpha-tocopherol inhibits aggregation of human platelets by a protein kinase C-dependent mechanism. *Circulation* 94, 2434–2440.

Garg, U.C. and Hassid, A. (1989) Nitric oxide generation vasodilators and 8-bromo-cyclic guanosine monophosphate inhibit mitogenesis and proliferation of cultured rat vascular smooth muscle cells. *Journal of Clinical Investigation* 83, 1774–1777.

Gladyshev, V.N. and Hatfield, D.L. (1999) Selenocysteine-containing proteins in mammals. *Journal of Biomedical Sciences* 6, 151–160.

Halliwell, B. (1996) Antioxidants in human health and disease. *Annual Review of Nutrition* 16, 33–50.

Halliwell, B. (2000) Why and how should we measure oxidative DNA damage in nutritional studies? How far have we come? *American Journal of Clinical Nutrition* 72, 1082–1087.

Hayashi, T., Ueno, Y. and Okamoto, T. (1993) Oxidoreductive regulation of nuclear factor kappa B. Involvement of a cellular reducing catalyst thioredoxin. *Journal of Biological Chemistry* 268, 11380–11388.

Hehenberger, K. and Hansson, A. (1997) High glucose-induced growth factor resistance in human fibroblasts can be reversed by antioxidants and protein kinase C-inhibitors. *Cell Biochemistry and Function* 15, 197–201.

Herrera, E. and Barbas, C. (2001) Vitamin E: action, metabolism and perspectives. *Journal of Physiological Biochemistry* 57, 43–56.

Ho, Y.S., Magnenat, J.L., Bronson, R.T., Cao, J., Gargano, M., Sugawara, M. and Funk, C.D. (1997) Mice deficient in cellular glutathione peroxidase develop normally and show no increased sensitivity to hyperoxia. *Journal of Biological Chemistry* 272, 16644–16651.

Hosomi, A., Arita, M., Sato, Y., Kiyose, C., Ueda, T., Igarashi, O., Arai, H. and Inoue, K. (1997) Affinity for alpha-tocopherol transfer protein as a determinant of the biological activities of vitamin E analogs. *FEBS Letters* 409, 105–108.

Imai, H., Narashima, K., Arai, M., Sakamoto, H., Chiba, N. and Nakagawa, Y. (1998) Suppression of leukotriene formation in RBL-2H3 cells that overexpressed phospholipid hydroperoxide glutathione peroxidase. *Journal of Biological Chemistry* 273, 1990–1997.

Ingold, K.U., Webb, A.C., Witter, D., Burton, G.W., Metcalfe, T.A. and Muller, D.P. (1987) Vitamin E remains the major lipid-soluble, chain-breaking antioxidant in human plasma even in individuals suffering severe vitamin E deficiency. *Archives of Biochemistry and Biophysics* 259, 224–225.

Jiang, Q., Elson-Schwab, I., Courtemanche, C. and Ames, B.N. (2000) γ-Tocopherol and its major metabolite, in contrast to α-tocopherol, inhibit cyclooxygenase activity in macrophages and epithelial cells. *Proceedings of the National Academy of Sciences USA* 97, 11494–11499.

Kagan, V.E. and Tyurina, Y.Y. (1998) Recycling and redox cycling of phenolic antioxidants. *Annals of the New York Academy of Sciences* 854, 425–434.

Kanno, T., Utsumi, T., Kobuchi, H., Takehara, Y., Akiyama, J., Yoshioka, T., Horton, A.A. and Utsumi, K. (1995) Inhibition of stimulus-specific neutrophil superoxide generation by alpha-tocopherol. *Free Radical Research* 22, 431–440.

Kayanoki, Y., Fujii, J., Islam, K.N., Suzuki, K., Kawata, S., Matsuzawa, Y. and Taniguchi, N. (1996) The protective role of glutathione peroxidase in apoptosis induced by reactive oxygen species. *Journal of Biochemistry* 119, 817–822.

Kayden, H.J. and Traber, M.G. (1993) Absorption, lipoprotein transport, and regulation of plasma concentrations of vitamin E in humans. *Journal of Lipid Research* 34, 343–358.

Knight, S.A. and Sunde, R.A. (1987) The effect of progressive selenium deficiency on anti-glutathione peroxidase antibody reactive protein in rat liver. *Journal of Nutrition* 117, 732–738.

Knight, S.A. and Sunde, R.A. (1988) Effect of selenium repletion on glutathione peroxidase protein level in rat liver. *Journal of Nutrition* 118, 853–858.

Kocher, O., Gabbiani, F., Gabbiani, G., Reidy, M.A., Cokay, M.S., Peters, H. and Huttner, I. (1991) Phenotypic features of smooth muscle cells during the evolution of experimental carotid artery intimal thickening. Biochemical and morphologic studies. *Laboratory Investigation* 65, 459–470.

Köhrle, J., Brigelius-Flohé, R., Böck, A., Gartner, R., Meyer, O. and Flohé, L. (2000) Selenium in biology: facts and medical perspectives. *Biological Chemistry* 381, 849–864.

Kretz-Remy, C., Mehlen, P., Mirault, M.E. and Arrigo, A.P. (1996) Inhibition of IκB-α phosphorylation and degradation and subsequent NF-κB activation by glutathione peroxidase overexpression. *Journal of Cell Biology* 133, 1083–1093.

Lagrost, L., Desrumaux, C., Masson, D., Deckert, V. and Gambert, P. (1998) Structure and function of the plasma phospholipid transfer protein. *Current Opinion in Lipidology* 9, 203–209.

Lanfear, J., Fleming, J., Wu, L., Webster, G. and Harrison, P.R. (1994) The selenium metabolite selenodiglutathione induces p53 and apoptosis: relevance to the chemopreventive effects of selenium? *Carcinogenesis* 15, 1387–1392.

Leeson, C.P.M., Whincup, P.H., Cook, D.G., Donald, A.E., Papacosta, O., Lucas, A. and Deanfield, J.E. (1997) Flow mediated dilation in 9- to 11-year old children – the influence of intrauterine and childhood factors. *Circulation* 96, 2233–2238.

Leinfelder, W., Zehelein, E., Mandrand-Berthelot, M.A. and Böck, A. (1988) Gene for a novel tRNA species that accepts L-serine and cotranslationally inserts selenocysteine. *Nature* 331, 723–725.

Li, D., Saldeen, T., Romeo, F. and Mehta, J.L. (2001) Different isoforms of tocopherols enhance nitric oxide synthase phosphorylation and inhibit human platelet aggregation and lipid peroxidation: implications in therapy with vitamin E. *Journal of Cardiovascular and Pharmacological Therapy* 6, 155–161.

Li, N.Q., Reddy, P.S., Thyagaraju, K., Reddy, A.P., Hsu, B.L., Scholz, R.W., Tu, C.P. and Reddy, C.C. (1990) Elevation of rat liver mRNA for selenium-dependent glutathione peroxidase by selenium deficiency. *Journal of Biological Chemistry* 265, 108–113.

Liao, F., Andalibi, A., deBeer, F.C., Fogelman, A.M. and Lusis, A.J. (1993) Genetic control of inflammatory gene induction and NF-kappa B-like transcription factor activation in response to an atherogenic diet in mice. *Journal of Clinical Investigation* 91, 2572–5279.

Lindner, V. and Collins, T. (1996) Expression of NF-κB and IκB-α by aortic endothelium in an arterial injury model. *American Journal of Pathology* 148, 427–438.

Low, S.C. and Berry, M.J. (1996) Knowing when not to stop: selenocysteine incorporation in eukaryotes. *Trends in Biochemical Sciences* 21, 203–208.

Maiorino, M., Wissing, J.B., Brigelius-Flohé, R., Calabrese, F., Roveri, A., Steinert, P., Ursini, F. and Flohé, L. (1998) Testosterone mediates expression of the selenoprotein PHGPx by induction of spermatogenesis and not by direct transcriptional gene activation. *FASEB Journal* 12, 1359–1370.

Malins, D.C., Johnson, P.M., Wheeler, T.M., Barker, E.A., Polissar, N.L. and Vinson, M.A. (2001) Age-related radical-induced DNA damage is linked to prostate cancer. *Cancer Research* 61, 6025–6028.

Mansell, J.B. and Berry, M.J. (2001) Towards a mechanism for selenocysteine incorporation in eukaryotes. In: Hatfield, D.L. (ed.) *Selenium. Its Molecular Biology and Role in Human Health.* Kluwer Academic Publishers, Boston, pp. 69–80.

Matthews, J.R., Botting, C.H., Panico, M., Morris, H.R. and Hay, R.T. (1996) Inhibition of NF-κB DNA binding by nitric oxide. *Nucleic Acids Research* 24, 2236–2242.

Matz, R.L., Schott, C., Stoclet, J.C. and Andriantsitohaina, R. (2000) Age-related endothelial dysfunction with respect to nitric oxide, endothelium-derived hyperpolarizing factor and cyclooxygenase products. *Physiological Research* 49, 11–18.

May, J.M., Qu, Z.C. and Mendiratta, S. (1998) Protection and recycling of alpha-tocopherol in human erythrocytes by intracellular ascorbic acid. *Archives of Biochemistry and Biophysics* 349, 281–289.

Müller, A.S. and Pallauf, J. (2002) Downregulation of GPx1 mRNA and the loss of GPx1 activity causes cellular damage in the liver of selenium deficient rabbits. *Journal of Animal Physiology and Animal Nutrition* 86, 273–287.

Nakamura, T., Goto, M., Matsumoto, A. and Tanaka, I. (1998) Inhibition of NF-kappa B transcriptional activity by alpha-tocopheryl succinate. *Biofactors* 7, 21–30.

Naruse, K., Shimuzu, K., Maramatsu, M., Toky, Y., Miyazaki, Y., Okamura, K., Hashimoto, H. and Ito, T. (1994) Long-term inhibition on NO synthesis promoter atherosclerosis in the hypercholesterolemic rabbit thoracic aorta: PGH2 does not contribute to impaired endothelium-dependent relaxation. *Arteriosclerosis and Thrombosis* 14, 746–752.

Packham, G., Ashmun, R.A. and Cleveland, J.L. (1996) Cytokines suppress apoptosis independent of increases in reactive oxygen levels. *Journal of Immunology* 156, 2792–2800.

Panza, J.A., Garcia, C.E., Kilcoyne, C.M., Quyymi, A.A. and Cannon, R.O. (1995) Impaired endothelium-dependent vasodilatation in patients with essential hypertension: evidence that nitric oxide abnormality is not localized to a single signal transduction pathway. *Circulation* 91, 1732–1738.

Park, H.S., Huh, S.H., Kim, Y., Shim, J., Lee, S.H., Park, I.S., Jung, Y.K., Kim, I.Y. and Choi, E.J. (2000) Selenite negatively regulates caspase-3 through a redox mechanism. *Journal of Biological Chemistry* 275, 8487–8491.

Parthasarathy, S., Santanam, N., Ramachandran, S. and Meilhac, O. (1999) Oxidants and antioxidants in atherogenesis. An appraisal. *Journal of Lipid Research* 40, 2143–2157.

Patel, R.P., Levonen, A.L., Crawford, J.H. and Darley-Usma, V.M. (2000) Mechanisms of the pro- and antioxidant actions of nitric oxide in atherosclerosis. *Cardiovascular Research* 47, 465–474.

Pence, B.C., Delver, E. and Dunn, D.M. (1994) Effects of dietary selenium on UVB-induced skin carcinogenesis and epidermal antioxidant status. *Journal of Investigative Dermatology* 102, 759–761.

Raisanen, S.R., Lehenkari, P., Tasanen, M., Rahkila, P., Harkonen, P.L. and Vaananen, H.K. (1999) Carbonic anhydrase III protects cells from hydrogen peroxide-induced apoptosis. *FASEB Journal* 13, 513–522.

Ricciarelli, R., Tasinato, A., Clement, S., Ozer, N.K., Boscoboinik, D. and Azzi, A. (1998) Alpha-tocopherol specifically inactivates cellular protein kinase C alpha by changing its phosphorylation state. *Biochemical Journal* 334, 243–249.

Ricciarelli, R., Maroni, P., Ozer, N., Zingg, J.M. and Azzi, A. (1999) Age-dependent increase of collagenase expression can be reduced by alpha-tocopherol via protein kinase C inhibition. *Free Radical Biology and Medicine* 27, 729–737.

Ricciarelli, R., Zingg, J.M. and Azzi, A. (2000) Vitamin E reduces the uptake of oxidized LDL by inhibiting CD36 scavenger receptor expression in cultured aortic smooth muscle cells. *Circulation* 102, 82–87.

Rimbach, G., Valacchi, G., Canali, R. and Virgili, F. (2000) Macrophages stimulated with IFN-γ activate NF-κB and induce MCP-1 gene expression in primary human endothelial cells. *Molecular Cell Biology Research Communications* 3, 238–242.

Rubanyi, G.M. (1993) The role of endothelium in cardiovascular homeostasis and diseases. *Cardiovascular Pharmacology* 22, S1–S14.

Sandstrom, P.A., Tebbey, P.W., Van Cleave, S. and Buttke, T.M. (1994) Lipid hydroperoxides induce apoptosis in T cells displaying a HIV-associated glutathione peroxidase deficiency. *Journal of Biological Chemistry* 269, 798–801.

Schnurr, K., Borchert, A. and Kuhn, H. (1999) Inverse regulation of lipid-peroxidizing and hydroperoxyl lipid-reducing enzymes by interleukins 4 and 13. *FASEB Journal* 13, 143–154.

Schütze, N., Bachthaler, M., Lechner, A., Köhrle, J. and Jakob, F. (1998) Identification by differential display PCR of the selenoprotein thioredoxin reductase as a 1 alpha,25(OH)2-vitamin D3-responsive gene in human osteoblasts – regulation by selenite. *Biofactors* 7, 299–310.

Schütze, N., Fritsche, J., Ebert-Dumig, R., Schneider, D., Köhrle, J., Andreesen, R., Kreutz, M. and Jakob, F. (1999) The selenoprotein thioredoxin reductase is expressed in peripheral blood monocytes and THP1 human myeloid leukemia cells – regulation by 1,25-dihydroxyvitamin D3 and selenite. *Biofactors* 10, 329–338.

Serbinova, E., Kagan, V., Han, D. and Packer, L. (1991) Free radical recycling and intramembrane mobility in the antioxidant properties of alpha-tocopherol and alpha-tocotrienol. *Free Radical Biology and Medicine* 10, 263–275.

Sies, H., Klotz, L.O., Sharov, V.S., Assmann, A. and Briviba, K. (1998) Protection against peroxynitrite by selenoproteins. *Zeitschrift für Naturforschung* 53, 228–232.

Simpson-Haidaris, P.J., Wright, T.W., Earnest, B.J., Hui, Z., Neroni, L.A. and Courtney, M.A. (1995) Cloning and characterization of a lung-specific cDNA corresponding to the gamma chain of hepatic fibrinogen. *Gene* 167, 273–278.

Sinha, R. and Medina, D. (1997) Inhibition of cdk2 kinase activity by methylselenocysteine in synchronized mouse mammary epithelial tumor cells. *Carcinogenesis* 18, 1541–1547.

Stocker, A., Zimmer, S., Spycher, S.E. and Azzi, A. (1999) Identification of a novel cytosolic tocopherol-binding protein: structure, specificity, and tissue distribution. *IUBMB Life* 48, 49–55.

Stroes, E.S., Koomans, H.A., de Bruin, T.W. and Rabelink, T.J. (1995) Vascular function in the forearm of vasodilation in patients with non-insulin-dependent diabetes mellitus. *Journal of the American College of Cardiology* 27, 567–574.

Sunde, R.A. (1997) Selenium. In: O'Dell, B.L. and Sunde, R.A. (eds) *Handbook of Nutritionally Essential Mineral Elements*. Marcel Dekker, New York, pp. 493–556.

Suzuki, Y.J. and Packer, L. (1993) Inhibition of NF-kappa B activation by vitamin E derivatives. *Biochemical and Biophysical Research Communications* 193, 277–283.

Suzuki, Y.J., Tsuchiya, M., Wassall, S.R., Choo, Y.M., Govil, G., Kagan, V.E. and Packer, L. (1993) Structural and dynamic membrane properties of alpha-tocopherol and alpha-tocotrienol: implication to the molecular mechanism of their antioxidant potency. *Biochemistry* 32, 10692–10699.

Tada, H., Ishii, H. and Isogai, S. (1997) Protective effect of D-alpha-tocopherol on the function of human mesangial cells exposed to high glucose concentrations. *Metabolism* 46, 779–784.

Terasawa, Y., Ladha, Z., Leonard, S.W., Morrow, J.D., Newland, D., Sanan, D., Packer, L., Traber, M.G. and Farese, R.V. Jr (2000) Increased atherosclerosis in hyperlipidemic mice deficient in alpha-tocopherol transfer protein and vitamin E. *Proceedings of the National Academy of Sciences USA* 97, 13830–13834.

Thiery, J., Teupser, D., Walli, A.K., Ivandic, B., Nebendahl, K., Stein, O., Stein, Y. and Seidel, D. (1996) Study of causes underlying the low

atherosclerotic response to dietary hypercholesterolemia in a selected strain of rabbits. *Atherosclerosis* 121, 63–73.
Thompson, K.M., Haibach, H. and Sunde, R.A. (1995) Growth and plasma triiodothyronine concentrations are modified by selenium deficiency and repletion in second-generation selenium-deficient rats. *Journal of Nutrition* 125, 864–873.
Traber, M.G. and Kayden, H.J. (1989) Preferential incorporation of alpha-tocopherol vs gamma-tocopherol in human lipoproteins. *American Journal of Clinical Nutrition* 49, 517–526.
Traber, M.G., Sokol, R.J., Burton, G.W., Ingold, K.U., Papas, A.M., Huffaker, J.E. and Kayden, H.J. (1990) Impaired ability of patients with familial isolated vitamin E deficiency to incorporate alpha-tocopherol into lipoproteins secreted by the liver. *Journal of Clinical Investigation* 85, 397–407.
Viita, H., Sen, C.K., Roy, S., Siljamaki, T., Nikkari, T. and Yla-Herttuala, S. (1999) High expression of human 15-lipoxygenase induces NF-κB-mediated expression of vascular cell adhesion molecule 1, intercellular adhesion molecule 1, and T-cell adhesion on human endothelial cells. *Antioxidants and Redox Signalling* 1, 83–96.
Weitzel, F., Ursini, F. and Wendel, A. (1990) Phospholipid hydroperoxide glutathione peroxidase in various mouse organs during selenium deficiency and repletion. *Biochimica et Biophysica Acta* 1036, 88–94.
Welty, F.K., Lichtenstein, A.H., Barrett, P.H., Jenner, J.L., Dolnikowski, G.G. and Schaefer, E.J. (2000) Effects of ApoE genotype on ApoB-48 and ApoB-100 kinetics with stable isotopes in humans. *Arteriosclerosis, Thrombosis and Vascular Biology* 20, 1807–1810.
Williams, S.B., Cusco, J.A., Roddy, M.A., Johnstone, M.T. and Creager, M.A. (1996) Impaired nitric oxide-mediated vasodilation in patients with non-insulin-dependent diabetes mellitus. *Journal of the American College of Cardiology* 27, 567–574.
Wingler, K., Müller, C., Schmehl, K., Florian, S. and Brigelius-Flohé, R. (2000) Gastrointestinal glutathione peroxidase prevents transport of lipid hydroperoxides in CaCo-2 cells. *Gastroenterology* 119, 420–430.
Wu, D., Hayek, M.G. and Meydani, S. (2001) Vitamin E and macrophage cyclooxygenase regulation in the aged. *Journal of Nutrition* 131, 382S–388S.
Yamamoto, Y. and Takahashi, K. (1993) Glutathione peroxidase isolated from plasma reduces phospholipid hydroperoxides. *Archives of Biochemistry and Biophysics* 305, 541–545.
Yamauchi, J., Iwamoto, T., Kida, S., Masushige, S., Yamada, K. and Esashi, T. (2001) Tocopherol-associated protein is a ligand-dependent transcriptional activator. *Biochemical and Biophysical Research Communications* 285, 295–299.
Yan, S.D., Schmidt, A.M., Anderson, G.M., Zhang, J., Brett, J., Zou, Y.S., Pinsky, D. and Stern, D. (1994) Enhanced cellular oxidant stress by the interaction of advanced glycation end products with their receptors/binding proteins. *Journal of Biological Chemistry* 269, 9889–9897.
Yoshida, N., Manabe, H., Terasawa, Y., Nishimura, H., Enjo, F., Nishino, H. and Yoshikawa, T. (2000) Inhibitory effects of vitamin E on endothelial-dependent adhesive interactions with leukocytes induced by oxidized low density lipoprotein. *Biofactors* 13, 279–288.
Zapolska-Downar, D., Zapolski-Downar, A., Markiewski, M., Ciechanowicz, A., Kaczmarczyk, M. and Naruszewicz, M. (2000) Selective inhibition by alpha-tocopherol of vascular cell adhesion molecule-1 expression in human vascular endothelial cells. *Biochemical and Biophysical Research Communications* 274, 609–615.
Zimmer, S., Stocker, A., Sarbolouki, M.N., Spycher, S.E., Sassoon, J. and Azzi, A. (2000) A novel human tocopherol-associated protein: cloning, *in vitro* expression, and characterization. *Journal of Biological Chemistry* 275, 25672–25680.

# 13 鞘脂类：癌症治疗和预防的新战略

Eva M. Schmelz
（卡马诺斯癌症研究所，韦恩州立大学医学院
底特律，密歇根，美国）

## 引　言

鞘脂类是所有真核及一些原核细胞中的重要结构成分。它们主要存在于膜中，影响膜的稳定性和流动性，并改变膜中蛋白质的行为作用。在几十年前，鞘脂代谢物作为脂类第二信使的作用才被发现。现已表明，鞘脂类参与细胞生长和细胞死亡广泛过程的调节。本文将对食物中的鞘脂类对细胞调节尤其是对结肠癌细胞调节的已知作用进行总结，并指出在结肠癌抑制中食物鞘脂类利用的信号途径。

## 鞘脂的结构与代谢

鞘脂类是由一个鞘氨脂碱（哺乳动物细胞中大多为鞘氨醇）、一个酰胺结合的脂肪酸和一个头基组成。如图 13.1 所示，鞘脂类的重新合成是在丝氨酸-棕榈酰转移酶的催化下，由丝氨酸和棕榈酰 CoA 经酮鞘氨醇合成鞘氨醇开始的，神经酰胺合成酶使氨基酰化形成二氢神经酰胺，再经脱饱和酶催化形成 4，5-反式双键形成神经酰胺，或加上一个头基合成更复杂的鞘脂类，该复杂的鞘脂类中可再引入双键（更详细内容见 Huwiler *et al.*，2000）。这些成分的变化使鞘脂类成为结构上最多变的一类膜脂类。鞘脂类的降解需要鞘氨醇（由复杂的鞘脂类水解而产生）经鞘氨醇激酶磷酸化为鞘氨醇-1-磷酸，且需要鞘氨醇裂解酶的活性以产生乙醇胺和六癸烯。细胞鞘脂代谢的中间产物对细胞行为有显著影响，这将在下面的部分中进行讨论。

## 鞘脂类作为第二信使

最先表明鞘脂代谢物参与信号转导的报告之一，是一篇鞘氨醇抑制蛋白激酶 C（PKC）活性的报道（Hannun *et al.*，1986）。随后，发现肿瘤坏死因子-α（TNF-α）或干扰素-γ（IFN-γ）处理细胞通过激活鞘磷脂酶水解鞘磷脂，从而引起神经酰胺积累。现今，激活该途径的介质或事件则包括生长因子、诱导分化或细胞凋亡的复合物（如维生素 D3、毒素、化疗制剂），以及细胞应激如低氧和 γ-辐射。一些介质不仅可激活鞘磷脂酶还可激活水解神经酰胺，并释放游离鞘氨脂碱的神经酰胺酶。这两个代谢物一般是生长抑制性的和有细胞毒性的，且已表明可诱导许多细胞系中的细胞凋亡（Spiegel and Merrill，1996；Mathias *et al.*，1998）。而且，两者的结合似乎更有效

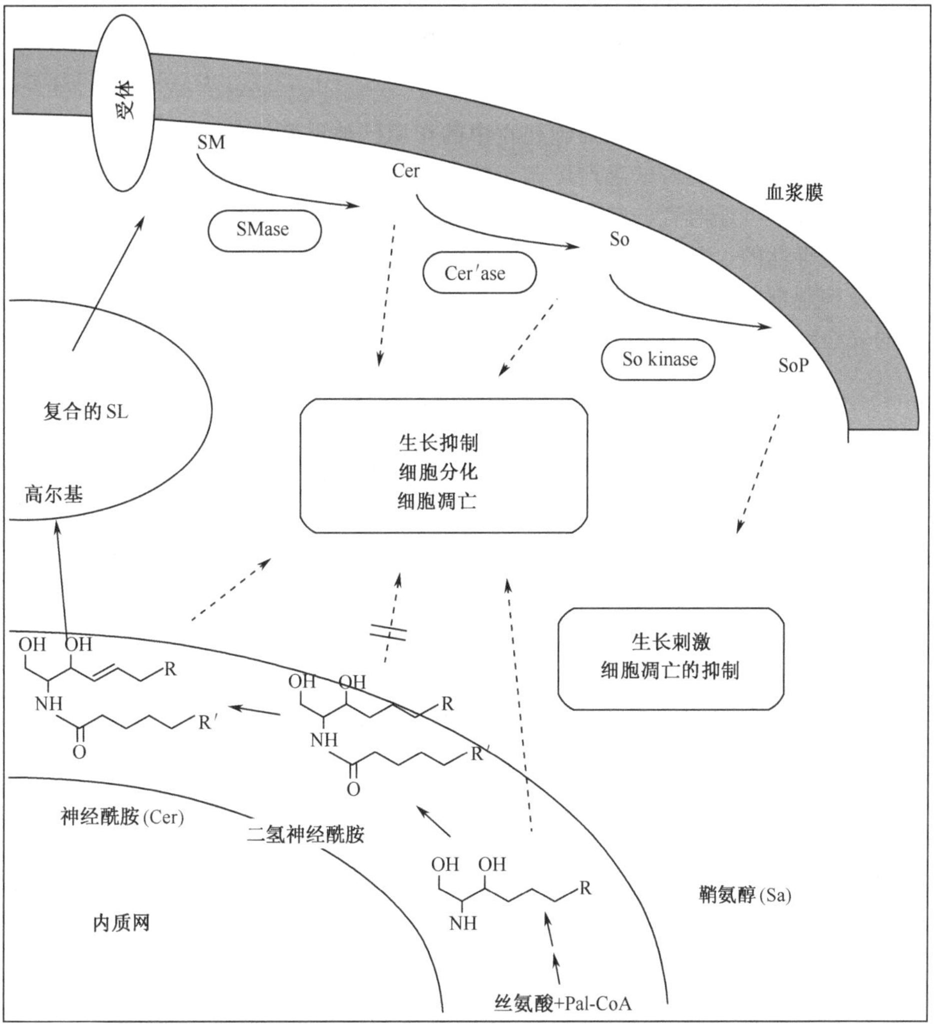


图 13.1 通过细胞外剂或应激激活受体或重新合成使鞘脂周转中的酶激活（鞘磷脂酶，SMase；神经酰胺酶，Cer'ase；鞘氨醇激酶，So kinase），可能有助于细胞内生物活性鞘脂代谢物的累积

（Jarvis *et al.*，1996）。生长因子如血小板衍生的生长因子（platelet derived growth factor，PDGF）也可激活鞘氨醇激酶（图 13.1），且与神经酰胺和鞘氨脂碱不同，鞘氨醇-1-磷酸的作用是参与有丝分裂，也能阻碍神经酰胺诱导的细胞凋亡（Spiegel，1999）。因此，细胞对外源和内源鞘脂类的反应取决于细胞中积累的代谢物或由细胞外刺激产生的代谢物的组合。一些介质不是通过复合膜鞘脂类的降解，而是通过激活重新合成来增加细胞内神经酰胺的浓度。对道诺红菌素（Bose *et al.*，1995）、TNF-α（Xu *et al.*，1998）、表鬼臼毒吡喃葡糖苷（Perry *et al.*，2000）和光动力学疗法（Separovic *et al.*，1997）的实验已表明了这一点。重要的是，内源代谢物的作用可被细胞通透的外源鞘脂类模拟。目前这已被广泛应用于细胞培养中，在体内试验中也用食物鞘脂类来治疗化学诱导的结肠癌。

# 癌症中的鞘脂类

鞘脂类在细胞生长、分化和细胞死亡中的作用已被许多体外研究所表明，人们认为外源和（或）内源鞘脂类对缺乏严厉调节的细胞如癌细胞可能有影响。癌的生长是超增生或细胞凋亡减少（或两者的结合）以及分化减少的结果。鞘脂类对大多数细胞系是生长抑制和细胞毒性的，并可诱导细胞凋亡或分化。此外，鞘脂类已表明可抑制或避免多药物抗性及抑制细胞游动性和血管发生。这些功能再加上食物中鞘脂类的可用性和其在膳食中的安全性，使鞘脂类成为癌症干预和（或）癌症预防研究中的良好候选者。

## 食物中的鞘脂类

特异性直接导向食物中鞘脂类的识别及其定量化的分析尚有待进行。然而，奶、肉产品、鸡蛋和大豆特别富含鞘脂类（Vesper *et al.*，1999）。在动物产品中，主要的鞘脂是鞘磷脂，通常含有鞘氨醇骨架和大多饱和且含 16 或 22～24 碳原子的酰胺结合的脂肪酸。在植物中，主要的鞘脂种类是含各种不同糖头基（葡萄糖、半乳糖和甘露糖），鞘氨脂碱骨架（不同数量和位置的双键）和脂肪酸（也是饱和的，但常为 *α*-羟基化的）的脑苷脂。

鞘脂类是食物中的微量成分，对食物中的能量含量没有显著贡献。尽管它们是脂类，但也被发现于“脱脂”食物如脱脂乳中，因为它们位于乳脂肪小球的膜中，而这些乳脂肪小球在脱脂后仍部分保留于乳中（Jensen，1995）。目前没有已知的对鞘脂类的营养需要量，但通过它们对啮齿类动物中的试验性结肠癌的作用表明，它们应被归类为“功能性食物”。Kobayashi 等（1997）研究表明，膳食鞘脂类用作化学预防性或保护性的物质是安全的，且食物中即便鞘脂类含量高（重量的 1%或超过平均消耗量的 100 多倍），对成年大鼠或它们后代的体重或血脂水平也没有影响。这可能是由复杂鞘脂类的有限消化造成的（Nyberg *et al.*，1997）。而且，将食物中的鞘脂类饲喂给啮齿动物并未永久增加其血中鞘脂代谢物的水平（Schmelz *et al.*，2001）。

## 富含鞘脂食物改变癌症发生率的流行病学证据

已做了许多研究来评估食物对癌症发生的作用。一些研究表明，乳及乳制品的消耗量与减少结肠癌发生率相关（Jarvinen *et al.*，2001），而其他研究则没有发现该作用。然而，由于不同的设计和一些变量如总脂或热量摄入的未知影响，故很难将这些研究结果加以比较。在降低癌症发生的研究中，癌症发生的降低通常归因于乳中钙含量（Lipkin，1999），尽管乳中的其他成分，如共轭亚油酸、丁酸、维生素 D 和乳铁传递蛋白也被证明可抑制结肠癌和其他癌症（Holt，1999；Parodi，1999）。因为目前还没有食物中鞘脂含量与（结肠）癌症联系的可利用的数据，因此，食物中鞘脂类的作用尚有待于在流行病学研究中加以阐明。

## 皮肤癌中的鞘脂类

鞘脂代谢物的理化特性及其相当大的毒性，使得鞘脂类到达体内目标的转运成了问题。这已把外源鞘脂类在癌症治疗和预防中的应用限制到了很少的试验模型中。已使用局部应用来测定鞘氨醇、甲基鞘氨醇和 *N*-乙酰鞘氨醇对 Sencar 鼠皮肤癌的影响（Enkvetchakul *et al.*，1989，1992；Birt *et al.*，1998）。鞘氨醇并不抑制二甲苯-［α］-蒽诱导的乳头状瘤的发育（Enkvetchakul *et al.*，1989，1992），在高剂量时甚至会促进乳头状瘤的形成（Enkvetchakul *et al.*，1992）。在后来测定鞘氨醇、甲基鞘氨醇和 *N*-乙酰鞘氨醇的有效性的研究中也观测到了这种现象，且鞘氨醇没有改变乳头状瘤的发生率（Birt *et al.*，1998）。然而，*N*-甲基鞘氨醇和 *N*-乙酰鞘氨醇都增加了无癌存活率。而且，将鼠用佛波酯处理后，每周应用鞘氨醇和 *N*-乙酰鞘氨醇处理，连续 10 周，抑制了这些鼠肿瘤的发展（Birt *et al.*，1998）。

另一个鞘脂衍生物沙芬戈（safingol）为鞘氨醇的 L-苏型异构体和蛋白激酶 C（PKC）的强力抑制剂，已被研制用来治疗皮肤病和癌症。局部应用 safingol 6 周后，雌鼠血清酶和肝损坏的增加明显高于雄鼠，尽管它们起始的血浆水平相似。这表明，局部使用的 safingol 能被大鼠吸收，并能诱导雌鼠肝脏中更高的细胞凋亡速率（以缺乏炎症和 DNA 序列梯为标志），因为雌鼠不能通过其细胞色素 P450 同工酶而充分清除 safingol（Carfagna *et al.*，1996）。

## 食物中鞘磷脂对 CF1 鼠化学诱导结肠癌早期阶段的抑制作用

食物中的复合鞘脂类可在小肠和结肠的所有部位被消化，且其生物活性代谢物神经酰胺和鞘氨醇可被肠细胞摄取（Schmelz *et al.*，1994，2001）。大多数代谢物可被小肠吸收，但只有约 10％到达结肠（Schmelz *et al.*，1994）。因而饲喂复合鞘脂类，其生物

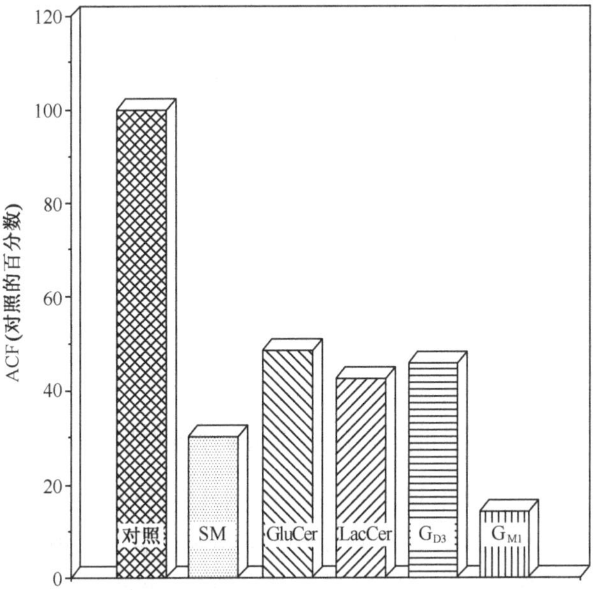


图 13.2　食物中鞘脂类对在结肠癌中的畸变隐窝病灶（ACF）形成的抑制

CF1 鼠经用结肠致癌剂处理后，饲喂基本无鞘脂（对照）或添补鞘磷脂（SM）、葡糖神经酰胺（GluCer）、乳糖苷神经酰胺（LacCer）或神经节苷脂 $G_{D3}$ 或 $G_{M1}$ 的饲粮。4 周后，测定了结肠癌中发生最早的一个形态学变化的数量，即 ACF 指标

活性代谢物可直接被运送到结肠细胞。所以，开展食物中鞘脂类对结肠癌发生的有效性的研究是很贴切的。在这些研究中，评价了食物中的鞘脂类对结肠癌发生的早期形态学标志，即在结肠癌中的畸变隐窝病灶（aberrant crypt foci，ACF）出现的影响，认为这些早期的损害经过一段时间后便会发展成腺瘤和腺癌，从而使 ACF 成为评价食物中化合物对结肠癌发生影响研究中的一个可以广泛应用的成本与时间有效性终点指标。

给 CF1 鼠注射 1，2-二甲基肼（dimethylhydrazine，DMH）用以诱导结肠肿瘤，肿瘤启动后，给小鼠饲喂半纯合且基本无鞘脂的 AIN 76A 饲粮（对照组）或补加 0.025%～0.1%（占饲粮重）鞘磷脂类的饲粮。饲喂从脱脂奶粉中提纯的鞘磷脂使 ACF 的形成减少了 70%（图 13.2），而长期饲喂后减少了肿瘤的形成（Dillehay *et al.*，1994）。在后来的一较大型研究中，饲粮鞘磷脂对 ACF 的抑制作用得到了证实，且可抑制良性瘤向恶性瘤发展（Schmelz *et al.*，1996）。这确实要归因于食物中的鞘磷脂，而不是纯化物中的其他杂质所致，因为合成的鞘磷脂与乳中的“天然”化合物对 ACF 的抑制作用相同（Schmelz *et al.*，1997）。

## 食物鞘脂类减少微型鼠的肿瘤形成

用于评价食物中鞘脂类对人结肠癌作用的临床试验尚有待进行。因啮齿类动物模型与人类疾病很相似，所以鼠的试验结果依然是可利用的。C57/B6J$^{Min/+}$ 小鼠（多种小肠瘤形成的微型小鼠）携带着腺瘤息肉大肠菌（adenomatous polyposis coli，APC）基因的突变，而这种突变几乎发现于所有患家族性腺瘤息肉病（familial adenomatous polyposis，FAP）的患者中。这些患者在结肠里自然产生数以千计的息肉，其中的一些不可避免地会发展成腺癌。在 40%～80%散发性结肠癌中也发现了 *APC* 基因突变（Nagase and Nakamura，1993；Sparks *et al.*，1998）。尽管微型鼠上肿瘤的位置与 FAP 患者的不同，FAP 患者的大多数位于结肠中，但二者都有相同的表型（完全常染色体显性遗传，多种小肠肿瘤发生，由 APC 基因突变诱导）（Kinzler and Vogelstein，1996）。微型鼠断奶后，在其膳食中补加 0.1%的复合鞘脂类的混合物，此混合物的成分与乳成分一致（70%鞘磷脂、5%乳糖苷神经酰胺、7.5%葡糖神经酰胺和 7.5%神经节苷脂 $G_{D3}$）。这是再一次晚期干涉研究，因为在这个时候肿瘤已在整个肠道发展。饲喂鞘脂类 65d 整个肠道的肿瘤发生率降低 40%（图 13.3）。在维持占膳食 0.1%的情况下，在此混合物中加入神经酰胺可将肿瘤抑制率提高到 50%。这在肠道的所有区域都可被观测到（Schmelz *et al.*，2001）。

## 鞘脂作用的结构依赖性

正如上面所提到的，食物中复合鞘脂类可被水解为神经酰胺和游离鞘氨脂碱。神经酰胺主链的结构已表明神经酰胺可决定其生物学作用。含鞘氨醇（无 4，5-反式双键的二氢神经酰胺）的神经酰胺是无生物活性的化合物（Bielawska *et al.*，1993），如果抑制 ACF 的主要代谢物是神经酰胺，那么含有二氢神经酰胺主链的食物复合鞘脂类便是无活性的。然而，给 DMH-处理的 CF1 鼠饲喂有鞘氨醇主链的二氢神经酰胺鞘磷脂的鞘磷脂，发现更进一步减少了 ACF 的形成（Schmelz *et al.*，1997）。这提示，在 ACF 抑制中最重要的生物活性代谢产物可能是游离的鞘氨脂碱而不是神经酰胺，因为鞘氨醇与

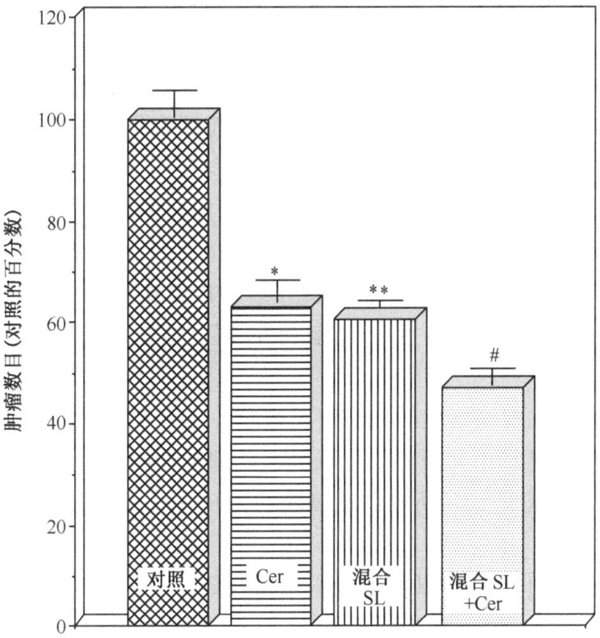


图 13.3 食物中鞘脂类对微型小鼠肿瘤形成的抑制

给微型小鼠单独饲喂无鞘脂饲粮（对照）或添补 0.1%神经酰胺（Cer），复合鞘脂类（SL）（正如在乳中的一样，混合 SL）或两者的结合（40%神经酰胺，60%混合 SL）的饲粮。65d 后，测定了肿瘤的数量和定位（$*P<0.05$；$**P<0.01$；$\#P<0.001$）

神经鞘氨醇的生物活性无差异。再则，由细胞内脱饱和酶引入双键（正如二氢神经酰胺摄入后内源鞘脂类的重新合成中所见到的一样）也不能被排除在此点之外。

鞘脂头基的结构及其复杂性可能影响消化，因而影响对 ACF 或肿瘤的抑制所必需的生物活性分子的释放。所以，我们比较了更复杂的鞘脂类对 ACF 发生的影响。结果表明，乳中数量可观的复合鞘脂类、葡糖神经酰胺、乳糖苷神经酰胺和神经节苷脂也减少了 CF1 鼠中 ACF 的形成（Schmelz *et al.*，2000）。如图 13.2 所示，脑中神经节苷脂 $G_{M1}$ 甚至比鞘磷脂更有效（Dillehay *et al.*，1994）。这大概归因于（或许是无特异性）所有这些复合鞘脂类中头基和脂肪酸的除去（Schmelz *et al.*，2000）以及这些化合物中鞘氨脂碱的释放。因此，食物中大多数或所有的复合鞘脂类都可能是可消化的，且在结肠癌的抑制中发挥作用。这得到大豆中葡糖神经酰胺［鞘氨醇中双键的位置和（或）数量与乳中的不同］也减少了微型鼠中肿瘤的形成（Schmelz 和 Merrill，未发表结果）的研究结果的支持。

## 鞘脂代谢物在化疗中的应用

根据我们的知识，食物中的鞘脂类除了上面提到的那些外，在研究中还没有被用作化学治疗剂，但在静脉应用被用来测定鞘脂代谢物的药物动力学和毒性。将 safingol（L-苏型-鞘氨醇，见上面）的剂量增加到 5mg/kg 体重，在鼠或狗上都未引起有害影响（Kedderis *et al.*，1995）。全身用药比成比例的程度更多地增加了血浆 safingol 浓度，并引起溶血、肾变性和坏死，且在更高的剂量时引起肝中毒。而且，safingol 在输入过程中会引起注射部位静脉血管内膜及中层的变性和坏死增加以及雄性动物输精管的黏着变性和坏死。阿霉素和 safingol 的联合使用没有改变阿霉素的药物动力学（Kedderis *et al.*，1995；Schwartz *et al.*，1997），说明如果引起器官损坏的起始溶血可被抑制的话，那么鞘脂类作为化疗药是可以静脉使用的。

在另一项研究中，先将溶解在大豆油中的亲脂神经酰胺衍生物以 1mg/kg 体重的剂量静脉注射到鼠体内，然后再注射高转移性的 Meth A-T 肿瘤细胞。只此一次注射未改变肺转移的发生率，然而重复注射（共 7 次）后即以结构依赖方式显著减少了肺小结的数量（Takenaga *et al.*，1999）。由于溶解度问题，短链神经酰胺在本研究中未能被测试。

给接种过 MKN74 人胃癌细胞的鼠注射鞘氨醇、二甲基鞘氨醇或三甲基鞘氨醇后，通过甲基化鞘氨醇衍生物显示出其对肿瘤生长的显著抑制作用，即使在停止鞘脂给药后还能继续维持。将肿瘤生长减少 50%所需的浓度大约为 8μmol/L（Endo *et al.*，1991）。

## 外源鞘脂类对肿瘤抑制作用的可能机制

许多体内和体外研究都已报道了癌细胞中鞘脂信号传导的缺陷。在用结肠致癌物处理的鼠结肠细胞中可观测到鞘磷脂酶活性的降低（Dudeja *et al.*，1986）。同样的降低也可在人结肠瘤中观测到（Hertervig *et al.*，1997），且这种降低在最近的单核苷酸阵列分析中得到了证实，而在该单核苷酸阵列分析中，与正常组织相比，结肠瘤中鞘磷脂酶的表达降低了 4 倍（Notterman *et al.*，2001）。由于这种低活性/表达，作为细胞外刺激时的适当结果，细胞内没有产生足够的神经酰胺鞘氨醇，造成细胞不随细胞行为的调节而做出反应的结果。这种现象已在经过 TNF-α（Wiegman *et al.*，1994）、喜树碱（Wang *et al.*，1999）或 γ 辐射（Wright *et al.*，1996；Chmura *et al.*，1997）处理而没有经历细胞凋亡的细胞系中观测到。另一个减少细胞内生物活性鞘脂代谢物和抑制随后的细胞生长调节的机制是鞘磷脂合成的增加（通过鞘磷脂合酶的激活；Luberto and Hannun，1998）或葡糖神经酰胺合成的增加（通过葡糖神经酰胺合酶的激活；Lavie *et al.*，1996）。这也与细胞系（Lavie *et al.*，1997）、人乳腺肿瘤以及黑素瘤（Lucci *et al.*，1998）中多重抗药性的发展有关。所有这些鞘脂信号途径的异常使细胞中生物活性代谢物的量减少，并取消了细胞凋亡和分化的诱导作用。通过饲喂鞘脂类，我们可直接将生物活性代谢物运送到细胞，这样或许可绕过癌细胞的缺陷（图 13.4）。

### 结肠增生的调节

用结肠致癌物如 DMH 处理可增加啮齿动物中结肠增生的速率。已报道在注射后其存在一个细胞凋亡的起始“波动”（Hirose *et al.*，1996）。在我们用 DMH-处理 CF1 鼠的研究中，饲喂对照饲粮或鞘脂添加物饲粮 4 周后都没有发现细胞凋亡的增加（Schmelz *et al.*，2000），表明细胞凋亡的诱导或许在食物中的鞘脂类对 ACF 的抑制中不起作用，但这并不排除鞘脂影响早期凋亡性的可能性。注射致癌物的鼠结肠增生速率增加，且增生面向肠腔（在正常结肠中只看到分化的非增生细胞）扩展。在我们的研究

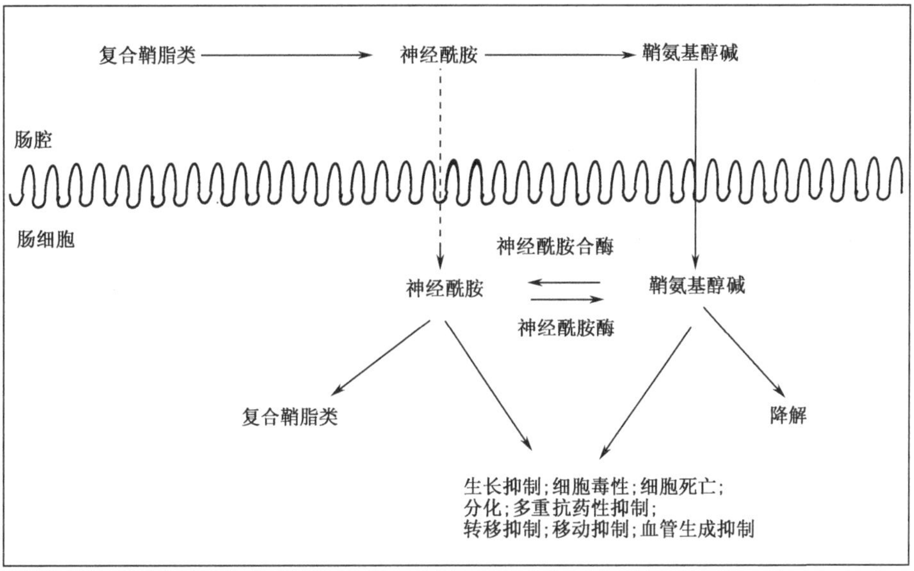


图 13.4 食物中鞘脂类的消化与吸收。复合鞘脂类在整个肠道中被消化为神经酰胺，但更可能是被消化为游离鞘氨基醇碱，并被肠细胞摄取。这两个代谢物的作用主要是生长抑制性和细胞毒性的

中应用的所有复合鞘脂类使增生约减少到未进行致癌物处理时对照动物的水平。而且，发现增生面被局限于结肠隐窝的下半部（Schmelz *et al.*，2000）。

## 外源鞘脂类对β-连环蛋白的调节

APC 基因产物位于细胞内代谢的主要“交叉点”，可对不同过程进行调节。被 APC 调节且对结肠癌特别重要的一种蛋白质为 β-连环蛋白。β-连环蛋白是一种细胞粘连蛋白，这种蛋白质通过 α-连环蛋白将 E-钙粘连蛋白连接到肌动蛋白细胞骨架（Pfeifer，1995；Sacco *et al.*，1995）以及其他跨膜和外周细胞质蛋白上（McNeill *et al.*，1990）。β-连环蛋白的这种功能与其作为发育系统中的信号分子的功能（Fagotto *et al.*，1996；Orsulic and Pfeifer，1996）以及其在生长刺激的细胞反应中的作用有显著区别（如 Wnt/Wingless 途径，表皮生长因子和肝细胞生长因子受体激活；Hoschuetzky *et al.*，1994；Shibamoto *et al.*，1994；Pfeifer，1995）。结肠癌中，APC 突变似乎是异常 ACF 的发育及向腺瘤和腺癌发展的关键现象（Jen *et al.*，1994）。作为此途径突变的结果，稳定的 β-连环蛋白在胞液中积累，随之被转运到核中，同 Tcf/Lef 家族的转录因子（T 细胞因子/淋巴细胞增值子因子）一起形成复合物。此复合物能调节基因转录过程并激活转录（Behrens，1999）。

正如上面提到的，在食物中发现的鞘脂类的数量抑制了微型鼠肿瘤的形成。在此模型中，肿瘤的形成发生在 APC 野生型等位基因丢失及相伴随的胞液 β-连环蛋白的蓄积之前。因此，在仅饲喂对照 AIN76A 饲粮并形成大量肠肿瘤的微型鼠中用荧光免疫组化法测得胞液 β-连环蛋白在各肠段中高度表达。饲喂补充鞘脂饲粮并只形成少量肿瘤的微型鼠大部分出现膜连 β-连环蛋白。这是在遗传背景鼠中发现的定位，表明食物鞘脂类

对β-连环蛋白表达/定位的正常化作用（Schmelz *et al.*，2001）。外源鞘脂类的相同作用可在体外通过应用人结肠癌细胞系而观测到，这种细胞系也携带APC突变并能稳定地过量表达胞液β-连环蛋白。鞘氨醇和神经酰胺在以无毒的浓度培育6h后，胞液中β-连环蛋白的数量都减少了，但要使核β-连环蛋白明显减少则需进行更长时间的培育（Schmelz *et al.*，2001）。从这些数据中并不能得出这是食物鞘脂类下调鼠中肿瘤形成的唯一途径的结论，因为参与其中的所有蛋白质和脂类都有多种功能。而且，假设增加的胞液β-连环蛋白不仅在早期结肠癌，而且在乳房、皮肤、前列腺和肾癌中都很重要，那么胞液β-连环蛋白的下调很明显是一个重要事件，且可直接参与鞘脂类的抗结肠癌作用。

## 鞘脂类与转移

趋药性细胞的迁移在胚胎生成、繁殖、伤口愈合、炎症以及血管发生中都起着重要作用。细胞迁移也是转移中的一个关键过程。已表明鞘氨醇-1-磷酸在体外能抑制肿瘤细胞（Wang *et al.*，1999）和非转化细胞（Kawa *et al.*，1997）的运动性和侵袭力，应用神经酰胺也得到了同样的结论（Metz *et al.*，1991）。对这种抑制来说，只需要纳摩尔浓度的鞘氨醇-1-磷酸（Wang *et al.*，1999）即可进行，但目前还不清楚食物中的鞘脂类是否在鞘氨醇-1-磷酸的循环池中发挥作用。

鞘糖脂α-半乳糖苷神经酰胺与肝和肺转移的预防有关（Toura *et al.*，1999；Nakagawa *et al.*，2001）。将从经体外半乳糖苷神经酰胺处理的鼠脾中制备的树突状细胞静脉注射到接种了黑素瘤或路易斯肺癌细胞的鼠后，在该鼠肝和肺中已形成的转移病灶几乎全部被根除（Toura *et al.*，1999）。这种抗肿瘤和抗转移作用是由半乳糖苷神经酰胺激活的NKT细胞介导的微转移中引发细胞凋亡所致，而被激活的NKT细胞产生IFN-γ，反过来通过激活Fas-Fas配体信号系统而诱导细胞凋亡（Nakagawa *et al.*，2001）。食物中半乳糖苷神经酰胺是否能激活肠道相关的抗肿瘤免疫，目前还不清楚，但可能是一个很有趣的概念。

## 肿瘤神经节苷脂

细胞表面神经节苷脂（含神经氨酸的糖鞘脂类）在癌细胞中的巨大变化已被注意到。正如在黑素瘤和成神经细胞瘤中所观察到的，这可能要归因于神经节苷脂前体（GM3、GM2、GD3和GD2）的不完全合成和伴随的积累。同样这也可能是由于新的神经节苷脂种类被表达，或在其头基或神经酰胺骨架上发生了结构改变，进而引起膜中的构象变化，产生或掩盖抗原决定基（Ritter and Livingston，1991），并改变神经节苷脂的活性（Ladisch *et al.*，1994）。在增殖迅速的细胞如癌细胞中，将神经节苷脂以微团、单体和膜小泡的形式排入细胞微环境，作为其神经节苷脂代谢的一个重要部分（Kong *et al.*，1998）。排出的神经节苷脂已表明其显著干扰抗肿瘤宿主反应的几个步骤，如对IL-2依赖的淋巴细胞增殖的抑制、T淋巴细胞中$CD4^+$的抑制（Valentino *et al.*，1990），以及巨噬细胞和$CD8^+$ T淋巴细胞数量的减少（Chu and Sharom，1993）。这就创造了一个有利于肿瘤形成与发展的肿瘤微环境。而且，神经节苷脂能直接或通过粘连受体的调节促进肿瘤细胞与内皮细胞的结合（Taki et al.，1997）。神经节苷脂与血细胞的结合导致了大团块的聚集和形成，可引起肺血栓，因此而促进转移（Ito *et al.*，2001）。另一方面，这些排出的神经节苷脂是通过抗-神经节苷脂单克隆抗体或神经节苷脂疫苗进行抗癌治疗的目标。然而，食物中的神经节苷脂是否也能增加血液中的神经节苷脂，由此产生相同的宿主和肿瘤反应，目前还不清楚。口服鞘脂类后，研究者虽未能观测到血液、淋巴或组织中的复合鞘脂类含量增加（Nilsson，1968，1969；Schmelz *et al.*，1994，2001；Kobayashi *et al.*，1997），但在缺失肠内源神经酰胺酶的情况下，较小的神经节苷脂也可能被吸收并直接同与肠相关的免疫系统接触（Ravindra *et al.*，2000），引起已见于生命早期经乳 $G_{D3}$ 的免疫（Takamizawa *et al.*，1986）。而且，老年人（大于 50 岁）的抗神经节苷脂抗体效价较低，这可能具有生理学意义，因为抗体能中和来自血清的循环肿瘤神经节苷脂（Ravindra *et al.*，2000）。利用营养手段提高抗肿瘤神经节苷脂水平将成为未来研究热点。

## 鞘脂类在癌症预防和治疗中的未来方向

我们用啮齿动物所进行的研究清楚表明，食物中鞘脂类对结肠癌的发生起着有益的作用。在这方面非常重要的是，鞘脂可以以非毒性的数量获得，且可在与西方膳食有密切关系的食品中被发现。然而，还需进行更多的研究，以测定食物中鞘脂类对人结肠癌及可能对其他器官癌的影响。此外，这些早期体内研究是很有前景的，且与大量体外研究所获得的可利用的数据一起，已鉴定了几条信号途径及外源鞘脂类的下游目标（图

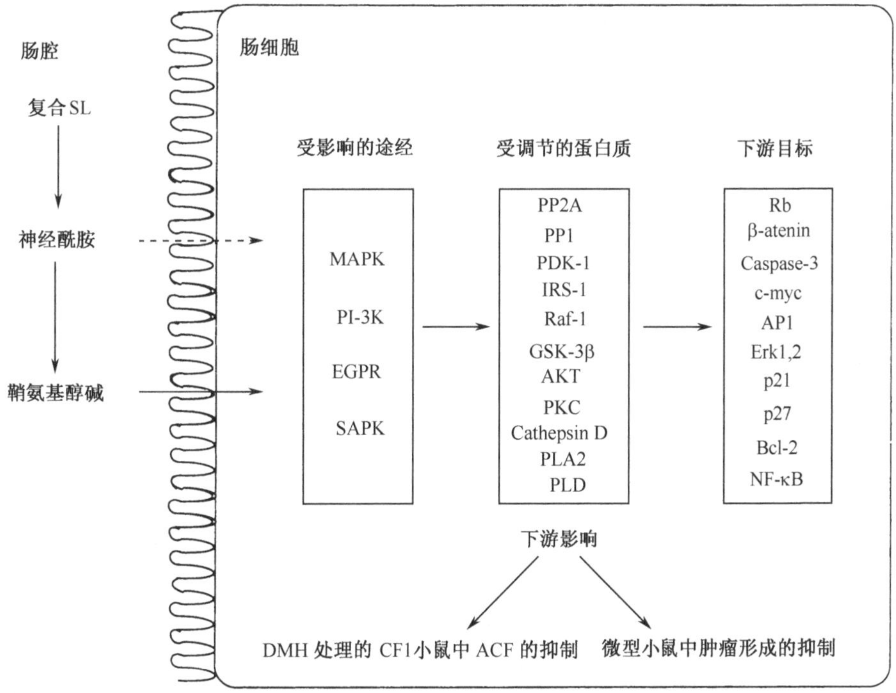


图 13.5　鞘脂代谢物的已知细胞内途径和目标的总结

这个到目前为止还不完全的列表阐明了鞘脂类可调节的几条可能的途径以调节可能参与啮齿动物 ACF 的下调和肿瘤形成的细胞生长和细胞死亡

13.5)。细胞内不断发现的鞘脂目标把未来研究导向到鞘脂类影响癌发生的机制上，且可能有助于将食物中鞘脂类确定为机制驱动性化疗剂及可能的化学预防剂。

## 参 考 文 献


Behrens, J. (1999) Cadherins and catenins: role in signal transduction and tumor progression. *Cancer and Metastasis Reviews* 18, 15–30.

Bielawska, A., Crane, H., Liotta, D.C., Obeid, L.M. and Hannun, Y.A. (1993) Selectivity of ceramide-mediated biology. Lack of activity of erythro-dihydro-ceramide. *Journal of Biological Chemistry* 268, 26226–26232.

Birt, D.F., Merrill, A.H. Jr, Barnett, T., Enkvetchakul, B., Pour, P.M., Liotta, D.C., Geisler, V., Menaldino, D.S. and Schwartzbauer, J. (1998) Inhibition of skin papillomas by sphingosine, *N*-methyl sphingosine, and *N*-acetyl sphingosine. *Nutrition and Cancer* 31, 119–126.

Bose, R., Verheil, M., Haimovitz-Friedman, A., Scotto, K., Fuks, Z. and Kolesnick, R. (1995) Ceramide synthase mediates daunorubicin-induced apoptosis: an alternative mechanism for generating death signals. *Cell* 82, 405–414.

Carfagna, M.A., Young, K.M. and Susick, R.L. (1996) Sex differences in rat hepatic cytolethality of the protein kinase C inhibitor safingol: role of biotransformation. *Toxicology and Applied Pharmacology* 137, 173–181.

Chmura, S.J., Nodzenski, E., Beckett, M.A., Kufe, D.W., Quintans, J. and Weichselbaum, R.R. (1997) Loss of ceramide production confers resistance to radiation-induced apoptosis. *Cancer Research* 57, 1270–1275.

Chu, J.W. and Sharom, F.J. (1993) Gangliosides inhibit T-lymphocyte proliferation by preventing the interaction of interleukin-2 with its cell surface receptors. *Immunology* 79, 10–17.

Dillehay, D.L., Webb, S.K., Schmelz, E.M. and Merrill, A.H. Jr. (1994) Dietary sphingomyelin inhibits 1,2-dimethylhydrazine-induced colon cancer in CF1 mice. *Journal of Nutrition* 124, 615–620.

Dudeja, P.K., Dahiya, R. and Brasitus, T.A. (1986) The role of sphingomyelin synthase and sphingomyelinase in 1,2-dimethylhydrazine-induced lipid alterations of rat colonic membranes. *Biochimica et Biophysica Acta* 863, 309–312.

Endo, K., Igarashi, Y., Nisar, M., Zhou, Q. and Hakomori, S.-I. (1991) Cell membrane signaling as target in cancer therapy: inhibitory effect of *N,N*-dimethyl and *N,N,N*-trimethyl sphingosine derivatives on *in vitro* and *in vivo* growth of human tumor cells in nude mice. *Cancer Research* 51, 1613–1618.

Enkvetchakul, B., Merrill, A.H. Jr and Birt, D.F. (1989) Inhibition of the induction of ornithine decarboxylase activity by 12-*O*-tetradecanoylphorbol-13-acetate in the mouse skin by sphingosine sulfate. *Carcinogenesis* 10, 379–381.

Enkvetchakul, B., Barnett, T., Liotta, D.C., Geisler, V., Menaldino, D.S., Merrill, A.H. Jr. and Birt, D.F. (1992) Influences of sphingosine on two-stage skin tumorigenesis in SENCAR mice. *Cancer Letters* 62, 35–42.

Fagotto, F., Funayama, N., Gluck, U. and Gumbiner, B.M. (1996) Binding to cadherins antagonizes the signaling activity of beta-catenin during axis formation in *Xenopus*. *Journal of Cell Biology* 132, 1105–1114

Hannun, Y.A., Loomis, C.R., Merrill, A.H. Jr and Bell, R.M. (1986) Sphingosine inhibition of protein kinase C activity and of phorbol dibutyrate binding *in vitro* and in human platelets. *Journal of Biological Chemistry* 261, 12604–12609.

Hertervig, E., Nilsson, Å., Nyberg, L. and Duan, R.D. (1997) Alkaline sphingomyelinase activity is decreased in human colorectal carcinoma. *Cancer* 79, 448–453.

Hirose, Y., Yoshimi, N., Makita, H., Hara, A., Tanaka, T. and Mori, H. (1996) Early apoptosis and cell proliferation in azoxymethane-initiated rat colonic epithelium. *Japanese Journal of Cancer Research* 87, 575–582.

Holt, P.R. (1999) Dairy foods and prevention of colon cancer: human studies. *Journal of the American College of Nutrition* 18 (Supplement 5), 379S–391S.

Hoschuetzky, H., Aberle, H. and Kemler, R. (1994) β-Catenin mediates the interaction of the cadherin–catenin complex with epidermal growth factor receptor. *Journal of Cell Biology* 127, 1375–1380.

Huwiler, A., Kolter, T., Pfeilschifter, J. and Sandhoff, K. (2000) Physiology and pathophysiology of sphingolipid metabolism and signaling. *Biochimica et Biophysica Acta* 1485, 63–99.

Ito, A., Handa, K., Withers, D.A., Satoh, M. and Hakomori, S. (2001) Binding specificity of siglec7 to disialogangliosides of renal cell carcinoma: possible role of disialogangliosides in tumor progression. *FEBS Letters* 504, 82–86.

Jarvinen, R., Knekt, P., Hakulinen, T. and Aromaa, A. (2001) Prospective study on milk products, calcium and cancers of the colon and rectum. *European Journal of Clinical Nutrition* 55, 1000–1007.

Jarvis, W.D., Fornari, F.A., Taylor, R.S., Martin, H.A., Kramer, L.B., Erukulla, R.K., Bittman, R. and Grant, S. (1996) Induction of apoptosis and potentiation of ceramide-mediated cytotoxicity by sphingoid bases in human myeloid leukemia cells. *Journal of Biological Chemistry* 271, 8275–8284.

Jen, J., Powell, S.M., Papdopoulos, N., Smith, K.J., Hamilton, S.R., Vogelstein, B. and Kinzler, K.W. (1994) Molecular determinants of dysplasia in colorectal lesions. *Cancer Research* 54, 5523–5526.

Jensen, R.G. (ed.) (1995) *Handbook of Milk Composition.* Academic Press, New York.

Kawa, S., Kimura, S., Hakomori, S.-I. and Igarashi, Y. (1997) Inhibition of chemotactic and trans-endothelial migration of human neutrophils by sphingosine-1-phosphate. *FEBS Letters* 420, 196–200.

Kedderis, L.B., Bozigan, H.P., Kleeman, J.M., Hall, R.L., Palmer, T.E., Harrison, S.D. Jr and Susick, R.L. Jr (1995) Toxicity of the protein kinase C inhibitor safingol administered alone and in combination with chemotherapeutic agents. *Fundamental and Applied Toxicology* 25, 201–217.

Kinzler, K.W. and Vogelstein, B. (1996) Lessons from hereditary colon cancer. *Cell* 87, 159–170.

Kobayashi, T., Shimizugawa, T., Osakabe, T., Watanabe, S. and Okuyama, H. (1997) A long-term feeding of sphingolipids affected the level of plasma cholesterol and hepatic triacylglycerol but not tissue phospholipids and sphingolipids. *Nutrition Research* 17, 111–114.

Kong, Y., Li, R. and Ladisch, S. (1998) Natural forms of shed tumor gangliosides. *Biochimica et Biophysica Acta* 1394, 43–56.

Ladisch, S., Li, R. and Olson, E. (1994) Ceramide structure predicts tumor ganglioside immunosuppressive activity. *Proceedings of the National Academy of Sciences USA* 91, 1974–1978.

Lavie, Y., Cao, H., Bursten, S.L., Giuliano, A.E. and Cabot, M.C. (1996) Accumulation of glucosylceramides in multi-drug resistant cancer cells. *Journal of Biological Chemistry* 27, 19530–19536.

Lavie, Y., Cao, H.T., Volner, A., Lucci, A., Han, T.Y., Geffen, V., Giuliano, A.E. and Cabot, M.C. (1997) Agents that reverse multidrug resistance, tamoxifen, verapamil, and cyclosporin A, block glycosphingolipid metabolism by inhibiting ceramide glycosylation in human cancer cells. *Journal of Biological Chemistry* 272, 1682–1687.

Lipkin, M. (1999) Preclinical and early human studies of calcium and colon cancer prevention. *Annals of the New York Academy of Sciences* 889, 120–127.

Luberto, C. and Hannun, Y.A. (1998) Sphingomyelin synthase, a potential regulator of intracellular levels of ceramide and diacylglycerol during SV40 transformation. Does sphingomyelin synthase account for the putative phosphatidylcholine-specific phospholipase C? *Journal of Biological Chemistry* 273, 14550–14559.

Lucci, A.E., Cho, W.I., Han, T.Y., Guiliani, A.E., Morton, D.L. and Cabot, M.C. (1998) Glucosylceramide: a marker for multiple-drug resistant cancers. *Anticancer Research* 18, 475–480.

Mathias, S., Peña, L.A. and Kolesnick, R.N. (1998) Signal transduction of stress via ceramide. *Biochemical Journal* 335, 465–480.

McNeill, H., Ozawa, M., Kemler, R. and Nelson, W.J. (1990) Novel function of the cell adhesion molecule uvomorulin as an inducer of cell surface polarity. *Cell* 62, 309–316.

Metz, R.J., Vellody, K., Patel, S., Bergstrom, R., Meisinger, J., Jackson, J., Wright, M.A. and Young, M.R. (1991) Vitamin D3 and ceramide reduce the invasion of tumor cells through extracellular matrix components by elevating protein phosphatase-2A. *Invasion and Metastasis* 16, 280–290.

Nagase, H. and Nakamura, Y. (1993) Mutations of the *APC* (adenomatous polyposis coli) gene. *Human Mutations* 2, 425–434.

Nakagawa, R., Nagafune, I., Tazunoki, Y., Ehara, H., Tomura, H., Iijima, R., Motoki, K., Kamishohara, M. and Seki, S. (2001) Mechanisms of the antimetastatic effect in the liver and of the hepatocyte injury induced by α-galactosylceramide in mice. *Journal of Immunology* 166, 6578–6584.

Nilsson, Å. (1968) Metabolism of sphingomyelin in the intestinal tract of the rat. *Biochimica et Biophysica Acta* 76, 575–584.

Nilsson, Å. (1969) Metabolism of cerebrosides in the intestinal tract of the rat. *Biochimica et Biophysica Acta* 187, 113–121.

Notterman, D.R., Alon, U., Sierk, A.J. and Levine, A.J. (2001) Transcriptional gene expression profiles of colorectal adenoma, adenocarcinoma, and normal tissue examined by oligonucleotide arrays. *Cancer Research* 61, 3124–3130.

Nyberg, L., Nilsson, Å., Lundgren, P. and Duan, R.-D. (1997) Localization and capacity of sphingomyelin digestion in the rat intestinal tract. *Journal of Nutritional Biochemistry* 8, 112–118.

Orsulic, S. and Pfeifer, M. (1996) An *in vitro* structure–function study of armadillo, the β-catenin homologue, reveals both separate and overlapping regions of the protein required for cell adhesion and for wingless signaling. *Journal of Cell Biology* 134, 1283–1300.

Parodi, P.W. (1999) Conjugated linoleic acid and other anticarcinogenic agents of bovine milk fat. *Journal of Dairy Sciences* 82, 1339–1349.

Perry, D.K., Carton, J., Shah, A.K., Meredith, F., Uhlinger, D.J. and Hannun, Y.A. (2000) Serine palmitoyl transferase regulates *de novo* ceramide generation during etoposide-induced apoptosis. *Journal of Biological Chemistry* 275, 9078–9084.

Pfeifer, M. (1995) Cell adhesion and signal transduction: the armadillo connection. *Trends in Cell Biology* 5, 224–229.

Ravindra, M.H., Gonzales, A.M., Nishimoto, K., Tam, W.-Y., Soh, D. and Morton, D.L. (2000) Immunology of gangliosides. *Indian Journal of Experimental Biology* 38, 301–312.

Ritter, G. and Livingston, P.O. (1991) Ganglioside antigens expressed by human cancer cells. *Seminars in Cancer Biology* 2, 401–409.

Sacco, P.A., McGranahan, M.J., Wheelock, M.J. and Johnson, K.R. (1995) Identification of plakoglobin domains required for association with N-cadherin and α-catenin. *Journal of Biological Chemistry* 270, 20201–20206.

Schmelz, E.M., Crall, K.L., LaRocque, R., Dillehay, D.L. and Merrill, A.H. Jr (1994) Uptake and metabolism of sphingolipids in isolated intestinal loops of mice. *Journal of Nutrition* 124, 702–712.

Schmelz, E.M., Dillehay, D.L., Webb, S.K., Reiter, A., Adams, J. and Merrill, A.H. Jr (1996) Sphingomyelin consumption suppresses aberrant colonic crypt foci and increases the proportion of adenomas versus adenocarcinomas in CF1 mice treated with 1,2-dimethylhydrazine: implications for dietary sphingolipids and colon carcinogenesis. *Cancer Research* 56, 4936–4941.

Schmelz, E.M., Bushnev, A.B., Dillehay, D.L., Liotta, D.C. and Merrill, A.H. Jr (1997) Suppression of aberrant colonic crypt foci by synthetic sphingomyelins with saturated or unsaturated sphingoid base backbones. *Nutrition and Cancer* 28, 81–85.

Schmelz, E.M., Sullards, M.C., Dillehay, D.L. and Merrill, A.H. Jr (2000) Inhibition of colonic cell proliferation and aberrant crypt foci formation by dairy glycosphingolipids in 1,2-dimethylhydrazine-treated CF1 mice. *Journal of Nutrition* 130, 522–527.

Schmelz, E.M.. Roberts, P.C., Kustin, E.M., Lemonnier, L.A., Sullards, M.C., Dillehay, D.L. and Merrill, A.H. Jr (2001) Modulation of β-catenin localization and intestinal tumorigenesis *in vitro* and *in vivo* by sphingolipids. *Cancer Research* 61, 6723–6729.

Schwartz, G.K., Ward, D., Saltz, L., Casper, E.S., Spiess, T., Mullen, E., Woodworth, J., Venuti, R., Zervos, P., Storniolo, A.M. and Kelsen, D.P. (1997) A pilot clinical/pharmacological study of the protein kinase C-specific inhibitor safingol alone and in combination with doxorubicin. *Clinical Cancer Research* 3, 537–543.

Separovic, D., He, J. and Oleinick, N.L. (1997) Ceramide generation in response to photodynamic treatment of L5178Y mouse lymphoma cells. *Cancer Research* 57, 1717–1721.

Shibamoto, S., Hayakawa, M., Takeuchi, K., Hori, T., Oku, N., Miyazawa, K., Kitamura, N., Takeichi, M. and Ito, F. (1994) Tyrosine phosphorylation of β-catenin and plakoglobin enhanced by hepatocyte growth factor and epidermal growth factor in human carcinoma cells. *Cellular Adhesion and Communication* 1, 295–305.

Sparks, A.B., Morin, P.J. and Kinzler, K.W. (1998) Mutational analysis of the APC/β-catenin/TCF pathway in colorectal cancer. *Cancer Research* 58, 1130–1134.

Spiegel, S. (1999) Sphingosine-1-phosphate: a prototype of a new class of second messengers. *Journal of Leukocyte Biology* 65, 341–344.

Spiegel, S. and Merrill, A.H. Jr (1996) Sphingolipid metabolism and growth regulation: a state-of-the-art review. *FASEB Journal* 10, 1388–1397.

Takamizawa, K., Iwamori, M., Mutai, M. and Nagai, Y. (1986) Selective changes in gangliosides of human milk during lactation: a molecular indicator for the period of lactation. *Biochimica el Biophysica Acta* 879, 73–77.

Takenaga, M., Igarashi, R., Matsumoto, K., Takeuchi, J., Mizushima, N., Nakayama, T., Morizawa, Y. and Mizushima, Y. (1999) Lipid microsphere preparation of a lipophilic ceramide derivative suppresses colony formation in a murine experimental metastasis model. *Journal of Drug Targeting* 7, 187–195.

Taki, T., Ishikawa, D., Ogura, M., Kakajima, M. and Handa, S. (1997) Ganglioside GD1a functions in the adhesion of metastatic tumor cells to endothelial cells of the target tissue. *Cancer Research* 57, 1882–1888.

Toura, I., Kawano, T., Akutsu, Y., Nkayama, T., Ochiai, T. and Taniguchi, M. (1999) Cutting edge: inhibition of experimental tumor metastasis by dendritic cells pulsed with α-galactosylceramide. *Journal of Immunology* 163, 2387–2391.

Valentino, L., Moss, T., Olson, E., Wang, H.J., Elashoff, R. and Ladisch, S. (1990) Shed tumor gangliosides and progression of human neuroblastoma. *Blood* 75, 1564–1567.

Vesper, H., Schmelz, E.M., Nikolova-Karakashian, M., Dillehay, D.L., Lynch, D.V. and Merrill, A.H. Jr (1999) Sphingolipids in food and the emerging importance of sphingolipids to nutrition. *Journal of Nutrition* 129, 1239–1250.

Wiegmann, K., Schütze, S., Machleid, T., Witte, D. and Krönke, M. (1994) Functional dichotomy of neutral and acidic sphingomyelinases in tumor necrosis factor signaling. *Cell* 78, 1005–1015.

Wang, F., Van Brocklyn, J.R., Edsall, L., Nava, V.E. and Spiegel, S. (1999) Sphingosine-1-phosphate inhibits motility of human breast cancer cells independently of cell surface receptors. *Cancer Research* 59, 6185–6191.

Wang, X.Z., Beebe, J.R., Pwiti, L., Bielawska, A. and Smyth, M.J. (1999) Aberrant sphingolipid signaling is involved in the resistance of prostate cancer cell lines to chemotherapy. *Cancer Research* 59, 5842–5848.

Wright, S.C., Zheng, H. and Zhong, J. (1996) Tumor cell resistance to apoptosis due to a defect in the activation of sphingomyelinase and the 24 kDa apoptotic protease (AP24). *FASEB Journal* 10, 325–332.

Xu, J., Yeh, C.H., Chen, S., He, L., Sensi, S.L., Canzoniero, L.M., Choi, D.W. and Hsu, C.Y. (1998) Involvement of *de novo* ceramide biosynthesis in tumor necrosis factor-alpha/cycloheximide-induced cerebral endothelial cell death. *Journal of Biological Chemistry* 1273, 16521–16526.

# 14 食物中异黄酮的保健作用

Thomas M. Badger，Martin J. J. Ronis 和 Nianbai Fang
（阿肯色州儿童营养中心，小石城，阿肯色州，美国）

## 引 言

本章将在异黄酮和健康方面做一介绍，集中给读者提供一个可理解的水平，使读者足以用批判的眼光来评价目前大量而且快速增长的异黄酮作用的相关文献，特别是与异黄酮营养和健康状态有关的文献。本文对植物和人体内异黄酮的分子结构以及目前广泛应用于异黄酮的鉴别及定量的方法予以描述，大豆蛋白分离物将被用作大豆食物模型来阐述大豆内含量最丰富的两种异黄酮在人尿与血浆中的药物动力学。动物模型将被用来阐述靶组织中异黄酮的分子形式及浓度。本文还将讨论人类与异黄酮在整个生命循环过程中的接触，研究异黄酮对健康的一些潜在的有利和不利的影响及这些作用可能的分子机制，并且对关于用大豆作为婴儿代乳品是否安全这一有争议的问题进行探讨。

大豆是高质量食物蛋白的一个极好来源，它的几种成分被用在食品加工及烹饪过程中。大豆可被加工为酱豆腐醅、日本豆面酱、面粉、油、纺织化纤维、蛋白质浓缩物和分离物，并可被提取为几种植物化学物如胰蛋白酶/胰凝乳蛋白酶的 Bowman-Birk 抑制剂，皂角苷和异黄酮类化合物。异黄酮类化合物是类黄酮的一个亚类。类黄酮是自然存在的植物成分中的一大组，其基本结构是由 3 个碳原子连接的两个芳香环。图 14.1 列出了类黄酮的这种结构及异黄酮的结构（其 A 环 9 位的碳原子上有一氧桥），以及它们的 C 环还原代谢物异黄烷醇和异黄烷。大豆中异黄酮以 A 环 7 位碳原子及 B 环 4′碳原子的氢化为特点。含量最丰富的三种大豆异黄酮的代表是黄苷元、染料木黄酮和黄豆黄素。它们通常以糖苷染料木苷、大豆苷和黄豆苷的形式，以及几种乙酰化和丙二酰化的衍生物存在（图 14.2）。应当注意还有其他已熟知的自然存在的异黄酮类化合物，如香豆素、鱼藤酮类化合物、紫檀烷类及其它们的异黄酮类化合物低聚物。本章我们将只讨论主要的大豆异黄酮。

异黄酮糖苷隐藏在大豆蛋白基质中，糖苷配基（染料木黄酮、黄苷元和黄豆黄素）在β-糖苷酶的作用下在小肠和结肠中释放，而β-糖苷酶存在于肠细胞和细菌酶中。第四个非常有潜力且对健康有重要影响的糖苷为牛尿酚，是在细菌酶的作用下由黄苷元衍生而来的。因为在尿中分别只有约 10％和 50％的总摄食的染料木黄酮和黄豆苷（Shelnutt *et al*.,2000），大豆蛋白分离物（soy protein isolate，$SPI^{+}$）中的异黄酮表观利用率非常低。在最初代谢中，大多被吸收的糖苷配基在葡萄糖苷酸化和硫酸化的作用下被结合。未知百分数的糖苷配基被吸收后通过门脉循环直接到达肝脏并在此被结合，留下一小部分（＜5％）的糖苷配基进入全身循环。而且，几种还原和（或）氧化的异黄酮

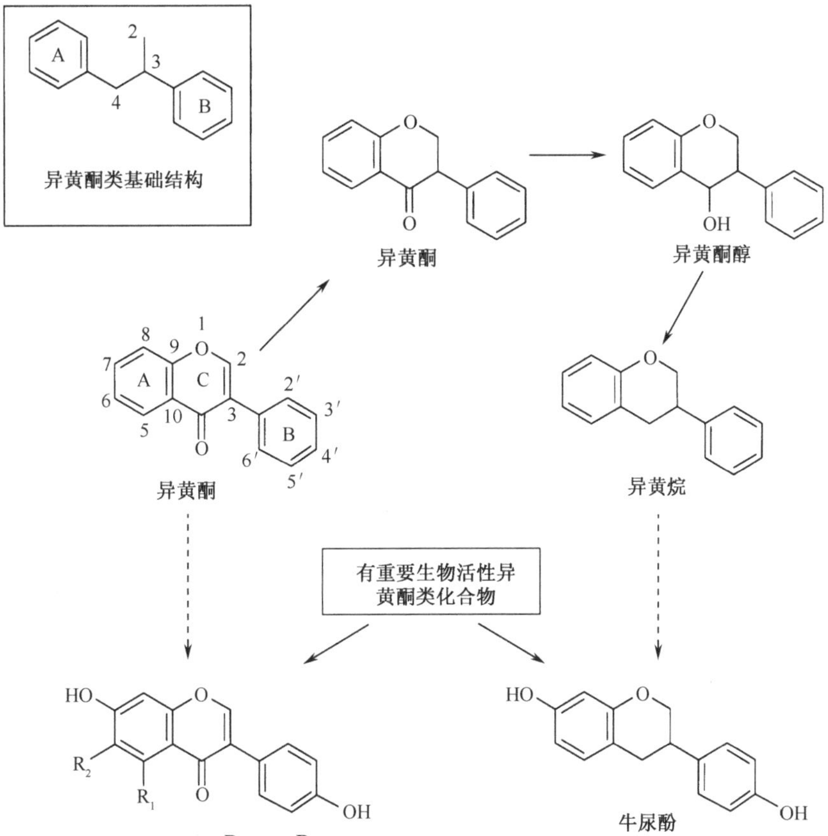


| | $R_1$ | $R_2$ |
|---|---|---|
| 大豆黄酮 | H | H |
| 染料木黄酮 | OH | H |
| 黄豆黄素 | H | OMe |

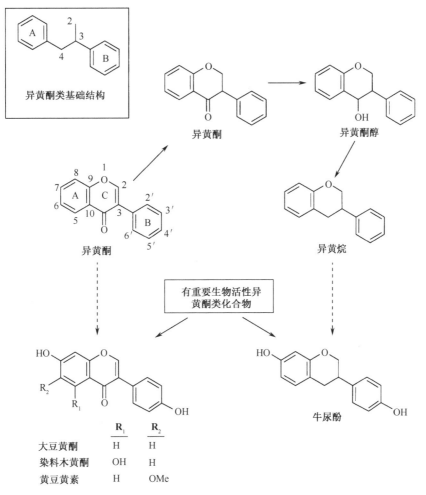

图 14.1　插图中显示了异黄酮的基础分子结构。通过 A 环 5 位和 6 位碳原子的改变形成三种主要的大豆异黄酮（大豆黄酮，染料木黄酮和黄豆黄素），牛尿酚是大豆黄酮的主要代谢物。异黄酮，isoflavanols 和 isoflavans 之间的关系包括 A 环和 B 环间链的 4 位碳原子

代谢物也被结合。因此，大豆异黄酮循环物以葡萄糖苷酸和硫酸盐结合物，自由循环糖苷配基和蛋白结合糖苷配基等几种分子形式排泄（Setchell，1988，1995；Coward *et al*.,1993；Barnes *et al*.,1996a）。饲喂 $SPI^+$ 饲粮的雌性大鼠的主要的尿异黄酮代谢物在图 14.3 和图 14.4（Fang *et al*.,2002）中列出。

染料木黄酮或黄苷元上有两个主要的结合位点（位点 7 和 4′，图 14.2），每一个都能被硫酸化和（或）葡糖苷酸化，因此存在单葡糖苷酸、单硫酸盐，二葡糖苷酸、二硫酸盐，和一个位点葡糖苷酸化，另一个位点硫酸化的混合结合物。大部分被吸收的异黄酮以结合物的形式分泌到尿中，但有一小部分进入肠肝循环（Setchell and Adlercruetz，1988；Setchell，1995）。在结合并分泌入尿方面，异黄酮与内源性腺和肾上腺类固醇相似。因为异黄酮大多以细胞内共轭的形式存在，人们认为这些结合物能够以与结合的内源类固醇相似的方式转运入和转运出细胞。这种转运为单葡糖苷酸或单硫酸盐在肝脏中

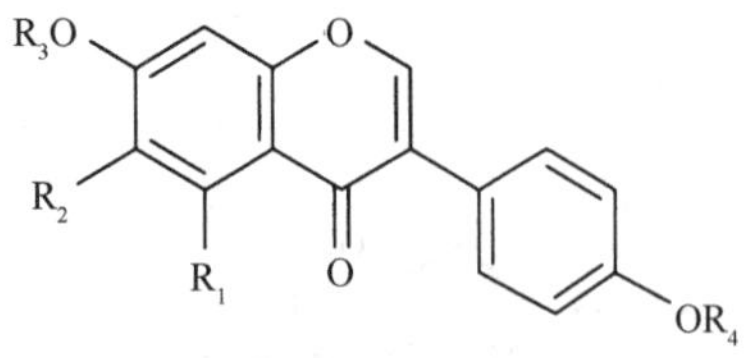

异黄酮糖苷

| | $R_1$ | $R_2$ | $R_3$ 或(和) $R_4$ |
|---|---|---|---|
| 黄豆苷 | | HH | 葡萄糖 |
| 染料木苷 | OH | H | 葡萄糖 |
| 黄豆黄素 | H | OMe | 葡萄糖 |
| 6″-*O*-乙酰黄豆苷 | | HH | 6″-*O*-乙酰葡萄糖 |
| 6″-*O*-乙酰染料木苷 | OH | H | 6″-*O*-乙酰葡萄糖 |
| 6″-*O*-乙酰黄豆苷 | H | OMe | 6″-*O*-乙酰葡萄糖 |
| 6″-*O*-丙乙酰黄豆苷 | | HH | 6″-*O*-丙乙酰葡萄糖 |
| 6″-*O*-丙乙酰染料木苷 | OH | H | 6″-*O*-丙乙酰葡萄糖 |
| 6″-*O*-丙乙酰黄豆苷 | H | OMe | 6″-*O*-丙乙酰葡萄糖 |

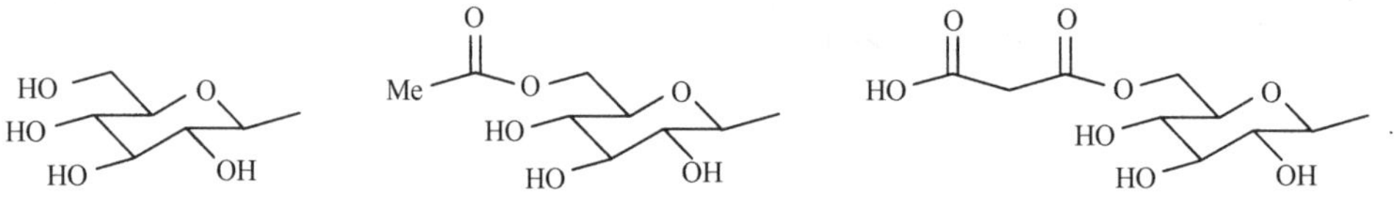

葡萄糖　　6″-*O*-乙酰葡萄糖　　6″-*O*-丙乙酰葡萄糖

图 14.2　已知的异黄酮糖苷的分子结构，通过 5，6，7 碳原子的改变形成

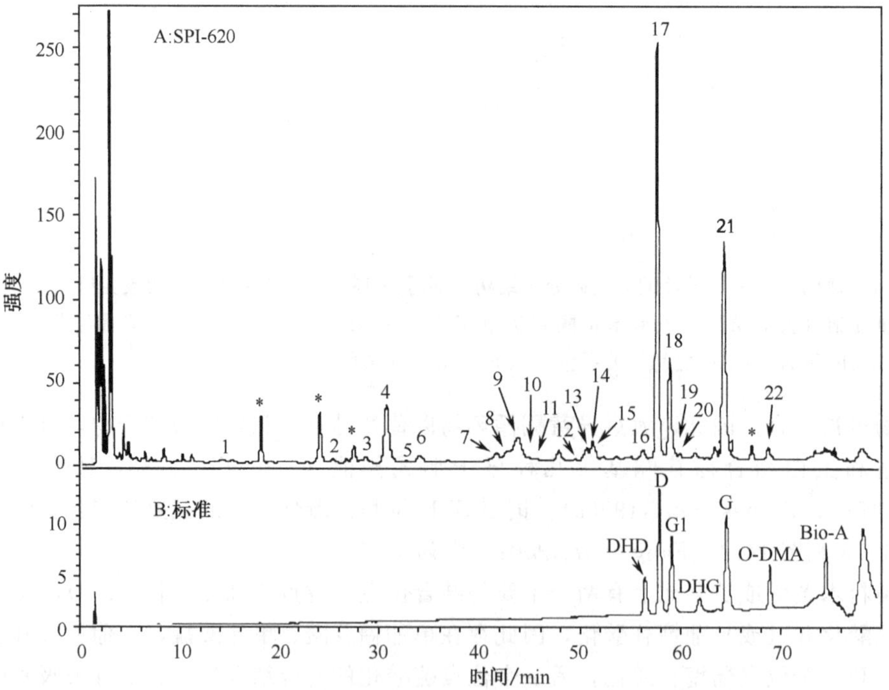


图 14.3　顶部显示了饲喂富含异黄酮饲粮大鼠尿中异黄酮代谢物的 MS-MS 分布图

SPI-620 是一含大豆蛋白分离物的饲粮，底部显示的是用于鉴定主要代谢物的市售可利用标准物。每一种代谢物的鉴定都在图 14.4 列出，Fang 等（2002）中描述了此过程的细节。* 表示尿和对照（在这儿没有显示）分布图的最高峰

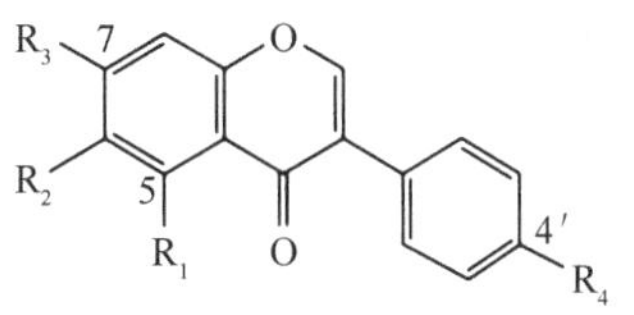

| No. | 代谢物 | $R_1$ | $R_2$ | $R_3$ | $R_4$ |
|---|---|---|---|---|---|
| **17** | 大豆黄酮 | H | H | OH | OH |
| **4** | 大豆黄酮 7-葡糖苷酸 | H | H | *OGlcUA | OH |
| **8** | 大豆黄酮 4′-葡糖苷酸 | H | H | OH | OGlcUA |
| **7** | 黄豆苯 | H | H | *OGlc | OH |
| **1** | 黄豆苷-葡糖苷酸 | H | H | OGlc | OGlcUA |
| **12** | 6″-*O*-丙乙酸黄豆苷 | H | H | Omalonyl | H |
| **21** | 染料木黄酮 | OH | H | OH | OH |
| **14** | 染料木黄酮-5-葡糖苷酸 | OGlcUA | H | OH | OH |
| **3** | 染料木黄酮 7-葡糖苷酸 | OH | H | OGlcUA | OH |
| **9** | 染料木黄酮 4′-葡糖苷酸 | OH | H | OH | OGlcUA |
| **2** | 染料木黄酮 7，4′-去葡糖苷酸 | OH | H | OGlcUA | OGlcUA |
| **18** | 黄豆黄素 | H | OMe | OH | OH |
| **6** | 黄豆黄素 7-葡糖苷酸 | H | OMe | OGlcUA | OH |
| **11** | 黄豆黄素 4′-葡糖苷酸 | H | OMe | OH | OGlcUA |
| 10 | 黄豆苷 | H | OMe | OGlc | OH |

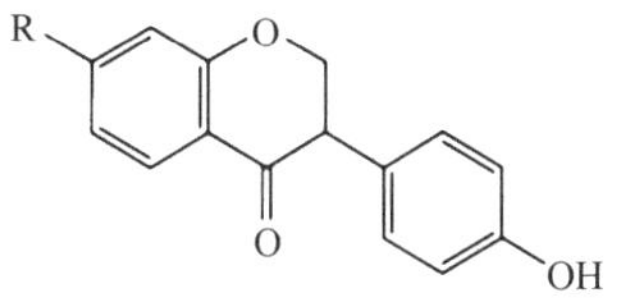

| No. | 代谢产物 | R |
|---|---|---|
| **16** | 二氢大豆黄酮 | OH |
| **5** | 二氢大豆酮 7-葡糖苷酸 | OGlcUA |

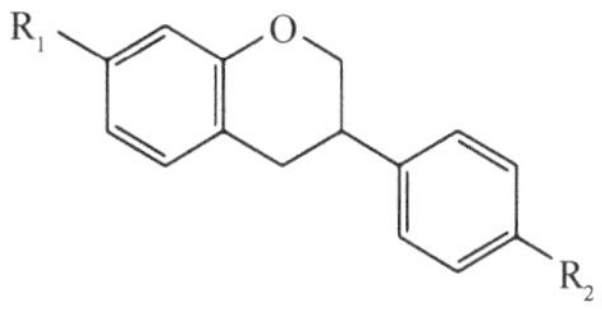

| No. | 代谢产物 | $R_1$ | $R_2$ |
|---|---|---|---|
| | 中尿酸 | OH | OH |
| ** **13** | 中尿酸葡糖苷酸 | OGlcUA | OH |
| ** **15** | 中尿酸葡糖苷酸 | OGlcUA | OH |

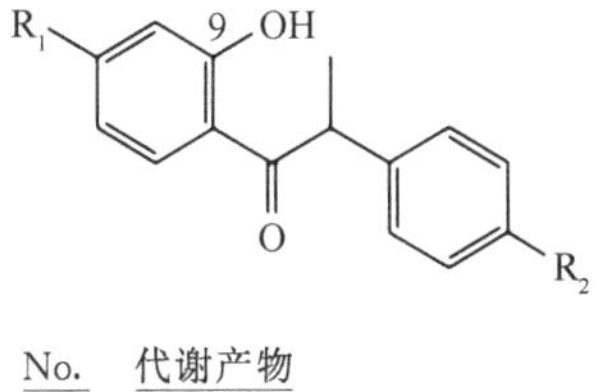

| No. | 代谢产物 | $R_1$ | $R_2$ |
|---|---|---|---|
| **22** | *O*-去甲基安哥拉紫檀素 | OH | OH |
| ** **19** | *O*-去甲基安哥拉紫檀素葡糖苷酸 | OGlcUA | OH |
| ** **20** | *O*-去甲基安哥拉紫檀素葡糖苷酸 | OGlcUA | OH |

* β-葡糖醛酸

β-葡萄糖

** 不能由 LC-MS-MS 确认共轭位置

图 14.4 尿代谢物的鉴定

粗体的数字与图 14.3 的吸收峰的数据相对应，Fang 等（2002）描述了此过程的细节

第二位点的结合提供了途径，从而为肾分泌做准备，同样也为大多数没有丰富结合酶的靶组织如乳腺、附属性器官、肾和脑等如何拥有丰富的结合形式的异黄酮做出了解释（Chang *et al.*,2002）。人们根据这些复杂的分子异黄酮种类在细胞内的存在，提出了几个重要的问题：①哪一个是生物活性最高的分子？②这些结合物是作为生物活性异黄酮

糖苷配基的靶组织池吗？③这些结合物确实是被转运入细胞了吗？或它们是在细胞内由糖苷配基形成的吗？

图 14.5 和表 14.1 列出了 12 个健康男性和女性在摄取含 SPI+ 的饮料后平均血清异黄酮（染料木黄酮加上黄豆苷等价物）的浓度，此饮料提供了 1.0mg/kg 的染料木黄酮（糖苷配基）等价物和 0.6mg/kg 的黄豆苷（糖苷配基）等价物。这些数据表明成年人在服用一种单一的市售可利用大豆饮料后（类似于大豆粉），血液中的异黄酮浓度很高，此浓度有一特定的时间过程。异黄酮在摄入后可很快被吸收，并在摄取后的 3～5.4h 染料木黄酮和大豆黄素达到高峰，3～8h 回复到基础水平（Cimino *et al.*,1999；Shelnutt *et al.*,2000，2002）。

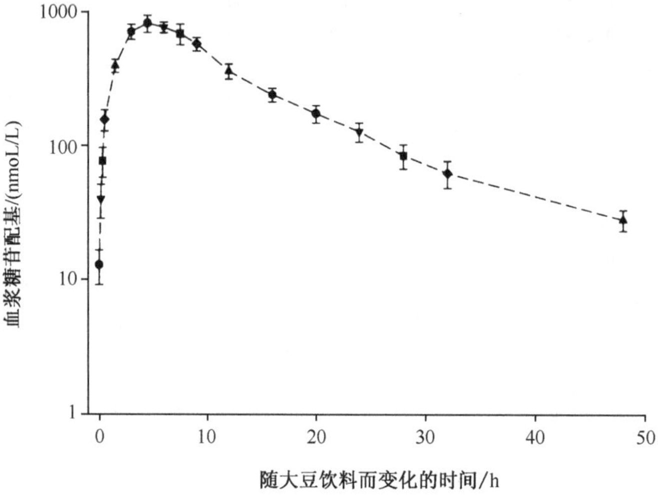


图 14.5　服用含大豆蛋白分离物的饮料后血浆总异黄酮的分布图

数值是依赖于硫酸酯酶和葡糖苷酸酶引起的解离而释放的总糖苷配基的平均值（±标准误）。详见 Shelnutt 等（2002）

**表 14.1　血浆和尿中异黄酮的药物动力学**

| 异黄酮 | 样品 | $T_{max}$/h[1] | $t_{1/2}$/h[2] |
|---|---|---|---|
| 染料木黄酮 | 血浆 | 5.2±0.4 | 8.2±0.7 |
| 葡糖苷酸 | 尿 | 4.3±0.5 | 6.0±0.4 |
| 硫酸盐 | 血浆 | 4.5±0.3 | 5.7±0.4 |
| | 尿 | 5.4±1.1 | 4.5±0.7 |
| 糖苷配基 | 尿 | 3.1±0.8 | ND[3] |
| 黄豆苷 | 血浆 | 5.0±0.5 | 3.3±0.3 |
| 葡糖苷酸 | 尿 | 5.2±0.4 | 3.8±0.4 |
| 硫酸盐 | 血浆 | 4.5±0.3 | 3.1±0.4 |
| | 尿 | 5.4±1.1 | 3.9±0.5 |
| 糖苷配基 | 尿 | 4.9±1.9 | ND[3] |

注：1 $T_{max}$为达到尿最高速度和血浆最大浓度的时间；2 $t_{1/2}$为表观半衰期；3 ND，因为浓度很低没有测定。

异黄酮类化合物的鉴定主要应用光谱方法，包括紫外光谱法（ultraviolet，UV）（Mabry *et al.*，1970）、核磁共振（nuclear magnetic resonance，NMR）光谱法（Markham and Geiger，1986）以及质谱法（mass spectrometry，MS）（Harbone *et al.*，1975）。异黄酮的紫外光谱有两个吸收峰，一个为 245～270nm（带Ⅱ），另一个为 300～340nm（带Ⅰ）（Harbone *et al.*，1975）。目前已报道的几种测量体液和食物中异黄酮及其代谢物的方法，包括高效液相色谱法（high pressure liquid chromatography，HPLC）（Franke *et al.*，1995；Supko and Phillips，1995；Kulling *et al.*，2001）、气相色谱法/质谱法（gas chromatography/mass spectrometry，GC/MS）（Adlercruetz，1993b）、高压液相色谱法/质谱法（high pressure liquid chromatography/mass spectrometry，LC/MS）（Cimino *et al.*，1999）、LC/MS/MS（Coward *et al.*，1996；Fang *et al.*，2002）、放射性免疫测定（radioimmunoassay，RIA）（Lapick *et al.*，1998）以及带有比色分析检测的 HPLC（Gamache and Acworth，1998）。因为生物样品中大多数异黄酮都是被结合的，经酶消化成糖苷配基后，LC/MS、GC/MS 或库仑电极分析技术（coulometric electrode array technology，CEAT）都是可选用的检测方法（Holder *et al.*，1999；Shelnutt *et al.*，2002）。随着更多可利用的结合标准物的出现，酶消化将变得再不必要。

用体外法已对提纯的异黄酮进行了广泛的研究，发现它能影响几个重要的细胞内系统（Polkowski and Mazyrek，2000），其中的一些列在了表 14.2 中。大多数发生在相对较高的微摩尔浓度下的体外作用，在体内并不会发生。许多体内和大多数体外的研究都是以提纯的异黄酮为材料进行的（Chang *et al.*，2000）。因为大多数被研究的细胞没有第一和第二阶段的酶，这些体外的作用都要归因于亲本复合物，因此，对异黄酮代谢物的生物学活性还了解很少。造成食物中异黄酮、提纯染料木黄酮或提纯黄苷元在体内真正起作用的生物学活性分子种类还未被确定。与之相似，对于牛尿酚代谢物的研究更少。因此，仍需要回答的问题是：异黄酮的生物学活性分子的种类有哪些；它们起作用所需的浓度是多少；以及食入含不同水平异黄酮的食物后，这些分子的细胞内浓度是多少。

**表 14.2　异黄酮的体外和体内生物学作用**

| | |
|---|---|
| 抑制利用 ATP 酶、酪氨酸特定蛋白激酶，拓扑异构酶Ⅱ | 改变雌激素代谢 |
| 抑制包含在磷酸肌醇转化过程中的酶 | 调节信号传递途径 |
| 诱导癌细胞中的细胞程序性死亡和分化 | EGF，TGF-α，TGF-β |
| 抑制细胞增殖 | 不同方式激活 ERα 和 ERβ |
| 抑制血管发生和转移 | 改变性腺类固醇代谢 |
| 发挥抗氧化作用 | 增加免疫反应并增强宿主抗性 |
| 调节细胞周期 | |

异黄酮类是植物化学物质中一个极好的范例，这类物质不是正常生长、正常发育或维持成体正常体内稳衡所必需的，但对哺乳动物却具有生物学活性并能够潜在的影响动物的健康。在这方面，异黄酮不是营养物质而是影响健康细胞的功能的重要潜在调节子。在动物和人类上的研究已证明异黄酮对健康存在有利和不利两方面的影响（Gilani and Anderson，2002)。大多数关于异黄酮的研究都表明它们对人类健康具有有利的影响，包括心脏病、癌症、肾病、骨质疏松症以及更年期综合征的减轻，后两个作用又引起了人们对大豆和异黄酮类化合物的一系列研究，研究其能否代替激素进行治疗。然而，也有几个研究指出了异黄酮类化合物对健康的潜在不利影响。要了解异黄酮在各种生理和环境条件下对健康的作用尚需不同生理和环境条件下进行更多的研究。三个最近举行的关于大豆在预防和治疗慢性疾病中作用的国际研讨会的资料（Messina，1998，2000；Messina *et al.*，2002）为读者获得异黄酮的健康作用的历史研究观点提供了很好的资源。

## 人类与异黄酮的接触

市场数据和医院出院记录表明，美国每年 400 万新生儿中大概有 25%的婴儿使用大豆婴儿代乳品（美国儿科营养研究会，1998）。美国市场上所出售的商品化可利用大豆婴儿食品是以 $SPI^+$ 为原料制作的，有几个植物化学物质与 $SPI^+$ 结合在一起，其中包括异黄酮。喂婴儿大豆代乳品的原因不同，从医学指示（如腹泻后乳糖不耐性、半乳糖血症和原发性乳糖酶缺乏）到想要维持素食生活方式（Hill and Stuart，1929；Setchell *et al.*，1997，1998）。尽管美国儿科研究会认为母乳喂养好于婴儿食品喂养，仍建议以分离大豆蛋白为基础的婴儿食品是母乳安全有效的替代品，因为它能够为婴儿的正常生长和发育提供合适的营养，而以母乳或牛奶为基础的婴儿食品却不能提供。

美国人的饮食习惯，特别是与大豆食品有关的习惯，明显与亚洲人不同。与大豆异黄酮的接触在一些国家可能真正开始于胎儿阶段，在这些国家，人们每天摄取大豆。脐血和羊水的数据表明，亚洲胎儿接触异黄酮的浓度高达 0.8μmol/L，该水平与所报道的怀孕期间母体血液循环水平（0.7μmol/L）接近（Adlercreutz *et al.*，1999）。然而，由于亚洲新生婴儿出生后的前几个月很可能食用牛奶或母乳，其异黄酮浓度降到几乎接近零。在牛奶和来自食用大豆的妇女的母乳中，异黄酮浓度都很低（5～50nmol/L），因此食用母乳或牛奶代乳品的亚洲新生儿与异黄酮的接触量很低（Franke and Caster，1996；Setchell *et al.*，1997，1998）。循环和组织中异黄酮的浓度一直会保持在低水平直到孩子开始食用大豆食品，如果孩子养成食用大豆食品（或其他含异黄酮的食品）的习惯，异黄酮就会增加到母体水平。其后，那些继续食用含有高水平异黄酮的典型亚洲食品的人体内保持一定的异黄酮含量，此含量相对要高于无大豆消耗国家人的异黄酮含量。

在美国，大多数怀孕的妇女并不食用含有异黄酮的食物，胎儿与异黄酮的接触很少，而且 75%的美国婴儿不食用大豆婴儿食品，因此他们的血浆异黄酮水平很低，在整个从婴儿到成年的发育过程中都保持着这种相同的低大豆消耗。除了加工食品中含有的大豆蛋白外，大多数美国人并不摄食含足够水平的大豆食品，因此他们的血浆异黄酮浓度在整个生命过程中都保持在低水平。

但是，25%的美国婴儿食用了大豆婴儿食品（美国儿科营养研究会，1998），此食品中含有大约 45μg/ml（166μmol/L）总异黄酮或大于 3000 倍人乳浓度的异黄酮（Setchell *et al.*，1997，1998）。食用大豆婴儿食品的婴儿（有时从出生就开始食用直到断奶）血浆异黄酮浓为 1～10μmol/L（Setchell *et al.*，1997），这也是药物动力学研究中，服用大豆饮料的成年人血浆异黄酮浓度的一般浓度范围（图 14.5 和表 14.1）（Shelnutt *et al.*，2002）。这些人的总异黄酮摄入量为 1.64mg/kg 体重，平均血浆总异黄酮浓度为 1.2μmol/L。食用含有与成年人药物动力学研究中相同 $SPI^+$ 的大豆婴儿食品的婴儿共食入 6～9mg/kg 体重的异黄酮，因此，大概 5 倍于成年人的消耗量，血浆总异黄酮浓度也比成年人高出 5 倍（Setchell *et al.*，1997）。这与从日本传统食物中摄取的平均每天 0.7mg/kg 的总异黄酮量（Messina，1995），及其所导致的约 0.9μmol/L 的血清总异黄酮水平相差不大（Patisaul *et al.*，2001）。因此，食用大豆婴儿代乳品的婴儿可能是人群中与大豆食品异黄酮接触最多的一部分人。食用大豆代乳品的美国孩子在断奶后可能摄入不到足够水平的大豆食品，他们的血浆异黄酮浓度下降至低于 20nmol/L，并且在以后的生命中一直保持这种低水平，因此，典型的亚洲人和美国人在他们的整个生命过程中与异黄酮的接触方式及异黄酮日摄入量都显著不同，这些接触方式对健康的潜在影响还没有完全阐明。

## 雌激素介导的异黄酮作用

尽管异黄酮的作用有雌激素受体介导和无雌激素介导两种，但前者已吸引了科学家和公众的注意力。大豆异黄酮通常也称为植物雌激素，要与雌激素受体（oestrogen receptor，ER）结合后，才发挥雌激素激动剂、拮抗剂或选择性雌激素受体调节者（selective oestrogen receptor modulator，SERM）的作用（Kupier *et al.*，1997；Barkhem *et al.*，1998；An *et al.*，2001；Setchell，2001）。其作用根据组织、细胞类型、异黄酮浓度及其他如激素水平、年龄等条件的不同而异（表 14.37）。

类固醇雌激素如 17β-雌二醇（$E_2$），通过影响分化和许多靶组织包括雄性和雌性生殖道、脑、骨和心血管系统的功能而影响生长和发育。雌激素是高亲脂性的，可以很容易地通过扩散作用穿过细胞和细胞内膜与细胞核内的雌激素受体（ER）结合。当雌激素配体与 ER 结合后，ER 的结构形态发生变化，这种形态的变化使受体-配体复合物与染色质相互作用从而调节靶基因的转录。尽管对 ER 的研究已进行了几十年，但是直到 1987 年才克隆出了 ERα（Koike *et al.*，1987），并且又花了十多年的时间发现并克隆了 ERβ（Kupier *et al.*，1996）。ERα 和 ERβ 蛋白有高度的同源性，但它们的靶细胞的分布和密度不同。尽管大多数配体都以很相似的亲和力与这两个受体相结合，但一些合成雌激素与自然类固醇激素的配体与受体结合的相对亲和力相差很大（Koike *et al.*，1987；Kupier *et al.*，1996）。人类食物中包含着一些非类固醇，弱雌激素性的植物化学物质（包括异黄酮、香豆素和木聚素），这些物质可以同 $E_2$ 竞争结合 ERα 和 ERβ 蛋白（Markiewicz，1993；Miksicek，1993；Korach，1994；Kupier *et al.*，1998）。

**表 14.3　雌激素与异黄酮功能的差异**

| |
|---|
| 募集转录因子 |
| （辅助调节子） |
| 激活或抑制基因转录 |
| 异黄酮激发抑制比激发激活途径优先 |
| 异黄酮比雌激素潜力小 |
| ER-配体构造 |

食用大豆婴儿食品的婴儿血液（或组织）中异黄酮浓度升高之所以会引起人们的兴

趣，是因为人类通常直到青春期后才会有高浓度的雌激素。这就提出了“激素”作用对婴儿发育的短期和长期的健康影响的问题。有报道称染料木黄酮的效力比 $E_2$ 低 3000～4000 倍（Farmakalidis and Murphy，1985；Kupier *et al.*，1997；An *et al.*，2001），这种较大差异可能要归因于方法学，目前经常使用 1∶1000 这个值。然而，对效能的研究主要在体外系统或动物中进行，至今尚无在人类婴儿靶细胞中进行真效能的研究报道。由于出生后雌激素因子通常很低而且异黄酮在婴儿中的效力及高血浆异黄酮浓度不能确定，因此科学家推测大豆婴儿食品中的异类酮可能存在副作用（Irvine *et al.*，1995；Robertson，1995；Sheehan，1997）。因为这个题目以前已经在一定程度上被讨论过（Badger *et al.*，2002a，b），在此只简单的予以总结。经常提到的两个副作用是对癌症和生殖的影响，这些将在下面进行简单的讨论。

## 癌　症

流行病学和实验动物数据表明不论是大豆食品还是在食物中补充的异黄酮，都能降低癌症的发生率（Adlercreutz and Mazur，1997；Barnes，1997）。如在食用高水平大豆国家中，乳腺癌和结肠癌的发生率低于大豆摄入量低的国家（Nomura *et al.*，1978；Hirayama，1985；Lee *et al.*，1991）。注射（Lamartiniere *et al.*，1995）、食用异黄酮（Fritz *et al.*，1998；Lamartiniere *et al.*，2001）或食用含大豆饲粮（Hakkak *et al.*，2000，2001）的动物，用化学方法诱导的乳腺和结肠癌的发生率较低。另一方面，给无胸腺小鼠饲喂异黄酮或大豆饲粮后促进了人类乳腺癌 MCF-7 细胞系的发展（Hsieh *et al.*，1998；Santell *et al.*，2000；Alfred *et al.*，2001），此结果对以前治疗过乳癌、有潜在乳瘤或绝经的妇女同样适用。这些研究中一个有趣的自相矛盾的特点是，最初一系列的研究证明异黄酮可减少癌症的发生（或许是通过减少肿瘤的起始和发展）（Nomura *et al.*，1978；Hirayama，1985；Lee *et al.*，1991；Lamartiniere *et al.*，1995，2001；Adlercreutz and Mazur，1997；Fritz *et al.*，1998；Hakkak *et al.*，2000，2001），而后来的研究却发现异黄酮增加了癌症发展的危险性（Hsieh *et al.*，1998；Santell *et al.*，2000；Alfred *et al.*，2001）。一种单一的膳食因素是如何具有癌促进和癌预防两种特性的呢？

考虑到以下的信息：①异黄酮结合并激活 ER，因此称为植物雌激素；②含有高水平植物雌激素的食物在以人类所做的研究中会减少癌症发生的危险，而在以动物所进行的研究中减少了实验诱导性癌的发生，但与之相矛盾的是，促进了与雌激素正相关的人类乳腺癌细胞的发展；③在激素替代治疗中应用的雌激素可能增加某些癌症发生的危险性。异黄酮能引起人们的兴趣是因为它不仅与雌激素有某些相同的生化特性，而且还具有明显的非雌激素的生物活性。如果雌激素和异黄酮的作用机制都包含 ER，那什么是它们独特性质的基础？答案是很复杂的，但可能在于 ER 配体的特异性，剂量依赖能力以及 ERα 和 ERβ 对基因激活或抑制的不同调节。

An 等（2001）有趣的研究证明 $E_2$ 有效地激发了 ERα 和 ERβ 介导的转录激活或抑制途径，而异黄酮也可有效地激活这些途径，但它主要是通过 ERβ 起作用的。因此，由异黄酮引起的转录活动与由雌激素通过独特基因的激活或抑制引起的转录活动不同，这能够为上面提到的相互矛盾作用做出解释。由此，即使雌激素疗法看起来与某些癌症发生的增加相关，食物中植物雌激素却与癌症发生的减少相关。

以下是异黄酮与雌激素在 ER 激活/抑制方面的几个不同：

(1) 异黄酮对 ERα 和 ERβ 的效力不同。如染料木黄酮对引发 ERα 介导的基因激活或抑制，是一个弱的激动剂，但却是 ERβ 介导的基因转录活性的强激动剂。实际上，在所报道的相关基因的研究中，ERβ 介导的基因转录的激活或抑制比 ERα 强 1000 倍 (An *et al.*, 2001)，其原因或根本的作用机制虽是复杂且多方面的，但部分可能要归因于 ERβ 与染料木黄酮的亲和力比 ERα 大近 30 倍。

(2) 现在认为 ER 引起的转录激活或抑制都需要辅调节子蛋白的参与 (An *et al.*, 1999; McKenna *et al.*, 1999a, b; Glass and Rosenfeld, 2000; Katzenellbogen *et al.*, 2000; Klinge, 2000)，因此，配体"补充"这些辅调节子蛋白的能力将极大地影响转录的活性。辅调节子蛋白与 $E_2$-ER 复合物结合，结合物不同于异黄酮-ER 复合物，这种结合与不同基因的转录激活或抑制有关。例如，在一定的试验条件下，染料木黄酮只补充 ERβ 的辅调节子，激发 ERβ 介导的途径，并引发与 ERβ 而不是 ERα 相关的临床作用。这将意味着 ERβ 介导途径中的有益作用将被实现，而 ERα 途径中经常的负面作用（如增加癌的危险性）不被实现。

(3) 配体剂量在调节作用中是一个很重要的因子。在靶细胞中配体的浓度可以很容易的通过膳食得到 (0.5～1μmol/L)。染料木黄酮只是在为 ERα 补充 GRIP1 时能力很弱，而为 ERβ 补充这种转录因子则很容易 (Barnes *et al.*, 1996b)。这些辅调节子有几种作用，包括减少激动剂从 ER-辅调节子复合物上的脱离，增强了转录作用。作为结果，这种稳定的作用可能是异黄酮作用中的一个重要因素。

(4) 异黄酮对基因转录抑制比转录激活更有效，大概为转录激活的 10～300 倍 (An *et al.*, 2001)。因此，异黄酮的有益作用可能要归因于"关闭"了导致不利作用的途径而不是激活其他途径，这为大豆异黄酮摄入量高的亚洲国家中较低的更年期症以及一些疾病如骨质疏松症、心血管病、乳房和子宫癌的发生率提供了概念框架 (Ingram *et al.*, 1997; Adlercreutz and Baillieres, 1998; Davis *et al.*, 1999)。

(5) 染料木黄酮结合在 ERβ 上时，ERβ 的构象与 $E_2$ 与之结合时的构象不同，异黄酮与 SERM 结合时呈现异黄酮-ERβ 构象。人们认为这种构象是上面第 (2) 点所提到的辅激活子和辅抑制子的选择性和特异性吸引的主要原因 (McKenna *et al.*, 1999a, b; Klinge, 2000)。异黄酮可能通过选择性的补充 ERβ 辅调节蛋白，从而引发一些与内源雌激素显著不同的临床作用，并且激发了特定的转录途径 (An *et al.*, 2001)。因此，该物理基础可以帮助解释雌激素和异黄酮在 ER 依赖的基因激活和抑制方面生物学作用不同，并能帮助解释看起来相矛盾的作用：①增加与雌激素相关的乳腺癌的危险，减少与大豆异黄酮高摄入量有关的危险；②在没有雌激素作用于子宫内膜的情况下，减轻绝经期妇女的更年期症状 (Albertazzi *et al.*, 1998; Duncan *et al.*, 1999; Upmalis *et al.*, 2000)，在短尾猿中也存在这种现象 (Foth and Cline, 1998)。

(6) 人们对内源的性激素和肾上腺激素的研究已进行了很多年，很好的阐明了"亲本"复合物及其代谢物的作用以及这些内源激素之间的互作。然而，对亲本大豆异黄酮及其代谢物对特定靶组织的生物学作用还没有进行过仔细地研究。这可能是应该考虑的很重要的营养因素，因为包含异黄酮的食物的组成成分是复杂的。例如，在大豆中有三种主要的异黄酮，染料木黄酮、黄豆苷和黄豆黄素以及几种代谢物。这些异黄酮并不是以相同的与 ER 结合的方式发挥作用，也没有其他的生物学作用。异黄酮混合物的联合作用可能与一个单独的异黄酮或性激素完全不同，而且，在大豆和复杂饮食的其他食物中还含有其他的植物化学物质，可能会改变对异黄酮的反应。因此，将来主要的挑战是测定这些异黄酮及其代谢物（包括结合物）和其他植物化学物质的组织特异性生物学作用，以及它们是如何相互作用而影响健康的。

(7) 最近的研究表明胞质 ERα 配体和 ERβ 配体复合物能够激活激酶级联反应 (Manolagas and Kousteni，2001)，这些是与胞质 ER 相连的不同配体的非基因组的、雌激素样的信号作用。异黄酮很明显是这些途径的潜在重要配体。尽管还没有进行很好的研究，但可以预见异黄酮在相似于其他配体（如男性激素）的一些方面，这与雌激素不同，于是提出了另外一种作用机制，通过此机制，雌激素和异黄酮能产生不同有时甚至相反的作用。

## 繁　殖

由于异黄酮-性腺雌激素链的存在，人们在异黄酮对繁殖系统的影响方面产生了很大的兴趣。关于异黄酮对繁殖影响的唯一的最有影响力的报道的是澳大利亚季节性繁殖绵羊采食富含有效异黄酮的苜蓿后，繁殖性能受到损坏 (Bennetts *et al.*，1946)。这是一个很重要的发现，因为它表明了在特定的环境下异黄酮对繁殖性能有着深刻的影响。虽然，从过去到现在这一发现仍是绵羊业中一个重要的因素 (Adams，1990)，但正如以前讨论过的一样，异黄酮对消费大豆食品的人的生殖性能的影响很小 (Badger *et al.*，2002)。这些发生在消费水平上的毒性作用不可能通过食物在人体上发生，但随着市场异黄酮供应量的增加，提纯的异黄酮可能会达到毒性水平并最终对人类的生殖产生毒性作用。

已有报道证明大豆蛋白的消费可改变月经周期中几种激素相，最显著的是在没有延长黄体期的情况下延长了卵泡期 (Cassidy *et al.*，1994)。对绝经前和绝经后妇女进行的相似研究都表明大豆蛋白的摄入减少了尿雌激素的排泄并降低了推定基因毒性的雌激素代谢物 (Xu *et al.*，1998，2000)。因此，人类消费的大豆蛋白食品中所包含的异黄酮的作用与较小的内分泌作用一致，它不会降低繁殖能力，但能通过以下方式防止癌的发生：①使接触雌激素最少的月经周期变长；②减少雌激素的合成并改变雌激素的代谢使其远离潜在的基因毒性代谢物。这些数据与该协会许多有关数据是一致的。这些有关数据内容是月经周期延长、雌激素接触与西方妇女月经周期动力学的不同之间的关系，这些西方妇女未摄入可观水平的大豆，且与每天摄入显著数量大豆食品的亚洲妇女相比，有较高的乳癌发病率。(Treolar *et al.*，1970；Olsson *et al.*，1983；世界卫生组织，1983；Henderson *et al.*，1985；Key *et al.*，1990；Lee *et al.*，1991；Munster *et al.*，1992)。

目前尚没有关于美国市场上出售的大豆食品对生殖的毒性特别是对繁殖能力的毒性的报道。与之相似，同样没有关于大豆食品对日本新生儿健康负面影响的报道，这表明围产期通过大豆婴儿代乳品（Setchell *et al.*，1997）或在子宫接触高浓度异黄酮（Adlercreutz *et al.*，1999）不可能会产生有害健康的作用，特别是不会引起归因于二乙醛己烯雌酚或胎儿酒精综合征的畸形。摄入大豆食品的妇女可产生高循环浓度的异黄酮，可以受孕，怀孕到期，产出正常的婴儿，正常哺乳并照顾孩子。这些妇女在怀孕前摄入大豆食品，并在怀孕和泌乳期间继续吃大豆食品，而大豆食品中的异黄酮并没有表现出对早期人的发育和后来的繁殖能力的不利作用。

### 婴儿生长和发育

在过去的 30 年中，数百万的美国婴儿食用大豆婴儿食品。几项研究表明大豆婴儿食品可支持婴儿正常的生长和发育（Graham *et al.*，1970；Kohler *et al.*，1984；Businco *et al.*，1992；Churella *et al.*，1994）。婴儿不论是食用大豆食品或牛奶为基础的食品或母乳，其体增重和体长都是一样的（Lasekan *et al.*，1999）。据报道，有关婴儿时期食用过大豆食品的青年人的研究也显现出相似的结果（Strom *et al.*，2001）。因此，可利用的数据表明大豆婴儿食品可以安全而有效地促进婴儿正常生长和发育。

### 动物的生长和发育

与大豆婴儿食品相关的一个问题是早期食用这些婴儿食品后所带来的长期的健康后果。亚洲人的一生中都摄入高水平大豆，为了建立一个与亚洲人相似的状况，将在婴儿食品中应用相同的 $SPI^{+}$ 饲喂给几代大鼠并对其作用进行研究（Badger *et al.*，2001）。在这些研究中，整个生命过程中都饲喂 AIN-93G 饲粮（由 $SPI^{+}$ 制成）的大鼠与饲喂商品饲粮或由酪蛋白制成的 AIN-93G 饲粮的大鼠相比，雄性和雌性大鼠有相同的生殖效率，饲喂大豆的大鼠后代的数目、性别比例、出生体重、出生体长、健康和一般外观都与饲喂酪蛋白的大鼠后代相同，并且雌性指标如第二性器官的重量、血浆雌激素浓度和乳腺发育都正常，唯一的影响是饲喂大豆的大鼠阴道开放时间早 1 天，目前对这种发现的实践意义还不清楚，因为在亚洲较早的青春期还未被认为是一个健康和生殖的问题。然而，对后面这一观点还没有进行很好的研究。

## 非雌激素介导的异黄酮作用

所谓的第一阶段酶（细胞色素 P450 酶，CYP）是异生素类（外源复合物，如食物因素、药物、杀虫剂和致癌物质）生物转化的主要催化剂，它们在内源复合物（内生的）如性腺和肾上腺类固醇、脂类和脂溶性维生素的代谢中也起到重要的作用。所提到的这些酶底物中的每一种物质对生理、内分泌、新陈代谢或药理都是很重要的，因此这些酶的调节对健康可能很重要。这些酶通常由它们的底物进行调节，从而将底物的作用引申到了细胞功能。食物因素、药物和其他外源及潜在毒性化学物质对第一阶段酶的诱导和抑制方面的例子很多（Guengerich，1995）。第一阶段代谢的调节是食物对健康产生有利或不利影响的潜在的主要作用机制。

第二阶段酶有两个主要的作用，都能够促进其分泌：①促进外生物质和内生物质的结合；②破坏复合物的反应性（Talalay *et al.*，1988；Prestera *et al.*，1993）。这些酶包括谷胱甘肽转移酶（glutathione S-transferase，GST）、UDP-葡糖醛酸基转移酶（uridine diphosphate glucuronosyltransferase，UDP-GT）、磺基转移酶（sulphotransferase，ST）、环氧化物酶和醌还原酶（quinone reductase，QR）。许多用啮齿动物进行的体内体外研究表明，大豆中提纯的异黄酮和香豆素可能对第二阶段酶 QR、GST 和 UDP-GT 有诱导作用（Appelt and Reicks，1997；Wang *et al.*，1998）。

已有报道称异黄酮和（或）异黄酮含量高的大豆食品通过改变 CYP 基因家族 1、2 和 3 中酶的表达对第一阶段代谢有显著的调节作用（Backlund *et al.*，1997；Heisby *et al.*，1998；Ronis *et al.*，1999，2001）。Rowlands 等（2001）对包含在前致癌剂 7，12 二甲苯-[*α*] 蒽（7，12-dimethylbenz-[α]anthracene，DMBA）激活过程中的第一阶段酶进行了研究，并报道了异黄酮含量丰富的食物是如何起保护作用从而降低乳腺癌发生率的。大豆食物起保护作用抵抗有毒复合物（如前致癌剂）的一个可能机制是通过抑制前致癌剂转化为致癌剂过程中所必需的酶，从而阻止致癌剂被激活，这依次被预测减少了加合物和后来导致乳腺癌基因突变的数目。对饲喂 $SPI^+$ 饲粮的大鼠用 DMBA 进行处理后，降低了肝 CYP1A1 和乳腺 CYP1B1 和 CYP1A1 的水平（Ronis *et al.*，2001）。CYP1A1 和 CYP1B1 的表达是由 DMBA 激活的芳基烃受体（AhR）调节的。AhR（aryl hydrocarbon receptor）与配体结合并与 AhR 核转运蛋白（AhR-nuclear translocator，ARNT）相互作用形成一核 AhR-ANRT 异二聚体复合物，此复合物作为配体激活转录因子结合在 CYP1 基因调节区域的异生素反应元件上，从而调节了 CYP1 基因的表达（Evans，1988）。用 DMBA 处理期间，胞液和核中 AhR 及 ARNT 蛋白水平较低，DMBA 激活作用降低后的下游结果是靶组织中致癌剂的浓度降低及 DNA 加合物的减少。

例如，饲喂 $SPI^+$ 的大鼠中，原本被报道的与其他器官相比，CYPIBI 水平非常高的卵巢和肾上腺中的 DMBA-DNA 加合物数量显著减少（$P<0.05$）（Bhattacharyya *et al.*，1995；Brake *et al.*，1999；McFadyen *et al.*，2001；Muskhelishvili *et al.*，2001）。这些数据补充了 Upadhyaya 和 el-Bayoumy（1998）的报道。Upadhyaya 和 el-Bayoumy（1998）的报道中指出，饲喂包含 $SPI^+$ 饲粮的大鼠乳腺 DMBA-DNA 加合物减少了。综上所述，现有数据表明，$SPI^+$ 通过负调节 CYP1A1 和 CYP1B1 基因的表达，减少 DMBA-DNA 加合物的数量和癌症发生率，进而减少了由 DMBA 诱导的乳腺癌的发生。

因此，摄食含高水平异黄酮的食物如含大豆丰富的食物，将会降低第一阶段的代谢，增加第二阶段代谢，从而增加清除作用和降低被吸收的致癌剂的致癌力。通过抑制第一阶段酶而抑制毒性异生素的激活及诱导第二阶段酶而增加脱毒作用，预计能将源于食物的有益于健康的作用，如对接触致癌剂前体或致癌剂所致的癌症的预防作用等最大化。

## 致　谢

本章中的许多工作都由 USDA ARS 阿肯色州儿童营养中心提供支持。

# 参 考 文 献


Adams, N.R. (1990) Permanent infertility in ewes exposed to plant oestrogens. *Australian Veterinary Journal* 67, 197–201.

Adlercreutz, H. and Baillieres, J.H. (1998) Epidemiology of phytoestrogens. *Clinical Endocrinology and Metabolism* 12, 605–623.

Adlercreutz, H. and Mazur, W. (1997) Phyto-oestrogens and Western diseases. *Annals of Medicine* 29, 95–120.

Adlercreutz, H., van der Widt, J. and Lampe, J. (1993) Quantitative determination of ligands and isoflavones in plasma of omivorous and vegetarian women by isotope dilution gas-chromatography–mass spectrometry. *Scandinavian Journal of Clinical Laboratory Investigation* 52, 97–102.

Adlercreutz, H., Yamad, T., Wahala, K. and Watanabe, S. (1999) Maternal and neonatal phytoestrogens in Japanese women during birth. *American Journal of Obstetrics Gynecology* 180, 737–743.

Albertazzi, P., Pansini, F., Bonaccorsi, G., Zanotti, L., Forini, E. and De Aloysio, D. (1998) The effect of dietary soy supplementation on hot flushes. *Obstetrics and Gynecology* 91, 6–11.

Allred, C.D., Allred, K.F., Young, H.J., Suzanne, M.V. and Helferich, W.G. (2001) Soy diets containing varying amounts of genistein stimulate growth of estrogen-dependent (MCF-7) tumors in a dose-dependent manner. *Cancer Research* 61, 5045–5050.

American Academy of Pediatrics Committee on Nutrition (1998) Soy protein-based formulas: recommendations for use in infant feeding. *Pediatrics* 101, 148–153.

An, J., Ribeiro, R.C., Webb, P., Gustafsson, J.A., Kushner, P.J., Baxter, J.D. and Leitman, D.C. (1999) Estradiol repression of tumor necrosis factor-alpha transcription requires estrogen receptor activation function-2 and is enhanced by coactivators. *Proceedings of the National Academy of Sciences USA* 96, 15161–15166.

An, J., Tzagarakis-Foster, C., Scharschmidt, T.C., Lomri, N. and Leitman, D.C. (2001) Estrogen receptor β-selective transcriptional activity and recruitment of coregulators by phytoestrogens. *Journal of Biological Chemistry* 276, 17808–17814.

Appelt, L.C. and Reicks, M.M. (1997) Soy feeding induces phase II enzymes in rat tissues. *Nutrition and Cancer* 28, 270–275.

Backlund, M., Johansson, I., Mkrtchian, S. and Ingelman-Sundberg, M. (1997) Signal transduction-mediated activation of the aryl hydrocarbon receptor in rat hepatoma H4IIE cells. *Journal of Biological Chemistry* 272, 31755–31763.

Badger, T.M., Ronis, M.J.J. and Hakkak, R. (2001) Developmental effects and health aspects of soy protein isolate, casein, and whey in male and female rats. *International Journal of Toxicology* 20, 165–174.

Badger, T.M., Ronis, M.J.J., Hakkak, R. and Korourian, S. (2002a) The health consequences of early soy consumption. *Journal of Nutrition* 132, 5595–5655.

Badger, T.M., Ronis, M.J.J., Hakkak, R. and Korourian, S. (2002b) The health consequences of soy infant formula, soy protein and isoflavones. In: Gilani, C.S. and Anderson, J.J.B. (eds) *Phytoestrogens and Health.* AOCS Press, Champaign, Illinois, pp. 586–605.

Barkhem, T., Carlson, B., Nilsson, Y., Enmark, E., Gustafsson, J. and Nilsson, S. (1998) Differential response of estrogen receptor α and estrogen receptor β to partial estrogen agonist/antagonists. *Molecular Pharmacology* 54, 105–112.

Barnes, S. (1997) The chemopreventive properties of soy isoflavonoids in animal models of breast cancer. *Breast Cancer Research and Treatment* 46, 169–179.

Barnes, S., Kirk, M. and Coward, L. (1996a) Isoflavones and their conjugates in soy foods – extraction conditions and analysis by HPLC mass spectrometry. *Journal of Agriculture and Food Chemistry* 442, 2466–2474.

Barnes, S., Sfakianos, J., Coward, L. and Kirk, M. (1996b) Soy isoflavonoids and cancer prevention. Underlying biochemical and pharmacological issues. *Advances in Experimental Medicine and Biology* 401, 87–100.

Bennetts, H.W., Underwood, E.J. and Shier, F.L. (1946) A specific breeding problem of sheep on subterranean clover pastures in Western Australia. *Australian Veterinary Journal* 22, 2–12.

Bhattacharyya, K.K., Brake, P.B., Eltom, S.E., Otto, S.A. and Jefcoate, C.R. (1995) Identification of a rat adrenal cytochrome P450 active in polycyclic hydrocarbon metabolism as rat CYP1B1. *Journal of Biological Chemistry* 270, 11595–11602.

Brake, P.B., Arai, M., As-Sanie, S., Jecoate, C.R. and Widmaier, E.P. (1999) Developmental expression and regulation of adrenocortical cytochrome P4501B1 in the rat. *Endocrinology* 140, 1672–1680.

Businco, I., Bruno, G. and Giampieto, P.G. (1992) Allergenicity and nutritional adequacy of soy protein formulas. *Journal of Pediatrics* 121, 821–828.

Cassidy, A., Bingham, S. and Setchell, K.D.R. (1994) Biological effects of a diet of soy protein rich in isoflavones on the menstrual cycle of premenopausal women. *American Journal of Clinical Nutrition* 60, 333–340.

Chang, H.C., Churchwell, M.I., Delclos, K.B., Newbold, R.R. and Doerge, D.R. (2000) Mass spectrometric determination of genistein tissue distribution in diet-exposed Sprague Dawley rats.

*Journal of Nutrition* 130, 1963–1970.

Chang, H.C., Fletcher, T., Ferguson, M., Hale, K., Fang, N., Ronis, M., Prior, R. and Badger, T.M. (2002) Serum and tissue profiles of isoflavone aglycones and conjugates in rats fed diets containing soy protein isolate (SPI). *FASEB Journal* 16, A1008.

Churella, H.R., Borschel, M.W., Thomas, M.R., Breem, M. and Jacobs, J. (1994) Growth and protein status of term infants fed soy protein formulas differing in protein content. *Journal of the American College of Nutrition* 13, 262–267.

Cimino, C.O., Shelnutt, S.R., Ronis, M.J. and Badger, T.M. (1999) An LC-MS method to determine concentrations of isoflavones and their sulfate and glucuronide conjugates in urine. *Clinica Chimica Acta* 287, 69–82.

Collaborative Group on Hormonal Factors in Breast Cancer (1997) Breast cancer and hormone replacement therapy: collaborative reanalysis of data from 51 epidemiological studies of 52,705 women with breast cancer and 108,411 women without breast cancer. *Lancet* 350, 1047–1059.

Coward, L., Barnes, N.C., Setchell, K.D.R. and Barnes, S. (1993) Genistein and daidzein and their β-glycosides conjugates: anti-tumor isoflavones in soybean foods from American and Asian diets. *Journal of Agriculture and Food Chemistry* 41, 1961–1967.

Coward, L., Kirk, M., Albin, N. and Barnes, S. (1996) Analysis of plasma isoflavones by reversed-phase HPLC-multiple reaction ion monitoring-mass spectrometry. *Clinica Chimica Acta* 247, 121–142.

Davis, S.R., Dalais, F.S., Simpson, E.R. and Murkies, A.L. (1999) Phytoestrogens in health and disease. *Recent Progress in Hormone Research* 54, 185–210.

Duncan, A.M., Underhill, K.E., Xu, X., Lavalleur, J., Phipps, W.R. and Kurzer, M.S. (1999) Modest hormonal effects of soy isoflavones in post menopausal women. *Journal of Clinical Endocrinology and Metabolism* 84, 3479–3484.

Evans, R.M. (1988) The steroid and thyroid hormone super-family. *Science* 240, 889–895.

Fang, N., Yu, S. and Badger, T.M. (2002) Characterization of isoflavones and their conjugates in female rat urine using LC/MS/MS. *Journal of Agriculture and Food Chemistry* 50, 2700–2707.

Farmakalidis, E. and Murphy, P.A. (1985) Isolation of 6″-*O*-acetylgenistein from toasted defatted soyflakes. *Journal of Agriculture and Food Chemistry* 33, 385–389.

Foth, D. and Cline, J.M. (1998) Effects of mammalian and plant estrogens on mammary glands and uteri of macques. *American Journal of Clinical Nutrition* 6S, 1413–1417.

Franke, A.A. and Custer, L.J. (1996) Daidzein and genistein concentrations in human milk after soy consumption. *Clinical Chemistry* 42, 955–964.

Franke, A.A., Custer, L.J., Cerna, C.M. and Narala, K. (1995) Rapid HPLC analysis of dietary phytoestrogens from legumes and from human urine. *Proceedings of the Society for Experimental Biology and Medicine* 208, 18–26.

Fritz, W., Coward, L., Wang, J. and Lamartiniere, C.A. (1998) Genistein: perinatal mammary cancer prevention, bioavailability and toxicity testing in the rat. *Carcinogenesis* 19, 2151–2158.

Gamache, P.H. and Acworth, I.N. (1998) Analysis of phytoestrogens and polyphenols in plasma, tissue, and urine using HPLC with coulometric array detection. *Proceedings of the Society for Experimental Biology and Medicine* 217, 274–280.

Gilani, G. and Anderson, J.J.B. (eds) (2002) *Phytoestrogens and Health.* AOCS Press, Champaign, Illinois, 660 pp.

Glass, C.K. and Rosenfeld, M.G. (2000) The coregulator exchange in transcriptional functions of nuclear receptors. *Genes and Development* 14, 121–141.

Graham, G.G., Placko, R.P. and Morals, E. (1970) Dietary protein quality in infants and children. VI. Isolated soy protein milk. *American Journal of Disabled Children* 120, 419–423.

Guengerich, F.G. (1995) Influence of nutrients and other dietary materials on cytochrome P-450 enzymes. *Journal of Nutrition* 61S, 651S–658S.

Hakkak, R., Korourian, S., Shelnutt, S.R., Lensing, S., Ronis, M.J.J. and Badger, T.M. (2000) Diets containing whey proteins or soy protein isolate protect against 7,12-dimethylbenz(a)anthracene-induced mammary tumors in female rats. *Cancer Epidemiology Biomarkers and Prevention* 9, 113–117.

Hakkak, R., Korourian, S., Ronis, M.J.J., Johnson, J. and Badger, T.M. (2001) Soy protein isolate consumption protects against azoxymethane-induced colon tumors in male rats. *Cancer Letters* 166, 27–32.

Hakkak, R., Korourian, S., Fletcher, T., Ferguson, M., Hale, K., Holder, D., Parker, J., Ronis, M., Treadaway, P. and Badger, T. (2002) The effects of soy protein containing negligible isoflavones and casein on DMBA-induced mammary tumors in rats. *American Association of Cancer Research* 43, 823.

Harborne, J.B., Mabry, T.J. and Mabry, H. (1975) *The Flavonoids.* Academic Press, New York.

Heisby, N.A., Chipman, J.K., Gescher, A. and Kerr, D. (1998) Inhibition of mouse and human CYP1A- and CYP2E1-dependent substrate metabolism by the isoflavonoids genistein and equol. *Food Chemistry and Toxicology* 36, 375–382.

Henderson, B.E., Ross, R.K., Judd, H.L., Frailo, M.D. and Pike, M.C. (1985) Do regulatory ovulatory cycles increase breast cancer risk? *Cancer* 56, 1206–1208.

Hill, L.W. and Stuart, H.C. (1929) A soy bean food preparation for feeding infants with milk idiosyncrasy. *Journal of the American Medical Association* 93, 985–987.

Hirayama, T. (1985) A large scale cohort study on cancer

risks by diet, with special reference to the risk reducing effects of green–yellow vegetable consumption. *Princess Takamatsu Symposium* 160, 41–53.

Holder, C.L., Churchwell, M.I. and Doerge, D.R. (1999) Quantitation of soy isoflavones, genestein and daidzein, and conjugates in rat blood using LC/ES-MS. *Journal of Agriculture and Food Chemistry* 47, 3764–3770.

Hsieh, C.Y., Santell, R.C., Haslam, S.Z. and Helferich, W.G. (1998) Estrogenic effects of genistein on the growth of estrogen receptor-positive human breast cancer (MCF-7) cells *in vitro* and *in vivo*. *Cancer Research* 58, 3833–3838.

Ingram, D., Sanders, K., Kolybaba, M. and Lopez, D. (1997) Case–control study of phyto-oestrogens and breast cancer. *Lancet* 350, 990–994.

Irvine, C., Fitzpatrick, M., Robertson, I. and Woodhams, D. (1995) The potential adverse effects of soybean phytoestrogens in infant feeding. *New Zealand Medical Journal* 108, 208–209.

Katzenellebogen, B.S., Montano, M.M., Ediger, T.R., Sun, J., Wlkena, K., Lazzennec, G., Martini, P.G., Mclnerney, E.M., Delage-Mourroux, R., Weis, K. and Katzenellebogen, J.A. (2000) Estrogen receptors: selective ligands, partners, and distinctive pharmacology. *Recent Progress in Hormone Research* 55, 163–193.

Key, T.J.A., Chen, D.Y., Wang, D.Y., Pike, M.C. and Boreham, J. (1990) Sex hormones in rural China and Britain. *British Journal of Cancer* 37, 467–480.

Klinge, C.M. (2000) Estrogen receptor interaction with co-activators and co-repressors. *Steroids* 65, 227–251.

Kohler, L., Meeuwisse, G. and Mortensson, W. (1984) Food intake and growth of infants between six and twenty-six weeks of age on breast milk, cow's milk formula, or soy formula. *Acta Paediatrica Scandinavica* 73, 40–48.

Koike, S., Sakai, M. and Muramatsu, M. (1987) Molecular cloning and characterization of rat estrogen receptor cDNA. *Nucleic Acids Research* 15, 2499–2513.

Kulling, S.E., Honig, D.M. and Metzler, M. (2001) Oxidative metabolism of the soy isoflavone daidzein and genistein in humans *in vitro* and *in vivo*. *Journal of Agriculture and Food Chemistry* 49, 3024–3033.

Kupier, G.G., Enmark, E., Pelton-Huikko, M.-H., Nilsson, S. and Gustafsson, J.-A. (1996) Cloning of a novel estrogen receptor expressed in rat prostate and ovary. *Proceedings of the National Academy of Sciences USA* 93, 5925–5930.

Kupier, G.G., Carlsson, B., Grandien, K., Enmark, E., Haggblad, J., Nilsson, S. and Gustafsson, J.-A. (1997) Comparison of the ligand binding specificity and transcript tissue distribution of estrogen receptor α and β. *Endocrinology* 138, 863–870.

Kupier, G.G., Lemmen, J.G., Carlsson, B., Corton, J.C., Safe, S.H., Van Der Saag, P.T., Van Der Burg, B. and Gustafsson, J.-A. (1998) Interaction of estrogenic chemicals and phytoestrogens with estrogen receptor β. *Endocrinology* 139, 4252–4263.

Lamartiniere, C.A., Moore, J., Holland, M. and Barnes, S. (1995) Genistein and chemoprotection of breast cancer. *Proceedings of the Society for Experimental Biology and Medicine* 208, 120–123.

Lamartiniere, C.A., Cotroneo, M.S., Frizt, W.A., Wang, J., Mentor-Marcel, R. and Elgavish, A. (2001) Genistein chemoprotection: timing and mechanisms of action in murine mammary and prostate. *Journal of Nutrition* 132, 552S–585S.

Lapcik, O., Hampl, R., Hill, M., Wahala, K., Maharik, N.A. and Adlercreutz, H. (1998) Radioimmunoassay of free genistein in human serum. *Journal of Steroid Biochemistry and Molecular Biology* 64, 261–268.

Lasekan, J.B., Ostrom, K.M., Jacobs, J.R., Blatter, M.M., Ndife, L.I., Gooch, W.M. and Cho, S. (1999) Growth of newborn, term infants fed soy formulas for 1 year. *Clinical Pediatrics* 38, 563–571.

Lee, H.P., Gourley, L., Diffy, S.W., Esteve, J., Lee, J. and Day, N.E. (1991) Dietary effects on breast cancer risk in Singapore. *Lancet* 337, 1197–1200.

Mabry, T.J., Markham, K.R. and Thomas, M.B. (1970) *The Systematic Identification of Flavonoids*. Springer, New York.

Manolagas, S.C. and Kousteni, S. (2001) Perspective: nonreproductive sites of action of reproductive hormones. *Endocrinology* 142, 2200–2204.

Markham, K.R. and Geiger, H. (1986) $^1$H nuclear magnetic resonance spectroscopy of flavonoids and their glycosides in hexadeuterodimethylsulfoxide. In: Harborne, J.B. (ed.) *Flavonoids*. Chapman & Hall, London, pp. 321–335.

Markiewicz, L., Garvey, J., Adlercreutz, H. and Gurpide, E. (1993) *In vitro* bioassays of non steroidal phytoestrogens. *Journal of Steroid Biochemistry and Molecular Biology* 45, 399–405.

McFadyen, M.C., Crickshank, M.E., Miller, I.D., McLeod, H.L., Melvin, W.T., Haites, N., Parin, D. and Murray, G.I. (2001) Cytochrome P450 CYP1B1 over-expression in primary and metastatic ovarian cancer. *British Journal of Cancer* 85, 242–246.

McKenna, N., Xu, J., Nawaz, Z., Tsai, S., Tsai, M.-J. and O'Malley, B. (1999a) Nuclear receptor coactivators: multiple functions. *Journal of Steroid Biochemistry and Molecular Biology* 69, 3–12.

McKenna, N.J., Lanz, R.B. and O'Malley, B.W. (1999b) Nuclear receptor coregulators: cellular and molecular biology. *Endocrine Review* 20, 321–344.

Messina, M. (1995) Isoflavone intakes by Japanese were overestimated. *American Journal of Clinical Nutrition* 62, 645.

Messina, M. (1998) The role of soy in preventing and treating chronic disease. *American Journal of Clinical Nutrition* 68, 1329S–1544S.

Messina, M. (2000) The role of soy in preventing and treating chronic disease. *American Journal of Clinical Nutrition* 70, 1329S–1541S.

Messina, M., Gardner, C. and Barnes, S. (2002) Gaining insight into the health effects of soy but a long way still to go: commentary on the Fourth International Symposium on the Role of Soy in Preventing and Treating Chronic Disease. *Journal of Nutrition* 132, 547S–551S.

Miksicek, R.J. (1993) Commonly occurring plant flavonoids have estrogenic activity. *Molecular Pharmacology* 44, 37–43.

Munster, K., Schmidt, L. and Helm, P. (1992) Length and variation in the menstrual cycle – a cross sectional study from Danish county. *British Journal of Obstetrics and Gynaecology* 99, 422–429.

Muskhelishvili, L., Thompson, P.A., Kusewitt, D.F., Wang, C. and Kadlubar, F.F. (2001) *In situ* hybridization and immunohistochemical analysis of cytochrome analysis of cytochrome P4501B1 expression in human normal tissues. *Journal of Histochemistry and Cytochemistry* 49, 229–236.

Nomura, A,. Henderson, B.E. and Lee, J. (1978) Breast cancer and diet among the Japanese in Hawaii. *American Journal of Clinical Nutrition* 31, 2020–2025.

Olsson, H., Landin-Olsson, M. and Gullberg, B. (1983) Retrospective assessment of menstrual cycle length in patients with breast cancer, in patients with benign breast disease and in women without breast disease. *Journal of the National Cancer Institute* 70, 17–20.

Patisaul, H.B., Dindo, M., Whitten, P.L. and Young, L.J. (2001) Soy isoflavone supplements antagonize reproductive behavior and estrogen receptor alpha and beta-dependent gene expression in the brain. *Endocrinology* 142, 2946–2952.

Pike, A.C., Brzozowski, A.M., Hubbard, R.E., Bonn, T., Thorsell, A.G., Engstrom, O., Ljunggren, J., Gustafsson, J.A. and Carlqyuist, M. (1999) Structure of the ligand-binding domain of oestrogen receptor β in the presence of a partial agonist and a full agonist. *EMBO Journal* 18, 4608–4618.

Polkowski, K. and Mazyrek, A.P. (2000) Biological properties of genistein. A review of *in vitro* and *in vivo* data. *Acta Polica Pharmica* 57, 135–155.

Prestera, Y., Holtzclaw, Y., Zhang, Y. and Talalay, P. (1993) Chemical and molecular regulation of enzymes that detoxify carcinogens. *Proceedings of the National Academy of Sciences USA* 92, 8965–8969.

Robertson, I.G.C. (1995) Phytoestrogens: toxicity and regulatory recommendations. *Proceedings of the Nutrition Society of New Zealand* 20, 35–42.

Ronis, M.J., Rowlands, J.C., Hakkak, R. and Badger, T.M. (1999) Altered expression and glucocorticoid-inducibility of hepatic CYP3A and CYP2B enzymes in male rats fed diets containing soy protein isolate. *Journal of Nutrition* 129, 1958–1965.

Ronis, M.J., Rowlands, J.C., Hakkak, R. and Badger, T.M. (2001) Inducibility of hepatic CYP1A enzymes by 3-methylcholanthrene and isosafrole differs in male rats fed diets containing casein, soy protein isolate or whey from conception to adulthood. *Journal of Nutrition* 131, 1180–1188.

Rowlands, J.C., He, J., Hakkak, R., Ronis, M.J.J. and Badger, T.M. (2001) Soy and whey proteins downregulate DMBA-induced liver and mammary gland CYP1 expression in female rats. *Journal of Nutrition* 131, 3281–3287.

Santell, R.C., Kieu, N. and Helferich, W.G. (2000) Genistein inhibits growth of estrogen-independent human breast cancer cells in culture but not in athymic mice. *Journal of Nutrition* 30, 1665–1669.

Setchell, K.D.R. (1995) Non-steroidal estrogens of dietary origin: possible roles in health and disease, metabolism and physiological effects. *Proceedings of the Nutrition Society of New Zealand* 20, 1–21.

Setchell, K.D.R. (2001) Soy isoflavones-benefits and risks from nature's selective estrogen receptor modulators (SERMs). *Journal of the American College of Nutrition* 20, 354S–362S.

Setchell, K.D.R. and Aldercreutz, H. (1988) Mammalian ligands and phytochemicals: recent studies on their formation, metabolism and biological role in health and disease. In: Rowland I.A. (ed.) *The Role of Gut Microflora in Toxicity and Cancer*. Academic Press, New York, pp. 315–345.

Setchell, K.D.R., Zimmer-Nechemias, L., Cai, J. and Heubi, J.E. (1997) Exposure of infants to phyto-oestrogens from soy-based infant formula. *Lancet* 350, 23–27.

Setchell, K.D.R., Zimmer-Nechemias, L., Cai, J. and Heubi, J.E. (1998) Isoflavone content of infant formulas and the metabolic fate of these phytoestrogens in early life. *American Journal of Clinical Nutrition* 68, 1453S–1461S.

Sheehan, D.M. (1997) Isoflavone content of breast milk and soy formula: benefits and risks. *Clinical Chemistry* 43, 850–852.

Shelnutt, S.R., Cimino, C.O., Wiggins, P.A. and Badger, T.M. (2000) Urinary pharmacokinetics of the glucuronide and sulfate conjugates of genistein and daidzein. *Cancer Epidemology Biomarkers and Prevention* 9, 413–419.

Shelnutt, S.R., Cimino, C.O., Wiggins, P.A., Ronis, M.J.J. and Badger, T.M. (2002) Pharmacokinetics of the glucuronide and sulfate conjugates of genistein and daidzein following a soy beverage in men and women. *American Journal of Clinical Nutrition* 76, 588–594.

Strom, B.L., Schinnar, R., Ziegler, E.E., Barnhart, K., Sammel, M., Macones, G., Stallings, V., Hanson, S.A. and Nelson, S.E. (2001) Follow-up study of a cohort fed soybased formula during infancy. *Journal of the American Medical Association*

286, 807–814.
Supko, J.G. and Phillips, L.R. (1995) High performance liquid chromatography assay for genistein in biological fluids. *Journal of Chromatography, Biomedical Applications* 666, 157–167.
Talalay, P., DeLong, M.J. and Prochaska, H.J. (1988) Identification of a common chemical signal regulating the induction of enzymes that protect against chemical carcinogenesis. *Proceedings of the National Academy of Sciences USA* 85, 8261–8265.
Treolar, A.E., Boynton, R.E., Behn, B.G. and Brown, B.W. (1970) Variation of the human menstrual cycle through reproductive life. *International Fertility* 12, 77–126.
Upahyaya, P. and el-Bayoumy, K. (1998) Effect of dietary soy protein isolate, genistein and 1,4-phenylenebis (methylene)selenocyanate on DNA binding of 7,12-dimethylbenz[a]anthracene in mammary glands of CD rats. *Oncology Reports* 5, 1541–1545.
Upmalis, D.H., Lobo, R., Bradley, L., Warren, M., Conne, F.L. and Lamia, C.A. (2000) Vasomotor symptom relief by soy isoflavone extract tablets in postmenopausal women: a multicenter, double-blind, randomized, placebo-controlled study. *Menopause* 7, 236–242.
Wang, W., Liu, L.Q., Higuchi, C.M., and Chen, H.W. (1998) Induction of NADPH-quinone reductase by dietary phytoestrogens in colonic colo205 cells. *Biochemical Pharmacology* 56, 189–195.
World Health Organization (1983) A prospective multicenter trial of the ovulation method of nature family planning. III Characteristics of the menstrual cycle and of the fertile phase. *Fertility and Sterility* 40, 773–778.
Xu, X., Duncan, A.M., Merz, B.E. and Kurzer, M.S. (1998) Effects of soy isoflavones on estrogen and phytoestrogen metabolism in premenopausal women. *Cancer Epidemiology Biomarkers and Prevention* 7, 1101–1108.
Xu, X., Duncan, A.M., Wangen, K.E. and Kurzer, M.S. (2000) Soy consumption alters endogenous estrogen metabolism in postmenopausal women. *Cancer Epidemiology Biomarkers and Prevention* 9, 781–786.

# 15　骨骼肌中遍在蛋白化和依赖于蛋白酶体的蛋白酶解机制

Didier Attaix，Lydie Combaret，Anthony J. Kee[1]
和 Daniel Taillandier
（法国克莱蒙费朗和法国国家农业研究院人类营养研究中心
营养与蛋白质代谢室，Ceyrat，法国）

## 引　　言

骨骼肌中的蛋白质，如同所有哺乳动物组织中的蛋白质一样，要经历一个连续合成和降解的过程，这个过程能调节所有肌肉蛋白的总量和特异蛋白的水平（Waterlow *et al*.，1978）。蛋白水解的增加导致肌肉损耗常见于某些病理条件下（如癌症、脓血症、糖尿病、烧伤、外伤等）。骨骼肌是体内主要的蛋白质库，骨骼肌中蛋白酶解的增加是一种关键的代谢性适应，这种适应通过游离氨基酸经糖异生和直接氧化而为生物体提供能量。在病理条件下肌肉蛋白的分解也提供游离氨基酸，以供给肝脏的急性期蛋白质合成和生命器官（如心脏和大脑）的蛋白质合成。

溶酶体、$Ca^{2+}$激活和依赖于遍在蛋白-蛋白酶体的途径是负责大量骨骼肌蛋白分解最重要的三个过程（Attaix and Taillandier，1998）。然而，肌肉中也有许多其他的活性蛋白酶，如半胱氨酸蛋白酶（Belizario *et al*.，2001）或者基质金属蛋白酶（Balcerzak *et al*.，2001）。溶酶体和$Ca^{2+}$激活蛋白酶（各自的代表性蛋白酶分别为组织蛋白酶、钙蛋白酶）在骨骼肌蛋白的水解中不发挥主要作用。在对照和极瘦弱的动物中这些酶所能水解的蛋白质不足肌肉中总蛋白质的15%～20%，这两种酶不直接参与肌原纤维蛋白质的分解（Lowell *et al*.，1986；Tiao *et al*.，1994），而且无论是组织蛋白酶还是钙蛋白酶的系统活化在肌肉消耗时都不能观察到（Attaix and Tailandier，1998）。相比之下，遍在蛋白-蛋白酶体途径的激活是主要应对出现在多种体质较差的动物模型的肌肉消耗（Attaix and Tailandier，1998；Attaix *et al*.，1998；Jagoe and Goldberg，2001；Hasselgren *et al*.，2002）。这个途径也在肌肉消耗迅速增加的患者中被激活（Mansoor *et al*.，1996；Tiao *et al*.，1997），但并不出现在慢性病条件下如肌肉营养失调（Combaret *et al*.，1996）或库欣综合征（又称皮质醇增多症）（Rallière *et al*.，1997）。在本章，我们将首先审慎地综述肌肉中遍在蛋白在和依赖于蛋白酶体的蛋白酶解机制，接着讨论在肌肉中蛋白酶解体系的调节，尤其强调营养和激素调节。

# 依赖于遍在蛋白-蛋白酶体的蛋白酶解机制

遍在蛋白-蛋白酶体系统是主要的非溶酶体过程，控制着哺乳动物细胞中大多数短寿命和长寿命蛋白的分解（Rock *et al.*，1994）。这个复杂的蛋白水解体系在数百个基因的调节下，其中包括调控遍在蛋白化和去遍在蛋白化系统的基因（200～300）、编码不同蛋白酶体亚基的基因（约 50）和内源蛋白酶体激活物（如下）及抑制物。这个途径在基础蛋白周转和错误编码、错误折叠或错误定位的异常蛋白去除中发挥持家的作用。另外，该系统也调节许多主要的生物功能如一级抗原呈递、细胞周期和转录、信号转导、蛋白质分类等的调节（Glickman and Giechanover，2002）。

在这个途径中有两个主要步骤具有严格的计划性：①在遍在蛋白化和去遍在蛋白化酶的调节下多聚遍在蛋白链共价结合到蛋白质底物上；②多聚遍在蛋白链的特异识别和通过 26S 蛋白酶体降解靶蛋白（图 15.1）。

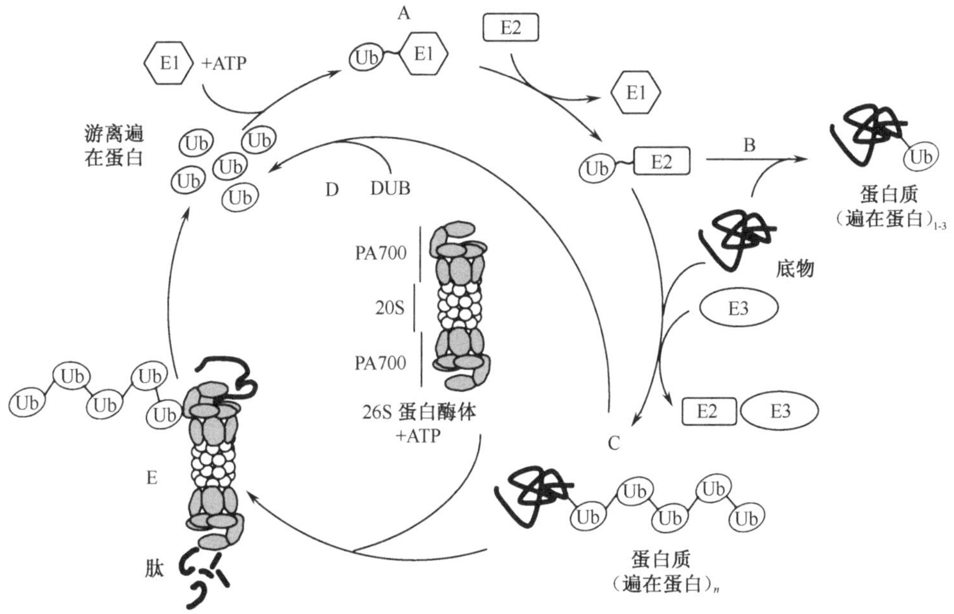


图 15.1 遍在蛋白-蛋白酶体途径的图示

A. 遍在蛋白（Ub）首先通过遍在蛋白激活酶（E1）激活，然后转移到遍在蛋白结合酶（E2）上。B. 不是为了分解的药物被 E2 单体、二聚体和三聚体遍在蛋白化［蛋白-（Ub）$_{1\sim3}$］。C. 相比之下，E2 通过或不通过 E3 的帮助形成一个多聚遍在蛋白化蛋白［蛋白-（Ub）$_n$］ D. 底物通过去遍在蛋白化酶（DUB）的作用脱遍在蛋白化，或者 E. 被识别、折叠，进入 26S 蛋白酶体，降解成肽。值得注意的是步骤 A 和 E 需要 STP 水解

## 遍在蛋白化

遍在蛋白是真核生物细胞中由 76 个氨基酸残基组成的一种高度保守的蛋白质。遍在蛋白化定义为遍在蛋白共价结合到蛋白质底物上，它是广泛存在的类似磷酸化具有翻译后修饰功能，同时又具有多种蛋白水解和非蛋白水解的重要功能。其作用底物为了通过蛋白酶体系统降解，必须伴随由至少 4 个遍在蛋白配基部分组成的多聚遍在蛋白降解

信号（Thrower *et al.*，2000）。单、二、三聚遍在蛋白化蛋白通常不被降解，而是具有其他作用。例如，单遍在蛋白化是大量受体细胞内吞的信号（Shih *et al.*，2000）。

多聚遍在蛋白化是一个复杂的多级过程（Pickart，2001）。简而言之，遍在蛋白首先被单遍在蛋白激活酶 E1 激活（图 15.1）。E1 然后把激活的遍在蛋白转运到遍在蛋白结合酶之一 E2，通过遍在蛋白的 C 端甘氨酸残基和蛋白底物中赖氨酸残基的 ε-氨基之间形成的异肽键，E2 把最初的遍在蛋白分子转移到蛋白底物上，形成单遍在蛋白化蛋白。一些 E2 能催化遍在蛋白分子的连续结合形成多聚遍在蛋白结合物，这一过程通常被待结合的活化遍在蛋白配基与前面结合了遍在蛋白分子的 Lys48 结合所激化。然而，大多数情况下多聚遍在蛋白链的形成需要遍在蛋白连接酶（E3）的参与，它在识别底物方面也具有重要的作用。

## 遍在蛋白激活酶 E1

E1 的序列在多种生物物种中是高度保守的。人的 E1 基因具有一个可选择的起始密码子，诱导一个 110kDa 核型 E1 和一个 117kDa 细胞质型 E1 表达（Haas and Siepmann，1997）。E1 催化的反应起始于 ATP-$Mg^{2+}$ 的结合，然后遍在蛋白结合到酶上，导致遍在蛋白腺苷酸中间产物的形成，该中间产物作为遍在蛋白的供体供与 E1 活性位点中关键的半胱氨酸残基。当满负荷时，E1 携带两个活化的遍在蛋白分子（分别作为硫羟酸酯和腺苷酸），以便连接硫醇的遍在蛋白转运到一个 E2 上。这个高效的反应为整个下游遍在蛋白结合途径提供了活化的遍在蛋白。

## 遍在蛋白结合酶 E2s

遍在蛋白结合酶 E2 是一个相关蛋白的超家族，分子质量为 14～35kDa。E2 分 4 类，均拥有一个约 150 个氨基酸残基的中心催化结构域，该结构中含有一个也能与遍在蛋白形成硫羟酸酯中间产物的关键的半胱氨酸残基和可变的 N 端和（或）C 端延伸（Scheffner *et al.*，1998）。在哺乳动物中，至少有 20～30 个 E2（Scheffner *et al.*，1998）。尽管它们结构相似，但每个 E2 具有独特的生物功能，使只有少数的 E2（如在酵母中 11 个 E2 中仅有 3 个）在多聚遍在蛋白降解信号形成中发挥作用。在 E1 和 ATP 存在时，一些 E2 具有不确定的底物识别位点并形成多聚遍在蛋白降解信号。然而，通常在 E3 存在时才形成多聚遍在蛋白链（Pickart，2001）。多数 E2 与许多 E3 相互作用（反之亦然），它能依次识别它们的特异蛋白质底物。而且，特定的蛋白质底物能通过 E2 和 E3 的不同组合被遍在蛋白化多可选择的遍在蛋白化途径（Glickman and Ciechanover，2002）。

## 遍在蛋白连接酶 E3

E3 在多聚遍在蛋白化中具有决定性作用，负责蛋白质底物的选择性识别。所有已知 E3 分三类：含有 HECT（homologous to E6-AP C-terminus）结构域的 E3、含有 RING（really interesting new gene）指的 E3（Pichart，2001）或含有 U 形框的 E3。

E3 的第一个主要类群是 HECT 结构域家族的单体酶。HECT 域（约 350 个氨基酸）位于酶的 C 端，调节 E2 的结合和靶蛋白经硫羟酸酯与遍在蛋白连接形成的遍在蛋白化。HECT 域 E3 的 N 端与特异底物相结合。人基因组编码至少 20 种 HECT 结构域蛋白。然而，哺乳动物基因组序列技术已经识别大量的具有潜在的非特征性 HECT E3（Pickart，2001）。

8 个半胱氨酸和组氨酸残基结合两个 $Zn^{2+}$ 离子被定义为 RING 指结构。在基因库中有数百个 cDNA 编码 RING 指蛋白，另外许多拥有未知功能的不相关 RING 指蛋白在体外起 E3 作用（Attaix *et al.*，2001）。

最简单的 RING 指 E3 是单体“N 端规则”酶 E3α 和 E3β。E3α 结合到蛋白质上产生碱性或大量疏水性 N 端氨基酸残基，而 E3β 结合到不带电荷的 N 端氨基酸上（Varshavsky，1996）。真核细胞识别各种特异底物的一个较复杂的途径是通过吸收大量特异的底物适配蛋白而实现的，此适配蛋白能将蛋白底物补充到中心遍在蛋白化复合物上（Attaix *et al.*，2001；Pichart，2001）。这些大分子质量多亚基复合物形成了最丰富的 E3 家族，有至少三类不同的 RING 指 E3 复合物：①循环体（cyclosome）或后期促进复合物（anaphase promoting complex，APC）；②SCF（Skpl-Cdc53-F-box 蛋白家族）E3 连接酶；③类VCB［遗传性斑痣性错构瘤抑制物——延伸素 C/B（von Hippel-Lindau tumour suppressor-eplongin C/B）］连接酶。这些复合物含有适配亚基，如 SCF 家族的 F 盒蛋白（与细胞周期蛋白 F 序列相类比）。F 盒蛋白通过特异的蛋白质与蛋白质相互作用结构域（如富含亮氨酸的重复序列或 WD40 结构域）识别不同底物。F 盒蛋白本身很快以依赖于遍在蛋白-蛋白酶体形式降解。这使 SCF 复合物转换而识别大量可选择性底物（图

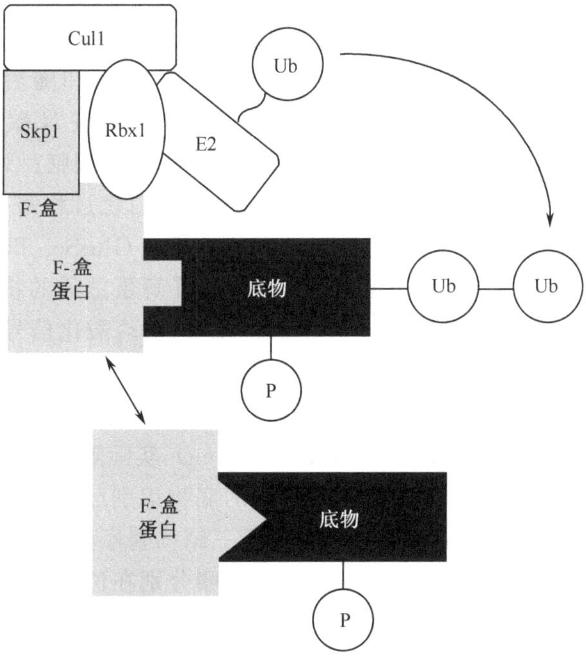


图 15.2 多亚基 RING 指 SCF E3 的图示（见正文）

催化中心的三个亚基涂成白色。Rbx1（Hrt1/Roc1）是一个 RING 指蛋白，Ubc3 家族的 E2 是催化中心的一部分。Cul1，cullin1。适配蛋白涂成灰色。Skp1 通过 F 盒基序结合到 F 盒蛋白上。F 盒蛋白的交换负责多种特异底物的识别。底物通过蛋白质-蛋白质的相互作用结合到各自的 F 盒蛋白上（见正文）。值得注意的是磷酸化底物通过 E3 家族遍在蛋白化

15.2)。最后，U 形框是一个由约 70 个氨基酸组成的结构域，具有类似于 RING 指的预测立体结构。只有很少几个 U 形框 E3 的性质近来被描述。

## 去遍在蛋白化

真核细胞也含有去遍在蛋白化酶（deubiquitinating enzyme，DUB），它由遍在蛋白羧基端水解酶（ubiquitin carboxyl terminal hydrolase，UCH）和遍在蛋白特异性加工蛋白酶（ubiquitin specific processing protease，UBP）基因家族编码。像 E3 连接酶一样，基于基因组序列工程资料；DUB 酶形成了一个由至少 90 个蛋白质成员组成的大家族（Chung and Baek，1999）。UCH 是相当小的蛋白质（小于 40kDa），在高等真核细胞中只有很少数的 UCH 异形体的性质被阐明。UCH 主要在遍在蛋白的 C 端水解小分子的氨基化合物和酯。比较之下，UBP 具有较大的结构（50～250kDa）和多种 DUB 的基团，去遍在蛋白化可以与去磷酸化过程相比较。因此，UBP 涉及许多生物过程，包括生长、分化和基因组完整的调节（Chung and Baek，1999），在依赖于蛋白酶体蛋白酶解中 DUB 的推测作用是：①通过加工多聚遍在蛋白降解信号和遍在蛋白前体进入自由单体来保持游离遍在蛋白水平；②使为降解而错误标记的底物去遍在蛋白化（校对）；③使 26S 蛋白酶体脱离多聚遍在蛋白化链。

## 遍在蛋白化和通过 26S 蛋白酶体底物识别的分子基础

为多聚遍在蛋白化和依赖于蛋白酶体的蛋白酶解而标记的蛋白质的一些信号已被识别。这些信号通常是通过在靶蛋白上主要序列的狭小区域确定的（Pichart，2001）。首先，蛋白质的 N 端氨基酸性质能确定它的遍在蛋白化速率和随后的降解速率（Varshavsky，1996）。在肌肉消耗增加可溶肌肉蛋白的分解时，该 N 端规则途径被正调节（Solomon *et al.*，1998），并且还控制许多生物功能（如肽运进细胞；Turner *et al.*，2000）。

其次，蛋白质通过 SCF E3 连接酶而有效的遍在蛋白化过程需要磷酸化（Pichart，2000）。$G_1$ 细胞周期蛋白的主要识别位点位于富含 Pro-Glu-Ser-Thr（PEST）基序的 100～200 氨基酸 C 端区域。PEST 基序作为能快速降解蛋白质的特征，几乎与多个蛋白激酶磷酸化位点不一致（Wilkinson，2000）。E3 识别磷酸化信号的能力可能是由于磷酸化氨基酸结合基序的存在，如 WW 或 WD40 结构域（Pickart，2001）。不幸的是，没有一个明确的磷酸化模式使靶底物遍在蛋白化。曾有报道磷酸化发生在单个或多个位点上。另外，一些蛋白质的磷酸化（如 c-Fos，c-Jun）实际阻止它们遍在蛋白化和降解（Attaix *et al.*，2001），而依赖于底物去磷酸化的底物识别已被报道（Pichart，2001）。一个在体内和体外似乎为靶蛋白遍在蛋白化的第三结构基序称为破坏盒（destruction）。这个退化的 9 个氨基酸基序（只有精氨酸和亮氨酸分别在位置 1 和位置 4 是不变的残基）是遍在蛋白化以及有丝分裂细胞周期蛋白和其他细胞周期调节物的降解的关键信号。破坏盒本身不是一个遍在蛋白化位点而是一个可转换的降解信号。据报道，含有这种基序的蛋白质以依赖细胞周期形式被迅速降解。破坏盒的特异的构造是为 E3 高效识别必需的（Pickart，2001）。

最后，26S 蛋白酶体也逐步降解部分非遍在蛋白化蛋白底物（Attaix *et al.*，2001；Benaroudj *et al.*，2001）。例如，一些骨骼肌特异蛋白（如肌钙蛋白 C）被 26S 蛋白酶体经非依赖于遍在蛋白的机制降解（Benaroudj *et al.*，2001）。另外，错误折叠、氧化、异常突变蛋白是 26S 蛋白酶体的很好的底物，它能识别非遍在蛋白化的错误折叠蛋白（Strickland *et al.*，2000）。疏水性通过 26S 蛋白酶体在多聚遍在蛋白链的识别上发挥主要作用（Thrower *et al.*，2000）。在一些 26S 蛋白酶体的非遍在蛋白化底物的主要序列中疏水氨基酸的延长［如钙调蛋白（calmodulin）和肌钙蛋白 C］可能取代遍在蛋白，而且对通过 26S 蛋白酶体的识别也是足够的（Benaroudj *et al.*，2001）。可选择的是，在底物识别时其他适配蛋白也会发挥作用。例如，鸟氨酸脱羧酶经抗酶（antizyme）与 26S 蛋白酶体结合，并以非依赖于遍在蛋白的形式被降解（Glickman and Ciechanover，2002）。

## 蛋白水解

在遍在蛋白–蛋白酶体途径中的次级主要步骤是多聚遍在蛋白化蛋白通过 26S 蛋白酶体复合物而降解（图 15.1）。26S 蛋白酶体是通过 20S 蛋白酶体与两个 19S 可调节复合物的连接而形成（图 15.3）（Voges *et al.*，1999；Attaix *et al.*，2001；Glickman and Ciechanover，2002）。

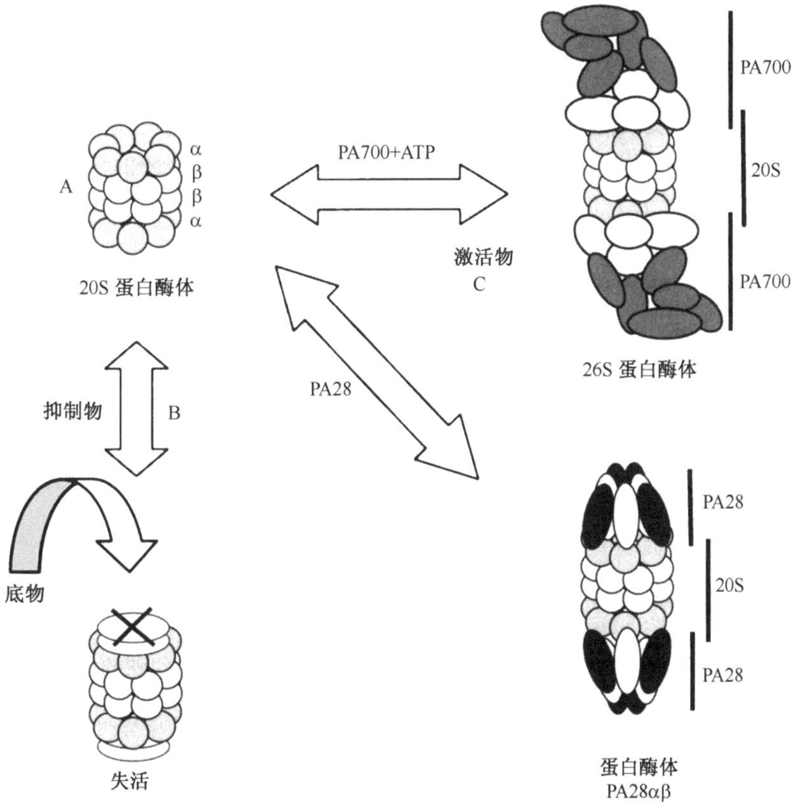


图 15.3　20S 蛋白酶体、26S 蛋白酶体和蛋白酶体-PA28αβ 的图示

A. α 和 β 分别代表 20S 蛋白酶体的 α 和 β 亚基。B. 蛋白酶体抑制物与蛋白酶体活化物（PA700 和 PA28）相竞争结合到 20S 蛋白酶体上；当蛋白酶体抑制物结合到 20S 蛋白酶体上，底物不能进入催化室。C. 相比之下，蛋白酶体活化物 PA700 或 PA28αβ 结合到 20S 蛋白酶体上，通过 α 环划界进入狭小的蛋白酶体通道。因而，26S 蛋白酶体表现为蛋白水解酶和肽酶活性增加，然而，蛋白酶体 PA28αβ 仅表现出活化肽酶活性。PA700 的底和盖分别用黑灰色和白色标记

## 20S 蛋白酶体

20S 蛋白酶体是蛋白水解体系的核心。这个桶形复合物形成相连的 4 个环，每个环有 7 个亚基（图 15.3A）。无催化活性的 α 亚基形成两个外部环，有催化活性的 β 亚基形成两个内部环。β 亚基活性位点位于圆桶内部，因此，蛋白酶体是自区分蛋白酶（Voges *et al.*，1999），因为底物必须进入被 β 环界定的催化室才能被降解成肽。底物在 α 环中往返穿过狭窄开口而接近活性位点，这个 α 环通过 α 亚基 N 端序列被阻挡在非配体游离 20S 蛋白酶体中。该通道的开放是由于与蛋白酶体激活物的结合（Kohler *et al.*，2000 和图 15.3）。真核细胞 20S 蛋白酶体含有至少两个糜蛋白酶、两个胰蛋白酶和两个类半胱氨酸蛋白酶激活位点（如下）。这些活性受到变构调节，已表明对蛋白质的降解有一个有序的循环机制（Kisselev *et al.*，1999）。多活性位点限制在极小的分隔间有另一优点。蛋白酶体水解大多数肽键，产生的肽具有典型 3～22 氨基酸长度，除了抗原肽外均没有保留生物活性。

## 26S 蛋白酶体

19S 复合物也称 PA700（蛋白酶体激活物 700），在 ATP 存在的条件下结合到 20S 蛋白酶体的两个 α 环上，形成 26S 蛋白酶体（图 15.3C）。PA700 含有至少 18 个不同的亚基，聚集在两个亚复合物内分别称为基底和盖子。在人的蛋白酶体中基底含有 6 个 ATP 酶和 3 个非 ATP 酶亚基。ATP 酶亚基是 AAA-ATP 酶家族的成员（ATP 酶与多种细胞活性有关），尽管它们的序列相似，但每一个成员都有许多特定的作用。它们能为 26S 蛋白酶体的装配提供能量、开放 20S 蛋白酶体通路、折叠并把蛋白底物注入蛋白酶体的催化室，另外也为肽的释放提供能量（Voges *et al.*，1999；Kohler *et al.*，2001；Navon and Goldberg，2001）。20S 蛋白酶体结合到基底上单独支持 ATP 依赖性肽的水解。相比之下，基底和盖子是依赖于遍在蛋白的蛋白水解所必需的。非 ATP 酶亚基 S5a 可以稳定基底和盖子亚复合物，然后紧密地连接到多聚遍在蛋白降解信号上（Deveraux *et al.*，1994）。然而，ATP 酶 S6′亚基在基底内也识别多聚遍在蛋白链，这种相互作用依赖于 ATP 的水解（Lam *et al.*，2002）。

## 其他蛋白酶体形式

免疫蛋白酶体是 20S 或者 26S 的蛋白酶体，在这些蛋白酶体中干扰素 γ 通过 3 个选择性亚基诱导具有不同催化活性的三个 β 亚基的取代反应，所生成的小微粒可有效地产生 Ⅰ 类抗原肽。

除了 PA700 以外，还有许多结合在 20S 核心颗粒上的蛋白酶体激活物已经被分离出来。例如，PA28（蛋白酶体激活物 28）以两个不同亚基的六聚体或七聚体复合物 PA28α 和 β 的形式存在，它能以非依赖于 ATP 的形式结合到 20S 蛋白酶体的任一 α 环上（图 15.3C）。该复合物与 20S 蛋白酶体结合从而激活肽酶活性，但不影响水解活性（表 15.1）。PA28-蛋白酶体复合物可能在 Ⅰ 类抗原呈递中发挥作用，也可能在连接 26S 免疫蛋白酶体过程中发挥作用（Rechseiner *et al.*，2000）。

PA28γ 与亚基 PA28α 和 β 密切相关，能形成结合到 20S 蛋白酶体颗粒上的同聚物。

由此所产生的蛋白酶体-PA28γ复合物参与生长调节（Muraata *et al.*，1999）。

**表 15.1　蛋白酶体的作用、底物和 ATP 依赖性**

| | 作用 | 亚基 | ATP[1] |
|---|---|---|---|
| 20S 蛋白酶体 | 蛋白质分解[2] 和肽酶活性 | 非遍在蛋白化蛋白和肽 | － |
| 26S 蛋白酶体 | 蛋白质分解 | 遍在蛋白化蛋白和非遍在蛋白化蛋白（如 ODC）[3] | ＋ |
| 20S 和 26S 免疫蛋白酶体 | Ⅰ类级抗原呈递 | 非遍在蛋白化（20S）或遍在蛋白化（26S）蛋白和肽 | －（20S）＋（26S） |
| 蛋白酶体 PA28αβ | Ⅰ类抗原呈递 | 肽 | － |
| 蛋白酶体 PA28γ | 生长 | 肽 | － |
| 杂合蛋白酶体 PA28αβ-20S-PA700 | Ⅰ类抗原呈递和蛋白水解（如 ODC） | 非遍在蛋白化（或遍在蛋白化?）蛋白和肽 | ＋ |
| PA200-20S-PA200 | DNA 修复 | 肽 | － |
| 杂合蛋白酶体 PA200-20S-PA700 | ? | ? | ? |

1 ATP 依赖性底物水解；2 20S 蛋白酶体降解氧化蛋白；3 ODC 鸟氨酸脱羧酶。

近来，第四蛋白酶体激活物（PA200）已被阐明（Ustrell *et al.*，2002）该核激活物参与 DNA 修复。

最后，还存在着一些杂合蛋白酶体，例如 PA28-20S 蛋白酶体-PA700，通过 PA28 和一个 PA700 结合到 20S 蛋白酶体的每一个 α 环上形成（Hendil *et al.*，1998）。这些杂合蛋白酶体复合物被干扰素 γ 诱导，作用于抗原呈递和一些蛋白质的分解（Tanahashi *et al.*，2000）。但是杂合 PA200-20S 蛋白酶体-PA700 的存在仍受到强烈质疑（Ustrell *et al.*，2002）。

因此，在哺乳动物细胞中多数蛋白酶体是共存的。蛋白酶体之所以存在不同形式，是因为不同的激活物结合到 20S 催化核心上，具有各自特定的蛋白水解和非蛋白水解功能（表 15.1）。另外，应该指出的是，19S 蛋白酶体亚复合物是 26S 蛋白酶体蛋白水解体系的前体，它也具有非蛋白水解的功能。例如，19S 蛋白具有核苷酸切除修复和转录延长的作用，可以列为活性启动子（Gonzalez *et al.*，2002）。

# 骨骼肌中遍在蛋白-蛋白酶体依赖性蛋白水解的调节

在本节中，我们依据有限的资料，不仅将激素和营养在肌肉中对遍在蛋白-蛋白酶体途径的调节做一综述，同时也会讨论其他因素对遍在蛋白-蛋白酶体途径的调节。

## 遍在蛋白共轭的调节

### 遍在蛋白

有研究称对啮齿动物（Attaix *et al.*，1998；Jagoe and Goldberg，2001）和人（Mansoor *et al.*，1996；Tao *et al.*，1997）而言，在许多分解代谢状态下的遍在蛋白可以增加骨骼肌中 mRNA 的水平。通过观察饥饿的大鼠（Medina *et al.*，1995）和其他

分解代谢模型Ⅰ和Ⅱ骨骼肌纤维发现，这种适应似乎是肌肉的通常反应。同时应当指出，增加的遍在蛋白 mRNA 水平并没有系统地反映肌肉蛋白水解的增加。例如，用地塞米松治疗老年大鼠表现为遍在蛋白表达增加，而其肌肉蛋白水解速率却没有任何改变（Dardevet *et al.*，1995）。在消瘦的条件下，增加遍在蛋白表达的一个关键调节途径似乎是增加血清糖（肾上腺）皮质激素水平。啮齿动物的皮质酮分泌量在多数分解代谢状态下会增加（Tiao *et al.*，1996；Marinovic *et al.*，2002），包括饥饿状态（Wing and Goldberg，1993）。给头部损伤体重减少的患者使用氢化可的松，患者表现为肌肉遍在蛋白和其他蛋白水解基因的表达增加（Mansoor *et al.*，1996）。增加的遍在蛋白 mRNA 经糖（肾上腺）皮质激素诱导，在酸中毒（Bailey *et al.*，1996）和糖尿病（Price *et al.*,1996）大鼠的肌肉中表现为转录增加。体外研究糖（肾上腺）皮质激素治疗 L6 肌肉细胞表明，遍在蛋白的转录受 Sp1 和 MEK1 调节（Marinovic *et al.*，2002）。据间接推测，遍在蛋白表达也受许多细胞因子包括肿瘤坏死因子-α（TNF-α）的正调节（Combaret *et al.*，2002）。

大鼠在饥饿一段时间后摄食，会导致其肌肉中遍在蛋白 mRNA 水平降低（Medina *et al.*，1995）。显然，胰岛素在此过程中发挥作用，因为在伴有高氨基酸血症存在（Larbaud *et al.*，1996）和缺失（Larbaud *et al.*，2001）的高胰岛素血症钳夹期间，遍在蛋白 mRNA 水平在Ⅱ型骨骼肌纤维中有一个快速（在 3～4h）和显著的转录负调节。这些结果表明氨基酸对体内遍在蛋白表达的调节较弱。在相关的实验中明确表明把氨基酸注入饥饿的大鼠，肌肉中遍在蛋白表达没有快速恢复到对照饲喂水平（Kee *et al.*，2003）。最后，胰岛素样生长因子-Ⅰ(IGF-Ⅰ）对体内遍在蛋白表达的调节作用还不清楚，至于负调节作用（Fang *et al.*，2000）或没有作用（Kee *et al.*，2002）均有文献报道。

## 遍在蛋白激活酶 E1

E1 在骨骼肌中的表达很低，且其 mRNA 水平在分解代谢状态下不受调节（Lecker *et al.*，1999）。这并不让人感到惊奇，因为：①E1 是一种活性非常高的酶，能够把遍在蛋白和过剩的 E2 连接起来（约 100pmol/L E2 的 $K_m$ 值）；②E1 是所有遍在蛋白共轭途径中的一个共同组分（Rajapurohitam *et al.*，2002）。因此，任何 E1 损伤都将影响整个下游遍在蛋白化级联反应。

## 遍在蛋白共轭酶 E2

在骨骼肌中已发现各种 E2，其中包括 14kDa、17kDa 和 20kDa（Attaix and Taillandier，1998）的 E2-F1（Gonen *et al.*，1996）、Ub-E2G（Chrysis and Underwood，1999）、UBC4（Rajapurohitam *et al.*，2002）和巨大的 230kDa 蛋白质（E2-230k），此蛋白质主要在骨骼肌和心脏中表达（Yokota *et al.*，2001）。在可溶性的肌肉提取物中，UBC2 家族 14kDa E2 或 UBC4 的免疫耗竭抑制了超过 50%的遍在蛋白化比率（Rajapurohitam *et al.*，2002）。在大多数肌肉消耗（Attaix and Taillandier，1998）（包括饥饿）（Wing and Banville，1994）的情况下，14kDa E2 中 1.2kb 转录物的表达受到正调节。Taillandier 等（1996）研究表明，在人们认为不重要的比目鱼肌中，增加的 14kDa E2 mRNA 进入了活性翻译过程。然而，在几个研究中，14kDa E2 表达的增加既不影响肌肉蛋白水解率（Temparis *et al.*，1994；Fang *et al.*，2000；Combaret *et al.*，2002），也不与改变的肌肉蛋白酶含量（Hobler *et al.*，1999a；Lecker *et al.*，1999）相关。在成年大鼠中，14kDa E2 的表达由糖皮质激素进行正调节（Dardevet *et al.*，1995）。在体外，胰岛素和 IGF-Ⅰ降低了 14kDa E2 mRNA 的稳定性（Wing and Bedard，1996）。然而，这样的负调节在注入胰岛素的动物中并不普遍（Larbaud *et al.*，1996，2001）。与之相似，在用 IGF-Ⅰ处理的大鼠肌肉中 14kDa E2 的表达并没有改变（Kee *et al.*，2002）。根据我们了解，只有 Chrysis 和 Underwood（1999）报道了大鼠肌肉中其他 E2 的 mRNA 水平发生改变。这些研究者证明地塞米松可提高 14kDa E2 和 Ub-E2G 的 mRNA 水平及 17kDa E2 mRNA 的 2E 异构体水平。他们还报道地塞米松处理动物，注射 IGF-Ⅰ后，14kDa E2 和 Ub-E2G 的高水平转录受到抑制，而 17kDa E2 的 2E 异构体只受到了中等程度的影响。

## 遍在蛋白-蛋白连接酶 E3

目前，只有一个报道描述了庞大的 E3（E3L)。E3（E3L）与 E2-F1 一道参与遍在蛋白化的肌动蛋白、肌钙蛋白 T 及肌源性决定因子（MyoD）体外降解（Gonen *et al.*，1996)。几个研究组（Lecker *et al.*，1999）已报道处于分解代谢的肌肉中的 E3α mRNA水平增加，E3α 即普遍存在的 N 端法则 RING 指连接酶，该酶可与 14kDa E2 一起发挥作用。然而，这样的变化并不与改变 E3α 的蛋白水平相关。最近的研究已证明了多种肌肉特异性 RING 指 E3 的存在。这些 E3 包括：肌肉 RING 指蛋白 MuRF-1、-2 和-3（muscle RING finger protein）（Centner *et al.*，2001)、横纹肌 RING 锌指 SMRZ（striated muscle RING zinc finger)（Dai and Liew（2001)、细胞分裂后期促进复合物 ANAPC11（anaphase-promoting complex)（Chan *et al.*，2001)、萎缩基因 atrogin-1/MAFbx（atrophy gene)（Gomes *et al.*，2001）也称作肌肉萎缩 F 盒 MAFbx)（muscle atrophy F-box)（Bodine *et al.*，2001)。Atrogin-1/MAFbx 和 MuRF-1 在肌肉萎缩过程中被不同程度地过量表达，其中对 Atrogin-1/MAFbx 来说在禁食和糖尿病情况下也被过量表达。这些发现都是非常重要的，因为 E3α（唯一一个目前报道过在肌肉消耗中发挥作用的 E3）可能有较小的生理作用：①E3α 参与可溶性肌肉蛋白而不是肌原纤维蛋白的遍在蛋白化（Lecker *et al.*，1999)；②缺乏 E3α 基因的小鼠能生育，能繁殖，只是其骨骼肌比对照组小（Kwon *et al.*，2001)。更加令人吃惊的是，缺乏 Atrogin-1/MAFbx 或 MURF-1 的小鼠对肌肉萎缩具有抗性（Bodine *et al.*，2001)。

## 遍在蛋白共轭物及遍在蛋白化/去遍在蛋白化比率

肌肉遍在蛋白共轭物的增加出现在一些消耗情况中（Attaix *et al.*，1998)，这表明 26S 蛋白酶体对遍在蛋白物的分解是限速反应，然而，另外几个研究却给出了相矛盾的结论。例如，尽管依赖遍在蛋白-蛋白酶体的蛋白酶解作用升高（Tiao *et al.*，1994；Voisin *et al.*，1996)，却没有观察到败血症动物肌肉中遍在蛋白共轭物的积累（Tiao *et al.*，1994)。目前对这种差异的原因还不清楚。然而，遍在蛋白共轭物的水平也依赖于遍在蛋白化比率、去遍在蛋白化活性及其分解率。Tilignac 等（2002）报道，在蛋白酶体活性低于基础对照水平的小鼠中，可溶性片段及肌原纤维片段中遍在蛋白共轭物的量不变，这表明遍在蛋白化比率也受损。虽然，有其他相矛盾的结论（Tiao *et al.*，1994），但是遍在蛋白化蛋白优先在肌原纤维片段中积累（Wing *et al.*，1995；Combaret *et al.*，2002），这些差异必须阐明，而且在骨骼肌中遍在蛋白化的收缩蛋白仍需进一步鉴定。

在不同分解代谢条件下，大鼠可溶性肌肉提取物中外源$^{125}$I标记的遍在蛋白及N端法则途径中的模型底物（如［$^{125}$I］α-乳白蛋白）的体外遍在蛋白化比率都有明显升高（Solomon *et al.*，1998；Lecker *et al.*，1999）。在一些动物的肌肉提取物中，因甲状腺激素的缺乏，蛋白质分解都会下降，用这些激素处理动物后，蛋白质分解又上升，二者之间存在着相同的遍在蛋白化比率变化，这证明了激素的调节作用（Solomon *et al.*，1998）。

在骨骼肌中存在很多未知功能的去遍在蛋白化酶（Woo *et al.*，1995）。如UBP45和UBP69的不同表达参与肌肉细胞分化调节（Park *et al.*，2002）。去遍在蛋白化酶在可溶性肌肉提取物中的活性很高（Rajapurohitam *et al.*，2002）。对于去遍在蛋白化是否在肌肉消耗中起着一定的作用仍需进一步研究。然而，禁食大鼠萎缩的肌肉中去遍在蛋白酶及遍在蛋白特异性蛋白酶（ubiquitin-specific protease，USPI）的mRNA水平增加，还能通过重新饲喂而降低（Combaret and Wing，未发表数据）。

## 蛋白酶体的调节

26S蛋白酶体的蛋白水解活性由多种精细机制调节，包括蛋白酶体亚基的合成与加工、20S和26S蛋白酶体的组装、翻译后修饰（如一些20S和PA700亚基的磷酸化）、蛋白酶体激活剂和抑制剂的结合等。在这些不同水平调节中，每一个调节的详细信息不在本章的范围内，读者可以参考一些好的而且广泛的文献来获得详细的信息（Voges *et al.*，1999；Glickman and Ciechanover，2002）。

### 蛋白水解率

在饥饿以及去神经支配肌肉萎缩过程中存在着肌肉消耗，Goldberg及他的同事们为遍在蛋白-蛋白酶体途径在肌肉消耗中的作用提供了最初的证据（Attaix and Taillandier，1998）。在这些条件下，大多数增加蛋白水解的作用都是依赖ATP的。在用饥饿和去神经大鼠培育的萎缩肌肉中，ATP耗竭［但不是钙蛋白酶和（或）组织蛋白酶抑制］抑制了蛋白水解率的升高，同样在其他啮齿动物分解代谢模型中也存在这种现象（Temparis *et al.*，1994；Price *et al.*，1996；Taillandier *et al.*，1996）。遍在蛋白通过E1的激活及通过26S蛋白酶体的蛋白水解都需要ATP（见上文）。

只有蛋白酶体抑制剂如乳胞素或MG132存在时，对消瘦情况下的全部蛋白水解率的升高有抑制作用的证明，可以为蛋白酶体在肌肉蛋白分解中的主要作用提供进一步有力的支持（Bailey *et al.*，1996；Price *et al.*，1996；Hobler *et al.*，1998；Combaret *et al.*，2002），并且只有蛋白酶体抑制剂抑制了培育萎缩肌肉中3-甲基组氨酸释放的升高（Hobler *et al.*，1998）。3-甲基组氨酸是由肌动蛋白和肌球蛋白的翻译后修饰形成的，其在培育介质中的出现频率反映了这些肌原纤维蛋白的分解程度（Attaix and Taillandier，1998）。Mykles（1993）对龙虾肌肉20S蛋白酶体的研究发现其具有5种肽酶的活性。体外研究已表明在收缩蛋白降解过程中包括支链氨基酸偏爱性、肽酰谷氨酰肽水解酶及胰蛋白酶样酶活性。同其他组织相比较，大鼠骨骼肌中的肽酶活性和蛋白酶活性相对较低（Farout *et al.*，2000）。在消瘦情况下，肌肉蛋白酶体中糜蛋白酶样肽酶活性增加（Hobler *et al.*，1999b；Ordway *et al.*，1999；Kee *et al.*，2002）。相反，这种肽酶活性在糖尿病例中没有变化（Liu *et al.*，2000）。骨骼肌中20S蛋白酶体家族至少有6种不同的亚型，包括组成型蛋白酶体、免疫蛋白酶体及它们的中间形式（Dahlmann *et al.*，2001）。因此，从一个特定组织分离出的所有20S蛋白酶体的性质代表了整个一套蛋白酶体亚型的普遍特性。这可能为总肌肉蛋白水解率和一些特异蛋白酶体活性的差异提供了解释。然而，与之相反，当用化学疗法减弱依赖蛋白酶体的蛋白水解作用时，体外测量的糜蛋白酶和胰蛋白酶样酶活性受到了抑制（Tilignac *et al.*，2002）。肌肉提取物中外源蛋白底物［如（甲基-$^{14}$C）酪蛋白］的ATP依赖性和遍在蛋白依赖性降解的增加，与取自运动后的大鼠（Kee *et al.*，2002）和慢性电刺激后的兔（Ordway *et al.*，1999）肌肉中的蛋白酶体的糜蛋白酶样酶活性的增加相关。

最后，据报道胰岛素能够通过胰岛素降解酶（Duckworth *et al.*，1998）和（或）蛋白酶体构象的改变（Hamel *et al.*，1998）对蛋白酶体的活性产生直接的抑制作用。

## 20S蛋白酶体亚基的表达和蛋白质含量

许多研究组已表明，在饥饿和饥饿/重新饲喂动物中，ATP依赖性的或蛋白酶体依赖性的肌肉蛋白水解率的升高与20S蛋白酶体中几个亚基mRNA水平的升高有关（Attaix *et al.*，1998；Jagoe and Goldberg，2001）。在癌症恶病质中，27kDa 20S蛋白酶体亚基中蛋白质丰度的增加与其他亚基表达的增加相关（Baracos *et al.*，1995）。相反，当用化学疗法抑制依赖蛋白酶体的蛋白水解作用时，20S蛋白酶体亚基的mRNA水平与两个亚基的减少量相关（Tilignac *et al.*，2002）。Taillandier等（1996）证明，萎缩肌肉中过量表达的RC9亚基能够进入活性翻译过程，同时Bailey等（1996）在酸毒症中观察到了转录的RC3蛋白酶体亚基mRNA增加。糖皮质激素（Price *et al.*，1994；Dardevet *et al.*，1995）和肿瘤坏死因子-α（TNF-α；Combaret *et al.*，2002）对20S蛋白酶体亚基的mRNA水平有正调节作用，在L6肌肉细胞中，糖皮质激素通过抗核因子（NF)-κB(Du *et al.*，2000）对蛋白酶体C3亚基转录的抑制诱导了其转录，而遍在蛋白的存在依赖于糖皮质激素增加的转录包括Sp1和MEK1（Marinovic *et al.*，2002）。因此，遍在蛋白-蛋白酶体系统中几个基因增加的同步转录激活了选择性信号途径。与糖皮质激素形成对比，胰岛素（Larbaud *et al.*，1996，2001）和IGF-Ⅰ（Kee *et al.*，2002）对20S肌肉蛋白酶体α或β亚基的体内表达没有调节作用。最后，饲喂蛋白缺乏饲粮的动物和甲状腺机能减退的动物的肌肉中蛋白酶体含量均有下降（Tawa and Goldberg，1993）。

## PA700亚基的表达和蛋白质含量

在肌肉消耗（Attaix *et al.*，1997）包括饥饿（Kee *et al.*，2003）的情况下，一些但不是全部PA700亚基的mRNA水平也被正调节。PA700中ATP酶亚基和非ATP

酶的表达调节不依赖于20S蛋白酶体亚基的表达，而是明显依赖于特定的分解代谢(Attaix *et al*., 1997；Combaret *et al*., 2002)。而且，个别调控亚基的mRNA水平和蛋白质含量的调节是独立的，与蛋白水解率之间没有系统地联系（Combaret *et al*., 2002；Tilignac *et al*., 2002)。这些数据与Dawson等（1995）在研究细胞程序性死亡期间对烟草蛾（*Manduca sexta*）腹部节间肌肉观察所得的数据相似。

目前已证明，糖皮质激素（Combaret et al.，未发表数据)、胰岛素（Attaix *et al*., 1997）和TNF-α（Combaret *et al*., 2002）都能调节PA700的几个ATP酶和非ATP酶亚基的mRNA水平。然而，对改变蛋白水解活性很关键的亚基表达和（或）蛋白含量的特异变化还没有被鉴定。

最近的两个报道表明二十碳五烯酸抑制了禁食（Whitehouse and Tisdale, 2001）及患肿瘤（Whitehouse *et al*., 2001）小鼠肌肉中蛋白酶体糜蛋白酶样酶活性的增加，并且也抑制了20S蛋白酶体α亚基和p42（PA700的一种ATP酶亚基）表达的增强。这种作用在添加相关的（*n*-3）脂肪酸二十二碳六烯酸或亚油酸的饲料喂养饥饿小鼠时没有发现。然而，5-、12-和15-脂肪氧化酶的抑制剂2,3,5-三甲基-6-(3-吡啶甲基）1,4-苯醌-1（CV-6504）同样也能减弱分解代谢反应。这些结果表明，在饥饿及癌症恶病质时蛋白分解代谢的介导途径相同，此途径被二十碳五烯酸抑制并可能包含一个脂肪氧化酶代谢物作为信号转导子。

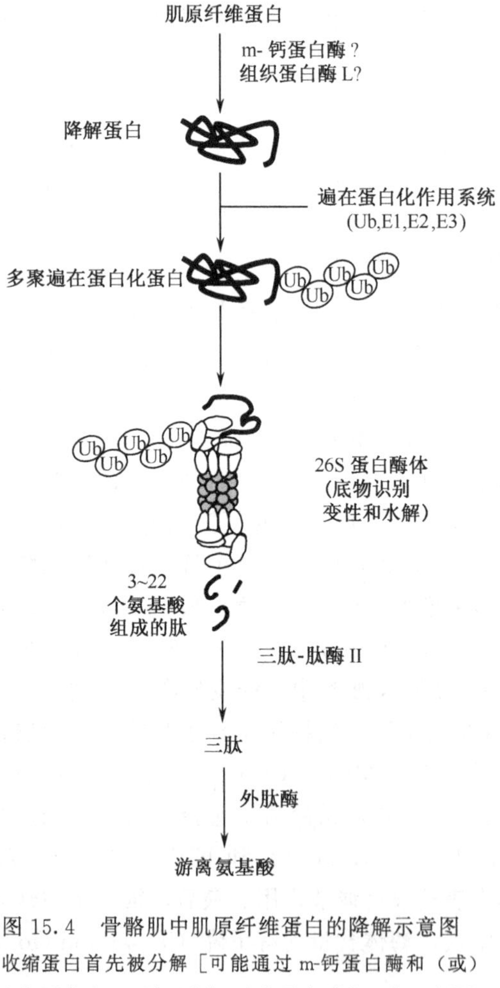


图15.4　骨骼肌中肌原纤维蛋白的降解示意图
收缩蛋白首先被分解［可能通过m-钙蛋白酶和（或）组织蛋白酶L，见正文］。这些蛋白［和（或）片段］然后被多聚遍在蛋白化并被26S蛋白酶体降解，产生3～22个氨基酸长的肽。大于三肽的肽通过三肽-肽酶Ⅱ水解，形成三肽，最后通过外肽酶的作用水解成游离氨基酸

## 与其他蛋白水解途径的功能联系

26S蛋白酶体只是将蛋白质降解为肽。除非作为主要组织相容性复合物（major histocompatibility complex, MHC）Ⅰ类分子出现，否则这些肽必须经过进一步水解成为游离氨基酸（Attaix *et al*., 2001)。近期研究结果显示，溶酶体外肽酶（extralysosomal peptidase）三肽-肽酶Ⅱ（tripeptidyl-peptidase Ⅱ，TPPⅡ）降解由蛋白酶体产生的肽（图15.4及Hasselgren *et al*., 2002)。TPPⅡ的表达、含量及活性在败血症的肌肉中有所增加，而且，糖皮质激素受体拮抗剂RU38486钝化了这些调节，表明糖皮质激素参与了TPPⅡ的正调节。

相反，其他蛋白酶可能作用于蛋白酶体的上游（图 15.4）。肌原纤维蛋白间特异的相互作用似乎可以保护它们免受依赖于遍在蛋白的降解，并且在其降解过程中的限速步骤可能是它们从肌原纤维上的分离过程（Solomon and Goldberg，1996）。钙蛋白酶在肌节蛋白分解及 Z 带的降解过程中起着关键的作用，从而导致肌丝的释放（Williams *et al.*，1999）。这些数据表明钙蛋白酶作用于蛋白酶体的上游（Hasselgren *et al.*，2002），而且 Deval 等（2001）最近发现，不同损耗情况对组织蛋白酶 L 的正调节作用不同，且其表达的改变紧密地跟随依赖遍在蛋白-蛋白酶体的蛋白质酶解调节。钙蛋白酶、组织蛋白酶 L 和遍在蛋白途径间是否存在功能上的联系还需要进一步研究。然而，在敲除肌肉特异性钙蛋白酶 p94 的小鼠中，几种蛋白水解基因（包括组织蛋白酶 L 和遍在蛋白-蛋白酶体系统中的几个成分）的表达被负调节（Combaret *et al.*，2003）。无论如何，阐明导致肌肉蛋白完全降解的蛋白酶体上游和下游的蛋白水解机制是很重要的。

## 小　　结

目前关于遍在蛋白-蛋白酶体途径如何降解肌肉蛋白特别是收缩蛋白还有很多地方不清楚。以肌原纤维蛋白分解为目标的信号及此途径的精确底物都没有被识别。很明显，为揭示营养素、激素及其信号传递途径的作用我们还需要做进一步详细的研究。遍在蛋白途径的复杂性无疑会妨碍肌肉蛋白水解调节中发挥重要作用的精确机制的识别。然而，对转录图的微阵列分析及对照和萎缩肌肉的蛋白质组分析应该能提供很有用的信息。最后，肌原纤维蛋白在蛋白酶体上游和下游降解的不同步骤的精确阐明对想出新的策略来预防肌肉萎缩是很重要的。

## 致　　谢

我们感谢 Daniel Bechet 博士对手稿的评论。作者所在实验室的研究得到了 the Association pour la Recherche surle Cancer、the Conseil Regional d Auvergne、法国国家农业研究院、国家健康与医学研究院和 the French Ministere de la Recherche and Nestle 的资助。同时，A. J. Kee 获得了法国国家农业研究院博士后奖学金的支持。在此一并表示感谢。

### 参 考 文 献

Attaix, D. and Taillandier, D. (1998) The critical role of the ubiquitin–proteasome pathway in muscle wasting in comparison to lysosomal and $Ca^{2+}$-dependent systems. In: Bittar, E.E. and Rivett, A.J. (eds) *Intracellular Protein Degradation.* JAI Press, Greenwich, Connecticut, pp. 235–266.

Attaix, D., Taillandier, D., Combaret, L., Rallière, C., Larbaud, D., Aurousseau, E. and Tanaka, K. (1997) Expression of subunits of the 19S complex and of the PA28 activator in rat skeletal muscle. *Molecular Biology Reports* 24, 95–98.

Attaix, D., Aurousseau, E., Combaret, L., Kee, A., Larbaud, D., Rallière, C., Souweine, B., Taillandier, D. and Tilignac, T. (1998) Ubiquitin–proteasome-dependent proteolysis in skeletal muscle. *Reproduction Nutrition Development* 38, 153–165.

Attaix, D., Combaret, L., Pouch, M.-N. and Taillandier,

D. (2001) Regulation of proteolysis. *Current Opinion in Clinical Nutrition and Metabolic Care* 4, 45–49.

Bailey, J.L., Wang, W., England, B.K., Price, S.R., Ding, X. and Mitch, W.E. (1996) The acidosis of chronic renal failure activates muscle proteolysis in rats by augmenting transcription of genes encoding proteins of the ATP-dependent ubiquitin–proteasome pathway. *Journal of Clinical Investigation* 97, 1447–1453.

Balcerzak, D., Querengesser, L., Dixon, W.T. and Baracos, V.E. (2001) Coordinate expression of matrix-degrading proteinases and their activators and inhibitors in bovine skeletal muscle. *Journal of Animal Science* 79, 94–107.

Baracos, V.E., DeVivo, C., Hoyle, D.H.R. and Goldberg, A.L. (1995) Activation of the ATP–ubiquitin–proteasome pathway in skeletal muscle of cachectic rats bearing a hepatoma. *American Journal of Physiology* 268, E996–E1006.

Belizario, J.E., Lorite, M.J. and Tisdale, M.J. (2001) Cleavage of caspases-1, -3, -6, -8 and -9 substrates by proteases in skeletal muscles from mice undergoing cancer cachexia. *British Journal of Cancer* 84, 1135–1140.

Benaroudj, N., Tarcsa, E., Cascio, P. and Goldberg, A.L. (2001) The unfolding of substrates and ubiquitin-independent protein degradation by proteasomes. *Biochimie* 83, 311–318.

Bodine, S.C., Latres, E., Baumhueter, S., Lai, V.K., Nunez, L., Clarke, B.A., Poueymirou, W.T., Panaro, F.J., Na, E., Dharmarajan, K., Pan, Z.Q., Valenzuela, D.M., DeChiara, T.M., Stitt, T.N., Yancopoulos, G.D. and Glass, D.J. (2001) Identification of ubiquitin ligases required for skeletal muscle atrophy. *Science* 294, 1704–1708.

Centner, T., Yano, J., Kimura, E., McElhinny, A.S., Pelin, K., Witt, C.C., Bang, M.L., Trombitas, K., Granzier, H., Gregorio, C.C., Sorimachi, H. and Labeit, S. (2001) Identification of muscle specific ring finger proteins as potential regulators of the titin kinase domain. *Journal of Molecular Biology* 306, 717–726.

Chan, A.H., Lee, S.M., Chim, S.S., Kok, L.D., Waye, M.M., Lee, C.Y., Fung, K.P. and Tsui, S.K. (2001) Molecular cloning and characterization of a RING-H2 finger protein, ANAPC11, the human homolog of yeast Apc11p. *Journal of Cellular Biochemistry* 83, 249–258.

Chrysis, D. and Underwood, L.E. (1999) Regulation of components of the ubiquitin system by insulin-like growth factor I and growth hormone in skeletal muscle of rats made catabolic with dexamethasone. *Endocrinology* 140, 5635–5641.

Chung, C.H. and Baek, S.H. (1999) Deubiquitinating enzymes: their diversity and emerging roles. *Biochemical and Biophysical Research Communications* 266, 633–640.

Combaret, L., Taillandier, D., Voisin, L., Samuels, S.E., Boespflug-Tanguy, O. and Attaix, D. (1996) No alteration in gene expression of components of the ubiquitin–proteasome proteolytic pathway in dystrophin-deficient muscles. *FEBS Letters* 393, 292–296.

Combaret, L., Taillandier, D. and Attaix, D. (2001) Nutritional and hormonal control of protein breakdown. *American Journal of Kidney Diseases* 37, Supplement 2, S108–S111.

Combaret, L., Béchet, D., Claustre, A., Taillandier, D., Richard, I. and Attaix, D. (2003) Down-regulation of genes in the lysosomal and ubiquitin-proteasome proteolytic pathway in calpain-3-deficient muscle. *International Journal of Biochemistry and Cell Biology* (in press).

Combaret, L., Tilignac, T., Claustre, A., Voisin, L., Taillandier, D., Obled, C., Tanaka, K. and Attaix, D. (2002) Torbafylline (HWA 448) inhibits enhanced skeletal muscle ubiquitin–proteasome-dependent proteolysis in cancer and septic rats. *Biochemical Journal* 361, 185–192.

Dahlmann, B., Ruppert, T., Kloetzel, P.M. and Kuehn, L. (2001) Subtypes of 20S proteasomes from skeletal muscle. *Biochimie* 83, 295–299.

Dai, K.S. and Liew, C.C. (2001) A novel human striated muscle RING zinc finger protein, SMRZ, interacts with SMT3b via its RING domain. *Journal of Biological Chemistry* 276, 23992–23999.

Dardevet, D., Sornet, C., Taillandier, D., Savary, I., Attaix, D. and Grizard, J. (1995) Sensitivity and protein turnover response to glucorticoids are different in skeletal muscle from adult and old rats. Lack of regulation of the ubiquitin–proteasome proteolytic pathway in aging. *Journal of Clinical Investigation* 96, 2113–2119.

Dawson, S.P., Arnold, J.E., Mayer, N.J., Reynolds, S.E., Billett, M.A., Gordon, C., Colleaux, L., Kloetzel, P.M., Tanaka, K. and Mayer, R.J. (1995) Developmental changes of the 26S proteasome in abdominal intersegmental muscles of *Manduca sexta* during programmed cell death. *Journal of Biological Chemistry* 270, 1850–1858.

Deval, C., Mordier, S., Obled, C., Béchet, D., Combaret, L., Attaix, D. and Ferrara, M. (2001) Identification of cathepsin L as a differentially-expressed message associated with skeletal muscle wasting. *Biochemical Journal* 360, 143–150.

Deveraux, Q., Ustrell, V., Pickart, C. and Rechsteiner, M. (1994) A 26S protease subunit that binds ubiquitin conjugates. *Journal of Biological Chemistry* 269, 7059–7061.

Du, J., Mitch, W.E., Wang, X. and Price, S.R. (2000) Glucocorticoids induce proteasome C3 subunit expression in L6 muscle cells by opposing the suppression of its transcription by NF-κB. *Journal of Biological Chemistry* 275, 19661–19666.

Duckworth, W.C., Bennett, R.G. and Hamel, F.G. (1998) Insulin acts intracellularly on proteasomes through insulin-degrading enzyme. *Biochemical and Biophysical Research Communications* 244, 390–394.

Fang, C.H., Li, B.G., Sun, X. and Hasselgren, P.O. (2000) Insulin-like growth factor I reduces ubiquitin and ubiquitin-conjugating enzyme gene expression but does not inhibit muscle proteolysis in septic rats. *Endocrinology* 141, 2743–2751.

Farout, L., Lamare, M.C., Cardozo, C., Harrisson, M., Briand, Y. and Briand, M. (2000) Distribution of proteasomes and of the five proteolytic activities in rat tissues. *Archives of Biochemistry and Biophysics* 374, 207–212.

Glickman, M.H. and Ciechanover, A. (2002) The ubiquitin–proteasome proteolytic pathway: destruction for the sake of construction. *Physiological Reviews* 82, 373–428.

Gomes, M.D., Lecker, S.H., Jagoe, R.T., Navon, A. and Goldberg, A.L. (2001) Atrogin-1, a muscle-specific F-box protein highly expressed during muscle atrophy. *Proceedings of the National Academy of Sciences USA* 98, 14440–14445.

Gonen, H., Stancovski, I., Shkedy, D., Hadari, T., Bercovich, B., Bengal, E., Mesilati, S., Abu-Atoum, O., Schwartz, A.L. and Ciechanover, A. (1996) Isolation, characterization, and partial purification of a novel ubiquitin-protein ligase, $E_3$. *Journal of Biological Chemistry* 271, 302–310.

Gonzalez, F., Delahodde, A., Kodadek, T. and Johnston, S.A. (2002) Recruitment of a 19S proteasome subcomplex to an activated promoter. *Science* 296, 548–550.

Haas, A.L. and Siepmann, T.J. (1997) Pathways of ubiquitin conjugation. *FASEB Journal* 11, 1257–1268.

Hamel, F.G., Bennett, R.G. and Duckworth, W.C. (1998) Regulation of multicatalytic enzyme activity by insulin and the insulin-degrading enzyme. *Endocrinology* 139, 4061–4066.

Hasselgren, P.O., Wray, C. and Mammen, J. (2002) Molecular regulation of muscle cachexia: it may be more than the proteasome. *Biochemical and Biophysical Research Communications* 290, 1–10.

Hendil, K.B., Khan, S. and Tanaka, K. (1998) Simultaneous binding of PA28 and PA700 activators to 20S proteasomes. *Biochemical Journal* 332, 749–754.

Hobler, S.C., Tiao, G., Fischer, J.E., Monaco, J. and Hasselgren, P.O. (1998) Sepsis-induced increase in muscle proteolysis is blocked by specific proteasome inhibitors. *American Journal of Physiology* 274, R30–R37.

Hobler, S.C., Wang, J.J., Williams, A.B., Melandri, F., Sun, X., Fischer, J.E. and Hasselgren, P.O. (1999a) Sepsis is associated with increased ubiquitin conjugating enzyme E2 14k mRNA in skeletal muscle. *American Journal of Physiology* 276, R468–R473.

Hobler, S.C., Williams, A., Fischer, D., Wang, J.J., Sun, X., Fischer, J.E., Monaco, J.J. and Hasselgren, P.O. (1999b) Activity and expression of the 20S proteasome are increased in skeletal muscle during sepsis. *American Journal of Physiology* 277, R434–R440.

Jagoe, R.T. and Goldberg, A.L. (2001) What do we really know about the ubiquitin–proteasome pathway in muscle atrophy? *Current Opinion in Clinical Nutrition and Metabolic Care* 4, 183–190.

Kee, A.J., Taylor, A.J., Carlsson, A.R., Sevette, A., Smith, R.C. and Thompson, M.W. (2002) IGF-I has no effect on postexercise suppression of the ubiquitin–proteasome system in rat skeletal muscle. *Journal of Applied Physiology* 92, 2277–2284.

Kee, A.J., Combaret, L., Tilignac, T., Souweine, B., Aurousseau, E., Dalle, M., Taillandier, D. and Attaix, D. (2003) Ubiquitin-proteasome-dependent muscle proteolysis responds slowly to insulin release and refeeding n starved rats. *Journal of Physiology* 546, 765–776.

Kisselev, A.F., Akopian, T.N., Castillo, V. and Goldberg, A.L. (1999) Proteasome active sites allosterically regulate each other, suggesting a cyclical bite–chew mechanism for protein breakdown. *Molecular Cell* 4, 395–402.

Kohler, A., Cascio, P., Leggett, D.S., Woo, K.M., Goldberg, A.L. and Finley, D. (2001) The axial channel of the proteasome core particle is gated by the Rpt2 ATPase and controls both substrate entry and product release. *Molecular Cell* 7, 1143–1152.

Kwon, Y.T., Xia, Z., Davydov, I.V., Lecker, S.H. and Varshavsky, A. (2001) Construction and analysis of mouse strains lacking the ubiquitin ligase UBR1 (E3α) of the N-end rule pathway. *Molecular and Cellular Biology* 21, 8007–8021.

Lam, Y.A., Lawson, T.G., Velayutham, M., Zweier, J.L. and Pickart, C.M. (2002) A proteasomal ATPase subunit recognizes the polyubiquitin degradation signal. *Nature* 416, 763–767.

Larbaud, D., Debras, E., Taillandier, D., Samuels, S.E., Temparis, S., Champredon, C., Grizard, J. and Attaix, D. (1996) Euglycemic hyperinsulinemia and hyperaminoacidemia decrease skeletal muscle ubiquitin mRNA in goats. *American Journal of Physiology* 271, E505–E512.

Larbaud, D., Balage, M., Taillandier, D., Combaret, L., Grizard, J. and Attaix, D. (2001) Differential regulation of the lysosomal, $Ca^{2+}$-dependent and ubiquitin–proteasome-dependent proteolytic pathways in fast-twitch and slow-twitch rat muscle following hyperinsulinemia. *Clinical Science* 101, 551–558.

Lecker, S.H., Solomon, V., Price, S.R., Kwon, Y.T., Mitch, W.E. and Goldberg, A.L. (1999) Ubiquitin conjugation by the N-end rule pathway and mRNAs for its components increase in muscles of diabetic rats. *Journal of Clinical Investigation* 104, 1411 1420.

Liu, Z., Miers, W.R., Wei, L. and Barrett, E.J. (2000) The ubiquitin–proteasome proteolytic pathway in heart vs. skeletal muscle: effects of acute diabetes. *Biochemical and Biophysical Research Communications* 276, 1255–1260.

Lowell, B.B., Ruderman, N.B. and Goodman, M.N. (1986) Evidence that lysosomes are not involved in the degradation of myofibrillar proteins in rat skeletal muscle. *Biochemical Journal* 234, 237–240.

Mansoor, O., Beaufrère, B., Boirie, Y., Rallière, C., Taillandier, D., Aurousseau, E., Schoeffler, P., Arnal, M. and Attaix, D. (1996) Increased mRNA levels for components of the lysosomal, $Ca^{2+}$-activated and ATP-ubiquitin-dependent proteolytic pathways in skeletal muscle from head trauma patients. *Proceedings of the National Academy of Sciences USA* 93, 2714–2718.

Marinovic, A.C., Zheng, B., Mitch, W.E. and Price, S.R. (2002) Ubiquitin (UbC) expression in muscle cells is increased by glucocorticoids through a mechanism involving Sp1 and MEK1. *Journal of Biological Chemistry* 277, 16673–16681.

Medina, R., Wing, S.S. and Goldberg, A.L. (1995) Increase in levels of polyubiquitin and proteasome mRNA in skeletal muscle during starvation and denervation atrophy. *Biochemical Journal* 307, 631–637.

Murata, S., Kawahara, H., Tohma, S., Yamamoto, K., Kasahara, M., Nabeshima, Y., Tanaka, K. and Chiba, T. (1999) Growth retardation in mice lacking the proteasome activator PA28γ. *Journal of Biological Chemistry* 274, 38211–38215.

Mykles, D.L. (1993) Lobster muscle proteasome and the degradation of myofibrillar proteins. *Enzyme Protein* 47, 220–231.

Navon, A. and Goldberg, A.L. (2001) Proteins are unfolded on the surface of the ATPase ring before transport into the proteasome. *Molecular Cell* 8, 1339–1349.

Ordway, G.A., Neufer, P.D., Chin, E.R. and DeMartino, G.N. (2000) Chronic contractile activity upregulates the proteasome system in rabbit skeletal muscle. *Journal of Applied Physiology* 88, 1134–1141.

Park, K.C., Kim, J.H., Choi, E.J., Min, S.W., Rhee, S., Baek, S.H., Chung, S.S., Bang, O., Park, D., Chiba, T., Tanaka, K. and Chung, C.H. (2002) Antagonistic regulation of myogenesis by two deubiquitinating enzymes, UBP45 and UBP69. *Proceedings of the National Academy of Sciences USA* 99, 9733–9738.

Pickart, C.M. (2001) Mechanisms underlying ubiquitination. *Annual Review of Biochemistry* 70, 503–533.

Price, S.R., England, B.K., Bailey, J.L., Van Vreede, K. and Mitch, W.E. (1994) Acidosis and glucocorticoids concomitantly increase ubiquitin and proteasome subunit mRNA levels in rat muscles. *American Journal of Physiology* 267, C955–C960.

Price, S.R., Bailey, J.L., Wang, X., Jurkovitz, C., England, B.K., Ding, X., Phillips, L.S. and Mitch, W.E. (1996) Muscle wasting in insulinopenic rats results from activation of the ATP-dependent, ubiquitin–proteasome proteolytic pathway by a mechanism including gene transcription. *Journal of Clinical Investigation* 98, 1703–1708.

Rajapurohitam, V., Bedard, N. and Wing, S.S. (2002) Control of ubiquitination of proteins in rat tissues by ubiquitin conjugating enzymes and isopeptidases. *American Journal of Physiology* 282, E739–E745.

Rallière, C., Tauveron, I., Taillandier, D., Guy, L., Boiteux, J.-P., Giraud, B., Attaix, D. and Thiéblot, P. (1997) Glucocorticoids do not regulate the expression of proteolytic genes in skeletal muscle from Cushing's syndrome patients. *Journal of Clinical Endocrinology and Metabolism* 82, 3161–3164.

Rechsteiner, M., Realini, C. and Ustrell, V. (2000) The proteasome activator 11S REG (PA28) and class I antigen presentation. *Biochemical Journal* 345, 1–15.

Rock, K.L., Gramm, C., Rothstein, L., Clark, K., Stein, R., Dick, L., Hwang, D. and Goldberg, A.L. (1994) Inhibitors of the proteasome block the degradation of most cell proteins and the generation of peptides presented on MHC class I molecules. *Cell* 78, 761–771.

Scheffner, M., Smith, S. and Jentsch, S. (1998) The ubiquitin-conjugation system. In: Peters, J.-M., Harris, J.R. and Finley, D. (eds) *Ubiquitin and the Biology of the Cell.* Plenum Press, New York, pp. 65–98.

Shih, S.C., Sloper-Mould, K.E. and Hicke, L. (2000) Monoubiquitin carries a novel internalization signal that is appended to activated receptors. *EMBO Journal* 19, 187–198.

Solomon, V. and Goldberg, A.L. (1996) Importance of the ATP–ubiquitin–proteasome pathway in the degradation of soluble and myofibrillar proteins in rabbit muscle extracts. *Journal of Biological Chemistry* 271, 26690–26697.

Solomon, V., Baracos, V., Sarraf, P. and Goldberg, A.L. (1998) Rates of ubiquitin conjugation increase when muscles atrophy, largely through activation of the N-end rule pathway. *Proceedings of the National Academy of Sciences USA* 95, 12602–12607.

Strickland, E., Hakala, K., Thomas, P.J. and DeMartino, G.N. (2000) Recognition of misfolding proteins by PA700, the regulatory subcomplex of the 26S proteasome. *Journal of Biological Chemistry* 275, 5565–5572.

Taillandier, D., Aurousseau, E., Meynial-Denis, D.,

Béchet, D., Ferrara, M., Cottin, P., Ducastaing, A., Bigard, X., Guezennec, C.-Y., Schmid, H.-P. and Attaix, D. (1996) Coordinate activation of lysosomal, $Ca^{2+}$-activated and ATP-ubiquitin-dependent proteinases in the unweighted rat soleus muscle. *Biochemical Journal* 316, 65–72.

Tanahashi, N., Murakami, Y., Minami, Y., Shimbara, N., Hendil, K.B. and Tanaka, K. (2000) Hybrid proteasomes. Induction by interferon-gamma and contribution to ATP-dependent proteolysis. *Journal of Biological Chemistry* 275, 14336–14345.

Tawa, N.E. Jr and Goldberg, A.L. (1993) Protein and amino acid metabolism in muscle. In: Engel, A.G. and Franzini-Armstrong, C. (eds) *Myology*. McGraw-Hill, New York, pp. 683–707.

Temparis, S., Asensi, M., Taillandier, D., Aurousseau, E., Larbaud, D., Obled, A., Béchet, D., Ferrara, M., Estrela, J.M. and Attaix, D. (1994) Increased ATP-ubiquitin-dependent proteolysis in skeletal muscles of tumor-bearing rats. *Cancer Research* 54, 5568–5573.

Thrower, J. S., Hoffman, L., Rechsteiner, M. and Pickart, C.M. (2000) Recognition of the polyubiquitin proteolytic signal. *EMBO Journal* 19, 94–102.

Tiao, G., Fagan, J.M., Samuels, N., James, J.H., Hudson, K., Lieberman, M., Fischer, J.E. and Hasselgren, P.O. (1994) Sepsis stimulates nonlysosomal, energy-dependent proteolysis and increases ubiquitin mRNA levels in rat skeletal muscle. *Journal of Clinical Investigation* 94, 2255–2264.

Tiao, G., Fagan, J.M., Roegner, V., Lieberman, M., Wang, J.J., Fischer, J.E. and Hasselgren, P.O. (1996) Energy-ubiquitin-dependent muscle proteolysis during sepsis is regulated by glucocorticoids. *Journal of Clinical Investigation* 97, 339–348.

Tiao, G., Hobler, S., Wang, J.J., Meyer, T.A., Luchette, F.A., Fischer, J.E. and Hasselgren, P.O. (1997) Sepsis is associated with increased mRNAs of the ubiquitin–proteasome proteolytic pathway in human skeletal muscle. *Journal of Clinical Investigation* 99, 163–168.

Tilignac, T., Temparis, S., Combaret, L., Taillandier, D., Pouch, M.-N., Cervek, M., Cardenas, D.M., Le Bricon, T., Debiton, E., Samuels, S.E., Madelmont, J.C. and Attaix, D. (2002) Chemotherapy inhibits skeletal muscle ubiquitin–proteasome-dependent proteolysis. *Cancer Research* 62, 2771–2777.

Turner, G.C., Du, F. and Varshavsky, A. (2000) Peptides accelerate their uptake by activating a ubiquitin-dependent proteolytic pathway. *Nature* 405, 579–583.

Ustrell, V., Hoffman, L., Pratt, G. and Rechsteiner, M. (2002) PA200, a nuclear proteasome activator involved in DNA repair. *EMBO Journal* 21, 3516–3525.

Varshavsky, A. (1996) The N-end rule: functions, mysteries, uses. *Proceedings of the National Academy of Sciences USA* 93, 12142–12149.

Voges, D., Zwickl, P. and Baumeister, W. (1999) The 26S proteasome: a molecular machine designed for controlled proteolysis. *Annual Review of Biochemistry* 68, 1015–1068.

Voisin, L., Breuillé, D., Combaret, L., Pouyet, C., Taillandier, D., Aurousseau, E., Obled, C. and Attaix, D. (1996) Muscle wasting in a rat model of long lasting sepsis results from the activation of lysosomal, $Ca^{2+}$-activated and ubiquitin–proteasome proteolytic pathways. *Journal of Clinical Investigation* 97, 1610–1617.

Waterlow, J.C., Garlick, P.J. and Millward, D.J. (1978) *Protein Turnover in Mammalian Tissues and in the Whole Body*. Elsevier-North Holland, Amsterdam.

Whitehouse, A.S. and Tisdale, M.J. (2001) Down-regulation of ubiquitin-dependent proteolysis by eicosapentaenoic acid in acute starvation. *Biochemical and Biophysical Research Communications* 285, 598–602.

Whitehouse, A.S., Smith, H.J., Drake, J.L. and Tisdale, M.J. (2001) Mechanism of attenuation of skeletal muscle protein catabolism in cancer cachexia by eicosapentaenoic acid. *Cancer Research* 61, 3604–3609.

Wilkinson, K.D. (2000) Ubiquitination and deubiquitination: targeting of proteins for degradation by the proteasome. *Seminars in Cell and Developmental Biology* 11, 141–148.

Williams, A.B., Decourten-Myers, G.M., Fischer, J.E., Luo, G., Sun, X. and Hasselgren, P.O. (1999) Sepsis stimulates release of myofilaments in skeletal muscle by a calcium-dependent mechanism. *FASEB Journal* 13, 1435–1443.

Wing, S.S. and Banville, D. (1994) 14-kDa ubiquitin-conjugating enzyme: structure of the rat gene and regulation upon fasting and by insulin. *American Journal of Physiology* 267, E39–E48.

Wing, S.S. and Bedard, N. (1996) Insulin-like growth factor I stimulates degradation of a mRNA transcript encoding the 14 kDa ubiquitin-conjugating enzyme. *Biochemical Journal* 319, 455–461.

Wing, S.S. and Goldberg, A.L. (1993) Glucocorticoids activate the ATP-ubiquitin-dependent proteolytic system in skeletal muscle during fasting. *American Journal of Physiology* 264, E668–E676.

Wing, S.S., Haas, A.L. and Goldberg, A.L. (1995) Increase in ubiquitin–protein conjugates concomitant with the increase in proteolysis in rat skeletal muscle during starvation and atrophy denervation. *Biochemical Journal* 307, 639–645.

Woo, S.K., Lee, J.I., Park, I.K., Yoo, Y.J., Cho, C.M., Kang, M.S., Ha, D.B., Tanaka, K. and Chung, C.H. (1995) Multiple ubiquitin C-terminal hydrolases from chick skeletal muscle. *Journal of Biological Chemistry* 270, 18766–18773.

Wray, C.J., Tomkinson, B., Robb, B.W. and Hasselgren, P.O. (2002) Tripeptidyl-peptidase II expression and

activity are increased in skeletal muscle during sepsis. *Biochemical and Biophysical Research Communications* 296, 41–47.
Yokota, T., Nagai, H., Harada, H., Mine, N., Terada, Y., Fujiwara, H., Yabe, A., Miyazaki, K. and Emi, M. (2001) Identification, tissue expression, and chromosomal position of a novel gene encoding human ubiquitin-conjugating enzyme E2-230k. *Gene* 267, 95–100.

# 第 4 部分
# 核酸与核酸结合化合物

# 16 膳食、DNA 甲基化与癌症

Judith K. Christman
（内布拉斯加州立大学医学中心生物化学和分子生物学系和 Eppley 癌症中心，
奥马哈，内布拉斯加州，美国）

## 引　　言

限制甲基供体的利用及其他一些一碳单位代谢所需的因素可影响癌症的发生，这在 50 多年前首次在大鼠模型中发现，并被越来越多的以啮齿动物为模型的试验证实。最近的研究发现，胎儿先天性缺陷、免疫缺陷、心血管病、糖尿病和某些智力发育迟钝等许多疾病也可能与甲基供体缺乏有关（Maier and Olek，2002；Issa，2002a；Van den Veyver，2002）。而这当中的某些疾病、膳食甲基供体缺乏与机体 DNA 甲基化状况的改变有关系。

大量研究发现支持，DNA 中胞嘧啶的甲基化在基因表达调节中起着非常重要的作用（Singal and Ginder，1999；Leonhardt and Cardoso，2000）。从细菌的限制修饰系统到哺乳动物细胞的基因表达调控，许多生物品种的研究已很清楚地表明，细胞的基因表达调控中，DNA 中碱基的甲基化可抑制或促进蛋白质与 DNA 的结合。在哺乳动物细胞中，DNA 甲基化主要发生在 CpG 二核苷酸序列中胞嘧啶的第五个碳原子上。在不同细胞和组织的 DNA 中，5-甲基胞嘧啶（5mC）占脊椎动物胞嘧啶总量的 3%～4%；其中，CpG 二核苷酸序列中的胞嘧啶有大约 70%被甲基化（Razin and Riggs，1980；Ehrlich and Wang，1981）。对 DNA 胞嘧啶（C5）甲基转移酶基因敲除小鼠的研究表明，DNA 甲基化是哺乳动物胎儿正常生长发育的关键（Li et al.，1992，Okano et al.，1999），在 X 染色体的失活和印迹中起着至关重要的作用（Goto and Monk，1998；Mann *et al.*，2000）。

关于癌症发生机制的合理学说首先是由 Nowell 于 1976 年提出的。他指出，癌症是由一系列可导致生长异常的基因突变引起的，肿瘤的发展则是具有生长优势［抗药性、高产和（或）对生长信号高敏感性或对肿瘤抑制基因产生的生长抑制信号的低敏感性］的细胞快速特异性生长的结果。在肿瘤发生过程中，胚胎正常生长发育所形成的特异性 DNA 甲基化模式调节异常，并由此导致基因组 DNA 总体甲基化水平的降低以及某些单个基因（包括重复 DNA）部分位点的甲基化缺失。尽管在肿瘤细胞发展过程中 DNA 的 5-甲基胞嘧啶含量总体水平降低，但甲基化程度反而会升高。DNA 高甲基化和低甲基化在基因组 DNA 中同时存在，且分别定位于不同的区域。DNA 甲基化缺失往往发生在卫星 DNA、反转录因子、内源性逆转录病毒的长末端单拷贝重复基因等重复序列中，而这些序列的甲基化缺失往往与基因的不稳定性和肿瘤发生的相关基因的激活有关（Ehrlich，2002）。DNA 高甲基化主要发生在基因组中富含 CpG 二核苷酸序列的区域

(CpG 岛)，而 CpG 二核苷酸序列在脊椎动物基因组中 CpG 岛以外的其他区域通常分布较少（Bird，1986）。CpG 岛的高甲基化往往与一系列肿瘤抑制基因沉默有关。现已确切证实：甲基化介导的后天性基因沉默不仅是先天性癌症发生的“先兆”，也是后天性癌症中肿瘤抑制因子正常等位基因失活的产生原因。

如果 DNA 甲基化调节异常是癌症产生发展的原因，那么以下问题仍有待证实：

(1) 在前肿瘤细胞中已经发生了甲基化变化和（或）在癌症的早期阶段甲基化变化已经存在；

(2) 在癌症发生过程中，当 DNA 甲基化的早期变化不足以引起肿瘤基因表型最佳的表达，则出现额外的甲基化变化和（或）甲基化调控缺失；

(3) DNA 甲基化改变与癌症发生发展之间存在本质联系。DNA 甲基化状态的改变往往容易导致基因突变、染色体缺失等基因改变，或导致癌症发生发展（细胞生长调节、细胞凋亡、细胞迁移）过程中相关蛋白质表达的丧失或放大；

(4) 当单个或多个基因的沉默对肿瘤表型至关重要时，通过地西他滨（2-氮-2-脱氧胞苷，ZdCyd）等抑制 DNA 甲基转移酶活性的药物使用和恰当的外源基因的转入表达，可重新激活已沉默的基因，从而达到恢复细胞的正常生长调节、修复细胞对诱导细胞凋亡药物的敏感性和（或）恢复正常基因表型的表达；

(5) 当特异基因活化或者甲基化缺失过程对肿瘤表型至关重要，那么通过抑制这类基因的表达，将能恢复其他基因的正常表达。尽管目前还无法通过试验手段改变细胞或组织的基因中特定区域的甲基化丧失或放大的模式，但基因沉默可以通过反义 RNA 或反义 DNA 的使用来实现。

在已研究的所有癌症中，家族性腺瘤性息肉病（adenomatous polyposis coli，APC）的发生过程与以上 5 点情况最为接近。在肿瘤形成前的息肉细胞 DNA 中 5-甲基胞嘧啶总含量低于正常结肠上皮细胞，并且这些特定基因癌前期病灶区域表现出极低的甲基化程度。在 Vogelstein 及其同事建立的 APC 线性发生模型中，DNA 低甲基化产生之前的基因变化只有 APC 基因的缺失或突变（Fearon and Vogelstein，1990）。随着该病的发展，DNA 甲基化程度似乎并没有进一步降低（Feinberg *et al*.，1988），许多低甲基化基因的表达也未进一步增加，因而这些基因可能与肿瘤的发生无关（Feinberg and Vogelstein，1983；Goelz *et al*.，1985）；然而，有证据表明在该病的发生发展中有不少基因渐成沉默。本文引言部分的所有内容也在其他刊物上均有相关综述报道。这些报道的部分作者如下：Doerfler（1983）；Riggs 和 Jones（1983）；Christman（1984）；Zingg 和 Jones（1997）；Baylin 等（1998）；Tyeko（2000）；Issa（2000b）；Costello 和 Plass（2001）；Jones 和 Takai（2001）；Reik 等（2001）；Robertson（2001）；Rountree 等（2001）；Esteller 等（2002）；Feinberg 等（2002）。

本章的重点主要集中在研究甲基供体的膳食来源的有效性对 DNA 甲基化的影响机制。参与 DNA 甲基化的酶都能利用“通用的甲基供体”——S-腺苷甲硫氨酸（AdoMet）。AdoMet 既可以直接由甲硫氨酸和腺苷合成，也可以利用来源于一碳单位代谢的甲基来合成。叶酸、甲硫氨酸、胆碱及膳食中某些成分均可为 AdoMet 的合成提供甲基，这些甲基供体的摄入不足可导致机体中 AdoMet 水平及 *S*-腺苷甲硫氨酸与 *S*-腺苷高半胱氨酸（AdoHcy）的比值（AdoMet/AdoHcy）降低，从而抑制 DNA 甲基化进程，从而引起疾病，这种推论是合乎逻辑的。本章将对人类癌症（尤其是结肠直肠癌）的流行病学研究和目前对哺乳动物细胞中 DNA 甲基化相关酶性质的研究进行综述，并将通过已报道的动物试验研究进一步阐述动物甲基供体缺乏、DNA 甲基化异常与癌症（肝癌、结肠癌）之间的关系；讨论目前存在的有关 DNA 甲基转移酶（DNMT）与染色质蛋白之间的相互作用对该酶活性的调节方式以及膳食甲基供体缺乏对这些过程的影响。

## 膳食与细胞甲基化状况

*S*-腺苷甲硫氨酸（AdoMet）是直接由膳食中甲硫氨酸和腺苷在甲硫氨酸腺苷转移酶（methionine adenosyl transferase，MAT；EC 2.5.1.6；图 16.1）作用下合成。甲硫氨酸腺苷转移酶有三种：MATⅠ、MATⅡ和 MAT Ⅲ。MATⅠ和 MAT Ⅲ分别是由 α1 亚基组成的四聚体和二聚体，两者主要在肝脏中表达；而 MATⅡ是由催化亚基（α2）和调节亚基（β）组成的杂二聚体，主要在多数组织和胎儿肝脏中表达（Kotb *et al.*，1997）。AdoMet 合成所需的甲硫氨酸可通过另外两条相互关联的途径合成：由 5-甲基四氢

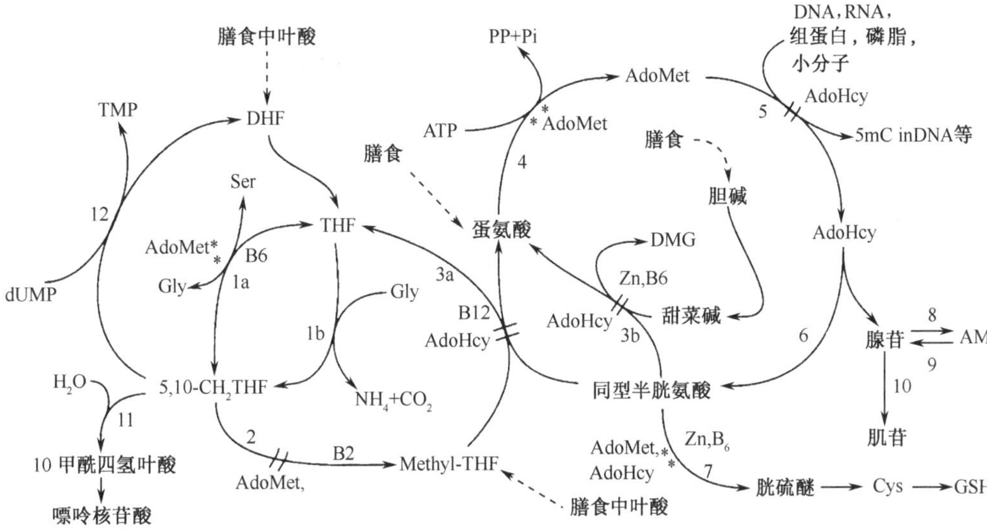


图 16.1 甲基代谢，参与甲基代谢的膳食因素、酶和反应底物

甲基代谢中的酶用数字区别表示：1a. 甘氨酸羟甲基转移酶（GHMT）；1b. 丝氨酸羟甲基转移酶；2. 亚甲基四氢叶酸还原酶（MTHFR）；3a. $N^5$-甲基四氢叶酸-高半胱氨酸-*S*-甲基转移酶（甲硫氨酸合成酶，MS）；3b. 甜菜碱-高半胱氨酸-*S*-甲基转移酶（BHMT）；4. 甲硫氨酸腺苷转移酶（MAT）；5. DNA 甲基转移酶（DNMT）等各种甲基转移酶；6. *S*-腺苷高半胱氨酸水解酶（SAHH）；7. 胱硫醚-*β*-合成酶（CBS）；8. 腺苷激酶（AK）；9. 5-核苷酸酶（5-NT）；10. 腺苷脱氨酶（ADA）；11. 亚甲基四氢叶酸环水解酶；12. 胸苷酸合成酶（TS）。缩写：DHF，二氢叶酸；Ser，丝氨酸；Gly，甘氨酸；Cys，半胱氨酸；THF，四氢叶酸；$B_6$，维生素 $B_6$；$B_{12}$，维生素 $B_{12}$或钴胺素；$B_2$，维生素 $B_2$ 或核黄素；5，10-$CH_2$ THF，$N^5$，$N^{10}$-亚甲基四氢叶酸；Methyl-THF，$N^5$-甲基四氢叶酸；Zn，锌；DMG，二甲基甘氨酸；AdoMet，*S*-腺苷甲硫氨酸；AdoHcy，*S*-腺苷高半胱氨酸；GSH，还原型谷胱甘肽；ATP，三磷酸腺苷；Pi，无机磷酸；PP，焦磷酸；TMP，一磷酸胸苷；dUMP，一磷酸脱氧尿苷；AMP，一磷酸腺苷；//，抑制；**，激活

叶酸（5-methyltetrahydrofolate，5MTHF）提供甲基，在叶酸依赖性的高半胱氨酸：甲硫氨酸合成酶（MTR，又称甲硫氨酸合成酶，MS；EC 2.1.1.13）作用下将高半胱氨酸甲基化而转变成甲硫氨酸；或是由胆碱氧化产物——甜菜碱提供甲基，在甜菜碱-高半胱氨酸 *S*-甲基转移酶（betaine-homocysteine *S*-methyltransferase，BHMT；EC 2.1.1.5）作用下，将高半胱氨酸甲基化而转变成甲硫氨酸。MTR 可在哺乳动物的所有组织中表达且具有活性。BHMT 有两种类型：BHMTⅠ和 BHMTⅡ，两者均可在动物肝脏和肾脏中表达。然而，不同于 BHMTⅠ，BHMTⅡ还可在成年动物的脑、心、骨骼肌以及胎儿组织（包括心脏、肝脏、肺、肾脏和眼）中低水平表达（Chadwick *et al.*，2000）。

机体所必需的甲基主要来源于膳食中的胆碱（约 30mmol 甲基/d）和蛋氨酸（约 10mmol 甲基/d），其余甲基（5～10mmol 甲基/d）可由提供一碳单位的丝氨酸和甘氨酸通过单碳代谢来提供（美国国家科学院和医学研究院，2000）。在后者反应过程中，甘氨酸羟甲基转移酶（glycine hydroxymethyltransferase，GHMT）可将丝氨酸中 β-碳原子转移至 5-亚胺甲基四氢叶酸中形成 $N^5$,$N^{10}$-二亚甲基四氢叶酸（5,10MTHF），而 GHMT 需要四氢叶酸（由叶酸转变）和维生素 $B_6$（吡哆醇）作为辅酶。此外，在机体的单碳代谢中，亚甲基四氢叶酸还原酶（methylenetetrahydrofolate reductase，MTHFR）需要维生素 $B_2$（核黄素）作为辅酶，甲硫氨酸合成酶需要维生素 $B_{12}$（钴胺素）作为甲基传递的甲基化中间体，甜菜碱-高半胱氨酸 *S*-甲基转移酶（BHMT）需要 $Zn^{2+}$ 作为辅酶。以上营养素（叶酸、维生素 $B_6$、维生素 $B_2$ 和维生素 $B_{12}$）均可从膳食中获取。

饮酒也会对一碳单位代谢产生不利影响（Finkelstein *et al.*，1974），长期过量饮酒往往会导致叶酸吸收不良，进而引发机体叶酸缺乏。同时，饮酒还会减少肝脏对叶酸的吸收和释放及增加尿中叶酸的排泄（Weir *et al.*，1985）。此外，乙醇的代谢产物——乙醛可使 5MTHF 失去活性并抑制 MTR 的活性（Kenyon *et al.*，1998；Homann *et al.*，2000）。MTR 受到抑制会导致叶酸以 5MTHF 的形式累积，且抑制 5,10MTHF 的重新合成，进而阻碍核苷酸的合成。同时，5MTHF 因不适宜作为多聚谷氨酸合成反应的底物而影响叶酸在细胞中的储留，最终导致机体细胞中叶酸的缺乏（Shane，1995）。

尽管本章节着重讨论膳食对机体 DNA 甲基化的影响，但膳食甲基供体缺乏对于癌症及其他人类疾病产生相关的其他三方面的影响也不容忽视：①对核苷酸代谢和 DNA 合成的影响；②对高半胱氨酸在体内蓄积的影响；③ AdoMet 缺乏对 DNA 以外的其他生物分子修饰的影响。

## 嘌呤代谢与 DNA 合成

5,10MTHF 参与嘌呤代谢和 DNA 合成。5,10MTHF 既可直接氧化成甲酰四氢叶酸参与嘌呤合成，又可将甲基直接传递给脱氧尿苷-磷酸（deoxyuridine monophosphate，dUMP）形成胸苷-磷酸（thymidine monophosphate，TMP）参与 DNA 合成。而后者是哺乳动物细胞 DNA 合成过程中的限速反应（Blount *et al.*，1997）。若 dUMP 不能甲基化为 TMP，则 dUMP 也会直接参与 DNA 合成（Pogribny *et al.*，1997），其严重后果将导致由糖基化酶介导的 DNA 修复过程中 DNA 尿嘧啶的清除及通过 DNA 修复的重复循环重新引入尿嘧啶，而这一过程是一个反复的“无效循环”。DNA 修复过程中发生的 DNA 链瞬间断裂可能会导致不可修复的 DNA 损伤、染色体结构的不稳定以及DNA低甲基化（Ryan and Weir，2001）。此外，细胞中AdoMet水平降低，也可导致DNA修复过程中合成的DNA片段甲基化无效。目前，DNA链断裂和DNA低甲基化的同时增加已经从各种叶酸缺乏的培养组织和动物模型（Choi，1999；James and Yin，1989；Kim *et al.*，1995，1997）以及叶酸缺乏的人白细胞中观察到（Jacob *et al.*，1998）。

## 高半胱氨酸

膳食甲基供体和维生素（胆碱、甲硫氨酸、叶酸、维生素$B_6$和维生素$B_{12}$）缺乏可导致高半胱氨酸在机体内的大量累积。*S*-腺苷高半胱氨酸水解酶（*S*-adenosylhomocysteine hydrolase，SAHH）催化的反应是可逆的，既可将*S*-腺苷高半胱氨酸（AdoHcy）水解成高半胱氨酸和腺苷，也可将半胱氨酸和腺苷合成为AdoHcy。在正常生理条件下，机体倾向于AdoHcy的合成，但AdoHcy的继续合成可被以下三种不可逆反应所阻碍：腺苷被腺苷脱氨酶转化为肌苷；高半胱氨酸被甲基转移酶（MTR）甲基化为甲硫氨酸的过程再合成AdoMet；高半胱氨酸直接进入转硫途径。在转硫途径中，维生素$B_6$依赖性胱硫醚-$\beta$合成酶（cystathionine β synthase，CBS）可利用丝氨酸将高半胱氨酸转变成胱硫醚，而胱硫醚在维生素$B_6$依赖性$\gamma$-胱硫醚酶（γ-cystathionase，GCT）作用下可分解成2-酮戊二酸和半胱氨酸。由于MTR（0.1mmol/L）和CBS（1mmol/L）对高半胱氨酸的$K_m$值之间的存在差异，使得与MTR相关的再甲基化途径在高半胱氨酸水平低的条件下起主导作用，而与CBS相关的转硫途径则在高半胱氨酸高水平降解中起主导作用。正由于以上两种反应为不可逆过程，机体高半胱氨酸水平才被严格控制（Finkelstein，1998，2000）。而高半胱氨酸的调节至关重要，因为AdoHcy是大多数利用AdoMet的甲基转移酶的竞争性抑制剂。然而，膳食中叶酸和维生素$B_{12}$（可能还有维生素$B_6$）缺乏可导致机体高半胱氨酸水平的升高，而当细胞中腺苷充足时，SAHH可促进AdoHcy的合成，从而抑制许多重要生物底物（包括DNA）的甲基化过程。此外，高半胱氨酸可影响细胞的生长和凋亡，并且通过细胞内蛋白质的高半胱氨酸化、活性氧的产生和谷胱甘肽水平的降低介导细胞损伤（见Medina *et al.*，2001）。研究还发现，高半胱氨酸可作为*N*-甲基-D-天冬氨酸（*N*-methyl-D-asparate，NMDA）型谷氨酸受体的拮抗剂，从而改变细胞的信号传导途径（Rosenquist *et al.*，1999）。高半胱氨酸还能细胞特异性地影响转录因子AP-1与DNA识别位点的结合过程（Torres *et al.*，1999）。因此，高半胱氨酸水平的升高可能会诱发心血管疾病、阿尔茨海默病及其他神经退化性疾病，并且能提高后天性生长发育异常、妊娠紊乱和癌症的发生率。

高半胱氨酸水平的调节过程、AdoMet的合成过程以及膳食一碳单位供体在一碳单位代谢中的相互作用都非常复杂，使得膳食甲基供体缺乏导致人类疾病发生的致病机制更难于揭示。然而，研究已发现，AdoMet和AdoHcy在调节从一碳循环到甲硫氨酸的甲基传递过程以及调节高半胱氨酸代谢中起着至关重要的作用（Finkelstein，1990）。AdoMet是亚甲基四氢叶酸还原酶（MTHFR）的别构抑制剂，而AdoHcy可解除AdoMet的这种抑制作用；胱硫醚-$\beta$合成酶（CSB）可被AdoMet和AdoHcy激活；AdoHcy可通过与高半胱氨酸（而不是甜菜碱）的竞争来抑制甜菜碱-高半胱氨酸*S*-甲基转移酶（BH-MT）的活性。在低水平AdoMet情况下，5甲基四氢叶酸（5MTHF）的合成处于最优水平，高半胱氨酸水平升高，这主要是因为 CBS 活性的降低促进了甲硫氨酸和 AdoMet 的合成。而在高水平 AdoMet 情况下，5MTHF 的合成受到抑制，甲硫氨酸水平下降，过多的高半胱氨酸被活化的 CBS 所消除。由于 CBS 仅在肝、胰、肾和脑中表达，大多数组织中过量高半胱氨酸主要通过细胞将它们释放到血液再由肾等器官将其清除。

## 作为通用甲基供体的 *S*-腺苷甲硫氨酸

*S*-腺苷甲硫氨酸（AdoMet）是机体生化反应中最重要的甲基供体，其参与甲基代谢反应的数量远高于不同形式的叶酸（Fauman，1999），认识这一点非常重要。例如，在哺乳动物中，5MTHF 作为甲基供体仅被 MTR 利用，而 AdoMet 则可被约 100 种酶利用，这可能是因为 AdoMet 中带电硫原子的存在极大地降低了甲基传递至各种受体的热力学反应屏障（活化能）。据估计，AdoMet 作为甲基供体对可极化的亲核物质（N、O 和 S）的反应活性比甲基化的叶酸高出三个数量级。值得注意的是，AdoMet 也可作为腺苷基和氨基丁酰侧链的供体，这也表现出较高的反应活性。对于哺乳动物来讲，约 10％的 AdoMet 经脱羧反应后，为多胺的合成提供丙胺基（Eloranta and Raina，1977）。胍基乙酸甲基化成肌酸酐的过程在肝中需要消耗 80％以上由 AdoMet 提供的甲硫氨酸（Mudd and Poole，1975；Im *et al.*，1979）。AdoMet 的另一个主要用途是参与膜磷脂的合成，即通过磷酸乙醇胺甲基转移酶（PEMT；EC 2.1.1.17）将磷酸乙醇胺（phosphoethanolamine，PE）甲基化为磷脂酰胆碱（phosphatidylcholine，PC）。除此之外，AdoMet 还参与 DNA、tRNA、rRNA、mRNA 等核酸、组蛋白等核蛋白以及与信号转导和神经传导相关的各种蛋白质和小分子物质的甲基化过程（Fauman，1999）。因此，即使在与膳食甲基供体缺乏相关的特殊疾病中出现了 DNA 甲基化的明显改变，也不足以证实 DNA 甲基化的改变是主要的致病原因。若要证实 DNA 甲基化的改变是主要的致病原因，至少需要通过试验动物不断重现这类疾病，并且特异性地显现抑制 DNA 甲基转移酶活性而不是抑制总的甲基代谢。

## 哺乳动物 DNA 胞嘧啶（C5）甲基转移酶：结构、功能及其作用机制

DNA(C5) 甲基转移酶（MTase）的催化机制包括：MTase 活性位点中的半胱氨酸残基和 DNA 胞嘧啶中第 6 个碳原子之间形成共价键，促使电子传递至 DNA 胞嘧啶中第 5 个碳原子，再对 AdoMet 上的甲基进行攻击，随后 MTase 夺取 DNA 胞嘧啶中第 5 个碳原子上的质子，通过 β-消减反应导致 DNA 胞嘧啶中第 5、6 碳原子之间双键的重新形成和 DNA MTase 的释放（图 16.2）（Santi *et al.*，1984）。因为哺乳动物的 DNA MTase 与细菌的 DNA(C5) MTase（M. *Hha* Ⅰ）都含有一套相同的保守的催化结构域基序[①]，它们可能具有相同的甲基转移催化机制（Lauster *et al.*，1989；Posfai *et al.*，1989）。这些催化基序包括 DNA 甲基转移酶多肽链中Ⅳ、Ⅰ、Ⅹ和Ⅵ区等。Ⅳ区是由脯氨酰半胱氨酰二肽组成的活性位点，Ⅰ区和Ⅹ区是一群分开的结构域，它们共同组成

①基序：指 DNA 上普遍存在的、具有共同结构特征的一段保守序列。

AdoMet 的结合袋，Ⅵ区包含使胞嘧啶第 3 个氮原子发生质子化的谷氨酸残基。DNA 甲基转移酶多肽链中与 DNA 大沟碱基特异性结合的识别区域通常位于其Ⅷ区和Ⅸ区之间。对于 M. *Hha*Ⅰ晶体结构的研究发现，Ⅳ区的脯氨酰半胱氨酰二肽中半胱氨酸残基与 DNA 中甲基化的 5-氟胞嘧啶能发生不可逆的共价结合，结合后 5-氟胞嘧啶滑出 DNA 双螺旋结构进入甲基转移酶的催化部位。在胞嘧啶中，其第 5 位碳原子与 AdoMet（与 DNA 甲基转移酶结合的）相互作用，其第 6 位碳原子与 DNA 甲基转移酶活性位点的半胱氨酸中巯基相互作用，一起协同攻击使甲基发生转移（Santi *et al.*，1984）。

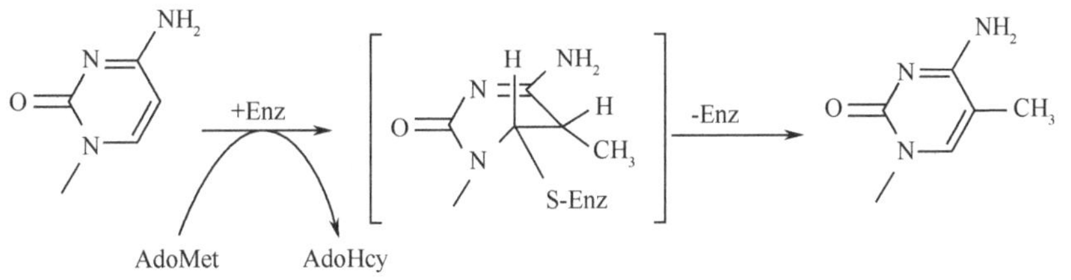


图 16.2　在含有胞嘧啶的目标 DNA 甲基化过程中中间产物 5,6-二氢嘧啶的形成

在哺乳动物细胞中，人们已鉴定出四种不同的编码蛋白（DNMT1，DNMT2，DNMT3a 和 DNMT3b），它们都含有 DNA(C5) MTase 所独有的保守基序。DNMT2 未被检测出 MTase 的活性，但是纯合的 *DnmtT*2 基因敲除的小鼠也未检测到任何生长发育缺陷（Okano *et al.*，1998b；Yoder and Bestor，1998）。此外，目前还发现一种与

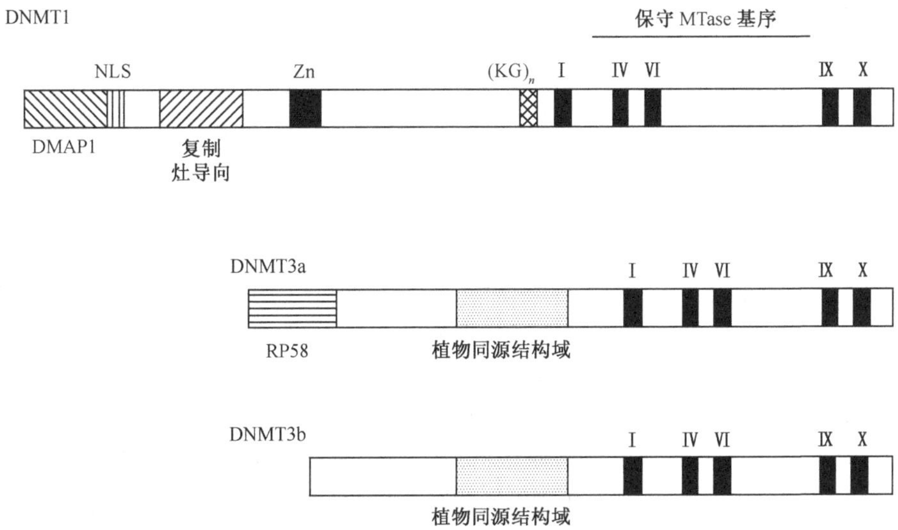


图 16.3　具有催化活性的哺乳动物 DNA 甲基转移酶的结构

图中所列三种酶均含有调节区域和催化区域。其中罗马数字表示的区域为催化区域中保守基序，且在正文中有说明。DNMT1 的另一结构特点是它含有核定位区域、复制灶和锌结合的导向区域。许多蛋白质结合区域也被发现，包括连接核基质的蛋白（p23，Zhang and Verdine，1996）、膜联蛋白Ⅴ（又称锚定蛋白）或转录抑制因子（HDAC1，Fuks *et al.*，2000；HDAC2-DMAP1，Rountree *et al.*，2000；pRb/DMAP1，Robertson *et al.*，2000）及导向 DNA 复制位点和修复位点的蛋白因子（PCNA，Chuang *et al.*，1997）的结合域。DNMT3a 和 DNMT3b 均具有富含半胱氨酸的植物同源结构域（PHD）（Aasland *et al.*，1995）及与 HDAC1 相互作用的转录抑制子结合的调节结构域（Bachman *et al.*，2001；Fuks *et al.*，2001）

DNA(C5) MTase 相似的蛋白质——DNMT3L，它与 DNMT3a 密切相关，具有相似的比较保守的Ⅰ、Ⅳ和Ⅵ区，但缺乏活性位点的结构域（Aapola *et al.*，2000）。然而，尽管 DNMT3L 缺乏 DNA（C5）MTase 活性，但 DNMTL 基因的分裂导致在生长发育过程中印迹区域的 DNA 甲基化异常（Bourc'his *et al.*，2001）。这可能暗示在基因印迹过程中需要 DNMT3a 与 DNMT3L 的相互作用来促进从头开始的 DNA 甲基化（Chedin *et al.*，2002）。以下对已知具有酶活性的哺乳动物 DNMT 的特性进行概述（图 16.3）。

## DNA 甲基转移酶 1（DNMT1）

DNMT1 是哺乳动物第一个被克隆的 DNA MTase（Bestor *et al.*，1988），也是哺乳动物细胞中含量最丰富、研究最透彻的 DNA MTase。DNMT1 是由定位于染色体 19p13.2 中单一基因编码的，其可供选择的外显子 1 中具有 3 个启动子，且都可整合到同一基因下游外显子 2 中。最远端启动子——1o 仅在卵母细胞中表达，由第四外显子中 AUG 翻译产生 DNMT1 的最短活性片段（170kDa）。而该酶主要定位于细胞质中，只有在细胞分裂的八细胞阶段才进入细胞核中，这暗示该酶对维持基因印迹过程中正常的 DNA 甲基化模式是非常必要的。第二个启动子——1s 引起 DNMT1 mRNA 在机体所有组织中表达（Mertineit *et al.*，1998）。虽然含有体细胞外显子 1 的 mRNA 具有 4 个潜在的 AUG 转录起始位点，但转录起始在第三个 AUG 效率最好，能翻译表达出 184kDa DNMT1 蛋白（Pradhan *et al.*，1999）。其他启动子位于基因第三个 AUG 上游的第 4、5 外显子之间（Bigey *et al.*，2000），含有分子质量更大的体细胞 DNMT1 的拼接变体。这些 DNMT1 的拼接变体含有 DNMT1 所没有的由 18 个氨基酸组成的序列，而该序列是由位于外显子 4、5 之间的 48 个核苷酸组成的 Alu 重复序列所编码（Hsu *et al.*，1999；Bonfils *et al.*，2000）。由第 4 个 AUG 上游基因编码的某些氨基酸的缺失对 DNMT1 催化活性没有显著影响，且 DNMT1 变体对其体内催化活性的影响机制还不清楚。此外，邻近于外显子 2 的启动子——1p 则仅仅在处于减数分裂的粗线期的初级精母细胞及分化的肌管细胞中表达（Mertinet *et al.*，1998）。

尽管由 500 个氨基酸组成的 DNMT1 C 端的结构域与细菌Ⅱ型 DNA(C5) MTase 具有相似的结构，但两者还存在以下几点不同：其一，DNMT1 的催化效率受 DNA 结构的影响（见下文），但其 DNA 特异性识别序列仍为 CpG 二核苷酸序列，而细菌Ⅱ型DNA(C5) MTase 至少需要由 4 个碱基组成的 DNA 识别序列；其二，DNMT1 多肽链 N 端有一个巨大的调节区域，其中含有与细胞复制体系相关的基序（图 16.3），然而，在细菌 DNMT1 中则缺乏这些基序。这些基序包括使蛋白质定位于复制灶的序列、与增殖细胞核抗原（proliferating cell nuclear antigen，PCNA）相互作用的区域及类似的多聚溴-1 蛋白区域，该多聚溴-1 蛋白也能在复制灶与蛋白质发生相互作用（Leonhardt *et al.*，1992；Chuang *et al.*，1997；Liu *et al.*，1998；Rountree *et al.*，2000）。此外，DNMT1 还存在一些区域，它们能通过与 pRB、E2F、DMAP1（新发现的抑制因子）、组蛋白脱乙酰基酶 HDAC1 和 HDAC2 等相互作用，以抑制转录（Robertson，2001）。在体外试验中，与 pRB 的互作可抑制 DNMT1 与 DNA 的结合，并降低其催化活性（Pradhan and Kim，2002）。DNMT1 调节区功能域的结构与维持 DNA 甲基化的主要功能相一致。DNMT1 的 N 端可与特异核蛋白相互作用，以保证其在适当的时间和地点组持 DNA 甲基化反应并直接作用于结合

HDAC 的蛋白质，以抑制 DNA 在该区域的转录（见下文）。

目前还不确定 DNMT1 是否也是一种从头开始的甲基转移酶。在体外试验中，纯化的 DNMT1 或富含 DNMT1 的核提取物作为底物对半甲基化的 DNA 表现出高度的亲合性(5～100 倍)。另外，纯化的完整 DNMT1 在体外可对未甲基化的 DNA 底物进行彻底甲基化，而这一功能在摩尔基础上与已知的从头开始的 DNMT（DNMT30 和 DNMT36）一致（Okano *et al.*，1998a)。然而，在体内尚未检测到 DNMT1 将从头开始的 DNMT 的底物——复制附加体（游离基因）甲基化（Hsieh，1999)。与各种用于体外试验研究的 DNA 底物之间的明显互作相比，这种差异极可能反映为 DNMT1 与存在于染色质中正在复制和没有复制的 DNA 区域中未甲基化的 CpG 位点之间相互作用的差异（Bestor，1992；Christman *et al.*，1995；Pradhan *et al.*，1999；Brank *et al.*，2002)。在这方面，目前的报道还表明，DNMT1 与甲基化的 DNA 结合蛋白（MBD2 和 MBD3）组成的复合体可特异性地与半甲基化的 DNA 结合（Tatematsu *et al.*，2000)；这一特性能进一步抑制 DNMT1 对体内完全未甲基化位点的重新甲基化，但能促进对新复制的半甲基化 DNA 进行甲基化。

## DNA 甲基转移酶 3a 和 DNA 甲基转移酶 3b

DNMT3a 和 DNMT3b 是对细菌Ⅱ型 DNA 甲基转移酶全序列进行 TBLASTN 研究时发现的（Okano *et al.*，1998a)。这两种酶基因分别定位于染色体 2p23 和 20q11.2。与 DNMT1 相似，DNMT3a 和 DNMT3b mRNA 存在大量的拼接变体和可变上游外显子（Chen *et al.*，2002；Weissenberger *et al.*，2002)。DNMT3a 和 DNMT3b mRNA 编码的最长蛋白质分别含有 908 个和 859 个氨基酸。mRNA 的可变剪接往往会导致比较保守的酶催化结构域的缺失，而由移码突变产生的新序列插入则可导致两个短链的 DNMT3b4 和 DNMT3b5 的产生。其中，DNMT3b4 可能是 DNA 甲基化的负性抑制因子（Saito *et al.*，2002)。在体外，DNMT3a、3b1 和 3b2 对半甲基化位点和完全未甲基化位点的亲和力没有差异，但都具有与 DNMT1 对非甲基化位点相似的催化效率。这暗示着，DNMT3a 和 DNMT3b 在体外对 DNA 半甲基化位点的甲基化效率要比 DNMT1 至少低一个数量级，因此，这两种酶不可能对维持基因组整个 DNA 甲基化模式起重大作用。此外，DNMT3a 和 DNMT3b 还可对 DNA 中除 CpG 以外的其他胞嘧啶（如 CpA）进行甲基化，但它们对 CpA 的甲基化效率低于 CpG（Gowher and Jertsch，2001；Aoki *et al.*，2001)。DNMT1 不能使 DNA 中除 CpG 外的胞嘧啶进行甲基化，也不能维持其甲基化状态（Wigler *et al.*，1981)，所以这些位点的甲基化状态若得不到其他 DNA 甲基转移酶的维持则会消失。然而，非 CpG 位点的甲基化可能在甲基化的扩展中发挥重要作用，因为形成链内 5mCpA/CpG 位点的环化 DNA 结构是 DNMT1 和 DNMT3b 的最佳底物，可与远处的 5mCpA 位点形成半甲基化的 CpG（Christman *et al.*，1995；Pogribny *et al.*，2000；Farrell and Christman，未发表数据)。研究已表明，DNMT1 和 DNMT3a 与 DNMT3b 之间具有协同效应，它们通过 N 端结构域发生相互作用（Fatemi *et al.*，2001；Liang *et al.*，2002；Rhee *et al.*，2002)。然而，在一种结肠直肠肿瘤细胞系中，*Dnmt1* 或 *Dnmt3b* 基因的敲除并未导致基因组总体 DNA 甲基化水平显著降低，但同时敲除 *Dnmt1* 和 *Dnmt3b* 基因则导致重复 DNA 序列的甲基化水平降低及 p16$^{INK4}$基因沉默或印迹的缺失（Rhee *et al.*，2002）。

DNMT3a 和 DNMT3b 的 N 端不具有任何与 DNMT1 所含有的复制点导向结构域同源的结构域（Leondardt *et al.*，1992；Liu *et al.*，1998），因而它们不会与 DNMT1 和 PCNA 发生作用而定位于增殖细胞核的复制灶中（Chuang *et al.*，1997）。然而，DNMT3a 的编码基因定位于外周着丝粒 DNA 中一些分散的复制灶，而这些复制灶随后将开始异染色质的复制（Bachman *et al.*，2001）。同时 DNMT3b 的编码基因也定位于 DNA 中一些分散的位点，虽然 DNMT3b 中指导其定位的结构域尚未被发现，但研究已证实，在 DNMT3a 和 DNMT3b 中存在的 ATRX 类似的富含半胱氨酸的结构域可能通过结合 HDAC 来抑制基因转录（见下文）。

## 被动与主动的 DNA 去甲基化作用

目前，我们对生长发育期间 DNA 甲基化的建立和改变机制的研究才刚刚起步，虽然参与从头甲基化反应的候选蛋白质已被分离鉴定，但基础理论正在形成中，尚不成熟。这些基础理论的形成将促进人们对导致基因激活或沉默的 DNA 甲基转移酶与核蛋白之间相互作用进行进一步研究。然而，即使 DNA 胞嘧啶残基的甲基化大量损耗已被确认是发生在哺乳动物胚胎发育和肿瘤发生的早期，但 DNA 5-甲基胞嘧啶（5-methyl-

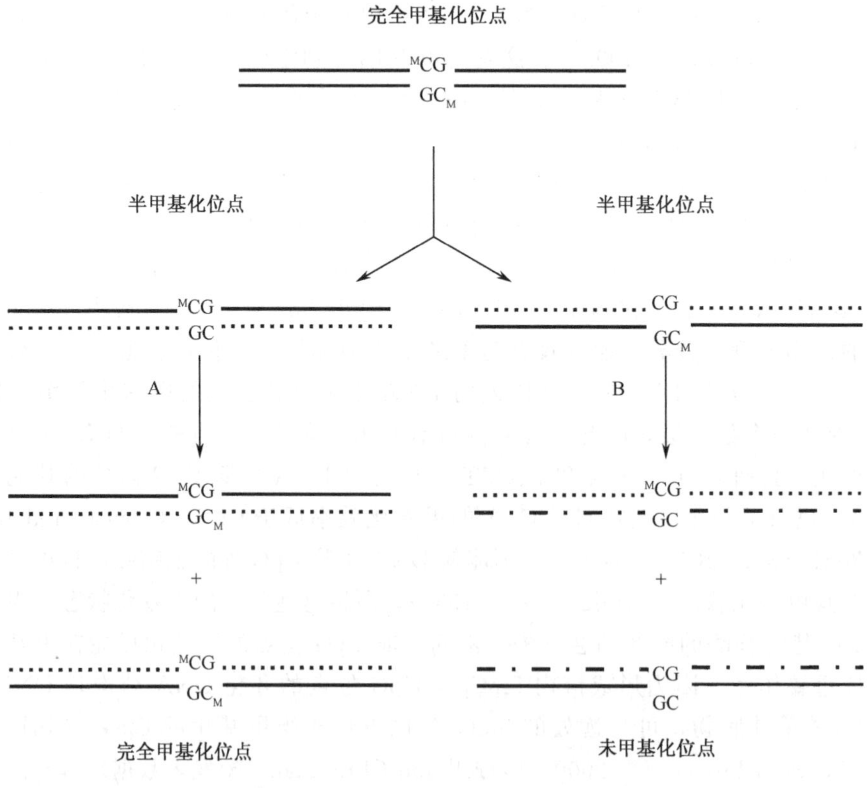


图 16.4　在 DNA 复制过程中 DNA 上的完全甲基化位点的变化情况

A. DNA 复制后，半甲基化位点被维持性 DNA 甲基转移酶 1（DNMT1）重新甲基化，以恢复原有的 DNA 甲基化形式；B. 一条子代 DNA 链的被动去甲基化。这是因为 DNMT1 未能对第一次 DNA 复制后产生的半甲基化位点进行重新甲基化。这种现象往往发生于 ZCyd 或 ZdCyd 处理后的细胞中，由于在 DNA 的 CpG 位点 ZCyt 残基与 DNMT1 共价结合使 DNMT1 失活

cytosine，5mC）的去除机制仍尚不清楚。当 DNMT1 的数量不足以维持 DNA 甲基化状态，或其活性被外源性或内源性抑制剂抑制时，DNA 5mC 的损耗可能是一种被动机制（图 16.4）。正如图 16.4 中 b 途径所示，在没有维持甲基化的情况下至少需经两次 DNA 复制才能保证在 50％新合成的 DNA 链中特定位点完成去甲基化。因此，由被动机制介导的 DNA 广泛去甲基化过程需不断进行 DNA 复制及 DNA 修复，且该过程不能维持原有 DNA 的甲基化状态。如果 DNA 合成速率超过有效 DNMT1 的甲基传递能力，那么被动去甲基化可能是快速生长的肿瘤细胞中 DNA 甲基化模式改变的原因。然而，在大多数处于细胞周期 S 时段的肿瘤细胞中，DNMT1 的表达仅轻微的升高（2～3 倍），且其表达速度与其他基因的表达相一致（Lee *et al.*，1996；Eads *et al.*，1999），因此，DNA 的被动去甲基化更可能是由于细胞中 AdoHcy 和 AdoMet 绝对或相对水平的改变造成的。此外，DNMT1 与 pRB 或其他蛋白质的相互作用可抑制其与 DNA 结合，阻碍其定位于复制灶，降低其催化活性，最终导致分裂细胞中 DNA 的被动去甲基化（Pradhan and Kim，2002）。

目前对小鼠胚胎发育早期 DNA 去甲基化现象的研究表明，母源染色体中 DNA 去甲基化是在发育中的受精卵经多次分裂后才逐步发生的。Mayer 等（2000）在小鼠受精卵第一次分裂期间就观察到父源染色体中 5mC 是部分（而不是完全）去甲基化，并认为机体中存在 DNA 中 5mC 主动去甲基化机制，而这也被未分裂细胞中 DNA 去甲基化的研究所证实（Jost *et al.*，2001；Pradhan and Kim，2002）。目前对完全甲基化 DNA 和半甲基化 DNA 的主动去甲基化酶的性质已经清楚（Gallinari and Jiricny，1996；Bird and Wolffe，1999；Jost *et al.*，2001），此外，DNA 中 5mC 也可通过氧化脱氨基转变成胸苷并经 G∶T 错配修复而去甲基化（Hardeland *et al.*，2001）。但目前尚未有证据表明这些酶参与父源染色体 DNA 的主动去甲基化过程。另外，尽管有研究表明甲基化 DNA 结合蛋白可参与 DNA 胞嘧啶残基上 C5 的主动去甲基化（Bhattacharya *et al.*，1999），但这仍有待于进一步证实（Magdinier and Wolffe，2001）。

# 膳食甲基供体缺乏与人类癌症关系流行病学表征

大量的例证表明，富含水果蔬菜的膳食能减少癌症（乳腺癌、子宫癌、卵巢癌、前列腺癌、睾丸癌和肾癌）的发生率（美国癌症研究所，AICR，1997）。由于豆类、十字花科蔬菜、一些非十字花科蔬菜、某些水果和谷物等食物也含有丰富的叶酸，叶酸与癌症预防之间可能存在某些联系。然而，根据 1995 年以前的研究资料，美国癌症研究所（AICR）的研究小组在《食物、营养与癌症预防》报告中得出结论：仅有结肠癌才体现出膳食甲基供体缺乏与癌症发生之间的关系。

## 膳食中的叶酸与结肠癌

正如许多综述中详细说到的，膳食叶酸缺乏可增加结肠腺瘤的发生率，而这往往是结肠癌发生的早期症状（Giovannucci and Willett，1994；Giovannucci *et al.*，1995；Garay and Engstrom，1999；Tomeo *et al.*，1999；Potter，1999；Fuchs *et al.*，2002）。在“医药自由取业者追踪调查项目”（the Health Professionals Follow-up）和“护士健

康研究项”（the Nurses Health study）中，通过摄入高水平和低水平叶酸人群的对比，并经过对年龄、体重指数、结肠癌家族病史以及摄食的总能、饱和脂肪、纤维等水平的矫正，女性和男性的腺瘤的相对发生率分别为 0.66 和 0.44（Giovannucci *et al.*，1993）。尽管男女的相对发生率发生了一些变化，但在西班牙、法国和美国进行的另外 4 个试验均表明，叶酸摄入量与结肠癌腺瘤的发生存在显著的负相关（Benito *et al.*，1993；Bird *et al.*，1995；Boutron-Ruault *et al.*，1996；Tseng *et al.*，1996）。

研究已证实，叶酸的摄入量与结肠癌的发生率之间存在负相关。五项个案控制研究项目表明，高水平叶酸降低了女性结肠癌的发生率（Benito *et al.*，1991；Freudenheim *et al.*，1991；Ferraroni *et al.*，1994；Slattery *et al.*，1997；White *et al.*，1997），但其中两项研究对男性的病例对照研究未发现类似的结果（Slattery *et al.*，1997；White *et al.*，1997）。在第六项研究中未发现叶酸摄入与结肠癌发生之间存在着必然联系（Boutron-Ruault *et al.*，1996）。另外五项前瞻性研究中，有两项研究表明膳食中叶酸摄入与男性结肠癌发生之间存在显著负相关，但这种关系在女性中尚未表现出来（Glynn *et al.*，1996；Su and Arab，2001）。第三项研究发现，长期摄入多种维生素降低了男性结肠癌的发生率，但是叶酸的总摄入量与男性结肠癌的发生率之间不存在显著的相关关系（Giovannucci *et al.*，1995）。第四项研究表明，每天摄入至少 400mg 叶酸或长期摄入多种维生素显著降低了结肠癌的发生率。另外，Kato 等（1999）在另一研究中发现，出现症状前的血浆叶酸水平与女性结肠癌的发生率之间存在显著负相关，同时，Ma 等（1997）也发现基准的血浆叶酸水平与男性结肠癌的发生率之间的相关关系。

有趣的是，许多研究表明，与高叶酸低乙醇相比，摄入低叶酸高乙醇大大提高了人结肠癌的相对发生率（Benito *et al.*，1991；Giovannucci *et al.*，1993，1995；Ferraroni *et al.*，1994；Boutron-Ruault *et al.*，1996；White *et al.*，1997）。高乙醇摄入导致男性和女性结肠癌的相对发生率范围分别为：2.67～5.07 和 1.2～4。如前所述，其主要原因可能是乙醇影响了叶酸吸收和甲基代谢。此外，饮酒者结肠黏膜中大量的乙醇经结肠细菌发酵后往往会产生大量的乙醛，这也可能是乙醇诱导结肠癌发生的另一重要原因（Salaspuro，1996）。研究已表明，乙醛的摄入使大鼠结肠黏膜中叶酸水平下降了 48%（Homan *et al.*，2000），而乙醇的长期摄入导致了机体 DNA 低甲基化（Choi *et al.*，1999）。

## 膳食中的叶酸与其他癌症

有关叶酸与宫颈癌之间的关系，人们已进行了大量的研究，但 AICR 专家组总结认为，这些研究结果表明叶酸与宫颈癌之间不存在相关关系（AICR，1997）。例如，最有说服力的例证是一项庞大的涉及多个民族案例对照的研究。该项目检测了 245 位已从组织学上确诊为入侵性宫颈癌的妇女和 545 位正常妇女的血液样品，并通过问卷调查对致病因素进行了评估。结果表明，与叶酸的最高摄入量组相比，叶酸最低的摄入量组提高了入侵性宫颈癌发生的可能性但提高幅度不显著（Weinstein *et al.*，2001a）。

目前对膳食甲基供体缺乏与癌症发生相关性研究的缺陷之一是，叶酸摄入水平一般是通过对人膳食的回想进行问卷调查来评定，这种评定方法可能相当不准确。由于膳食中叶酸存在形式的不同、叶酸的不稳定性以及叶酸生物利用率的差异，被调查者的叶酸摄入水平难以准确定量。同时，以血浆或红细胞的叶酸水平作为膳食叶酸摄入量的评定指标也受各种可用的测定方法固有缺陷的影响，而且它们不能真实反映敏感组织的叶酸水平（Gunter，1996）。最近研究发现，血清高半胱氨酸水平作为机体叶酸状况的评定指标更为敏感。Weinstein 等（2001b）通过对 183 名宫颈癌患者和 540 名正常人的调查研究表明，血清高半胱氨酸水平与宫颈癌的发生率呈显著正相关（相对比例 RR 为 2.4～3.2，$P=0.01$）。

目前对膳食维生素摄入与吸烟者肺部鳞状细胞癌（squamous cell carcinoma，SCC）发生之间关系的研究表明，肿瘤组织中局部的叶酸和维生素 $B_{12}$ 水平降低，同时 SCC DNA 胞嘧啶的甲基化水平也降低，此时 *Sss*I（一种对 CpG 位点特异的并向完全未被甲基化的位点和半甲基化的位点导入甲基的酶）释放的甲基数量增加（见尾注；Piyathilake *et al.*，2000）。Piyathilake 等（1992）对吸烟者与不吸烟者口腔颊部组织和血清中叶酸水平的研究表明，颊部组织叶酸水平降低可能与长期吸烟直接有关。后来，对 14 位被调查者的研究表明，在原发性非小细胞性肺癌和迁移性肺癌的口腔黏膜和恶性肺病灶组织中分离出的 DNA 的甲基化水平显著高于非病灶组织或淋巴细胞（$P<0.05$）（Piyathilake and Johanning，2002）。

目前关于叶酸与乳腺癌的几项病案与对照组的研究表明，叶酸高水平摄入显著降低了绝经前后乳腺癌的发生率（Graham *et al.*，1991；Freudenheim *et al.*，1996；Ronco *et al.*，1999；Negri et al.，2000）。然而，其中有两项研究发现，矫正蔬菜中叶酸摄入量后，乳腺癌的发生率并没有显著降低（Freudenheim *et al.*，1996；Ronco *et al.*，1999）。Freudenheim 等（1996）和 Potischman 等（1999）在一项研究中未发现叶酸添加能降低绝经期乳腺癌发生的风险，在另一项研究中，也未发现膳食叶酸或补充叶酸与绝经前和绝经后乳腺癌发生率的降低之间存在必然的联系。同样，在护士健康研究项目中（3483 个病例）和加拿大军队的试验（1469 个病例）中，对总叶酸摄入量与乳腺癌的各种风险走向也未显示出相关联系（Zhang *et al.*，1999；Rohan *et al.*，2000）。有趣的是，这两个试验以及 Negri 等（2000）的研究表明，叶酸摄入显著降低了饮酒较多的妇女乳腺癌的发生率，这与前面所述乙醇影响甲基代谢的事实相符合。

另外，也有一个试验发现男性吸烟者血清叶酸水平与胰腺癌的发生率呈显著负相关。

## 甲基四氢叶酸还原酶的作用对结肠癌发生率的影响

甲基四氢叶酸还原酶（MTHFR）基因的多态性（C677T：编码区第 677 位的胞嘧啶被胸腺嘧啶取代）可降低 MTHFR 的热稳定性，使其活性下降（Todesco *et al.*，1999）。与 MTHFR 基因多态性相关的基因型主要有：TT、CC、CT，大约 10%的人中存在 MTHFR 的 TT 纯合子基因。TT 基因型的人 MTHFR 活性仅为 CC 基因型的 30%，并且其血浆高半胱氨酸水平表现出中等程度的升高（Frosst *et al.*，1995）。男性的 TT 基因型比 CC 或 CT 基因型具有更低的结肠癌发生率，但是，当摄入高乙醇、低叶酸、低蛋氨酸的膳食后，TT 基因型男性的结肠癌发生率反而比 CC 基因型的高（Chen *et al.*，1996；Ma *et al.*，1997；Slattery *et al.*，1999）。目前多项研究报道表明，饮酒是导致 TT 基因型的人结肠癌发生的统计学显著因素（Chen *et al.*，1996；Ulric *et al.*，1999；Levine *et al.*，2000）。TT 基因型人的外周血白细胞 DNA 甲基化水平也比 CC 基因型人的低（Stern *et al.*，2000）；同时采用质谱分光光度测定方法直接测定 DNA 的 5-甲基胞嘧啶（5mC）含量，发现白细胞 DNA 的甲基化水平与叶酸摄入状况直接相关，并与高半胱氨酸水平呈负相关（Friso *et al.*，2002）。另外，也有一些证据表明 TT 基因型的人也具有较高的子宫癌、食道癌、胃癌、乳腺癌、卵巢癌及骨髓癌等癌症发生率（Estelle *et al.*，1997；Gershoni-Baruc *et al.*，2000；Shen *et al.*，2001；Song *et al.*，2001；Wiemels *et al.*，2001；Sharp *et al.*，2002）。然而，TT 基因型对肺癌的发生率的影响较小（Skibola *et al.*，1999；Franco *et al.*，2001），并对一些白血病有防护性作用。

总之，许多研究表明，叶酸缺乏人群的结肠癌发生率呈中度升高，而过量饮酒或者膳食中甲基供体缺乏同时还伴随 MTHFR 活性的降低时则会显著增加结肠癌的发生。这是否暗示着甲基供体缺乏在癌症发生中起着非常重要的作用？如前所述，大多数肿瘤组织中 DNA 甲基化水平都显著低于正常组织，并在肿瘤发生前的结肠损伤中就能检测到 DNA 低甲基化。对于其他组织（乳腺、卵巢、子宫、前列腺和脑）而言，随着肿瘤的进一步发展，DNA 低甲基化程度也逐步增强（Kim *et al.*，1994；Fowler *et al.*，1998；Narayan *et al.*，1998；Qu，1999；Shen *et al.*，1998；Santourlidis *et al.*，1999）。目前对膳食甲基供体缺乏与结肠癌发生因果关系的研究已经开展。膳食中叶酸缺乏可在白细胞中降低叶酸水平、诱导 DNA 链断裂和降低 DNA 甲基化（Jacob *et al.*，1998）。正常结肠组织中叶酸水平可反映膳食中叶酸摄入状况（Kim *et al.*，2001），有研究发现膳食中添加叶酸可增加直肠黏膜中 DNA 甲基化水平（Cravo *et al.*，1998）。对于家族性腺瘤性息肉病，肿瘤恶化程度是按时间顺序与甲基化状况相关的，而在肿瘤形成前的结肠组织中 5 甲基胞嘧啶的损失总量并未随病情的发展而进一步增加（Feinberg *et al.*，1988）。由于人们对结肠癌的发生发展过程已认识得比较清楚，目前的研究主要集中在该过程中肿瘤抑制基因的后天性沉默上（Issa，2000a）。而且，膳食甲基供体缺乏、DNA 甲基化的抑制和关键基因激活之间的关系还有待研究。膳食中足量甲基的供应可能在抑制肿瘤形成前细胞中 DNA 甲基化的异常调节中起到主要的作用。相反，膳食甲基供体缺乏可能是已经发生基因改变的细胞进一步发展的促进因素，而不是癌症发生的诱发因素。

## 膳食中甲基缺乏可通过抑制 DNA 甲基化引起或促进癌症发生的动物模型实验结果

### 甲基供体缺乏作为致癌物质诱导的啮齿动物肝癌的启动子

Copeland 和 Salmon（1946）报道，长期饲喂胆碱缺乏的膳食增加了大鼠肝脏及其他组织的肿瘤发生率。随后采用被黄曲霉毒素污染的花生粕配制的饲粮进行的试验表明，胆碱缺乏促进了肿瘤的生长但没有直接的致癌作用（Newberne *et al.*，1964）。利用各种不同的致癌物质进行的试验也逐渐证实，胆碱和（或）其他抗脂肪肝物质（蛋氨酸、叶酸和维生素 $B_{12}$）缺乏促进了肿瘤的发生（Christman，1995b）。研究表明，不添加致癌物质的膳食也会发生肿瘤，如饲喂缺乏蛋氨酸和胆碱但补充叶酸、维生素 $B_{12}$ 和高半胱氨酸的一种合成氨基酸（mino acid-defined，AAD）膳食增加了肝脏肿瘤的发生率（Mikol *et al.*，1983）。这些研究有两方面值得注意：①与通常用于肿瘤研究的低蛋氨酸或叶酸缺乏的膳食不同，单独缺乏叶酸或维生素 $B_{12}$ 并不能导致肿瘤的发生；②蛋氨酸和胆碱缺乏实际上“对肝外组织的肿瘤自然产生还有一定的抑制作用”。上述表明，甲基供体的严重缺乏与本模型中的肿瘤发生有因果关系，上述第二点表明甲基供体缺乏对肿瘤的诱导作用还存在一定的组织特异性。

正如 Christman（1995a，b）在前面综述中认为的，膳食甲基缺乏与大鼠必需的肝脏特异性致癌因素之间的联系也许取决于肝脏对甲基缺乏的特异性反应。肝脏中磷脂酰胆碱的利用率可调节肝脏中极低密度脂蛋白的分泌。由于血浆中脂蛋白大部分在肝脏中合成，且磷脂酰胆碱约占大鼠脂蛋白中磷脂总量的 75%，所以大鼠肝脏中胆碱缺失将导致甘油三酯在肝脏而非其他组织中快速、大量的沉积（Yao and Vance，1988）。尽管甲基缺乏导致肝癌发生的机制尚待进一步研究（Christman，1995a，b），但已有几项研究表明，当机体缺乏甲基时，体内脂肪数天内就在肝脏中迅速沉积，DNA 合成和细胞增殖也迅速加快（≤1 周）；同时，肝脏中 AdoMet 水平降低和 AdoHcy/AdoMet 比值增加（表 16.1）。DNA 复制加快的同时伴随着 AdoMet 水平降低和 AdoHcy（AdoHcy 是甲基转移反应的竞争性抑制剂）水平相对提高，这不仅表明供所有细胞甲基转移酶利用的 AdoMet 数量下降，同时在新复制出的肝细胞 DNA 中维持 DNA 甲基化模式所需的 AdoMet 增加。正如图 16.4b 表明的，随着 DNA 复制的不断循环发生，甲基转移的抑制很容易导致 DNA 的被动去甲基化。

**表 16.1　短期饲喂氨基酸抗脂肪肝物质缺乏的膳食对大鼠肝脏的影响**

| 效果 | 出现时间 | 参考文献[1] |
|---|---|---|
| 增加 DNA 合成速度 | ≤1 周 | Wainfan 等（1989）；Christman 等（1993b） |
| 增加有丝分裂细胞数量 | ≤1 周 | Christman 等（1993b） |
| | <6 周 | Hoover 等（1984） |
| 增加 DNA 甲基转移酶的活性 | 1～2 周 | Christman 等（1993a）；Dizik 等（1991） |
| 增加 tRNA 甲基转移酶活性 | 1～2 周 | Christman 等（1993a）；Wainfan 等（1984，1988） |
| 增加特异性 mRNA 的水平 | ≤1 周 | Dizik 等（1991） |
| 降低 AdoMet 水平或 AdoMet/AdoHcy 比值 | ≤1 周 | Shivapurkar 和 Poirier（1983） |
| 降低特异性 mRNA 的水平 | ≤1 周 | Dizik 等（1991） |
| 降低 DNA 甲基化 | 1 周 | Wainfan 等（1989） |
| 降低 tRNA 甲基化 | 5 天 | Wainfan 等（1986，1988，1989） |

1 日粮按照 Shivapurkar 和 Poirier（1983）及 Wainfan 等（1986）要求配制。

Poirier 及其同事的研究表明，甲基缺乏导致肝脏肿瘤的 DNA 5-甲基胞嘧啶含量显著降低（Wilson *et al.*，1984）。然而，对于喂给含充足甲基膳食的动物，致癌物质诱导的肝脏肿瘤的 DNA 5-甲基胞嘧啶含量也比正常肝脏的低，即使其 DNMT 水平是升高的（Lapeyre and Becker，1979）。我们采用 DNA 通过 DNMT1 发生甲基化的能力来确定DNA去甲基化，结果表明，DNA去甲基化发生在甲基供体总量缺乏后一周内，而在*c-myc*、*c-fos*和*c-Ha-ras*等致癌基因中特定位点的去甲基化则发生在4周内，且伴随着这些基因表达的提高（Wainfan *et al.*，1989；Dizik *et al.*，1991）。*p53*基因的编码区也发生了去甲基化，但*p53* mRNA表达量并未增加（Christman *et al.*，1993a）。研究发现，tRNA的甲基化受到抑制，而tRNA甲基化酶的活性却升高。采用胆碱和蛋氨酸缺乏但含叶酸和维生素$B_{12}$的膳食进行试验可得到同样的结果，单独缺乏叶酸并不产生此结果（Chen *et al.*，1993）。在甲基供体缺乏4～8周后补充充足的甲基供体，DNA合成和有丝分裂细胞数量在一周内降至正常对照组水平，新合成DNA（半甲基化位点）的甲基化也在1～2周内恢复到正常。然而，对于两条链均未甲基化的*c-myc*基因，只有在补充充足甲基膳食后至少10个月，采用限制性内切核酸酶降解后，其CpG位点从头开始的甲基化才能被检测到（表16.2）。

**表16.2 短期饲喂氨基酸抗脂肪肝物质缺乏的膳食的效应的可逆性变化**

| 效果 | 达到对照组水平的时间[1] |
|---|---|
| 增加DNA合成 | ≤1周 |
| 增加有丝分裂细胞数量 | ≤1周 |
| 增加脂肪沉积 | 8～10周 |
| 增加DNA甲基转移酶活性 | 1～2周 |
| 增加tRNA甲基转移酶活性 | 1～2周 |
| 增加或减少特殊mRNA水平 | 1～3周 |
| 降低DNA和tRNA的甲基化（作为体外甲基化的底物—见正文） | 1～2周 |
| 降低特殊基因中CCGG位点的甲基化 | >12周～<10月 |

1所有指标都与采食甲基供体充足膳食的年龄配对的动物肝脏的结果进行比较。数据引用于Christman等（1993a，b）和Chen等（1993）。

根据前面的研究结果，在癌症发生、DNA甲基化与膳食甲基供体缺乏之间的关系方面，可得到以下结论：①膳食中甲基缺乏可降低AdoMet水平和AdoMe/AdoHcy的值，进而影响肝脏中许多甲基转移过程（DNA甲基化、tRNA甲基化、由磷酸乙醇胺合成磷脂酰胆碱以及很可能也影响组蛋白及其他关键底物的甲基化）；②由短期甲基缺乏所引起的各种变化都将在膳食中补充甲基供体后1～2周内得到恢复，但两条链的CpG位点DNA甲基化缺失时除外，这与体细胞中无效的从头开始的甲基化过程是一致的；③至少对于被测的基因，一旦DNA合成和细胞分裂恢复正常（低）水平，DNA甲基化的缺失将不足以导致基因表达的稳定增加；④完全不含叶酸但其他甲基供体充足的膳食不能改变肝脏中DNA甲基化状态，这可能是由于肠道微生物合成的叶酸的同时可满足机体需要的缘故。

对饲喂含胆碱但甲基缺乏膳食的大鼠，研究肝细胞*p53*基因DNA甲基化程度随时间变化的规律表明，*p53*基因外显子6～7中CpG位点的甲基化是随时间逐步减少的。这些位点的再甲基化非常缓慢，但在肿瘤DNA中这些位点全部被甲基化。双链均失去甲基化作用位点的再甲基化与肝脏和肿瘤中从头开始的甲基转移酶活性增加有关。有趣的是，肿瘤中 *p53* 基因表达的缺失（Pogribny *et al.*，1997）与非 CpG 岛启动子中 CpG 位点从头开始的甲基化有关（Pogribny and James，2002）。然而，有关膳食多种甲基供体的缺乏导致肝外组织中 DNA 合成的增加和 AdoMet 水平降低的研究，目前尚未见报道。Chen 和 Christman 未发表的资料表明，膳食中胆碱、叶酸、蛋氨酸和维生素 $B_{12}$ 的同时缺乏确实抑制大鼠肺、睾丸、肾脏和结肠中 DNA 的合成。这与叶酸缺乏对动物或人的造血组织的影响结果一致。而且，甲基供体的缺乏不能使多数组织（包括结肠）中 AdoMet 水平发生显著变化和（或）AdoMet/AdoHcy 比值发生显著提高，但肺和睾丸除外（Shivapurkar and Poirier，1983）。值得注意的是，目前有两种非膳食成分的 DNA 甲基化的抑制剂：L-乙硫氨酸和胞嘧啶类似物 5-氮胞嘧啶核苷（5-azacytidine，ZCyd），两者均为致肝癌物质（Cox and Irving，1977；Carr *et al.*，1984）。其中，L-乙硫氨酸对肝脏无基因毒性，作为 AdoMet 的类似物抑制 DNA 甲基化（Cox and Irving，1977）。5-氮胞嘧啶核苷（ZCyd）是一种胞嘧啶的类似物，当与半甲基化 CpG 位点结合后，能与 DNMT1 结合形成抑制性复合体（Christman，2002），而且，5-氮胞嘧啶还能形成潜在的 DNA 突变分解产物（Jackson-Grosby *et al.*，1997）。

## 叶酸缺乏与肝脏和结肠中 DNA 甲基化抑制

试验采用缺乏叶酸的纯合膳食，同时在膳食中添加抗生素以抑制肠道中的微生物合成叶酸，从而使动物处于完全缺失叶酸的状态。结果表明，肝脏中 AdoMet 水平降低、AdoHcy 水平升高，同时在 4 周内可检测到 DNA 甲基化缺失，但是在癌症末期不会发生（Wagner，1982）。未经抗生素处理的叶酸缺乏膳食导致中等程度的叶酸缺乏症状，但没有出现由叶酸严重缺乏导致的生长完全受阻。中度叶酸缺乏也引起肝脏中 AdoMet/AdoHcy 比值的变化，但在结肠中未发现显著变化。试验采用 DNA 被 *Sss*I 基因催化进行甲基化的能力来检测 DNA 甲基化缺失情况，结果表明，即使在中等程度叶酸缺乏 24 周后，也未在肝脏中检测到 DNA 甲基化的缺失。对于结肠，中度叶酸缺乏对 AdoMet 或 AdoHcy 水平没有产生显著影响，也未发现所有 DNA 甲基化缺失或 *c-myc* 特殊位点甲基化缺失（Kim *et al.*，1995）。然而，在这些叶酸缺失的大鼠肝脏和结肠中，DNA 链断裂增强，*p53* 基因甲基化模式也发生了外显子特异性变化（Kim *et al.*，1997）。此外，给大鼠饲喂中等叶酸缺乏膳食 20 周也不足以诱导其结肠发生肿瘤前病变，但能促进由二甲基肼（dimethylhydrazine，DMH）诱导的肿瘤的恶化（Cravo *et al.*，1996）。相反，DMH 单独处理就可导致结肠 *p53* 基因特异外显子的甲基化缺失。可见，增加或减少膳食叶酸水平对提高或降低结肠 DNA 中这些特定位点的甲基化缺失的影响是不定的，但 DMH 处理后补充叶酸对肉眼可见的结肠肿瘤发生发展具有一定的抑制作用（Kim *et al.*，1996）。这些结果与流行病学研究的结果一致，即叶酸缺失是结肠癌发生的致病因素，而补充叶酸可预防结肠癌的发生，同时也说明，叶酸缺失对结肠组织中肿瘤形成的影响不只是由于 AdoMet 和 AdoHcy 水平或比值的变化所致的 DNA 甲基化改变的缘故，在不添加抗生素的情况下，中等程度的叶酸缺乏对啮齿类动物肝外组织的影响可能与甲基供体严重缺乏时的影响相类似。如果这种膳食不能促进细胞分裂，或者实际上降低正常细胞的分裂速度，则中度甲基缺乏将抑制 DNA 甲基化和促进已发生基因突变的细胞增殖，从而增强细胞程序性死亡的抗性，提高生长速度，最后诱导肿瘤的发生。

## $Apc^{min}$ 小鼠模型中的肠癌是一种不同的范例

上述研究有力地证实 DNA 甲基化的缺失在癌症发生中起着非常重要的作用，然而与此相反，研究发现，一个 *Dnmt1* 等位基因被敲除后，当 DNMT1 水平降低约 50% 时，$Apc^{min}$ 杂合小鼠自发性肠道肿瘤形成受到抑制，这种抑制作用经 ZdCyd 处理后会进一步加强。由于 $Apc^{min}$ 杂合小鼠的完整 *Apc* 等位基因的缺失频率在 DNMT1 正常水平与较低水平情况下相同，由此得出结论：杂合子的缺失（loss of heterozygosity，LOH）不可能是肿瘤受抑制的主要原因，而是另有原因（Laird *et al.*，1995）。

这一结果证实了以下的假设：CpG 岛的甲基化使肿瘤抑制基因失活是肿瘤发生的原因之一；且 DNMT1 的抑制也能降低肿瘤的发生或延缓肿瘤的发展。然而，ZdCyd 的作用仅局限在动物出生后一个相当窄的时间阶段内。这也将出现如下的可能，即这一药物可能诱导某些细胞的缺失，而这些细胞只存在于肠道发育成熟过程的某一特定阶段，尤其是在息肉形成和发展过程中起着重要作用。然而，随后采用 DNMT1 基因表达受调节的 $Apc^{min}$ 小鼠的试验表明，其 DNMT1 水平比 *Dnmt1* 敲除杂合鼠的低，并完全抑制了息肉的形成，这进一步证实了抑癌作用完全是由于 DNMT1 水平较低和（或）降低了 CpG 岛甲基化所致（Eads *et al.*，2002）。

膳食甲基供体对 $Apc^{min}$ 小鼠的作用并不明显。一项补充叶酸对断奶 $Apc^{min}$ 鼠的研究表明，膳食中补充叶酸减少了回肠中息肉的产生。在补充叶酸 3 个月时，与叶酸缺乏鼠相比，补充组的异常隐窝病灶（aberrant crypt foci）比叶酸缺乏组的降低了 75%～100%。但是，在随后的时间点，补充叶酸却可能增加了回肠息肉的数量（Song *et al.*，2000a，b）。对饲喂充足叶酸的 $Apc^{min}$ 小鼠的另一项研究发现，癌前期肠道中 AdoMet 和 AdoHcy 的水平与肿瘤的数量呈线性正相关。该研究还发现，这些动物的总体 DNA 低甲基化与肿瘤多样性之间存在正相关。然而，尽管 AdoMet 水平的降低和 DNA 低甲基化使 $Apc^{min}$ 小鼠肿瘤数量增加，但叶酸/胆碱缺乏对这些动物的影响结果目前还不一致（Sibani *et al.*，2002）。这两项研究提示，膳食甲基供体对癌症发生的调节作用可能受细胞转化状态的影响。

# DNMT 功能：更复杂的观点

长期以来，人们一直认为哺乳动物的 DNA 甲基转移酶（DNMT）不与裸露的 DNA 发生相互作用，而与染色质发生相互作用；而且与染色质结合的组蛋白和其他非组蛋白可能在 DNA 甲基化的时间和空间调节中起着重要作用。研究表明，即使大量的细胞核 DNMT 能够与无活性的染色质相结合，但是无活性的染色质中的非甲基化 CpG 位点只有在染色质蛋白去除之后才能与 DNMT 相结合（Creusot and Christman，1981），这也进一步证实了上面的观点。最近研究发现，一些非组蛋白的核蛋白可与 DNMT1、DNMT3a、DNMT3b 和 DNMT3L 的 N 端结合域相互作用，而对这些核蛋白的鉴定大大加强了我们对 DNA 甲基化和染色质压缩如何使基因沉默的认识。现在形成的一种观点是，DNMT 的功能是作为巨大多蛋白复合体的组成部分，这些复合体中的蛋白质能决定 DNMT 在染色质中的位点，有助于增强 CpG 甲基化的基因沉默功能，并且影响 DNMT1 维持 DNA 甲基化的能力。越来越多的试验表明，DNMT 还可能抑制基因的表达，而这种抑制并不依赖其作为甲基转移酶的功能（Bachman *et al.*，2001；Fuks *et al.*，2001）。

## DNMT、组蛋白脱乙酰酶与甲基化的 DNA 结合蛋白

含甲基结合区域（methyl-binding domain，MBD）的蛋白质结合到甲基化的 DNA 上，并且富聚到含组蛋白脱乙酰酶（histone deacetylase，HDAC）的蛋白质复合体上，或者作为该复合体的一部分。反过来，组蛋白的脱乙酰化在染色质压缩中起着关键作用，而致密染色质使 DNA 无法进入转录机制。组蛋白脱乙酰化的抑制剂能促进由 Zd-Cyd 诱导的基因表达，并常常能解除由 DNMT1 介导的基因沉默（Robertson，2001；Burgers *et al.*，2002）。有关膳食与癌症之间相互作用的研究发现，由结肠微生物产生的丁酸是 HDAC 的抑制剂，至少对于一些培养细胞而言，丁酸诱导了细胞的分化（Sealy and Chalkley，1978；Kruh，1982），并且伴随着 DNA 甲基化的降低（Christman *et al.*，1980）。

DNMT1、DNMT3a 和 DNMT3b 也可直接与 HDAC 结合（图 16.3），并能抑制基因表达（Robertson，2001）。在此过程中，DNMT1 也可直接与含 HDAC 的 MeCP1 抑制复合体中的 MBD2 和 MBD3（Tatematsu *et al.*，2000）相互作用（Feng and Zhang，2001）。由于在细胞分裂 S 时段后期，DNMT1 与 HDAC2 都定位于 DNA 复制位点，这些复合体均能通过确保组蛋白脱乙酰化和随后复制基因中基因沉默的传播来维持基因沉默，与此同时，核小体被重新组装并重新定位于新复制的 DNA 中（Rountree *et al.*，2000）。

此外，DNMT1 也能与 DMAP1 和 HDAC2 结合形成复合体，并与 DNMT1 在细胞分裂 S 时段共同定位于 DNA 复制点并执行类似的功能（Rountree *et al.*，2000）。

## DNMT 和染色质相关蛋白

除了影响 DNMT1 活性外，Rb 蛋白还可与 DNMT1 共同抑制由 E2F 应答启动子开始的转录过程（Robertson *et al.*，2000）。Rb 使 DNMT1、HDAC 和组蛋白甲基转移酶（histone methyltransferase，HMT）聚集在这些启动子上，随后，与 HP1 蛋白质家族中甲基化的赖氨酸结合蛋白相结合。

DNMT3a 可与异染色质结合的转录抑制因子 RP58 结合（Fuks *et al.*，2001）。DNMT3a 的 N 端部分也可与异染色质相关蛋白（heterochromatin-associated protein）HP12 和单核细胞趋药性蛋白（monocyte chemotactic protein）MCP2 协同定位，这对致密的、已甲基化的、着丝点外的染色质中 DNA 甲基化的定位以及特定 DNA 序列从头开始的甲基化起着非常重要的作用（Bachman *et al.*，2001）。然而，目前还没有直接证据说明 DNMT3a 在这些蛋白质复合体中的功能，因为催化活性失活也能导致基因沉默。

DNA 甲基转移酶定位的蛋白质复合体中包括早幼粒细胞白血病基因-视黄酸受体复合体（promyelocytic leukaemia gene-retinoic acid receptor，PML-RAR），它是一种由

染色体 17 上的视黄酸受体 α（RARα）基因与染色体 15 上的早幼粒细胞白血病基因（promyelocytic leukaemia gene，PML）易位形成的融合蛋白。PML-RAR 可与 DNMT1 和 DNMT3a 发生免疫沉淀反应，同时，PML-RAR 可与 DNMT1 和 DNMT3a 一起定位在 PML-RAR 过量表达的细胞中，还可介导 RARβ2 启动子从头开始的甲基化和基因沉默（Di Croce *et al.*，2002）。此外，转录抑制因子 Daxx 也可与 DNMT1 结合并与 PML 基因相互定位，这是使 RARβ2 启动子从头开始的甲基化和基因沉默的另一可能途径。

## DNMT 与染色质重建

如果染色质压缩限制了 DNMT 结合到 DNA 中剩余的未甲基化位点，那么染色质重建蛋白能够在维持和（或）启动 DNMT 对 DNA 中特定位点的甲基化作用方面发挥一定的作用，这似乎是合乎逻辑的。

哺乳动物细胞含有大量的酵母蔗糖非发酵基因（sucrose non-fermenter gene，SNF2）家族的同源物。SNF2 家族成员可利用 ATP 水解释放出的能量抑制组蛋白与 DNA 之间的相互作用，从而使核小体沿着 DNA 滑至下一个新位点（Wolffe and Pruss *et al.*，1996；Vignali *et al.*，2000）。人类 SNF2 结构性同系物的植物同源域（plant homeodomain，PHD）ATRX 的突变可导致 X 染色体相关的 *α*-珠蛋白生成障碍性贫血智力发育不良（X-linked α-thalassaemia mental retardation，ATR-X）综合征（Gibbons *et al.*，1997）。尽管人们还不能确定 ATRX 在染色质重建中的实际作用，但它能导致核糖体的 DNA 重复序列的甲基化缺失。而且，正常 ATRX 通常定位于核糖体 DNA 重复序列中。然而，研究发现 Y 染色体特异性的重复序列表现出 DNA 高甲基化（Gibbons *et al.*，2000）。

ATRX 的作用可能对基因组中某些区域具有极高的选择性，且很可能在这些位点与各种的 DNMT-HDAC 复合体发生相互作用。ATRX 的 PHD 结构域与 DNMT3a 的一致性暗示，ATRX 能够阻碍在特异性的着丝粒外位点中 DNMT3a-RP58-HDAC 复合体的压缩。然而，ATRX 在 rRNA 重复序列中的作用仅限于染色质重建中，从而使含有 DNMT1 但不具有 PHD 结构域的复合体能够维持甲基化状况。

另一个类似 SNF2 的染色质重建蛋白是淋巴细胞特异性解旋酶（lymphocyte-specific helicase，LSH），它能影响哺乳动物 DNA 的甲基化。尽管 LSH 首先在淋巴细胞中被发现（Jarvis *et al.*，1996），但它实际上广泛分布于哺乳动物的各种细胞中，而且其表达与细胞增殖紧密相关（Geiman and Muegge，2000；Raabe *et al.*，2001）。LSH 基因敲除纯合鼠在出生时是活着的，但是其出生体重低、肾脏病变、淋巴细胞数量下降，仅能存活数天；基因组 5-甲基胞嘧啶总量在所有组织中均下降 50%～60%；在正常条件下均可高度甲基化的各种类型的重复 DNA 序列的甲基化状况均受到显著影响；与正常组织相比，所测的全部单拷贝基因（包括组织特异性基因、持家基因和印迹基因）均表现出更低的甲基化水平（Dennis *et al.*，2001）。以上结果不仅说明了染色质结构在建立和维持 DNA 甲基化模式中的重要性，而且还暗示了至少在胚胎发育期间 DNA 甲基化的严重受抑并不能阻止细胞的大量分裂。尽管在 $Lsh^{-/-}$ 鼠胚胎 DNA 中是否存在一些区域其 DNA 甲基化被有选择性地保留还有待证实，但是 DNA 甲基化的大量缺失并不能抑制复合体的正常合成，这些复合体对细胞命运具有决定性作用，并参与建立胚胎形成过程中基因表达的组织特异性模型。

# 膳食甲基缺乏与癌症：染色质、染色质相关蛋白与DNA甲基转移酶之间的相互调节作用

采用$Apc^{min}$杂合小鼠和*Dnmt1*基因敲除小鼠来降低DNA甲基化水平的研究结果与采用叶酸缺乏或叶酸和胆碱同时缺乏的膳食进行研究的结果之间存在显著的差异，DNA甲基转移酶、赖氨酸甲基化组蛋白、定位靶蛋白与转录抑制之间的相互关系可能为产生此差异的原因提供重要线索。*Dnmt1*基因敲除可降低具有催化活性的DNMT1水平，但不能推测其是否影响细胞中除DNA外其他细胞组分的甲基化状况。

我们的研究表明，L-乙硫氨酸诱导了人HL-60细胞和鼠红白血友病（friend erythroleukaemia，FLC）细胞产生不同的细胞分化（Christman *et al.*，1977；Mendelsohn *et al.*，1980）。在一些细胞（包括肝细胞）中，L-乙硫氨酸可转变成*S*-腺苷乙硫氨酸（Brada and Bulba，1987），同时它也是AdoMet结合蛋白的有效抑制剂（Cox and Irving，1997），因此，利用L-乙硫氨酸可模拟膳食甲基供体缺乏对AdoMet依赖性甲基化反应的不利影响。FLC细胞产生不同在含L-乙硫氨酸的培养基中生长时可降低其DNA甲基化总体水平，并且抑制了其RNA和组蛋白的甲基化（Christman *et al.*，1997）。有意思的是，在组蛋白中，由L-乙硫氨酸导致的赖氨酸甲基化抑制程度远高于精氨酸的甲基化抑制程度（Copp，纽约大学博士学位论文，1981）。

这些研究结果表明，膳食甲基供体缺乏不仅可通过干扰DNA甲基化来影响基因的表达，还可通过干扰抑制因子复合体的组装来影响基因的表达。这些抑制因子复合体与失活的DNMT1相互作用，但是其组装需要甲基化组蛋白和其他一些未知的甲基化核蛋白的参与。

然而，对于$Dnmt1^{+/-}$基因敲除小鼠，活性DNMT1水平降低，但其他蛋白质和RNA等的甲基化不受影响；同时，形成一些抑制因子复合体。

$Lsh^{-/-}$小鼠和$Dnmt1^{-/-}$小鼠之间的差异也可以得到类似的解释：$Lsh^{-/-}$小鼠细胞存在大量的DNA甲基化缺失，但仍能保持野生型DNMT1蛋白的正常水平及RNA、组蛋白和其他蛋白质的正常甲基化，同时也使胚胎发育所必需的基因表达发生变化。而$Dnmt1^{-/-}$小鼠不能表达野生型DNMT1蛋白，也不能使基因沉默。这些小鼠往往在胚胎发育的第10.5天死亡（Li *et al.*，1992）。

尽管目前人们对DNMT与核蛋白之间的相互作用、DNA甲基化模式的建立以及基因表达调控的认识仍有待进一步深入，但是很明显，在未来几年，人们应该集中揭示膳食叶酸/甲基供体的缺乏与先天性发育缺陷、癌症及其他人类疾病之间的关系。

# 致　谢

在此首先感谢我的良师益友Elsie Wainfan在膳食甲基供体缺乏对癌发生的重要性及其在我们对膳食甲基供体缺乏与核酸甲基化之间关系进行的研究中所做出的不可估量的贡献；感谢James Finkelstein对AdoMet和AdoHcy在甲基代谢中的调节作用以及机体AdoHcy和高半胱氨酸水平的内稳衡调节等方面给予的帮助；同时，感谢我们研究

组成员 Klinkebiel、Uzvolgyi、Boland、Farrell 和 Van Bemmel 等在制作图表和数据处理方面给予的帮助。

此外，本文中对膳食、DNA 甲基化和癌症的科学研究的经费由美国癌症研究所（94A37-REN 和 96A115-REV）资助；本文中对 DNA 甲基化、DNA 甲基转移酶抑制剂和癌症的研究由美国国防部（DAMD-17-98-1-8215）和美国国家卫生研究院（NIH）（R21 CA91315）资助。

# 尾　注

在阅读 DNA 甲基化相关的资料时，认真评定 DNA 甲基化状况的检测方法显得尤为重要。这些方法能够决定一个基因是“过甲基化”了或者是甲基化偏低、还是总 DNA 甲基化已经发生并被保留。确定这些变化的 DNA 甲基化检测方法仅能对 DNA 中少数区域进行检测，而且（或者）主要针对富含 CpG 的区域和含有甲基化敏感性核酸内切酶酶切位点的区域的甲基化增强或减少进行检测。评价 DNA 中特定区域 DNA 甲基化的理想方法是亚硫酸盐测序（Warnecke *et al.*，2002）。该方法可对约 300bp 长的单分子 DNA 片段中每个胞嘧啶甲基化状况进行检测，同时，通过显微解剖和激光捕获的方法保证细胞类型的一致性，人们可获得关于肿瘤发生前或肿瘤细胞中 DNA 特定区域相对于正常组织是否处于 DNA 高甲基化或低甲基化等有价值的结论，同时也能确定这些通常的变化是否与肿瘤的发生发展有关，并且还能得出这些变化与基因功能增减的关系。此外，许多其他 DNA 甲基化检测方法可对大量的样品进行快速的检测。这些方法通常用来检测特定 DNA 序列是否被高甲基化或低甲基化，同时也能作为癌症分期的有价值的诊断标示物。然而，大多数 DNA 甲基化检测方法主要集中在一些可能的位置如 CpG 岛的少量位点上，而这些位点的甲基化状况是否与基因活性或细胞功能有关目前尚不能确定。

DNA 全局甲基化缺失是指样品 DNA 中 5-甲基胞嘧啶水平低于对照样品。尽管 DNA 中 5-甲基胞嘧啶可通过目前相当灵敏的质谱法进行测定，但一些其他方法也已经被开发。例如，通过 *Sss*Ⅰ（CpG 位点的 DNA 甲基转移酶），利用［$^3$H］-甲基-AdoMet 对 DNA 进行甲基化修饰后检测被修饰 DNA 中$^3$H 的放射性。由于未对 DNA 或染色质片段结构进行分析，以上两种方法均不能对基因中特定位点的甲基化状态进行研究。同时，这两种方法也不能区分出 DNA 甲基化缺失是发生在一条 DNA 上还是发生在两条互补的 DNA 上（图 16.4）。DNMT1 目前已用于对 DNA 不完全甲基化的评定（Christman *et al.*，1993b）。应用甲基化敏感性限制性内切核酸酶的 DNA 甲基化检测方法可对先前未甲基化的 DNA 位点的甲基化变化进行检测，但在大多数研究中该方法仅限于对少数位点的检测研究（Christman 于 1984 年提供了此领域的一篇很好的综述）。作为检测大量 CpG 岛 DNA 序列的甲基化状态的新方法，微列阵或二维凝胶分析（Yan *et al.*，2001）已极大地提高了我们检测 DNA 总体甲基化变化的能力，但可能低估了 DNA 甲基化变化对 CpG 位点较少的 DNA 区域的影响（Rein *et al.*，1998）。而以上各方面的考虑与对经历干细胞分化、末期分化、凋亡的一个正常成熟周期的细胞 DNA 甲基化的研究尤为相关。甲基供体缺乏导致了细胞 DNA 低甲基化，但补充甲基供体后可恢复DNA甲基化。DNA低甲基化是由于对新复制DNA甲基化不完全，还是反映出DNA甲基化模式的改变；DNA再甲基化是维持原有DNA甲基化状态、恢复原有DNA甲基化模式、用在甲基供体充足时分化产生的细胞替代DNA低甲基化细胞，还是对保持稳定DNA甲基化模式的细胞的选择性生长，这些方面的研究非常重要。

## 参 考 文 献


Aapola, U., Kawasaki, K., Scott, H.S., Ollila, J., Vihinen, M., Heino, M., Shintani, A., Minoshima, S., Krohn, K., Antonarakis, S.E., Shimizu, N., Kudoh, J. and Peterson, P. (2000) Isolation and initial characterization of a novel zinc finger gene, DNMT3L, on 21q22.3, related to the cytosine-5-methyltransferase 3 gene family. *Genomics* 65, 293–298.

Aasland, R., Gibson, T.J. and Stewart, A.F. (1995) The PHD finger: implications for chromatin-mediated transcriptional regulation. PG-56-9. *Trends in Biochemical Science* 20(2).

AICR (1997) *Food, Nutrition and the Prevention of Cancer: a Global Perspective.* American Institute for Cancer Research, Washington, DC.

Aoki, K., Meng, G., Suzuki, K., Takashi, T., Kameoka, Y., Nakahara, K., Ishida, R. and Kasai, M. (1998) RP58 associates with condensed chromatin and mediates a sequence-specific transcriptional repression. *Journal of Biological Chemistry* 273, 26698–26704.

Aoki, A., Svetake, I., Miyagawa, J., Fujio, T., Chijiwa, T., Sasaki, H. and Tajima, S. (2001) Enzymatic properties of denovotype mouse (cystosine-5) methyltransferases. *Nucleic Acids Research* 29, 2506–3512.

Bachman, K.E., Rountree, M.R. and Baylin, S.B. (2001) Dnmt3a and Dnmt3b are transcriptional repressors that exhibit unique localization properties to heterochromatin. *Journal of Biological Chemistry* 276, 32282–32287.

Baylin, S.B., Herman, J.G., Graff, J.R., Vertino, P.M. and Issa, J.P. (1998) Alterations in DNA methylation: a fundamental aspect of neoplasia. *Advances in Cancer Research* 72, 141–196.

Benito, E., Stiggelbout, A., Bosch, F.X., Obrador, A., Kaldor, J., Mulet, M. and Munoz, N. (1991) Nutritional factors in colorectal cancer risk: a case-control study in Majorca. *International Journal of Cancer* 49, 161–167.

Benito, E., Cabeza, E., Moreno, V., Obrador, A. and Bosch, F.X. (1993) Diet and colorectal adenomas: a case-control study in Majorca. *International Journal of Cancer* 55, 213–219.

Bestor, T., Laudano, A., Mattaliano, R. and Ingram, V. (1988) Cloning and sequencing of a cDNA encoding DNA methyltransferase of mouse cells. The carboxyl-terminal domain of the mammalian enzymes is related to bacterial restriction methyltransferases. *Journal of Molecular Biology* 203, 971–983.

Bestor, T.H. (1992) Activation of mammalian DNA methyltransferase by cleavage of a Zn binding regulatory domain. *EMBO Journal* 11, 2611–2617.

Bhattacharya, S.K., Ramchandani, S., Cervoni, N. and Szyf, M. (1999) A mammalian protein with specific demethylase activity for mCpG DNA. *Nature* 397, 579–583.

Bigey, P., Ramchandani, S., Theberge, J., Araujo, F.D. and Szyf, M. (2000) Transcriptional regulation of the human DNA methyltransferase (dnmt1) gene. *Gene* 242, 407–418.

Bird, A.P. (1986) CpG-rich islands and the function of DNA methylation. *Nature* 321, 209–213.

Bird, A.P. and Wolffe, A.P. (1999) Methylation-induced repression – belts, braces, and chromatin. *Cell* 99, 451–454.

Bird, C.L., Swendseid, M.E., Witte, J.S., Shikany, J.M., Hunt, I.F., Frankl, H.D., Lee, E.R., Longnecker, M.P. and Haile, R.W. (1995) Red cell and plasma folate, folate consumption, and the risk of colorectal adenomatous polyps. *Cancer Epidemiology Biomarkers Prev* 4, 709–714.

Blount, B.C., Mack, M.M., Wehr, C.M., MacGregor, J.T., Hiatt, R.A., Wang, G., Wickramasinghe, S.N., Everson, R.B. and Ames, B.N. (1997) Folate deficiency causes uracil misincorporation into human DNA and chromosome breakage: implications for cancer and neuronal damage. *Proceedings of the National Academy of Sciences USA* 94, 3290–3295.

Bonfils, C., Beaulieu, N., Chan, E., Cotton-Montpetit, J. and MacLeod, A.R. (2000) Characterization of the human DNA methyltransferase splice variant Dnmt1b. *Journal of Biological Chemistry* 275, 10754–10760.

Bourc'his, D., Xu, G.L., Lin, C.S., Bollman, B. and Bestor, T.H. (2001) Dnmt3L and the establishment of maternal genomic imprints. *Science* 294, 2536–2539.

Boutron-Ruault, M.C., Senesse, P., Faivre, J., Couillault, C. and Belghiti, C. (1996) Folate and alcohol intakes: related or independent roles in the adenoma carcinoma sequence? *Nutrition and Cancer* 26, 337–346.

Brada, Z. and Bulba, S. (1987) The synthesis of the adenosyl-moiety of S-adenosylethionine in liver of rats fed DL-ethionine. *Research Communications in Chemical Pathology and Pharmacology* 56, 133–136.

Brank, A.S., Van Bemmel, D.M. and Christman, J.K. (2002) Optimization of baculovirus-mediated expression and purification of hexahistidine-tagged murine DNA (cytosine-C5)-methyltransferase-1 in *Spodoptera frugiperda* 9 cells. *Protein Expression and Purification* 25, 31–40.

Burgers, W.A., Fuks, F. and Kouzarides, T. (2002) DNA methyltransferases get connected to chromatin. *Trends in Genetics* 18, 275–277.

Carr, B.I., Reilly, J.G., Smith, S.S., Winberg, C. and Riggs, A. (1984) The tumorigenicity of 5-azacytidine in the male Fischer rat. *Carcinogenesis* 5, 1583–1590.

Chadwick, L.H., McCandless, S.E., Silverman, G.L., Schwartz, S., Westaway, D. and Nadeau, J.H. (2000) Betaine-homocysteine methyltransferase-2: cDNA cloning, gene sequence, physical mapping, and expression of the human and mouse genes. *Genomics* 70, 66–73.

Chedin, F., Lieber, M.R. and Hsieh, C.L. (2002) The DNA methyltransferase-like protein DNMT3L stimulates de novo methylation by Dnmt3a. *Proceedings of the National Academy of Sciences USA* 99, 16916–16921.

Chen, J., Giovannucci, E., Kelsey, K., Rimm, E.B., Stampfer, M.J., Colditz, G.A., Spiegelman, D., Willett, W.C. and Hunter, D.J. (1996) A methylenetetrahydrofolate reductase polymorphism and the risk of colorectal cancer. *Cancer Research* 56, 4862–4864.

Chen, M.-L., Abileah, S., Wanfan, E. and Christman, J.K. (1993) Influence of folate and vitamin B12 on the effects of dietary lipotrope deficiency. *Proceedings of the American Association for Cancer Research* 34, 131.

Chen, T., Ueda, Y., Xie, S. and Li, E. (2002) A novel Dnmt3a isoform produced from an alternative promoter localizes to euchromatin and its expression correlates with active de novo methylation. *Journal of Biological Chemistry*.

Choi, S.-W., Stickel, F., Baik, H., Kim, Y., Seitz, H., amd Mason, J. (1999) Chronic alcohol consumption induces genomic but not p53-specific DNA hypomethylation in rat colon. *Journal of Nutrition* 129, 1945–1950.

Christman, J.K. (1984) DNA methylation in Friend erythroleukemia cells: the effects of chemically induced differentiation and of treatment with inhibitors of DNA methylation. *Current Topics in Microbiology and Immunology* 108, 49–78.

Christman, J.K. (1995a) Dietary effects on DNA methylation: Do they account for the hepatocarcinogenic properties of lipotrope deficient diets? *Advances in Experimental Medicine and Biology* 369, 141–154.

Christman, J.K. (1995b) Lipotrope deficiency and persistent changes in DNA methylation. *Advances in Experimental Medicine and Biology* 375, 97–106.

Christman, J.K. (2002) 5-Azacytidine and 5-aza-2′-deoxycytidine as inhibitors of DNA methylation: mechanistic studies and their implications for cancer therapy. *Oncogene* 21, 5483–5495.

Christman, J.K., Price, P., Pedrinan, L. and Acs, G. (1977) Correlation between hypomethylation of DNA and expression of globin genes in Friend erythroleukemia cells. *European Journal of Biochemistry* 81, 53–61.

Christman, J.K., Weich, N., Schoenbrun, B., Schneiderman, N. and Acs, G. (1980) Hypomethylation of DNA during differentiation of Friend erythroleukemia cells. *Journal of Cell Biology* 86, 366–370.

Christman, J.K., Sheikhnejad, G., Dizik, M., Abileah, S. and Wainfan, E. (1993a) Reversibility of changes in nucleic acid methylation and gene expression induced in rat liver by severe dietary methyl deficiency. *Carcinogenesis* 14, 551–557.

Christman, J.K., Cheng, M.-L., Sheikhnejad, G., Dizik, M., Alileah, S., and Wainfan, E. (1993b) Methyl deficiency, DNA methylation, and cancer: studies on the reversibility of the effects of lipotrope-deficient diet. *Journal of Nutrition and Biochemistry* 4, 672–680.

Christman, J.K., Sheikhnejad, G., Marasco, C.J. and Sufrin, J.R. (1995) 5-Methyl-2′-deoxycytidine in single-stranded DNA can act in cis to signal de novo DNA methylation. *Proceedings of the National Academy of Sciences USA* 92, 7347–7351.

Chuang, L.S., Ian, H.I., Koh, T.W., Ng, H.H., Xu, G. and Li, B.F. (1997) Human DNA-(cytosine-5) methyltransferase-PCNA complex as a target for p21WAF1. *Science* 277, 1996–2000.

Copeland, D.H. and Salmon, W.D. (1946) The occurrence of enoplasms in the liver, lungs, and other tissues of rats as a result of prolonged choline deficiency. *American Journal of Pathology* 22, 1059.

Costello, J.F. and Plass, C. (2001) Methylation matters. *Journal of Medical Genetics* 38, 285–303.

Costello, J.F., Smiraglia, D.J. and Plass, C. (2002) Restriction landmark genome scanning. *Methods* 27, 144–149.

Cox, R. and Irving, C.C. (1977) Inhibition of DNA methylation by S-adenosylethionine with the production of methyl-deficient DNA in regenerating rat liver. *Cancer Research* 37, 222–225.

Cravo, M.L., Mason, J.B., Dayal, Y., Hutchinson, M., Smith, D., Selhub, J. and Rosenberg, I.H. (1992) Folate deficiency enhances the development of colonic neoplasia in dimethylhydrazine-treated rats. *Cancer Research* 52, 5002–5006.

Cravo, M.L., Pinto, A.G., Chaves, P., Cruz, J.A., Lage, P., Nobre Leitao, C. and Costa Mira, F. (1998) Effect of folate supplementation on DNA methylation of rectal mucosa in patients with colonic adenomas: correlation with nutrient intake. *Clinical Nutrition* 17, 45–49.

Creusot, F. and Christman, J.K. (1981) Localization of DNA methyltransferase in the chromatin of Friend erythroleukemia cells. *Nucleic Acids Research* 9, 5359–5381.

Dennis, K., Fan, T., Geiman, T., Yan, Q. and Muegge, K. (2001) Lsh, a member of the SNF2 family, is required for genome-wide methylation. *Genes and Development* 15, 2940–2944.

Di Croce, L., Raker, V.A., Corsaro, M., Fazi, F., Fanelli, M., Faretta, M., Fuks, F., Lo Coco, F., Kouzarides, T., Nervi, C., Minucci, S. and Pelicci, P.G. (2002) Methyltransferase recruitment and DNA hypermethylation of target promoters by an oncogenic transcription factor. *Science* 295, 1079–1082.

Dizik, M., Christman, J.K. and Wainfan, E. (1991) Alterations in expression and methylation of specific genes in livers of rats fed a cancer promoting methyl-deficient diet. *Carcinogenesis* 12, 1307–1312.

Doerfler, W. (1983) DNA methylation and gene activity. *Annals of Biochemistry* 52, 93–124.

Eads, C.A., Danenberg, K.D., Kawakami, K., Saltz, L.B., Danenberg, P.V. and Laird, P.W. (1999) CpG island hypermethylation in human colorectal tumors is not associated with DNA methyltransferase overexpression. *Cancer Research* 59, 2302–2306.

Eads, C.A., Nickel, A.E. and Laird, P.W. (2002) Complete genetic suppression of polyp formation and reduction of CpG-island hypermethylation in Apc(Min/+) Dnmt1-hypomorphic mice. *Cancer Research* 62, 1296–1299.

Ehrlich, M. (2002) DNA methylation in cancer: Too much, but also too little. *Oncogene* 21, 5400–5413.

Ehrlich, M. and Wang, R.Y. (1981) 5-Methylcytosine in eukaryotic DNA. *Science* 212, 1350–1357.

Eloranta, T.O. and Raina, A.M. (1977) S-adenosylmethionine metabolism and its relation to polyamine synthesis in rat liver. Effect of nutritional state, adrenal function, some drugs and partial hepatectomy. *Biochemistry Journal* 168, 179–185.

Esteller, M., Garcia, A., Martinez-Palones, J.M., Xercavins, J. and Reventos, J. (1997) Germ line polymorphisms in cytochrome-P450 1A1 (C4887 CYP1A1) and methylenetetrahydrofolate reductase (MTHFR) genes and endometrial cancer susceptibility. *Carcinogenesis* 18, 2307–2311.

Esteller, M., Fraga, M.F., Paz, M.F., Campo, E., Colomer, D., Novo, F.J., Calasanz, M.J., Galm, O., Guo, M., Benitez, J. and Herman, J.G. (2002) Cancer epigenetics and methylation. *Science* 297, 1807–1808.

Fatemi, M., Hermann, A., Pradhan, S. and Jeltsch, A. (2001) The activity of the murine DNA methyltransferase Dnmt1 is controlled by interaction of the catalytic domain with the N-terminal part of the enzyme leading to an allosteric activation of the enzyme after binding to methylated DNA. *Journal of Molecular Biology* 309, 1189–1199.

Fauman, E.B. (1999) *S-Adenosylmethionine-Dependent Methyltransferases: Structures and Functions.* World Scientific Publishing.

Fearon, E.R. and Vogelstein, B. (1990) A genetic model for colorectal tumorigenesis. *Cell* 61, 759–767.

Feinberg, A.P. and Vogelstein, B. (1983) Hypomethylation of *ras* oncogenes in primary human cancers. *Biochemistry and Biophysics Research Communications* 111, 47–54.

Feinberg, A.P., Gehrke, C.W., Kuo, K.C. and Ehrlich, M. (1988) Reduced genomic 5-methylcytosine content in human colonic neoplasia. *Cancer Research* 48, 1159–1161.

Feinberg, A., Cui, H. and Ohlsson, R. (2002) DNA methylation and genomic imprinting: insights from cancer into epigenetic mechanisms. *Seminars in Cancer Biology* 12, 389–395.

Fenech, M. (2001) The role of folic acid and Vitamin B12 in genomic stability of human cells. *Mutation Research* 475, 57–67.

Feng, Q. and Zhang, Y. (2001) The MeCP1 complex represses transcription through preferential binding, remodeling, and deacetylating methylated nucleosomes. *Genes and Development* 15, 827–832.

Ferraroni, M., La Vecchia, C., D'Avanzo, B., Negri, E., Franceschi, S. and Decarli, A. (1994) Selected micronutrient intake and the risk of colorectal cancer. *British Journal of Cancer* 70, 1150–1155.

Finkelstein, J.D. (1990) Methionine metabolism in mammals. *Journal of Nutritional Biochemistry* 1, 228–237.

Finkelstein, J.D. (1998) The metabolism of homocysteine: pathways and regulation. *European Journal of Pediatrics* 157 (Supplement 2), S40–S44.

Finkelstein, J.D. (2000) Pathways and regulation of homocysteine metabolism in mammals. *Seminars in Thrombosis and Hemostasis* 26, 219–225.

Finkelstein, J.D., Cello, J.P. and Kyle, W.E. (1974) Ethanol-induced changes in methionine metabolism in rat liver. *Biochemistry and Biophysics Research Communications* 61, 525–531.

Fowler, B.M., Giuliano, A.R., Piyathilake, C., Nour, M. and Hatch, K. (1998) Hypomethylation in cervical tissue: is there a correlation with folate status?' *Cancer Epidemiology Biomarkers Prev* 7, 901–906.

Franco, R.F., Simoes, B.P., Tone, L.G., Gabellini, S.M., Zago, M.A. and Falcao, R.P. (2001) The methylenetetrahydrofolate reductase C677T gene polymorphism decreases the risk of childhood acute lymphocytic leukaemia. *British Journal of Haematology* 115, 616–618.

Freudenheim, J.L., Graham, S., Marshall, J.R., Haughey, B.P., Cholewinski, S. and Wilkinson, G. (1991) Folate intake and carcinogenesis of the colon and rectum. *International Journal of Epidemiology* 20, 368–374.

Freudenheim, J.L., Marshall, J.R., Vena, J.E., Laughlin, R., Brasure, J.R., Swanson, M.K., Nemoto, T. and Graham, S. (1996) Premenopausal breast cancer risk and intake of vegetables, fruits, and related nutrients. *Journal of National Cancer Institute* 88, 340–348.

Friso, S., Choi, S.W., Girelli, D., Mason, J.B., Dolnikowski, G.G., Bagley, P.J., Olivieri, O., Jacques, P.F., Rosenberg, I.H., Corrocher, R. and Selhub, J. (2002) A common mutation in the 5,10-methylenetetrahydrofolate reductase gene affects genomic DNA methylation through an interaction with folate status. *Proceedings of the National Academy of Sciences USA* 99, 5606–5611.

Frosst, P., Blom, H.J., Milos, R., Goyette, P., Sheppard, C.A., Matthews, R.G., Boers, G.J., den Heijer, M., Kluijtmans, L.A., van den Heuvel, L.P. and Rozen, R. (1995) A candidate genetic risk factor for vascular disease: a common mutation in methylenetetrahydrofolate reductase. *Nature Genetics* 10, 111–113.

Fuchs, C.S., Willett, W.C., Colditz, G.A., Hunter, D.J., Stampfer, M.J., Speizer, F.E. and Giovannucci, E.L. (2002) The influence of folate and multivitamin use on the familial risk of colon cancer in women. *Cancer Epidemiology Biomarkers Prev* 11, 227–234.

Fuks, F., Burgers, W.A., Brehm, A., Hughes-Davies, L. and Kouzarides, T. (2000) DNA methyltransferase Dnmt1 associates with histone deacetylase activity. *Nature Genetics* 24, 88–91.

Fuks, F., Burgers, W.A., Godin, N., Kasai, M. and Kouzarides, T. (2001) Dnmt3a binds deacetylases and is recruited by a sequence-specific repressor to silence transcription. *EMBO Journal* 20, 2536–2544.

Gallinari, P. and Jiricny, J. (1996) A new class of uracil-DNA glycosylases related to human thymine-DNA glycosylase. *Nature* 383, 735–738.

Garay, C.A. and Engstrom, P.F. (1999) Chemoprevention of colorectal cancer: dietary and pharmacologic approaches. *Oncology (Huntingt)* 13, 89–97.

Geiman, T.M. and Muegge, K. (2000) Lsh, an SNF2/helicase family member, is required for proliferation of mature T lymphocytes. *Proceedings of the National Academy of Sciences USA* 97, 4772–4777.

Gershoni-Baruch, R., Dagan, E., Israeli, D., Kasinetz, L., Kadouri, E. and Friedman, E. (2000) Association of the C677T polymorphism in the MTHFR gene with breast and/or ovarian cancer risk in Jewish women. *European Journal of Cancer* 36, 2313–2316.

Gibbons, R.J., Bachoo, S., Picketts, D.J., Aftimos, S., Asenbauer, B., Bergoffen, J., Berry, S.A., Dahl, N., Fryer, A., Keppler, K., Kurosawa, K., Levin, M.L., Masuno, M., Neri, G., Pierpont, M.E., Slaney, S.F. and Higgs, D.R. (1997) Mutations in transcriptional regulator ATRX establish the functional significance of a PHD-like domain. *Nature Genetics* 17, 146–148.

Gibbons, R.J., McDowell, T.L., Raman, S., O'Rourke, D.M., Garrick, D., Ayyub, H. and Higgs, D.R. (2000) Mutations in ATRX, encoding a SWI/SNF-like protein, cause diverse changes in the pattern of DNA methylation. *Nature Genetics* 24, 368–371.

Giovannucci, E. and Willett, W.C. (1994) Dietary factors and risk of colon cancer. *Annals of Medicine* 26, 443–452.

Giovannucci, E., Stampfer, M.J., Colditz, G.A., Rimm, E.B., Trichopoulos, D., Rosner, B.A., Speizer, F.E. and Willett, W.C. (1993) Folate, methionine, and alcohol intake and risk of colorectal adenoma. *Journal of the National Cancer Institute* 85, 875–884.

Giovannucci, E., Rimm, E.B., Ascherio, A., Stampfer, M.J., Colditz, G.A. and Willett, W.C. (1995) Alcohol, low-methionine–low-folate diets, and risk of colon cancer in men. *Journal of the National Cancer Institute* 87, 265–273.

Giovannucci, E., Stampfer, M.J., Colditz, G.A., Hunter, D.J., Fuchs, C., Rosner, B.A., Speizer, F.E. and Willett, W.C. (1998) Multivitamin use, folate, and colon cancer in women in the Nurses' Health Study. *Annals of Internal Medicine* 129, 517–524.

Glynn, S.A., Albanes, D., Pietinen, P., Brown, C.C., Rautalahti, M., Tangrea, J.A., Gunter, E.W., Barrett, M.J., Virtamo, J. and Taylor, P.R. (1996) Colorectal cancer and folate status: a nested case-control study among male smokers. *Cancer Epidemiological Biomarkers Prev* 5, 487–494.

Goelz, S.E., Vogelstein, B., Hamilton, S.R. and Feinberg, A.P. (1985) Hypomethylation of DNA from benign and malignant human colon neoplasms. *Science* 228, 187–190.

Goto, T. and Monk, M. (1998) Regulation of X-chromosome inactivation in development in mice and humans. *Microbiology and Molecular Biology Reviews* 62, 362–378.

Gowher, H. and Jeltsch, A. (2001) Enzymatic properties of recombinant Dnmt3a DNA methyltransferase from mouse: the enzyme modifies DNA in a non-processive manner and also methylates non-CpG sites. *Journal of Molecular Biology* 309, 1201–1208.

Graham, S., Hellmann, R., Marshall, J., Freudenheim, J., Vena, J., Swanson, M., Zielezny, M., Nemoto, T., Stubbe, N. and Raimondo, T. (1991) Nutritional epidemiology of postmenopausal breast cancer in western New York. *American Journal of Epidemiology* 134, 552–566.

Gunter, E., Bowman, B., Caudill, S., Twite, D., Adams, M. and Sampson, E. (1996) Results of an international round robin for serum and whole-blood folate. *Clinical Chemistry* 42, 1689–1694.

Hardeland, U., Bentele, M., Lettieri, T., Steinacher, R., Jiricny, J. and Schar, P. (2001) Thymine DNA glycosylase. *Progress in Nucleic Acid Research and Molecular Biology* 68, 235–253.

Homann, N., Tillonen, J. and Salaspuro, M. (2000) Microbially produced acetaldehyde from ethanol may increase the risk of colon cancer via folate deficiency. *International Journal of Cancer* 86, 169–173.

Hoover, K.L., Lynch, P.H. and Poirier, L.A. (1984) Profound postinitiation enhancement by short-term severe methionine, choline, vitamin B12, and folate deficiency of hepatocarcinogenesis in F344 rats given a single low-dose diethylnitrosamine injection. *Journal of the National Cancer Institute* 73, 1327–1336.

Hsieh, C.L. (1999) In vivo activity of murine de novo methyltransferases, Dnmt3a and Dnmt3b. *Molecular Cell Biology* 19, 8211–8218.

Hsu, D.W., Lin, M.J., Lee, T.L., Wen, S.C., Chen, X. and Shen, C.K. (1999) Two major forms of DNA (cytosine-5) methyltransferase in human somatic tissues. *Proceedings of the National Academy of Sciences USA* 96, 9751–9756.

Im, Y.S., Chiang, P.K. and Cantoni, G.L. (1979) Guanidoacetate methyltransferase. Purification and molecular properties. *Journal of Biological Chemistry* 254, 11047–11050.

Institute of Medicine and National Academy of Science, USA (2000) *Dietary Reference Intakes for Thiamin, Riboflavin, Niacin, Vitamin B6, Folate, Vitamin B12, Pantothenic Acid, Biotin and Choline.* National Academy Press, Washington, DC.

Issa, J.P. (2000a) CpG-island methylation in aging and cancer. *Current Topics in Microbiology and Immunology* 249, 101–118.

Issa, J.P. (2000b) The epigenetics of colorectal cancer. *Annals of the New York Academy of Science* 910, 140–53; discussion 153–155.

Jackson-Grusby, L., Laird, P.W., Magge, S.N., Moeller, B.J. and Jaenisch, R. (1997) Mutagenicity of 5-aza-2′-deoxycytidine is mediated by the mammalian DNA methyltransferase. *Proceedings of the National Academy of Sciences USA* 94, 4681–4685.

Jacob, R.A., Gretz, D.M., Taylor, P.C., James, S.J., Pogribny, I.P., Miller, B.J., Henning, S.M. and Swendseid, M.E. (1998) Moderate folate depletion increases plasma homocysteine and decreases lymphocyte DNA methylation in postmenopausal women. *Journal of Nutrition* 128, 1204–1212.

James, S.J. and Yin, L. (1989) Diet-induced DNA damage and altered nucleotide metabolism in lymphocytes from methyl-donor-deficient rats. *Carcinogenesis* 10, 1209–1214.

Jarvis, C.D., Geiman, T., Vila-Storm, M.P., Osipovich, O., Akella, U., Candeias, S., Nathan, I., Durum, S.K. and Muegge, K. (1996) A novel putative helicase produced in early murine lymphocytes. *Gene* 169, 203–207.

Jones, P.A. and Takai, D. (2001) The role of DNA methylation in mammalian epigenetics. *Science* 293, 1068–1070.

Jost, J.P., Oakeley, E.J., Zhu, B., Benjamin, D., Thiry, S., Siegmann, M. and Jost, Y.C. (2001) 5-Methylcytosine DNA glycosylase participates in the genome-wide loss of DNA methylation occurring during mouse myoblast differentiation. *Nucleic Acids Research* 29, 4452–4461.

Kato, I., Dnistrian, A.M., Schwartz, M., Toniolo, P., Koenig, K., Shore, R.E., Akhmedkhanov, A., Zeleniuch-Jacquotte, A. and Riboli, E. (1999) Serum folate, homocysteine and colorectal cancer risk in women: a nested case-control study. *British Journal of Cancer* 79, 1917–1922.

Kenyon, S.H., Nicolaou, A. and Gibbons, W.A. (1998) The effect of ethanol and its metabolites upon methionine synthase activity in vitro. *Alcohol* 15, 305–309.

Kim, Y.I., Giuliano, A., Hatch, K.D., Schneider, A., Nour, M.A., Dallal, G.E., Selhub, J. and Mason, J.B. (1994) Global DNA hypomethylation increases progressively in cervical dysplasia and carcinoma. *Cancer* 74, 893–899.

Kim, Y.I., Christman, J.K., Fleet, J.C., Cravo, M.L., Salomon, R.N., Smith, D., Ordovas, J., Selhub, J. and Mason, J.B. (1995) Moderate folate deficiency does not cause global hypomethylation of hepatic and colonic DNA or c-myc-specific hypomethylation of colonic DNA in rats. *American Journal of Clinical Nutrition* 61, 1083–1090.

Kim, Y.I., Salomon, R.N., Graeme-Cook, F., Choi, S.W., Smith, D.E., Dallal, G.E. and Mason, J.B. (1996) Dietary folate protects against the development of macroscopic colonic neoplasia in a dose responsive manner in rats. *Gut* 39, 732–740.

Kim, Y.-I., Pogribny, I., Basnakian, A., Miller, J., Selhub, J., James, S., and Mason, J. (1997) Folate deficiency in rats induces DNA strand breaks and hypomethylation within the p53 tumor suppressor gene. *American Journal of Clinical Nutrition* 65, 46–52.

Kim, Y.I., Fawaz, K., Knox, T., Lee, Y.M., Norton, R., Libby, E. and Mason, J.B. (2001) Colonic mucosal concentrations of folate are accurately predicted by blood measurements of folate status among individuals ingesting physiologic quantities of folate. *Cancer Epidemiology Biomarkers Prev* 10, 715–719.

Knudson, A.G. Jr (1971) Mutation and cancer: statistical study of retinoblastoma. *Proceedings of the National Academy of Sciences USA* 68, 820–823.

Kotb, M., Mudd, S.H., Mato, J.M., Geller, A.M., Kredich, N.M., Chou, J.Y. and Cantoni, G.L. (1997) Consensus nomenclature for the mammalian methionine adenosyltransferase genes and gene products. *Trends in Genetics* 13, 51–52.

Kruh, J. (1982) Effects of sodium butyrate, a new pharmacological agent, on cells in culture. *Molecular*

*Cell Biochemistry* 42, 65–82.
Laird, P.W., Jackson-Grusby, L., Fazeli, A., Dickinson, S.L., Jung, W.E., Li, E., Weinberg, R.A. and Jaenisch, R. (1995) Suppression of intestinal neoplasia by DNA hypomethylation. *Cell* 81, 197–205.
Lapeyre, J.N. and Becker, F.F. (1979) 5-Methylcytosine content of nuclear DNA during chemical hepatocarcinogenesis and in carcinomas which result. *Biochemical and Biophysical Research Communications* 87, 698–705.
Lauster, R., Trautner, T.A. and Noyer-Weidner, M. (1989) Cytosine-specific type II DNA methyltransferases. A conserved enzyme core with variable target-recognizing domains. *Journal of Molecular Biology* 206, 305–312.
Lee, P.J., Washer, L.L., Law, D.J., Boland, C.R., Horon, I.L. and Feinberg, A.P. (1996) Limited up-regulation of DNA methyltransferase in human colon cancer reflecting increased cell proliferation. *Proceedings of the National Academy of Sciences USA* 93, 10366–10370.
Leonhardt, H. and Cardoso, M.C. (2000) DNA methylation, nuclear structure, gene expression and cancer. *Journal of Cell Biochemistry* (Supplement 35), 78–83.
Leonhardt, H., Page, A.W., Weier, H.U. and Bestor, T.H. (1992) A targeting sequence directs DNA methyltransferase to sites of DNA replication in mammalian nuclei. *Cell* 71, 865–873.
Levine, A.J., Siegmund, K.D., Ervin, C.M., Diep, A., Lee, E.R., Frankl, H.D. and Haile, R.W. (2000) The methylenetetrahydrofolate reductase 677C→T polymorphism and distal colorectal adenoma risk. *Cancer Epidemiology Biomarkers Prev* 9, 657–663.
Li, E., Bestor, T.H. and Jaenisch, R. (1992) Targeted mutation of the DNA methyltransferase gene results in embryonic lethality. *Cell* 69, 915–926.
Liang, G., Chan, M.F., Tomigahara, Y., Tsai, Y.C., Gonzales, F.A., Li, E., Laird, P.W. and Jones, P.A. (2002) Cooperativity between DNA methyltransferases in the maintenance methylation of repetitive elements. *Molecular Cellular Biology* 22, 480–491.
Liu, Y., Oakeley, E.J., Sun, L. and Jost, J.P. (1998) Multiple domains are involved in the targeting of the mouse DNA methyltransferase to the DNA replication foci. *Nucleic Acids Research* 26, 1038–1045.
Ma, J., Stampfer, M.J., Giovannucci, E., Artigas, C., Hunter, D.J., Fuchs, C., Willett, W.C., Selhub, J., Hennekens, C.H. and Rozen, R. (1997) Methylenetetrahydrofolate reductase polymorphism, dietary interactions, and risk of colorectal cancer. *Cancer Research* 57, 1098–1102.
Magdinier, F. and Wolffe, A.P. (2001) Selective association of the methyl-CpG binding protein MBD2 with the silent p14/p16 locus in human neoplasia. *Proceedings of the National Academy of Sciences USA* 98, 4990–4995.
Maier, S. and Olek, A. (2002) Diabetes: a candidate disease for efficient DNA methylation profiling. *Journal of Nutrition* 132 (Supplement 8), 2440S–2443S.
Mann, J.R., Szabo, P.E., Reed, M.R. and Singer-Sam, J. (2000) Methylated DNA sequences in genomic imprinting. *Critical Reviews in Eukaryotic Gene Expression* 10, 241–257.
Mayer, W., Niveleau, A., Walter, J., Fundele, R. and Haaf, T. (2000) Demethylation of the zygotic paternal genome. *Nature* 403, 501–502.
Medina, M., Urdiales, J.L. and Amores-Sanchez, M.I. (2001) Roles of homocysteine in cell metabolism: old and new functions. *European Journal of Biochemistry* 268, 3871–3882.
Mendelsohn, N., Michl, J., Gilbert, H.S., Acs, G. and Christman, J.K. (1980) L-Ethionine as an inducer of differentiation in human promyelocytic leukemia cells (HL-60). *Cancer Research* 40, 3206–3210.
Mertineit, C., Yoder, J.A., Taketo, T., Laird, D.W., Trasler, J.M. and Bestor, T.H. (1998) Sex-specific exons control DNA methyltransferase in mammalian germ cells. *Development* 125, 889–897.
Mikol, Y.B., Hoover, K.L., Creasia, D. and Poirier, L.A. (1983) Hepatocarcinogenesis in rats fed methyl-deficient, amino acid-defined diets. *Carcinogenesis* 4, 1619–1629.
Mudd, S.H. and Poole, J.R. (1975) Labile methyl balances for normal humans on various dietary regimens. *Metabolism* 24, 721–735.
Narayan, A., Ji, W., Zhang, X.Y., Marrogi, A., Graff, J.R., Baylin, S.B. and Ehrlich, M. (1998) Hypomethylation of pericentromeric DNA in breast adenocarcinomas. *International Journal of Cancer* 77, 833–838.
Negri, E., La Vecchia, C. and Franceschi, S. (2000) Re: dietary folate consumption and breast cancer risk. *Journal of the National Cancer Institute* 92, 1270–1271.
Newberne, P.M., Carltob, W.W. and Wong, G.N. (1964) Hepatomas in rats and hepatorenal injury induced by peanut meal in *Aspergillus flavus* extract. *Path Vet* 1, 105–132.
Nowell, P. (1976) The clonal evolution of tumor cell populations. *Science* 194, 23–28.
Okano, M., Xie, S. and Li, E. (1998a) Cloning and characterization of a family of novel mammalian DNA (cytosine-5) methyltransferases. *Nature Genetics* 19, 219–220.
Okano, M., Xie, S. and Li, E. (1998b) Dnmt2 is not required for de novo and maintenance methylation of viral DNA in embryonic stem cells. *Nucleic Acids Research* 26, 2536–2540.
Okano, M., Bell, D.W., Haber, D.A. and Li, E. (1999) DNA methyltransferases Dnmt3a and Dnmt3b are essential for de novo methylation and mammalian development. *Cell* 99, 247–257.
Piyathilake, C.J. and Johanning, G.L. (2002) Cellular

vitamins, DNA methylation and cancer risk. *Journal of Nutrition* 132 (Supplement 8), 2340S–2344S.

Piyathilake, C.J., Hine, R.J., Dasanayake, A.P., Richards, E.W., Freeberg, L.E., Vaughn, W.H. and Krumdieck, C.L. (1992) Effect of smoking on folate levels in buccal mucosal cells. *International Journal of Cancer* 52, 566–569.

Piyathilake, C.J., Johanning, G.L., Macaluso, M., Whiteside, M., Oelschlager, D.K., Heimburger, D.C. and Grizzle, W.E. (2000) Localized folate and vitamin B-12 deficiency in squamous cell lung cancer is associated with global DNA hypomethylation. *Nutrition and Cancer* 37, 99–107.

Pogribny, I P. and James, S.J. (2002) Reduction of p53 gene expression in human primary hepatocellular carcinoma is associated with promoter region methylation without coding region mutation. *Cancer Letters* 176, 169–174.

Pogribny, I.P., Miller, B.J. and James, S.J. (1997a) Alterations in hepatic p53 gene methylation patterns during tumor progression with folate/methyl deficiency in the rat. *Cancer Letters* 115, 31–38.

Pogribny, I.P., Muskhelishvili, L., Miller, B.J. and James, S.J. (1997b) Presence and consequence of uracil in preneoplastic DNA from folate/methyl-deficient rats. *Carcinogenesis* 18, 2071–2076.

Pogribny, I.P., Pogribna, M., Christman, J.K. and James, S.J. (2000) Single-site methylation within the p53 promoter region reduces gene expression in a reporter gene construct: possible in vivo relevance during tumorigenesis. *Cancer Research* 60, 588–594.

Posfai, J., Bhagwat, A.S., Posfai, G. and Roberts, R.J. (1989) Predictive motifs derived from cytosine methyltransferases. *Nucleic Acids Research* 17, 2421–2435.

Potischman, N., Swanson, C.A., Coates, R.J., Gammon, M.D., Brogan, D.R., Curtin, J. and Brinton, L.A. (1999) Intake of food groups and associated micronutrients in relation to risk of early-stage breast cancer. *International Journal of Cancer* 82, 315–321.

Potter, J.D. (1999) Colorectal cancer: molecules and populations. *Journal of the National Cancer Institute* 91, 916–932.

Pradhan, S., Bacolla, A., Wells, R.D. and Roberts, R.J. (1999) Recombinant human DNA (cytosine-5) methyltransferase. I. Expression, purification, and comparison of de novo and maintenance methylation. *Journal of Biological Chemistry* 274, 33002–33010.

Pradhan, S. and Kim, G.D. (2002) The retinoblastoma gene product interacts with maintenance human DNA (cytosine-5) methyltransferase and modulates its activity. *EMBO Journal* 21, 779–788.

Qu, G., Dubeau, L., Narayan, A., Yu, M.C., Ehrlich, M. (1999) Satellite DNA hypomethylation vs overall genomic hypomethylation in ovarian epithelial tumors of different malignant potential. *Mutation Research Fund Molecular Mech Mutagens* 423, 91–101.

Raabe, E.H., Abdurrahman, L., Behbehani, G. and Arceci, R.J. (2001) An SNF2 factor involved in mammalian development and cellular proliferation. *Developmental Dynamics* 221, 92–105.

Razin, A. and Riggs, A.D. (1980) DNA methylation and gene function. *Science* 210, 604–610.

Reik, W., Dean, W. and Walter, J. (2001) Epigenetic reprogramming in mammalian development. *Science* 293, 1089–1093.

Rein, T., DePamphilis, M.L. and Zorbas, H. (1998) Identifying 5-methylcytosine and related modifications in DNA genomes. *Nucleic Acids Research* 26, 2255–2264.

Rhee, I., Bachman, K.E., Park, B.H., Jair, K.W., Yen, R.W., Schuebel, K.E., Cui, H., Feinberg, A.P., Lengauer, C., Kinzler, K.W., Baylin, S.B. and Vogelstein, B. (2002) DNMT1 and DNMT3b cooperate to silence genes in human cancer cells. *Nature* 416, 552–556.

Riggs, A.D. and Jones, P.A. (1983) 5-Methylcytosine, gene regulation, and cancer. *Advances in Cancer Research* 40, 1–30.

Robertson, K.D. (2001) DNA methylation, methyltransferases, and cancer. *Oncogene* 20, 3139–3155.

Robertson, K.D. (2002) DNA methylation and chromatin – unraveling the tangled web. *Oncogene* 21, 5361–5379.

Robertson, K.D., Ait-Si-Ali, S., Yokochi, T., Wade, P.A., Jones, P.L. and Wolffe, A.P. (2000) DNMT1 forms a complex with Rb, E2F1 and HDAC1 and represses transcription from E2F-responsive promoters. *Nature Genetics* 25, 338–342.

Rohan, T.E., Jain, M.G., Howe, G.R. and Miller, A.B. (2000) Dietary folate consumption and breast cancer risk. *Journal of the National Cancer Institute* 92, 266–269.

Ronco, A., De Stefani, E., Boffetta, P., Deneo-Pellegrini, H., Mendilaharsu, M. and Leborgne, F. (1999) Vegetables, fruits, and related nutrients and risk of breast cancer: a case-control study in Uruguay. *Nutrition and Cancer* 35, 111–119.

Rosenquist, T.H., Schneider, A.M. and Monaghan, D.T. (1999) *N*-methyl-D-aspartate receptor agonists modulate homocysteine-induced developmental abnormalities. *FASEB Journal* 13, 1523–1531.

Rountree, M.R., Bachman, K.E. and Baylin, S.B. (2000) DNMT1 binds HDAC2 and a new co-repressor, DMAP1, to form a complex at replication foci. *Nature Genetics* 25, 269–277.

Rountree, M.R., Bachman, K.E., Herman, J.G. and Baylin, S.B. (2001) DNA methylation, chromatin inheritance, and cancer. *Oncogene* 20, 3156–3165.

Ryan, B.M. and Weir, D.G. (2001) Relevance of folate metabolism in the pathogenesis of colorectal cancer. *Journal of Laboratory and Clinical Medicine* 138,

164–176.

Saito, Y., Kanai, Y., Sakamoto, M., Saito, H., Ishii, H. and Hirohashi, S. (2002) Overexpression of a splice variant of DNA methyltransferase 3b, DNMT3b4, associated with DNA hypomethylation on pericentromeric satellite regions during human hepatocarcinogenesis. *Proceedings of the National Academy of Sciences USA* 99, 10060–10065.

Salaspuro, M. (1996) Bacteriocolonic pathway for ethanol oxidation: characteristics and implications. *Annals of Medicine* 28, 195–200.

Santi, D.V., Norment, A. and Garrett, C. E. (1984) Covalent bond formation between a DNA-cytosine methyltransferase and DNA containing 5-azacytosine. *Proceedings of the National Academy of Sciences USA* 81, 6993–6997.

Santourlidis, S., Florl, A., Ackermann, R., Wirtz, H.C. and Schulz, W.A. (1999) High frequency of alterations in DNA methylation in adenocarcinoma of the prostate. *Prostate* 39, 166–174.

Sealy, L. and Chalkley, R. (1978) The effect of sodium butyrate on histone modification. *Cell* 14, 115–121.

Shane, B. (1995) Folate chemistry and metabolism. In: *Folate in Health and Disease.* Marcel Dekker, New York, pp. 1–22.

Sharp, L., Little, J., Schofield, A.C., Pavlidou, E., Cotton, S.C., Miedzybrodzka, Z., Baird, J.O., Haites, N.E., Heys, S.D. and Grubb, D.A. (2002) Folate and breast cancer: the role of polymorphisms in methylenetetrahydrofolate reductase (MTHFR). *Cancer Letters* 181, 65–71.

Shen, L., Fang, J., Qiu, D., Zhang, T., Yang, J., Chen, S. and Xiao, S. (1998) Correlation between DNA methylation and pathological changes in human hepatocellular carcinoma. *Hepatogastroenterology* 45, 1753–1759.

Shen, H., Spitz, M.R., Wang, L.E., Hong, W.K. and Wei, Q. (2001) Polymorphisms of methylenetetrahydrofolate reductase and risk of lung cancer: a case-control study. *Cancer Epidemiology Biomarkers Prev* 10, 397–401.

Shivapurkar, N. and Poirier, L.A. (1983) Tissue levels of S-adenosylmethionine and S-adenosylhomocysteine in rats fed methyl-deficient, amino acid-defined diets for one to five weeks. *Carcinogenesis* 4, 1051–1057.

Sibani, S., Melnyk, S., Pogribny, I.P., Wang, W., Hiou-Tim, F., Deng, L., Trasler, J., James, S.J. and Rozen, R. (2002) Studies of methionine cycle intermediates (SAM, SAH), DNA methylation and the impact of folate deficiency on tumor numbers in Min mice. *Carcinogenesis* 23, 61–65.

Singal, R. and Ginder, G.D. (1999) DNA methylation. *Blood* 93, 4059–4070.

Skibola, C.F., Smith, M.T., Kane, E., Roman, E., Rollinson, S., Cartwright, R.A. and Morgan, G. (1999) Polymorphisms in the methylenetetrahydrofolate reductase gene are associated with susceptibility to acute leukemia in adults. *Proceedings of the National Academy of Sciences USA* 96, 12810–12815.

Slattery, M.L., Schaffer, D., Edwards, S.L., Ma, K.N. and Potter, J.D. (1997) Are dietary factors involved in DNA methylation associated with colon cancer?' *Nutrition and Cancer* 28, 52–62.

Slattery, M.L., Potter, J.D., Samowitz, W., Schaffer, D. and Leppert, M. (1999) Methylenetetrahydrofolate reductase, diet, and risk of colon cancer. *Cancer Epidemiology Biomarkers Prev* 8, 513–518.

Song, C., Xing, D., Tan, W., Wei, Q. and Lin, D. (2001) Methylenetetrahydrofolate reductase polymorphisms increase risk of esophageal squamous cell carcinoma in a Chinese population. *Cancer Research* 61, 3272–3275.

Song, J., Sohn, K.-J., Medline, A., Ash, C., Gallinger, S. and Kim, Y.-I. (2000b) Chemopreventive effects of dietary folate on intestinal polyps in Apc+/– Msh2–/– mice. *Cancer Research* 60, 3191–3199.

Song, J., Medline, A., Mason, J.B., Gallinger, S. and Kim, Y.I. (2000a) Effects of dietary folate on intestinal tumorigenesis in the apcMin mouse. *Cancer Research* 60, 5434–5440.

Stern, L.L., Mason, J.B., Selhub, J. and Choi, S.W. (2000) Genomic DNA hypomethylation, a characteristic of most cancers, is present in peripheral leukocytes of individuals who are homozygous for the C677T polymorphism in the methylenetetrahydrofolate reductase gene. *Cancer Epidemiology Biomarkers Prev* 9, 849–853.

Stolzenberg-Solomon, R.Z., Albanes, D., Nieto, F.J., Hartman, T.J., Tangrea, J.A., Rautalahti, M., Sehlub, J., Virtamo, J. and Taylor, P.R. (1999) Pancreatic cancer risk and nutrition-related methyl-group availability indicators in male smokers. *Journal of the National Cancer Institute* 91, 535–541.

Su, L.J. and Arab, L. (2001) Nutritional status of folate and colon cancer risk: evidence from NHANES I epidemiologic follow-up study. *Annals of Epidemiology* 11, 65–72.

Tatematsu, K.I., Yamazaki, T. and Ishikawa, F. (2000) MBD2-MBD3 complex binds to hemi-methylated DNA and forms a complex containing DNMT1 at the replication foci in late S phase. *Genes and Cells* 5, 677–688.

Todesco, L., Angst, C., Litynski, P., Loehrer, F., Fowler, B. and Haefeli, W.E. (1999) Methylenetetrahydrofolate reductase polymorphism, plasma homocysteine and age. *European Journal of Clinical Investigations* 29, 1003–1009.

Tomeo, C.A., Colditz, G.A., Willett, W.C., Giovannucci, E., Platz, E., Rockhill, B., Dart, H. and Hunter, D.J. (1999) *Harvard Report on Cancer Prevention,* Vol. 3, *Prevention of Colon Cancer in the United States. Cancer*

*Causes Control* 10, 167–180.

Torres, L., Garcia-Trevijano, E.R., Rodriguez, J.A., Carretero, M.V., Bustos, M., Fernandez, E., Eguinoa, E., Mato, J.M. and Avila, M.A. (1999) Induction of TIMP-1 expression in rat hepatic stellate cells and hepatocytes: a new role for homocysteine in liver fibrosis. *Biochimica Biophysica Acta* 1455, 12–22.

Tseng, M., Murray, S.C., Kupper, L.L. and Sandler, R.S. (1996) Micronutrients and the risk of colorectal adenomas. *American Journal of Epidemiology* 144, 1005–1014.

Tycko, B. (2000) Epigenetic gene silencing in cancer. *Journal of Clinical Investigations* 105, 401–407.

Ulrich, C.M., Kampman, E., Bigler, J., Schwartz, S.M., Chen, C., Bostick, R., Fosdick, L., Beresford, S.A., Yasui, Y. and Potter, J.D. (1999) Colorectal adenomas and the C677T MTHFR polymorphism: evidence for gene–environment interaction? *Cancer Epidemiology Biomarkers Prev* 8, 659–668.

Van den Veyver, I.B. (2002) Genetic effects of methylation diets. *Annual Review of Nutrition* 22, 255–282.

Vignali, M., Hassan, A.H., Neely, K.E. and Workman, J.L. (2000) ATP-dependent chromatin-remodeling complexes. *Molecular Cell Biology* 20, 1899–1910.

Wagner, C. (1982) Cellular folate binding proteins; function and significance. *Annual Review of Nutrition* 2, 229–248.

Wainfan, E., Dizik, M. and Balis, M.E. (1984) Increased activity of rat liver N2-guanine tRNA methyltransferase II in response to liver damage. *Biochimica et Biophysica Acta* 799, 282–290.

Wainfan, E., Dizik, M., Hluboky, M. and Balis, M.E. (1986) Altered tRNA methylation in rats and mice fed lipotrope-deficient diets. *Carcinogenesis* 7, 473–476.

Wainfan, E., Kilkenny, M. and Dizik, M. (1988) Comparison of methyltransferase activities of pair-fed rats given adequate or methyl-deficient diets. *Carcinogenesis* 9, 861–863.

Wainfan, E., Dizik, M., Stender, M. and Christman, J.K. (1989) Rapid appearance of hypomethylated DNA in livers of rats fed cancer-promoting, methyl-deficient diets. *Cancer Research* 49, 4094–4097.

Warnecke, P.M., Stirzaker, C., Song, J., Grunau, C., Melki, J.R. and Clark, S.J. (2002) Identification and resolution of artifacts in bisulfite sequencing. *Methods* 27, 101–107.

Weinstein, S.J., Ziegler, R.G., Frongillo, E.A.Jr, Colman, N., Sauberlich, H.E., Brinton, L.A., Hamman, R.F., Levine, R.S., Mallin, K., Stolley, P.D. and Bisogni, C.A. (2001a) Low serum and red blood cell folate are moderately, but nonsignificantly associated with increased risk of invasive cervical cancer in U.S. women. *Journal of Nutrition* 131, 2040–2048.

Weinstein, S.J., Ziegler, R.G., Selhub, J., Fears, T.R., Strickler, H.D., Brinton, L.A., Hamman, R.F., R.S., L., Malin, K. and Stolley, P.D. (2001b) Elevated serum homocysteine levels and increased risk of invasive cervical cancer in U.S. women. *Cancer Causes Control* 12, 317–324.

Weir, D.G., McGing, P.G. and Scott, J.M. (1985) Folate metabolism, the enterohepatic circulation and alcohol. *Biochemical Pharmacology* 34, 1–7.

Weisenberger, D.J., Velicescu, M., Preciado-Lopez, M.A., Gonzales, F.A., Tsai, Y.C., Liang, G. and Jones, P.A. (2002) Identification and characterization of alternatively spliced variants of DNA methyltransferase 3a in mammalian cells. *Gene* 298, 91–99.

White, E., Shannon, J.S. and Patterson, R.E. (1997) Relationship between vitamin and calcium supplement use and colon cancer. *Cancer Epidemiology Biomarkers Prev* 6, 769–774.

Wiemels, J.L., Smith, R.N., Taylor, G.M., Eden, O.B., Alexander, F.E. and Greaves, M.F. (2001) Methylenetetrahydrofolate reductase (MTHFR) polymorphisms and risk of molecularly defined subtypes of childhood acute leukemia. *Proceedings of the National Academy of Sciences USA* 98, 4004–4009.

Wigler, M., Levy, D. and Perucho, M. (1981) The somatic replication of DNA methylation. *Cell* 24, 33–40.

# 17　人类细胞中组蛋白的生物素化

Janos Zempleni
（内布拉斯加州-林肯大学营养科学与营养学系，林肯，内布拉斯加州，美国）

## 前　　言

### 组蛋白的翻译后修饰

组蛋白是介导 DNA 折叠形成染色质的主要蛋白质（Wolffe，1998）。染色质由可复制的核蛋白复合物（即核小体）组成。每个核小体由 DNA 缠绕在核心组蛋白的八聚体（组蛋白 H2A、H2B、H3 和 H4 各 2 分子）上形成；核小体的组装是由连接蛋白（组蛋白 H1）来完成。DNA 与组蛋白之间的结合依赖其静电作用，由 DNA 中带负电荷的磷酸基与组蛋白中带正电荷的 ε-氨基和胍基结合而成。

体内试验表明，组蛋白能进行翻译后的乙酰化修饰（Ausio and Holde，1986；Hebbes *et al.*，1988；Lee *et al.*，1993）、甲基化修饰（Wolffe，1998）、磷酸化修饰（Wolffe，1998）、遍在蛋白化修饰（ubiquitination）（Wolffe，1998）和多聚酶（ADP-核糖基化）修饰（Chambon *et al.*，1966；Boulikas，1988；Boulikas *et al.*，1990），这些基团与组蛋白的氨基酸残基（如赖氨酸残基的 ε-氨基和精氨酸残基的胍基）共价结合。一些修饰作用［如乙酰化和多聚（ADP-核糖基化）］使组蛋白失去一个正电荷，从而削弱了 DNA 与组蛋白之间的结合力。

### DNA 的转录活性

研究表明，染色质的转录活性受组蛋白共价修饰的调节，特别是对于组蛋白的乙酰化修饰在 DNA 转录中的作用，人们已进行了广泛的研究。这些研究结果表明，转录激活因子具有组蛋白乙酰转移酶的活性（Brownell *et al.*，1996），并且组蛋白去乙酰酶是转录抑制因子（Taunton *et al.*，1996）。组蛋白的乙酰化和染色质的转录活性具有非常强的相关性（Allfrey *et al.*，1964；Gorovsky，1973；Mathis *et al.*，1978），同时，组蛋白的遍在蛋白化与染色质的转录活性也有相关性（Pham and Sauer，2000）。然而，位于核心组蛋白中的赖氨酸残基的甲基化可能与转录抑制有关（Wolffe，1998）。

### DNA 修复

研究已表明，组蛋白的多聚（ADP-核糖基化）与 DNA 的修复机制有关（Durkacz *et al.*，1980；Althaus，1992）。由化学治疗或 UV 照射诱导的哺乳动物 DNA 损伤能显著提高组蛋白多聚（ADP-核糖基化），即提高 ADP-核糖基的共价结合力（Juarez-Salinas *et al.*，1979）。同样，淋巴样细胞受到有丝分裂原刺激后多聚（ADP-核糖基化）的增加，也是由于植物凝集素诱导的 DNA 链断裂增多所引起的（Boulikas *et al.*，1990）。

多聚（ADP-核糖化）作用是由多聚（ADP-核糖）聚合酶（EC2.4.2.30）催化完成的（Chambon *et al*.，1966；Nishizuka *et al*.，1971；Boulikas *et al*.，1990）。多聚（ADP-核糖）聚合酶的抑制剂能促进被 UV 照射过的人淋巴细胞异常的 DNA 合成（Sims *et al*.，1982）。

在核心组蛋白中，H2B 能最大限度地被多聚（ADP-核糖基化）（Boulikas *et al*.，1990），H2A、H3 和 H4 及连接组蛋白 H1 也能被多聚（ADP-核糖基化）（Boulikas *et al*.，1990）。而且，多聚（ADP-核糖）聚合酶还能催化 DNA 损伤后的自动 ADP-核糖基化（Althaus，1992）。

组蛋白的多聚（ADP-核糖基化作用）促使组蛋白与 DNA 之间发生临时的分离和再结合（Althaus，1992），人们认为该机制可把酶导入 DNA 修复位点（Althaus，1992）；组蛋白的多聚（ADP-核糖基化）可能还参与 DNA 切除修复中染色质的核小体解链过程，因为组蛋白 H1 的多聚（ADP-核糖基化）还可能参与早期的细胞凋亡过程（Yoon *et al*.，1996）。

营养状况对 DNA 修复很重要。细胞中的辅酶Ⅰ（nicotinamide adenine dinucleotide，NAD）是多聚（ADP-核糖基化作用）的底物，它的营养性耗竭可阻止 DNA 链断裂片段的再连接（Durkacz *et al*.，1980）。食物中烟酸和色氨酸是 NAD 的前体物质，即使烟酸轻度缺乏，也能降低大鼠肝脏的多聚（ADP-核糖）浓度（Rawling *et al*.，1994）。

### 细胞的增殖

组蛋白的翻译后修饰是保证细胞正常增殖所必需的。例如，用精氨酸或谷氨酸替换酿酒酵母（*Saccharomyces cerevisiae*）中组蛋白 H4 的 4 个可乙酰化的赖氨酸，会严重减缓细胞生长，或细胞 $G_2$/M 期阻滞（Megee *et al*.，1995）。在真核细胞的细胞周期中，组蛋白的修饰方式发生了显著的变化（Wolffe，1998），这与组蛋白在细胞增殖过程中发挥的作用一致如核心组蛋白在有丝分裂期进行去乙酰化。在细胞周期中，组蛋白的翻译后修饰可能调节参与细胞周期的基因转录（Wolffe，1998）。

### 精子发生

精子发生需要 DNA 组装成惰性的染色质结构。在精子发生过程中，组蛋白发生比较强的乙酰化反应，并临时被“过渡蛋白”取代（Wolffe，1998），随后过渡蛋白被含碱性强的精蛋白所取代，从而使细胞核发生最大限度的压缩。精蛋白与 DNA 之间的结合受到精蛋白的翻译后修饰（如磷酸化）的调节。压缩使精子的核 DNA 对酶解、物理和化学降解具有更大的抵抗能力。同时，这些研究结果表明，组蛋白和其他核酸结合蛋白的翻译后修饰在精子的形成过程中占有重要的地位。

## 组蛋白的酶性生物素化

近些年，研究发现组蛋白还存在另一种翻译后的修饰方式：赖氨酸残基的生物素化。Hymes 等（1995）已提出了生物素酶（EC 3.5.1.12）催化生物素与组蛋白进行共价结合的反应机制。他们还指出生物素酶裂解生物胞素（生物素-ε-赖氨酸）时，在酶活性中心及附近形成一种中间产物——生物素酰硫酯（半胱氨酸结合的生物素）（图

17.1）（Hymes *et al.*，1995；Hymes and Wolf，1999），随后，生物素酰基从硫酯转移到组蛋白中赖氨酸残基的 ε-氨基（或其他氨基）上（图 17.2）。生物素酶普遍存在于哺乳动物细胞，约 26％有活性的细胞生物素酶存在于细胞核中（Pispa，1965）。

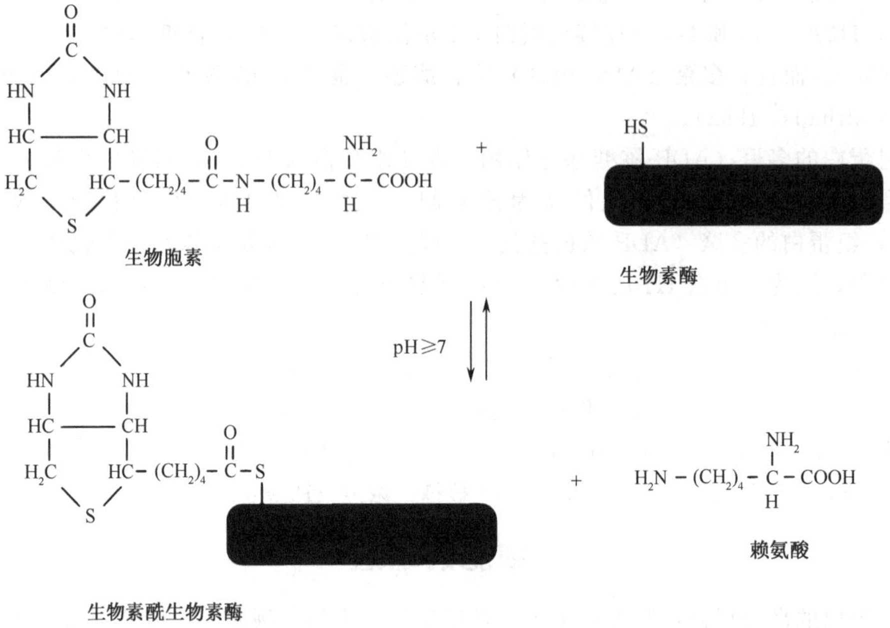


图 17.1　生物素酶裂解生物胞素（生物素-ε-赖氨酸）时，在酶活性中心及附近形成一种中间产物——生物素酰硫酯（半胱氨酸结合的生物素、生物素酰生物素酶）（Hymes and Wolf，1999）

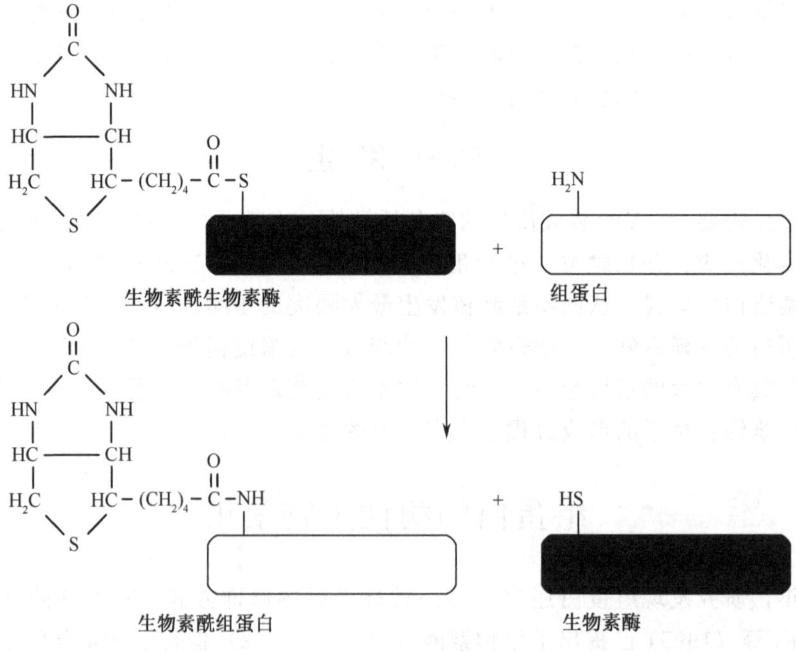


图 17.2　来自中间产物生物素酰生物素酶的生物素酰基转移到组蛋白中赖氨酸的 ε-氨基（或其他氨基）上（Hymes and Wolf，1999）

# 人类细胞的生物素化组蛋白

## 生物素化组蛋白的鉴定

我们实验室研究了人类细胞是否含有生物素化组蛋白，采用酸提取法，从人外周血单核细胞（PBMC）的细胞核中分离出组蛋白（Stanley *et al*.，2001）。根据商品组蛋白在聚丙烯酰胺凝胶上的共同迁移情况和考马斯蓝染色后的非常重要的非组蛋白条带的缺少情况，该项研究分为 5 类主要组蛋白（H1、H2A、H2B、H3 和 H4）的纯化制备（图 17.3，泳道 1 和 2）。取相同的 PBMC 细胞核的提取物，检测组蛋白是否共价结合了生物素。采用组蛋白的 Western 印迹法，用链霉抗生物素蛋白结合的过氧化物酶来检测生物素，过氧化物酶的活性采用化学发光法来检测。组蛋白 H1、H3 和 H4 含有链霉抗生物素蛋白结合的部位，这表明这些组蛋白可被生物素化。组蛋白 H2A 和 H2B 电泳成一条带，这样就无法确定链霉抗生物素蛋白是与 H2A 或 H2B 结合，还是同时与两者结合。

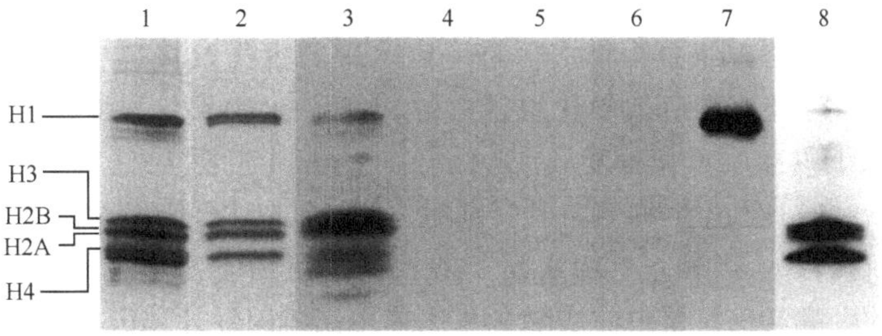


图 17.3　人 PBMC 的核含有生物素化组蛋白

组蛋白从 PBMC 核中提取，采用 SDS 凝胶电泳进行色谱分离。非组蛋白和合成的多肽（多聚-L-精氨酸和多聚-L-赖氨酸）从 Sigma 公司购买，采用组蛋白的分离方法来分离。样品用 Gelcode 蓝进行染色（泳道 1 和 2）、用链霉抗生物素蛋白结合的过氧化物酶来检测（泳道 3～7）或用抗生物素的单克隆抗体来检测（泳道 8）。泳道 1，人 PBMC 核的组蛋白；泳道 2，犊牛胸腺提取的商品组蛋白；泳道 3，人 PBMC 核的组蛋白；泳道 4，抑肽酶、α-乳白蛋白、β-乳球蛋白、胰蛋白酶抑制因子和胰蛋白酶原的混合物；泳道 5，多聚-L-精氨酸；泳道 6，多聚-L-赖氨酸；泳道 7，化学生物素化的组蛋白 H1；泳道 8，人 PBMC 核的组蛋白。除化学合成的生物素化的组蛋白 H1 大约稀释 20 000 倍外，试验采用摩尔数相等的组蛋白、多聚-L-精氨酸、多聚-L-赖氨酸和非组蛋白（Stanley *et al*.，2001）

随后的试验表明，除了与非生物素化蛋白质发生非特异性的结合外，链霉抗生物素蛋白还能特异性地结合生物素化组蛋白。非生物素化蛋白质（抑肽酶、α-乳白蛋白、β-乳球蛋白、胰蛋白酶抑制因子和胰蛋白酶原）的分子质量与这些组蛋白的分子质量相近，当用链霉抗生物素蛋白结合的过氧化物酶来检测非生物素化蛋白质时（图 17.3，泳道 4），没有观察到条带；试验用合成的多肽（多聚-L-赖氨酸和多聚-L-精氨酸）来模拟富含赖氨酸和精氨酸的组蛋白，再用链霉抗生物素蛋白结合的过氧化物酶来检测时，也没观察到条带（图 17.3，泳道 5 和 6），随后，对组蛋白 H1 则采用前面叙述的化学方法进行生物素化（Zempleni and Mock，1999），这种合成的复合物用作正对照（Stanley *et al*.，2001）；若用链霉抗生物素蛋白结合的过氧化物酶来检测，这种生物素化组蛋白 H1 形成一较强的条带（图 17.3，泳道 7）；最后，PBMC 细胞核的提取物采用抗生物素的单克隆抗体来检测（图 17.3，泳道 8），抗体与组蛋白之间的结合进一步证明了组蛋白含有生物素。

## 细胞的增殖作用

提高不同基因的表达是确保细胞正常增殖必不可少的。从理论上讲，组蛋白的生物素化能提高 DNA 的转录，以此类推，组蛋白的乙酰化和遍在蛋白化均能增强 DNA 的转录活性（Lee *et al.*，1993；Sommerville *et al.*，1993；Pham and Sauer，2000）。那么组蛋白的生物素化是否与细胞增殖有关呢？为此，本实验室采用刀豆素 A 诱导 PBMC 的增殖，而对照组不添加刀豆素 A（Stanley *et al.*，2001）。培养基含有生理浓度的 [$^3$H] 生物素（475pmol/L），采用链霉抗生物素蛋白结合的过氧化物酶的 Western 印迹法和流体闪烁计数法，研究组蛋白的生物素化。

PBMC 进行增殖的同时组蛋白的生物素化也提高了。PBMC 组蛋白的凝胶电泳分析结果证实，所有组蛋白（即 H1、H2A、H2B、H3 和 H4）的生物素化程度均得到了提高（图 17.4A）。研究还发现，静止对照组也发生了组蛋白的生物素化，这表明不管细胞处于哪种增殖状态，组蛋白都应保持一定水平的生物素化。与 Western 印迹法相比，流体闪烁计数法能更准确地量化组蛋白的 [$^3$H] 生物素化程度。PBMC 进行增殖时，组蛋白的 [$^3$H] 生物素化程度从（3.9±1.3）amol 生物素/$10^6$ 个细胞（静止状态）提高到（20.1±2.9）amol 生物素/$10^6$ 个细胞（图 17.4B）。值得注意的是流体闪烁计数法只能检测新结合的 [$^3$H] 生物素而不能检测组蛋白本身含有的未标记的生物素，导致每个细胞的生物素化组蛋白的实际含量比采用流体闪烁计数法检测的要高。

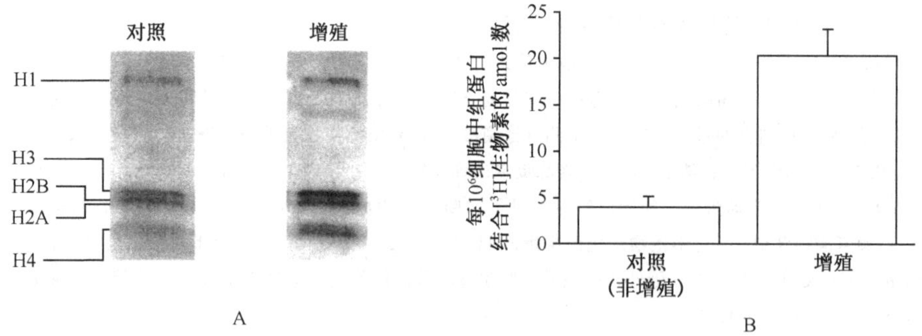


图 17.4 人类 PBMC 对细胞增殖（伴随组蛋白生物素化增加）的反应
A. 用刀豆素 A 培养 PBMC 以诱导细胞增殖；非增殖对照组没用刀豆素 A 培养。从增殖和未增殖 PBMC（$5.1\times10^6$PBMC）中分离等量的组蛋白，用 SDS-PAGE 和化学荧光法进行分析。B. 用液体闪烁计数法检测在增殖和未增殖 PBMC 中结合到组蛋白中的 [$^3$H] 生物素的量。从增殖和未增殖（对照）PBMC（$20.48\times10^6$PBMC）中分离等量的组蛋白；PBMC 数量根据（A）中使用的进行确定。$P<0.05$；$n=3$（Stanley *et al.*，2001）

## 细胞增殖期间组蛋白的生物素化

从理论上讲，在增殖的 PBMC 中，组蛋白的生物素化的提高可能局限在细胞周期的某个特定的时段。用处于停滞期的 PBMC 来检测不同时段的组蛋白生物素化（Stanley *et al.*, 2001)。PBMC 采用 475pmol/L [$^3$H] 生物素和 20μg/ml 伴刀豆素 37℃培养 30h，用以诱导细胞增殖，静止对照组的 PBMC 在不添加刀豆素 A 情况下进行培养。然后，把下面的化学试剂之一加入到增殖的 PBMC 中，而且，为了使细胞被阻滞在细胞周期的某一特定时段，继续培养 16h：100nmol/L（最终的浓度）渥曼青霉素使细胞被阻滞在 $G_1$ 期（Yano *et al.*, 1993），118μmol/L 阿非迪霉素（一种 DNA 聚合阻抑剂——蚜肠霉素）使细胞被阻滞在 S 期（Dasso and Newport，1990），1μmol/L 冈田酸使细胞被阻滞在 $G_2$ 期（Cohen，1989；Cohen *et al.*, 1989；Haavik *et al.*, 1989），5 μmol/L 秋水仙碱使细胞被阻滞在 M 期（Chalifour and Dakshinamurti，1982）。组蛋白是从 PBMC 的细胞核中提取的，[$^3$H] 生物素化的组蛋白采用流体闪烁计数法来检测，与静止对照组（$G_0$ 期）相比，组蛋白的 [$^3$H] 生物素化在细胞周期早期（$G_1$ 期）就得到了提高，并在后面的阶段仍保持上升（S、$G_2$ 和 M 期）；若采用 $1/10^6$ 细胞来进行标准化，提高幅度达 4 倍以上（图 17.5A）。

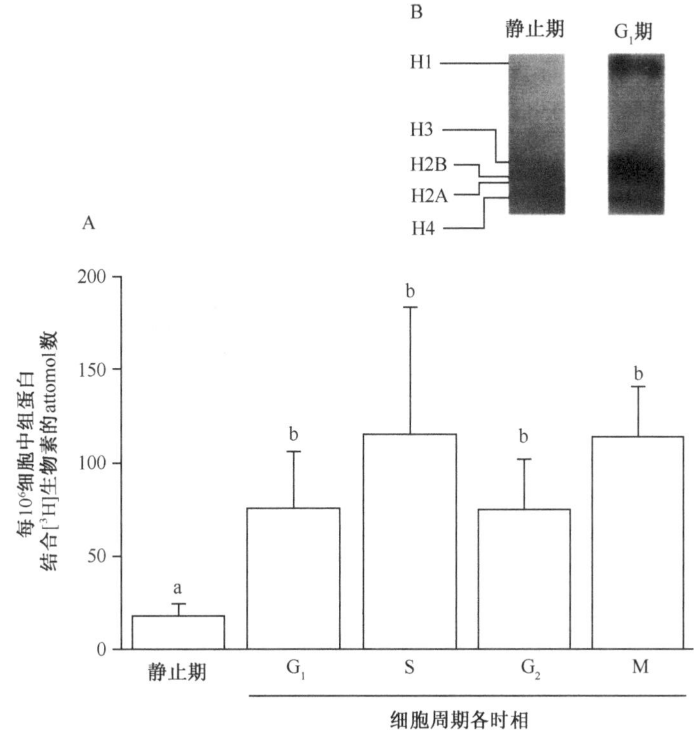


图 17.5　细胞周期各时相中人 PBMC 的组蛋白生物素化

PBMC 按本文所述的方法和 [$^3$H] 生物素一起培养；采用化学方法对细胞周期的各个时期进行阻滞。A. 组蛋白的 $^3$H-生物素化采用流体闪烁计数法进行检测。静止的 PBMC 用作对照。不含相同字母的柱形图之间差异显著；$P<0.01$；$n=4$。B. 组蛋白的生物素化采用 SDS-PAGE 和化学发光法来分析。研究 $G_0$ 和 $G_1$ 期的 PBMC；阻滞在 $G_1$ 期的 PBMC 发生的生物素化方式，代表了 $G_1$、S、$G_2$ 和 M 期的细胞的生物素化方式（Stanley *et al.*, 2001）

是细胞周期的不同时相的所有组蛋白的生物素化增加还是只有部分组蛋白的生物素化增加？采用 SDS-PAGE 和化学发光法所得到的分析结果表明，与 $G_0$ 期相比，细胞周

期的 $G_1$、S、$G_2$ 和 M 期的所有组蛋白的生物素化均得到了提高（图 17.5B），这可能暗示组蛋白的生物素化是所有组蛋白以相似的程度进行的一般反应，其具体机制仍不清楚。

总之，这些试验结果与原本的假设一致，该假设认为，与静止态的 PBMC（$G_0$ 期）相比，在细胞周期的各个时相（$G_1$、S、$G_2$ 和 M 期），所有组蛋白的生物素化均有提高。因此，组蛋白的生物素化在整个细胞周期里均得到提高，而不是在细胞周期的某个特定时期才提高，但其具体原因目前仍不清楚。

## 生物素酶基因的表达

Hymes 等的研究证明，组蛋白的生物素化是由生物素酶来催化完成的（Hymes *et al.*，1995；Hymes and Wolf，1999）。为了研究增殖的 PBMC 的组蛋白的生物素化的提高是否与生物素酶活性的增强一致，我们对增殖和非增殖的 PBMC 的生物素酶的基因表达进行了检测（Stanley *et al.*，2001）。生物素酶 mRNA 的细胞水平在增殖和非增殖的 PBMC 中类似（图 17.6）。生物素酶 mRNA 的轻微提高（约 0.6 倍）显著低于参与生物素代谢的其他蛋白质的 mRNA 的提高，生物素酶 mRNA 的提高很可能不具有显著的生理意义。例如，编码钠离子依赖性的多种维生素转运蛋白（能转运生物素）的 mRNA 在增殖的 PBMC 中提高了约 10 倍（Zempleni *et al.*，2001），同样，转铁蛋白受体 mRNA（对照组）提高了 5.5 倍（图 17.6）。

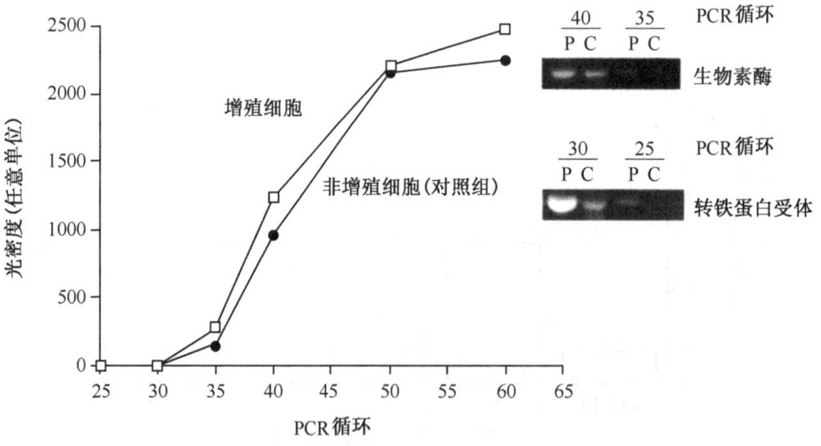


图 17.6　PBMC 中生物素酶的基因表达不受细胞增殖的影响

增殖（P）和非增殖（C）的 PBMC 中生物素酶 mRNA 采用定量的 PCR 方法进行检测。右上角的插入图表示编码生物素酶和转铁蛋白受体的 mRNA 的琼脂糖凝胶图（Stanley *et al.*，2001）

对 PBMC 进行的表型（即生物素酶活性）研究也表明，生物素酶不受细胞增殖的影响（Zempleni *et al.*，2001）。试验检测了静止和增殖的 PBMC 的生物素酶活性，对于这两种细胞，其生物素酶活性没有低于最低检测限（$<30U/10^{-9}$ PBMC），从而无法得到可靠的数据［1U 为每分钟 *N*-D-生物素酰基-*p*-氨基苯甲酸释放出的对氨基苯甲酸（*p*-aminobenzoic acid，PABA）的 nmol 数］。该结果与 Pispa（1965）早期的报道一致，

表明人白细胞具有非常低的生物素酶活性，相对而言，血浆中生物素酶的活性要高得多。本实验室的检测结果表明，5 位健康成年人的血浆中生物素酶的活性平均为（7980±1324）U/L。总之，PBMC 中生物素酶活性非常低，并且增殖细胞中生物素酶基因表达的提高也不明显。虽然生物素酶催化组蛋白进行生物素化反应已被确定（Hymes *et al.*，1995），我们猜测其他的酶也可能具有催化组蛋白进行生物素化反应的活性，或组蛋白生物素化的水平受到其降解速率的调节，或两者均存在。

## 组蛋白的去生物素化

最近的研究表明，生物素酶既能催化组蛋白进行生物素化，也能催化组蛋白进行去生物素化（Ballard *et al.*，2002），这与早期的假设一致。若组蛋白的生物素化是调节细胞过程的一个重要机制，既然组蛋白的生物素化和去生物素化均由同一酶来完成，那么细胞又是如何调节组蛋白的生物素化状态？下面对此进行解释。

（1）除了生物素酶，其他的酶也可能催化组蛋白进行生物素化或去生物素化。例如，羧化全酶合成酶（EC 6.3.4.10）可以催化生物素与 4 种哺乳动物羧化酶中赖氨酸残基的 ε-氨基发生共价结合（Zempleni，2001），因此，羧化全酶合成酶也可能具有催化生物素与组蛋白发生反应的活性，但该推论还没得到证实。

（2）生物素酶的共价修饰既可能有利于组蛋白的生物素化，也可能是有利于组蛋白的去生物素化。糖基化是目前唯一被证实的生物素酶的翻译后修饰方式（Cole *et al.*，1994）。

（3）辅助因子可能促进组蛋白的生物素化，也可能促进组蛋白的去生物素化。例如，高浓度的酶作用底物生物胞素能提高组蛋白生物素化的速率。同样，生物素酰多肽通过竞争结合生物素酶能抑制组蛋白的去生物素化。相反，生物素酶的生物素化活性的最适 pH 与去生物素化活性的相同（pH8）（Hymes *et al.*，1995）。这样，组蛋白微环境的 pH 变化不会影响生物素化状态。

## 致　　谢

感谢国家卫生协会（DK 60447）和 USDA/CSREES 项目基金 2001-35200-10187 给予的大力支持。

## 参 考 文 献

Allfrey, V., Faulkner, R.M. and Mirsky, A.E. (1964) Acetylation and methylation of histones and their possible role in the regulation of RNA synthesis. *Proceedings of the National Academy of Sciences USA* 51, 786–794.

Althaus, F. (1992) Poly ADP-ribosylation: a histone shuttle mechanism in DNA excision repair. *Journal of Cell Science* 102, 663–670.

Ausio, J. and van Holde, K.E. (1986) Histone hyperacetylation: its effect on nucleosome conformation and stability. *Biochemistry* 25, 1421–1428.

Ballard, T.D., Wolffe, J., Griffin, J.B., Stanley, J. S., van Calcar, S. and Zempleni, J. (2002) Biotinidase catalyzes debiotinylation of histones. *European Journal of Biochemistry* 41, 78–84.

Boulikas, T. (1988) At least 60 ADP-ribosylated variant histones are present in nuclei from dimethylsulfate-treated and untreated cells. *EMBO Journal* 7, 57–67.

Boulikas, T., Bastin, B., Boulikas, P. and Dupuis, G. (1990) Increase in histone poly(ADP-ribosylation) in mitogen-activated lymphoid cells. *Experimental Cell*

*Research* 187, 77–84.

Brownell, J.E., Zhou, J., Ranalli, T., Kobayashi, R., Edmondson, D.G., Roth, S.Y. and Allis, C.D. (1996) *Tetrahymena* histone acetyltransferase A: a homolog to yeast Gcn5p linking histone acetylation to gene activation. *Cell* 84, 843–851.

Chalifour, L.E. and Dakshinamurti, K. (1982) The biotin requirement of human fibroblasts in culture. *Biochemical and Biophysical Research Communications* 104, 1047–1053.

Chambon, P., Weill, J.D., Doly, J., Strosser, M.T. and Mandel, P. (1966) On the formation of a novel adenylic compound by enzymatic extracts of liver nuclei. *Biochemical and Biophysical Research Communications* 25, 638–643.

Cohen, P. (1989) The structure and regulation of protein phosphatases. *Annual Review of Biochemistry* 58, 453–508.

Cohen, P., Klumpp, S. and Schelling, D.L. (1989) An improved procedure for identifying and quantitating protein phosphatases in mammalian tissues. *FEBS Letters* 250, 596–600.

Cole, H., Reynolds, T.R., Lockyer, J.M., Buck, G.A., Denson, T., Spence, J.E., Hymes, J. and Wolf, B. (1994) Human serum biotinidase cDNA cloning, sequence, and characterization. *Journal of Biological Chemistry* 269, 6566–6570.

Dasso, M. and Newport, J.W. (1990) Completion of DNA replication is monitored by a feedback system that controls the initiation of mitosis *in vitro*: studies in *Xenopus*. *Cell* 61, 811–823.

Durkacz, B.W., Omidiji, O., Gray, D.A. and Shall, S. (1980) (ADP-ribose)$_n$ participates in DNA excision repair. *Nature* 283, 593–596.

Gorovsky, M.A. (1973) Macro- and micronuclei of *Tetrahymena pyriformis*: a model system for studying the structure of eukaryotic nuclei. *Journal of Protozoology* 20, 19–25.

Haavik, J., Schelling, D.L., Campbell, D.G., Andersson, K.K., Flatmark, T. and Cohen, P. (1989) Identification of protein phosphatase 2A as the major tyrosine hydroxylase phosphatase in adrenal medulla and corpus striatum: evidence from the effects of okadaic acid. *FEBS Letters* 251, 36–42.

Hebbes, T.R., Thorne, A.W. and Crane-Robinson, C. (1988) A direct link between core histone acetylation and transcriptionally active chromatin. *EMBO Journal* 7, 1395–1402.

Hymes, J. and Wolf, B. (1999) Human biotinidase isn't just for recycling biotin. *Journal of Nutrition* 129, 485S-489S.

Hymes, J., Fleischhauer, K. and Wolf, B. (1995) Biotinylation of histones by human serum biotinidase: assessment of biotinyl-transferase activity in sera from normal individuals and children with biotinidase deficiency. *Biochemical and Molecular Medicine* 56, 76–83.

Juarez-Salinas, H., Sims, J.L. and Jacobson, M.K. (1979) Poly(ADP-ribose) levels in carcinogen-treated cells. *Nature* 282, 740–741.

Lee, D.Y., Hayes, J.J., Pruss, D. and Wolffe, A.P. (1993) A positive role for histone acetylation in transcription factor access to nucleosomal DNA. *Cell* 72, 73–84.

Mathis, D.J., Oudet, P., Waslyk, B. and Chambon, P. (1978) Effect of histone acetylation on structure and *in vitro* transcription of chromatin. *Nucleic Acids Research* 5, 3523–3547.

Megee, P.C., Morgan, B.A. and Smith, M.M. (1995) Histone H4 and the maintenance of genome integrity. *Genes and Development* 9, 1716–1727.

Nishizuka, Y., Ueda, K. and Hayaishi, O. (1971) Adenosine diphosphoribosyltransferase in chromatin. In: McCormick, D.B. and Wright, L.D. (eds) *Vitamins and Coenzymes*. Academic Press, New York, pp. 230–233.

Pham, A.-D. and Sauer, F. (2000) Ubiquitin-activating/conjugating activity of $TAF_{II}250$, a mediator of activation of gene expression in *Drosophila*. *Science* 289, 2357–2360.

Pispa, J. (1965) Animal biotinidase. *Annals of Medicine and Experimental Biology Fenniae* 43, 4–39.

Rawling, J.M., Jackson, T.M., Driscoll, E.R. and Kirkland, J.B. (1994) Dietary niacin deficiency lowers tissue poly(ADP-ribose) and $NAD^+$ concentrations in Fischer-344 rats. *Journal of Nutrition* 124, 1597–1603.

Sims, J.L., Sikorski, G.W., Catino, D.M., Berger, S.J. and Berger, N.A. (1982) Poly(adenosinediphosphoribose) polymerase inhibitors stimulate unscheduled deoxyribonucleic acid synthesis in normal human lymphocytes. *Biochemistry* 21, 1813–1821.

Sommerville, J., Baird, J. and Turner, B.M. (1993) Histone H4 acetylation and transcription in amphibian chromatin. *Journal of Cell Biology* 120, 277–290.

Stanley, J.S., Griffin, J.B. and Zempleni, J. (2001) Biotinylation of histones in human cells: effects of cell proliferation. *European Journal of Biochemistry* 268, 5424–5429.

Taunton, J., Hassig, C.A. and Schreiber, S.L. (1996) A mammalian histone deacetylase related to a yeast transcriptional regulator Rpd3. *Science* 272, 408–411.

Wolffe, A. (1998) *Chromatin*. Academic Press, San Diego, California.

Yano, H., Nakanishi, S., Kimura, K., Hanai, N., Saitoh, Y., Fukui, Y., Nonomura, Y. and Matsuda, Y. (1993) Inhibition of histamine secretion by wortmannin through the blockade of phosphatidylinositol 3-kinase in RBL-2H3 cells. *Journal of Biological Chemistry* 268, 25846–25856.

Yoon, Y.S., Kim, J.W., Kang, K.W., Kim, Y.S., Choi, K.H. and Joe, C.O. (1996) Poly(ADP-ribosyl) ation of histone H1 correlates with internucleosomal DNA fragmentation during apoptosis. *Journal*

*of Biological Chemistry* 271, 9129–9134.
Zempleni, J. (2001) Biotin. In: Bowman, B.A. and Russell, R.M. (eds) *Present Knowledge in Nutrition*, 8th edn. International Life Sciences Institute, Washington, DC, pp. 241–252.
Zempleni, J. and Mock, D.M. (1999) Chemical synthesis of biotinylated histones and analysis by sodium dodecyl sulfate–polyacrylamide gel electrophoresis/streptavidin-peroxidase. *Archives of Biochemistry and Biophysics* 371, 83–88.
Zempleni, J., Stanley, J.S. and Mock, D.M. (2001) Proliferation of peripheral blood mononuclear cells causes increased expression of the sodium-dependent multivitamin transporter gene and increased uptake of pantothenic acid. *FASEB Journal* 15, A964 [abstract].

# 18　烟酸营养状态、多聚体（ADP-核糖）的代谢及基因组的不稳定性

Jennifer C. Spronck 和 James B. Kirkland
（圭尔夫大学人类生物与营养学系，安大略，加拿大）

## 前　　言

细胞周期的检查点和 DNA 修复途径可以对 DNA 缺失进行检测和修复，防止在细胞分裂期内发生永久性的突变（Kaufmann and Paules，1996）。DNA 损伤的识别和修复途径是确保遗传信息准确传递的关键，若 DNA 损伤严重，细胞凋亡有助于消除遗传信息发生变化的细胞，这些过程的相互协调是维持机体自主分裂细胞的基因组稳定性所必需的。看管基因被认为是基因组的监护者，它确保基因组基本的组成的稳定和复制的准确。看管基因（caretaker）包括参与错配修复、核苷酸和碱基切除及双链断裂修复的基因。持家基因表达的蛋白质对细胞的命运具有决定性作用，这些蛋白质可以决定细胞的存活或凋亡。在含持家基因多态性的家族中，肿瘤发生率的提高表明，大量的细胞应答在维持基因组的稳定性方面具有重要作用。

除了基因型外，这些细胞过程还可能受环境因素的影响。流行病学、实验室和临床研究表明，许多养分与基因组的不稳定性或肿瘤的发生有关。人缺乏烟酸，会导致糙皮病，该病的主要特征是“4 D 症状”，即腹泻、皮炎、痴呆和死亡。皮炎局限于暴露于阳光的皮肤区域，这可能是由 DNA 修复途径出现了问题所致。过去，认为该病的发生是与以玉米作为主食有关，但是以玉米为主食的北美和南美的土著美洲人不发生糙皮病，因为土著美洲人采用壁炉灰末和石灰水来加工玉米产品，现在人们已经知道可以用碱来提高玉米中烟酸的生物学利用率（Carpenter，1981）。在工业化国家，烟酸的来源包括肉类如鱼肉、禽肉、猪肉和牛肉。谷物、种子和豆类也含有烟酸，其形式主要是烟酸，但它可能与复杂的碳水化合物和蛋白质结合，从而限制了其吸收，因此，在绝大多数发达国家，食用谷类食物时均要补加烟酸。肝脏还能利用色氨酸来合成烟酰胺腺嘌呤二核苷酸（nicotinamide ademine dinucleotide，NAD），在肝脏中 1mg 过剩的食物色氨酸可合成 1/60mg 烟酸（Kirkland and Rawling，2001）。在发达国家，虽然真正的糙皮病几乎未见报道，但是由于症状不明显，轻度的烟酸缺乏可能不易被发现，尽管摄食了补加烟酸的食物，部分人群仍表现出极低的烟酸症状。在瑞典马尔摩膳食与癌症研究中一项关于女性的研究表明，15%的女性有烟酸的亚临床缺乏症（Jacobson，1993），而在美国老年人群中也观察到类似的现象，22%老年人有低烟酸症状（Knebl and Jacobson，1992）。

## NAD 的功能和代谢

人们已知，NAD 和 NADP 的主要代谢功能是氧化还原作用。它们的氧化形式分别

是 $NAD^+$ 和 $NADP^+$，还原形式分别是 NADH 和 NADPH。NADH 的主要氧化还原作用是把代谢中间产物的电子转移到电子传递链上，产生 ATP；而 NADPH 在脂肪酸合成和其他生物合成途径中作为还原剂，同时也能作为氧化防护的还原剂（Kirkland and Rawling，2001）。

## 多聚体（ADP-核糖）

最近几年，人们还发现了 $NAD^+$ 的非氧化还原功能，包括 $NAD^+$ 用于合成多聚体（ADP-核糖）（poly ADP-rebose，pADPr）、单 ADP-核糖和环 ADP-核糖。大部分的细

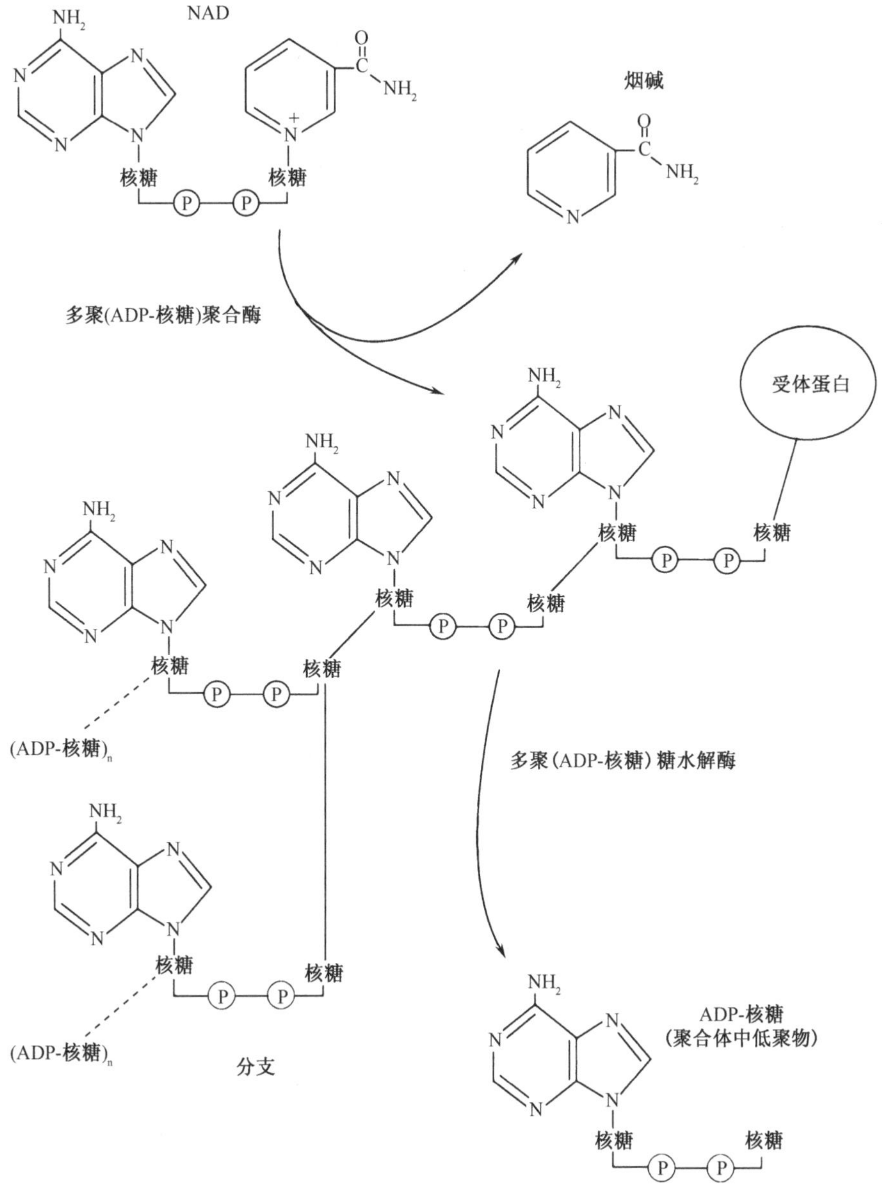


图 18.1 多聚体（ADP-核糖）（pADPr）的合成和降解

胞 pADPr 是由核酶——多聚（ADP-核糖）聚合酶（poly ADP-rebose polymerase，PARP）催化而成的。PARP 与 DNA 链裂口结合，并被特异性地激活，这是细胞对 DNA 损伤的最早反应之一。激活后，PARP 用 $NAD^+$ 来合成 pADPr，pADPr（自动修饰）和许多其他受体蛋白（翻译后修饰）均参与维持染色质结构和 DNA 代谢的稳定（D′Amours *et al.*，1999）（图 18.1）。各种 pADPr 的长度不一，复杂性差异也较大，含有大于 200 个残基的直链和支链部分（D′Amours *et al.*，1999）。PARP 蛋白及相关的多聚（ADP-核糖基化）反应，在许多生化过程（如 DNA 修复、重组、细胞凋亡和维持基因组的稳定）中发挥重要作用（Le Rhun *et al.*，1998；D′Amours *et al.*，1999）。

在所有较高级的真核生物中，PARP 含量丰富，并具有高度保守性（D′Amours *et al.*，1999）。锌指蛋白作为损伤的检测器，不管 DNA 的序列如何，它都能识别并结合 DNA 链裂口（Ikejima *et al.*，1990）。除了与 DNA 结合外，N 端结构域还能作为蛋白质-蛋白质相互作用的界面，但该作用需要包括组蛋白、DNA 聚合酶 α、X 射线交互辅助因子-1（X-ray cross-complementing factor-1，XRCCF1）和多种转录因子（Niedergang *et al.*，2000）的辅助。该结构域还含有一个核定位信号和一个半胱氨酸蛋白酶切割位点（DEVD），该位点是发生细胞凋亡的靶点（D′Amours *et al.*，1999）。

当与带负电荷的 pADPr 结合时，PARP 的自动修饰使它失去对 DNA 链裂口的亲和力，因而失去催化活性。PARP 自动修饰区域（蛋白质的中间区域）还含有一个 BRCT（BRCA1 C 端）序列，许多细胞周期检查点和 DNA 修复蛋白也含有这种序列（Niedergang *et al.*，2000），该序列使 PARP 和 XRCC1 与 DNA 连接酶Ⅲ之间发生特异性的蛋白质-蛋白质反应成为可能（D′Amours *et al.*，1999）。

PARP 的 C 端区域含有 $NAD^+$ 结合位点，PARP 的催化作用包括降解 $NAD^+$ 及参与 pADPr 合成的起始、延长、分支和终止（de Murcia and Menissier，1994）。pADPr 的合成是瞬间的、可逆的，且 DNA 损伤后，pADPr 半衰期小于 1min（Alvarez-Gonzalez and Althaus，1989）。多聚（ADP-核糖）糖水解酶（glycohydrolase）降解 pADPr，释放出游离的 ADP-核糖单体和游离的 pADPr 的短链（D′Amours *et al.*，1999）。

## PARP 与 DNA 修复

DNA 切除修复途径包括碱基切除修复和核苷酸切除修复。碱基切除指的是切除单一受损的碱基（如烷化剂、氧化损伤的碱基等），核苷酸切除指的是切除大的 DNA 扭曲损伤（如 UV 光诱导的二聚体）。PARP 在切除修复中发挥着重要的作用。早期研究发现，PARP 催化 pADPr 的合成过程（Durkacz *et al.*，1980；Jacobson *et al.*，1980），PARP 的抑制、NAD 的营养缺失或突变的 PARP DNA 结合区域的过度表达都会提高基因毒性（Nduka *et al.*，1980；Chatterjee *et al.*，1991；Molinete *et al.*，1993）或阻止链断裂的修复/再连接（Durkacz *et al.*，1980；Nduka *et al.*，1980）。后来的试验也发现，补充烟酸的人淋巴细胞用氧自由基处理后，DNA 损伤减少（Weitberg，1989），并且在人淋巴细胞中添加烟酸可降低腺苷脱氨酶抑制剂诱导的 DNA 链断裂（Weitberg and Corvese，1990）。以 PARP 无效小鼠为模型的试验结果表明，PARP 蛋白是细胞发挥最佳修复功能不可缺少的（Dantzer *et al.*，1999）。最近研究表明，PARP 与切除修复系统的各种组分之间存在物理的相互作用（Masson *et al.*，1998）。PARP 通过 BRCT 序列与 XRCC1 发生相互作用，而 XRCC1 是一种调节蛋白，能调节两种重要的切除修复蛋白（DNA 连接酶Ⅲ和 DNA 聚合酶 β）之间的相互作用。有人提出，PARP 可能通过直接补充 DNA 修复系统到 DNA 损伤位点来促进机体 DNA 修复（D′Amours *et al.*，1999），若合成 pADPr 的能力受损且没有更改，则 DNA 结合的 PARP 将干扰修复过程，并且 DNA 链裂口不能被修复（Satoh and Lindahl，1992）。

然而，当 PARP 从无细胞系统中去除时，DNA 修复就不能依靠 NAD，但是其修复速度与对照组的相同（Lindahl *et al.*，1995），这暗示 PARP 本身并不是切除修复和 DNA 链裂口再连接必不可少的（D′Amours *et al.*，1999），但当它失去催化活性，PARP 则抑制修复。为了证明这个结论，PARP 缺失小鼠的细胞用烷化剂或 γ 射线处理后，仍有效地进行了切除修复（Vodenicharov *et al.*，2000）。在体外，用不含 PARP 的纯化人类蛋白质来重建这些修复途径，其结果也表明切除修复并不直接需要 PARP（Vodenicharov *et al.*，2000），但与体内试验相比，这些修复系统不含复杂的染色质结构。尽管人们对 PARP 去除效应的观点还存在争议，但是无催化活性的 PARP 干扰了切除修复，这一点已很清楚，它可能发生在烟酸缺乏期间。

## PARP 与细胞凋亡

细胞凋亡是胚胎形成、组织平衡和免疫系统的专门化过程中调节细胞数量的不可缺少的生理过程。细胞凋亡的抑制造成 DNA 损伤的细胞不能被清除，这通常与癌症的发生有关（Schmitt and Lowe，1999）。细胞凋亡的一些生化特点包括早期 pADPr 合成的瞬间增加（Simbulan-Rosenthal *et al.*，1998）、半胱氨酸蛋白酶的激活及 PARP 特异性地裂解成 24kDa 和 89kDa 两个多肽片段（Kaufmann *et al.*，1993）。关于 PARP 裂解对细胞凋亡信号传递的影响机制，现已提出了几种观点。由溶蛋白性裂解导致的 PARP 失活通过抑制 $NAD^+$ 过度消耗和有害坏疽来确保正常的细胞凋亡，$NAD^+$ 过度消耗会引起 ATP 消耗。也有人提出，PARP 的裂解片段直接在细胞凋亡中发挥作用，或完整的 PARP 抑制细胞凋亡，这一观点已被试验证实。研究发现，不被半胱氨酸蛋白酶裂解的突变 PARP 能保护细胞免受 DNA 损伤诱导的凋亡，使细胞长期存活下去并进行细胞分裂（Halappanavar *et al.*，1999），受损细胞的残存提高了基因组的不稳定性概率，最后发生癌变。

对于 PARP 在凋亡过程中发挥的具体作用，人们进行了大量的研究，但仍存在争议。采用各种 PARP 无效小鼠进行的试验也不能阐明 PARP 对细胞凋亡的作用（de Murcia *et al.*，1997；Leist *et al.*，1997；Wang *et al.*，1997；Simbulan-Rosenthal *et al.*，1998；Masutani *et al.*，1999）。大部分争论可能是由于所使用的凋亡诱导物的不同、所研究的细胞类型的多样性和使 PARP 功能失活的方法的差异所致。

## PARP 与 p53

p53 是一种多功能蛋白质，在调节细胞过程中发挥关键性作用，如细胞周期的阻滞和DNA损伤导致的凋亡（Gottlieb and Oren，1998）。受基因毒性的应激（如化疗药物、UV和离子辐射）后，p53蛋白被各种翻译后的修饰作用［包括磷酸化、去磷酸化、乙酰化和多聚化（ADP-核糖基化）］所稳定激活（Steegenga *et al*.，1996；Colman *et al*.，2000）。作为转录因子，p53与DNA上的特殊序列结合，并使参与执行细胞周期阻滞的目标基因被反式激活（transactivation）（如p21$^{WAF1}$、GADD45和B99）（el Deiry，1998）和介导细胞凋亡（如Bax、Fas/APO1和Killer/DR5）（el Deiry，1998；Gottlieb and Oren，1998）；p53还能通过与DNA非序列特异性的结合来抑制许多基因（包括抗凋亡基因Bcl-2）的转录（el Deiry，1998）。

DNA损伤后，PARP和p53均被激活。PARP对DNA损伤的反应和pADPr合成在数秒钟到数分钟内被激活（Berger，1985），这表明多聚化（ADP-核糖基化）可能作为DNA损伤的信号产生器，产生下游效应分子。研究表明，DNA损伤后，PARP的竞争性抑制导致基础p53蛋白水平和稳定性降低。pADPr合成能力的缺乏或$NAD^+$的细胞系的缺失抑制了对鬼臼乙叉苷（etoposide）诱导的正调节不敏感的基础p53蛋白水平的降低（Whitacre *et al*.，1995）。另外，对于人胶质母细胞瘤（glioblastoma）细胞，采用3-氨基苯甲酰胺对PARP进行的化学抑制降低了射线诱导的p53的积聚，表明p53的多聚化（ADP-核糖基化）是蛋白质稳定和积聚所必需的（Wang *et al*.，1998）。PARP无效的小鼠细胞中基础p53蛋白降低了约50%，p53的正调节也受损（Agarwal *et al*.，1997）。然而，不同来源的PARP缺失细胞含有较低的p53基础水平，但当DNA损伤时，这些细胞发生了正常的正调节（Wesierska-Gadek *et al*.，1999）。有趣的是，PARP无效细胞还含有一种p53的拼接形式（p53AS），p53AS的功能与正常的p53不同，因为它在组成上具有序列特异性的DNA结合区域（Bayle *et al*.，1995）并能够降低细胞凋亡（Almog *et al*.，1997）。本研究表明，PARP缺失细胞在组成上表达p53AS，并且正常形式p53在这些细胞中极不稳定（Wesierska-Gadek *et al*.，1999）。DNA损伤对PARP缺失细胞和野生型PARP细胞中的两种p53拼接形式的诱导作用不同，正常拼接p53的诱导只发生在野生型PARP细胞，而p53AS的诱导在PARP缺失细胞和野生型PARP细胞之间的动力学不同。

PARP对p53的表达及功能的调节机制有多种。p53 DNA结合区域和低聚化区域含有pADPr结合基序，低聚化区域可使p53与游离pADPr或pADPr结合的PARP之间产生较强的非共价结合（Malanga *et al*.，1998）。另外，前期的研究已表明，在体外p53利用pADPr进行了翻译后修饰（Wesierska-Gadek *et al*.，1996a；Kumari *et al*.，1998），同样，体内试验也发现，细胞凋亡的早期阶段也如此（Simbulan-Rosenthal *et al*.，1999a；Smulson *et al*.，2000）。在完整细胞和体外环境中，p53-PARP复合物已被分离出来（Wesierska-Gadek *et al*.，1996a；Vaziri *et al*.，1997；Kumari *et al*.，1998）。研究表明，这种与pADPr共价和非共价结合的p53修饰能调节p53的表达（基础的和诱导的），也能调节p53的DNA结合特征（Wesierska-Gadek *et al*.，1996b；Wang *et al*.，1998；Malanga *et al*.，1998）。从功能上讲，DNA损伤后，化学抑制剂导致的PARP失活，在蛋白质和mRNA水平上抑制了p53应答基因产物的积聚（Vaziri *et al*.，1997；Wang *et al*.，1998）。

根据细胞耐受细胞凋亡和（或）细胞周期阻滞的能力，人们对p53功能损伤的生物相关性进行了研究，但文献报道不一。一些文献报道 DNA 损伤（Whitacre *et al*.，1995）和细胞周期阻滞（Nozaki *et al*.，1994；Masutani *et al*.，1995）使细胞凋亡降低，而也有一些文献报道 p53 的激活和功能不依赖于 PARP（Agarwal *et al*.，1997）。

# PARP 同系物

多年来，人们都认为只有一种蛋白质具有 PARP 的活性。最近的研究发现，PARP 无效的细胞也能合成少量的 pADPr（Shieh *et al*.，1998），并发现多种蛋白质具有 PARP 类似的活性。与 PARP（现在有时也被称为 PARP-1）相同，PARP-2 也能被基因毒性的应激激活，即使它们的 DNA 结合区域的结构不一样（Ame *et al*.，1999）。PARP-2 位于细胞核内，具有特殊的催化活性或作为 PARP 的后备（Oliver *et al*.，1999）。另一同系物 PARP-3 也被发现，但其功能还不清楚（Johansson，1999）。

端锚聚合酶是一种含有锚蛋白和 PARP 催化区域同源的蛋白质，已被分离鉴定，位于人端粒复合物中（Smith *et al*.，1998）。在体外，端锚聚合酶与端粒重复结合因子-1（telomere repeat-binding factor-1，TRF-1）结合，TRF-1 是一种对端粒长度进行负调节的蛋白质（Smith *et al*.，1998）。端锚聚合酶合成 pADPr，使 TRF-1 发生多聚化（ADP-核糖基化），并将 TRF-1 从端粒上释放从而增强端粒酶的活性。另外，在细胞周期的不同时段，核孔复合物和中心体中也都发现含有端锚聚合酶（Smith and de Lange，1999）。

第四种 PARP 的同系物是 vault-PARP（vPARP），其活性的发挥不需要 DNA 的参与，它能够在细胞质中与穹隆体颗粒结合，它还能与有丝分裂纺锤体和端粒酶结合蛋白进行结合（Kickhoefer *et al*.，1999a，b），尽管 vPARP 的功能还不清楚，但是有丝分裂器和 DNA 中的 PARP 同系物的鉴定及定位暗示 vPARP 能够维持基因组完整性（Jacobson，1999）。

# 单 ADP-核糖基化作用和环 ADP-核糖的形成

大多数细胞均含大量的单 ADP-核糖转移酶，但其功能还不清楚。这些翻译后修饰大多发生在 G 蛋白上，可能在与细胞分裂有关的信号转导调控中发挥作用（Okazaki and Moss，1999）。合成环 ADP-核糖需要烟酸，最近发现的分子尼克酸腺嘌呤二核苷酸磷酸（nicotinic acid adenine dinucleotide phosphate，NAADP）的合成也需要烟酸，这两种物质能调节细胞内钙信号传递途径（Lee，2001）。钙信号传递途径的变化，使细胞凋亡和细胞周期的调节也发生改变。

# 烟酸的非-ADP-核糖基化作用对基因组稳定性的影响

## 氧化防御/应激

NADPH 由戊糖磷酸途径合成，能为谷胱甘肽过氧化物酶消除氧自由基提供动力。谷胱甘肽过氧化物酶能催化过氧化氢生成水，防止损伤力高的氢氧根自由基的形成，因为氢氧根自由基能与细胞内的所有大分子（包括 DNA）发生反应，所以烟酸缺乏可能提高诱变 DNA 损伤的物质，如 8-羟基-脱氧鸟苷（8-OHdG），但这还没被试验证实。

### 沉默信号调控因子（SIR2）

在维持基因组稳定性中 NAD 的另一功能可能是通过 SIR2——NAD-依赖性的组蛋白去乙酰化酶来实现的。去乙酰化作用使染色质结构更紧凑、使基因沉默，也能阻止染色质的敏感位点（如端粒）发生易位，从而延长与能量限制有关的生命周期（Lin *et al.*，2000）。SIR2 还能通过去乙酰化作用实现对 p53 的调节（Vaziri *et al.*，2001）。

## PARP、烟酸营养状况与基因组的稳定性

直接研究烟酸营养状况和基因组的不稳定性或癌变的资料还不多。在本章节，由于在机制上可能存在相同点，我们已把烟酸缺乏和 PARP 损坏结合在一起讨论。我们将从 PARP 着手，然后再概述目前对烟酸缺乏的认识。

PARP 现已被认为是基因组的护卫者，具有看管基因（caretaker）和持家基因（gatekeeper）的功能，是一种高度保守的酶（Tsutsumi *et al.*，2001）。在各种研究 PARP 功能的试验模型中，PARP 无效小鼠是最新发现的试验模型，每个模型都有其优点，也存在缺点。一般来说，以前采用竞争性抑制剂和无催化活性 PARP 片段基因进行试验的结果与用 PARP 无效小鼠试验模型进行研究的结果一致，这些结果虽然也存在矛盾点，但可通过理论进行较好的解释。例如，由于不能自动修复，化学抑制剂和 PARP 片段模型能产生与 DNA 链裂口紧密结合的无活性 PARP 形式，这将阻止随后的修复步骤。相反，PARP 无效模型不含影响损伤位点修复的 PARP，在损伤位点，PARP 无效模型真正消除了由 PARP 引起的蛋白质-蛋白质之间的相互作用，而且两种模型均不能进行 PARP-1 依赖性的 pADPr 合成。对 PARP 同系物（PARP-2、PARP-3、端锚聚合酶和 vPARP）的活性和功能的进一步认识也有利于阐述 PARP 无效小鼠模型的不同点。

在比较这些试验模型与烟酸缺乏的可能后果时，必须注意的是烟酸缺乏能降低 PARP 的活性，但不影响任何被检测组织中 PARP 的表达水平（未发表资料，图 18.2）。如前所述，无催化活性的 PARP 分子与 DNA 链裂口结合，阻止修复（Satoh and Lindahl，1992；Satoh *et al.*，1994）。另外，其他的 PARP（如 PARP-2 和-3、端锚聚合酶-1 和-2 及 vPARP）的活性直接受到烟酸营养状况和 $NAD^+$ 的利用率影响。单 ADP-核糖基化反应和钙信号分子 cADPr 和 NAADP 的形成也可能受到烟酸缺乏的影响，同时还受到组蛋白和 p53 的 NAD 依赖性去乙酰化作用的影响。因此，烟酸缺乏比研究 PARP 功能的特定试验模型要复杂得多。

基因组不稳定性的量化方法有多种。染色体的结构畸变和数目畸变可采用细胞遗传学的方法进行检测（如传统的染色体畸变分析法或微核形成分析法），而 DNA 损伤（DNA 加合物、链断裂、碱不稳定位点和链交叉）能采用各种方法进行分析，包括彗星分析法（Albertini *et al.*，2000）。姐妹染色单体互换（sister chromatid exchange，SCE）被认为是由 DNA 双链断裂和同源重组来诱导的（Tucker and Preston，1996）。SCE 形

成的确切机制目前还不清楚，但常作为基因毒性的指标，反映 DNA 损伤的情况（Tucker and Preston，1996；Albertini *et al*.，2000）。

采用化学抑制剂、底物限制或通过分子遗传方法来抑制 PARP，总能提高 SCE 的发生频率。PARP 的竞争性抑制剂使 SCE 的基础发生频率提高了 3～10 倍（Oikawa *et al*.，1980；Hori，1981；Morgan and Cleaver，1982），并能增强烷化剂对 SCE 的提高幅度（Morgan and Cleaver，1982；Park *et al*.，1983）。一些细胞含有能表达无催化活性的 PARP 的 DNA 结合片段，这些细胞对烷化剂极其敏感，并能显著提高自然产生的 SCE（Schreiber *et al*.，1995）。

另外，所有的 PARP 无效小鼠对基因毒性的应激因子（如离子辐射和烷化剂）极其敏感（Wang *et al*.，1997；Trucco *et al*.，1999；Masutani *et al*.，2000）。受基因毒性应激后，PARP 无效小鼠表现为自然产生的 SCE 提高了 2～3 倍，且增强了 SCE 和微核的形成概率（de Murcia *et al*.，1997；Wang *et al*.，1997）。烷化剂或离子辐射诱导的损伤导致 PARP 无效小鼠的染色单体裂片提高了 33～36 倍（de Murcia *et al*.，1997）。最近的研究发现，PARP 无效小鼠对亚硝胺诱导的肿瘤发生更敏感（Tsutsumi *et al*.，2001），这进一步证明了 PARP 在基因组稳定性和癌发生中发挥着重要作用。

DNA 双链断裂的修复可能由同源重组和非同源重组来完成（Alberts *et al*.，1994）。SCE 是通过同源重组方式实现的，研究发现 PARP 可作为断裂重组的抑制因子，降低断裂位点不恰当的同源重组（Satoh and Lindahl，1992）。抑制不恰当的重组可能存在两种机制：①PARP 迅速与 DNA 链裂口紧密结合，从而阻止了外切核酸酶与 DNA 的接触；②带负电荷的 pADPr 在链裂口处聚集，通过静电排斥来抵制游离 DNA 末端，从而防止染色体不正常的易位（Satoh and Lindahl，1992；Lindahl *et al*.，1995）。

基因组不稳定性的其他起因采用不同的试验体系来进行评估。对野生型 PARP 小鼠、PARP（$PARP^{+/-}$）杂合小鼠和 PARP 无效小鼠的端粒长度进行分析发现，PARP 活性对端粒长度具有负面影响（d'Adda *et al*.，1999）。端粒抑制了染色体的融合，因而参与了基因组稳定性的调节过程（McClintock，1941）。PARP 无效小鼠与野生型 PARP 小鼠相比，染色体的端对端融合（end-to-end fusion）、非整倍体细胞和染色体片段的数量显著增加，而 PARP（$PARP^{+/-}$）杂合小鼠的基因组稳定性居中（d'Adda *et al*.，1999）。相反，另一试验表明，PARP 无效的初级细胞与野生型 PARP 细胞的端粒长度之间没有差异，并且 PARP 无效细胞提高了端对端融合的频率，但提高幅度比前人报道的低（Samper *et al*.，2001）。

PARP 无效小鼠的永生化纤维原细胞含有不稳定的四倍体，再引入 PARP cDNA 后，四倍体消失（Simbulan-Rosenthal *et al*.，1999b）。Simbulan-Rosenthal 等（1999b）采用比较基因组杂交分析法来研究遗传变异，结果表明，PARP 无效小鼠和 PARP 无效的永生化纤维原细胞含有 4 号、5 号、14 号染色体的部分区域，同时缺失 14 号染色体。由于非整倍体和多倍体通常与癌的发生有关（Halloway *et al*.，1999），这些研究结果也提示 PARP 在维持基因组稳定和预防肿瘤方面具有重要作用。

为了更深入地认识 PARP 对基因组稳定性的影响机制，利用寡核苷酸微阵列技术来研究基因表达的变化情况。PARP 缺失小鼠的初级纤维原细胞抑制参与细胞周期和有丝分裂的调节过程中数个基因的表达，而促进参与肿瘤的启动和发展过程的基因表达

(Simbulan-Rosenthal *et al.*, 2000)。

因此，PARP 活性的降低可能通过影响 DNA 修复、重组、凋亡及 p53 表达和功能的调节来增强基因组的不稳定性，从而提高了癌症的发生率。

## 烟酸、基因组的不稳定性与肿瘤

我们已经讨论了烟酸对基因不稳定性的影响机制。最后这部分将概述与基因不稳定性更直接相关的研究。在这个领域中，绝大多数试验都重点研究了最后发生癌变的某些指标，这是基因组不稳定性的间接但敏感的反映指标。

流行病学研究报道，在南非和中国某些特殊省份，食道癌的发生与边缘性烟酸缺乏有关，这些人群摄食低蛋白质以及以玉米为基础的食物（Hageman and Stierum，2001；Kirkland and Rawling，2001），在意大利，过量的饮酒进一步加剧了与玉米摄入有关的食道癌的发病率（Franceschi *et al.*，1990）；在美国，低摄入量的烟酸也与口腔癌（oral cancer）发病率的提高有关（Marshall *et al.*，1992）。在中国的临县地区的研究中，边缘性烟酸缺乏与食道癌的发生有关，然而，随后的干预研究发现，补充烟酸和核黄素既不降低食道癌的发生率，也不降低癌症的死亡率（Blot *et al.*，1995）。在这项研究中，5 年的补添时间不能说明烟酸在癌症发病早期的作用。有关高剂量的烟酸或烟酰胺在人类癌症发生方面作用的报道非常少。然而，为了治疗高血脂，服用药理剂量的烟酸也不能降低癌症的总发病率（Kirkland and Rawling，2001）。

动物的烟酸营养与其他癌症之间的关系非常复杂（Bryan，1986）。对于饲喂玉米的大鼠，用烷化剂来诱导食道癌和营养不良后，添加烟酸可以降低肿瘤的发生概率，减小肿瘤的体积和减慢肿瘤恶化的进程（Van Rensburg *et al.*，1986）。相反，以患肾癌和（或）肝癌的动物为试验模型，其结果不能完全证明烟酸在癌发生中的作用。烟酸缺乏或添加药理剂量的烟酸或烟酰胺对二乙基亚硝胺诱导的大鼠肝病变无影响（Jackson *et al.*，1995；Rawling *et al.*，1995），而烟酸缺乏提高了亚硝基二甲胺诱导的大鼠肾癌的发生（Miller and Burns，1984）。

对不同致癌物质的敏感性差异及组织特异性癌的发生都受到靶组织对膳食烟酸含量变化的敏感性的影响。我们的研究结果表明，烟酸缺乏时，不同组织对 NAD 耗竭的反应差异非常大。肝脏、肺脏和肾脏等一些组织本身含较高水平的 NAD，因此烟酸缺乏时，它们对 NAD 耗竭具有一定的抵抗力（Rawling *et al.*，1994，1995，1996）。与之相反，骨髓和皮肤对 NAD 耗竭非常敏感，若分别用烷化剂或 UV 光处理，这两种组织癌发生的概率均有提高（Boyonoski *et al.*，2002a；Shah *et al.*，2002）。这两种组织的共同特点是：由于细胞的损失（皮肤）或输出（骨髓），细胞分裂的速度都非常高，新细胞不断地产生导致这些组织的烟酸营养必须得到保证。

以大鼠为试验动物，我们发现骨髓对膳食的烟酸含量变化最为敏感（图 18.2），而且烟酸缺乏提高了乙基亚硝基脲诱导的癌症的发生率，这些癌症大部分是骨髓病变所引起的白血病（Boyonoski *et al.*，2002a）。我们也发现，药理剂量的烟酸或烟酰胺延缓了乙基亚硝基脲诱导的癌症的发展（Boyonoski *et al.*，2002b），在这些试验中，烟酸缺乏和药理剂量的烟酸，都使骨髓 $NAD^+$ 发生显著变化（图 18.2A），并使基础 pADPr 和

DNA 损伤诱导的 pADPr 水平发生明显变化（图 18.2B）（Boyonoski *et al.*，2002a，b）。

皮肤也是对烟酸营养状况非常敏感的组织，若烟酸缺乏，人易发生阳光敏感性的损伤。同骨髓试验相似，对皮肤的试验也进行了烟酸缺乏和药理剂量的研究。研究发现，由于大鼠体内的色氨酸能有效地转化为烟酸，使大鼠缺乏烟酸的程度相对较轻，即便如此，大鼠缺乏烟酸也能增加 UV 诱导的皮肤癌的发生概率（Shah *et al.*，2002）。对于通过膳食已经摄入规定剂量烟酸的大鼠，口服或局部注射高剂量的烟酸可抑制 UV 诱导的光致癌发生（Gensler，1997；Gensler *et al.*，1999），这证实了烟酸营养具有防癌的作用。这些试验表明，添加烟酸至少可以通过增强机体免疫来部分地发挥其作用（Gensler *et al.*，1999）。

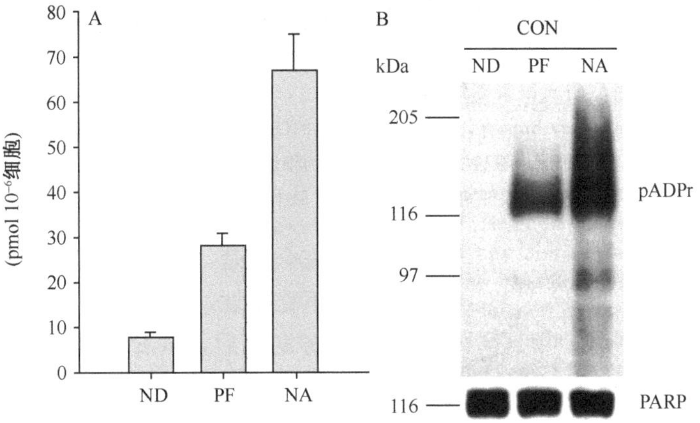


图 18.2　A. 与配对饲喂（paired fed，PF）的对照组相比，骨髓细胞对烟酸缺乏（niacin deficiency，ND）和药理剂量的烟酸（nicotinic acid，NA，4g/kg 膳食）的高敏感性。对照组是摄入正常需要量的烟酸 0.03g/kg 膳食。在本试验中，骨髓 $NAD^+$ 的变化约 10 倍。B. 蛋白质结合的 pADPr（上）和 PARR 蛋白（下）的 Western 印迹图。采用快绿染色法（fast green staining）检测表明，每个泳道的总蛋白质和分布形式相同，但是与配对饲喂对照（PF）比，烟酸缺乏（ND）和药理剂量（NA）添加显著影响 pADPr 修饰方式。pADPr 水平上的差异不是由 PARR 蛋白表达的变化所致（这些试验的详细资料见 Boyonoski *et al.*，2002a，b）

最近，我们对基因组的不稳定性进行了更直接地研究。单独缺乏烟酸使大鼠骨髓多染性红细胞的微核率提高了 6 倍（Spronck and Kirkland，2002）。图 18.3 显示烟酸缺乏大鼠的骨髓细胞用吖啶橙染色后的荧光图，包括有微核的多染性红细胞。微核指的是不能正确进入子代细胞核的整条染色体（含着丝粒的微核）或无着丝粒的染色体断片（无着丝粒的微核），它们的形成机制不同。含着丝粒的微核是由纺锤丝或着丝粒功能失调所致，而无着丝粒的微核是由 DNA 在损伤位点断裂产生。我们尽管没有对烟酸缺乏时增加的微核类型进行研究，但推测相当大部分的微核是含着丝粒的，因为我们没发现单独烟酸缺乏对 DNA 链断裂有任何影响（Boyonoski *et al.*，2002a），这暗示无着丝粒的微核不是主要原因。烟酸缺乏也提高基因毒性药物诱导产生的微核和 DNA 损伤试剂诱导产生的链断片（资料未发表），这表明无着丝粒的微核对烟酸缺乏具有一定的敏感性。

图 18.4 是烟酸缺乏的大鼠骨髓细胞的分裂中期图，用溴脱氧尿苷进行染色，以分辨染色单体（箭头）之间的同源重组体，被称为 SCE。尽管与 SCE 有关的机制目前还

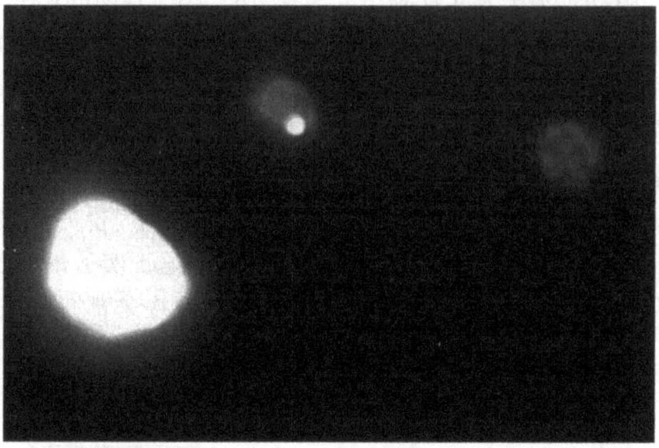

图 18.3 用吖啶橙染色后的大鼠骨髓细胞

多染性红细胞（polychromatic erythrocyte，PCE）是不成熟的红细胞，不含细胞核，但含 RNA，显现红色荧光。微核被染成亮绿色，只在 PCE 细胞中被检测到，因为有核的细胞中微核难以检测，含微核的成熟红细胞易被脾脏清除。左边可以看到一个含完整的大细胞核的淋巴细胞（原书为黑白印刷，故无法显示彩色）

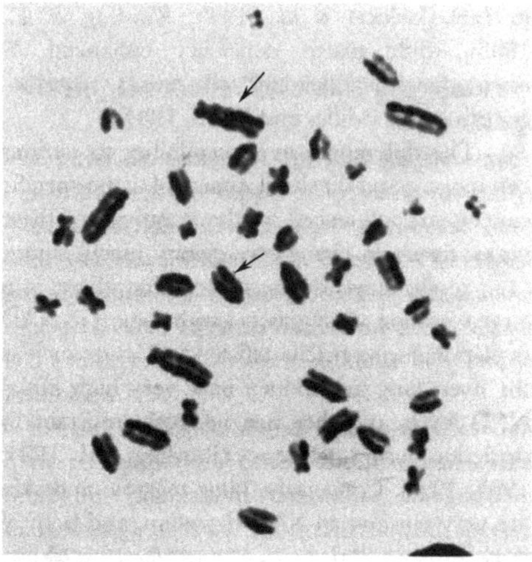

图 18.4 烟酸缺乏的大鼠骨髓细胞的分裂中期图

用溴脱氧尿苷进行前处理，新合成的染色单体被染色。新合成的和原来的染色单体之间的交换称为姐妹染色单体交换，属于同源重组，增强了 DNA 损伤和基因组的不稳定性

不完全清楚，但它们反映了各种类型的基因组不稳定性，并且几乎所有类型的基因毒性损伤发生后，SCE 水平都提高，如前所述，所有的 PARP 缺失模型都会提高 SCE 频率。我们发现单独烟酸缺乏使大鼠骨髓细胞的 SCE 频率提高了 3 倍（Spronck and Kirkland，2002）。

另外，我们还发现烟酸缺乏可能提高骨髓细胞中染色体畸变的频率（图 18.5），改变细胞周期中期的染色体结构（资料未发表）。初步试验还表明，烟酸缺乏导致 p53 异常表达并增加鬼臼乙叉苷处理后的 DNA 链断裂（资料未发表）。

图 18.5 表明单独烟酸缺乏提高了总染色体的不稳定性。在我们的试验中，烟酸缺乏对基因组不稳定性的诱导机制基本上还不清楚。pADPr 积聚的明显变化提示，pADPr 合成是重要的致病因素，然而，许多 PARP 同系物也能合成 pADPr，同时 pADPr 对基因组稳定性的调节还存在许多其他机制。p53 信号转导的改变也能破坏细胞周期的阻滞，降低 DNA 修复的效率，并通过破坏细胞凋亡信号的传递使严重受损的细胞不发生凋亡。对这些作用机制的全面认识将增强人们对烟酸代谢、PARP 的功能及癌症发生的认识。

烟酸缺乏导致基因组不稳定的影响是多方面的，尽管在西方国家，真正的烟酸缺乏

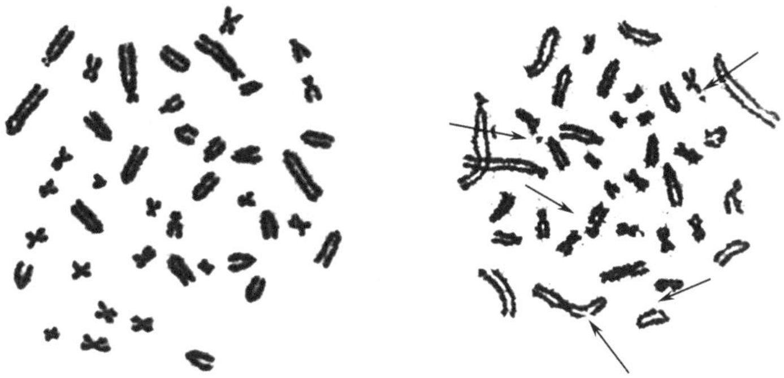


图 18.5　对照和烟酸缺乏的大鼠骨髓细胞的分裂中期图

烟酸缺乏组表现失常并且染色体有粗糙的外观

非常罕见，但还是有部分人群（妇女和老年人）处于亚临床的烟酸缺乏状态，而且，接受化疗的癌症患者中有 40%出现烟酸缺乏（Inculet *et al.*，1987）。我们认为烟酸缺乏可能会提高与治疗有关的继发性恶性疾病的发生概率。大多数白血病的特点是染色体易位，而 pADPr 合成受损可能增加这些异常的重组。皮肤癌是人类另一个重要的健康问题。在动物试验中，皮肤癌会受到烟酸营养状况的严重影响。由于过去对烟酸的研究较少，烟酸营养与癌症的发病率之间的重要关系没能引起我们的重视。本文综述的基础研究表明，烟酸营养状况能影响人类的基因组稳定性和癌症的发病率。

## 参　考　文　献


Agarwal, M.L., Agarwal, A., Taylor, W.R., Wang, Z.Q., Wagner, E.F. and Stark, G.R. (1997) Defective induction but normal activation and function of p53 in mouse cells lacking poly-ADP-ribose polymerase. *Oncogene* 15, 1035–1041.

Albertini, R.J., Anderson, D., Douglas, G.R., Hagmar, L., Hemminki, K., Merlo, F., Natarajan, A.T., Norppa, H., Shuker, D.E., Tice, R., Waters, M.D. and Aitio, A. (2000) IPCS guidelines for the monitoring of genotoxic effects of carcinogens in humans. International Programme on Chemical Safety. *Mutation Research* 463, 111–172.

Alberts, B., Bray, D., Lewis, J., Raff, M., Roberts, K. and Watson, J.D. (1994) Cancer. In: Robertson, M., Adams, R., Cobert, S.M. and Goertzen, D. (eds) *Molecular Biology of the Cell.* Garland Publishing, New York, pp. 1255–1294.

Almog, N., Li, R., Peled, A., Schwartz, D., Wolkowicz, R., Goldfinger, N., Pei, H. and Rotter, V. (1997) The murine C′-terminally alternatively spliced form of p53 induces attenuated apoptosis in myeloid cells. *Molecular and Cellular Biology* 17, 713–722.

Alvarez-Gonzalez, R. and Althaus, F.R. (1989) Poly(ADP-ribose) catabolism in mammalian cells exposed to DNA-damaging agents. *Mutation Research* 218, 67–74.

Ame, J.C., Rolli, V., Schreiber, V., Niedergang, C., Apiou, F., Decker, P., Muller, S., Hoger, T., Menissier-de Murcia, J. and de Murcia, G. (1999) PARP-2, a novel mammalian DNA damage-dependent poly(ADP-ribose) polymerase. *Journal of Biological Chemistry* 274, 17860–17868.

Bayle, J.H., Elenbaas, B. and Levine, A.J. (1995) The carboxyl-terminal domain of the p53 protein regulates sequence-specific DNA binding through its nonspecific nucleic acid-binding activity. *Proceedings of the National Academy of Sciences USA* 92, 5729–5733.

Berger, N.A. (1985) Poly(ADP-ribose) in the cellular response to DNA damage. *Radiation Research* 101, 4–15.

Blot, W.J., Li, J.Y., Taylor, P.R., Guo, W., Dawsey, S.M. and Li, B. (1995) The Linxian trials: mortality rates by vitamin–mineral intervention group. *American Journal of Clinical Nutrition* 62 (Supplement),

1424S–1426S.
Boyonoski, A.C., Spronck, J.C., Gallacher, L.M., Jacobs, R.M., Shah, G.M., Poirier, G.G. and Kirkland, J.B. (2002a) Niacin deficiency decreases bone marrow poly(ADP-ribose) and the latency of ethylnitrosourea-induced carcinogenesis in rats. *Journal of Nutrition* 132, 108–114.
Boyonoski, A.C., Spronck, J.C., Jacobs, R.M., Shah, G.M., Poirier G.G. and Kirkland, J.B. (2002b) Pharmacological intakes of niacin increase bone marrow poly(ADP-ribose) and the latency of ethylnitrosourea-induced carcinogenesis in rats. *Journal of Nutrition* 132, 115–120.
Bryan, G.T. (1986) The influence of niacin and nicotinamide on *in vivo* carcinogenesis. *Advances in Experimental Medicine and Biology* 206, 331–338.
Carpenter, K.J. (1981) *Pellagra.* Hutchinson Ross, Stroudsburg.
Chatterjee, S., Cheng, M.F., Berger, S.J. and Berger, N.A. (1991) Alkylating agent hypersensitivity in poly(adenosine diphosphate-ribose) polymerase deficient cell lines. *Cancer Communications* 3, 71–75.
Colman, M.S., Afshari, C.A. and Baffett, J.C. (2000) Regulation of p53 stability and activity in response to genotoxic stress. *Mutation Research* 462, 179–188.
d'Adda, D.F., Hande, M.P., Tong, W.M., Lansdorp, P.M., Wang, Z.Q. and Jackson, S.P. (1999) Functions of poly(ADP-ribose) polymerase in controlling telomere length and chromosomal stability. *Nature Genetics* 23, 76–80.
D'Amours, D., Desnoyers, S., D'Silva, I. and Poirier, G.G. (1999) Poly(ADP-ribosyl)ation reactions in the regulation of nuclear functions. *Biochemical Journal* 342, 249–268.
Dantzer, F., Schreiber, V., Niedergang, C., Trucco, C., Flatter, E., De La Rubia, G., Oliver, J., Rolli, V., Menissier-de Murcia, J. and de Murcia, G. (1999) Involvement of poly(ADP-ribose) polymerase in base excision repair. *Biochimie* 81, 69–75.
de Murcia, G. and Menissier, D.M. (1994) Poly(ADP-ribose) polymerase: a molecular nick-sensor. *Trends in Biochemical Sciences* 19, 172–176.
de Murcia, J.M., Niedergang, C., Trucco, C., Ricoul, M., Dutrillaux, B., Mark, M., Oliver, F.J., Masson, M., Dierich, A., LeMeur, M., Walztinger, C., Chambon, P. and de Murcia, G. (1997) Requirement of poly(ADP-ribose) polymerase in recovery from DNA damage in mice and in cells. *Proceedings of the National Academy of Sciences USA* 94, 7303–7307.
Durkacz, B.W., Omidiji, O., Gray, D.A. and Shall, S. (1980) $(ADP\text{-ribose})_n$ participates in DNA excision repair. *Nature* 283, 593–596.
el Deiry, W.S. (1998) Regulation of p53 downstream genes. *Seminars in Cancer Biology* 8, 345–357.
Franceschi, S., Bidoli, E., Baron, A.E. and La Vecchia, C. (1990) Maize and risk of cancers of the oral cavity, pharynx, and esophagus in northeastern Italy. *Journal of the National Cancer Institute* 82, 1407–1411.
Gensler, H.L. (1997) Prevention of photoimmunosuppression and photocarcinogenesis by topical nicotinamide. *Nutrition and Cancer* 29, 157–162.
Gensler, H.L., Williams, T., Huang, A.C. and Jacobson, E.L. (1999) Oral niacin prevents photocarcinogenesis and photoimmunosuppression in mice. *Nutrition and Cancer* 34, 36–41.
Gottlieb, T.M. and Oren, M. (1998) p53 and apoptosis. *Seminars in Cancer Biology* 8, 359–368.
Hageman, G.J. and Stierum, R.H. (2001) Niacin, poly (ADP-ribose) polymerase-1 and genomic stability. *Mutation Research* 475, 45–56.
Halappanavar, S.S., Rhun, Y.L., Mounir, S., Martins, L.M., Huot, J., Earnshaw, W.C. and Shah, G.M. (1999) Survival and proliferation of cells expressing caspase-uncleavable poly(ADP-ribose) polymerase in response to death-inducing DNA damage by an alkylating agent. *Journal of Biological Chemistry* 274, 37097–37104.
Halloway, S.L., Poruthu, J. and Scata, K. (1999) Chromosome segregation and cancer. *Experimental Cell Research* 253, 308–314.
Hori, T. (1981) High incidence of sister chromatid exchanges and chromatid interchanges in the conditions of lowered activity of poly(ADP-ribose) polymerase. *Biochemical and Biophysical Research Communications* 102, 38–45.
Ikejima, M., Noguchi, S., Yamashita, R., Ogura, T., Sugimura, T., Gill, D.M. and Miwa, M. (1990) The zinc fingers of human poly(ADP-ribose) polymerase are differentially required for the recognition of DNA breaks and nicks and the consequent enzyme activation. *Journal of Biological Chemistry* 265, 21907–21913.
Inculet, R.I., Norton, J.A., Nichoalds, G.E., Maher, M.M., White, D.E. and Brennan, M.F. (1987) Water-soluble vitamins in cancer patients on parenteral nutrition: a prospective study. *Journal of Parenteral and Enteral Nutrition* 11, 243–249.
Jackson, T.M., Rawling, J.M., Roebuck, B.D. and Kirkland, J.B. (1995) Large supplements of nicotinic acid and nicotinamide increase tissue $NAD^+$ and poly(ADP-ribose) levels but do not affect diethylnitrosamine-induced altered hepatic foci in Fischer-344 rats. *Journal of Nutrition* 125, 1455–1461.
Jacobson, E.L. (1993) Niacin deficiency and cancer in women. *Journal of the American College of Nutrition* 12, 412–416.
Jacobson, M.K. and Jacobson, E.L. (1999) Discovering new ADP-ribose polymer cycles: protecting the genome and more. *Trends in Biochemical Sciences* 24, 415–417.
Jacobson, M.K., Levi, V., Juarez-Salinas, H., Barton, R.A. and Jacobson, E.L. (1980) Effect of carcinogenic *N*-alkyl-*N*-nitroso compounds on nicotinamide adenine dinucleotide metabolism. *Cancer*

*Research* 40, 1797–1802.
Johansson, M. (1999) A human poly(ADP-ribose) polymerase gene family (ADPRTL): cDNA cloning of two novel poly(ADP-ribose) polymerase homologues. *Genomics* 57, 442–445.
Kaufmann, W.K. and Paules, R.S. (1996) DNA damage and cell cycle checkpoints. *FASEB Journal* 10, 238–247.
Kaufmann, S.H., Desnoyers, S., Ottaviano, Y., Davidson, N.E. and Poirier, G.G. (1993) Specific proteolytic cleavage of poly(ADP-ribose) polymerase: an early marker of chemotherapy-induced apoptosis. *Cancer Research* 53, 3976–3985.
Kickhoefer, V.A., Siva, A.C., Kedersha, N.L., Inman, E.M., Ruland, C., Streuli, M. and Rome, L.H. (1999a) The 193-kD vault protein, VPARP, is a novel poly(ADP-ribose) polymerase. *Journal of Cell Biology* 146, 917–928.
Kickhoefer, V.A., Stephen, A.G., Harrington, L., Robinson, M.O. and Rome, L.H. (1999b) Vaults and telomerase share a common subunit, TEP1. *Journal of Biological Chemistry* 274, 32712–32717.
Kirkland, J.B. and Rawling, J.M. (2001) Niacin. In: Rucker R, Suttie J.W., McCormick D.M. and Machlin L.J. (eds) *Handbook of Vitamins*. Marcel Dekker, New York, pp. 211–252.
Knebl, J.A. and Jacobson, E.L. (1992) Assessment of niacin status in an elderly population. *Gerontology* 32, 247A.
Kumari, S.R., Mendoza-Alvarez, H. and Alvarez-Gonzalez, R. (1998) Functional interactions of p53 with poly(ADP-ribose) polymerase (PARP) during apoptosis following DNA damage: covalent poly (ADP-ribosyl)ation of p53 by exogenous PARP and noncovalent binding of p53 to the M(r) 85,000 proteolytic fragment. *Cancer Research* 58, 5075–5078.
Le Rhun, Y., Kirkland, J.B. and Shah, G.M. (1998) Cellular responses to DNA damage in the absence of poly(ADP-ribose) polymerase. *Biochemical and Biophysical Research Communications* 245, 1–10.
Lee, H.C. (2001) Physiological functions of cyclic ADP-ribose and NAADP as calcium messengers. *Annual Reviews of Pharmacology and Toxicology* 41, 317–345.
Leist, M., Single, B., Kunstle, G., Volbracht, C., Hentze, H. and Nicotera, P. (1997) Apoptosis in the absence of poly-(ADP-ribose) polymerase. *Biochemical and Biophysical Research Communications* 233, 518–522.
Lin, S.J., Defossez, P.A. and Guarente, L. (2000) Requirement of NAD and SIR2 for life-span extension by calorie restriction in *Saccharomyces cerevisiae*. *Science* 289, 2126–2128.
Lindahl, T., Satoh, M.S., Poirier, G.G. and Klungland, A. (1995) Post-translational modification of poly (ADP-ribose) polymerase induced by DNA strand breaks. *Trends in Biochemical Sciences* 20, 405–411.
Malanga, M., Pleschke, J.M., Kleczkowska, H.E. and Althaus, F.R. (1998) Poly(ADP-ribose) binds to specific domains of p53 and alters its DNA binding functions. *Journal of Biological Chemistry* 273, 11839–11843.
Marshall, J.R., Graham, S., Haughey, B.P., Shedd, D., O'Shea, R., Brasure, J., Wilkinson, G.S. and West, D. (1992) Smoking, alcohol, dentition and diet in the epidemiology of oral cancer. *European Journal of Cancer B Oral Oncology* 28B, 9–15.
Masson, M., Niedergang, C., Schreiber, V., Muller, S., Menissier-de Murcia, J. and de Murcia, G. (1998) XRCC1 is specifically associated with poly(ADP-ribose) polymerase and negatively regulates its activity following DNA damage. *Molecular and Cellular Biology* 18, 3563–3571.
Masutani, M., Nozaki, T., Wakabayashi, K. and Sugimura, T. (1995) Role of poly(ADP-ribose) polymerase in cell-cycle checkpoint mechanisms following gamma-irradiation. *Biochimie* 77, 462–465.
Masutani, M., Suzuki, H., Kamada, N., Watanabe, M., Ueda, O., Nozaki, T., Jishage, K., Watanabe, T., Sugimoto, T., Nakagama, H., Ochiya, T. and Sugimura, T. (1999) Poly(ADP-ribose) polymerase gene disruption conferred mice resistant to streptozotocin-induced diabetes. *Proceedings of the National Academy of Sciences USA* 96, 2301–2304.
Masutani, M., Nozaki, T., Nakamoto, K., Nakagama, H., Suzuki, H., Kusuoka, O., Tsutsumi, M. and Sugimura, T. (2000) The response of Parp knockout mice against DNA damaging agents. *Mutation Research* 462, 159–166.
McClintock, B. (1941) The stability of broken ends of chromosomes in *Zea mays*. *Genetics* 25, 234–282.
Miller, E.G. and Burns, H. Jr (1984) *N*-Nitrosodimethylamine carcinogenesis in nicotinamide-deficient rats. *Cancer Research* 44, 1478–1482.
Molinete, M., Vermeulen, W., Burkle, A., Menissier-de Murcia, J., Kupper, J.H., Hoeijmakers, J.H. and de Murcia, G. (1993) Overproduction of the poly (ADP-ribose) polymerase DNA-binding domain blocks alkylation-induced DNA repair synthesis in mammalian cells. *EMBO Journal* 12, 2109–2117.
Morgan, W.F. and Cleaver, J.E. (1982) 3-Aminobenzamide synergistically increases sister-chromatid exchanges in cells exposed to methyl methanesulfonate but not to ultraviolet light. *Mutation Research* 104, 361–366.
Nduka, N., Skidmore, C.J. and Shall, S. (1980) The enhancement of cytotoxicity of *N*-methyl-*N*-nitrosourea and of gamma-radiation by inhibitors of poly(ADP-ribose) polymerase. *European Journal of Biochemistry* 105, 525–530.
Niedergang, C., Oliver, F.J., Menissier-de-Murcia, J. and de Murcia, G. (2000) Involvement of poly(ADP-ribose) polymerase in the cellular response to DNA

damage. In: Szabo, C. (ed.) *Cell Death: the Role of PARP.* CRC Press, Boca Raton, Florida, pp. 183–207.

Nozaki, T., Masutani, M., Akagawa, T., Sugimura, T. and Esumi, H. (1994) Suppression of G1 arrest and enhancement of G2 arrest by inhibitors of poly (ADP-ribose) polymerase: possible involvement of poly(ADP- ribosyl)ation in cell cycle arrest following gamma-irradiation. *Japanese Journal of Cancer Research* 85, 1094–1098.

Oikawa, A., Tohda, H., Kanai, M., Miwa, M. and Sugimura, T. (1980) Inhibitors of poly(adenosine diphosphate ribose) polymerase induce sister chromatid exchanges. *Biochemical and Biophysical Research Communications* 97, 1311–1316.

Okazaki, I.J. and Moss, J. (1999) Characterization of glycosylphosphatidylinositol-anchored, secreted, and intracellular vertebrate mono-ADP-ribosyltransferases. *Annual Review of Nutrition* 19, 485–509.

Oliver, F.J., Menissier-de Murcia, J. and de Murcia, G. (1999) Poly(ADP-ribose) polymerase in the cellular response to DNA damage, apoptosis, and disease. *American Journal of Human Genetics* 64, 1282–1288.

Park, S.D., Kim, C.G. and Kim, M.G. (1983) Inhibitors of poly(ADP-ribose) polymerase enhance DNA strand breaks, excision repair, and sister chromatid exchanges induced by alkylating agents. *Environmental Mutagenesis* 5, 515–525.

Rawling, J.M., Jackson, T.M., Driscoll, E.R. and Kirkland, J.B. (1994) Dietary niacin deficiency lowers tissue poly(ADP-ribose) and $NAD^+$ concentrations in Fischer-344 rats. *Journal of Nutrition* 124, 1597–1603.

Rawling, J.M., Jackson, T.M., Roebuck, B.D., Poirier, G.G. and Kirkland, J.B. (1995) The effect of niacin deficiency on diethylnitrosamine-induced hepatic poly(ADP-ribose) levels and altered hepatic foci in the Fischer-344 rat. *Nutrition and Cancer* 24, 111–119.

Rawling, J.M., ApSimon, M.M. and Kirkland, J.B. (1996) Lung poly(ADP-ribose) and $NAD^+$ concentrations during hyperoxia and niacin deficiency in the Fischer-344 rat. *Free Radicals in Biology and Medicine* 20, 865–871.

Samper, E., Goytisolo, F.A., Menissier-de Murcia, J., Gonzalez-Suarez, E., Cigudosa, J.C., de Murcia, G. and Blasco, M.A. (2001) Normal telomere length and chromosomal end capping in poly(ADP-ribose) polymerase-deficient mice and primary cells despite increased chromosomal instability. *Journal of Cell Biology* 154, 49–60.

Satoh, M.S. and Lindahl, T. (1992) Role of poly (ADP-ribose) formation in DNA repair. *Nature* 356, 356–358.

Satoh, M.S., Poirier, G.G. and Lindahl, T. (1994) Dual function for poly(ADP-ribose) synthesis in response to DNA strand breakage. *Biochemistry* 33, 7099–7106.

Schmitt, C.A. and Lowe, S.W. (1999) Apoptosis and therapy. *Journal of Pathology* 187, 127–137.

Schreiber, V., Hunting, D., Trucco, C., Gowans, B., Grunwald, D., de Murcia, G. and de Murcia, J.M. (1995) A dominant-negative mutant of human poly(ADP-ribose) polymerase affects cell recovery, apoptosis, and sister chromatid exchange following DNA damage. *Proceedings of the National Academy of Sciences USA* 92, 4753–4757.

Shah, G.M., Le Rhun, Y., Sutarjono, I. and Kirkland, J.B. (2002) Niacin deficient SKH-1 mice are more susceptible to ultraviolet B radiation-induced skin carcinogenesis. *Cancer Research* 131, 3150S.

Shieh, W.M., Ame, J.C., Wilson, M.V., Wang, Z.Q., Koh, D.W., Jacobson, M.K. and Jacobson, E.L. (1998) Poly(ADP-ribose) polymerase null mouse cells synthesize ADP-ribose polymers. *Journal of Biological Chemistry* 273, 30069–30072.

Simbulan-Rosenthal, C.M., Rosenthal, D.S., Iyer, S., Boulares, A.H. and Smulson, M.E. (1998) Transient poly(ADP-ribosyl)ation of nuclear proteins and role of poly(ADP-ribose) polymerase in the early stages of apoptosis. *Journal of Biological Chemistry* 273, 13703–13712.

Simbulan-Rosenthal, C.M., Rosenthal, D.S., Luo, R. and Smulson, M.E. (1999a) Poly(ADP-ribosyl)ation of p53 during apoptosis in human osteosarcoma cells. *Cancer Research* 59, 2190–2194.

Simbulan-Rosenthal, C.M., Haddad, B.R., Rosenthal, D.S., Weaver, Z., Coleman, A., Luo, R., Young, H.M., Wang, Z.Q., Ried, T. and Smulson, M.E. (1999b) Chromosomal aberrations in PARP(−/−) mice: genome stabilization in immortalized cells by reintroduction of poly(ADP-ribose) polymerase cDNA. *Proceedings of the National Academy of Sciences USA* 96, 13191–13196.

Simbulan-Rosenthal, C.M., Ly, D.H., Rosenthal, D.S., Konopka, G., Luo, R., Wang, Z.Q., Schultz, P.G. and Smulson, M.E. (2000) Misregulation of gene expression in primary fibroblasts lacking poly (ADP-ribose) polymerase. *Proceedings of the National Academy of Sciences USA* 97, 11274–11279.

Smith, S. and de Lange, T. (1999) Cell cycle dependent localization of the telomeric PARP, tankyrase, to nuclear pore complexes and centrosomes. *Journal of Cell Science* 112, 3649–3656.

Smith, S., Giriat, I., Schmitt, A. and de Lange, T. (1998) Tankyrase, a poly(ADP-ribose) polymerase at human telomeres. *Science* 282, 1484–1487.

Smulson, M.E., Simbulan-Rosenthal, C.M., Boulares, A.H., Yakovlev, A., Stoica, B., Iyer, S., Luo, R., Haddad, B., Wang, Z.Q., Pang, T., Jung, M., Dritschilo, A. and Rosenthal, D.S. (2000) Roles of poly(ADP-ribosyl)ation and PARP in apoptosis, DNA repair, genomic stability and functions of p53 and E2F-1. *Advances in Enzyme Regulation* 40, 183–215.

Spronck, J.C. and Kirkland, J.B. (2002) Niacin deficiency increases spontaneous and etoposide-induced chromosomal instability in rat bone marrow cells *in vivo*. *Mutation Research* 508, 83–97.

Steegenga, W.T., van der Eb, A.J. and Jochemsen, A.G. (1996) How phosphorylation regulates the activity of p53. *Journal of Molecular Biology* 263, 103–113.

Trucco, C., Rolli, V., Oliver, F.J., Flatter, E., Masson, M., Dantzer, F., Niedergang, C., Dutrillaux, B., Menissier-de Murcia, J. and de Murcia, G. (1999) A dual approach in the study of poly (ADP-ribose) polymerase: *in vitro* random mutagenesis and generation of deficient mice. *Molecular and Cellular Biochemistry* 193, 53–60.

Tsutsumi, M., Masutani, M., Nozaki, T., Kusuoka, O., Tsujiuchi, T., Nakagama, H., Suzuki, H., Konishi, Y. and Sugimura, T. (2001) Increased susceptibility of poly(ADP-ribose) polymerase-1 knockout mice to nitrosamine carcinogenicity. *Carcinogenesis* 22, 1–3.

Tucker, J.D. and Preston, R.J. (1996) Chromosome aberrations, micronuclei, aneuploidy, sister chromatid exchanges, and cancer risk assessment. *Mutation Research* 365, 147–159.

Van Rensburg, S.J., Hall, J.M. and Gathercole, P.S. (1986) Inhibition of esophageal carcinogenesis in corn-fed rats by riboflavin, nicotinic acid, selenium, molybdenum, zinc, and magnesium. *Nutrition and Cancer* 8, 163–170.

Vaziri, H., West, M.D., Allsopp, R.C., Davison, T.S., Wu, Y.S., Arrowsmith, C.H., Poirier, G.G. and Benchimol, S. (1997) ATM-dependent telomere loss in aging human diploid fibroblasts and DNA damage lead to the post-translational activation of p53 protein involving poly(ADP-ribose) polymerase. *EMBO Journal* 16, 6018–6033.

Vaziri, H., Dessain, S.K., Eaton, E.N., Imai, S.I., Frye, R.A., Pandita, T.K., Guarente, L. and Weinberg, R.A. (2001) hSir2(SIRT1) functions as an NAD-dependent p53 deacetylase. *Cell* 107, 149–159.

Vodenicharov, M.D., Sallmann, F.R., Satoh, M.S. and Poirier, G.G. (2000) Base excision repair is efficient in cells lacking poly(ADP-ribose) polymerase 1. *Nucleic Acids Research* 28, 3887–3896.

Wang, X., Ohnishi, K., Takahashi, A. and Ohnishi, T. (1998) Poly(ADP-ribosyl)ation is required for p53-dependent signal transduction induced by radiation. *Oncogene* 17, 2819–2825.

Wang, Z.Q., Stingl, L., Morrison, C., Jantsch, M., Los, M., Schulze-Osthoff, K. and Wagner, E.F. (1997) PARP is important for genomic stability but dispensable in apoptosis. *Genes and Development* 11, 2347–2358.

Weitberg, A.B. (1989) Effect of nicotinic acid supplementation *in vivo* on oxygen radical-induced genetic damage in human lymphocytes. *Mutation Research* 216, 197–201.

Weitberg, A.B. and Corvese, D. (1990) Niacin prevents DNA strand breakage by adenosine deaminase inhibitors. *Biochemical and Biophysical Research Communications* 167, 514–519.

Wesierska-Gadek, J., Bugajska-Schretter, A. and Cerni, C. (1996a) ADP-ribosylation of p53 tumor suppressor protein: mutant but not wild-type p53 is modified. *Journal of Cellular Biochemistry* 62, 90–101.

Wesierska-Gadek, J., Schmid, G. and Cerni, C. (1996b) ADP-ribosylation of wild-type p53 *in vitro*: binding of p53 protein to specific p53 consensus sequence prevents its modification. *Biochemical and Biophysical Research Communications* 224, 96–102.

Wesierska-Gadek, J., Wang, Z.Q. and Schmid, G. (1999) Reduced stability of regularly spliced but not alternatively spliced p53 protein in PARP-deficient mouse fibroblasts. *Cancer Research* 59, 28–34.

Whitacre, C.M., Hashimoto, H., Tsai, M.L., Chatterjee, S., Berger, S.J. and Berger, N.A. (1995) Involvement of NAD-poly(ADP-ribose) metabolism in p53 regulation and its consequences. *Cancer Research* 55, 3697–3701.

# 第 5 部分
# 分子活动对生理的影响

# 19 转运甘油三酯的血浆脂蛋白的装配

Joan A. Higgins
（舍菲尔德大学，分子生物学与生化学院，英国）

## 前　　言

脂肪具有重要的生理作用，如储存在脂肪组织中的甘油三酯是生物体的重要能源，磷脂是组成生物膜的必需成分，胆固醇也是生物膜的组分之一，并且是包括类固醇激素和胆汁盐等生物活性物质的前体。脂类之所以具有上述生理作用是由于它在水溶液中具有低溶解度特性，因此，由于磷脂的亲水性和疏水性，它参与组成生物膜基本的双层结构。此外，不以碳水化合物，而以甘油三酯作为储存能量的重要形式，有其生物学优势，脂肪不与水结合，所占体积最节省，增加的体重也最轻。然而，正是脂类发挥生理作用的水不溶性，使得其在血液或淋巴器官中的转运发生困难，解决的办法是以脂类与蛋白质复合物的形式（即血浆脂蛋白）进行转运。

## 血浆脂蛋白

脂类以多种不同的血浆脂蛋白形式进行转运，不同的脂蛋白是根据其密度、组成和电泳活性进行定义和区分的（表 19.1）。四类主要的血浆脂蛋白分别为：乳糜微粒（chylomicron）、极低密度脂蛋白（very low-density lipoprotein，VLDL）、低密度脂蛋白（low-density lipoprotein，LDL）和高密度脂蛋白（high-density lipoprotein，HDL），每一类又包括若干特性和功能不同的亚类。不同种类和亚类的脂蛋白之间的关系非常复杂，各种血浆脂蛋白以及它们的亚类和相互关系的详细论述不在本章范围之内，下面仅进行简单说明。

### 转运甘油三酯的脂蛋白

甘油三酯是以乳糜微粒的形式运输的，乳糜微粒携带来自小肠的膳食脂肪，经淋巴系统进入血浆，同时 VLDL 携带来自肝脏的内源性甘油三酯进入血浆，乳糜微粒是体积最大、密度最小的血浆脂蛋白。非极性脂类主要指甘油三酯和一些胆固醇酯，它们占脂类的 90%以上（$m/m$）。双亲性的磷脂和胆固醇及若干称为载脂蛋白的蛋白质组成脂蛋白的外层，使脂滴保持稳定。外层的载脂蛋白对乳糜微粒在肠上皮细胞内的合成及其在血管内的代谢均有重要作用（表 19.1 和表 19.2）。VLDL 的体积比乳糜微粒小，却比乳糜微粒重。然而，它们的基本结构相似，均由外层为两性的脂类和载脂蛋白的非极性脂滴组成（图 19.1）。

乳糜微粒和 VLDL 分别向细胞和器官提供膳食中的或内源性的甘油三酯脂肪酸。甘油三酯在靶器官毛细血管内皮细胞表面的脂蛋白脂酶的作用下发生水解，生成脂肪酸，并被细胞所摄取。随着甘油三酯内核逐渐分解，脂蛋白颗粒也逐渐变小，而且一些表面组分被转移给其他脂蛋白。乳糜微粒成为乳糜微粒残滴后，被肝脏清除，而 VLDL 先被转换为中间密度脂蛋白（intermediate-density lipoprotein，IDL），再变为 LDL。IDL 被肝脏清除，而 LDL 为各类细胞提供胆固醇，多余的 LDL 也由肝脏清除。

**表 19.1　血浆脂蛋白**

A. 一般特性

| 脂蛋白 | 直径/(nmol/L) | 密度/(g/ml) | 电泳活动性[1] | 悬浮率/(Sf) | 主要组织来源 | 主要功能 |
|---|---|---|---|---|---|---|
| 乳糜微粒 | 80～500 | <0.94 | 初始 | >400 | 小肠 | 转运外源性脂肪 |
| VLDL | 30～100 | 0.94～1.006 | β | 20～400 | 肝脏 | 转运内源性脂肪 |
| LDL | 19～25 | 1.006～1.063 | 前 β | 0～20 | 肝脏（由 VLDL 转变而成） | 转运内源性胆固醇至组织中 |
| HDL | 4～12 | 1.063～1.21 | α | 0～9 | 肝脏和小肠 | 转运组织中的胆固醇至肝脏 |

B. 组成

| 脂蛋白 | 蛋白质/% | 甘油三酯/% | 胆固醇/% | 磷脂/% | 载脂蛋白 |
|---|---|---|---|---|---|
| 乳糜微粒 | 2 | 89 | 5 | 4 | A1，A11 和 A1V<br>B48<br>C1，C11 和 C111<br>E |
| VLDL | 10 | 57 | 17 | 16 | B100<br>C1，C11 和 C111<br>E |
| LDL | 23 | 6 | 47 | 24 | B100 |
| HDL | 48 | 4 | 17 | 30 | A1，A11 和 A111<br>E |

1 在琼脂糖凝胶上的移动性，此处 α 向阴极方向移动的距离最远。

**表 19.2　人血浆脂蛋白**

| 载脂蛋白 | 分子质量/kDa | 合成部位 | 脂蛋白 | 血浆中的含量/(g/L) | 主要功能 |
|---|---|---|---|---|---|
| Apo-A1 | 28.1 | 肝脏和小肠 | 乳糜微粒、HDL | 1.3～1.8 | LCAT 激活剂、HDL 结构蛋白 |
| Apo-A11 | 17.4 | 肝脏和小肠 | 乳糜微粒、HDL | 0.4～0.65 | LCAT 抑制剂、HDL 结构蛋白 |
| Apo-A1V | 46.0 | 肝脏和小肠 | 乳糜微粒 | 0.4 | |
| Apo-B48 | 260.0 | 小肠 | 乳糜微粒 | 0.005～0.2 | 乳糜微粒结构蛋白 |
| Apo-B100 | 550.0 | 肝脏 | VLDL | 0.6～1.2 | VLDL 结构蛋白 |
| Apo-C1 | 6.3 | 肝脏 | 乳糜微粒、VLDL、HDL | 0.1～0.18 | LCAT 激活剂（?） |
| Apo-C11 | 8.8 | 肝脏 | 乳糜微粒、VLDL、HDL | 0.78～2.0 | 脂蛋白脂酶激活剂 |
| Apo-C111 | 8.8 | 肝脏 | 乳糜微粒、VLDL、HDL | 0.2～0.4 | 脂蛋白脂酶抑制剂 |
| Apo-E | 36.5 | 肝脏 | 乳糜微粒、VLDL、HDL | 0.78～1.5 | 参与清除乳糜微粒残滴和 VLDL 的合成 |

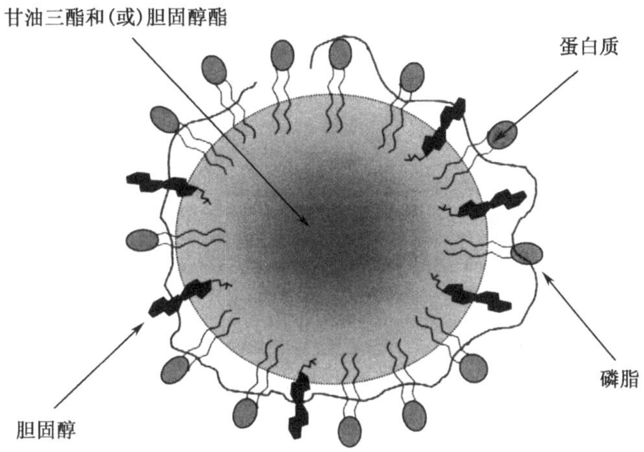


图 19.1 血浆脂蛋白的基本结构

## 胆固醇转运的脂蛋白

LDL 是一类体积较小的脂蛋白颗粒，其非极性脂核主要为胆固醇酯，外层为磷脂和胆固醇以及单一的载脂蛋白 B（apolipoprotein B，apo-B）。经与 apo-B 结合的 LDL 受体介导的胞吞作用，LDL 颗粒被需要胆固醇的所细胞吸收。胆固醇的摄取会导致 LDL 受体的副调控和胆固醇的合成降低，从而调节细胞内的总胆固醇水平。当血浆中 LDL 的浓度增加，LDL apo-B 有被破坏或修饰（如氧化）的倾向。被破坏的蛋白质成为动脉管壁中巨噬细胞上的清除受体的一个配基，这些受体不会发生负调节作用，因此胆固醇累积在巨噬细胞中成为泡沫细胞，这是动脉硬化症发生的第一步。

HDL 是体积最小、密度最大的一类脂蛋白，是肝脏和肠道产生的新生形态。HDL 的作用似乎是将胆固醇从包括动脉管壁在内的外围组织，直接地或通过其他脂蛋白间接地运输到肝脏，胆固醇不被降解，体内胆固醇的清除主要途径是随肝脏分泌的胆汁排出，因此 HDL 运输胆固醇的作用对于预防动脉硬化症非常重要。

# 甘油三酯转运的脂蛋白的合成

## 乳糜微粒和 VLDL 的总特征

乳糜微粒和 VLDL 具有很多共同的特征。两类脂蛋白均由非极性脂滴、甘油三酯和少量的胆固醇酯构成，外层由磷脂、胆固醇和载脂蛋白组成（表 19.1 和表 19.2）。构成乳糜微粒和 VLDL 的主要结构性载脂蛋白是 apo-B，它对于乳糜微粒和 VLDL 的合成和分泌均是必需的。乳糜微粒和 VLDL 中还存在其他的载脂蛋白，它们在血管内的代谢作用与 apo-B 相似（表 19.2）。

尽管乳糜微粒和 VLDL 具有许多相似之处，二者在结构和功能上也存在着明显差异。乳糜微粒体积较大，含有更多的甘油三酯分子，而密度却小于 VLDL（表 19.1）。Apo-B 是两类脂蛋白都必需的结构蛋白，然而 VLDL 中含有完整的多肽 apo-B100，乳糜微粒中含有 apo-B48，为一截短的形式，只有 apo-B100 N 端 48%的长度（Hussain *et al.*，1996；Davidson and Shelness，2000；Hussain，2000）。LDL 的受体结合区域处于 apo-B100 C 端一半长度的位置，这说明乳糜微粒残留物在循环系统的清除方式不同于 LDL（Hussain *et al.*，1996）。乳糜微粒残留物通过 apo-E 与 LDL 受体相关蛋白质的相互作用被肝脏摄取（Hussain *et al.*，1996）。脂蛋白脂酶对乳糜微粒的作用比对 VLDL 更迅速，因而，乳糜微粒在循环系统中的半衰期为若干分钟，而 VLDL 的半衰期为若干天。摄食后，随着循环系统中乳糜微粒水平的升高，VLDL 的浓度也会增加，这将导致患动脉硬化症的概率变大（Bjorkegren *et al.*，1996；Packard *et al.*，2000）。

小肠中来自膳食的脂肪数量和组成常发生大幅度的波动，且膳食中的脂肪必须被有效吸收以合成乳糜微粒。与之相反的是，肝脏可摄取脂肪酸作为底物，以合成浓度范围相当窄的 VLDL。因此，乳糜微粒合成的调节必然不同于 VLDL。乳糜微粒和 VLDL 的合成对普郎尼克酸抗原具有不同的敏感性也说明了这一点，普郎尼克酸可以抑制乳糜微粒的合成，却不影响 VLDL 的合成（Tso *et al.*，1981）。此外，在遗传性障碍乳糜微粒滞留病中，乳糜微粒的合成被特异性抑制，而 VLDL 的合成却不受影响（Anderson *et al.*，1961；Hussain *et al.*，2000），因此，至少有一种未知蛋白可特异性地参与乳糜微粒的合成。

## Apo-B mRNA 的编译

Apo-B 48 和 apo-B 100 是同一基因的产物（图 19.2），但是在小肠中，apo-B 100 mRNA 进行转录后修饰。由于胞嘧啶在锌-依赖性 mRNA 编辑脱氨酶 apo-bec-1 的作用下脱氨生成尿嘧啶，apo-B 100 mRNA 的核苷酸 6666 被改变，从而在谷氨酰胺编码密码子（CAA）的位置生成止动子（UAA），因此，apo-B 合成到 48%时就被终止了（Chen *et al.*，1987；Powell *et al.*，1987；Davidson and Shelness，2000）。

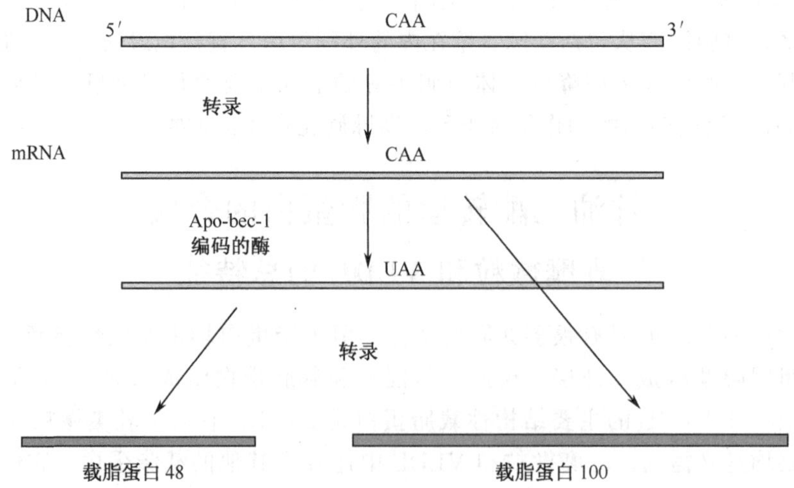


图 19.2　Apo-B mRNA 的编译

Apo-B 100 是完整的 apo-B mRNA 的产物。Apo-B 100 mRNA 被编译，在 LCAT（卵磷脂-胆固醇酰基转移酶）核苷酸 6666 的位置生成终止子，编译后所得到的截短蛋白质即 apo-B 48

Apo-B mRNA 的编译修饰只发生于所有被研究的哺乳类动物的小肠中，这表明 apo-B 48 的生成在乳糜微粒的合成和脂肪的吸收过程中具有重要作用。在人及几种动物如仓鼠及兔中，编译只发生于小肠。然而，对于某些种类的动物，尤其是常用的实验动物小鼠和大鼠，一些 apo-B mRNA 的转录后修饰也发生于肝脏，导致分泌既含有 apo-B 48 又含有 apo-B 100 的 VLDL（Davidson and Shelness，2000；Hussain *et al.*，2000）。

小肠中 apo-B 的修饰随着发育的过程而被调节。胎儿的肠道不表达 apo-bec-1 编译酶，只生成 apo-B 100，而在婴儿和成人阶段，apo-bec-1 的表达增加，生成 apo-B 48（Teng *et al.*，1990；Patterson *et al.*，1992）。因此，编译是非常特殊的，被精确调节。近期研究已表明，编译可以使肠道的上皮细胞随着膳食变化发生轻微的改变，使得肠道在营养物质的可获得性被限制的条件下仍能充分发挥作用（见下文）（Kendrick *et al.*，2001）。

## 肝脏中极低密度脂蛋白合成的分子基础

极低密度脂蛋白（VLDL）在肝脏中由肝细胞合成并被分泌到肝淋巴间隙处，再由此进入血浆。VLDL 分泌的速度是决定血浆中 LDL 水平的因素之一。由于血浆中 LDL 胆固醇水平的增加，易引发动脉粥样硬化，人们对弄清 VLDL 在肝脏中合成和分泌的调节机制产生了浓厚的兴趣。电子显微镜、超微结构细胞化学和放射自显影等形态学研究已经表明，VLDL 的合成和分泌是通过典型的分泌途径进行的。Apo-B 100 由粗面内质网（rough endoplasmic reticulum，RER）上的结合核糖体合成，脂滴（VLDL 及其前体）则出现于光面内质网（smooth endoplasmic reticulum，SER）和高尔基体腔内。

### 研究 VLDL 合成的细胞和动物实验模型

#### 培养细胞系

研究 VLDL 合成的最终目的是将试验结果外推至人，因此选择合适的试验模型很重要。有许多试验是利用培养细胞系进行研究，其中主要为人的肝癌细胞系 Hep-G2，而这类细胞不能进行甘油三酯合成，原因可能是这些细胞中光面内质网（SER）的数量比正常肝细胞的数量少，但 Hep-G2 可合成和分泌与 LDL 大小和密度相同的颗粒（Gibbons，1994）。对 Hep-G2 细胞的研究已经为了解 apo-B 100 在细胞内运输的早期分子基础提供了许多信息，但是对了解脂质在细胞内的运输帮助不大。一种大鼠肝瘤细胞系 McArdle 细胞可分泌体积更大、更轻的含 apo-B 的微粒，所以也被用作试验模型（Boren *et al.*，1994；Gordon *et al.*，1996）。然而，和人肝细胞不同的是，这些细胞既分泌 apo-B 100，又分泌 apo-B 48，而且这两种 apo-B 组成脂蛋白的方式不同。与制备供体动物的成年肝细胞相比，人的肝瘤细胞系更容易获得。然而，该细胞系来自于肿瘤，常有某些生化特性发生了变化，这就很难对结果做出解释和外推至正常生理状态。

#### 大鼠、仓鼠和兔的分离肝细胞

悬浮液或原代培养物中新鲜制备的肝细胞也已用于 VLDL 合成的研究（Dixon and

Ginsberg, 1993; Gibbons, 1994; Cartwright and Higgins, 1995, 1996; Cartwright *et al.*, 1997; Kendrick and Higgins, 1999)。尽管从技术上来说应该使用成年细胞，但此试验方法的优势是在分离肝细胞之前，可用不同的方式处理供体动物，如影响分泌的 VLDL 的特性及其分泌速度的短期和（或）长期的膳食改变，而且，仓鼠和兔的肝细胞在分泌仅含 apo-B 100 的 VLDL 方面与人的肝脏相似。

**亚细胞成分的分级分离**

VLDL 是由粗面内质网（RER）、色氨酸（SER）、顺面高尔基体和反面高尔基体组成，主要组分的分离方法已经形成（Higgins and Hutson, 1984）。最近本试验室已研发出几种方法，即用单一自生成梯度法制备分泌室亚细胞组分（Plonne *et al.*, 1999）。结合分子组分的分析，亚细胞成分的应用已为研究 VLDL 合成的细胞内剖析和分子学基础提供了一种方法。

## Apo-B 100 和 VLDL 分泌的翻译后调节

当肝细胞中 apo-B 100 分泌水平的变化达到 7 倍时，apo-B 100 mRNA 的量并无明显改变（Kosykh *et al.*, 1988; Pullinger *et al.*, 1989）。对肝癌细胞和分离肝细胞的研究已表明，apo-B 100 的合成量多于分泌量，多余的 apo-B 100 蛋白在细胞内被降解（Borchardt and Davis, 1987; Boren *et al.*, 1990; Dixon and Ginsberg, 1991; White *et al.*, 1992; Cartwright and Higgins, 1997; Yao *et al.*, 1997）。尽管表面上看来是个效率很低的过程，但这意味着机体可立即提供 VLDL 合成所需要的 apo-B 100。因此脂肪的有效性促使 VLDL 合成，apo-B 100 在细胞内的翻译后转运对 VLDL 合成是很重要的。

## VLDL 合成的两步法模型

许多研究室的工作表明，VLDL 在肝细胞内的合成包括一系列过程（图 19.3）。将新合成的 apo-B 100 加入到 RER 膜中，以单克隆抗体和（或）蛋白酶作为探针的试验结果表明组合到 RER 膜上的 apo-B 100，部分位于此膜的细胞质一侧，部分位于胞腔一侧（Davis *et al.*, 1989; Dixon *et al.*, 1992; Furakawa *et al.*, 1992; Wilkinson *et al.*, 1993），这与完全进入内质网（ER）腔的正常分泌蛋白不同。RER 膜中的 apo-B 100 主要是在 RER 中被降解，如果降解被抑制，apo-B 100 将积聚在 SER 和高尔基体膜上。结合到膜上的 apo-B 100 在遍在蛋白化作用后通过蛋白体途径降解（Yeung *et al.*, 1996）。部分 apo-B 100 和少量存在于 VLDL 前体颗粒中的脂质一起进入 RER 腔，这些脂质的密度与 HDL 相同。当脂质化不完全或 apo-B 100 的折叠不正常时，上述颗粒中的 apo-B 100 可能在腔内被降解（Cartwright and Higgins, 1996; Wu *et al.*, 1997; Shelness *et al.*, 1999; Davidson and Shelness, 2000; Fisher *et al.*, 2001）。大部分最终参与形成 VLDL 的脂质（甘油三酯、胆固醇酯和胆固醇）以轻的大颗粒形式被运送进入 SER 腔，大颗粒再与含有 apo-B 100 的前体颗粒融合形成 VLDL。因此，VLDL 合成的第一步是形成密度较大的含 apo-B 100 的前体颗粒，第二步是这些颗粒与富含甘油三酯的大颗粒的小滴融合。

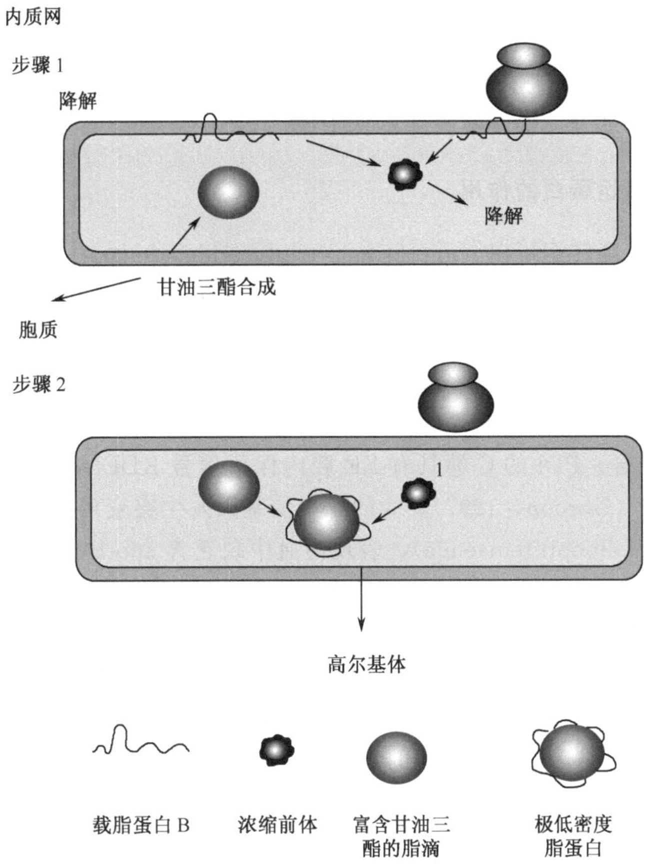


图 19.3　肝细胞中 VLDL 合成的两步法模型

在内质网腔中由两个步骤合成 VLDL。步骤 1，apo-B 100 由结合核糖体合成，并以颗粒形式转移进入 RER 腔，此颗粒的密度与 HDL 相同，由磷脂、胆固醇和少量的非极性脂类组成。甘油三酯（triglyceride，TAG）在 SER 中由酶催化合成，并以缺乏 apo-B 100 但富含脂质的颗粒形式转移进入腔内。步骤 2，上述两种颗粒结合形成 VLDL，所形成的 VLDL 在 SER 和（或）高尔基体中被进一步修饰。Apo-B 100 被翻译后摄取并组合于膜上，部分结构位于膜的细胞质一侧。同时有证据表明，apo-B 100 能从 ER 的腔侧反运输至细胞质一侧。膜中的 apo-B 100 可在 RER、SER 和高尔基体中通过蛋白质降解体的路径被降解。当甘油三酯含量低或者高密度颗粒的脂质化不完全或 apo-B 100 的折叠不正常时，腔中的 apo-B 100 也被降解

人们对 VLDL 合成的第一步即高密度的含 apo-B 100 前体颗粒在分子水平上的生成过程已有了相当的认识，而对第二步即大量脂质颗粒参与的情况了解还很少，因为第二步决定了分泌的 VLDL 颗粒的大小、密度和组分而显得十分重要。

**新合成的 apo-B 100 通过粗面内质网膜的迁移**

VLDL 合成调节的第一步是 apo-B 100 通过粗面内质网膜迁移并加入到前体颗粒中。当只有小部分 apo-B 100 蛋白位于膜腔侧时，apo-B 100 显然会发生翻译抑制作用（Du *et al*.，1994；Zhou *et al*.，1995；Davidson and Shelness，2000）。蛋白质的合成完成后，大部分的多肽位于膜的细胞质一侧。在此阶段，蛋白质与膜并不结合，而被认为是存在于转

运蛋白复合体中（Shelness *et al.*，1999；Davidson and Shelness，2000）。位于细胞质一侧的蛋白质与胞质中的热激蛋白 70 结合。当存在 ATP 时，热激蛋白 70 分解，且 apo-B 100 在蛋白质降解体的作用下被降解（Zhou *et al.*，1995；Yeung *et al.*，1996）。在有油酸盐存在时，apo-B 100 似乎不被降解就进入 RER 腔。

**微粒体甘油三酯转运蛋白的作用**

微粒体甘油三酯转运蛋白（microsomal triglyceride transport protein，MTP）是一种异源二聚体，由较大的 MTP 亚单位（97kDa）与蛋白二硫化物异构酶（protein disulphide isomerase，PDI）以非共价键的形式结合而成（Wetterau *et al.*，1997）。PDI 是一种 ER 常驻蛋白，可催化二硫键的形成和异构化，对新合成蛋白质的折叠具有重要作用。PDI 蛋白在腔内含量丰富，多数 PDI 以同源二聚体的形式存在，且与较大的 MTP 亚单位相结合。PDI 的 C 端具有 ER 腔内保留信号 KDEL，它的作用可能是保持 MTP 于 ER 腔内（Gordon，1997）。大的亚单位基因发生突变可导致人类遗传疾病无 β 脂蛋白血症（abetalipoproteinaemia），表现为血中缺乏含 apo-B 白浆脂蛋白，且 VLDL 和乳糜微粒的分泌极少（Sharp *et al.*，1993；Gordon，1997）。体外试验结果表明，MTP 催化胆固醇酯和甘油三酯在血浆脂蛋白和（或）膜囊微粒之间的转运，说明 MTP 可能在非极性脂类转移到 apo-B 中的富含甘油三酯脂蛋白合成中起作用。与此观点一致的是，免疫沉淀研究表明，MTP 与 apo-B 的结合是短暂的，且此结合被甘油三酯的合成所促进（Wu *et al.*，1996）。不同的实验室已对 MTP 抑制体进行了各种研究，这些抑制体在 VLDL 合成的第一步，即高密度含 apo-B 100 的颗粒的生成过程中与 MTP 相关（Gordon *et al.*，1996；Jamil *et al.*，1996）。也有证据表明，MTP 可参与第二步，即富含甘油三酯的颗粒的形成（Tietge *et al.*，1999）。甘油三酯从 RER 膜的转运机制尚不清楚，然而，推测其模型是，当膜上甘油三酯的浓度增加，将出现相的分离，膜上就萌发出甘油三酯的颗粒。这些颗粒可以进入细胞质以形成甘油三酯的储存颗粒，也可以进入 RER 腔。

## 营养素对 VLDL 合成的影响

**VLDL 亚类**

通过超速离心发现人的血浆中有两种 VLDL 亚类：一类为体积较大、重量较轻的富含甘油三酯的 VLDL1；另一类为体积较小、较重的 VLDL2（Demant *et al.*，1993，1996）。肝脏可分泌这两类 VLDL，且两类 VLDL 合成的调节是相互独立的。对于通常的实验对象，胰岛素可立即抑制体内 VLDL1 的产生，但是在抗胰岛素的实验对象的体内，胰岛素不具有此作用，而 VLDL2 的产生不受胰岛素的影响（Malmstrom *et al.*，1997）。雌激素可促进 VLDL1 的生成，而不影响 VLDL2 的产生。两类 VLDL 均是诱发动脉硬化症的危险因素，VLDL2 比 VLDL1 能更多地转换为 LDL，从而导致高胆固醇血症（Gaw *et al.*，1995）。摄食后，可能由于乳糜微粒和 VLDL1 在清除时存在竞争，血浆中 VLDL1 的水平升高（Karpe *et al.*，1993；Bjorkegren *et al.*，1996，1997），VLDL1 的增加将促进甘油三酯向 LDL 转移，产生的富含甘油三酯的 LDL 导致更能发生动脉粥样化的体积较小但较重的 LDL 前体的产生。

**VLDL 生成速率及其分泌微粒特性的调节**

对人和实验动物的研究结果表明，底物运输进入肝脏是 VLDL 分泌速率的主要调节因素（Sniderman and Cianflone，1993）。因此，富含碳水化合物的膳食将导致大而轻的 VLDL1 的生成，而富含甘油三酯的膳食将导致小而重的 VLDL2 的生成。膳食中脂肪酸的特性也是 VLDL 分泌的决定因素。关于膳食脂肪酸在体内作用的最佳报道是，膳食中鱼油（富含 *n*-3 不饱和脂肪酸）可抑制体内外 VLDL 的合成和分泌（Wilkinson *et al.*，1998；Kendrick and Higgins，1999），这是因为 VLDL 合成的第二步被抑制，从而导致 VLDL 分泌的减少和 apo-B 100 前体颗粒在 RER 腔中的降解（Wilkinson *et al.*，1998；Kendrick and Higgins，1999）。在上述研究中，人们还发现增加膳食中的向日葵油（富含 *n*-6 不饱和脂肪酸）可以促进 VLDL 的合成和增加 apo-B 100 通过分泌路径的转移。

## 肠道细胞中乳糜微粒合成的分子基础

乳糜微粒由肠道细胞（吸收性上皮细胞）合成和分泌，这些肠道细胞排列于小肠肠腔表面的微绒毛上。与 VLDL 在肝脏中的分泌不同，乳糜微粒在肠道中的分泌为一相对连续的过程，而且是在摄食脂肪后发生。人的血浆中乳糜微粒的浓度在摄食脂肪后的 2～4h 最高。然而，有许多因素能调节血浆中乳糜微粒的出现和清除速率。其中包括膳食中脂肪数量和脂肪酸组成的差异，以及与遗传和生活方式相关的个体和种群的差异。

脂肪的吸收包括如下步骤：①肠腔中的消化；②消化产物通过肠细胞刷状缘的转运；③脂质的再合成及合成的脂质组合为乳糜微粒，这些乳糜微粒在肠细胞间隙释放，并由此进入肠壁固有层中的乳糜管。乳糜管汇入淋巴管，乳糜微粒流入血液。甘油三酯的主要消化产物是单甘油酯和游离脂肪酸，这些产物和胆盐以及其他脂质（磷脂和胆固醇）均匀地散布于组成肠细胞生理底物的脂质微团中。消化产物进入肠细胞后，通过单甘油酯的酰化途径重新合成甘油三酯。

单甘油酯路径：

$$\text{脂酰辅酶 A} + \text{单甘油酯} \longrightarrow \text{甘油二酯} \quad ①$$

$$\text{脂酰辅酶 A} + \text{甘油二酯} \longrightarrow \text{甘油三酯} \quad ②$$

这与肝脏中通过 α-磷酸甘油的酰化途径合成甘油三酯不同。

α-甘油磷酸路径：

$$\alpha\text{-甘油磷酸} + 2\,\text{脂酰辅酶 A} \longrightarrow \text{磷脂酸} \quad ①$$

$$\text{磷脂酸} \longrightarrow \text{甘油二酯} + \text{磷酸} \quad ②$$

$$\text{甘油二酯} + \text{脂酰辅酶 A} \longrightarrow \text{甘油三酯} \quad ③$$

单甘油酯的酰化途径只发生于肠细胞中，通过此途径可以直接利用消化的主要产物。

### 研究乳糜微粒合成的实验细胞和动物模型

**CACO-2 细胞**

对乳糜微粒合成的研究比 VLDL 的研究少。其部分原因是缺乏适宜的培养细胞模型。一些研究者所采用的是Caco-2细胞（Hussain，2000），这些细胞来源于结肠，但是可以被诱导分化为具有与肠细胞相似特性的细胞。Caco-2细胞可分泌类似于LDL的具有漂浮特性的脂蛋白，然而，油酸盐会导致Caco-2细胞除了分泌LDL大小的颗粒，还分泌VLDL大小的颗粒。Hussain及其同事通过改变Caco-2细胞的培养条件，使其能够分泌小的乳糜微粒（Luchoomun and Hussain，1999），然而，这些乳糜微粒既含有apo-B 48，又含有apo-B 100，这与只分泌apo-B 48的成年肠细胞不同。目前看来，供给含牛磺胆酸的生理物质且在适宜条件下培养的Caco-2细胞系是最佳的培养细胞模型。但是用Caco-2细胞进行研究也同样存在上述用培养细胞进行VLDL合成研究的缺陷，即细胞系来源于肿瘤细胞，因此某些生物化学特性可能已改变，这将导致其结果被外推至正常生理状态的动物模型时可能出现问题，而且用细胞系进行生理试验和不同膳食的试验也是不可能的。

**分离的成年肠细胞**

我们近来已建立了分离兔和大鼠肠细胞的方法，该方法是将小肠的肠腔表面暴露于低浓度的螯合剂——柠檬酸钠和EDTA中（图19.4）（Cartwright and Higgins，1999a，b）。使用充氧的等渗平衡液对制备有活力的细胞非常重要，细胞的活力在Dulbecco氏改良Eagle培养基（Dulbecco's modified eagle medium，DMEM）的悬浮液中可保持2h。当供给生理性的微团底物时，分离的肠细胞可以合成和分泌apo-B 48及甘油三酯。90%以上被分泌的apo-B 48和甘油三酯共同沉淀于颗粒中，这些颗粒具有乳糜微粒的悬浮特性。

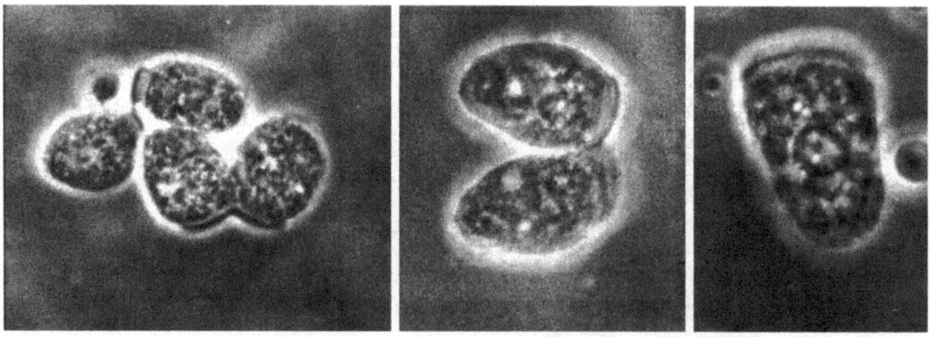

图19.4　分离的兔肠绒毛上皮细胞

**肠绒毛和腺窝梯度处的细胞**

小肠黏膜的上皮细胞排列成指状并凸向肠腔的微绒毛，在微绒毛的底部上皮细胞内陷，形成管状的腺窝（图19.5）。位于微绒毛顶端的细胞吸收脂肪和合成乳糜微粒最活跃。这些细胞的半衰期约为2天，它们被从腺窝底部向绒毛顶部移动的细胞所代替，而腺窝正是细胞分裂的场所。随着上皮细胞向绒毛部位的移动，它们便获得成年肠细胞的功能性和形态学的特征。这种独特的组织方式和细胞快速的更新速率保证了小肠通过改变吸收细胞的特性来适应膳食条件的变化。

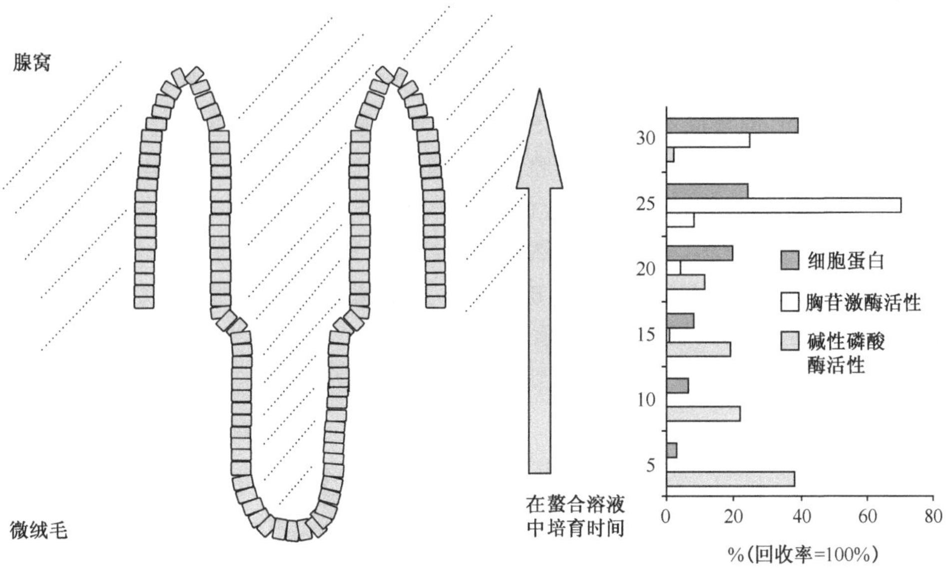


图 19.5　来自于绒毛/腺窝梯度不同位置处上皮细胞的分离

将肠腔表面暴露于螯合剂中以分离肠细胞。暴露时间的长短决定了从腔面上获得的细胞种类。首先获得的是绒毛细胞（碱性磷酸酶标记），接着获得的是腺窝细胞（胸苷激酶标记）

通过将小肠肠腔暴露于螯合剂中 5～30min，便可制备出源于微绒毛和腺窝梯度处不同位置的上皮细胞群（Cartwright and Higgins，1999a）。首先获得位于绒毛顶端的分化肠细胞（用碱性磷酸酶活性标记），接着获得的是腺窝细胞（用胸苷激酶活性标记）（图 19.5）。制备这些细胞对于研究小肠是如何适应包括膳食改变在内的生理变化具有重要作用。

**亚细胞分级分离**

近来，我们已研究出在单一自生成梯度中制备分离肠细胞中分泌腔亚细胞组分的方法（Cartwright *et al.*，2000）。与分子组分分析相结合，亚细胞的分级分离为研究乳糜微粒合成的细胞内及分子学基础提供了方法。

## 肠细胞中乳糜微粒合成的两步法模型

显微生物化学和放射自显影等电镜研究结果表明，乳糜微粒合成的基本步骤类似于 VLDL 的合成（图 19.6）。Apo-B48 由 RER 中的核糖体合成；被认为是乳糜微粒和（或）乳糜微粒前体的脂滴存在于 SER 腔、高尔基体腔及上皮细胞的细胞间隙中。然而，除了这一基本相似之处外，乳糜微粒和 VLDL 合成的细胞内步骤及其调节均存在差异。

分离肠细胞和亚细胞组分的研究结果表明，乳糜微粒的合成包括下述几步（图 19.6）（Kumar and Mansbach，1997；Mansbach and Nevin，1998；Cartwright and Higgins，2000，2001；Mansbach and Dowell，2000）：由 RER 上结合核糖体合成的 Apo-B48，与 RER 膜结合，并与一些磷脂一起转移进入 SER 腔，形成与 HDL 密度相

同的颗粒（步骤1）；在SER中合成的甘油三酯，转移入腔体（步骤2）。从SER腔中只分离到两种乳糜微粒的前体：高密度的含apo-B48的颗粒（与HDL的密度相同）、低密度的富含甘油三酯的apo-B48颗粒（与VLDL或乳糜微粒的密度相同）。这表明，与肝细胞中VLDL的合成不同的是，SER腔中不形成富含甘油三酯且缺乏apo-B48的颗粒，或者是当它们从膜上分离进入腔体时，能够与高密度的含apo-B48的颗粒快速结合。MTP在乳糜微粒合成中起到一定作用，其抑制作用可以减少甘油三酯从SER膜进入腔体（Cartwright *et al.*，2000），然而，此过程中其他因素如乳糜微粒存留疾病中受影响的蛋白质也具有作用。和VLDL的合成一样，调节甘油三酯是转移至细胞质储备还是转移至ER腔中的机制和影响因素尚不清楚。

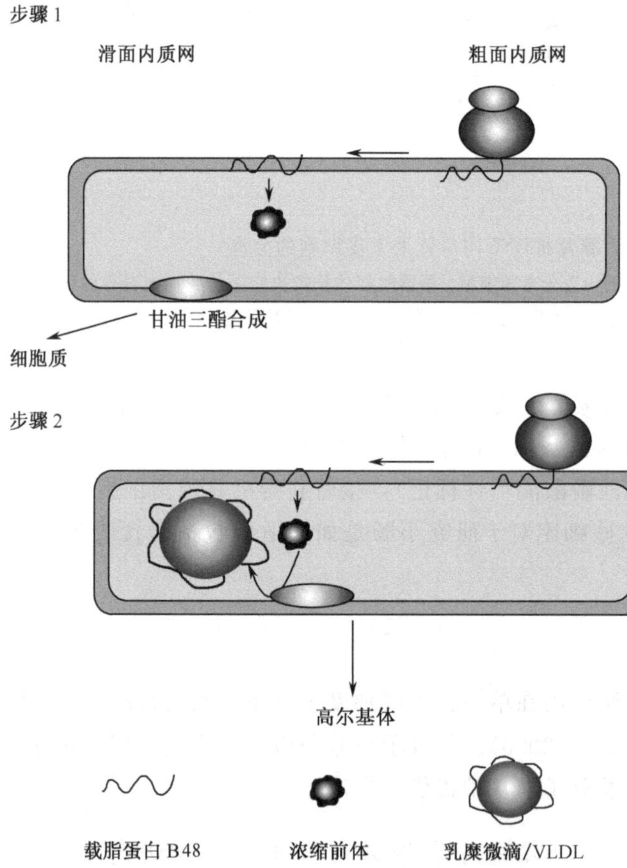


图19.6　肠绒毛上皮细胞内乳糜微粒合成的两步法模型

乳糜微粒是在肠细胞间隙中以两步法合成的。步骤1，由结合核糖体合成apo-B48，并转移进入SER膜中，参与形成与HDL密度相同的体积较小且富含磷脂的颗粒，甘油三酯（TAG）在SER中合成。步骤2，当腔体中没有去除apo-B48的甘油三酯脂滴时，含apo-B48的颗粒获得甘油三酯的过程与从膜上分离出甘油三酯脂滴的过程同时发生或紧随其后

### 肠细胞对脂肪吸收的适应

在饲喂低脂膳食（chow-fed）兔的肠细胞中，大多数细胞内甘油三酯与 SER 膜结合，且乳糜微粒合成过程中的限速步骤似乎是膜上的甘油三酯向含 apo-B48 的颗粒中转移（Cartwright *et al.*，2000；Cartwright and Higgins，2001）。饲喂富含脂肪的饲粮后，甘油三酯积聚在细胞质中，在 SER 腔中出现富含甘油三酯且含 apo-B48 的颗粒。这些发现表明乳糜微粒前体从 SER 转移进高尔基体是一个饱和步骤，当膳食中富含脂肪时，此步骤就成了限速步骤。

饲喂高脂肪膳食两周，将对分离的兔微绒毛细胞合成和分泌乳糜微粒的能力产生明显影响。与低脂膳食 chow（其中脂肪提供的能量占总能量的 7%）相比，向日葵油（其中脂肪提供的能量占总能量的 21%）使细胞质中甘油三酯的储存量大幅度增加，而且 apo-B48 和甘油三酯的分泌量分别增加 20 和 50 倍（Cartwright 和 Higgins，1999a）。不同的膳食脂肪其促进效果也不同，其中向日葵油＞典型西方膳食中的脂肪＞鱼油。与饲喂 chow 低脂膳食相比，饲喂 18h 的脂肪对分离肠细胞分泌乳糜微粒的能力仅有较小的影响，说明脂肪的影响比膳食要缓慢。这些结果说明，当各种肠细胞从腺窝处向微绒毛顶部移动时，将受到摄食脂肪的调节，细胞对脂肪的吸收变得“敏感”，而且不同的脂肪酸具有不同的促进效果。

### Apo-B 蛋白的编译在乳糜微粒合成中的作用

肠细胞中 apo-B 100 mRNA 编译产生 apo-B 48 的过程是存在时间特异性的，随着发育阶段而被调节。然而，敲除了编译酶（apo-bec-1）的小鼠能吸收脂质，并生成含 apo-B 100 的乳糜微粒（Hirano *et al.*，1996；Morrison *et al.*，1996；Nakumuta *et al.*，1996；Kendrick *et al.*，2001），这些乳糜微粒的直径比野生型小鼠在同样条件下产生的 apo-B 100 大约大 50%（Kendrick *et al.*，2001），因此，乳糜微粒的合成是肠细胞特异性的活动，并非 apo-B 48 的特性。这引起人们对 apo-B mRNA 的编译作用产生疑问：为什么要通过进化形成一个如此特异性的、准确而且随着发育阶段而调节的过程？近来研究表明，在禁食条件下，apo-B48 在乳糜微粒合成中的作用比 apo-B 100 更有效。饲喂低脂 chow 膳食的 apo-bec-1 小鼠的肠细胞要比野生型小鼠能蓄积更多的胞质甘油三酯，一晚的禁食可耗竭所储存的甘油三酯。在禁食条件下，与野生型小鼠相比，apo-bec-1 小鼠的体外乳糜微粒合成和体内脂肪吸收被大幅度降低。因此，当限制摄食量并且摄入低脂膳食时，apo-B mRNA 的编译可以提高脂肪吸收效率。

## VLDL 和乳糜微粒合成的比较

运输甘油三酯的两类主要脂蛋白即乳糜微粒和 VLDL 分别由小肠和肝脏生成，它们的生成过程大致相似。然而，二者合成及调节的详细过程均存在差异。肠细胞吸收乳糜微粒的能力相对于肝脏是无限大的，而且因为肠上皮组织的更新速率很快，参与吸收的细胞群能迅速地适应膳食的改变。肝细胞和小肠中脂蛋白合成的细胞内步骤也有所不同，然而，迄今尚不了解存在的差异大小。

# 参 考 文 献


Anderson, C.M., Townley, R.R. and Freeman, J.P. (1961) Unusual causes of steatorrhea in infancy and childhood. *Medical Journal of Australia* 11, 617–621.

Bjorkegren, J., Packard, C.J., Hamsten, A., Bedford, D., Caslake, M., Foster, L., Shepherd, J., Stewart, P. and Karpe, F. (1996) Accumulation of large very low density lipoprotein in plasma during intravenous infusion of a chylomicron-like triglyceride emulsion reflects competition for a common lipolytic pathway. *Journal of Lipid Research* 37, 76–86.

Bjorkegren, J., Hamsten, A., Milne, R.W. and Karpe, F. (1997) Alterations of VLDL composition during alimentary lipemia. *Journal of Lipid Research* 38, 301–314.

Borchardt, R.A. and Davis, R.A. (1987) Intracellular assembly of very low density lipoproteins. *Journal of Biological Chemistry* 262, 16394–16402.

Boren, J., Wettesten, M., Sjoberg, A., Thorin, T., Bondjer, G., Wiklund, A., Carlsson, P. and Olofsson, S.O. (1990) Studies on the assembly of apo-B100 containing lipoproteins in Hep-G2 cells. *Journal of Biological Chemistry* 263, 4434–4442.

Cartwright I.J. and Higgins, J.A. (1995) Intracellular events in the assembly of very low density lipoprotein lipids with apolipoprotein B in rabbit hepatocytes. *Biochemical Journal* 310, 897–907.

Cartwright I.J. and Higgins, J.A. (1996) Intracellular degradation in the regulation of apolipoprotein B 100 by rabbit hepatocytes. *Biochemical Journal* 314, 977–984.

Cartwright, I.J. and Higgins, J.A. (1999a) Isolated enterocytes as a model cell system for investigations of chylomicron assembly and secretion. *Journal of Lipid Research* 40, 1357–1365.

Cartwright, I.J. and Higgins, J.A. (1999b) Increased dietary triacylglycerol markedly enhances the ability of isolated rabbit enterocytes to secrete chylomicrons: an effect related to dietary fatty acid composition. *Journal of Lipid Research* 40, 1858–1866.

Cartwright, I.J. and Higgins, J.A. (2001) Direct evidence for a two-step assembly of apo-B48 containing lipoproteins in the lumen of the smooth endoplasmic reticulum of rabbit enterocytes. *Journal of Biological Chemistry* 276, 48048–48057.

Cartwright, I.J., Higgins, J.A., Wilkinson, J. Bellevia, S., Kendrick, J.S, and Graham, J.M. (1997) Investigation of the role of lipids in the assembly of VLDL in rabbit hepatocytes. *Journal of Lipid Research* 38, 531–545.

Cartwright, I.J., Plonne, D. and Higgins, J.A. (2000) Intracellular events in the assembly of chylomicrons in rabbit enterocytes. *Journal of Lipid Research* 41, 1728–1739.

Chen, S.H, Habib, G., Yang, C.Y., Gu, Z.W., Lee, B.T., Weng, S.A., Siberman, S.R., Cai, S.J., Desylpere, J.P., Rossenau, M. and Chan, L. (1987) Apolipoprotein B48 is the product of a messenger RNA with an organ specific in-frame stop codon. *Science* 238, 363–366.

Davidson, N.O. and Shelness, G.S. (2000) Apolipoprotein B: mRNA editing, lipoprotein assembly, and presecretory degradation. *Annual Review of Nutrition* 20, 169–193.

Davis, R.A., Prewett, D.C., Chan, D.C.F. Thompson, J.J., Borchardt, R.A. and Gallagher, W.R. (1989) Intraheptic assembly of very low density lipoproteins; immunological characterisation of apolipoprotein B in lipoproteins and hepatic membrane fractions and its intracellular distribution. *Journal of Lipid Research* 30, 1185–1196.

Demant, T., Gaw, A., Watts, G.F., Durrington, P., Buckley, B., Imrie, C.W., Wilson, C., Packard, C.J. and Shepherd, J. (1993) Metabolism of apoB-100-containing lipoproteins in familial hyperchylomicronemia. *Journal of Lipid Research* 34, 147–156.

Demant, T., Packard, C.J., Demmelmair, H., Stewart, P., Bedynek, A., Bedford, D., Seidel, D. and Shepherd, J. (1996) Sensitive methods to study human apolipoprotein B metabolism using stable isotope-labeled amino acids. *American Journal of Physiology* 270, E1022–E1036.

Dixon J.L. and Ginsberg, H.N. (1991) Oleate stimulates secretion of apo-B containing lipoproteins from Hep-G2 cells by inhibiting early intracellular degradation of apo-B. *Journal of Biological Chemistry* 266, 5080–5086.

Dixon, J.L. and Ginsberg, H.N. (1993) Regulation of hepatic secretion of apolipoprotein B containing lipoproteins: information obtained from cultured cells. *Journal of Lipid Research* 34, 167–178.

Dixon, J.L. Chattapadhyay, R., Hulma, T., Redman, C.N. and Banjeree, D. (1992) Biosynthesis of lipoprotein: location of the nascent A1 and apo-B in the rough endoplasmic reticulum of chicken hepatocytes. *Journal of Cell Biology* 117, 1161–1169.

Du, E.Z., Kurth, J., Wang, S.L., Humiston, P. and Davis, R.A. (1994) Proteolysis-coupled secretion of the N terminus of apolipoprotein B. Characterization of a transient, translocation arrested intermediate. *Journal of Biological Chemistry* 269, 24169–24176.

Fisher, E.A., Pan, M., Chen, X., Wu, X., Wang, H., Jamil, H., Sparks, J.D. and Williams, K.J. (2001) The triple threat to nascent apolipoprotein B. *Journal of Biological Chemistry* 276, 27855–27863.

Furukawa, S., Sakata, N. and Ginsberg, H.N. (1992) Studies of the sites of intracellular degradation of

apolipoprotein B in Hep-G2 cells. *Journal of Biological Chemistry* 271, 18445–18455.
Gaw, A. Packard, C.J., Lindsay, G.M., Griffin, B.A., Caslake, M.J., Lorimer, A.R. and Sheperd, J. (1995) Overproduction of small very low density lipoproteins (Sf 20–60) in moderate hypercholesterolaemia: relationships between apolipoprotein B kinetics and plasma lipoproteins. *Journal of Lipid Research* 36, 158–171.
Gibbons, G.F. (1994) A comparison of *in vitro* models to study hepatic lipid and lipoprotein metabolism. *Current Opinion in Lipidology* 23, 465–500.
Gordon, D.A. (1997) Recent advances in elucidation of the role of the microsomal triglyceride transfer protein in apolipoprotein B lipoprotein assembly. *Current Opinion in Lipidology* 8, 136–150.
Gordon, D.A., Jamil, H., Gregg, R.E., Olofsson, S.-V. and Boren, J. (1996) Inhibition of microsomal triglyceride transfer protein blocks the first step of apolipoprotein B lipoprotein assembly but not the addition of the core bulk lipids in the second step. *Journal of Biological Chemistry* 271, 33047–33053.
Higgins, J.A. and Hutson, J.L. (1984) The role of Golgi and endoplasmic reticulum in the synthesis and assembly of lipoprotein lipids in rat hepatocytes *Journal of Lipid Research* 25, 1295–1305.
Hirano, K.I., Young, S.G., Farese, R.V. Jr, Ng, J., Sande, E., Warburton, C., Powell-Braxton, L.M. and Davidson, N.O. (1996) Targeted distruption of the mouse apo-bec-gene-1 abolished apolipoprotein B mRNA editing and eliminates apolipoprotein apoB48. *Journal of Biological Chemistry* 271, 9887–9890.
Hussain, M.M. (2000) A proposed model for the assembly of chylomicrons. *Atherosclerosis* 14, 1–15.
Hussain, M.M., Kancha, R.K., Zhou, Z., Luchoomun, J., Zu, H. and Bakillah, A. (1996) Chylomicron assembly and catabolism: role of apolipoproteins and receptors. *Biochimica et Biophysica Acta* 1300, 151–170.
Jamil, H., Gordon, D.A., Eustice, D., Brooks, C.M., Dickson, J.K. Jr, Chen, Y., Ricci, B., Chu, C.H., Harrity, T.W., Ciosek, C.P. Jr, Biller, S.A., Gregg, R.E. and Wetterau, J.R. (1996) An inhbibitor of the microsomal transfer protein inhbibits apo-B secretion from Hep-G2 cells. *Proceedings of the National Academy of Sciences USA* 93, 11991–11995.
Karpe, F., Steiner, G., Olivecrona, T., Carlson, L.A. and Hamsten, A. (1993) Metabolism of triglyceride-rich lipoproteins during alimentary lipemia. *Journal of Clinical Investigations* 91, 748–758.
Kendrick, J.S. and Higgins, J.A. (1999) Dietary fish oils inhibit early events in the assembly of very low density lipoproteins and target apolipoprotein B for degradation within the lumen of the rough endoplasmic reticulum of hamster hepatocytes. *Journal of Lipid Research* 40, 504–514.
Kendrick, J.S., Chan, L. and Higgins, J.A. (2001) Superior role of apolipoprotein B48 over apolipoprotein B100 in chylomicron assembly and fat absorption: an investigation of apobec-1 knock-out and wild-type mice. *Biochemical Journal* 356, 821–827.
Kosykh, V.A., Surguchow, A.P., Podres, E.A., Novikov, D.K. and Sudarickoc, N. (1988) VLDL apoprotein secretion and apo-B mRNA levels in primary cultures of cholesterol loaded rabbit hepatocytes. *FEBS Letters* 232, 103–106.
Kumar, N.S. and Mansbach, C.M. (1997) Determinants of triglyceride transport from the endoplasmic reticulum to the Golgi in intestine. *American Journal of Physiology* 273, G18–G30.
Luchoomun, J. and Hussain, M.M. (1999) Assembly and secretion of chylomicrons by differentiated Caco-2 cells. *Journal of Biological Chemistry* 274 19565–19572.
Malmstrom, R., Packard, C.J., Watson, T.G., Ranniko, S., Caslake, M., Bedford, D., Stewart, P., Yid-Jarvinen, H., Shepherd, J. and Taskinen, M.R. (1997) Metabolic basis of hypertriglyceridaemic effect of insulin in normal men. *Arteriosclerosis, Thrombosis and Vascular Biology* 17, 1454–1464.
Mansbach, C.M. and Dowell, R. (2000) Effect of increasing lipid load on the ability of the endoplasmic reticulum to tranport lipid to the Golgi. *Journal of Lipid Research* 41, 605–612.
Mansbach, C.M. and Nevin, P. (1998) Intracellular movement of triacylglycerols in the intestine. *Journal of Lipid Research* 39, 963–968.
Morrison, J.R., Pasty, Ch., Stevens, M.E., Hughes, S.D., Forte, T., Scott, J. and Rubin, E.M. (1996) Apolipoprotein B RNA editing enzyme deficient mice are viable despite alterations in lipoprotein metabolism. *Proceedings of the National Academy of Sciences USA* 93, 7154–7159.
Nakamuta, M., Chang, B.H.J., Zsigmond, E., Kobayashi, K., Lei, H., Ishida, B.Y., Oka, K., Li, E. and Chan. L. (1996) Complete phenotypic characterisation of apobec-1 knock-out mice with a wild type genetic background and restoration of mRNA editing by somatic transfer of apobec-1. *Journal of Biological Chemistry* 271, 25981–25988.
Packard, C.J., Demant, T., Stewart, J.P., Bedford, D., Caslake, M.J., Schwertfeger, G., Bedynek, A., Shepherd, J. and Seidel, D. (2000) Apolipoprotein B metabolism and the distribution of VLDL and LDL subfractions. *Journal of Lipid Research* 41, 305–317.
Patterson, A.P., Tennyson, G.E., Hoeg, J.M., Sviridov, D.D. and Brewer, H.B. (1992) Ontogenic regulation of apo-lipoprotein B mRNA editing during human and rat development *in vivo*. *Arteriosclerosis, Thrombosis and Vascular Biology* 12, 463–473.
Plonne, D., Cartwright, I.J., Graham, J.M., Dargel, R. and Higgins, J.A. (1999) Separation of the components of the secretory compartment from rat liver and isolated rat heptocytes in a single step in

self generating gradients of iodixanol. *Analytical Biochemistry* 279, 88–89.

Powell, L.M., Wallis, S.C., Pease, R.J. Edwards, Y.H., Knott, T.J. and Scott, J. (1987) A novel form of tissue specific RNA processing produces apolipoprotein B48 in intestine. *Cell* 50, 831–846.

Pullinger, C.R., North, J.D., Teng, B., Rifici, V.A. Ronhild de Brito, A.E. and Scott, J. (1989) The apolipoprotein B gene is constitutively expressed in Hep-G2 cells. *Journal of Lipid Research* 38, 1065–1076

Sharp, D.L., Blinderman. K.A., Combe, R.A., Kienzie, B., Ricci, B., Wager-Smith, K., Gil, C.M., Turck, C.W., Bourma, M.E., Rader, D.J. *et al.* (1993) Cloning and gene defects in microsomal triglyceride transfer proteins associated with abetalipoproteinaemia. *Nature* 356, 65–69.

Shelness, G.S., Ingram, M., Huand, X.F. and DeLozier, J.A. (1999) Apolipoprotein B in the rough endoplasmic reticulum: translation, translocation and initiation of lipoprotein assembly. *Journal of Nutrition.* 129, 456S–462S.

Sniderman, A.D. and Cianflone, K. (1993) Substrate delivery as a determinant of hepatic apo-B secretion. *Arteriosclerosis, Thrombosis and Vascular Biology* 13, 629–636.

Teng, B., Verp, M., Salomon, J. and Davidson, N.O. (1990) Apolipoprotein B messenger RNA editing is developmentally regulated and widely expressed in human tissues. *Journal of Biological Chemistry* 265, 20616–20620.

Tietge, U.J., Bakillah, A., Maugeatis, C.M., Tsukamoto, K., Hussain, M. and Rader, D.J. (1999) Hepatic overexpression of microsomal triglyceride transfer protein (MTP) results in increased *in vivo* secretion of VLDL triglycerides and apolipoprotein B. *Journal of Lipid Research* 40, 2134–2138.

Tso, P., Balint, J.A., Bishop. M.B. and Rodgers, J.B. (1981) Acute inhibition of intestinal lipid transport by pluronic L-81 in the rat. *American Journal of Physiology* 241, G487–G497.

Wetterau, J.R., Lin, M.C. and Jamil, H. (1997) Microsomal triglyceride transfer protein. *Biochemica et Biophysica Acta* 1345, 136–150.

White A.L., Graham, D.L., LeGros, J., Pease, R.J. and Scott, J. (1992) Oleate mediated stimulation of apolipoprotein B secretion from rat hepatoma cells. *Journal of Biological Chemistry* 267, 15657–15684.

Wilkinson, J. Higgins, J.A., Groot, P.H.E., Gherardi, E. and Bowyer, D.E. (1993) Topography of apolipoprotein B in subcellular fractions of rabbit liver probed with a panel of monoclonal antibodies. *Journal of Lipid Research* 34, 815–825.

Wilkinson, J., Fitzsimmons, C., Higgins, J.A. and Bowyer, D.E (1998) Dietary fish-oils modify the assembly of very low density lipoproteins and expression of the low density lipoprotein receptor in rabbit liver. *Arteriosclerosis, Thrombosis and Vascular Biology* 18, 1490–1497.

Wu, X., Zhou, L.-S., Wetterau, J. and Ginsberg. H.N. (1996) Demonstration of a physical interaction between MTP protein and apo-B during assembly of apo-B containing lipoproteins. *Journal of Biological Chemistry* 271 10277–10281.

Wu, X., Sakata, N., Lele, K.M., Zhou, M., Jiang, H and Ginsberg, H.N. (1997) A two site model for apo-B degradation in Hep G2 cells. *Journal of Biological Chemistry* 272, 11575–11580.

Yao, Z., Tran, K. and McCleod, R.S. (1997) Intracellular degradation of newly synthesised apolipoprotein B. *Journal of Lipid Research* 38, 1937–1953.

Yeung, S.J., Chen, S.-W. and Chan, L. (1996) Ubiquitin-proteasome mediated pathway for the degradation of apolipoprotein B. *Biochemistry* 35, 13843–13848.

Zhou, M. Wu, X., Huang, L.-S. and Ginsberg, H.N. (1995) Apolipoprotein B 100 an inefficiently translocated secretory protein is bound to cytosolic chaperone heat shock protein. *Journal of Biological Chemistry* 270, 25220–25224.

# 20 细胞胆固醇的调节

Ji-Young Lee，Susan H. Mitmesser 和 Timothy P. Carr
（营养科学与营养学系，
内布拉斯加州立大学，林肯，内布拉斯加州，美国）

## 前　　言

胆固醇在两个世纪以前作为胆石的一种可溶于乙醇的成分被发现，后来证实，胆固醇是大多数动物组织的成分，更具体地说，是作为细胞膜的必需组分。在哺乳动物中，胆固醇还是胆汁酸和类固醇激素的前体物。虽然主要来源于动物，但是在真菌和绿色植物中也发现了微量的胆固醇。

人类的正常代谢功能需要胆固醇。实际上，所有细胞在有需要的时候都可以合成胆固醇，食物来源的胆固醇同样也对胆固醇的总体池有贡献。胆固醇在体内大多存在于细胞膜中，而一小部分同血浆脂蛋白结合作为血管内精细转运系统的一部分。体内过多的胆固醇通过胆汁分泌以游离胆固醇或胆汁酸的形式排出体外。因为肝脏决定着胆固醇从体内排出的基本路线，而且也是血浆胆固醇浓度的主要决定者，所以肝脏是胆固醇代谢中的一个决定性器官。如果肝细胞内胆固醇不能维持动态平衡，将会导致血浆中胆固醇的累积，从而增加了患冠心病和中风的概率。

细胞内胆固醇浓度必须保持在一个恒定的水平以免胆固醇积累过多而达到细胞毒素水平。许多调节蛋白也都参与肝胆固醇动态平衡的维持，其中在科学文献中经常提到的是：①3-羟基-3-甲基戊二酸单酰-辅酶 A（3-hydroxy-3-methylglutaryl-CoA，HMG-CoA）还原酶，是胆固醇合成中的限速酶；②低密度脂蛋白（LDL）受体，是血浆 LDL 清除的主要装置；③胆固醇 7α-羟化酶，是经典胆汁酸合成途径的限速酶。它们中的每一个蛋白质都受转录因子的调节，这些转录因子包括固醇调节元件结合蛋白（sterol regulatory element binding protein，SREBP）、肝脏 X 受体（liver X receptor，LXR）和法尼醇 X 受体（farnesoid X receptor，FXR）。

如图 20.1 所示，本章集中讨论了转录因子和主要调节蛋白的内在联系。对大多数人而言，尽管可以通过外在因素（如食物）来调节胆固醇代谢，从而影响动脉粥样硬化疾病的发展，但在人类中，基因突变是胆固醇不平衡的最显著证据（Goldstein and Brown，2001），因此，为了与“分子营养”的主题保持一致，我们也将讨论特定营养素在细胞内胆固醇调节中的作用。

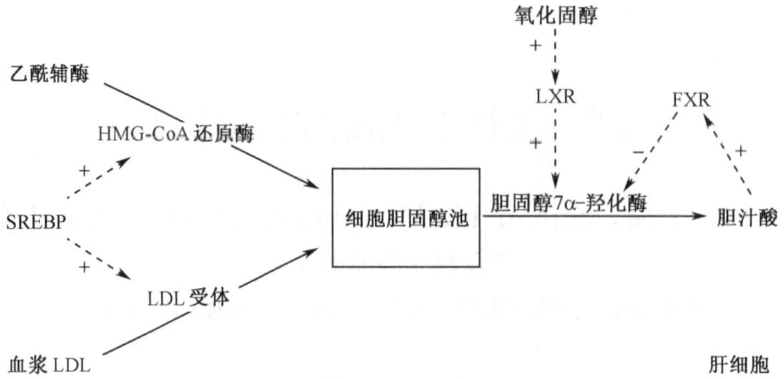


图 20.1 SREBP、LXR 和 FXR 在肝胆固醇代谢中的作用

SREBP 通过激活 HMG-CoA 还原酶和 LDL 受体的转录增加细胞胆固醇池，分别引起胆固醇的合成和循环中 LDL 的摄取。细胞胆固醇浓度增加引起氧化固醇浓度的增加，增加的氧化固醇与 LXR 结合，激活了编码胆固醇 7α-羟化酶［传统（中性）胆汁酸合成中限速酶］的 Cyp7a1 基因的转录。胆汁酸作为 FXR 的配体对胆汁酸合成起负反馈作用

# 转 录 因 子

## SREBP

SREBP 是通过一发夹结构域与内质网和核被膜相结合的完整蛋白，此发夹结构域由两个跨膜结构域组成，这两个跨膜结构域是被一个朝向内质网膜和核被膜的由 31 个氨基酸组成的短环所隔开的（Hua *et al.*，1995）。N 端结构域的大约 480 个氨基酸和 C 端结构域的大约 590 个氨基酸朝向胞质。SREBP 的 N 端结构域包含一个碱性螺旋-环-螺旋亮氨酸拉链（basic helix-loop-helix leucine zipper，bHLH-ZIP）基元，该基元提供一个 DNA 结合结构域。为发挥转录因子的作用，SREBP 需要被激活，激活后其 N 端结构域才能转运到细胞核内，在核内它激活参与胆固醇和脂肪酸代谢的固醇响应基因的转录（Brown and Goldstein，1999）。

应对如细胞胆固醇衰竭反应的信号，需要两步连续性的蛋白水解切除反应以释放一个 SREBP 的 N 端转录因子结构域。第一个切除反应通过 1 位蛋白酶（site-1 protease，S1P）的作用发生在两个跨膜结构域之间的腔环上（位点 1 裂解），第二个切除反应通过 2 位蛋白酶（S2P）的作用发生在第一个跨膜结构域内（位点 2 裂解），释放出 SREBP 的 N 端片段（Sakai *et al.*，1996）。成熟 SREBP 的 N 端片段进入细胞核并发挥转录因子的作用。图 20.2 阐明了 SREBP 的蛋白水解切除反应以及成熟的 SREBP 进入细胞核的转运过程。固醇通过选择性的阻碍 1 位蛋白酶（site-1 protease，S1P）引发的切除反应来抑制 SREBP 的激活。因为固醇只在位点 1 裂解后才对 SREBP 起作用，所以 S2P 受固醇的间接调节。

SREBP 裂解-激活蛋白（SREBP cleavage-activating protein，SCAP）是另外一个参与 SREBP 激活的因子。SCAP 是一种内质网膜结合糖蛋白，是通过对中国仓鼠卵巢（Chinese hamster ovary，CHO）细胞的一个突变系的研究而被发现的。Nohturfft 等

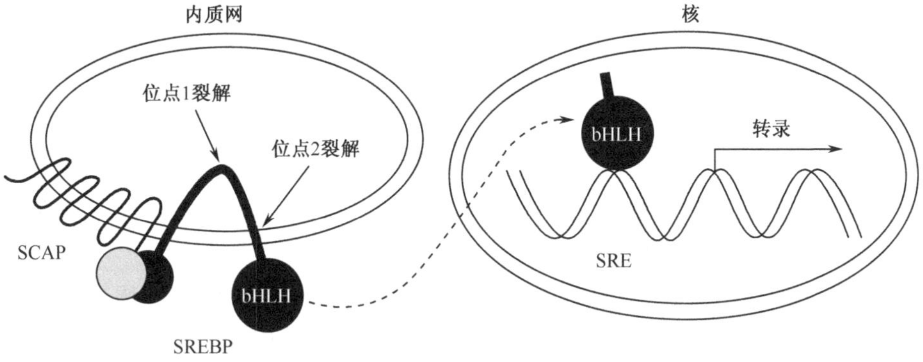


图 20.2　SREBP 的成熟及其反式激活作用

S1P 和 S2P 分别在位点 1 及后来在位点 2 对 SREBP 的顺序蛋白水解作用从内质网膜上释放了 SREBP 的 N 端结构域。这个结构域包含一个转移到核中的碱性螺旋-环-螺旋亮氨酸拉链 DNA-结合基序（bHLH）。成熟 SREBP 与固醇调节元件（SRE）的结合激活了固醇-响应基因及编码 HMG-CoA 还原酶和 LDL 受体基因的转录 。SREBP 水解激活蛋白（SCAP）发挥固醇传感器的作用。SCAP 与 SREBP C 端结构域的相互作用对位点 1 的水解是很必要的

（1996）发现，在固醇存在的情况下，SCAP 基因突变的 CHO 细胞不能抑制 SREBP 的裂解作用。研究者认为，SCAP 的 N 端结构域具有固醇感受功能。SCAP 和 SREBP 通过其胞质 C 端结构域之间的相互作用在内质网膜上形成复合物，该复合物引导 S1P 到达其位于 SREBP 腔结构域的目标上，并且此复合物的分裂可以阻止 1 位裂解反应（Sakai *et al.*，1998）。Brown 和 Goldstein（1999）对固醇介导的 SREBP N 端结构域的蛋白水解的作用机制进行了充分的阐述。

SREBP 最先以特异转录因子被发现，该转录因子结合于对 LDL 受体和 HMG-CoA 合成酶进行编码的基因启动子内 10bp 固醇调节元件（SRE）上（Briggs *et al.*，1993；Wang *et al.*，1993）。SREBP 的遍在表达说明它们在细胞胆固醇和脂肪酸水平调节中的重要作用，此调节作用通过其激活一系列固醇响应基因而实现。当细胞内胆固醇减少时，SREBP 被激活，从而增加编码胆固醇生物合成途径中酶的基因转录。LDL 受体可通过受体介导的血浆 LDL 的细胞内吞作用提供胆固醇，LDL 受体的转录同样也受 SREBP 的调节（Wang *et al.*，1993）。

SREBP 的 3 个同工型已被鉴定：SREBP-1a、SREBP-1c 和 SREBP-2。SREBP-1a 和 SREBP-1c 由单一基因通过可变的转录起始位点形成，该起始位点能编码可变的第一外显子，这些外显子被剪接成一个共同的第二外显子（Shimomura *et al.*，1997）。SREBP-1c 与 SREBP-1a 转录的比率变化很大，在肝脏（9∶1）、肾上腺和脂肪组织中有较高的比率，而在脾中比率低（1∶10）。这些差异表明 SREBP-1a 和 SREBP-1c 的转录受到不同的调节，从而对胆固醇和脂肪酸代谢相关的器官特异性因子产生不同的反应（Shimomura *et al.*，1997）。SREBP-2 与 SREBP-1a 的 47％是相同的，但编码 SREBP-2 的基因为单独基因（Hua *et al.*，1993）。虽然 SREBP-1 与 SREBP-2 共用一些氨基酸序列并且在结构上有一定的相似性，但它们的功能不一定相同。SREBP-1 是在脂肪酸代谢中发挥重要作用，而 SREBP-2 在胆固醇调节中发挥着更大的调节作用。而且，

SREBP-1 对脂肪酸代谢的调节还包含肝 X 受体（liver X receptor，LXR），这将在下面的章节进行讨论。

## LXR

长久以来，人们一直认为胆固醇或其代谢对细胞过多胆固醇的反应是激活它们自己的分解代谢（如转化成胆汁酸）。近来 LXR 及其自然配体（图 20.1）的发现在理解胆固醇如何影响其自身分解代谢的作用机制方面有了突破。

LXR 最先是作为孤儿核受体被发现的。它们由中心 DNA 结合结构域和疏水的 C 端结构域组成，此 C 端结构域介导配体识别、受体二聚化和依赖于配体的激活（Peet *et al.*，1998a）作用。LXR 被自然存在的氧化固醇激活，氧化固醇是胆固醇被氧化而形成的衍生物。22（R）-羟基胆固醇、24（S）-羟基胆固醇和 24（S），25-双环氧胆固醇都是 LXR 的高亲和性配体（Janowski *et al.*，1996；Lehmann *et al.*，1997）。要发挥转录因子的作用，LXR 必须与类维生素 A X 受体（retinoid X receptor，RXR）形成异源二聚体。LXR-RXR 异源二聚体能被 RXR 的配体（如 9-顺视黄酸）或 LXR 的配体激活。配体的结合诱导了 LXR-RXR 异源二聚体构象的改变，导致靶基因转录活性的提高。

哺乳动物中，两种 LXR 已被识别：LXRα 和 LXRβ。LXRα 主要在肝脏中表达，而在肾、肠、脾和肾上腺中有少量的表达；LXRβ 在大多数组织中都普遍表达（Willy *et al.*，1995）。LXRα 和 LXRβ 对氧化固醇配体有相似的特征。然而，在 LXRα 敲除的小鼠中，LXRβ 不能替代 LXRα 在胆汁酸合成和分泌中的作用（Peet *et al.*，1998b）。目前已知，LXR-RXR 异源二聚体与 DR-4 激素反应元件结合，DR-4 激素反应元件为由 4 个核苷酸间隔的两个相似六核苷酸的直接重复（Lehmann *et al.*，1997）。

以 LXR 的表达模式及其配体为氧化固醇的论据表明 LXR 参与胆固醇代谢。Lehmann 等（1997）在编码胆固醇 7α-羟化酶的 *Cyp7a1* 基因启动子区域对 LXR 反应元件（LXR response element，LXRE）的识别为 LXR 作为胆汁酸代谢中关键转录调节子的作用提供了强有力的证据。Peet 等（1998b）证明，当饲喂低胆固醇膳食时，LXRα 敲除的小鼠表现正常，然而当饲喂高胆固醇膳食时，LXRα 敲除小鼠的 *Cyp7a1* 基因的转录不会增加，结果引起了肝胆固醇酯的显著增加。与此相反，饲喂高胆固醇膳食的野生型小鼠会增加 *Cyp7a1* 基因的转录。

LXRα 可能参与脂肪酸和胆固醇代谢。同野生型小鼠相比，饲喂高胆固醇膳食的 LXRα 敲除小鼠中，表达参与脂肪酸代谢的一些蛋白质（包括 SREBP-1、硬脂酰-CoA 脱饱和酶 1 和脂肪酸合成酶）的基因受到负调节（Peet *et al.*，1998b）。LXR-RXR 在对 SREBP-1c 基因启动子内的结合位点的识别说明，LXR 引起的脂肪生成是通过 SREBP-1c 介导的（Repa *et al.*，2000）。这些研究表明胆固醇和脂肪酸代谢以协同的方式进行调节，此方式由 LXR 介导。

## FXR

胆固醇转化为胆汁酸具有重要的生理作用，因为它是体内胆固醇清除的主要途径。通过胆固醇分解为胆汁酸途径清除的胆固醇约占胆固醇日排除量的 50%（Vlahcevic *et al.*，1999)。然而，不同于 LXR 对 *Cyp7a1* 的正调节作用，FXR 通过减少 *Cyp7a1* 基因的转录对胆汁酸的合成起负调节作用（图 20.1)，这种作用由胆汁酸通过肠肝循环到肝脏的回流介导。最近研究证明，胆汁酸为 FXR 的配体，这一发现为理解胆汁酸的调节作用增加了一个新的方向。

FXR 为核激素受体，主要在肝、肠、肾和肾上腺皮质中表达（Forman *et al.*，1995)。FXR 需要与 RXR 形成一个异源二聚体 FXR-RXR，并与由一个单核苷酸（IR-1 基序）分离开的反向六核苷酸重复序列结合（Forman *et al.*，1995)。体外试验证明，鹅脱氧胆酸（chenodeoxycholic acid，CDCA)、脱氧胆酸（deoxycholic acid，DCA）和胆汁酸石胆酸（litocholic acid，LCA）可以强烈地激活 FXR，此激活作用发生在细胞生理浓度范围内，表明胆汁酸是 FXR 的自然配体（Makishima *et al.*，1999；Parks *et al.*，1999；Wang *et al.*，1999)。

胆固醇 7α-羟化酶的活性在转录水平通过肠肝循环内的胆汁酸/盐的流出进行负调节。在大鼠肝细胞中，相对疏水的胆汁酸（如牛磺胆酸和脱氧牛磺胆酸）可抑制胆固醇 7α-羟化酶的 mRNA，而亲水的牛磺熊脱氧胆酸和牛磺猪脱氧胆酸没有这种作用（Stravitz *et al.*，1993)。内源胆汁酸激活的 FXR 的识别促使研究者将 *Cyp7a1* 作为 FXR 的靶基因进行进一步的研究。

*Cyp7a1* 基因启动子能够被胆汁酸和 FXR 抑制，这表明胆汁酸和 FXR 参与了 *Cyp7a1* 基因转录的调节（Makishima *et al.*，1999)。有两个胆汁酸反应元件：位于 *Cyp7a1* 基因启动子－75～－54 核酸的 BARE-I（bile acid response element，B・ARE-I）和位于 *Cyp7a1* 基因启动－149～－118 核苷酸的 BARE-Ⅱ（bie acid response element，BARE-Ⅱ）已被识别（Chiang and Stroup，1994)。将 BARE-Ⅱ从 *Cyp7a1* 启动子中敲除的试验证明，BARE-Ⅱ负责由胆汁酸介导的 *Cyp7a1* 基因表达的负调节（Stroup *et al.*，1997)。尽管已经证明 BARE-Ⅱ为 FXR 的反应元件，但是 FXR-RXR 异源二聚体好像并不与此序列相结合。此后的研究表明，FXR 在被胆汁酸激活后通过间接的方式抑制 *Cyp7a1* 基因的转录（Chiang *et al.*，2000)。

肝脏受体同系物-1（liver receptor homologue，LRH-1，也称作 CPF）和小的异源二聚体配偶体（small heterodimer partner，SHP）好像也在 FXR 的间接抑制作用中发挥着一定的作用。LRH-1 为单体核孤儿受体，发挥组织特异性转录因子作用，且只在肝脏、肠和胰中表达（Becker-Andre *et al.*，1993)。LXRα 对 *Cyp7a1* 的反式激活作用需要 LRH-1 在 *Cyp7a1* 启动子部位与 BARE-Ⅱ结合（Lu *et al.*，2000)。SHP 为与众不同的核受体，因为它缺乏 DNA 结合域，并与其他几个核激素受体形成一个异源二聚体（Seol *et al.*，1996)。SHP 通常被认为降低其配偶体的反式激活作用。SHP 启动子含有一个胰岛素受体-1（insulin receptor，IR-1)，该受体为 FXR-RXR 异源二聚体的结合位点。胆汁酸激活的 FXR 增加了 SHP 基因的转录（Lu *et al.*，2000)，因此，增加了 SHP 蛋白含量，SHP 蛋白的增加通过形成异源二聚体使 LRH-1 失去活性，抑制了 *Cyp7a1* 的转录。图 20.3 阐明了 LXR 和 FXR 诱导或抑制 *Cyp7a1* 基因转录的作用机制。

FXR 在胆汁酸代谢中的作用不仅是负调节胆汁酸的合成，还参与胆汁酸在肝肠循环中的转化。回肠胆汁酸结合蛋白（ileal bile acid-binding protein，IBABP）是一种高亲和性与胆汁酸结合的胞质蛋白，并被认为在细胞胆汁酸摄取和运输中发挥着一定的作

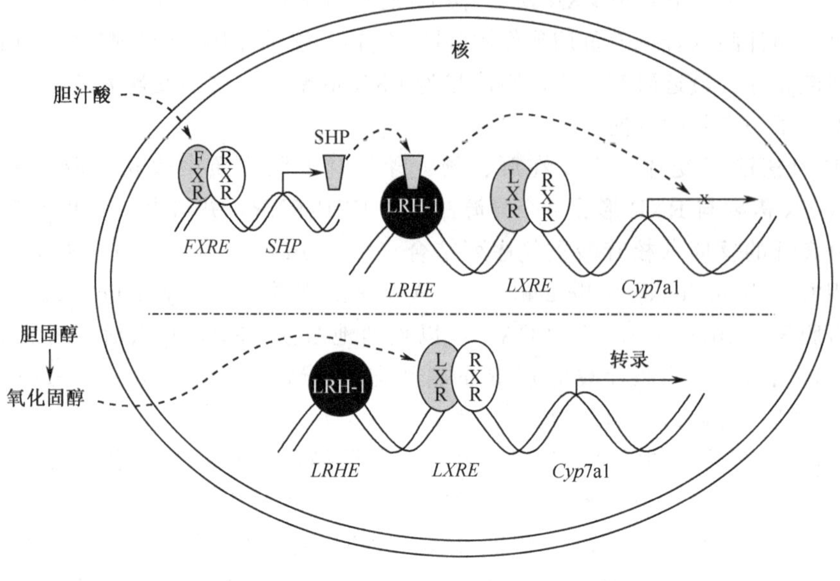


图 20.3 LXR 和 FXR 对 *Cyp7a1* 基因转录的调节

胆汁酸与 FXR 的结合诱导了 *SHP* 基因的表达（图的上半部分），SHP 蛋白与已连接到位于 *Cyp7a1* 基因启动子中 LRH 反应元件（LRHE）上的 LRH-1 结合，减少了 *Cyp7a1* 基因的转录。当胆固醇转化为 LXR 的配体氧化固醇时，*Cyp7a1* 基因的转录增加（图的下半部分）。细胞氧化固醇浓度的增加代表了胆固醇浓度的增加。LXRE 代表 LXR 反应元件；FXRE 代表 FXR 反应元件

用（Kramer *et al.*，1993）。在人类 IBABP 启动子中，位于－160～－148 核苷酸的 IR-1 为 FXR-RXR 异源二聚体的结合位点，这表明胆汁酸能够通过的激活 FXR 诱导 IBABP 的转录（Grober *et al.*，1999；Makishima *et al.*，1999）。胆汁盐输出泵（bile soft export pump，BSEP）在胆汁酸跨肝细胞微管膜转运过程中发挥着重要的作用。BSEP 基因启动子包含一个 IR-1 元件，并且胆汁酸对 BSEP 的反式激活作用依赖于 FXR（Ananthanarayanan *et al.*，2001）。

FXR 为胆汁酸代谢乃至胆固醇代谢中的一个重要的调节子，它有正负两种方式的调节功能。与 FXR 在胆汁酸合成中的间接负调节作用相比，FXR 通过与 *IBABP* 和 *BSEP* 基因启动子的结合，对这些基因的表达起正调节作用。通过对 FXR 敲除小鼠的研究也阐明了 FXR 其他可能的作用（Sinal *et al.*，2000）：肝基底钠牛磺酸协同转运蛋白（natrium taurocholate cotransporter protein，NTCP）和肝脂肪酸结合蛋白（liver fatty acid-binding protein，L-FABP）受 FXR 的调节；FXR 还可作为一种胆汁酸传感器，它通过改变几个基因的表达在胆固醇、胆汁酸和脂肪酸代谢中发挥着重要的作用。

# 调节蛋白质和食物的作用

## HMG-CoA 还原酶

胆固醇在一个多酶途径中合成，在此途径中 HMG-CoA 还原酶为限速酶。HMG-CoA 还原酶是一种 ER 的内在膜蛋白，它催化甲羟戊酸的合成，甲羟戊酸为固醇和非固

醇类异戊二烯复合物形成过程中的关键中间产物。胆固醇的生物合成因细胞胆固醇水平的基因调节而被调节（Goldstein and Brown，1990）。在缺少固醇的细胞中转录水平高，在含有过多固醇的细胞中转录水平低。胆固醇的摄入、吸收和转运到细胞的增多，会通过 SREBP 介导的基因表达调控降低 HMG-CoA 还原酶的活性。

当细胞胆固醇耗尽时，成熟的 SREBP 通过蛋白水解作用被转运到核中，然后与 HMG-CoA 还原酶启动子部位的固醇调节元件（SRE）相结合，增加其转录水平。相反，当存在过多的胆固醇时，胆固醇代谢物（如羟基胆固醇）阻止 SREBP 通过蛋白水解作用转化为其活性形式，引起对 HMG-CoA 还原酶的负调节作用（Edwards *et al.*，2000）。

不像食物中的胆固醇，单独食物中的脂肪酸对 HMG-CoA 还原酶的作用是多样的，目前还没有彻底弄明白。一般而言，饱和脂肪酸（SFA）与多不饱和脂肪酸（PUFA）和单不饱和脂肪酸（MUFA）相比，有增加 HMG-CoA 活性的趋势。对 SREBP 成熟的抑制可通过油酸对氧化固醇的作用实现，食物中 MUFA（如油酸）通过抑制 SREBP 的成熟进而影响氧化固醇，从而对 HMG-CoA 还原酶的活性进行间接负调节（Thewke *et al.*，1998）。MUFA 和 PUFA 都通过减少 SREBP 的成熟形式降低表达 HMG-CoA 还原酶的 mRNA，而 SFA 的这种影响很小（Worgall *et al.*，1998）。

食物中纤维——主要指水溶性纤维，趋向于降低血清胆固醇浓度。研究者已提出若干种机制来解释食物中纤维的降胆固醇作用，包括肠胆固醇吸收、脂蛋白代谢、胆汁酸代谢、副产物发酵的改变以及它们在肝胆固醇合成中的作用（Kay，1982）。在动物研究中，可溶性食物纤维（如胶质、瓜尔豆胶和蚤草）增加了 HMG-CoA 还原酶的活性（Fernandez *et al.*，1995；Moundras *et al.*，1997）。这些作用主要归因于肝胆固醇浓度的减少，而采食食物纤维可使肝胆固醇浓度减少。

## LDL 受体

极低密度脂蛋白（VLDL）是在肝脏形成的富含三酰甘油的颗粒，其作用是将内源脂质转运到外周组织。一旦三酰甘油耗尽后，VLDL 转换成胆固醇含量丰富的残留颗粒，称作 LDL。尽管血浆中过多的 LDL 与冠心病和中风有关，LDL 仍没有明显的生物学功能。因此，研究者主要将注意力集中在 LDL 的细胞摄取途径方面。

LDL 可以通过受体依赖和非受体依赖途径从血浆中去除。大多发生在肝脏的受体依赖型摄取占 LDL 清除总量的 60%～80%（Spady *et al.*，1986）。LDL 受体存在于细胞表面，对 LDL 的唯一蛋白成分——载脂蛋白 $B_{100}$ 进行识别。LDL 受体对食物中胆固醇、脂肪酸和纤维的变化高度敏感，而 LDL 的非受体依赖型摄取不受食物的影响。

LDL 受体的活性随细胞游离胆固醇浓度的变化在转录水平上受到调节。正如在 SREBP 章节中描述的一样，LDL 受体基因的启动子包含一个 SRE，SRE 为 SREBP 的结合位点。细胞内低胆固醇浓度诱导 SREBP 的成熟，然后 SREBP 的 N 端转录因子结构域被释放，引起 LDL 受体基因的反式激活（Hua *et al.*，1995）。相反，当细胞内含有过量的胆固醇时，胆固醇和（或）胆固醇代谢物如氧化固醇会阻止 SREBP 的成熟，导致对 LDL 受体基因的负调节（Towle，1995）。

食物中胆固醇可影响肝脏中胆固醇的浓度，肝脏胆固醇浓度又可影响 LDL 受体的活性。尽管对于肝胆固醇浓度的增加，细胞内存在数个稳态作用途径来调节，但 LDL 受体活性的降低是主要的补救途径。食用胆固醇含量丰富的食物能显著增加肝胆固醇浓度，从而抑制 LDL 受体的 mRNA 丰度（Boucher *et al.*，1998）。食物中高含量的胆固醇会超过 LDL 受体将 LDL 从血浆中充分清除的负调节能力，导致血浆 LDL 浓度的增加（Applebaum-Bowden *et al.*，1984；Spady and Dietschy，1988）。因此，人们普遍认为，食物中胆固醇能通过影响肝胆固醇池，随后影响胆固醇或其代谢物对 SREBP 激活的抑制而减少 LDL 受体的基因表达。

食入的脂肪酸类型可以影响肝脂肪酸成分，肝脂肪酸成分又可影响 LDL 受体的活性。通常，食物中 SFA 降低 LDL 受体活性并因此可引起与血浆 LDL 有关的高胆固醇症。食物中 MUFA 及 PUFA 相对而言是降胆固醇的，有一个例外是食物中的硬脂酸。硬脂酸是一种 SFA，但它并不降低 LDL 受体的活性或 mRNA 丰度，也不增加血浆 LDL 浓度（Nicolosi，1997），其确切的作用机制目前还不清楚，但有可能包含转录和翻译后作用机制。

可溶性食物纤维的降胆固醇作用主要归因于纤维能降低胆固醇的吸收和胆汁酸在小肠中的重吸收。动物研究表明，可溶性纤维的摄食能增加 LDL 受体活性及 mRNA 水平，并同时降低血浆胆固醇浓度（Fernandez *et al.*，1995；Fukushima *et al.*，2000）。在细胞水平，可溶性纤维的降胆固醇作用是由肝胆固醇调节池和 SREBP 信号途径的减少介导的。

## 胆固醇 7α-羟化酶

胆汁酸在脂类食物的消化和吸收中是很重要的。近来的研究已阐明了胆汁酸在调节参与胆固醇和胆汁酸代谢的基因转录中的作用（Makishima *et al.*，1999；Parks *et al.*，1999；Wang *et al.*，1999）。胆汁酸合成有两条截然不同的途径：经典（中性的）途径和选择性（酸性的）途径。微粒体胆固醇 7α-羟化酶为经典胆汁酸合成过程中的限速酶，线粒体固醇 27-羟化酶被认为是选择性胆汁酸合成过程中的限速酶。选择性途径对总胆汁酸合成的作用目前还不清楚。在抑制胆固醇 7α-羟化酶活性的情况下，胆汁酸合成仍保持着相对较高的水平，这表明在胆固醇 7α-羟化酶活性持续抑制时，选择性途径成为胆汁酸合成的主要途径（Xu *et al.*，1999）。胆汁酸在分泌到胆囊中进行暂时储存之前，是与肝脏中的甘氨酸或牛磺酸相结合的。大约 95%的胆汁酸/盐在小肠中通过被动或主动转运机制重吸收，并返回肝脏。胆汁酸/盐的肠肝循环在脂类吸收、胆汁酸合成和胆固醇动态平衡中发挥着重要作用。

食物中胆固醇增加时，肝胆固醇池主要靠增加胆固醇转化成胆汁酸或将游离胆固醇分泌到胆汁中的方式来维持稳定。胆汁酸合成多少与食物中胆固醇的含量相关，这种相关性随物种不同而不同。作为维持大鼠（Pandak *et al.*，1991；Shefer *et al.*，1992）和小鼠（Deuland *et al.*，1993；Torchia *et al.*，1996）胆固醇动态平衡的补偿性反应，食物中胆固醇会增加胆固醇 7α-羟化酶的 mRNA 丰度及其活性，然而在新西兰兔（Xu *et al.*，1995）、非洲绿猴（Rudel *et al.*，1994）和仓鼠（Horton *et al.*，1995）中，饲喂胆固醇抑制了胆固醇 7α-羟化酶的 mRNA 丰度及其活性。对兔的进一步研究表明，饲喂胆固醇在短期内增加了胆固醇 7α-羟化酶的活性，但随之而来的就是酶活性的降低

（Xu *et al.*，1999）。由于选择性胆汁酸合成途径中的固醇 27-羟化酶活性的增加，胆汁酸池就显著的增大。因为 *Cyp7a1* 基因的表达受 LXR 的正调节或 FXR 的负调节，因此研究者猜测，开始时饲喂胆固醇也许可以激活 LXR，但是，肝中胆汁酸流出量的增加，导致 FXR 可能被激活并抑制 *Cyp7a1* 基因的反式转录。通过 FXR 而对 *Cyp7a1* 基因抑制敏感的物种中，选择性胆汁酸合成可能是胆汁酸合成的主要途径。

胆固醇 7α-羟化酶的活性也受到食物中脂肪酸的调节，确切的途径还没有被识别。总之，在动物研究中，PUFA 和 MUFA 趋向于增加胆固醇 7α-羟化酶的活性，而 SFA 却限制其活性（Bravo *et al.*，1996）。例如，仓鼠膳食中包含的油酸会增加肝胆固醇 7α-羟化酶的活性，而棕榈酸则抑制其活性（Kurushima *et al.*，1995）。有关特定的脂肪酸在细胞和分子水平的特定作用方面的资料还很少。

水溶性纤维的降胆固醇作用至少部分归因于它们形成黏性基质的能力，此黏性基质能够抑制胆汁酸/盐和固醇从小肠的吸收（Eastwood and Morris，1992）。结果胆汁酸分泌和总固醇的增加对胆汁酸合成有正调节作用，以补充肝胆汁酸池。血脂正常的青年人连续摄食含燕麦麸的食物两个月，胆汁酸合成和经粪排泄均会增加（Marlett *et al.*，1994）。饲喂果胶膳食的大鼠血清和肝脏胆固醇浓度较低，粪胆汁酸排泄以及胆固醇 7α-羟化酶的活性增加（Garcia-Diez *et al.*，1996）。摄食亚麻增加了雄性荷兰猪肝胆固醇 7α-羟化酶的活性以及大鼠和仓鼠的胆固醇 7α-羟化酶 mRNA（Horton *et al.*，1994；Fernandez *et al.*，1995；Buhman *et al.*，1998）。在这些研究的每一项观察中，当肠源性胆固醇经转化而从体内排出时，动物机体都需要合成更多的胆固醇并将其转化成胆汁酸。

## 小　　结

在肝胆固醇代谢中有 3 个主要的调节蛋白，即 HMG-CoA 还原酶、LDL 受体和胆固醇 7α-羟化酶，它们在转录水平受 SREBP、LXR 和（或）FXR 的协同调节，以维持胆固醇的稳衡。通过调节编码 HMG-CoA 还原酶和 LDL 受体基因的表达，SREBP 在胆固醇合成和细胞摄取中发挥着重要的作用。LXR 和 FXR 通过胆固醇 7α-羟化酶调节细胞胆固醇到胆汁酸的转化。因为胆汁酸分泌是胆固醇从体内排出的主要途径，LXR 和 FXR 在调节全身胆固醇周转和平衡中也有一定的作用。而且，所有这些过程都受食物特别是食物中胆固醇、脂肪酸和纤维的影响。其他食物成分也可能影响细胞胆固醇代谢并无疑会成为将来研究的方向。

### 参 考 文 献


Ananthanarayanan, M., Balasubramanian, N., Makishima, M., Mangelsdorf, J.J. and Suchy, F.J. (2001) Human bile salt export pump (BSEP) promoter is transactivated by the farnesoid X receptor/bile acid receptor (FXR/BAR). *Journal of Biological Chemistry* 276, 28857–28865.

Applebaum-Bowden, D., Haffner, S.M., Hartsook, E., Luk, K.H., Albers, J.J. and Hazzard, W.R. (1984) Down-regulation of the low-density lipoprotein receptor by dietary cholesterol. *American Journal of Clinical Nutrition* 39, 360–367.

Becker-Andre, M., Andre, E. and DeLamarter, J.F. (1993) Identification of nuclear receptor mRNAs by RT-PCR amplication of conserved zinc-finger motif sequences. *Biochemical and Biophysical Research Communications* 194, 1371–1379.

Boucher, P., de Lorgeril, M., Salen, P., Crozier, P., Delaye, J., Vallor, J.J., Geyssant, A. and Dante, R.

(1998) Effect of dietary cholesterol on low density lipoprotein-receptor, 3-hydroxy-3-methylglutaryl-CoA reductase, and low density lipoprotein receptor-related protein mRNA expression in healthy humans. *Lipids* 33, 1177–1186.

Bravo, E., Cantafora, A., Marinelli, T., Avella, M., Mayes, P.A. and Botham, K.M. (1996) Differential effects of chylomicron remnants derived from corn oil or palm oil on bile acid synthesis and very low density lipoprotein secretion in cultured rat hepatocytes. *Life Sciences* 59, 331–337.

Briggs, M.R., Yokoyama, C., Wang, X., Brown, M.S. and Goldstein, J.L. (1993) Nuclear protein that binds sterol regulatory element of low density lipoprotein receptor promoter. *Journal of Biological Chemistry* 268, 14490–14496.

Brown, M.S. and Goldstein, J.L. (1999) A proteolytic pathway that controls the cholesterol content of membranes, cells and blood. *Proceedings of the National Academy of Sciences USA* 96, 11041–11048.

Buhman, K.K., Furumoto, E.J., Donkin, S.S. and Story, J.A. (1998) Dietary psyllium increases fecal bile acid excretion, total steroid excretion and bile acid biosynthesis in rats. *Journal of Nutrition* 128, 1199–1203.

Chiang, J.Y.L. and Stroup, D. (1994) Identification and characterization of a putative bile acid-response element in cholesterol 7α-hydroxylase gene promoter. *Journal of Biological Chemistry* 269, 17502–17507.

Chiang, J.Y.L., Kimmel, R., Weinberger, C. and Stroup, D. (2000) Farnesoid X receptor responds to bile acids and repressed cholesterol 7α-hydroxylase gene (CYP7A1) transcription. *Journal of Biological Chemistry* 275, 10918–10924.

Deuland, S., Drisko, J., Graf, L., Machleder, D., Lusis, A.J. and Davis, R.A. (1993) Effect of dietary cholesterol and taurocholate on cholesterol 7 alpha-hydroxylase and hepatic LDL receptors in inbred mice. *Journal of Lipid Research* 34, 923–931.

Eastwood, M.A. and Morris, E.R. (1992) Physical properties of dietary fibre that influence physiological function: a model for polymers along the gastrointestinal tract. *American Journal of Clinical Nutrition* 55, 436–442.

Edwards, P.A., Tabor, D., Kast, H.R., Venkateswaran, A. (2000) Regulation of gene expression by SREBP and SCAP. *Biochimica et Biophysica Acta* 1529, 103–113.

Fernandez, M.L., Ruiz, L.R., Conde, A.K., Sun, D.M., Erickson, S.K. and McNamara, D.J. (1995) Psyllium reduces plasma LDL in guinea pigs by altering hepatic cholesterol homeostasis. *Journal of Lipid Research* 36, 1128–1138.

Forman, B.M., Goode, E., Chen, J., Oro, A.E., Bradley, D.J., Perimann, T., Noonan, D.J., Burka, L.T., McMorris, T., Lamph, W.W., Evans, R.M. and Weinberger, C. (1995) Identification of a nuclear receptor that is activated by farnesol metabolites. *Cell* 81, 687–693.

Fukushima, M., Nakano, M., Morii, Y., Ohashi, T., Fujiwara, Y. and Sonoyama, K. (2000) Hepatic LDL receptor mRNA in rats is increased by dietary mushroom (*Agaricus bisporus*) fibre and sugar beet fibre. *Journal of Nutrition* 130, 2151–2156.

Garcia-Diez, F., Garcia-Mediavilla, V., Bayon, J.E. and Gonzalez-Gallego, J. (1996) Pectin feeding influences fecal bile acid excretion, hepatic bile acid and cholesterol synthesis and serum cholesterol in rats. *Journal of Nutrition* 126, 1766–1771.

Goldstein, J.L. and Brown, M.S. (1990) Regulation of the mevalonate pathway. *Nature* 343, 425–430.

Goldstein, J.L. and Brown, M.S. (2001) The cholesterol quartet. *Science* 292, 1310–1312.

Grober, J., Zaghin, I., Fujii, H., Jones, S.A., Kliewer, S.A., Willson, T.M., Ono, T. and Besnard, P. (1999) Identification of a bile acid-responsive element in the human ileal bile acid-binding protein gene. *Journal of Biological Chemistry* 274, 29749–29754.

Horton, J.D., Cutbert, J.A. and Spady, D.K. (1995) Regulation of hepatic 7α-hydroxylase expression and response to dietary cholesterol in the rat and hamster. *Journal of Biological Chemistry* 270, 5381–5387.

Hua, X., Yokoyama, C., Wu, J., Briggs, M.R., Brown, M.S., Goldstein, J.L and Wang, X. (1993) SREBP-2, a second basic-helix–loop–helix-leucine zipper protein that stimulates transcription by binding to a sterol regulatory element. *Proceedings of the National Academy of Sciences USA* 90, 11603–11607.

Hua, X., Wu, J., Goldstein, J.L., Brown, M.S. and Hobbs, H.H. (1995) Structure of the human gene encoding sterol regulatory element binding protein-1 (SREBF-1) and localization of SREBF-1 and SREBF-2 to chromosomes 17p11.2 and 22q13. *Genomics* 25, 667–673.

Janowski, B.A., Willy, P.J., Devi, T.R., Falck, J.R. and Mangelsdorf, D.J. (1996) An oxysterol signaling pathway mediated by the nuclear receptor LXRα. *Cell* 383, 728–731.

Kay, R.M. (1982) Dietary fibre. *Journal of Lipid Research* 23, 221–242.

Kramer, W., Birbig, F., Gutjahr, U., Kowalewski, S., Jouvenal, K., Muller, G., Tripier, D. and Wess, G. (1993) Intestinal bile acid absorption. Na(+)-dependent bile acid transport activity in rabbit small intestine correlates with the coexpression of an integral 93-kDa and a peripheral 14-kDa bile acid-binding membrane protein along the duodenum–ileum axis. *Journal of Biological Chemistry* 268, 18035–18046.

Kurushima, H., Hayashi, K., Shingu, T., Kuga, Y., Ohtani, H., Okura, Y., Tanaka, K., Yasunobu, Y., Nomura, K. and Kajiyama, G. (1995) Opposite effects on cholesterol metabolism and their mechanisms induced by dietary oleic acid and palmitic

acid in hamsters. *Biochimica et Biophysica Acta* 1258, 251–256.
Lehmann, J.M., Kliewer, S.A., Moore, L.B., Smith-Oliver, T.A., Oliver, B.B., Su, J.L., Sundseth, S.S., Winegar, D.A., Blanchard, D.E., Spencer, T.A. and Willson, T.M. (1997) Activation of the nuclear receptor LXR by oxysterols defines a new hormone response pathway. *Journal of Biological Chemistry* 272, 3137–3140.
Lu, T.T., Makishima, M., Repa, J.J., Schoonjans, K., Kerr, T.A., Auwerx, J. and Mangelsdorf, D.D. (2000) Molecular basis for feedback regulation of bile acid synthesis by nuclear receptors. *Molecular Cell* 6, 507–515.
Makishima, M., Lkamoto, A.Y., Repa, J.J., Tu, H., Learned, M., Luk, A., Hull, M.V., Lustig, K.D., Mangelsdorf, D.J. and Shan, B. (1999) Identification of a nuclear receptor for bile acids. *Science* 284, 1362–1365.
Marlett, J.A., Hosig, K.B., Vollendorf, N.W., Shinnick, F.L., Haack, V.S. and Story, J.A. (1994) Mechanism of serum cholesterol reduction by oat bran. *Hepatology* 20, 1450–1457.
Moundras, C., Behr, S.R., Remesy, C. and Demigne, C. (1997) Fecal losses of sterols and bile acids induced by feeding rats guar gum are due to greater pool size and liver bile acid secretion. *Journal of Nutrition* 127, 1068–1076.
Nicolosi, R.J. (1997) Dietary fat saturation effects on low-density-lipoprotein concentrations and metabolism in various animal models. *American Journal of Clinical Nutrition* 65 (Supplement 5), 1617S–1627S.
Nohturfft, A., Hua, X., Brown, M.S. and Goldstein, J.L. (1996) Recurrent G-to-A substitution in a single codon of SREBP cleavage-activating protein causes sterol resistance in three mutant Chinese hamster ovary cell lines. *Proceedings of the National Academy of Sciences USA* 93, 13709–13714.
Pandak, W.M., Li, Y.C., Chiang, J.Y.L., Studer, E.J., Gurley, E.C., Heuman, D.M., Vlahcevic, Z.R. and Hylemon, P.B. (1991) Regulation of cholesterol 7α-hydroxylase mRNA and transcriptional activity by taurocholate and cholesterol in the chronic biliary diverted rat. *Journal of Biological Chemistry* 266, 3416–3421.
Parks, D.J., Blanchard, S.G., Bledsoe, R.K., Chandra, G., Consler, T.G., Kliewer, S.A., Stimmel, J.B., Wilson, T.M., Zavacke, A.M., Moore, D.D. and Lehmann, J.M. (1999) Bile acids: natural ligand for an orphan nuclear receptor. *Science* 284, 1365–1368.
Peet, D.J., Janowski, B.A. and Mangelsdorf, D.J. (1998a) The LXRs: a new class of oxysterol receptors. *Current Opinion in Genetics and Development* 8, 571–575.
Peet, D.J., Turley, S.D., Ma, W., Janowski, B.A., Lobaccaro, J.A., Hammer, R.E. and Mangelsdorf, D.J. (1998b) Cholesterol and bile acid metabolism are impaired in mice lacking the nuclear oxysterol receptor LXRα. *Cell* 93, 693–704.
Repa, J.J., Liang, G., Ou, J., Bashmakov, Y., Lobaccaro, J.M.A., Shimomura, I., Shan, B., Brown, M. S., Goldstein, J.L. and Mangelsdorf, D.J. (2000) Regulation of mouse sterol regulatory element-binding protein-1c gene (SREBP-1c) by oxysterol receptors, LXRalpha and LXRbeta. *Genes and Development* 14, 2819–2830.
Rudel, L., Deckelman, C., Wilson, M., Scobey, M. and Anderson, R. (1994) Dietary cholesterol and downregulation of cholesterol 7α-hydroxylase and cholesterol absorption in African green monkeys. *Journal of Clinical Investigation* 93, 2463–2472.
Sakai, J., Duncan, E.A., Rawson, R.B., Hua, X., Brown, M.S. and Goldstein, J.L. (1996) Sterol-regulated release of SREBP-2 from cell membranes requires two sequential cleavages, one within a transmembrane segment. *Cell* 85, 1037–1046.
Sakai, J., Nohturfft, A., Goldstein, J.L. and Brown, M.S. (1998) Cleavage of sterol regulatory element-binding proteins (SREBPs) at site-1 requires interaction with SREBP cleavage-activating protein. Evidence from *in vivo* competition studies. *Journal of Biological Chemistry* 273, 5785–5793.
Seol, W., Choi, H.-S. and Moore, D.D. (1996) An orphan nuclear hormone receptor that lacks a DNA binding domain and heterodimerizes with other receptors. *Science* 272, 1336–1339.
Shefer, S., Nguyen, L.B., Salen, G., Ness, G.C., Chowdhary, I.R., Lerner, S., Batta, A.K. and Tint, G.S. (1992) Differing effects of cholesterol and taurocholate on steady state hepatic HMG-CoA reductase and cholesterol 7 alpha-hydroxylase activities and mRNA levels in the rat. *Journal of Lipid Research* 33, 1193–1200.
Shimomura, I., Shimano, H., Horton, J.D., Goldstein, J.L. and Brown, M.S. (1997) Differential expression of exons 1a and 1c in mRNAs for sterol regulatory element binding protein-1 in human and mouse organs and cultured cells. *Journal of Biological Chemistry* 99, 838–845.
Sinal, C.J., Tohkin, M., Miyata, M., Ward, J.M., Lambert, G. and Gonzalez, F.J. (2000) Targeted disruption of the nuclear receptor FXR/BAR impairs bile acid and lipid homeostasis. *Cell* 102, 731–744.
Spady, D.K. and Dietschy, J.M. (1988) Interaction of dietary cholesterol and triglycerides in the regulation of hepatic low density lipoprotein transport in hamster. *Journal of Clinical Investigation* 81, 300–309.
Spady, D.K., Stange, E.F., Bilhartz, L.E. and Dietschy, J.M. (1986) Bile acids regulate hepatic low density lipoprotein receptor activity in the hamster by altering cholesterol flux across the liver. *Proceedings of the National Academy of Sciences USA* 83, 1916–1920.
Stravitz, R.T., Hylemon, P.B., Heuman, D.M., Hagey,

L.R., Schteingart, C.D., Ton-Nu, H.-T., Hofmann, A.F. and Vlahcevic, Z.R. (1993) Transcriptional regulation of cholesterol 7α-hydroxylase mRNA by conjugated bile acids in primary cultures of rat hepatocytes. *Journal of Biological Chemistry* 268, 13967 13993.

Stroup, D., Crestani, M. and Chiang, J. Y. L. (1997) Identification of a bile acid response element in the cholesterol 7 alpha-hydroxylase gene CYP7A. *American Journal of Physiology* 273, G508–G517.

Thewke, D.P., Panini, S.R. and Sinensky, M. (1998) Oleate potentiates oxysterols inhibition of transcription from sterol regulatory element-1-regulated promoters and maturation of sterol regulatory element-binding proteins. *Journal of Biological Chemistry* 173, 21402–21407.

Torchia, E.C., Cheema, S.K. and Agellon, L.B. (1996) Coordinate regulation of bile acid biosynthetic and recovery pathways. *Biochemical and Biophysical Research Communications* 225, 128–133.

Towle, H.C. (1995) Metabolic regulation of gene transcription in mammals. *Journal of Biological Chemistry* 270, 23235–23238.

Vlahcevic, Z.R., Pandak, W.M. and Stravitz, R.T. (1999) Regulation of bile acid biosynthesis. *Gastroenterology Clinics of North America* 28, 1–25.

Wang, X., Briggs, M.R., Hua, X., Yokoyama, C., Goldstein, J.L. and Brown, M.S. (1993) Nuclear protein that binds sterol regulatory element of low density lipoprotein promoter. II. Purification and characterization. *Journal of Biological Chemistry* 268, 14497–14504.

Wang, H., Chen, J., Hollister, K., Sowers, L.C. and Forman, B.M. (1999) Endogenous bile acids are ligands for the nuclear receptor FXR/BAR. *Molecular Cell* 3, 543–553.

Willy, P.J., Umesono, K., Ong, E.S., Evans, R.M., Heyman, R.A. and Mangelsdorf, D.J. (1995) LXR, a nuclear receptor that defines a distinct retinoid response pathway. *Genes and Development* 9, 1033–1045.

Worgall, T.S., Sturley, S.L., Seo, T., Osborne, T.F. and Deckelbaum, R.J. (1998) Polyunsaturated fatty acids decrease expression of promoters with sterol regulatory elements by decreasing levels of mature sterol regulatory element-binding protein. *Journal of Biological Chemistry* 273, 25537–25540.

Xu, G., Salen, G., Shefer, S., Ness, G.C., Nguyen, L.B., Parker, T.S., Chen, T.S., Zhao, Z., Donnelly, T.M. and Tint G.S. (1995) Unexpected inhibition of cholesterol 7 alpha-hydroxylase by cholesterol in New Zealand white and Watanabe heritable hyperlipidemic rabbits. *Journal of Clinical Investigation* 95, 1497–1504.

Xu, G., Salen, G., Shefer, S., Tint, G.S., Nguyen, L.B., Chen, T.S. and Greenblatt, D. (1999) Increasing dietary cholesterol induces different regulation of classic and alternative bile acid synthesis. *Journal of Clinical Investigation* 103, 89–95.

# 21　2002年度：营养对白内障发生率影响的评估

Allen Taylor 和 Mark Siegal
（美国农业部老年人营养研究中心营养与视力研究实验室，
塔福兹大学，波士顿，马萨诸塞州，美国）

## 白内障：公共卫生问题

人们已知氧化应激会使晶状体受到损伤。最近研究发现，抗氧化剂可降低晶状体的损伤并减少白内障的发病率。本章将简要回顾白内障的病原学及流行病学资料，重点阐述维生素C、维生素E和类胡萝卜素对白内障的作用。全面而较新的综述可参见Taylor（1999b，c）。

## 年龄相关性流行病的年度治疗费

白内障是世界上可预防性失明的主要原因之一（Kupfer，1985；Schwab，1990；世界卫生组织，1991）。在美国，白内障的流行情况是从65岁以上的约5%提高到75岁以上的将近50%（Leibowitz *et al.*，1980；Klein *et al.*，1992，1993）。在不发达国家，如印度（Chatterjee *et al.*，1982）、中国（Wang *et al.*，1990）和肯尼亚（Whitfield *et al.*，1990），白内障更普遍，发病的年龄也更早（Leibowitz *et al.*，1980；Chatterjee *et al.*，1982），而且对视力损伤的影响力也更大，其中有些国家的眼睛失明和视觉损伤的发病率超过90%（Taylor，1999a，c），同时在这些国家缺乏眼科专家来进行晶状体的摘除术。

据估计，在未来的10年内，致视觉失明的白内障发病率将降低到约45%（Kupfer，1985），这将改善世界上许多老年人的生活质量，大大减少白内障相关性残疾和白内障手术带来的经济损失（50亿～60亿美元）（Young，1993）。

## 年龄性损伤与氧化损伤和自身保护能力下降有关

眼睛晶状体的主要功能是将光线聚焦到视网膜上（图21.1），为此，在整个生命期，它都必须保持透明。晶状体组装精密，单层上皮细胞直接位于前囊膜下，在这里被压缩成胶囊状（图21.1b）。生发区上皮细胞发生分裂，然后转移并分化成晶状体纤维细胞。作为主要的基因产物，这些纤维主要是晶状体蛋白质，也称为晶状体球蛋白。在整个生命过程中，新细胞不断生成，但衰老的细胞通常不被清除，相反，它们被压缩到中心或晶状体核。随着晶状体的老化，或因光线（McCarty and Taylor，1999）或因吸烟（West，1999）产生的应激，这些晶状体蛋白质便发生光氧化损伤，然后凝集。高能辐射、各种活性氧、阳光和光暴露均可产生氧化损伤，且第二防御系统的失调也可产生氧化损伤（图 21.2）。随后晶状体蛋白质和晶状体本身一起发生脱水，从而使每毫升房水的蛋白质浓度提高至数百毫克（Taylor *et al.*，1981）。这些变化以及其他与年龄相关的蛋白质修饰［如糖基化、糖氧化（glycoxidation）］和蛋白质的脂化共同作用，降低晶状体的调节能力，导致晶状体更不灵活。最后，晶状体蛋白质的大量凝集使晶状体发生混浊，甚至形成白内障。

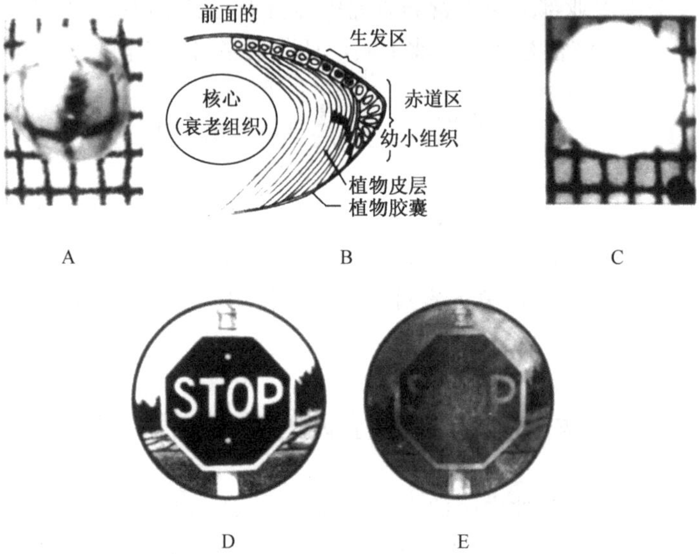


图 21.1　透明的晶状体和患白内障的晶状体

A. 透明晶状体能在后面线栅上形成自由视野。B. 晶状体的结构图。晶状体前表面是单层上皮细胞（最新生成的组织）。当它们被未成熟细胞覆盖后，前赤道部的细胞发生分裂并转移到皮质层。大部分晶状体球蛋白是由这些细胞产生的。随着细胞的生长和成熟，细胞发生脱核和伸长。早在胚胎期产生的晶状体组织现位于晶状体中心或核（最老的组织处）。C. 患白内障的晶状体不能在后面线栅上形成图像。D. 通过透明的、未染色的、年幼的晶状体投影的图标图像：清晰明亮。E. 通过白内障晶状体投影的图标图像：部分图像模糊，黑色区是由衰老过程中形成的晶状体褐斑所致（Taylor，1999a）

年龄相关性白内障与老年人的晶状体混浊不同于其他因素（如先天性和代谢性失调或外伤）导致的混浊。目前白内障的评估和分类方法有多种，大多数用于评估混浊的程度、密度及发生混浊的位置（Chylack *et al.*，1993；Chylack，1999）。一般情况下，可对后囊膜下、核性、皮质性和混合性的混浊度进行评估，还能对着色进行评估（Wolfe *et al.*，1993；Chylack *et al.*，1993）。然而，目前人们还不能确定各种白内障是否具有完全不同的病原学。

## 抗氧化剂是晶状体损伤的第一道防线

晶状体的保护系统主要有两道防线，它们相互关联、共同作用，从而实现对光氧化损伤的预防与保护（图 21.2）。第一道防线是抗氧化剂和抗氧化酶，它们通过减轻氧化损伤或修复蛋白质来保护晶状体；第二道防线包括蛋白质降解和修复过程，可及时降解

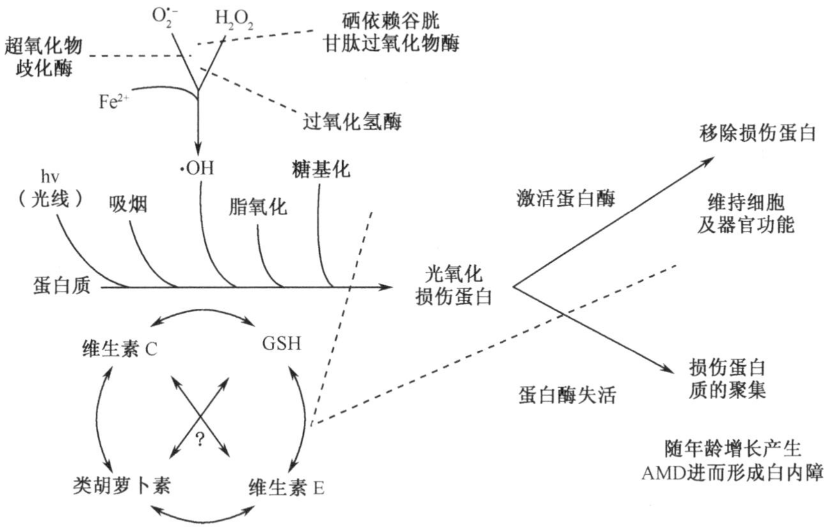


图 21.2　晶状体蛋白质、氧化剂、光线、吸烟、抗氧化酶和蛋白酶之间可能的相互作用

晶状体蛋白质存活时间相当长，故易受光线和各种形式氧的损伤。具有间接保护晶状体蛋白质作用的抗氧化酶：超氧化物歧化酶、过氧化氢酶、GSH 还原酶/过氧化物酶，能把活性氧转化成毒性更小的物质。而抗氧化剂：GSH、抗坏血酸（维生素 C）、生育酚（维生素 E）和类胡萝卜素具有直接的保护作用。部分但不是所有物质的氧化形式和还原形式的水平取决于这些物质之间或与环境之间的相互作用（Taylor and Davies，1987；Taylor *et al.*，1991a；Chylack *et al.*，1993，1994；Wolfe *et al.*，1993；Chylack，1999）。在很多研究系统中，GSH 与抗坏血酸水平相关，但是对抗坏血酸必需的大鼠补足后，我们没发现这种关系（Bunce *et al.*，1990；Taylor and Jacques，1997）。若晶状体具有足够的蛋白质降解能力，退化蛋白质和受损蛋白质就可能被降解成各种氨基酸。随着年龄的增长，眼部的一些抗氧化剂逐渐减少，抗氧化酶和蛋白酶的活性也逐渐降低。这可能使受损蛋白质发生聚集、凝集，最后沉淀形成白内障。GSH：谷胱甘肽，hv：光线，AMD：年龄相关性斑点退化

和消除受损蛋白质和其他生物分子（Taylor and Davies，1987）。

维生素 C 和谷胱甘肽（GSH）是晶状体中主要的水溶性抗氧化剂（Bunce *et al.*，1990；Reddy，1990；Taylor *et al.*，1991a，1995b，1997；Sastre *et al.*，1994；Mune *et al.*，1995；Taylor and Jacques，1997；Smith *et al.*，1987），二者在晶状体中的浓度非常低，只有几毫摩尔。

维生素 C 可能是哺乳动物中最有效且毒性最小的抗氧化剂（Levine，1986；Berger *et al.*，1988，1989；Frei *et al.*，1988；Taylor *et al.*，1991a；Taylor and Jacques，1997），眼部的维生素 C 水平与摄入量有关（Berger *et al.*，1988，1989）。然而，研究还发现，对于摄入推荐膳食允许量（70mg/d）2 倍多的人群，补充维生素 C 也能增加其在晶状体中的水平（Taylor *et al.*，1991a，1997）。

提高维生素 C 的饲喂量能延缓或预防豚鼠（Kosegarten and Maher，1978；Yokoyama *et al.*，1994）和大鼠（Vinson *et al.*，1986）的半乳糖性白内障、大鼠的亚硒酸诱导的白内障（Devamanoharan *et al.*，1991）及 GSH 缺失鸡胚的晶状体混浊（Nishigori *et al.*，1986），并延缓 UV 对豚鼠晶状体中蛋白质和蛋白质酶造成的损伤（Blondin *et al.*，1986，1987；Blondin and Taylor，1987；Taylor *et al.*，1995a）。晶状体维生素 C 浓度只要提高 2 倍，就能保护晶状体免受白内障类似的损伤（Blondin *et al.*，

1986）。

由于维生素C也是碳水化合物，因此，根据生化理论来推导，维生素C似乎也能诱导体内晶状体的损伤（Garland，1991；Nagaraj and Monnier，1992）。然而，目前该推论还未得到试验证实。尽管我们已发现，对于需要补充维生素C的大鼠，膳食维生素C的增加提高了糖血红蛋白水平（Smith *et al.*，1999），但是饲喂8%维生素C的小鼠并未发生白内障（Bensch *et al.*，1985）。

对于GSH，眼部的浓度比全血中的浓度高出数倍，比血浆中的浓度高出几个数量级。与维生素C相同，老年人和白内障患者的晶状体几乎不含GSH（Reddy，1990）。至于半乳糖诱导的白内障，初步结果也表明，维持较高水平的GSH对大鼠有一定的保护作用（Sastre *et al.*，1994）。然而，目前人们还不能确定饲喂GSH是否能提高眼睛GSH的水平（Sastre *et al.*，1994）。在体内很多器官中，GSH和维生素C的水平是相互关联的，但是，给需要补充维生素C的大鼠补足后，我们并未发现此种关系（Smith *et al.*，1999）。

维生素E和类胡萝卜素是脂溶性抗氧化剂（Schalch *et al.*，1999；Yeum *et al.*，1999），其作用可能是维持膜的完整性（Machlin and Bendich，1987）和GSH再生能力（Costagliola *et al.*，1986）。维生素E在整个晶状体中的浓度处于微摩尔级，而在新生组织中的浓度相对来说比较高（Yeum *et al.*，1995）。晶状体与膳食的维生素E水平可能没有相关性（Stephens *et al.*，1988），因为维生素E大部分存在于膜上，从而导致膜上的浓度可能高出几个数量级。除了晶状体中的分布情况（Yeum *et al.*，1999），维生素E和类胡萝卜素水平随年龄变化的情况目前还不清楚。现已报道，维生素E能有效地抑制动物的各种诱发性白内障，包括半乳糖诱导的白内障（Bhuyan *et al.*，1983；Creighton *et al.*，1985；Jacques and Taylor，1991）及氨基三唑诱导的兔白内障（Bhuyan，1984）。

最近人们对类胡萝卜素和眼睛健康之间的关系进行了综述（Schalch *et al.*，1999；Taylor，1999b）。人晶状体中$\beta$-胡萝卜素水平非常有限（Yeum *et al.*，1995，1999），而成年人的晶状体和黄斑中的类胡萝卜素（11～44ng/g湿重）为叶黄素和玉米黄素（Hammond *et al.*，1997；Schalch *et al.*，1999，Snodderly and Hammond，1999），同时，还含有视黄醇、视黄酯（21～50ng/g湿重）和维生素E（1232～2550ng/g湿重）。

晶状体还含有多种抗氧化酶：谷胱甘肽过氧化物酶/还原酶、过氧化氢酶、超氧化物歧化酶和谷胱甘肽氧化还原循环的酶类（Giblin *et al.*，1982；Fridovich，1984；Varma *et al.*，1984；Zigler and Goosey，1984；Rathbun *et al.*，1996）。这些抗氧化酶与各种形式的氧之间发生互作同时与抗氧化剂之间也发生相互作用，如GSH是谷胱甘肽过氧化物酶的作用底物。许多抗氧化酶的活性在发育、衰老和白内障形成过程中会受到抑制（Berman，1991）。

## 蛋白水解酶是保护晶状体的第二道防线

受损蛋白质的积聚具有细胞毒性，许多与年龄相关的神经退化性疾病（包括阿尔茨海默氏病和匹克氏病）中均发现这种细胞毒性，其中，由受损蛋白质的沉淀导致的白内障是最典型的例子。蛋白质降解系统被认为是第二道防线，它们能够清除晶状体及其他组织中的细胞毒素损伤的蛋白质或退化蛋白质（Kosegarten and Maher，1978；Jahngen *et al.*，1986，1990；Taylor and Davies，1987；Eisenhauer *et al.*，1988；Taylor *et al.*，1991b，1993；Jahngen-Hodge *et al.*，1992，1997；Huang *et al.*，1993；Obin *et al.*，1994，1996，1998，1999；Shang and Taylor，1995；Shang *et al.*，1997a，b）。这些蛋白质降解系统存在于新生晶状体组织和晶状体的外周组织，也可能受第一道防线即抗氧化系统的保护。然而，随着年龄的增长或受到氧化应激后，大部分蛋白降解酶的活性降低（Taylor and Davies，1987；Taylor *et al.*，1999b）。研究发现，老年人的晶状体中氧化蛋白质（还有其他方式修饰的蛋白质）发生积聚，这与蛋白质降解酶不能及时清除掉受损的晶状体蛋白质相符，其部分原因是组成第二防御系统的酶类与其他大分子蛋白质一样受到了光氧化损伤（Blondin and Taylor，1987；Taylor *et al.*，1993；Shang and Taylor，1995；Shang *et al.*，1995；Jahngen-Hodge *et al.*，1997；Taylor，1999b）。

一些研究表明，不同的抗氧化剂之间存在相互作用，以便一种抗氧化剂能节约另一种抗氧化剂（图 21.2）。研究证实，在氧化（或光氧化）诱导的蛋白质降解功能损伤方面，维生素 C 和 GSH 之间存在直接的节约效应（Shang and Taylor，1995；Jahngen-Hodge *et al.*，1997；Obin *et al.*，1998）。这些资料还表明，年幼的晶状体具有良好的第一、第二防御功能。然而，抗氧化酶活性、抗氧化剂浓度和第二道防线的年龄相关性损伤可能使晶状体失去对氧化损伤的防御作用。这种防御功能的失去使生命周期长的蛋白质及其他组分更易受到氧化损伤。而受损蛋白质的积聚和沉淀会使晶状体发生混浊。根据现有资料，人们预测增加抗氧化剂的摄入量能增强部分蛋白质降解酶的活动（Blondin and Taylor，1987）。

## 流行病学研究：抗氧化剂与白内障之间的关系

氧化应激是导致白内障发生的部分原因，因为氧化应激既可损伤晶状体蛋白质组分，也能损害蛋白质降解酶。在正常情况下，蛋白质降解酶能清除蛋白质受损部分，若这些酶受到氧化损伤，它们就不能及时清除掉蛋白质受损部分，从而形成白内障（图 21.2；Taylor，2000），所以，人们正致力于研究抗氧化剂是否能降低白内障的发生率。目前人们普遍认为，营养物质的摄入与白内障的发生率有关，且部分营养物质能降低白内障发生率。

下面将综述对一些特殊养分的研究资料。人们认为在晶状体中，代谢和发育上完全不同的区域发生的混浊具有不同的病原学，因此，白内障被划分为后囊下、皮质性和核性白内障。根据营养状况采用各种调查问卷或检测血液营养水平来进行评估。对于某一类型的白内障，我们将依次讨论补充养分、食物（包括补充物质）和血液水平之间的效应。

由于试验设计之间的差异，各研究结果常常很难进行直接的比较。各种试验设计的优缺点已在前面进行了讨论（Christen，1999）。过去，大多数试验采用追溯性的病例-对照或代表性的方法，将白内障患者与晶状体透明者的测定结果进行比较（Mohan *et al.*，1989；Robertson *et al.*，1989；Jacques and Chylack，1991；Leske *et al.*，1991，

1995；意大利-美国白内障研究，1991；Vitale *et al.*，1993；Mares-Perlman *et al.*，2000；Luthra et al.，1997；Cumming et al.，2000；Jacques *et al.*，2001；Taylor *et al.*，2002）。我们对这些追溯到的结果进行解释，而且仅限于把晶状体健康状况和营养水平联系起来进行的评价。优先对白内障进行诊断可能会影响患者的行为，包括饮食和偏好一般食物。

还有一些研究是评价补充一些特殊养分及其补充水平对白内障的作用，并对晶状体正常的人群进行了长达13年之久的跟踪试验（Hankinson *et al.*，1992；Knekt *et al.*，1992；Seddon *et al.*，1994；Mares-Perlman *et al.*，1996；Rouhianinen *et al.*，1996；Leske *et al.*，1998a，b；Brown *et al.*，1999；Chasan-Taber *et al.*，1999a，b；Lyle *et al.*，1999a，b；Christen，2001）。这些前瞻性研究不易产生偏差，因为在结果出来之前就对研究中各个处理组进行了评估。其中，部分研究采用白内障摘除术，或已报道的白内障诊断方法来评价白内障的发生率（Hankinson *et al.*，1992；Knekt *et al.*，1992；Seddon *et al.*，1994；Brown *et al.*，1999；Chasan-Taber *et al.*，1999a，b）。然而，一些研究采用多种潜在易混淆的变量来矫正白内障的发生率，而另一些研究则没有进行矫正。追溯性的和前瞻性的研究资料分别列在不同的图中。

同样，营养物质摄入的评定时间和检测频率也能影响这些结果的准确性，因为白内障的发生常常需要多年的时间。随着时间的推移，一个指标的频繁检测并不能准确评估一般摄入的情况，而多个指标的准确性要比单个的高。

除了试验设计的差异外，不同研究常常采用不同晶状体分类标准，而且对营养物质的高低水平的界定也不一样，研究对象的年龄也存在很大差异。

尽管，目前有关多种营养物质的流行病学已进行了大量的研究（Taylor，1999b）；但是本章只对一些重要的研究资料进行综述。

## 抗坏血酸

由于抗坏血酸的摄入量与眼组织中抗坏血酸水平有关，而且也有研究报道称维生素具有潜在的抑制或促进白内障发生的作用。因此，有关抗坏血酸摄入和白内障发生率的流行病学资料引起人们极大的关注。

在维生素C的摄入与白内障发生率之间的关系方面，人们进行了大量的研究（Mohan *et al.*，1989；Robertson *et al.*，1989；Jacques and Chylack，1991；，1998a；意大利-美国白内障研究，1991；Hankinson *et al.*，1992；Vitale *et al.*，1993；Jacques *et al.*，1994，1997，2001；Mares-Perlman *et al.*，1994，1995a，2000；Brown *et al.*，1999；Lyle *et al.*，1999b；Simon and Hudes，1999；Cumming *et al.*，2000；Taylor *et al.*，2002）。其中多数研究表明，维生素C，尤其是外源维生素C，至少与一种类型的白内障之间存在负相关（图21.3）。

在一项维生素C与白内障的研究中，对摄入高水平维生素C（平均为294mg/d）的165名女性和摄入较低水平（平均为77mg/d）的136名女性进行调查，所得数据用年龄矫正，结果表明，与低水平组相比，补充维生素C 10年以上组的早期晶状体混浊发生率的降低幅度超过70%［相对发生率（relative risk，RR）=0.23，95%的置信区间（confidence interval，CI）=0.09～0.60，图21.3A］，中期晶状体混浊发生率的降低

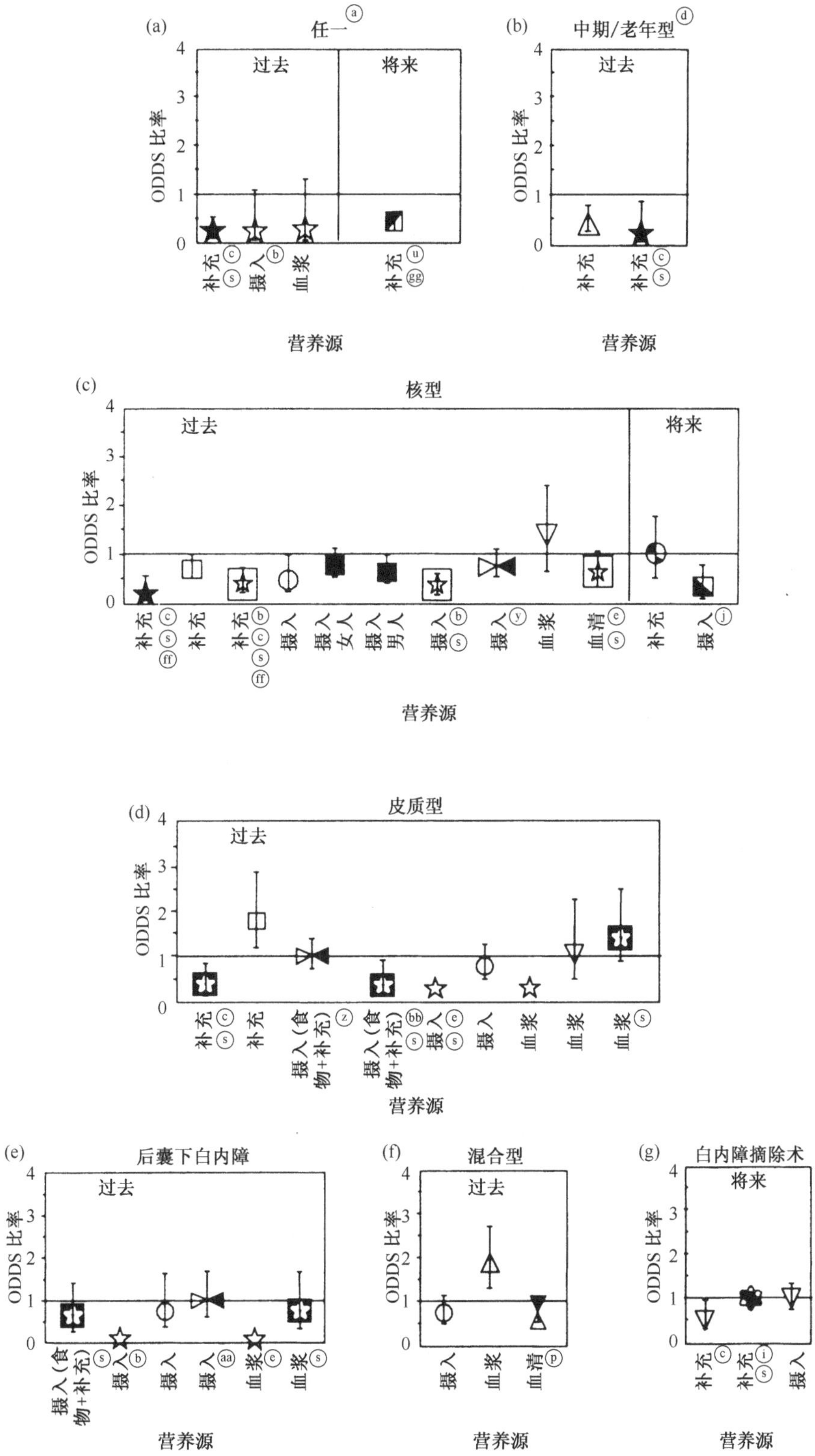
(a)
任一ⓐ
过去
将来
ODDS 比率
补充ⓒⓢ
摄入ⓑ
血浆
补充ⓤⓖⓖ
营养源
(b)
中期/老年型ⓓ
过去
ODDS 比率
补充
补充ⓒⓢ
营养源
(c)
核型
过去
将来
ODDS 比率
补充ⓒⓢⓕⓕ
补充
补充ⓑⓒⓢⓕⓕ
摄入
摄入女人
摄入男人
摄入ⓑⓢ
摄入ⓨ
血浆
血清ⓔⓢ
补充
摄入ⓙ
营养源
(d)
皮质型
过去
ODDS 比率
补充ⓒⓢ
补充
摄入(食物+补充)ⓩ
摄入(食物+补充)ⓑⓑⓢ
摄入ⓔⓢ
摄入
血浆
血浆
血浆ⓢ
营养源
(e)
后囊下白内障
过去
ODDS 比率
摄入(食物+补充)ⓢ
摄入ⓑ
摄入
摄入ⓐⓐ
血浆ⓒ
血浆ⓢ
营养源
(f)
混合型
过去
ODDS 比率
摄入
血浆
血清ⓟ
营养源
(g)
白内障摘除术
将来
ODDS 比率
补充ⓒ
补充ⓘⓢ
摄入
营养源

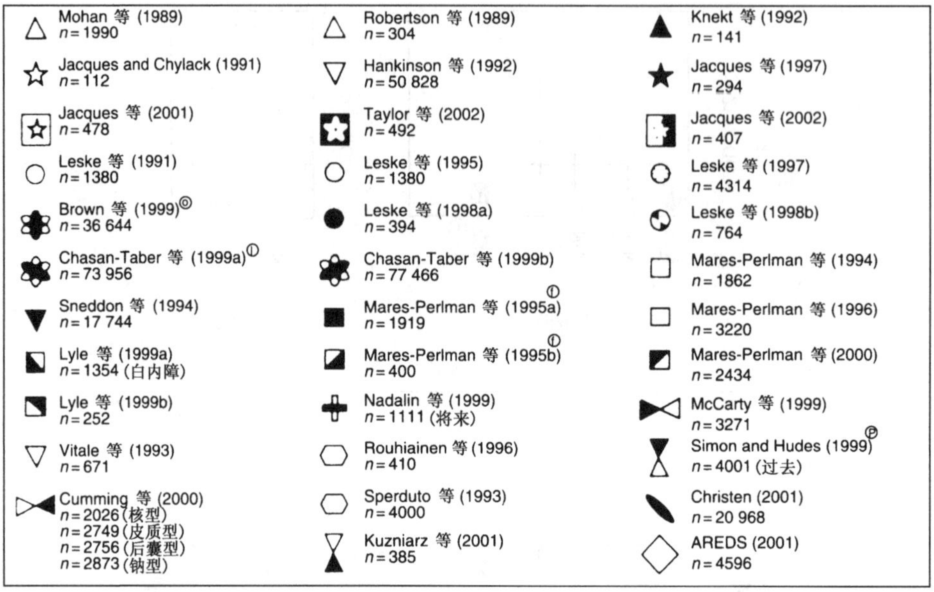


ⓐ"任一"表示 LOCSⅡ级数大于 1 的白内障；ⓑ$P \ll 0.05$；ⓒ补充者使用大于 10 年；ⓓ中期/老年型表示 LOCSⅡ级数大于 2 的白内障；ⓔ$P \ll 0.01$；ⓕ当数据可用于研究对象的比较组时，从最大样本中得到的最新数据可被用于这种情况；ⓖ应用对象年龄小于 70 岁；ⓗ叶黄素/玉米黄质，尤其是椰菜和甘蓝；ⓘ维生素 E、C、A 及多维使用超过 10 年；ⓙ有超过 25 年吸烟史的人（$n=80$）患眼睛混浊的概率，高血压患者（$n=90$）有相同的 OR，对于整个组来说，维生素 C 摄入与核型混浊发生率没有显著相关性；ⓚ通过对饮食评估（1978～1980）超过 65 岁患眼睛混浊的概率（$n=102$），1988～1990 饮食评估差异不显著，吃较多的蛋及菠菜的人的 OR 值降低；ⓛ（$n=143$），对所有组来说，维生素 E 摄入与核型混浊发生率没有显著相关性；ⓜ大于 5 年的眼睛混浊率所需 α-及 γ-生育酚数量；ⓝ5 年里递增的眼睛混浊风险；ⓞ尤其是椰菜和菠菜；ⓟ已报道的白内障在 60～74 岁每 mg/dl 血清抗坏血酸盐中的增加伴随着 26%白内障的降低（CI=0.56～0.97）；ⓠ多风险因素包括 UV-B 暴露；ⓡ每天摄入 5～10mg 与每天摄入小于 5mg 相比；ⓢ女人；ⓣ5 年的干预试验；ⓤ一批研究，补充超过 10 年；ⓥ调整年龄、性别、吸烟、糖尿病、高血压及类固醇使用；ⓦ维生素 A 补充摄入；ⓧ补充 β-胡萝卜素超过 13.2 年；ⓨ$n=2026$，摄入=食物+补充；ⓩ$n=2749$，摄入=食物+补充；ⓐⓐ$n=2756$，摄入=食物+补充；ⓑⓑ年龄小于 60 岁的女性；ⓒⓒ从不吸烟的女性，摄入=食物+补充；ⓓⓓ调整年龄、性别、能量、教育、吸烟、糖尿病、高血压及类固醇的应用；ⓔⓔ年龄及性别的调整；ⓕⓕ维生素 E 及多维添加剂使用的矫正；ⓖⓖ调整少年时期年龄、性别、吸烟、糖尿病、高血压、BMI、UV-B 暴露及营养使用；ⓗⓗ调整年龄、乙醇摄入、吸烟、BMI；ⓘⓘ调整年龄、性别、高血压、糖尿病、教育及口服或吸入类固醇的使用；ⓙⓙ60 像素密度单位混浊（提供手稿）；ⓚⓚ AREDS 使用的日补充物中包括 500mg 维生素 C、400IU 维生素 E、15mg β-胡萝卜素、80mg 氧化锌、2mg 氧化铜，平均间隔是 6.3 年，在曲线基线上的改变：核混浊度（以 0.9～6.1U 为数值范围增加 1.5U），皮质型混浊度（绝对增加混浊度面积 10%，此面积为一个标准的半径 5mm 圆环），PSC（绝对增加混浊度面积 5%，此面积为一个标准的半径 5mm 圆环）

图 21.3 （上页和上图）白内障的发生率、高摄入量与低摄入量（补充或不补充）的维生素 C 和血浆维生素 C 水平

白内障的类型：任一、中期/老年型、核型、皮质型、后囊下、混合型和白内障摘除术。追溯性和前瞻性的研究资料分开列出。摘自 Taylor 等《维生素 C 在健康与疾病中的作用》（纽约，Marcel Dekker，1997）

幅度超过 80%（RR=0.17，CI=0.03～0.87；图 21.3B）（Jacques *et al.*，1994）。

“营养与视力项目”基础阶段的研究进一步证实了该结果。其结果表明，根据晶状体混浊度分类系统Ⅲ（LOCSⅢ），补充维生素C超过10年的人，早期核性混浊的RR降低了64%（RR=0.36；CI=0.18～0.72；全面矫正后的结果，包括其他养分的矫正；图21.3C）（Jacques *et al.*，2001）。虽然，目前还不能确定皮质性和核性白内障是否具有相同的病原学，但是，对于年龄不到60岁的女性，补充维生素C至少10年使皮质性白内障的发生率降低60%［危险系数比（odds ratio，OR）=0.40，CI=0.18～0.87，图21.3D］（Taylor *et al.*，2002）。Mares-Perlman等（2000）研究也发现，补充维生素C超过10年者与未补充维生素C者相比，任何类型白内障的5年发生率均降低了60%。对血液指标的研究结果也表明，维生素C的营养状况与核性混浊的发生率之间存在负相关（Jacques *et al.*，2001）。与上面的研究结果类似，Hankinson等（1992）发现补充维生素C超过10年的女性白内障手术比率降低45%（RR=0.55，CI=0.32～0.96，图21.3G）。然而，对9种潜在的影响因子（包括年龄、糖尿病、吸烟和能量摄入等）进行矫正后，发现维生素C与白内障手术的比率之间没有相关性，而且，Chasan-Taber等（1999a）采用同一批人群进行试验，也发现延长维生素C的补充时间对白内障摘除术的危险性没有正面影响（图21.3G）。

与上面的研究结果不同，Mares-Perlman等（1994）研究发现，对年龄、性别、吸烟和过量饮酒的历史进行矫正后，补充维生素C降低了核性白内障的发生率（RR=0.7，CI=0.5～1.0，图21.3C），但却提高了皮质性白内障的发生率（矫正后的RR=1.8，CI=1.2～2.9）（图21.3D）。

其他研究也常常证实维生素C与白内障之间的负相关关系。Robertson等（1989）比较了视力受损的白内障患者与年龄和性别配对的、视力正常的未患白内障或轻度混浊的人，结果表明，每天补充超过300mg维生素C的人白内障的发生率约为未补充组的1/3（RR=0.30，CI=0.24～0.77，图21.3B），而且其他一些研究也发现，提高食物中维生素C含量对降低白内障发生率有利。Leske等（1991）发现，对年龄和性别进行矫正后，受试人群中维生素C摄入量最高的20%人群的核性白内障发生率比摄入量最低的20%人群要低52%（RR=0.48，CI=0.24～0.99，图21.3C），但是维生素C对其他类型的白内障的影响极其微小（图21.3f）。Jacques和Chylack（1991）也发现，维生素C摄入量高（大于490mg/d）的人群的白内障发生率是摄入量低（小于125mg/d)的人群的25%（RR=0.25，CI=0.06～1.09，图21.3A）。在“营养与视力项目”的基础阶段对早期晶状体混浊的研究中，我们发现，提高维生素C的摄入量使整个试验人群发生核性混浊的危险性降低了69%（RR=0.31，CI=0.16～0.58，图21.3C），使不到60岁的人群发生皮质性混浊的危险性降低了57%（RR=0.43，CI=0.2～0.93，图21.3D）（Taylor *et al.*，2002）。人们发现，每天摄入200～360mg维生素C可能最为有效（Taylor *et al.*，1991a；Taylor and Jacques，1997；Jacques *et al.*，2001），但是值得注意的是参照人群的维生素C摄入量已经是膳食推荐允许量的2倍。在海狸坝研究组（Beaver Dam Group）最近进行的一项前瞻性研究中，Lyle等（1999a）发现，摄入维生素C使严重吸烟者和患高血压者发生核性白内障的危险性降低了70%（RR=0.3，CI=0.1～0.8），但是对于整个人群来说，维生素C与核性混浊之间的关联并不显著（图 21.3C）。

然而，Vitale 等（1993）发现，维生素 C 高摄入量（大于 261mg/d）与低摄入量（小于115mg/d）的人群的白内障发生率几乎没有差异。意大利-美国白内障研究组（1991）也发现白内障发生率与维生素 C 摄入之间不存在显著关系。而且，在一项大型的前瞻性研究中，维生素 C 高摄入量（中值＝ 705mg/d）与低摄入量（中值＝ 70 mg/d）的女性在白内障摘除术的危险性方面没有显著的相关性（RR＝0.98，CI＝0.72～1.32，图 21.3G）（Hankinson *et al.*，1992）。

有关白内障发生的危险与血浆维生素 C 水平之间的关系，目前的报道不一。Jacques 和 Chylack（1991）发现，含血浆维生素 C 水平高（大于 90μmol/L）的人群的早期白内障发生率只有含血浆维生素 C 水平低（小于 40μmol/L）的人群的不足 1/3，但利用年龄、性别、种族和糖尿病病史进行矫正后，没有显著差异(RR＝0.29，CI＝0.06～1.32)（图 21.3A）。我们在“营养与视力项目”基础阶段的研究中也证实了该结果（图 21.3C）（Jacques *et al.*，2001）。但 Mohan 等（1989）发现，就标准差而言，随着血浆维生素 C 水平的提高，混合型白内障（后囊下和核型）的发生率提高了 87%（RR＝1.87，CI＝1.29～2.69，图 21.3F）。Vitale 等（1993）也发现，对年龄、性别和糖尿病进行矫正后，血浆维生素 C 水平超过 80μmol/L 与低于 60μmol/L 的人群的核性（RR＝1.31，CI＝0.61～2.39）和皮质性（RR＝1.01，CI＝0.45～2.26）白内障的发生率没有显著差异（图 21.3C 和 D）。

Sperduto 等（1993）的干预研究也表明，补充维生素 C 对试验人群没有正面影响。另一个大型（*n*＝4596）、加倍掩饰的、随机性的、安慰剂对照的干预研究即年龄相关性眼病研究（age-related eye disese study，AREDS）表明，补充维生素 E、β-胡萝卜素、锌和 500mg 维生素 C 6.3 年后，也没有产生任何效果（年龄相关性眼病研究专题组，2001）（图 21.4）。

然而，罗氏欧美白内障试验（Roche Europen American Cataract Trial，REACT）（*n*＝158）利用维生素 C（750mg/d）、维生素 E 和 β-胡萝卜素进行加倍掩饰的、随机性的、安慰剂对照的研究，3 年后，多种维生素在一定程度上能延缓 40 岁以上的美国人的早期浑浊的发展，但是对于营养状况更差的英国人来说，并没有出现类似情况（Chylack *et al.*，2002）（表 21.1）。

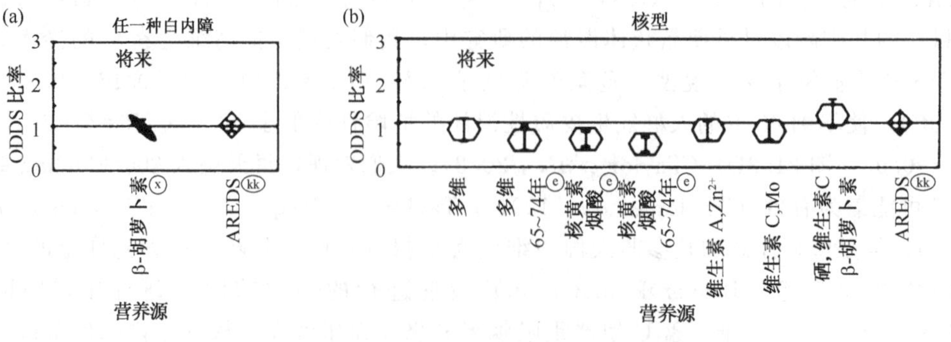

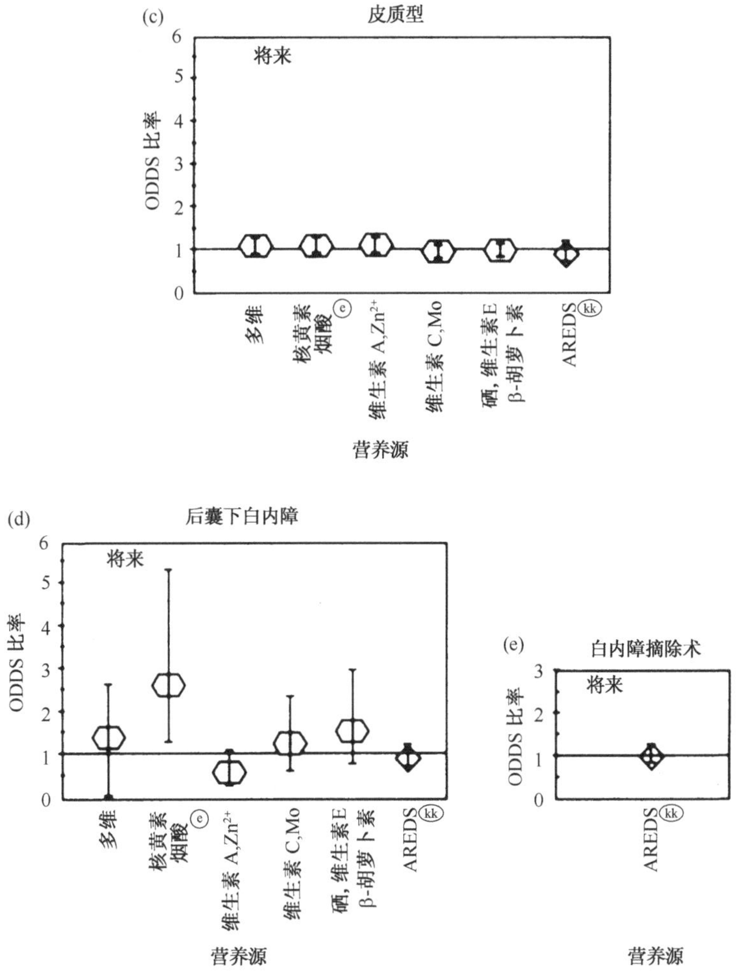


图 21.4　白内障的发生率、高低摄入量（补充或不补充）的维生素 C 与干预研究

白内障的类型：任一、核型、皮质型、后囊下和白内障摘除术。摘自 Taylor 等《维生素 C 在健康与疾病中的作用》（纽约，Marcel Dekker，1997）

**表 21.1　未采用固定终点的研究结果**

| 研究 | 设计 | 变化类型 | 变化 | | | *P* |
|---|---|---|---|---|---|---|
| | | | 不补充 | 补充 | 差值（不补充-补充） | |
| REACT[a] Chylack 等(2001) $n=158$ | 安慰剂-对照干预试验<br>每天补充：<br>18mg $\beta$-胡萝卜素<br>750mg 维生素 C<br>600mg 维生素 E | 超过 3 年的像素不透明度的变化 | 安慰剂组 3.3％像素不透明度 | 安慰剂组 1.7％像素不透明度 | 1.6％ | 0.05 |
| NVP Jacques 等（2002）$n=407$ | 补充维生素 E<br>补充时间≥10 年 | 超过 5 年的像素不透明度的变化 | 17 | 12 | 5 | 0.004 |

a 加倍掩饰的、随机性的、安慰剂对照的研究；研究对象的年龄超过 40 岁，基本上有一只眼患有早期白内障，英国人的白内障比美国人严重；有 2/3 英国人血浆维生素 C、维生素 E、番茄红素较低，有 1/3 英国人玉米黄质较低；英国有 23％人吸烟；美国有 15％人吸烟；3 年后美国像素不透明性差异为 0.0014，英国差异不显著。

# 维 生 素 E

经常补充维生素 E 比仅靠食物摄取能为机体提供更多的维生素 E。在维生素 E 的补充、膳食摄入和血浆水平与各种类型的白内障发生率之间的关系方面，人们进行了大量研究，但是各个研究报道不一。一些前瞻性研究表明，补充维生素 E 与白内障发生率之间存在负相关（图 21.5）。Robertson 等（1989）发现，对于年龄和性别配对的病例与对照人群，补充维生素 E（大于 400IU/d）的人，老年性白内障的发生率比未补充组的降低 56%（RR=44，CI=0.24～0.77，图 21.5b）。Jacques 和 Chylack（未发表的资料）也发现，经年龄、性别、种族和糖尿病矫正后的服用者，补充维生素 E 使白内障的发生率降低了 67%（RR=33，CI=0.12～0.96，图 21.5A）。在“营养与视力项目”基础研究中，我们发现，将维生素 E 的摄入超过 90mg/d 者与低于 6.7mg/d 者相比，女性核性白内障发生率降低了 55%（RR=0.45，CI=0.23～0.86），而对于补充时间超过 10 年的人群，其核性白内障发生率降低了 51%（RR=0.49，CI=0.22～1.09，图 21.5C）（Jacques *et al.*，2001）。对维生素 C 补充组的结果进行矫正后，其显著性降低。海狸坝研究组对维生素 E 的补充效果进行了一些研究。Mares-Perlman 等（1994）发现，补充维生素 E 与核性白内障之间只存在微弱的、不显著的差异（RR=0.9，CI=0.6～1.5，图 21.5C）。在相同的研究组中，Lyle 等（1999a）进行的一项前瞻性研究也发现，维生素 E 的摄入与核性白内障之间不存在显著的负相关（RR=0.5，CI=0.3～1.1，图 21.5C）。然而，Mares-Perlman 等（2000）最近的研究结果表明，补充维生素 E 使任何一种白内障发生率降低了 60%（RR=0.4，CI=0.3～0.6，图 21.5A）。在白内障的跟踪（纵向）研究中，维生素 E 的补充也降低了核性混浊的进一步发展（RR=0.43，CI=0.19～0.99，图 21.5C）（Leske *et al.*，1998a）。

与核性白内障的研究结果相反，Mares-Perlman 等（1994）研究发现，虽然差异不显著，但补充维生素 E 有提高皮质性白内障的发生率的趋势（RR=1.2，CI=0.6～2.3，图 21.5D），而且在蓝山组（Blue Mountain Group）工作的 McCarty 等（1999）发现，补充维生素 E 对核性白内障没有影响（图 21.5C），但提高后囊下白内障的发生率（RR=1.47，CI=1.04～2.09，图 21.5E）。在 Nadalin 等（1999）的前瞻性研究中，补充维生素 E 使皮质性白内障的发生率降低了 53%(RR=0.47，CI=0.28～0.83，图 21.5D)，但并没有降低核性白内障的发生率（图 21.5C）。

在国家健康与营养调查研究（National Health and Nutrition Examination Survey，NHANES）中，Simon 和 Hudes（1999）发现，补充维生素 E 与白内障之间不存在显著的相关性（RR=0.93，CI=0.52～1.67，图 21.5B），且 Chasan-Taber 等（1999a）研究也发现，即使补充维生素 E 10 年，也没有对白内障的摘除术产生任何影响（RR=0.99，CI=0.74～1.32，图 21.5G）。

Leske 等（1991）发现，年龄和性别矫正后，维生素 E 的摄入量最高的 20%人群的皮质性白内障（RR=0.59，CI=0.36～0.97，图 21.5D）和混合型白内障（RR=0.58，CI=0.37～0.93，图 21.5F）发生率比最低摄入量的 20%人群约低 40%。Jacques 和 Chylack（1991）发现，维生素 E 的总摄入量（食物本身含有的和补充的总和）与白内障的发生率之间不存在显著的负相关。每天摄入超过 35.7mg 维生素 E 的人早期

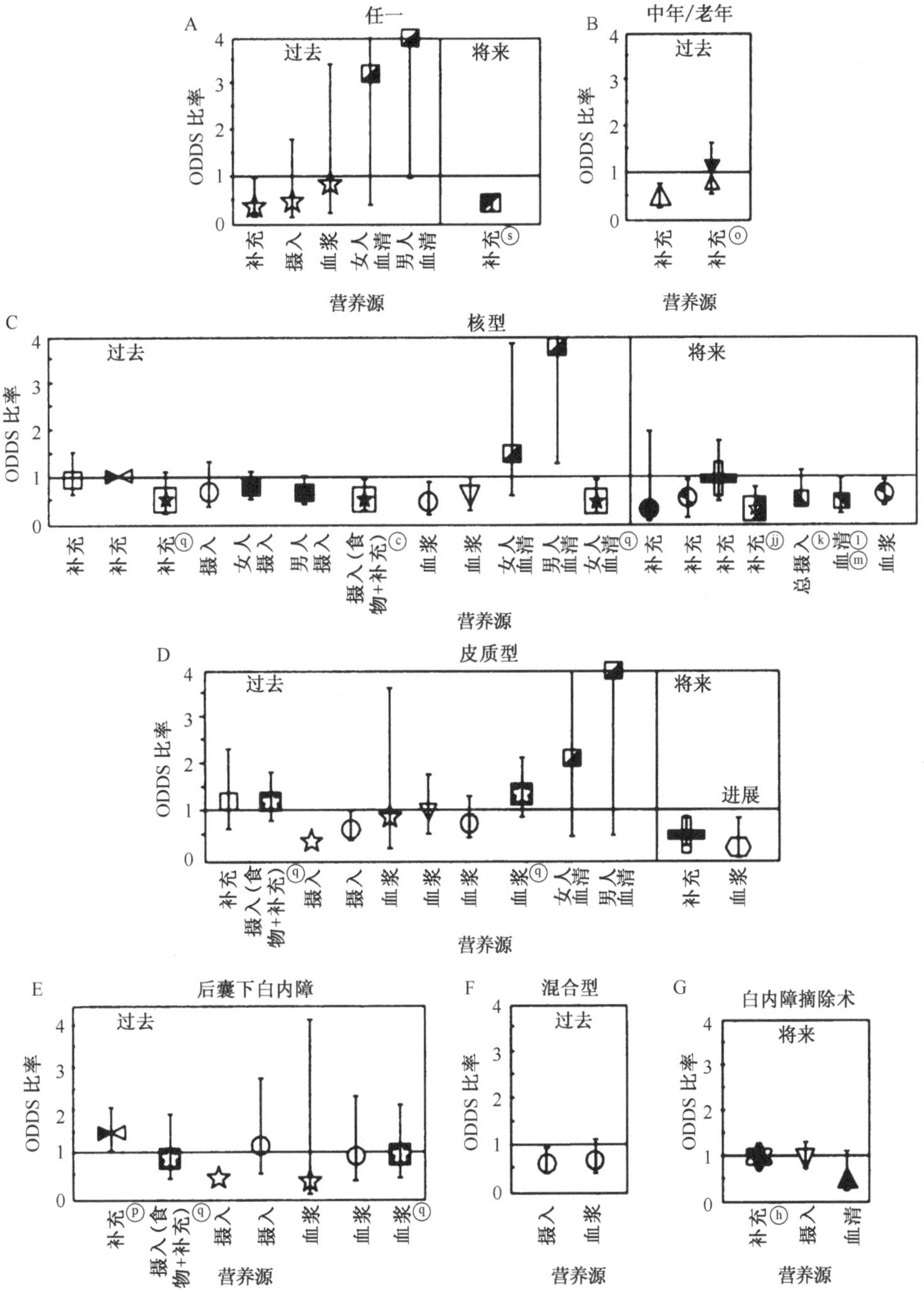


图 21.5　白内障的发生率、高低摄入量（补充或不补充）的维生素 E 和血浆维生素 E 水平（α-生育酚）

白内障的类型：任一、中期/老年型、核型、皮质型、后囊下、混合型和白内障摘除术。追溯性和前瞻性的研究资料分开列出。摘自 Taylor 等《维生素 C 在健康与疾病中的作用》（纽约，Marcel Dekker，1997）

白内障的发生率比每天摄入低于 8.4mg 维生素 E 的低 55%（RR＝0.45，CI＝0.12～

0.79，图 21.5A）(Jacques and Chylack，1991)。他们还发现摄入高水平的维生素 E 显著降低了皮质性白内障的发生率（图 21.5D，RR＝0.37，$P<0.1$）。

然而，Hankinson 等（1992）发现，维生素 E 的摄入对白内障手术无影响（图 21.5G）。维生素 E 的高摄入量组（中值＝210mg/d）与低摄入量组（中值＝3.3mg/d）的女性白内障手术率基本上相同（RR＝0.96，CI＝0.72～1.29）。该试验结果与 Chasan-Taber 等（1999a）的研究结果一致。但是 Mares-Perlman 等（1995a）研究发现，膳食维生素 E 降低（不显著）男性核性白内障的发生率，但是对女性没有此作用（图 21.5C），而且血清 γ-生育酚水平与白内障之间存在正相关关系（图 21.5A，C，D）。

Lyle 等（1999a，b）的前瞻性研究发现，血清维生素 E 水平的提高与白内障发病率的降低有关（图 21.5C，RR＝0.4，CI＝0.2～0.9）。其他一些研究也发现，血浆维生素 E 水平与白内障发生率之间存在显著的负相关（图 21.5）。在“营养与视力项目”基础阶段的研究中，我们发现，41μmol/L 与 23μmol/L 的女性血清维生素 E 水平相比，其白内障发生率降低了 52%（RR＝0.48，CI＝0.25～0.95）（图 21.5C）(Jacques *et al.*，2001)。Knekt 等（1992）对 1419 名芬兰人进行了 15 年的跟踪研究，发现 47 名白内障患者并转入眼科病房。每个患者选择两个年龄、性别和自治地区配对的人作为对照。研究结果表明，当血清维生素 E 浓度超过约 20μmol/L 时，与该浓度以下患者相比白内障的手术降低约一半（RR＝0.53，CI＝0.24～1.1，图 21.5G）。Vitale 等（1993）的追溯性研究发现，与血浆维生素 E 低于 18.6μmol/L 组相比，对年龄、性别和糖尿病进行矫正后，血浆维生素 E 超过 29.7μmol/L 组的核性白内障的发生率也降低了约一半（RR＝0.52，CI＝0.27～0.99，图 21.5C），但是对于皮质性白内障的发生率，血浆维生素 E 水平高组和低组几乎相同（RR＝0.96，CI＝0.52～1.78，图 21.5D）。Jacques 和 Chylack（1991）研究发现，对年龄、性别、种族和糖尿病矫正后，血浆维生素 E 水平超过 35μmol/L 的人群后囊下白内障的发生率比血浆维生素 E 水平低于 21μmol/L 组的低 67%（RR＝0.33，CI＝0.03～4.13，图 21.5E），但是该差异并无统计上的显著意义。对于高血浆维生素 E 水平和低血浆维生素 E 水平的人群，早期白内障（RR＝0.83，CI＝0.20～3.40，图 21.5A）或皮质性白内障（RR＝0.84，CI＝0.20～3.60，图 21.5D）的发生率之间没有显著差异。Rouhiainen 等（1996）的前瞻性研究发现，血浆维生素 E 水平较高的人群皮质性白内障的发生率降低了 73%（RR＝0.27，CI＝0.08～0.83，图 21.5D）。在意大利进行的一项大型研究发现，对年龄和性别矫正后，血浆维生素 E 水平也与白内障发生率呈负相关，但是对教育、光处理和白内障家族史等其他因素矫正后，它们之间的关系没有统计上的显著意义（意大利-美国白内障研究，1991）。Leske 等（1995）表明，血浆维生素 E 水平高的人群核性白内障发生率显著降低（RR＝0.44，CI＝0.21～0.90），但是维生素 E 与晶状体其他位点的白内障没有相关性（图 21.5C）。Leske 等（1998a）的前瞻性研究发现，血浆维生素 E 水平高的人群核性白内障的发生率降低了 42%（RR＝0.58，CI＝0.36～0.94，图 21.5C）。

但 Mares-Perlman 等（1995b）通过对 Lyle 等（1996b）研究相似的人群进行研究发现，高水平的血浆维生素 E 显著提高了男性核性白内障的发生率（RR＝3.74，CI＝1.25～11.2，图 21.5C），而且女性核性白内障的发生率也显著提高（RR＝1.47，CI＝0.57～3.82，图 21.5C）。而 Mohan 等（1989）未发现血浆维生素 E 与白内障发生率之间的相关性。

随着对 $\gamma$-生育酚抗氧化功能的认识不断深入，人们也对 $\gamma$-生育酚与白内障发生率之间的关系进行了一些研究。$\gamma$-生育酚的生物学活性比 $\alpha$-生育酚的低。Mares-Perlman 等（1995b）发现，血清 $\gamma$-生育酚的水平与核性硬化症的严重性之间存在负相关，但不显著（RR＝0.61，CI＝0.32～1.19），而提高血清 $\gamma$-生育酚的水平与男性和女性的皮质性白内障的严重性之间存在显著的正相关。

我们最近在“营养与视力项目”中对晶状体浑浊的发展进行了研究，其结果表明，维生素 E 补充超过 10 年，可以延缓核性白内障的发展（图 21.1）（P. F. Jacques *et al*.，未发表数据）。

分别补充 600mg（Chylack *et al*.，2002）和 400IU（年龄相关性眼病研究小组，2001）维生素 E 的两项掩饰性干预研究得到的结果差异较大。虽然 REACT 研究组的部分人员发现早期白内障的发展受到抑制，但在 AREDS 研究中，补充多种维生素 6.3 年对各种类型的混浊发病率没有影响（年龄相关性眼病研究小组，2001）。

## 类胡萝卜素、维生素 A 及其他营养成分

与维生素 E 一样，类胡萝卜素也是天然的脂溶性抗氧化剂（Machlin and Bendich，1987；Shang *et al*.，1997a；Curran Celentano *et al*.，2002；Krinsky，2002）。$\beta$-胡萝卜素是一种最常见的类胡萝卜素，其功能是作为维生素 A 的前体物质。晶状体中 $\beta$-胡萝卜素的含量非常低（Yeum *et al*.，1999）。除了 $\beta$-胡萝卜素外，$\alpha$-胡萝卜素、叶黄素、玉米黄质和番茄红素也是人类食物中重要的类胡萝卜素，这些物质在晶状体中的含量约为 10ng/g 湿重（Daicker *et al*.，1987；Yeum *et al*.，1995，1999）。有关补充类胡萝卜素对白内障发生率的影响的研究资料越来越多，而且多种“眼用维生素”制剂均含有类胡萝卜素。

Jacques 和 Chylack（1991）首先发现，每天的胡萝卜素摄入量超过 18 700IU 的人群与低于 5677IU 的人群具有相同的白内障发生率（RR＝0.91，CI＝0.23～3.78，图 21.6A）。随后，Hankinson 等（1992）也发现，摄入高剂量的胡萝卜素（中值＝14558 IU/d）的女性与摄入低剂量的胡萝卜素（中值＝2935IU/d，图 21.6e）的女性相比，多变量矫正后的白内障手术比率约低 30%（RR＝0.73，CI＝0.55～0.97）。“营养与视力项目”利用相同数量的人群来进行研究，结果表明，每天摄入 6.6mg $\beta$-胡萝卜素的女性核性白内障发生率比每天摄入 3.0mg 的低 48%（RR＝0.52，CI＝0.28～0.97，图 21.7B）（Jacques *et al*.，2001）。该试验还表明，叶酸（RR＝0.44，CI＝0.24～0.81）和核黄素（RR＝0.37，CI＝0.19～0.73，图 21.8A）也降低了核性白内障的 RR。对于未吸过烟的女性，我们发现后囊下白内障的 RR 降低情况：类胡萝卜素的总摄入水平 24mg/d 组比 12mg/d 组低 81%（RR＝0.19，CI＝0.08～0.68，图 21.6D）、$\alpha$-胡萝卜素的 1.2 mg/d 组比 0.44mg/d 组低 71%（RR＝0.29，CI＝0.08～1.05，不显著，图 21.9D）、$\beta$-胡萝卜素的 6.6 mg/d 组比 3.0mg/d 组低 72%（RR＝0.28，CI＝0.08～0.96，图 21.7D）（Taylor *et al*.，2002）。他们还发现，提高叶酸摄入量也降低了后囊下白内障的 RR：每天摄入量超过 548μg 组比低于 284μg 组低 74%（RR＝0.26，CI＝0.09～0.77）。但是试验结果被矫正后，只有总的类胡萝卜素摄入水平对后囊下白内障的影响显著（Taylor *et al*.，2002）。Mares-Perlman 等（1996）也发现，补充 $\beta$-胡萝卜

素使核性白内障的 RR 降低。

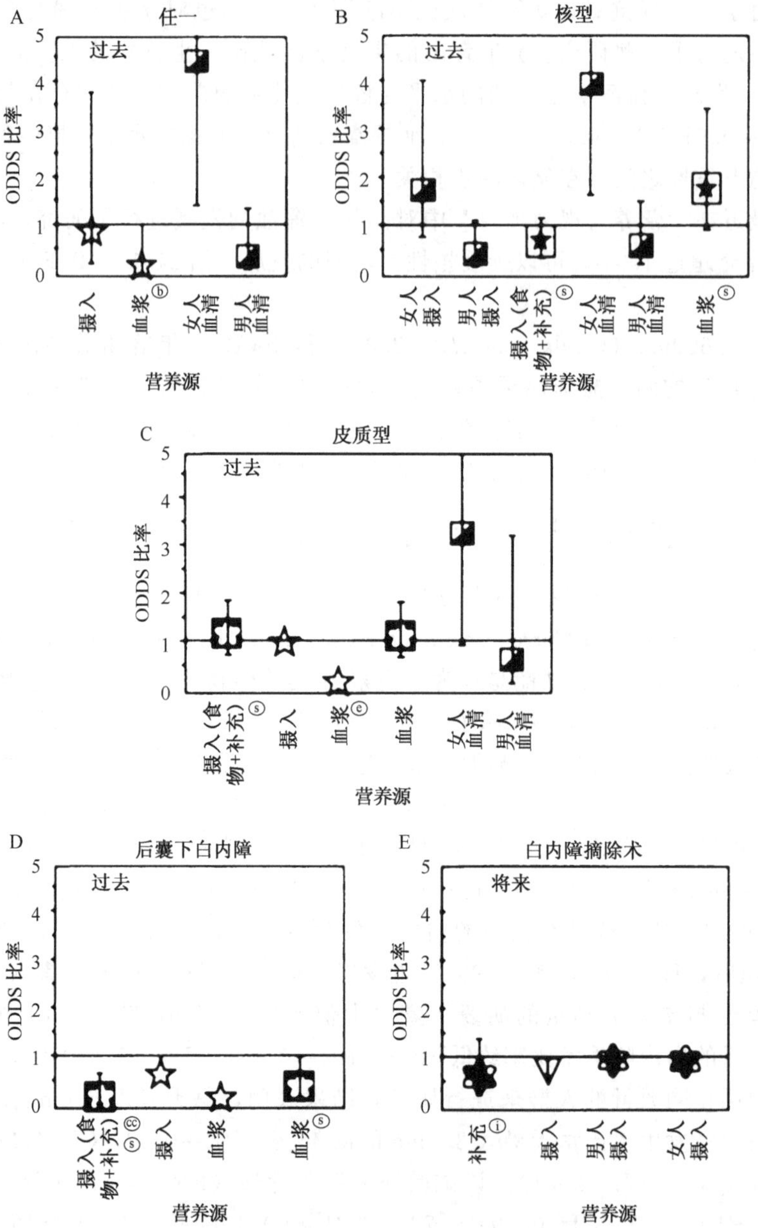


图 21.6　白内障的发生率、高低摄入量（补充或不补充）的类胡萝卜素和血浆类胡萝卜素水平

白内障的类型：任一、中期/老年型、核型、皮质型、后囊下、混合型和白内障摘除术。追溯性和前瞻性的研究资料分开列出。摘自 Taylor 等《维生素 C 在健康与疾病中的作用》(纽约，Marcel Dekker，1997)

另有试验发现，补充 $\beta$-胡萝卜素（Chasan-Taber *et al.*，1999a，图 21.6E）和提高总的类胡萝卜素摄入量均轻度降低了女性白内障摘除术的发生率（RR＝0.85，CI＝0.7～1.03）

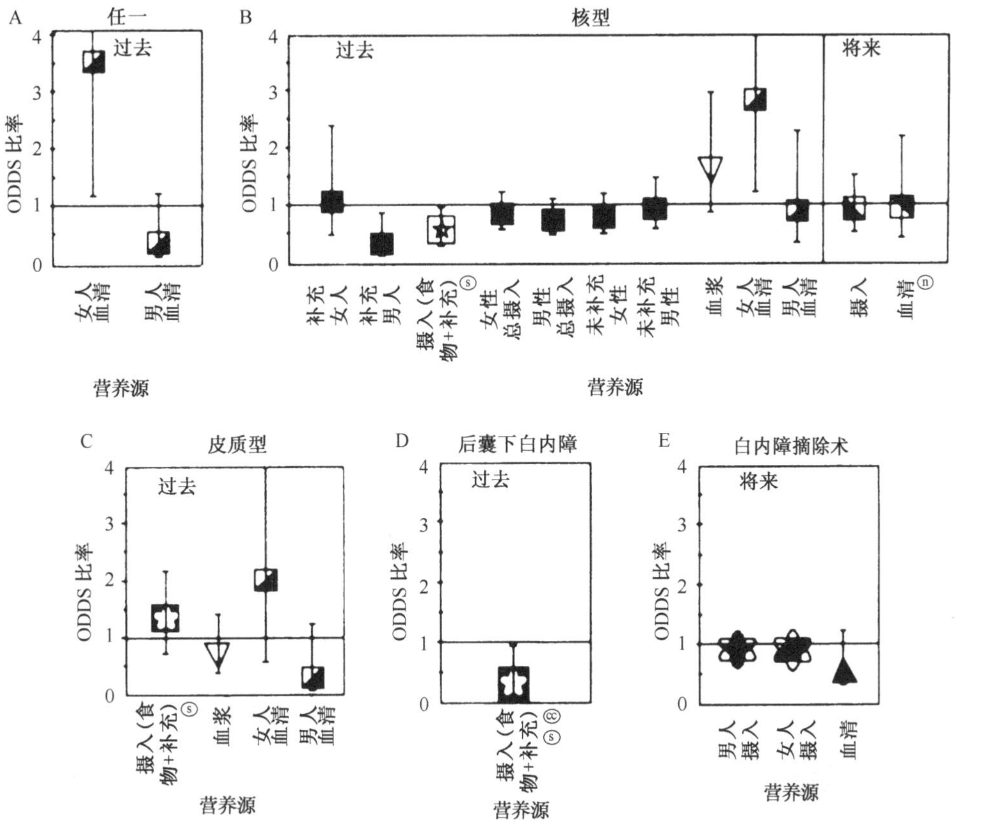


图 21.7 白内障的发生率、高低摄入量（补充或不补充）的 β-胡萝卜素和血浆 β-胡萝卜素水平

白内障的类型：任一、中期/老年型、核型、皮质型、后囊下、混合型和白内障摘除术。追溯性和前瞻性的研究资料分开列出。摘自 Taylor 等《维生素 C 在健康与疾病中的作用》（纽约，Marcel Dekker，1997）

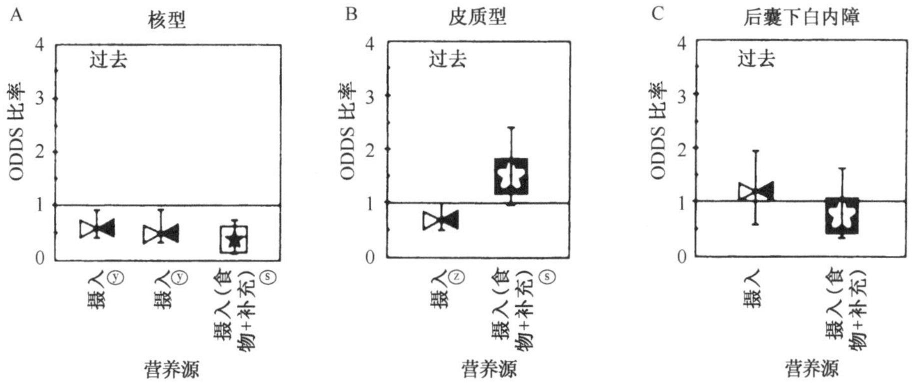


图 21.8 白内障的发生率、高低摄入量（补充或不补充）的核黄素和血浆核黄素水平

白内障的类型：核型、皮质型和后囊下。摘自 Taylor 等《维生素 C 在健康与疾病中的作用》（纽约，Marcel Dekker，1997）

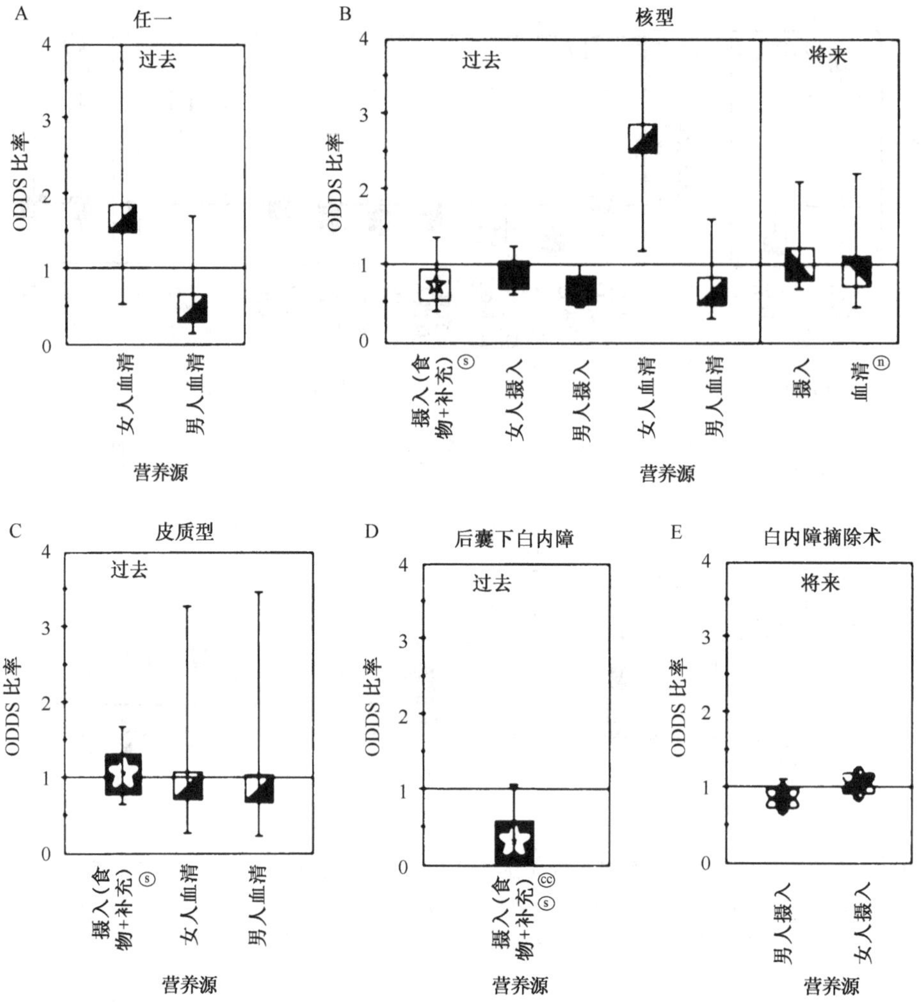


图 21.9　白内障的发生率、高低摄入量（补充或不补充）的 α-胡萝卜素和血浆 α-胡萝卜素水平

白内障的类型：任一、核型、皮质型、后囊下和白内障摘除术。追溯性和前瞻性的研究资料分开列出。摘自 Taylor 等《维生素 C 在健康与疾病中的作用》(纽约，Marcel Dekker，1997)

(Chasan-Taber *et al.*，1999b)。同样，Brown 等（1999）研究也表明，提高类胡萝卜素的摄入使男性白内障摘除术发生率也有所下降（RR＝0.85，CI＝0.68～1.07，图 21.6E)。然而这些试验表明，白内障手术与胡萝卜素的总摄入量之间存在负相关，但是与富含胡萝卜素的食物（如胡萝卜）不存在较高的相关性。另外，白内障手术与某些食物（如菠菜，它富含叶黄素和茜草黄质，而 β-胡萝卜素含量低）的低摄入量有关。该结果可能与本文作者发现人晶状体含有叶黄素和玉米黄质但不含 β-胡萝卜素，以及 Christen（2001）发现补充 β-胡萝卜素没有效果的结果相吻合（见下面）。而在其他一些试验中，白内障手术不是研究叶黄素（xanthophyll）的判定指标（Mares-Perlman *et al.*，1994，1995b)。Mares-Perlman 等（1995a）发现特殊类胡萝卜素与白内障发生之间没有显著相关性。

Chasan-Taber 等（1999b）的前瞻性研究也发现，α-胡萝卜素（RR＝0.95，CI＝

0.7～2.1，图 21.9B）和β-胡萝卜素（RR＝0.9，CI＝0.5～1.4，图 21.7B）与白内障的发生率之间没有显著关系。α-胡萝卜素、β-胡萝卜素（图 21.7E）、番茄红素和β-隐黄质（β-cryptoxanthin）与白内障摘除术发生的危险性都不相关。对于男性，α-胡萝卜素（RR＝0.89，CI＝0.72～1.1，图 21.9E）、β-胡萝卜素（RR＝0.92，CI＝0.73～1.16，图 21.7E）、番茄红素（RR＝1.10，CI＝0.88～1.36）或β-隐黄质（RR＝1.09，CI＝0.87～1.37）与白内障摘除术发生的危险性均没有显著的相关性（Brown *et al.*，1999），而 Mares-Perlman 等（1994，1995b）也没发现这些养分对白内障发生的危险性具有显著影响。

血浆类胡萝卜素水平也与白内障的发生率有关。Jacques 和 Chylack（1991）发现，利用年龄、性别、种族和糖尿病矫正后，血浆总的类胡萝卜素含量高（3.3μmol/L）的人群具有的白内障发病率只有血浆类胡萝卜素总含量低于 1.7μmol/L 人群的不足 1/5（RR＝0.18，CI＝0.03～1.03，图 21.6A），但是，他们并没有发现胡萝卜素的摄入量与白内障的发生率之间的相关性。Knekt 等（1992）报道，对于年龄和性别配对的病例和对照人群，血清β-胡萝卜素浓度超过 0.1μmol/L 的人群的白内障手术率比血清β-胡萝卜素浓度低于 0.1μmol/L 组的约低 40%（RR＝0.59，CI＝0.26～1.25，图 21.7E）。

Mares-Perlman 等（1995b）研究了血清类胡萝卜素水平与核性和皮质性混浊之间的关系，发现血清单种类胡萝卜素和总类胡萝卜素的浓度提高并不减少核性或皮质性白内障的发生率（图 21.6、图 21.7 和图 21.9）。有些类型的白内障发生率与营养状况（nutriture）之间的关系随男性和女性的不同而差异较大，如核性白内障与α-胡萝卜素的摄入之间的关系就是如此（Mares-Perlman *et al.*，1995a）。血清β-胡萝卜素（图 21.7a，b）和血清番茄红素等其他养分与白内障发生率之间的关系因性别而完全相反。对于男性，血清β-胡萝卜素水平的提高使皮质性混浊的危险系数降低，但是对于女性，并没有出现此现象。血清α-胡萝卜素、β-隐黄质和叶黄素水平的提高与核性硬化症的降低之间的关系只对吸烟的男性才显著相关。相反，部分类胡萝卜素水平的提高往往与核性硬化症和皮质性白内障的发生率提高直接相关（图 21.6B，C，21.7B，C 和 21.9B），对于女性更是如此。但是 Lyle 等（1999b）发现，血清α-胡萝卜素（RR＝0.9，CI＝0.4～2.2，图 21.9B）、β-胡萝卜素（RR＝0.9，CI＝0.4～2.2，图 21.7B）、番茄红素（RR＝1.1，CI＝0.5～2.6）、β-隐黄质（RR＝0.7，CI＝0.3～1.6）和叶黄素（RR＝0.7，CI＝0.3～1.6，图 21.10B）对白内障的发生率几乎没有影响。

Vitale 等（1993）也研究了血浆β-胡萝卜素水平与年龄、性别和糖尿病矫正后的皮质性和核性白内障发生率之间的关系（图 21.7B，C）。尽管资料表明血浆β-胡萝卜素水平与皮质性白内障之间存在弱负相关，而与核性白内障之间存在弱正相关，但它们均没有达到统计上的显著水平。血浆β-胡萝卜素水平超过 0.88μmol/L 的人群与低于 0.33μmol/L 的人群相比，其皮质性白内障的发生率降低了 28%（RR＝0.72，CI＝0.37～1.42，图 21.7C），而核性白内障的发生率提高了 57%（RR＝1.57，CI＝0.84～2.93，图 21.7B）。

叶黄素的摄入也与老年性黄斑部病变（age-related maculopathy）发生率相关，而最近在叶黄素对白内障发生率的影响方面，人们也进行了更全面的研究。与前面的β-胡萝卜素相同，一些研究发现叶黄素的摄入与白内障发生率之间不存在显著的负相关，然而，也有研究发现增加富含叶黄素食物的摄入与降低白内障发生的危险性有关。同时，叶黄素和玉米黄质的摄入与核性白内障的发生（RR=0.4，CI=0.2～0.8，图 21.10B）(Lyle *et al.*，1999a)、女性白内障摘除术（RR=0.78，CI=0.63～0.95，图 21.10E）(Chasan-Taber *et al.*，1999b）和男性白内障摘除术（RR=0.81，CI=0.65～1.01）(Brown *et al.*，1999）之间存在负相关。其中，花椰菜和煮熟的菠菜与白内障摘除术发生的危险性降低之间的关系最为显著，而这些食物富含叶黄素和玉米黄质（Brown *et al.*，1999）。我们最近还发现，每天至少摄入 5.6mg 叶黄素和玉米黄质的人核性白内障的发生率比每天摄入 2.4mg 的人低 51%（RR=0.49，CI=0.25～0.94，图 21.10B）(Jacques *et al.*，2001）。

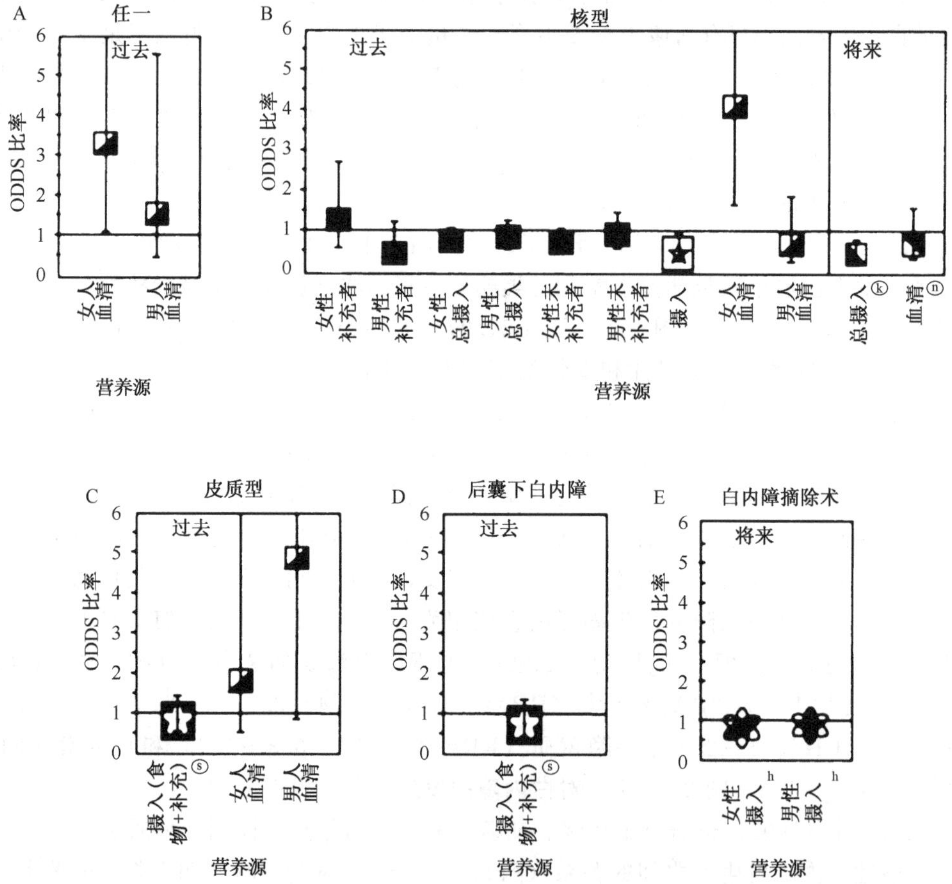


图 21.10　白内障的发生率、高低摄入量（补充或不补充）的叶黄素和血浆叶黄素水平

白内障的类型：任一、中期/老年型、核型、皮质型、后囊下、混合型和白内障摘除术。追溯性和前瞻性的研究资料分开列出。摘自 Taylor 等《维生素 C 在健康与疾病中的作用》（纽约，Marcel Dekker，1997）

在蓝山组研究中，Cumming 等（2000）发现，每天摄入 3.2mg 维生素 A 的人群与每天摄入 0.3mg 的人群相比，其核性白内障的发生率降低了 50%（RR=0.5，CI=0.3～0.8）。

然而，干预研究得到的结果并不一致。AREDS 干预研究表明，每天摄入 15mg 的 $\beta$-胡萝卜素，同时补充维生素 C、维生素 E 和锌，6.3 年后，其混浊的发生率或进一步

发展并未受到影响（年龄相关性眼病研究小组，2001）（图 21.4）。相反，REACT 研究中每天摄入 18mg 的 $\beta$-胡萝卜素，同时补充维生素 C 和 E 确实延缓了白内障的发展，至少延缓了美国人的白内障发展（Chylack *et al.*，2002）。

## 多种抗氧化剂的联合使用

为了评估食物含有的多种抗氧化剂联合使用对白内障发生率的影响，我们采用“抗氧化指标”来评判。然而，个别营养物质可能会显著影响这些指标，因此目前这些指标的有效性还存在很大的争议。有关多种维生素对白内障发生率的影响，更新的研究资料非常匮乏，读者可以主要参考以前的一些综述（Taylor，1999b）。人们对各种不同食物的摄入与白内障发生率之间的关系进行了一些试验研究，并进行了一些综述（Taylor，2000）。

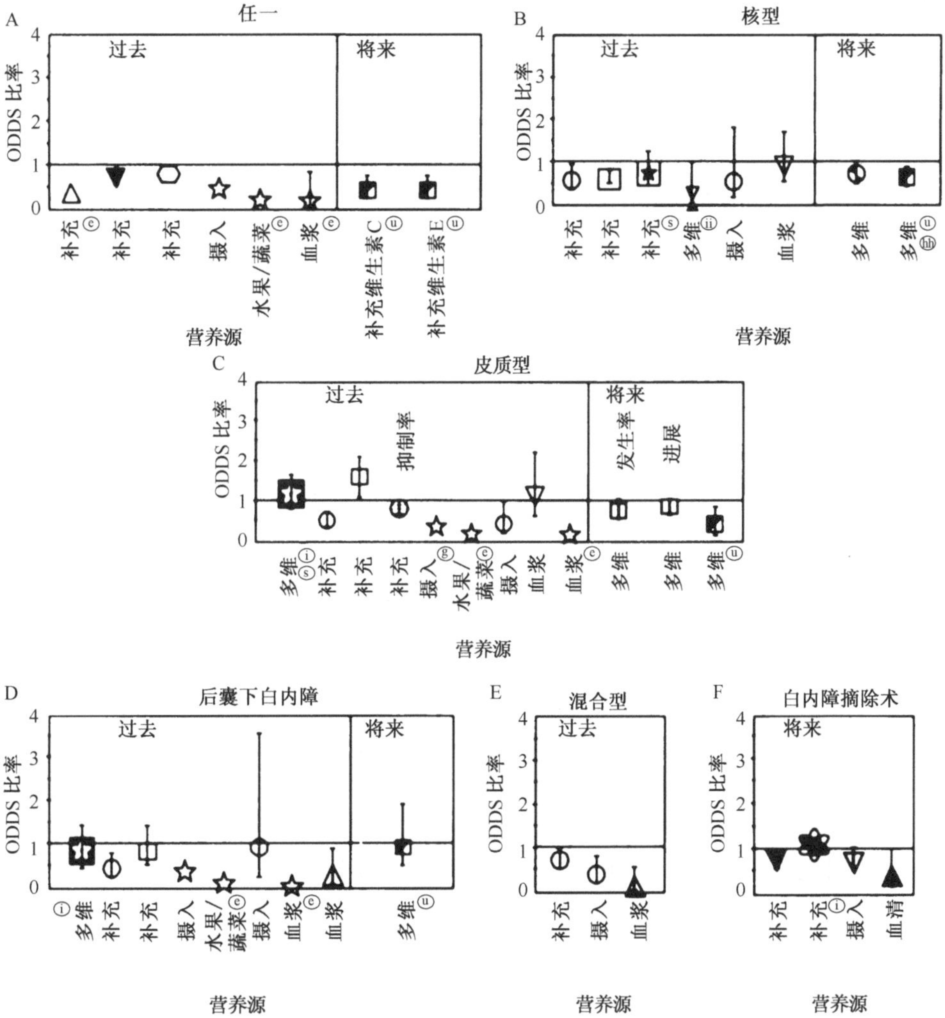


图 21.11　白内障的发生率、高低摄入量（补充或不补充）和抗氧化指标

白内障的类型：任一、核型、皮质型、后囊下、混合型和白内障摘除术。追溯性和前瞻性的研究资料分开列出。摘自 Taylor 等《维生素 C 在健康与疾病中的作用》（纽约，Marcel Dekker，1997）

多种维生素的抗氧化功能可能具有累加效应，这在以前阐述过。然而，最近的一项研究发现，对于多种维生素摄入超过 10 年的蓝山人（Blue Mountain population）（$n$=385），其核性和皮质性混浊的发生率分别为 0.1（CI=0.0～1.0）和 0.9（CI=0.3～1.1）（图 21.11B，C）（Kuzniarz *et al.*，2001）。这表明，多种维生素对白内障的发生具有一定的抑制作用，但是未达到统计上的显著水平。

## 干预研究

人们还进行了一些干预试验来研究补充维生素对白内障发生率的影响（Sperduto *et al.*，1993；年龄相关性眼病研究小组，2001；Christen，2001；Chylack *et al.*，2002）。Sperduto 等（1993）进行了两个试验来评价补充维生素对白内障发生率的影响。其试验对象为 4000 个中国临县农村的 45～74 岁的营养不良人群，第一个试验采用简单的试验设计：补充多种维生素或安慰剂；第二个试验采用更多因子处理来评估维生素和矿物元素的四种不同组合的效果：视黄醇（5000IU）和锌（22mg）、核黄素（3mg）和烟酸（40mg）、维生素 C（120mg）和钼（30μg）及维生素 E（30mg）、β-胡萝卜素（15mg）和硒（50μg）。试验时间为 5～6 年。试验结束时对所有的试验对象进行眼检，以确定白内障的发生率。

第一个试验的结果表明，补充多种维生素使 65～74 岁的人核性白内障的发生率显著降低了 43%（RR=0.57，CI=0.36～0.90；图 21.4A）。第二个试验也表明，与未补充组相比，补充核黄素/烟酸的人核性白内障发生率显著降低（RR=0.59，CI=0.45～0.79；图 21.4A），该效应对 65～74 岁的人群最明显（RR=0.45，CI=0.31～0.64；图 21.4A）。然而，补充核黄素/烟酸可能提高后囊下白内障的发生率（RR=2.64，CI=1.31～5.35；图 21.4D）。蓝山研究组对每天摄入 44mg 与 11.7mg 烟酸的人群进行研究，结果表明，利用能量矫正后，皮质性白内障（RR=0.7，CI=0.5～1.0）和核性白内障（RR=0.6，CI=0.4～0.9）发生率均降低（Cumming *et al.*，2000）。Sperduto 等（1993）的结果还表明，补充视黄醇/锌（RR=0.77，CI=0.58～1.02）或维生素 C/钼（RR=0.78，CI=0.59～1.04）对核性白内障的发生具有一定的预防作用（图 21.4D）。Cumming 等（2000）也发现，每天摄入 14.3mg 与 7.8mg 锌的人群相比，核性白内障的发生率也有所降低（RR=0.7，CI=0.5～1.1），但这还有待于进一步验证。

Christen（2001）的干预研究发现，每隔一天补充 β-胡萝卜素 50mg，持续 13 年以后，自报白内障（self-reported cataract）的发生率几乎未受到影响。最近一项大型的干预（$n$=4596）研究表明，每天摄入 500mg 维生素 C、400IU 维生素 E、15mg β-胡萝卜素和 80mg 锌，持续 6.3 年后，对所有类型的晶状体混浊也没有产生任何影响（年龄相关性眼病研究小组，2001）。相反，REACT 的干预研究（$n$=158）发现，在美国人群中，每天摄入 750mg 维生素 C、600mg 维生素 E 和 18mg β-胡萝卜素，持续 3 年后，延缓了人晶状体混浊的发展过程，但是在英国人群中，并未发现此现象（Chylack *et al.*，2002）（表 21.1）。

根据与自由基反应之间的关系，铁代谢也可能影响白内障的发生。然而，Cumming 等（2000）指出，提高铁的摄入量（15.4 mg/d 与 8.5mg/d）使核性白内障的发生率降低了 40%（RR=0.6，CI=0.4～0.9）。

# 总　　结

光和氧对人体既有益处也有坏处。尽管光和氧是机体发挥正常生理功能必不可少的因素，但是若在不加控制的条件下过量存在时，它们常常能导致白内障。随着年龄的增长，主要的抗氧化剂、抗氧化酶活性和第二道防御系统如蛋白酶逐渐减少或缺失，晶状体功能逐渐减退，其受损程度会不断加深。吸烟（West，1999）和光处理（McCarty and Taylor，1999）能产生氧化应激，提高白内障的发生率。研究发现，教育水平低和社会经济状况差也能显著提高白内障发生率（McLaren，1980；Harding and van Heyningen，1987；Mohan *et al.*，1989；Leske *et al.*，1991），这都与比较差的营养状况有关。资料表明，增加抗氧化剂的摄入量对降低白内障的发生率能产生一定的正效应。根据我们的研究结果和其他试验结果，维生素 C、E 和叶黄素均有利于降低核性白内障发病率，而且每天维生素 C、E 和叶黄素的摄入水平分别为 250mg、90mg 和 3mg 就能对身体产生有利影响（Taylor and Jacques，1997；Jacques *et al.*，2001）。这很可能是通过调整正常膳食行为来满足这些养分（Jacob *et al.*，1988；Taylor *et al.*，1991a；Hankinson *et al.*，1992；Jacques *et al.*，1997）。随着年龄的增长，维生素 C 的生物学利用率将会降低，因此，有必要适当提高维生素 C 的摄入量来满足老年人的需要。而其他养分的适宜摄入水平的确定，还有待于进一步研究。不同研究结果很难进行比较，其主要原因是：对于养分与某一类型白内障的相关性，各研究报道不一；各试验对白内障的分类方法不尽相同；各种白内障具有不同的一般病原学特点。本章综述的研究大部分采用病例-对照的试验设计，且只对各种营养状况评价一次。由于摄入量和营养指标的变异非常大，且食物各组分的作用很可能具有累积效应，因此有关摄入量的研究对象应该是有长期饮食记录的人群。在各种反映营养状况的指标中，血浆指标可能更适合于摄入量的研究。改善膳食结构和补充必要的养分能使人体获得最佳的营养状况，尤其对于那些营养状况差和患病的人群。因此，根据目前的研究资料，我们得出结论：补充营养物质，尤其是维生素类抗氧化剂，是一种延缓白内障发生的最经济实惠、最实用的手段。

# 致　　谢

感谢 Tom Nowell 和 Marisa Hobbs 在图表制作方面给予的大力帮助，同时感谢 Paul Jacques 在评价流行病学资料方面提供的宝贵帮助。

## 参 考 文 献

Age-Related Eye Disease Study Research Group (2001) A randomized, placebo-controlled, clinical trial of high-dose supplementation with vitamins C and E and beta carotene for age-related cataract and vision loss: AREDS report no. 9. *Archives of Ophthalmology* 119, 1439–1452.

Bensch, K.G., Fleming, J.E. and Lohmann, W. (1985) The role of ascorbic acid in senile cataract. *Proceedings of the National Academy of Sciences USA* 82, 7193–7196.

Berger, J., Shephard, D., Morrow, F., Sadowski, J., Haire, T. and Taylor, A. (1988) Reduced and total ascorbate in guinea pig eye tissues in response to dietary intake. *Current Eye Research* 7, 681–686.

Berger, J., Shepard, D., Morrow, F. and Taylor, A. (1989) Relationship between dietary intake and tissue levels of reduced and total vitamin C in the nonscorbutic guinea pig. *Journal of Nutrition* 119, 734–740.

Berman, E.R. (1991) *Biochemistry of the Eye*. Plenum Press, New York.

Bhuyan, K.C. and Bhuyan, D.K. (1984) Molecular mechanism of cataractogenesis: III. Toxic metabolites of oxygen as initiators of lipid peroxidation and cataract. *Current Eye Research* 3, 67–81.

Bhuyan, D.K., Podos, S.M., Machlin, L.T., Bhagavan, H.N., Chondhury, D.N., Soja, W.S. and Bhuyan, K.C. (1983) Antioxidant in therapy of cataract II: effect of all-roc-alpha-tocopherol (vitamin E) in sugar-induced cataract in rabbits (ARVO abstracts). *Investigative Ophthalmology and Visual Science* 24, 74.

Blondin, J. and Taylor, A. (1987) Measures of leucine aminopeptidase can be used to anticipate UV-induced age-related damage to lens proteins: ascorbate can delay this damage. *Mechanisms of Ageing and Development* 41, 39–46.

Blondin, J., Baragi, V., Schwartz, E., Sadowski, J.A. and Taylor, A. (1986) Delay of UV-induced eye lens protein damage in guinea pigs by dietary ascorbate. *Journal of Free Radicals in Biology and Medicine* 2, 275–281.

Blondin, J., Baragi, V.J., Schwartz, E., Sadowski, J. and Taylor, A. (1987) Dietary vitamin C delays UV-induced age-related eye lens protein damage. *Annals of the New York Academy of Sciences* 498, 460–463.

Brown, L., Rimm, E.B., Seddon, J.M., Giovannucci, E.L., Chasan-Taber, L., Spiegelman, D., Willett, W.C. and Hankinson, S.E. (1999) A prospective study of carotenoid intake and risk of cataract extraction in US men. *American Journal of Clinical Nutrition* 70, 517–524.

Bunce, G.E., Kinoshita, J. and Horwitz, J. (1990) Nutritional factors in cataract. *Annual Review of Nutrition* 10, 233–254.

Chasan-Taber, L., Willett, W.C., Seddon, J.M., Stampfer, M.J., Rosner, B., Colditz, G.A. and Hankinson, S.E. (1999a) A prospective study of vitamin supplement intake and cataract extraction among U.S. women. *Epidemiology* 10, 679–684.

Chasan-Taber, L., Willett, W.C., Seddon, J.M., Stampfer, M.J., Rosner, B., Colditz, G.A., Speizer, F.E. and Hankinson, S.E. (1999b) A prospective study of carotenoid and vitamin A intakes and risk of cataract extraction in US women. *American Journal of Clinical Nutrition* 70, 509–516.

Chatterjee, A., Milton, R.C. and Thyle, S. (1982) Prevalence and aetiology of cataract in Punjab. *British Journal of Ophthalmology* 66, 35–42.

Christen, W.G. (1999) Evaluation of epidemiologic studies of nutrition and cataract. In: Taylor, A. (ed.) *Nutritional and Environmental Influences on the Eye*. CRC Press, Boca Raton, Florida, pp. 95–104.

Christen, W.G. (2001) Beta-carotene and age-related cataract in a randomized trial of U.S. physicians. *Investigative Ophthalmology and Visual Science* 42, S518.

Chylack, L.T. Jr (1999) Function of the lens and methods of quantifying cataract. In: Taylor, A. (ed.) *Nutritional and Environmental Influences on the Eye*. CRC Press, Boca Raton, Florida, pp. 25–52.

Chylack, L.T. Jr, Wolfe, J.K., Singer, D.M., Leske, M.C., Bullimore, M.A., Bailey, I.L., Friend, J., McCarthy, D. and Wu, S.Y. (1993) The Lens Opacities Classification System III. The Longitudinal Study of Cataract Study Group. *Archives of Ophthalmology* 111, 831–836.

Chylack, L.T. Jr, Wolfe, J.K., Judith, F., Singer, D.M., Wu, S.Y. and Leske, M.C. (1994) Nuclear cataract: relative contributions to vision loss of opalescence and brunescence. *Investigative Ophthalmology and Visual Science* 35, 42632 [abstract].

Chylack, L.T. Jr, Brown, N.P., Bron, A., Hurst, M., Kopcke, W., Thien, U. and Schalch, W. (2002) The Roche European American Cataract Trial (REACT): a randomized clinical trial to investigate the efficacy of an oral antioxidant micronutrient mixture to slow progression of age-related cataract. *Ophthalmic Epidemiology* 9, 49–80.

Costagliola, C., Iuliano, G., Menzione, M., Rinaldi, E., Vito, P. and Auricchio, G. (1986) Effect of vitamin E on glutathione content in red blood cells, aqueous humor and lens of humans and other species. *Experimental Eye Research* 43, 905–914.

Creighton, M.O., Ross, W.M., Stewart-DeHaan, P.J., Sanwal, M. and Trevithick, J.R. (1985) Modelling cortical cataractogenesis VII: effects of vitamin E treatment on galactose-induced cataracts. *Experimental Eye Research* 40, 213–222.

Cumming, R.G., Mitchell, P. and Smith, W. (2000) Diet and cataract: the Blue Mountains Eye Study. *Ophthalmology* 107, 450–456.

Curran Celentano, J., Burke, J.D. and Hammond, B.R. Jr (2002) *In vivo* assessment of retinal carotenoids: macular pigment detection techniques and their impact on monitoring pigment status. *Journal of Nutrition* 132, 535S–539S.

Daicker, B., Schiedt, K., Adnet, J.J. and Bermond, P. (1987) Canthaxanthin retinopathy. An investigation by light and electron microscopy and physicochemical analysis. *Graefe's Archive for Clinical and Experimental Ophthalmology* 225, 189–197.

Devamanoharan, P.S., Henein, M., Morris, S., Ramachandran, S., Richards, R.D. and Varma, S.D. (1991) Prevention of selenite cataract by vitamin C. *Experimental Eye Research* 52, 563–568.

Eisenhauer, D.A., Berger, J.J., Peltier, C.Z. and Taylor, A. (1988) Protease activities in cultured beef lens epithelial cells peak and then decline upon progressive passage. *Experimental Eye Research* 46, 579–590.

Frei, B., Stocker, R. and Ames, B.N. (1988) Antioxidant defenses and lipid peroxidation in human blood plasma. *Proceedings of the National Academy of Sciences*

*USA* 85, 9748–9752.
Fridovich, I. (1984) Oxygen: aspects of its toxicity and elements of defense. *Current Eye Research* 3, 1–2.
Garland, D.L. (1991) Ascorbic acid and the eye. *American Journal of Clinical Nutrition* 54, 1198S–1202S.
Giblin, F.J., McCready, J.P. and Reddy, V.N. (1982) The role of glutathione metabolism in the detoxification of $H_2O_2$ in rabbit lens. *Investigative Ophthalmology and Visual Science* 22, 330–335.
Hammond, B.R. Jr, Wooten, B.R. and Snodderly, D.M. (1997) Density of the human crystalline lens is related to the macular pigment carotenoids, lutein and zeaxanthin. *Optometry and Vision Science* 74, 499–504.
Hankinson, S.E., Stampfer, M.J., Seddon, J.M., Colditz, G.A., Rosner, B., Speizer, F.E. and Willett, W.C. (1992) Nutrient intake and cataract extraction in women: a prospective study. *British Medical Journal* 305, 335–339.
Harding, J.J. and van Heyningen, R. (1987) Epidemiology and risk factors for cataract. *Eye* 1, 537–541.
Huang, L.L., Jahngen-Hodge, J. and Taylor, A. (1993) Bovine lens epithelial cells have a ubiquitin-dependent proteolysis system. *Biochimica et Biophysica Acta* 1175, 181–187.
Italian-American Cataract Study (1991) Risk factors for age-related cortical, nuclear, and posterior subcapsular cataracts. *American Journal of Epidemiology* 133, 541–553.
Jacob, R.A., Otradovec, C.L., Russell, R.M., Munro, H.N., Hartz, S.C., McGandy, R.B., Morrow, F.D. and Sadowski, J.A. (1988) Vitamin C status and nutrient interactions in a healthy elderly population. *American Journal of Clinical Nutrition* 48, 1436–1442.
Jacques, P.F. and Chylack, L.T. Jr (1991) Epidemiologic evidence of a role for the antioxidant vitamins and carotenoids in cataract prevention. *American Journal of Clinical Nutrition* 53, 352S–355S.
Jacques, P.F. and Taylor, A. (1991) Micronutrients and age-related cataracts. In: Bendich, A. and Butterworth, C.E. (eds) *Micronutrients in Health and in Disease Prevention.* Marcel Dekker, New York, pp. 359–379.
Jacques, P.F., Chylack, L.T. Jr and Taylor, A. (1994) Relationships between natural antioxidants and cataract formation. In: Frei, B. (ed.) *Natural Antioxidants in Human Health and Disease.* Academic Press, Orlando, Florida, pp. 513–533.
Jacques, P.F., Taylor, A., Hankinson, S.E., Willett, W.C., Mahnken, B., Lee, Y., Vaid, K. and Lahav, M. (1997) Long-term vitamin C supplement use and prevalence of early age-related lens opacities. *American Journal of Clinical Nutrition* 66, 911–916.
Jacques, P.F., Chylack, L.T. Jr, Hankinson, S.E., Khu, P.M., Rogers, G., Friend, J., Tung, W., Wolfe, J.K., Padhye, N., Willett, W.C. and Taylor, A. (2001) Long-term nutrient intake and early age-related nuclear lens opacities. *Archives of Ophthalmology* 119, 1009–1019.
Jahngen, J.H., Haas, A.L., Ciechanover, A., Blondin, J., Eisenhauer, D. and Taylor, A. (1986) The eye lens has an active ubiquitin–protein conjugation system. *Journal of Biological Chemistry* 261, 13760–13767.
Jahngen, J.H., Lipman, R.D., Eisenhauer, D.A., Jahngen, E.G. Jr and Taylor, A. (1990) Aging and cellular maturation cause changes in ubiquitin–eye lens protein conjugates. *Archives of Biochemistry and Biophysics* 276, 32–37.
Jahngen-Hodge, J., Cyr, D., Laxman, E. and Taylor, A. (1992) Ubiquitin and ubiquitin conjugates in human lens. *Experimental Eye Research* 55, 897–902.
Jahngen-Hodge, J., Obin, M.S., Gong, X., Shang, F., Nowell, T.R. Jr, Gong, J., Abasi, H., Blumberg, J. and Taylor, A. (1997) Regulation of ubiquitin-conjugating enzymes by glutathione following oxidative stress. *Journal of Biological Chemistry* 272, 28218–28226.
Klein, B.E., Klein, R. and Linton, K.L. (1992) Prevalence of age-related lens opacities in a population: the Beaver Dam Eye Study. *Ophthalmology* 99, 546–552.
Klein, R., Klein, B.E., Linton, K.L. and DeMets, D.L. (1993) The Beaver Dam Eye Study: the relation of age-related maculopathy to smoking. *American Journal of Epidemiology* 137, 190–200.
Knekt, P., Heliovaara, M., Rissanen, A., Aromaa, A. and Aaran, R.K. (1992) Serum antioxidant vitamins and risk of cataract. *British Medical Journal* 305, 1392–1394.
Kosegarten, D.C. and Maher, T.J. (1978) Use of guinea pigs as model to study galactose-induced cataract formation. *Journal of Pharmaceutical Sciences* 67, 1478–1479.
Krinsky, N.I. (2002) Possible biologic mechanisms for a protective role of xanthophylls. *Journal of Nutrition* 132, 540S–542S.
Kupfer, C. (1985) The conquest of cataract: a global challenge. *Transactions of the Ophthalmological Societies of the United Kingdom* 104, 1–10.
Kuzniarz, M., Mitchell, P., Cumming, R.G. and Flood, V.M. (2001) Use of vitamin supplements and cataract: the Blue Mountains Eye Study. *American Journal of Ophthalmology* 132, 19–26.
Leibowitz, H.M., Krueger, D.E., Maunder, L.R., Milton, R.C., Kini, M.M., Kahn, H.A., Nickerson, R.J., Pool, J., Colton, T.L., Ganley, J.P., Loewenstein, J.I. and Dawber, T.R. (1980) The Framingham Eye Study monograph: an ophthalmological and epidemiological study of cataract, glaucoma, diabetic retinopathy, macular degeneration, and visual acuity in a general population of 2631 adults, 1973–1975. *Survey of Ophthalmology* 24, 335–610.

Leske, M.C., Chylack, L.T. Jr and Wu, S.Y. (1991) The Lens Opacities Case–Control Study. Risk factors for cataract. *Archives of Ophthalmology* 109, 244–251.

Leske, M.C., Wu, S.Y., Hyman, L., Sperduto, R., Underwood, B., Chylack, L.T., Milton, R.C., Srivastava, S. and Ansari, N. (1995) Biochemical factors in the lens opacities. Case–control study. The Lens Opacities Case–Control Study Group. *Archives of Ophthalmology* 113, 1113–1119.

Leske, M.C., Wu, S.Y., Connel, A.M., Hyman, L. and Schachat, A.P. (1997) Lens opacities, demographic factors and nutritional suppplements in the Barbados Eye Study. *International Journal of Epidemiology* 26, 1314–1322.

Leske, M.C., Chylack, L.T. Jr, He, Q., Wu, S.Y., Schoenfeld, E., Friend, J. and Wolfe, J. (1998a) Antioxidant vitamins and nuclear opacities: the longitudinal study of cataract. *Ophthalmology* 105, 831–836.

Leske, M.C., Chylack, L.T. Jr, He, Q., Wu, S.Y., Schoenfeld, E., Friend, J. and Wolfe, J. (1998b) Risk factors for nuclear opalescence in a longitudinal study. LSC Group. Longitudinal Study of Cataract. *American Journal of Epidemiology* 147, 36–41.

Levine, M. (1986) New concepts in the biology and biochemistry of ascorbic acid. *New England Journal of Medicine* 314, 892–902.

Luthra, R., Wa, S. and Leske, M.C. (1997) Lens opacities and use of nutritional supplements: the Barbados Study. *Investigative Ophthalmology and Visual Science* 8, S450.

Lyle, B.J., Mares-Perlman, J.A., Klein, B.E., Klein, R. and Greger, J.L. (1999a) Antioxidant intake and risk of incident age-related nuclear cataracts in the Beaver Dam Eye Study. *American Journal of Epidemiology* 149, 801–809.

Lyle, B.J., Mares-Perlman, J.A., Klein, B.E., Klein, R., Palta, M., Bowen, P.E. and Greger, J.L. (1999b) Serum carotenoids and tocopherols and incidence of age-related nuclear cataract. *American Journal of Clinical Nutrition* 69, 272–277.

Machlin, L.J. and Bendich, A. (1987) Free radical tissue damage: protective role of antioxidant nutrients. *FASEB Journal* 1, 441–445.

Mares-Perlman, J.A., Klein, B.E., Klein, R. and Ritter, L.L. (1994) Relation between lens opacities and vitamin and mineral supplement use. *Ophthalmology* 101, 315–325.

Mares-Perlman, J.A., Brady, W.E., Klein, B.E., Klein, R., Haus, G.J., Palta, M., Ritter, L.L. and Shoff, S.M. (1995a) Diet and nuclear lens opacities. *American Journal of Epidemiology* 141, 322–334.

Mares-Perlman, J.A., Brady, W.E., Klein, B.E., Klein, R., Palta, M., Bowen, P. and Stacewicz-Sapuntzakis, M. (1995b) Serum carotenoids and tocopherols and severity of nuclear and cortical opacities. *Investigative Ophthalmology and Visual Science* 36, 276–288.

Mares-Perlman, J.A., Brady, W.E., Klein, R., Klein, B.E., Bowen, P., Stacewicz-Sapuntzakis, M. and Palta, M. (1995c) Serum antioxidants and age-related macular degeneration in a population-based case–control study. *Archives of Ophthalmology* 113, 1518–1523.

Mares-Perlman, J.A., Brady, W.E., Klein, B.E.K., Klein, R. and Palta, M. (1996) Supplement use and 5-year progression of cortical opacities. *Investigative Ophthalmology and Visual Science* 37, S237.

Mares-Perlman, J.A., Lyle, B.J., Klein, R., Fisher, A.I., Brady, W.E., VandenLangenberg, G.M., Trabulsi, J.N. and Palta, M. (2000) Vitamin supplement use and incident cataracts in a population-based study. *Archives of Ophthalmology* 118, 1556–1563.

McCarty, C. and Taylor, H.R. (1999) Light and risk for age-related diseases. In: Taylor, A. (ed.) *Nutritional and Environmental Influences on the Eye*. CRC Press, Boca Raton, Florida, pp. 135–150.

McCarty, C.A., Mukesh, B.N., Fu, C.L. and Taylor, H.R. (1999) The epidemiology of cataract in Australia. *American Journal of Ophthalmology* 128, 446–465.

McLaren, D. S. (1980) *Nutritional Ophthalmology*. Academic Press, London.

Mohan, M., Sperduto, R.D., Angra, S.K., Milton, R.C., Mathur, R.L., Underwood, B.A., Jaffery, N., Pandya, C.B., Chhabra, V.K., Vajpayee, R.B. *et al.* (1989) India–US case–control study of age-related cataracts. India–US Case–Control Study Group. *Archives of Ophthalmology* 107, 670–676.

Mune, M., Meydani, M., Jahngen-Hodge, J., Martin, A., Smith, D., Palmer, V., Blumberg, J.B. and Taylor, A. (1995) Effect of calorie restriction on liver and kidney glutathione in aging Emory mice. *Age* 18, 43–49.

Nadalin, G., Robman, L.D., McCarty, C.A., Garrett, S.K., McNeil, J.J. and Taylor, H.R. (1999) The role of past intake of vitamin E in early cataract changes. *Ophthalmic Epidemiology* 6, 105–112.

Nagaraj, R.H. and Monnier, V.M. (1992) Isolation and characterization of a blue fluorophore from human eye lens crystallins: *in vitro* formation from Maillard reaction with ascorbate and ribose. *Biochimica et Biophysica Acta* 1116, 34–42.

Nishigori, H., Lee, J.W., Yamauchi, Y. and Iwatsuru, M. (1986) The alteration of lipid peroxide in glucocorticoid-induced cataract of developing chick embryos and the effect of ascorbic acid. *Current Eye Research* 5, 37–40.

Obin, M., Nowell, T. and Taylor, A. (1994) The photoreceptor G-protein transducin (Gt) is a substrate for ubiquitin-dependent proteolysis. *Biochemical and Biophysical Research Communications* 200, 1169–1176.

Obin, M.S., Jahngen-Hodge, J., Nowell, T. and Taylor, A. (1996) Ubiquitinylation and ubiquitin-dependent proteolysis in vertebrate photoreceptors (rod outer segments). Evidence for ubiquitinylation of Gt and rhodopsin. *Journal of Biological Chemistry* 271, 14473–14484.

Obin, M., Shang, F., Gong, X., Handelman, G., Blumberg, J. and Taylor, A. (1998) Redox regulation of ubiquitin-conjugating enzymes: mechanistic insights using the thiol-specific oxidant diamide. *FASEB Journal* 12, 561–569.

Obin, M., Mesco, E., Gong, X., Haas, A.L., Joseph, J. and Taylor, A. (1999) Neurite outgrowth in PC12 cells. Distinguishing the roles of ubiquitylation and ubiquitin-dependent proteolysis. *Journal of Biological Chemistry* 274, 11789–11795.

Rathbun, W.B., Killen, C.E., Holleschau, A.M. and Nagasawa, H.T. (1996) Maintenance of hepatic glutathione homeostasis and prevention of acetaminophen-induced cataract in mice by L-cysteine prodrugs. *Biochemical Pharmacology* 51, 1111–1116.

Reddy, V.N. (1990) Glutathione and its function in the lens – an overview. *Experimental Eye Research* 50, 771–778.

Robertson, J.M., Donner, A.P. and Trevithick, J.R. (1989) Vitamin E intake and risk of cataracts in humans. *Annals of the New York Academy of Sciences* 570, 372–382.

Rouhiainen, P., Rouhiainen, H. and Salonen, J.T. (1996) Association between low plasma vitamin E concentration and progression of early cortical lens opacities. *American Journal of Epidemiology* 144, 496–500.

Sastre, J., Meydani, M., Martin, A., Biddle, L., Taylor, A. and Blumberg, J. (1994) Effect of glutathione monoethyl ester administration on galactose-induced cataract in the rat. *Life Chemistry Reports* 12, 89–95.

Schalch, W., Dayhaw-Barker, P. and Barker, F.M. II (1999) The carotenoids of the human retina. In: Taylor, A. (ed.) *Nutritional and Environmental Influences on the Eye*. CRC Press, Boca Raton, Florida, pp. 215–250.

Schwab, L. (1990) Cataract blindness in developing nations. *International Ophthalmology Clinics* 30, 16–18.

Seddon, J.M., Christen, W.G., Manson, J.E., LaMotte, F.S., Glynn, R.J., Buring, J.E. and Hennekens, C.H. (1994) The use of vitamin supplements and the risk of cataract among US male physicians. *American Journal of Public Health* 84, 788–792.

Shang, F. and Taylor, A. (1995) Oxidative stress and recovery from oxidative stress are associated with altered ubiquitin conjugating and proteolytic activities in bovine lens epithelial cells. *Biochemical Journal* 307, 297–303.

Shang, F., Gong, X. and Taylor, A. (1995) Changes in ubiquitin conjugation activities in young and old lenses in response to oxidative stress. *Investigative Ophthalmology and Visual Science* 36, S528.

Shang, F., Gong, X., Palmer, H.J., Nowell, T.R. Jr and Taylor, A. (1997a) Age-related decline in ubiquitin conjugation in response to oxidative stress in the lens. *Experimental Eye Research* 64, 21–30.

Shang, F., Gong, X. and Taylor, A. (1997b) Activity of ubiquitin-dependent pathway in response to oxidative stress. Ubiquitin-activating enzyme is transiently up-regulated. *Journal of Biological Chemistry* 272, 23086–23093.

Simon, J.A. and Hudes, E.S. (1999) Serum ascorbic acid and other correlates of self-reported cataract among older Americans. *Journal of Clinical Epidemiology* 52, 1207–1211.

Smith, D., Shang, F., Nowell, T.R., Asmundsson, G., Perrone, G., Dallal, G., Scott, L., Kelliher, M., Gindelsky, B. and Taylor, A. (1999) Decreasing ascorbate intake does not affect the levels of glutathione, tocopherol or retinol in the ascorbate-requiring osteogenic disorder shionogi rats. *Journal of Nutrition* 129, 1229–1232.

Snodderly, D.M. and Hammond, B.R. Jr (1999) *In vivo* psychophysical assessment of nutritional and environmental influences on human ocular tissues: lens and macular pigment. In: Taylor, A. (ed.) *Nutritional and Environmental Influences on the Eye*. CRC Press, Boca Raton, Florida, pp. 251–285.

Sperduto, R.D., Hu, T.S., Milton, R.C., Zhao, J.L., Everett, D.F., Cheng, Q.F., Blot, W.J., Bing, L., Taylor, P.R., Li, J.Y., Dawsey, S. and Guo, W. (1993) The Linxian cataract studies. Two nutrition intervention trials. *Archives of Ophthalmology* 111, 1246–1253.

Stephens, R.J., Negi, D.S., Short, S.M., van Kuijk, F.J., Dratz, E.A. and Thomas, D.W. (1988) Vitamin E distribution in ocular tissues following long-term dietary depletion and supplementation as determined by microdissection and gas chromatography–mass spectrometry. *Experimental Eye Research* 47, 237–245.

Taylor, A. (1999a) Lens and retina function: introduction and challenge. In: Taylor, A. (ed.) *Nutrition and Environmental Influences on the Eye*. CRC Press, Boca Raton, Florida, pp. 1–4.

Taylor, A. (1999b) Nutritional and environmental influences on risk for cataract. In: Taylor, A. (eds) *Nutritional and Environmental Influences on the Eye*. CRC Press, Boca Raton, Florida, pp. 53–93.

Taylor, A. (ed.) (1999c) *Nutritional and Environmental Influences on the Eye*. CRC Press, Boca Raton, Florida.

Taylor, A. (2000) Nutritional influences on risk for cataract. In: Fuchs, J. and Packer, L. (eds) *Environmental Stressors: Effects on Lung, Skin, Eye and Immune System Function*. Marcel Dekker, New York, pp. 457–487.

Taylor, A. and Davies, K.J. (1987) Protein oxidation and loss of protease activity may lead to cataract formation in the aged lens. *Free Radical Biology and Medicine*

3, 371–377.

Taylor, A. and Jacques, P. (1997) Antioxidant status and risk for cataract. In: Bendich, A. and Deckelbaum, R.J. (eds) *Preventive Nutrition: the Guide for Health Professionals.* Humana Press, Totowa, New Jersey.

Taylor, A., Tisdell, F.E. and Carpenter, F.H. (1981) Leucine aminopeptidase (bovine lens): synthesis and kinetic properties of ortho-, meta-, and para-substituted leucyl-anilides. *Archives of Biochemistry and Biophysics* 210, 90–97.

Taylor, A., Jacques, P.F., Nadler, D., Morrow, F., Sulsky, S.I. and Shepard, D. (1991a) Relationship in humans between ascorbic acid consumption and levels of total and reduced ascorbic acid in lens, aqueous humor, and plasma. *Current Eye Research* 10, 751–759.

Taylor, A., Jahngen-Hodge, J., Huang, L. and Jacques, P. (1991b) Aging in the eye lens: roles for proteolysis and nutrition in formation of cataract. *Age* 14, 65–71.

Taylor, A., Jacques, P.F. and Dorey, C.K. (1993) Oxidation and aging: impact on vision. *Toxicology and Industrial Health* 9, 349–371.

Taylor, A., Jacques, P.F. and Epstein, E.M. (1995a) Relations among aging, antioxidant status, and cataract. *American Journal of Clinical Nutrition* 62, 1439S–1447S.

Taylor, A., Jahngen-Hodge, J., Smith, D.E., Palmer, V.J., Dallal, G.E., Lipman, R.D., Padhye, N. and Frei, B. (1995b) Dietary restriction delays cataract and reduces ascorbate levels in Emory mice. *Experimental Eye Research* 61, 55–62.

Taylor, A., Lipman, R.D., Jahngen-Hodge, J., Palmer, V., Smith, D., Padhye, N., Dallal, G.E., Cyr, D.E., Laxman, E., Shepard, D., Morrow, F., Salomon, R., Perrone, G., Asmundsson, G., Meydani, M., Blumberg, J., Mune, M., Harrison, D.E., Archer, J.R. and Shigenaga, M. (1995c) Dietary calorie restriction in the Emory mouse: effects on lifespan, eye lens cataract prevalence and progression, levels of ascorbate, glutathione, glucose, and glycohemoglobin, tail collagen breaktime, DNA and RNA oxidation, skin integrity, fecundity, and cancer. *Mechanisms of Ageing and Development* 79, 33–57.

Taylor, A., Jacques, P.F., Nowell, T., Perrone, G., Blumberg, J., Handelman, G., Jozwiak, B. and Nadler, D. (1997) Vitamin C in human and guinea pig aqueous, lens and plasma in relation to intake. *Current Eye Research* 16, 857–864.

Taylor, A., Jacques, P.F., Chylack, L.T. Jr, Hankinson, S.E., Khu, P.M., Rogers, G., Friend, J., Tung, W., Wolfe, J.K., Padhye, N. and Willett, W.C. (2002) Long-term intake of vitamins and carotenoids and odds of early age-related cortical and posterior subcapsular lens opacities. *American Journal of Clinical Nutrition* 75, 540–549.

Varma, S.D., Chand, D., Sharma, Y.R., Kuck, J.F. Jr and Richards, R.D. (1984) Oxidative stress on lens and cataract formation: role of light and oxygen. *Current Eye Research* 3, 35–57.

Vinson, J.A., Possanza, C.J. and Drack, A.V. (1986) The effect of ascorbic acid on galactose-induced cataracts. *Nutrition Reports International* 33, 665–668.

Vitale, S., West, S., Hallfrisch, J., Alston, C., Wang, F., Moorman, C., Muller, D., Singh, V. and Taylor, H.R. (1993) Plasma antioxidants and risk of cortical and nuclear cataract. *Epidemiology* 4, 195–203.

Wang, G.M., Spector, A., Luo, C.Q., Tang, L.Q., Xu, L.H., Guo, W.Y. and Huang, Y.Q. (1990) Prevalence of age-related cataract in Ganzi and Shanghai. The Epidemiological Study Group. *Chinese Medical Journal (England)* 103, 945–951.

West, S.K. (1999) Smoking and the risk of eye disease. In: Taylor, A. (ed.) *Nutritional and Environmental Influences on the Eye.* CRC Press, Boca Raton, Florida, pp. 151–164.

Whitfield, R., Schwab, L., Ross-Degnan, D., Steinkuller, P. and Swartwood, J. (1990) Blindness and eye disease in Kenya: ocular status survey results from the Kenya Rural Blindness Prevention Project. *British Journal of Ophthalmology* 74, 333–340.

Wolfe, J.K., Chylack, L.T., Leske, M.C. and Wu, S.Y. (1993) Lens nuclear color and visual function. *Investigative Ophthalmology and Visual Science* 34 (Supplement), 2550.

World Health Organization (1991) Use of intraocular lenses in cataract surgery in developing countries: memorandum from a WHO meeting. *Bulletin of the World Health Organization* 69, 657–666.

Yeum, K.J., Taylor, A., Tang, G. and Russell, R.M. (1995) Measurement of carotenoids, retinoids, and tocopherols in human lenses. *Investigative Ophthalmology and Visual Science* 36, 2756–2761.

Yeum, K.J., Shang, F.M., Schalch, W.M., Russell, R.M. and Taylor, A. (1999) Fat-soluble nutrient concentrations in different layers of human cataractous lens. *Current Eye Research* 19, 502–505.

Yokoyama, T., Sasaki, H., Giblin, F.J. and Reddy, V.N. (1994) A physiological level of ascorbate inhibits galactose cataract in guinea pigs by decreasing polyol accumulation in the lens epithelium: a dehydroascorbate-linked mechanism. *Experimental Eye Research* 58, 207–218.

Young, R.W. (1993) The Charles F. Prentice Medal Award Lecture 1992: optometry and the preservation of visual health. *Optometry and Vision Science* 70, 255–262.

Zigler, J.S. Jr and Goosey, J.D. (1984) Singlet oxygen as a possible factor in human senile nuclear cataract development. *Current Eye Research* 3, 59–65.

# 22 营养与免疫功能

Parveen Yaqoob[1] 和 Philip C. Calder[2]
(1 食品生物科学学院，雷丁大学，怀特莱茨，雷丁，英国
2 南安普敦大学，南安普敦，英国)

## 前　言

关于饥荒和传染病的流行之间的关系已记入史册，早在公元前370年，希波克拉底就已认识到，营养不良的人更易感染传染病。一般来说，营养不良损害免疫系统，且抑制保护宿主免于病原微生物侵害所必需的免疫功能。导致免疫系统削弱的营养不良可能是由于能量和常量养分的摄入不足和（或）由于缺乏特定微量养分（维生素和矿物质）所致。以上情况常结合发生，这对蛋白质能量的营养不良和微量养分如维生素A、铁、锌和碘的缺乏尤其明显。显然，营养不良的影响在发展中国家是最大的，但在发达国家也是很明显，尤其是那些老年人、饮食性疾病者、酗酒者、具有某些疾病的患者、早熟的和妊娠时较小的婴儿中更是如此。众所周知，单个养分对免疫功能的不同方面的精确影响是很难研究的。然而，越来越清楚的是，在免疫反应中许多养分都有确切的作用，且每个养分都有支持最佳免疫功能的不同范围的摄入量。在这个范围以外降低养分的水平或增加超过这一范围时，都能削弱免疫功能。因此，免疫系统的功能发挥受到作为膳食正常成分摄入的养分的影响，且受合适的营养为宿主保持对细菌、病毒、真菌和寄生虫的完备免疫防御所必需。养分影响免疫功能的细胞及其分子机制才刚开始被理解。本章从综述免疫系统的关键组成开始，主体是致力于对单个常量和微量养分对免疫功能的影响及其建议的作用机制进行评价。

## 免疫系统

### 先天性免疫

免疫系统的作用是保护宿主免受存在于环境中的病原体（细菌、病毒、真菌和寄生虫）和其他有害损伤物的侵害。免疫系统有两个功能分区：先天性（或自然）免疫系统与获得性（也称特异性或适应性）免疫系统。先天性免疫由物理屏障、可溶性因子和包括粒细胞（嗜中性粒细胞、嗜碱性粒细胞和嗜酸性粒细胞）的噬菌细胞、单核细胞和巨噬细胞组成。先天性免疫没有记忆功能，因此，不受以前接触微生物的影响。噬菌细胞是这些先天性免疫的主要效应器，对细菌表面抗原表达特定表面受体。将抗原连接到受体上可诱发噬菌作用及随后通过补体或如超氧化物自由基和过氧化氢等有毒化学物质对病原微生物的破坏。自然杀伤者（natural killer，NK）细胞也拥有表面受体，且通过释放细胞毒素蛋白破坏病原菌。通过这种方式，先天性免疫形成抵抗病原体入侵的第一

道防线。然而，免疫反应通常需要先天性免疫与更加强大且灵活的获得性免疫二者的协调作用。

## 获得性免疫

获得性免疫特异性识别入侵的病原体上的分子（抗原），并将它辨认为宿主的外来物。淋巴细胞可细分为T淋巴细胞和B淋巴细胞，它们影响免疫的形成。所有淋巴细胞（免疫系统的所有细胞）都起源于骨髓，在释放进入循环系统前B淋巴细胞在骨髓中进一步发育成熟，而T淋巴细胞在胸腺中成熟。淋巴细胞能从血流中进入包括淋巴结、脾脏、黏膜淋巴组织、扁桃体和与肠有关的淋巴组织等外周淋巴器官。免疫反应大部分发生在这些淋巴器官上，这些器官能被高效地组织起来促进细胞和入侵病原体间的相互作用。

获得性免疫系统是高度特异性的，因为每个淋巴细胞携带针对每个抗原的表面受体。然而，获得性免疫又是极易变化的，据估计，人的淋巴细胞的所有组成成分能识别近 $10^{11}$ 个抗原，高度的特异性与巨大的淋巴细胞的所有组成成分结合意味着只有相对较少的淋巴细胞能识别一些特异抗原。获得性免疫已经提高了用克隆扩增来解决此问题的能力。一旦与特异抗原相互作用，克隆扩增就涉及一个淋巴细胞的增殖，于是单一淋巴细胞产生一个淋巴细胞克隆，每一个都有能力识别和破坏引起起始反应的抗原。获得性免疫的这个特征就像组成一个军队来抵抗外来入侵。获得性免疫反应在开始激活的几天内是有效的，但在抗原去除后也会持续一段时间。这种持续产生的免疫记忆也是获得性免疫的一个特征。它是对再次接触抗原（即用相同抗原再次感染）产生更强大更有效的免疫反应的基础，也是接种疫苗的基础。最后，通过包含细胞间通信的自我调节机制，免疫系统将重新建立动态平衡。

## B和T淋巴细胞

B淋巴细胞以能产生抗体或免疫球蛋白（immunoglobulin，Ig）为特征，抗体将抗原的特异性传到获得性免疫系统（即B淋巴细胞产生的抗体对单个抗原是特异的）。这种预防抗感染的形成被称为体液免疫，主要通过B淋巴细胞单独作用。在B淋巴细胞携带着能将抗原结合在细胞表面的免疫球蛋白，免疫球蛋白与抗原结合使B淋巴细胞增殖，随后转变成血浆细胞，它能分泌大量具有和双亲细胞一样特异性的抗体。

免疫球蛋白是由两个相同的重链和两个相同的轻链组成的蛋白质。重链的五种不同类型产生了免疫球蛋白的五个主要种类（IgA、IgD 、IgG、IgM和IgE），每一种都能产生体液免疫反应的不同成分。抗体以几种方式来抵抗入侵的病原体。它们通过与毒素或微生物结合，且通过阻止它们对宿主细胞的依附来中和它们，抗体能激活血浆中的补体蛋白，补体蛋白反过来通过噬菌细胞加强对细菌的消灭。既然补体上有抗原结合的位点和供噬菌细胞上的受体结合的位点，抗体也能通过形成物理桥的方式促进两个组分的相互作用，这个桥是一个通常所说的调理作用的过程。被抗体结合的噬菌细胞的类型由抗体种类决定，巨噬细胞和中性粒细胞对IgM和IgG是特异的，嗜酸性粒细胞对IgE是特异的。通过这种方式，抗体成为获得性免疫和先天性免疫反应之间的一种通信形式，它们是通过高度特异的机制引起的，但最终被翻译成能被先天免疫系统理解的形式，使它能破坏病原体。

体液免疫专门抵抗细胞外部病原体的入侵，但一些病原体，尤其是病毒及某些细菌仍能进入细胞内部使个体感染。这类病原体不受体液免疫的影响，但细胞免疫可作用于它们，细胞免疫是通过T淋巴细胞而进行的。T淋巴细胞表面表达特异抗原T细胞受体（T cell receptor，TCR)，TCR有大量抗原所有的组成成分。然而不像B淋巴细胞，T细胞只能识别呈现到细胞表面的抗原，这是体液免疫和细胞免疫明显的区别。因此，通过一个细胞内病原体引起的一个细胞的感染通过信号传递给T淋巴细胞，这种传递是通过来自该病原体肽片段的细胞表面的表达来实现的。这些片段被运送到受感染的细胞表面，并在那里与主要组织相容性复合物（major histocompatibility complex，MHC）协同表达。对人类而言，MHC被称为人白细胞抗原（human leukocyte antigen，HLA)，它是结合到MHC上的病原体肽片段结合体，可被T淋巴细胞识别。MHC有两类，MHCⅠ和MHCⅡ，结合到每一种MHC上的肽源都不同。MHCⅠ结合来自宿主细胞液中合成的抗原蛋白的胚，它们通常是来自病毒或某些细菌；结合到MHCⅡ的肽来自被巨噬细胞吞噬或被抗原呈递细胞（巨噬细胞、树突细胞和B淋巴细胞）内吞的抗原。MHC肽复合物在T淋巴细胞上被TCR识别。表达CD8的T淋巴细胞（细胞毒性T细胞，cytotoxic T cell）识别MHCⅠ，然而表达CD4的T淋巴细胞（辅助T细胞，helper T cell）识别MHCⅡ。因此，细胞内病原体激活细胞毒性T淋巴细胞以杀死受感染细胞，而细胞外病原体则激活辅助T细胞在延迟型超敏感（delayed type hypersensitivity，DTH）反应中，抗原激活的$CD4^+$ T淋巴细胞（辅助T细胞）分泌的细胞因子具有许多作用，包括补充嗜中性粒细胞和来自血液中的单核细胞到达单核细胞的抗原激发和激活位点来有效消灭抗原。人类的DTH反应可通过与化学物质和环境抗原接触致敏或通过微生物抗原的内皮注射来诱导，这已作为细胞介导的免疫的一种快速的体内标记而被广泛使用。通常是接触抗原48h后，通过测量皮肤厚度来估计DTH的反应程度。

## 免疫系统间的通信：细胞因子

在获得性免疫系统内以及在先天性免疫系统和获得性免疫系统间的通信是通过直接的包含连接分子的细胞与细胞的接触和通过把信号从一个细胞传送到另一个细胞的化学信使的生成而实现的（图22.1）。这些化学信使主要是被称为细胞因子的蛋白质，它们能调节产生细胞因子的细胞及其他细胞的活性。每个细胞因子对不同的细胞类型具有多种活性。细胞因子通过与细胞表面的特异受体结合而发挥作用，因而引起靶细胞生长、发育和活性的变化。

肿瘤坏死因子（tumour necrosis factor α，TNF-α)、白细胞介素（IL-1）和IL-6是单核细胞和巨噬细胞产生的最重要的细胞因子。这些细胞因子激活嗜中性粒细胞、单核细胞和巨噬细胞以引发细菌和肿瘤细胞的死亡，增加连接分子在嗜中性粒细胞和内皮细胞表面的表达，激活T、B淋巴细胞增殖和引起其他前炎症反应细胞因子的产生（如TNF诱导IL-1和IL-6的产生，IL-1诱导IL-6的产生）。因此，TNF、IL-1和IL-6是自然免疫和获得性免疫的调节因子，也是两种免疫之间最重要的连接（图22.1）。另外，这些细胞因子具有调节炎症如发烧、失重和急性期肝脏中蛋白质的合成等全身作用。炎症是机体对感染或损伤的直接反应，常表现出发红、肿胀、发热、疼痛等症状。这些症状是血流增加，通过血液毛细管使大分子（如补体、抗体和细胞因子）离开血流穿过内皮的渗透性增加，以及白细胞从血流向周围组织的移动增加的结果。因而，炎症是先天性免疫的一个完整部分，生成适当量的 TNF、IL-1 和 IL-6 对感染反应是重要的。然而，不适当的或过量的产生是危险的，这些细胞因子尤其是 TNF 可导致发生在慢性炎症反应条件下的某些病理反应的发生（如风湿性关节炎、牛皮癣）。

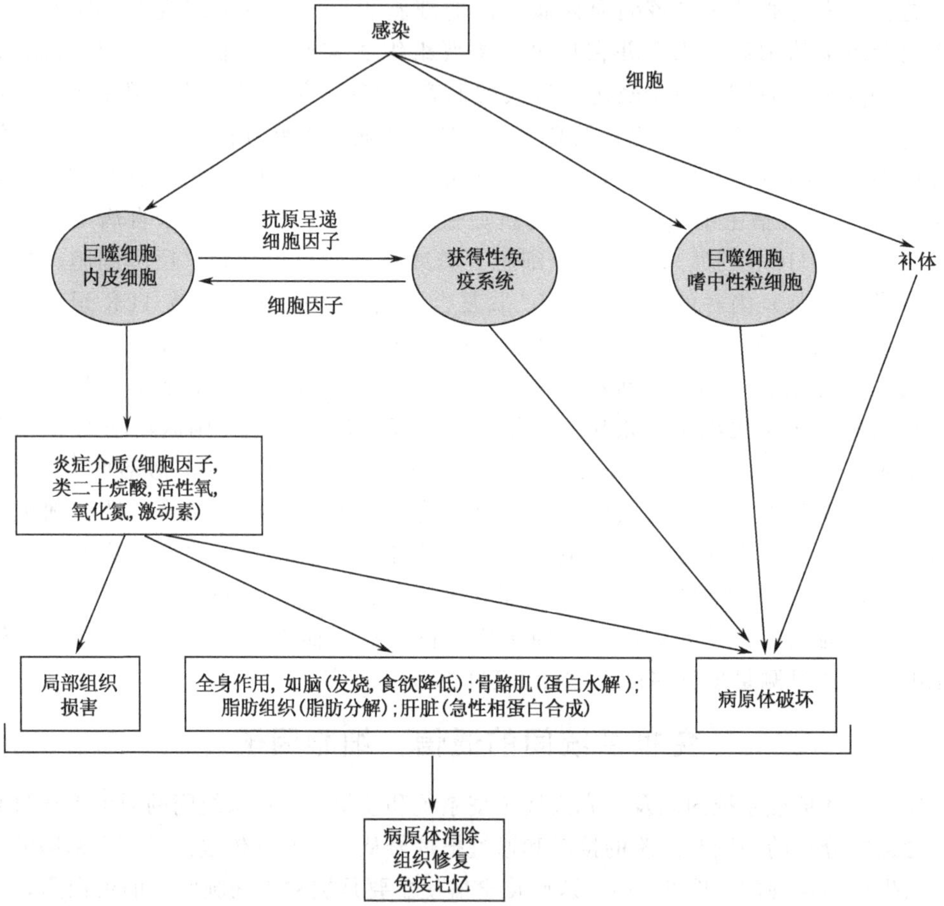


图 22.1　免疫系统概观

修改自 Calder (2002)

辅助 T 淋巴细胞根据它们产生的细胞因子的模式被细分为两大类（图 22.2）。没有遇到抗原以前的辅助 T 淋巴细胞在开始遇到抗原时主要产生 IL-2。这些细胞可能分化成一个有时被称为 Th0 细胞的群体，Th0 进一步分化成 Th1 或 Th2 细胞（图 22.2）。这种分化由细胞因子调节：IL-12 和干扰素 γ（IFN-γ）促进 Th1 细胞发育，而 IL-4 促进 Th2 细胞发育（图 22.2）。在细胞因子产生方面 Th1 和 Th2 细胞本身具有相当严格的限制：Th1 细胞产生 IL-2 和 IFN-γ，它能激活巨噬细胞、NK 细胞和细胞毒素 T 淋巴细胞，是细胞介导的主要免疫效应器；Th2 细胞产生 IL-4，它能激活免疫球蛋白 E

的产生，IL-5 是一个嗜酸性粒细胞激活因子，IL-10 与 IL-4 一起抑制细胞介导的免疫（图 22.2）。Th2 细胞负责防御蠕虫类寄生虫，这是由于柱状细胞和嗜碱性细胞的由 IgE 介导的激活作用所致。通过 Th1 和 Th2 淋巴细胞，细胞因子的分泌模式在小鼠上首次得到证明，人的辅助 T 淋巴细胞在细胞因子外形上确实显示出差异，而在小鼠上这种区分不明显。

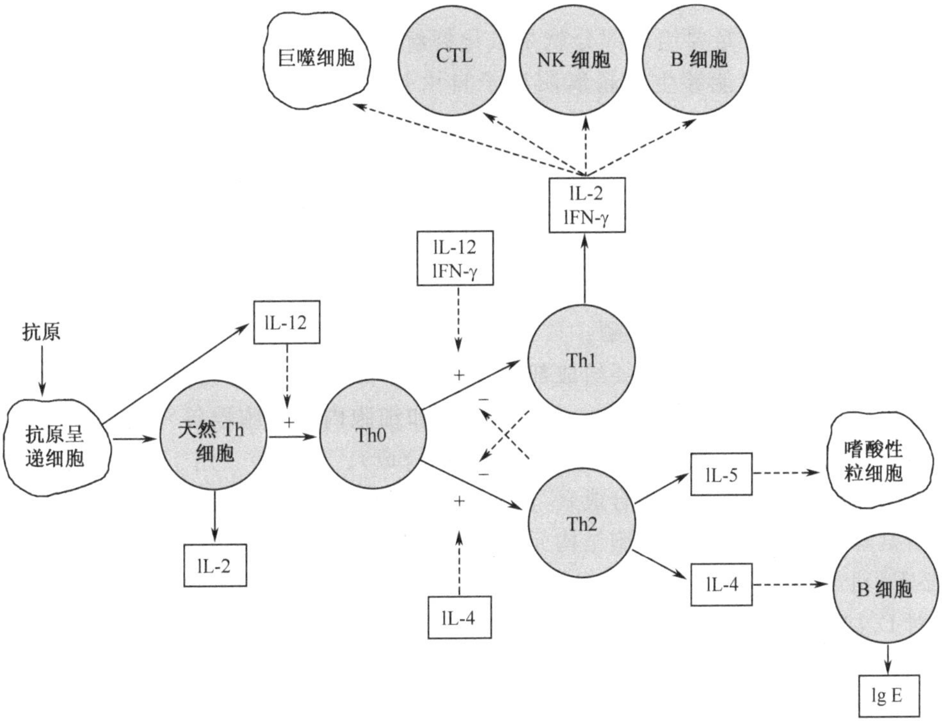


图 22.2　辅助性 T 淋巴细胞沿 Th1 和 Th2 途径的分化以及细胞因子 Th1 和 Th2 在引发效应器细胞功能中的作用

# 为什么养分能影响免疫功能

尽管免疫系统在任何时候都在发挥作用，但是特异性免疫在宿主被病原体刺激时才被激活。这种激活作用与免疫系统对底物和养分的需要显著增加以提供现成的能源相关，这一需要可由外源（如从食物中）和（或）由能源库提供。免疫系统的细胞有代谢活性，能利用葡萄糖、氨基酸和脂肪酸作为燃料。能量的产生包括电子载体，它们是核苷酸衍生物［如烟酰胺腺嘌呤二核苷酸（NAD）和黄素腺嘌呤二核苷酸（FAD）］和大量辅酶。电子载体和辅酶通常是维生素的衍生物：焦磷酸硫胺素来源于硫氨素（维生素 $B_1$），FAD 和黄素单核苷酸来自核黄素（维生素 $B_2$），NAD 来源于烟碱（烟酸），磷酸吡哆醛来源于维生素 $B_6$，辅酶 A 来源于泛酸盐、四氢叶酸来自叶酸、钴胺酰胺来自钴胺酸（维生素 $B_{12}$）。另外，某些酶的活性需要依靠生物素的存在。能量产生的途径的最终组成（线粒体电子转移链）也包括在其活性位点中含有铁或铜的电子载体。

免疫反应的激活诱导蛋白（免疫球蛋白、细胞因子、细胞因子受体、连接分子和急性期蛋白）和来源于脂类的调节因子（前列腺素和白三烯）的生成。为达到最佳的反应目的，显然在合适的位置要有合适的酶（用于 RNA 和蛋白质合成以及它们的调节）和足够的可利用底物用于合成 RNA 的核苷酸，用于合成蛋白质的所有氨基酸，用于合成类二十烷的。

免疫反应的一个重要组成部分就是氧化裂解，在此裂解期间，与 NADPH 的氧化相关联的反应中，氧能够生成超氧阴离子自由基。所生成的各种活性氧可能使宿主组织损伤，因而，抗氧化保护机制是必需的。抗氧化剂包括传统的抗氧化维生素［α-生育酚（维生素 E）和抗坏血酸（维生素 C）］、谷胱甘肽（谷氨酸、胱氨酸和甘氨酸组成的三肽）、抗氧化酶（过氧化物歧化酶和过氧化氢酶）和谷胱甘肽循环酶（过氧化物酶）。超氧化物歧化酶有两种存在形式：线粒体形式和细胞液形式，线粒体形式在其活性位点含有锰，细胞液形式则含有铜和锌。过氧化氢酶在其活性位点含有铁，而谷胱甘肽过氧化物酶则含有硒。

细胞增殖是免疫反应的重要组成部分，用于扩增和记忆：分裂前必须进行 DNA 的复制，然后是所有细胞组成成分（蛋白质、膜和细胞内的细胞器等）的复制。除能量外，显然需要提供核苷酸（用于 DNA 和 RNA 合成）、氨基酸（用于蛋白质合成）、脂肪酸、碱基和磷酸（用于磷脂合成）、其他脂类（如胆固醇）和细胞组分。尽管核苷酸主要由氨基酸合成，但一些细胞结构元件仍不能在哺乳动物细胞中合成，而必须从食物（如必需脂肪酸、必需氨基酸和矿物质）中摄取。氨基酸（如精氨酸）是多胺合成的前体，对 DNA 复制和细胞分裂的调节起作用。多种微量养分（如铁、叶酸、锌和镁）也参与核苷酸和核酸的合成。

总之，营养物质在免疫功能中的作用有很多，而且是多样的。很容易理解，如果要进行一个合适的免疫反应，营养物质的充足和均衡的供给是必需的，但是，我们对营养物质在免疫功能中作用的分子基础的理解尚不完全，即使已有证据表明它对表面受体/蛋白调节、调节生成和氧化还原作用条件有影响。

## 蛋白质能量营养不良和免疫功能

蛋白能量营养不良虽然常被认为只是发展中国家的一个问题，但在一些最富裕国家中也已经被述及。在发达国家，中等的营养不良常见于那些老年、厌食、食欲旺盛、早熟婴儿、医院患者和具有多种疾病的患者（如胆囊纤维化、获得性免疫缺乏症和某些癌症）中。在动物上的大量研究表明蛋白质缺乏对免疫具有不利作用（Woodward，1998），在多种人类环境中，这些作用已得到证实（Gross and Newberne，1980；Chandra，1991）。蛋白质缺乏对免疫反应的降低和对传染易感性的增加是不足为奇的，因为免疫防御依赖于细胞复制和具有生物活性的蛋白质的产生（如抗体、细胞因子和急性期蛋白）。但是，人们认为蛋白质能量营养不良不仅体现为摄入的能量和常量养分总量的不足，而且体现为微量养分的缺乏。

实际上蛋白质能量营养不良可能影响免疫的所有形式，但是它对非特异性防御和细胞介导的免疫的影响比对体液免疫（抗体）反应的影响严重得多（Chandra，1991；Woodward，1998）。蛋白质能量营养不良造成实验动物和人的淋巴器官萎缩（胸腺、脾、淋巴结和扁桃体），循环的白细胞量增加，但是这是由嗜中性粒细胞的数量增加，单核细胞、淋巴细胞、$CD4^+$ 细胞和 $CD8^+$ 细胞的绝对和相对数量减少，以及 $CD4^+/CD8^+$ 的比例也减少所导致的。循环的 T 淋巴细胞数量的下降与营养不良的程度是成比例的（Rivera *et al.*，1986）。营养不良使 T 淋巴细胞对有丝分裂原和抗原的增殖反应下降，细胞因子 IL-2 和 IFN-γ 的合成以及 NK 细胞的活性也会降低。单核细胞所产生的细胞因子（包括 TNF-α、IL-1 和 IL-6）也因营养不良而下降。在体内受特异记忆抗原刺激的 DTH 反应也因营养不良而下降（Rivera *et al.*，1986）。嗜中性粒细胞的杀细菌活性和在呼吸系统中的释放也因营养不良而下降，但嗜中性粒细胞和单核细胞的噬菌能力似乎不受影响。循环中的 B 淋巴细胞和免疫球蛋白水平似乎不受营养不良的影响，也有可能会增加（Chandra，1991），这种增加可能代表潜在感染。分泌型 IgA 在眼泪、腮腺和肠道洗液中的水平因营养不良而下降，这可能与负责 IgA 的跨上皮细胞转运的聚合 Ig 受体表达水平的下降相关，而与 IgA 合成的降低无关（Woodward，1998）。

# 微量养分和免疫功能

## 维生素 A

维生素 A 缺乏常伴随传染病和某些癌症的发病率和严重性增加，这与受损的免疫反应有关（Semba，1998，1999）。研究表明维生素 A 缺乏时几乎所有的免疫反应会受到削弱（Semba，1998，1999），而且维生素 A 对保持表皮和黏膜的完整性是必需的，维生素 A 缺乏的小鼠其肠道黏膜有组织病理学变化，这与消化道屏障的完整性受损和削弱的黏膜分泌相一致性，这两者都有利于病原体的入侵。饲喂低蛋白质食物的小鼠表现出黏膜 IgA 和细胞因子（IL-4 、IL-5 和 IL-6）浓度的下降，这能通过提高膳食中维生素 A 水平来预防（Nikawa *et al.*，1999）。维生素 A 可以调节角化细胞的分化，它的缺乏将会诱导皮肤角化的变化，这可以用来解释为什么维生素 A 缺乏会导致皮肤感染发生率的增加（Semba，1998）。

在发展中国家已广泛研究了维生素 A 缺乏对传染病的影响。维生素 A 缺乏与儿童的发病率和死亡率的增加有关，它们易引起呼吸感染、腹泻、囊虫病（Chandra，1991；Scrimshaw and SanGiovanni，1997；Semba，1999；Calder and Jackson，2000）。给耗竭或缺乏维生素 A 的动物或受试者补充维生素 A 会导致淋巴器官的发育、循环中的免疫细胞数量、免疫细胞的功能和 DTH 反应的恢复，从而能够增加病原体对感染的抵抗力（Semba，1998）。在服用维生素 A 与儿童的发病率和死亡率的关系方面已经进行了大量的研究。这些研究中大多数表明可以降低发病率，也可以降低与麻疹和与腹泻有关疾病的死亡率，但不降低呼吸道感染（Chandra，1991；Scrimshaw and SanGiovanni，1997；Semba，1999；Calder and Jackson，2000）。食物中高水平的维生素可使免疫反应的增加超过正常水平。相反的作用也曾观察到，即高水平的维生素 A 可增加、没有改变或降低免疫反应（Fridman and Sklan，1993）。动物试验表明维生素 A 摄入过量与维生素 A 缺乏均可将免疫功能和对疾病的抵抗力降低到相似的程度（Friedman and Sklan，1993）。维生素 A 经核视黄酸和类维生素 A 受体发挥作用，它们作用于许多特定靶基因的转录激活因子（见第 10 章）。

## 叶酸和 B 族维生素

在试验动物中叶酸的缺乏引起胸腺和脾脏萎缩，循环中 T 细胞数量、细胞毒素 T 淋巴细胞的活性和脾脏淋巴细胞的增殖的降低，但是没有改变嗜中性粒细胞的吞噬作用和杀菌作用（Gross and Newberne，1980）。相反，维生素 $B_{12}$ 缺乏降低嗜中性粒细胞的吞噬作用和杀菌作用（Gross and Newberne，1980）。在试验动物中，维生素 $B_6$ 的缺乏导致胸腺和脾脏萎缩，淋巴细胞增殖和 DTH 反应降低，同种异体移植的幸存增加（Gross and Newberne，1980）。在一个对健康老年人 21 天的研究中，维生素 $B_6$ 缺乏的食物［3μg/（kg 体重・d）或分别对男性和女性约 0.17 和 0.1mg/d］导致循环的淋巴细胞百分含量和总数量下降，对有丝分裂原的反应中 T 淋巴细胞和 B 淋巴细胞的增殖降低，IL-2 的生成减少（Meydani *et al.*，1991）。以 15 或 22.5μg/（kg 体重・d）经过 21 天的充分供应没有恢复免疫功能到起始值。然而，在每天以 33.75μg/kg 体重的供应（分别对男性和女性约 1.9 和 1.1mg/d）可恢复免疫参数到起始值。提供 41mg/d 的维生素 $B_6$ 经过 4 天，使淋巴细胞增殖和 IL-2 生成增加。这些数据表明维生素 $B_6$ 的缺乏削弱了人免疫功能（至少在老年），这种削弱通过充分供应可取消，在维生素 $B_6$ 水平超过典型的日常消费量时，淋巴细胞的功能提高。

## 维 生 素 C

维生素 C 是水溶性抗氧化剂，在循环的白细胞中具有较高的浓度，并在感染期间被利用。维生素 C 对免疫功能的影响非常依赖于剂量，人们在试图比较使用不同剂量维生素 C 的研究时产生了迷惑（Siegel，1993）。人维生素 C 缺乏没有削弱淋巴细胞的增殖，但可能增加循环中的 Ig 浓度（尽管一些研究不能表明这一点；Siegel，1993）。Jacob 等（1991）在研究不同维生素 C 水平对年轻健康的非吸烟者免疫功能的影响时发现：缺乏维生素 C 的食物降低单核细胞维生素 C 含量的 50%，降低 DTH 对 7 个回忆抗原的反应，但没有改变淋巴细胞增殖。即使单核细胞维生素 C 含量恢复到基础水平，添加维生素 C 10、20、60 或 250mg/d（每次持续 28 天）不能引起 DTH 反应的恢复。

## 维 生 素 D

维生素 D 受体在大多数免疫系统的细胞中是存在的，表明它具有免疫调节的特性（Lemire，1992）。它对免疫功能的影响包括对 T 淋巴细胞增殖、IL-2 和 IFN-γ 产生、单核细胞分化、NK 细胞活性和通过 B 淋巴细胞产生的抗体的生成的抑制作用（Lemire，1992；Overbergh *et al.*，2000）。活性的代谢物（1，25-二羟维生素 $D_3$）似乎对细胞因子的生成具有特异性的影响，推动免疫反应朝向 Th2 形式进行，这种作用部分通过抑制巨噬细胞/树突细胞 IL-12 的生成（诱导 Th1 反应的主要细胞因子），而且也通过直接抑制 IL-2 和 IFN-γ 的生成（Th1 细胞因子）及刺激 IL-4 的生成而进行（Alroy *et al.*，1995；D'Ambrosio *et al.*，1998；Takeuchi *et al.*，1998；Singh *et al.*，1999）。这可以用来解释为什么维生素D能延长移植物的存活（Lemire，1992；Overbergh *et al.*，2000），且对自身免疫疾病的动物模型也是有益的。例如，维生素D缺乏会增加小鼠对自身免疫试验性过敏脑脊髓炎（一种多样硬化症模型）的易感性（Cantorna *et al.*，1996）。在小鼠和大鼠的这种疾病开始之前或之后口服维生素D，便能阻止该病的发展和消除针对髓磷脂碱性蛋白的抗体的增加以及组织学变化（Lemire，1992；Cantorna *et al.*，1996）。腹膜内注入维生素D能预防皮肤损伤、蛋白尿以及在遗传易感小鼠品系中伴随狼疮发育的自身抗体的产生（Lemire，1992），减少遗传易感小鼠中胰岛炎的发生（Mathieu *et al.*，1992），在小鼠上防止胶原质诱导的关节炎的发生（Cantorna *et al.*，1998）。这些结果表明维生素D抑制Th1细胞的活性。

根据典型的维生素D基因组作用，1，25（$OH_2$）$D_3$ 是从血清结合蛋白中释放出来，穿过细胞膜扩散，在细胞核中结合到典型的含有锌指的受体上（维生素D受体，vitamin D receptor，VDR）（Macdonald *et al.*，1994）。在维生素A辅助受体（类维生素A X受体，RXR）和维生素D间有某种相互作用，甲状腺素也一样，在能影响靶基因转录的不同组合中受体形成异质二聚体（Haag，1999）。在过去的10年间，人们已经清楚认识到除了缓慢基因组作用的模式之外，经过快速非基因组作用，维生素D也能发挥作用，这涉及许多第二信使途径。

维生素D对IFN-γ启动子起负调节作用，这种作用通过RXR被增强（Cippitelli and Santoni，1998）。由1，25（$OH_2$）$D_3$ 通过直接干涉启动子的激活而对IL-2基因表达的抑制已被描述过，而这种抑制似乎是由于VDR-RXR异质二聚体使激活的T细胞/激活物蛋白-1（nuclear factor of activated T-cell/activator protein-1，NFAT-AP-1）复合物的核因子的形成受到阻止所致（Alroy *et al.*，1995）。1，25（$OH_2$）$D_3$ 对T细胞的核因子——卡巴B（nuclear factor-kappa B，NF-κB）活性的抑制性影响也已被描述，表明一个调节机制可能阻止过量淋巴细胞的活化（Yu *et al.*，1995）。此类作用可能是维生素D在治疗自身免疫疾病的动物模型中具有有益作用的原因。而且，已证明免疫抑制剂环孢霉素可增加负责1，25（$OH_2$）$D_3$ 的最终激活酶的表达，在自身免疫缺损的小鼠上已经发现该酶的表达不足（Overbergh *et al.*，2000）。尽管该酶的不足不一定是小鼠自身免疫缺损的一个原因，但它可能有助于小鼠自身免疫缺损的发生，并且发炎时的正调节［导致1，25（$OH_2$）$D_3$ 的合成］也可能为其发生提供一个负反馈环（Casteel *et al.*，1998a，b）。

## 维生素E

维生素E是体内主要的脂溶性抗氧化剂，对保护细胞膜脂类免于过氧化损伤是必需的。既然自由基和脂类的过氧化是抑制免疫的，人们认为维生素E应该在使免疫反应最优化、甚至是在增强免疫反应方面起作用（Bendich，1993；Meydani and Beharka，1998）。除了在早产婴儿和老年人中外，维生素E的临床缺乏在人类中是很少见的，尽管在许多国家有许多人的维生素E的摄入低于日常推荐的量。

对试验动物而言，维生素E缺乏降低脾脏淋巴细胞增殖、NK细胞活性、接种后特异抗体的产生和嗜中性粒细胞的吞噬作用并增加动物对传染病病原体的易感性（Bendich，1993；Meydani and Beharka，1998；Han and Meydani，1999）。当给动物饲喂含有高水平的多不饱和脂肪酸时，维生素 E 消耗是相当显著的，在试验动物膳食中添加维生素 E 能增加抗体的产生、淋巴细胞的增殖、IL-2 产生、NK 细胞活性、巨噬细胞的吞噬作用和 DTH 反应（Meydani and Beharka，1998）。膳食中维生素 E 促进肉鸡、火鸡、鼠、猪、绵羊和牛对病原体的抵抗力（Meydani and Beharka，1998；Han and Meydani，1999）。给小鼠饲喂过量的维生素 E 会阻止脾脏淋巴细胞 IL-2 和 IFN-γ 生成以及逆转录病毒诱导的 NK 细胞活性的下降（Wang *et al.*，1994），并降低肺中流感病毒的滴度（Hayek *et al.*，1997）。

血浆维生素 E 水平和 DTH 反应呈正相关，但是与 60 岁以上的健康成年人传染病的发生率呈负相关（Bendich，1993；Meydani and Beharka，1998；Han and Meydani，1999）。似乎在老年人中补充维生素 E 尤其有益。综合研究表明，给老年受试者每天补充 60、200 和 800mg 的维生素 E 可增加 DTH 反应，在剂量为 200mg/d 时具有最大的效果（Meydani *et al.*，1997）。这个剂量也会显著增加抗体对肝炎 B、破伤风类毒素和肺炎双球菌疫苗的反应。因此，在食物中补充超过正常可达到的水平的维生素 E 能提高某些免疫反应，而与许多其他微量养分一样，极大地超过正常需要的剂量可能抑制免疫反应，如 800mg/d 的维生素 E 组的抗体反应比空白对照组还低（Meydani *et al.*，1997）。

## 锌

动物的锌缺乏与很多免疫损伤相关（Fraker *et al.*，1993；Shankar and Prasad，1998）。锌缺乏对骨髓有明显的影响，它能够降低有核细胞的数量和作为淋巴细胞前体物的细胞的数量和比列（Fraker *et al.*，1993；Fraker and King，1998）。缺乏锌与镰刀型红细胞病相关，NK 细胞活性降低，但补充锌后可恢复到正常水平。肠致病性肢端皮炎的特征是肠道锌吸收降低、胸腺萎缩、淋巴细胞的发育减弱、淋巴细胞反应及 DTH 降低。人的中等程度的锌缺乏或试验性缺乏（被诱导以小于 3.5mg/d 的消耗量；在英国成年人日常摄入量是 9～12mg/d）降低了胸腺活性、NK 细胞活性、淋巴细胞增殖、IL-2、IFN-γ 和 TNF-α 产生和 DTH 反应，这些都可通过充足供应锌而得以纠正（Shankar and Prasad，1998）。

血浆低锌水平可用来预测随后在营养不良的群体中下呼吸道感染和腹泻的发生（Shankar and Prasad，1998；Calder and Jackson，2000）。事实上，腹泻被认为是一种缺锌症状。人们已证明，给营养不良的儿童口服锌[2mg/(kg 体重・d)]可使腹泻、呼吸和皮肤感染发生率降低 50%以上，与服用低剂量的锌（3.5mg/d）的儿童相比，还可促进生长（Calder and Jackson，2000）。大量研究表明补充锌降低了儿童腹泻和呼吸疾病的发生，尽管有些研究并未见到补锌对呼吸疾病有益（Chandra，1991；Scrimshaw and SanGiovanni，1997；Shankar and Prasad，1998；Calder and Jackson，2000）。给前期低体重的婴儿口服锌（每天 1mg/kg，经过 30 天）可增加循环 T 淋巴细胞数量和淋巴细胞的增殖（Chandra，1991）。给低出生重、妊娠时较小的婴儿每天提供 5mg 锌，6 个月便增加了细胞介导的免疫功能从而降低了胃肠道和上呼吸感染的发生（Calder and Jackson，2000），但是1mg/d锌的剂量没有效果。尽管增加锌摄入可提高免疫功能，过量的锌摄入却削弱免疫反应。例如，给年轻的成年人每天补锌300mg，6周后降低了淋巴细胞和巨噬细胞的功能（Chandra，1984）。高锌的摄入也导致铜耗竭，铜的缺乏削弱免疫功能（见下）。

锌是许多酶的辅助因子，包括铜锌超氧化物歧化酶（即细胞质的抗氧化剂）因此锌涉及保护宿主细胞不受免疫反应期间产生的各种自由基的细胞毒素的影响。锌是胸苷激酶、DNA聚合酶、RNA聚合酶和氨酰基-tRNA合成酶的活性所必需的元素，因此，锌在DNA、RNA和蛋白质合成以及整个细胞周期演化中都具有关键的作用。而且许多转录因子包括NF-κB的DNA结合区域都含有锌指，这也进一步强调了锌在基因转录上的作用。近来研究指出，锌通过对NF-κB途径的作用，在调节细胞因子产生方面具有关键作用。

## 铜

尽管人们相信人体中明显的铜缺乏是罕见的，但中等缺乏可能会在一些人群中出现。锌和铁拮抗铜的吸收，以至于高剂量锌和铁的摄入造成铜的中等缺乏。铜缺乏在早产婴儿和肠外营养的患者中已被述及。铜缺乏的典型症状是Menkes综合征，这是一种导致血中携带铜的铜蓝蛋白完全缺失的先天性疾病。Menkes综合征的孩子的细菌感染、腹泻和肺炎的发生会增加。试验和农场动物的铜缺乏会大大削弱试验动物的免疫功能并增加对细菌和寄生虫的易感性（Prohaska and Failla，1993；Failla and Hopkins，1998）。在人体所做的研究表明，低铜食物降低淋巴细胞的增殖和IL-2的产生，然而，口服铜将逆转这些作用（Kelley *et al.*，1995）。与许多其他微量养分一样，过量的铜可能抑制免疫。

## 铁

铁缺乏对试验动物和人的免疫功能有多种影响（Sherman and Spear，1993）。缺铁的个体具有正常的吞噬功能，且嗜中性粒细胞杀菌的能力受到削弱，这可能是呼吸脉冲发生改变所致。铁缺乏与肠道和呼吸感染相关（Calder and Jackson，2000）。尽管铁缺乏对免疫反应有抑制作用，负荷过多铁的疾病和过量铁的补充也与感染的危险性增加相关（Caldera and Jackson，2000）。由于微生物需要铁，故口服铁可能有利于病原体，至少部分如此。确实，伴随感染的循环，铁浓度的下降是宿主试图使感染物缺铁而死这一观点至今还存在争议。然而，铁过量也降低T淋巴细胞的数量、淋巴细胞增殖、IL-2产生、细胞毒素T细胞、NK细胞活性以及嗜中性粒细胞的吞噬作用（Sherman and Spear，1993）。铁过量的影响机制还不清楚，但是它可能与影响细胞运动和功能的淋巴组织中的铁的沉积或游离铁的水平的过氧化损伤的增加有关。

关于铁的营养状况和疟疾的情况关系很复杂，既然红色血细胞是疟原虫的宿主，宿主所储存的铁和入侵的病原体占同一空间。这可解释疟疾在铁丰富的个体比铁缺乏的个体更常见，疟疾的感染水平和疾病的严重性因补铁而增加（Calder and Jackson，2000）。实际上，人们对降低铁水平可能改善疟疾的结论还存在争议，产生争议主要因为螯合铁治疗能加强对寄生虫的清除和提高抗疟疾的效果（Calder and Jackson，2000）。

## 硒

硒在肝脏、脾脏和淋巴结中具有较高的浓度。试验动物缺硒将会降低一系列的免疫功能和增加对细菌、病毒、真菌和寄生虫攻击的易感性（Stabel and Spears，1993；McKenzie *et al.*，1998）。硒缺乏对嗜中性粒细胞或巨噬细胞的吞噬能力没有影响，但是一旦微生物被细胞摄入，缺硒确实会降低这些细胞杀死微生物的能力（Stabel and Spears，1993）。动物补硒研究证明，硒能提高免疫（接种）和抗原攻击反应的抗体的效价，使淋巴细胞增殖、增加 IFN-$\gamma$ 产生和提高 NK 细胞活性、增加 DTH 反应和皮肤异源移植排斥作用，并降低对感染的易感性（Stabel and Spears，1993；McKenzie *et al.*，1998）。人的硒缺乏导致循环中的 IgG 和 IgM 浓度下降。

研究表明，在削弱的宿主硒营养状况与对病毒传染的易感性之间，存在一个吸引人的关系。用库克萨基病毒 B3-诱导的心肌炎小鼠模型表明，缺硒的小鼠对病毒的易感性高于补硒小鼠（Beck，1999）。一个正常的无毒病毒株在小鼠缺硒时会变成有毒株，这表明增加的毒性是病毒基因组特定变化的结果（Beck，1999）。

依赖于硒的谷胱甘肽过氧化物酶和硫氧还蛋白还原酶是抗氧化酶的重要例子，它们可以保护细胞免受被释放的杀死微生物的自由基的细胞毒素的作用，这些酶在单核细胞分化期间是正调节，且受维生素 D 的影响。硫氧还蛋白还原酶减少巯基和涉及调节蛋白质与蛋白质之间的相互作用和蛋白质与 DNA 的相互作用（如转录因子的 DNA 连接），这种相互作用代表了硒在免疫系统中作用的一种潜在的分子连接。

## 微量养分的组合

一些研究探讨了微量养分组合在免疫方面的作用。Penn 等（1991）研究了补充维生素 A、C 和 E 对健康老人的影响，结果表明这种补充可以增加循环的 T 细胞数量，增加辅助细胞数量，增加辅助细胞对细胞毒素细胞的比率和增加淋巴细胞对有丝分裂原的反应。在另一个微量养分组合的研究中，健康老人接受无效剂或一种多种维生素/微量元素添补物达 12 个月，多种维生素的补充增加了 T 细胞的数量、NK 细胞的活性和 IL-2 的生成，促进了抗体反应（Chandra，1992），与无效剂组相比，该组与感染相关的疾病也较少（Chandra，1992）。

# 食物脂肪和免疫功能

## 食物脂肪的量和免疫功能

动物研究表明，与低脂肪食物相比较，高脂肪食物降低了淋巴细胞增殖和 NK 细胞活性（Calder，1998，2001a，b）。然而，这种精确影响还取决于在高脂肪食物中所用脂肪的确切水平和它的来源。在与有丝分裂原反应时，食物中总脂肪摄入的降低（从总能量的 40%到 25%）可使人的血淋巴细胞因有组分裂原的作用而大量增殖（Calder，1998，2001a，b）。当脂肪的摄入减少到小于 30%的能量时，人的 NK 细胞活性显著增加（Calder，1998，2001a，b）。当脂肪的摄入量降低，使能量从 36％降低到 27％时，来自老年人的通过脂多糖激活的单核细胞的 IL-1 生成量就会增加（Calder，1998）。总之，这些数据表明，高脂肪食物抑制了人天然免疫和细胞介导的免疫的细胞成分的活性。

## 必需脂肪酸和免疫功能

必需脂肪酸有两个主要家族，*n*-6（或 ω-6）和 *n*-3（ω-3）家族，它们在动物体内不能相互转化。这两个家族的最简单成员是亚麻酸（18∶2*n*-6）和 α-亚麻酸（18∶3*n*-3）。植物组织和植物油是亚麻酸和 α-亚麻酸丰富的来源。当亚油酸随食物被摄入时，能转化成花生四烯酸（20∶4*n*-6）（图 22.3）。食物中 α-亚麻酸通过相同路径能转化成二十碳五烯酸（20∶5*n*-3）和二十二碳六烯酸（22∶6*n*-3）（图 22-3）。二十碳五烯酸和二十二碳六烯酸在多油的鱼组织和鱼油制备物中含量丰富。当摄入不到多油鱼时，α-亚麻酸是主要的食物（*n*-3）脂肪酸。

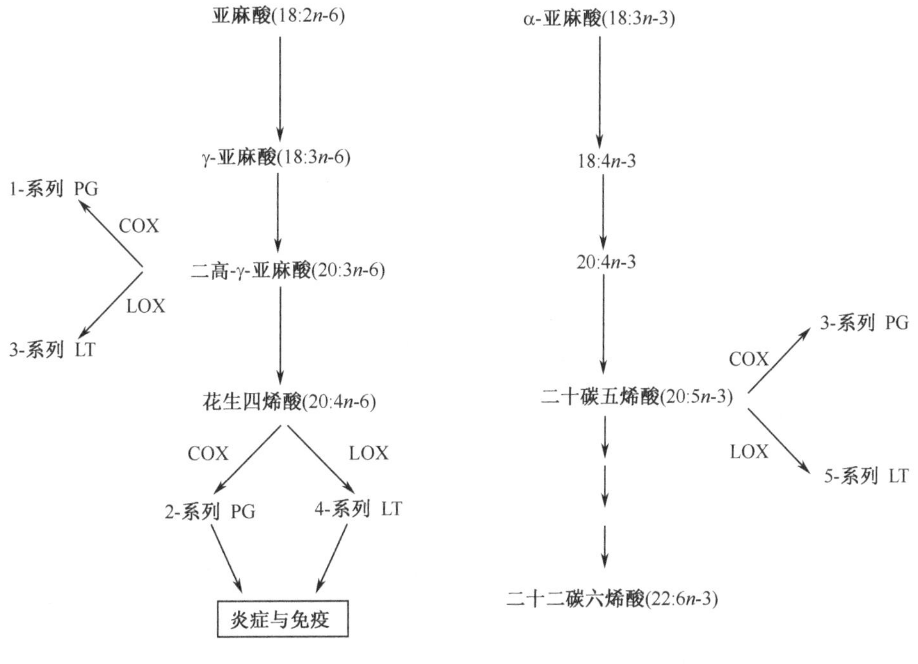


图 22.3　*n*-6 和 *n*-3 多不饱和脂肪酸的代谢途径

COX，环加氧酶，LOX，脂肪氧化酶 LT，白细胞三烯；PG，前列腺素

与饲喂有足够亚麻酸和 α-亚麻酸的食物的大鼠或小鼠相比较，饲喂缺乏这两种脂肪酸的膳食的大鼠或小鼠的胸腺和脾脏的重量、淋巴细胞的增殖、嗜中性粒细胞的趋化性、巨噬细胞调节的细胞毒素和 DTH 反应均会降低（Kelley and Daudu，1993）。这很可能是因为免疫细胞需要多不饱和脂肪酸用于膜的合成和作为类二十烷酸（指白细胞三烯、前列腺素和凝血烷等激素）合成的前体物。

## 类二十烷酸：脂肪酸和免疫系统的联系

在脂肪酸和免疫功能之间的关键联系是称为类二十烷酸的介质家族，它由脂肪酸尤其是二高-γ-亚麻酸、花生四烯酸和二十碳五烯酸组成（图 22.3）。既然大多数细胞膜含有大量的花生四烯酸，与二高-γ-亚麻酸和二十碳五烯酸相比较，花生四烯酸通常是类二十烷酸合成的主要前体。花生四烯酸在细胞膜中被多种磷酸化酶（大部分是磷酸化酶 $A_2$）所释放，游离花生四烯酸随后作为环加氧酶的底物，形成前列腺素和相关复合物，或者作为脂肪氧化酶形成白介素及其相关复合物。这些复合物涉及调节炎症和嗜中性粒细胞、单核细胞/巨噬细胞、T 细胞和 B 细胞的功能。通过增加对（*n*-3）脂肪酸尤其是来自鱼油的二十碳五烯酸的利用，降低了感染细胞产生来自于花生四烯酸的类二十烷酸的能力（Calder，2001a，b，c）。伴随鱼油的摄入，花生四烯酸来源的调节物的含量将会下降，因此，人们认为鱼油能抗炎症反应。有大量的证据支持这一观点。例如，（*n*-3）脂肪酸通过单核细胞/巨噬细胞影响细胞因子的产生模式（Calder，2001a，b，c）。然而，对感染的易感性的影响在动物中还存在争议，一些动物试验表明饲喂大量鱼油的动物对感染的易感性增加，其他动物试验则表明会降低易感性（Calder，2001a，b）。但很显然，鱼油具有免受内毒素的有害作用的保护功能（Calder，2001a，b），且对炎症疾病如风湿性关节炎的现在疗法是一个非常有用的辅助剂（Calder，2001a，b，c）。

## 作用机制

除了能够降低来自花生四烯酸的类二十烷酸的生成的直接诱导作用，近来研究表明（*n*－3）脂肪酸能影响细胞因子和其他免疫调节基因的表达（Renier *et al*.，1993；Curtis *et al*.，2000；Mikes *et al*.，2000；Wallace *et al*.，2001）。这可能是由于改变了类二十烷酸的生成，或者也可能是不依赖于这种改变（Miles and Calder，1998；Yaqoob，1998）。例如，不同脂肪酸可能与各种转录因子相互作用，大量转录因子已被发现具有前炎症反应的功能。其中的一部分是细胞特异性的，然而，其他如 NF-κB 是普遍性的。NF-κB 很重要，因为它在健康和慢性疾病中控制编码细胞因子、细胞连接分子、生长因子和一些急性期蛋白的基因的表达（Barnes and Karin，1997）。脂肪酸可直接或间接地影响 NF-κB 的活化和其 DNA 的结合。某些（*n*-3）脂肪酸的这种作用模式在最近得以证明（Lo *et al*.，1999；Xi *et al*.，2001）。

过氧化物酶体增殖物激活受体（peroxisome proliferator activated receptor，PPAR）是最近发现的一种转录因子，已表明可被脂肪酸及其代谢衍生物结合并激活，其中两种同质异构体 PPARα、PPARγ 存在于免疫系统的细胞内，具有抗炎性作用（Gelman *et al*.，1999）。尽管，这些作用已经用 PPAR 的合成激活物所证实，似乎对 PPARγ 至少这些复合物的一些抗炎性作用不依赖于它们转录因子本身（Thieringer *et al*.，2000）。而且，PPAR 的生理配体的性质还不清楚，因此，尽管认为 PPAR 可作为一些脂肪酸的抗炎性作用的分子调节物这一观点很引人注意，但 PPAR 在免疫和炎性反应中的作用还需要被阐明，脂肪酸真正的分子靶体尚需被证明。值得注意的是，脂肪酸对基因表达的影响不一定是直接对单一转录因子的作用。许多转录因子参与交互作用，导致转移活化或转移抑制，这明确证明 PPAR 可参与 NF-κB 和 AP-1 的交互作用（Delerive *et al.*，1999）。

# 食物中的氨基酸及其相关复合物和免疫功能

## 含硫氨基酸和谷胱甘肽

含硫氨基酸是人类必需的。缺乏蛋氨酸和半胱氨酸导致甲状腺、脾脏和淋巴结萎缩，阻止蛋白质能量营养不良的恢复（Gross and Newberne，1980）。当与作为必需氨基酸的异亮氨酸和缬氨酸的缺乏相结合时，含硫氨基酸的缺乏会导致肠道淋巴组织严重损伤，这与蛋白质缺乏的影响非常相似（Gross and Newberne，1980）。

谷胱甘肽是一个由甘氨酸、半胱氨酸和谷氨酸盐组成的抗氧化三肽。面对炎症反应的刺激，谷胱甘肽在肝脏、肺、小肠和免疫细胞中的浓度有所下降，通过在食物中补充半胱氨酸，从而在一些器官中能阻止这种下降（Hunter and Grimble，1997）。谷胱甘肽本身能提高人细胞毒素 T 细胞的活性，细胞内谷胱甘肽的损耗会减少淋巴细胞的增殖和细胞毒素 T 淋巴细胞的产生（Droge *et al.*，1994；Kinscherf *et al.*，1994）。谷胱甘肽损耗与 IFN-γ 的降低是相关的，但与 IL-2 或 IL-4 以及通过抗原激活的鼠类淋巴结淋巴细胞的产生无关（Peterson *et al.*，1998），这种作用是通过抗原呈递细胞进行调节的，作者认为谷胱甘肽通过诱导 IL-12 的生成，并通过这些细胞改变 Th1/Th2 的平衡以有利于 Th1 反应。

## 精　氨　酸

精氨酸是人的一种非必需氨基酸，参与蛋白质、尿素、核苷合成和 ATP 的生成。它也是含氮氧化物的前体，是一种有潜力的免疫调节介质，这种介质对肿瘤细胞和某些微生物有细胞毒性。精氨酸是多胺合成的前体，它在 DNA 复制、细胞周期和细胞分化的调节方面具有关键的作用。在试验动物上，发现精氨酸可以降低与损伤相关的胸腺萎缩，促进胸腺重新聚集和细胞构成，增加淋巴细胞增殖、NK 细胞活性和巨噬细胞的细胞毒性，增加 DTH，增加细菌感染的抗性和脓血症与烧伤的存活率，促进伤口愈合和皮肤移植的排斥（Redmond and Daly，1993；Evoy *et al.*，1998）。研究表明，给健康人添加精氨酸（30g/d；作为西方日常食物的一部分的摄入量是约 4g/d），可增加因有丝分裂原而产生的血液淋巴细胞的增殖以促进的伤口愈合（Redmond and Daly，1993；Evoy *et al.*，1998）。特别有趣的是将精氨酸包含在肠道处方中给予因手术、损伤和烧伤而住院的患者，似乎会降低感染并发症的严重性和住院时间（Evoy *et al.*，1998）。然而在对这些患者的许多临床研究中，肠道处方含有大量的具有潜在免疫作用的养分，因此把这些观察到的影响归因于某一特定养分是很困难的。

## 谷 胱 甘 肽

在血中和体内游离的氨基酸库中，谷胱甘肽是体内最丰富的氨基酸。骨骼肌被认为是体内最重要的谷胱甘肽的产生者，它提供谷胱甘肽作为器官间氮的转运蛋白。免疫系统被认为是谷胱甘肽的重要使用者。肌肉和血浆中谷胱甘肽浓度在应激情况下会下降，如脓血症和癌症、烧伤等情况（Wilmore and Shabert，1998；Calder and Yaqoob，1999），这些结果表明骨骼肌谷胱甘肽库的明显损耗是损伤的重要特征。血浆谷胱甘肽浓度的下降最可能是体内一些器官系统（如肝脏、肾脏、肠道和免疫系统）对谷胱甘肽供不应求的结果，在应激期间谷胱甘肽被认为是条件性必需氨基酸。表明在这种情况下血浆谷胱甘肽的较低浓度会削弱免疫功能，至少在部分上是如此。

动物研究表明谷胱甘肽丰富的食物促进 T 淋巴细胞增殖和 IL-2 产生，增加草食动物受到感染攻击后的幸存能力（Wilmore and Shabert，1998；Calder and Yaqoob，1999）。对多种患者（骨髓移植患者、低初生重婴儿、重症护理的患者）的研究揭示了谷胱甘肽有明显的临床益处，包括降低感染率和住院时间（Wilmore and Shabert，1998；Calder and Yaqoob，1999）。在这些研究中，增加疗效是与免疫功能的提高是连在一起的（其他的则不调节免疫功能）。除了这种直接的免疫影响外，即使肠胃外注射谷胱甘肽就能增加有感染危险的患者肠道的屏障功能。这种功能对减少细菌从肠道的转移和消除关键的感染源有益。

## 核　苷

核苷具有多种细胞功能，包括在 DNA 和 RNA 结构、能量代谢、信号传导、磷脂的生物合成和酶活性的调节等方面的作用，核苷主要来源于氨基酸，能被重新合成，另外，所有食物都含有核酸和核苷。淋巴细胞的活化使核苷合成迅速增加，这是增强代谢从而提供能量以及随后的核酸合成所需要的（RNA 用于蛋白质合成，DNA 用于细胞分裂）。因此，食物提供的外源核苷（或它们的前体）通过进入淋巴细胞和其他白细胞可利用的核苷库来增强免疫力，进而节约供完整的重新合成的需要。

无核苷的膳食削弱鼠科动物的免疫反应，而将 RNA 或核苷添加到小鼠膳食中会增加依赖于 T 淋巴细胞的抗体反应、淋巴细胞的增殖、IL-2 的产生、DTH 和增强小鼠在细菌和真菌攻击时的幸存能力（Boza，1998）。核苷通常包含在住院患者经肠或非经肠的食物中，但是因为有氨基酸的存在，把特定的作用归因于这些制备物中的单个组分是很困难的。因为人乳中含有丰富的核苷，故人们对在婴儿食品中补充核苷具有较大的兴趣。然而，就新生儿的抗感染而言，几乎没有证据表明这样的营养强化能产生益处（Boza，1998）。

## 总　结

总能量或必需养分中的一种或多种养分缺乏（包括维生素 A、维生素 $B_6$、维生素 $B_{12}$、维生素 C、维生素 E、叶酸、锌、铁、铜、硒、必需氨基酸和必需脂肪酸）能削弱免疫功能，增加对传染性病原体的易感性。这很可能是因为这些养分参与分子和细胞对免疫系统攻击的反应。提供这些养分给缺乏这些养分的患者能恢复免疫功能，增加抗感染性。对有些养分而言，使免疫功能提高到最强时的食物中的摄入量比推荐量要高，然而，有些养分的过量摄入会削弱免疫反应。因此，养分的摄入和免疫功能两者间似乎存在四种内在的关系（图 22.4）。这些不同类型的关系可能部分反应养分间的相互作用，如一种养分过量会负面地影响第二种养分的水平（如锌和铜）。通常认为应当确定免疫与养分摄入之间的关系，免疫系统的所有组件都对一种指定的养分以相同的剂量依赖的形式作出反应。至少就涉及的一些养分而言，这是不正确的，免疫系统的不同组件似乎可能对一种指定的养分的可利用性显示出一种特有的剂量—反应关系。

一方面在这一综述中不涉及激素在调节免疫功能中的作用。养分的供应不足可能给机体造成生理上的应激，导致循环中的肾上腺皮质激素和儿茶酚胺浓度升高。这类激素对免疫功能具有抑制作用，因此，当考虑养分供应和免疫学结果之间的关系时，它们可能是重要的因子。

最后，显然养分在免疫功能中的作用有许多，且是多种多样的，很容易理解，如果要获得适合的免疫反应，这些养分充足和平衡的供给是必需的，我们对大多数养分作用的分子基础的理解还不完善。

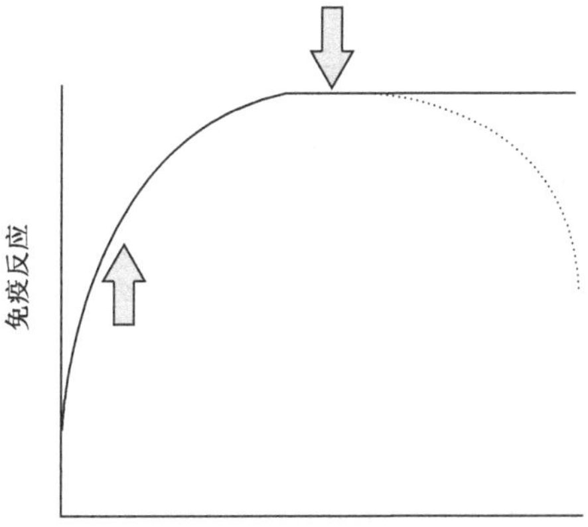


图 22.4　营养素摄入与免疫反应之间的潜在关系

假定特定营养素低摄入或无摄入损害免疫反应，因此，增加此营养素的摄入增加免疫反应。两种潜在的关系的结果为：营养素摄入量显著超过达到最大免疫反应所需的量时并不一定会导致免疫反应的进一步变化。或者是营养素摄入量超过达到最大免疫反应所需的量时可能损害免疫反应。营养素的惯常（或推荐）摄入量如图中的箭头所示。因此，最大免疫反应可能发生在惯常（或推荐）摄入量以上，或可能发生在与惯常（或推荐）摄入量一致的摄入量上

## 参 考 文 献

Alroy, I., Towers, T.L. and Freedman, L.P. (1995) Transcriptional repression of the interleukin-2 gene by vitamin D3: direct inhibition of NFAT/AP-1 complex formation by a nuclear hormone receptor. *Molecular and Cellular Biology* 15, 5789–5799.

Barnes, P.J. and Karin, M. (1997) Nuclear factor-κB: pivotal transcription factor in chronic inflammatory diseases. *New England Journal of Medicine* 336, 1066–1071.

Beck, M.A. (1999) Selenium and host defence towards viruses. *Proceedings of the Nutrition Society* 58, 707–711.

Bendich, A. (1993) Vitamin E and human immune functions. In: Klurfeld, D.M. (ed.) *Nutrition and Immunology*. Plenum Press, New York, pp. 217–228.

Boza, J. (1998) Nucleotides in infant nutrition. *Monatsschrift für Kinderheilkunde* 98, S39–S48.

Calder, P.C. (1998) Dietary fatty acids and the immune system. *Nutrition Reviews* 56, S70–S83.

Calder, P.C. (2001a) The effect of dietary fatty acids on the immune response and susceptibility to infection. In: Suskind, R. and Tontisirin, K (eds) *Nutrition, Immunity and Infectious Diseases in Infants and Children*. Karger, Basel, pp. 137–172.

Calder, P.C. (2001b) Polyunsaturated fatty acids, inflammation and immunity. *Lipids* 36, 1007–1024.

Calder, P.C. (2001c) *N*-3 polyunsaturated fatty acids, inflammation and immunity: pouring oil on troubled waters or another fishy tale? *Nutrition Research* 21, 309–341.

Calder, P.C. (2002) Dietary modification of inflamation with lipids. *Proceedings of the Nutrition Society* 61, 345–358.

Calder, P.C. and Jackson, A.A. (2000) Undernutrition, infection and immune function. *Nutrition Research Reviews* 13, 3–29.

Calder, P.C. and Yaqoob, P. (1999) Glutamine and the immune system. *Amino Acids* 17, 227–241.

Cantorna, M.T., Hayes, C.E. and de Luca, H.F. (1996) 1,25-Dihydroxyvitamin D3 reversibly blocks the progression of relapsing encephalomyelitis, a model of multiple sclerosis. *Proceedings of the National Academy* treatment on CD4+ and CD8+ cells. *FASEB Journal* 8, 448–451.

Lemire, J.M. (1992) Immunomodulatory role of 1.25-dihydroxyvitamin $D_3$. *Journal of Cellular Biochemistry* 49, 26–31.

Lo, C.J., Chiu, K.C., Fu, M., Lo, R. and Helton, S. (1999) Fish oil decreases macrophage tumour necrosis factor gene transcription by altering the NFkappaB activity. *Journal of Surgical Research* 82, 216–221.

MacDonald, P.N., Dowd, D.R. and Hayssler, M.R. (1994) New insight into the structure and functions of the vitamin D receptor. *Seminars in Nephrology* 14, 101–118.

Mathieu, C., Laureys, J., Sobis, H., van de Putte, M., Waer, M. and Bouillon, R. (1992) 1,25 Dihydroxyvitamin D3 prevents insultis in NOD mice. *Diabetes* 41, 1491–1495.

McKenzie, R.C., Rafferty, T.S. and Beckett, G.J. (1998) Selenium: an essential element for immune function. *Immunology Today* 19, 342–345.

Meydani, S.N. and Beharka, AA. (1998) Recent developments in vitamin E and immune response. *Nutrition Reviews* 56, S49–S58.

Meydani, S.N., Ribaya-Mercado, J.D., Russell, R.M., Sahyoun, N., Morrow, F.D. and Gershoff, S.N. (1991) Vitamin B6 deficiency impairs interleukin-2 production and lymphocyte proliferation in elderly adults. *American Journal of Clinical Nutrition* 53, 1275–1280.

Meydani, S.N., Meydani, M., Blumberg, J.B., Leka, L.S., Siber, G., Loszewski, R., Thompson, C., Pedrosa, M.C., Diamond, R.D. and Stollar, B.D. (1997) Vitamin E supplememtation and *in vivo* immune response in healthy subjects. *Journal of the American Medical Association* 277, 1380–1386.

Miles, E.A. and Calder, P.C. (1998) Modulation of immune function by dietary fatty acids. *Proceedings of the Nutrition Society* 57, 277–292.

Miles, E.A., Wallace, F.A. and Calder, P.C. (2000) Dietary fish oil reduces intercellular adhesion molecule 1 and scavenger receptor expression on murine macrophages. *Atherosclerosis* 152, 43–50.

Nikawa, T., Odahara, K., Koizumi, H., Kido, Y., Teshima, S., Rokutan, K. and Kishi, K. (1999) Vitamin A prevents the decline in immunoglobulin A and Th2 cytokine levels in small intestine mucosa of protein-malnourished mice. *Journal of Nutrition* 129, 934–941.

Overbergh, L., Decallone, B., Valckx, D., Verstuyf, A., Depovere, J., Laureys, J., Rutgeerts, O., Saint-Arnaud, R., Bouillon, R. and Mathieu, C. (2000) Identification and immune regulation of 25-hydroxyvitamin D-1-α-hydroxylase in murine macrophages. *Clinical and Experimental Immunology* 120, 139–146.

Penn, N.D., Purkins, I. and Kelleher, J. (1991) The effects of dietary supplementation with vitamins A, C and E on cell-mediated immune function in elderly long-stay patients: a randomised controlled study. *Age and Ageing* 20, 169–174.

Peterson, J.D., Herzenberg, L.A., Vasquez, K. and Waltenbaugh, C. (1998) Glutathione levels in antigen-presenting cells modulate Th1 versus Th2 response patterns. *Proceedings of the National Academy of Sciences USA* 95, 3071–3076.

Prasad, A.S., Bao, B., Beck, F.W.J. and Sarkar, F.H. (2001) Zinc activates NF-κB in HUT-78 cells. *Journal of Laboratory and Clinical Medicine* 138, 250–256.

Prohaska, J.R. and Failla, M.L. (1993) Copper and immunity. In: Klurfeld, D.M. (ed.) *Nutrition and Immunology*. Plenum Press, New York, pp. 309–332.

Redmond, H.P. and Daly, J.M. (1993) Arginine. In: Klurfeld, D.M. (ed.) *Nutrition and Immunology*. Plenum Press, New York, pp. 157–166.

Renier, G., Skamene, E., de Sanctis, J. and Radzioch, D. (1993) Dietary *n*-3 polyunsaturated fatty acids prevent the development of atherosclerotic lesions in mice: modulation of macrophage secretory activities. *Arteriosclerosis and Thombosis* 13, 1515–1524.

Rivera, J., Habicht, J.-P., Torres, N., Cossio, T., Utermohlen, V., Tovar, A., Robson, D.S. and Bourges, H. (1986) Decreased cellular immune response in wasted but not in stunted children. *Nutrition Research* 6, 1161–1170.

Scrimshaw, N.S. and SanGiovanni, J.P. (1997) Synergism of nutrition, infection and immunity: an overview. *American Journal of Clinical Nutrition* 66, 464S–477S.

Semba, R.D. (1998) The role of vitamin A and related retinoids in immmune function. *Nutrition Reviews* 56, S38–S48.

Semba, R.D. (1999) Vitamin A and immunity to viral, bacterial and protozoan infections. *Proceedings of the Nutrition Society* 58, 719–727.

Shankar, A.H. and Prasad, A.S. (1998) Zinc and immune function: the biological basis of altered resistance to infection. *American Journal of Clinical Nutrition* 68, 447S–463S.

Sherman, A.R. and Spear, A.T. (1993) Iron and immunity. In: Klurfeld, D.M. (ed.) *Nutrition and Immunology*. Plenum Press, New York, pp. 285–307.

Siegel, B.V. (1993) Vitamin C and the immune response in health and disease. In: Klurfeld, D.M. (ed.) *Nutrition and Immunology*. Plenum Press, New York, pp. 167–196.

Singh, S., Aiba, S., Manome, H. and Tagami, H. (1999) The effects of dexamethasone, cyclosporine, and vitamin D3 on the activation of dendritic cells stimulated by haptens. *Archives of Dermatological Disease* 291, 548–554.

Stabel, J.R. and Spears, J.W. (1993) Role of selenium in immune responsiveness and disease resistance. In: Klurfeld, D.M. (ed.) *Nutrition and Immunology*. Plenum Press, New York, pp. 333–356.

Takeuchi, A., Reddy, G.S., Kobayashi, T., Okani, T.,

Park, J. and Sharma, S. (1998) Nuclear factor of activated T cells (NFAT) as a molecular target for 1,25-dihydroxyvitamin D3-mediated effects. *Journal of Immunology* 160, 209–218.

Thieringer, R., Fenyk-Melody, J.E., Le Grand C.B., Shelton, B.A., Detmers, P.A., Somers, E.P., Carbin, L., Moller, D.E., Wright, S.D. and Berger, J. (2000) Activation of peroxisome proliferator-activated receptor γ does not inhibit IL-6 or TNF-α responses of macrophages to lipopolysaccharide *in vitro* or *in vivo*. *Journal of Immunology* 164, 1046–1054.

Wallace, F.A., Miles, E.A., Evans, C., Stock, T.E., Yaqoob, P. and Calder, P.C. (2001) Dietary fatty acids influence the production of Th1- but not Th2-type cytokines. *Journal of Leukocyte Biology* 69, 449–457.

Wang, Y., Huang, D.S., Eskelson, C.D. and Watson, R.R (1994) Long-term dietary vitamin E retards development of retrovirus-induced dysregulation in cytokine production. *Clinical Immunology and Immunopathology* 72, 70–75.

Wilmore, D.W. and Shabert, J.K. (1998) Role of glutamine in immunologic responses. *Nutrition* 14, 618–626.

Woodward, B. (1998) Protein, calories and immune defences. *Nutrition Reviews* 56, S84–S92.

Xi, S., Cohen, D., Barve, S. and Chen, L.H. (2001) Fish oil suppressed cytokines and nuclear factor kappaB induced by murine AIDS virus infection. *Nutrition Research* 21, 865–878.

Yaqoob, P. (1998) Lipids and the immune response. *Current Opinion in Clinical Nutrition and Metabolic Care* 1, 153–161.

Yu, X., Bellido, T. and Manolagas, S.C. (1995) Down-regulation of NFκB protein levels in activated human lymphocytes by 1,25-dihydroxyvitamin D3. *Proceedings of the National Academy of Sciences USA* 92, 10990–10994.

# 第 6 部分
# 食　　物

# 23 食物过敏的分子机制

J. Steven Stanley[1]和 Gary A. Bannon[2]
（1 小儿科和生化分子生物部，2 阿肯色州儿童医院研究所和
阿肯色州医科大学，小石城，阿肯色州，美国）

## 流 行

有关人们对食物产生有害反应的流行性的一项消费者调查显示：受访者中 30%的人认为他们或他们的一些家庭成员对某种食物产品过敏（Anderson and Sogn，1984）；22%的受访者避开某种食物仅仅因为这种食物可能含有一种过敏原。事实上，食物过敏反应只对儿童人口的 6%～8%和成年人口的 1%～2%产生影响（Burks and Sampson，1993；Jansen *et al.*，1994），且过敏是由我们所消费的食物的一小部分引起的（Hanson and Telemo，1997）。所有过敏性疾病的发生率在工业化社会中似乎在增加（Hanson and Telemo *et al.*，1997）。最常见的影响儿童的食物过敏反应是由免疫球蛋白 E（IgE）介导的对牛奶、鸡蛋、花生、大豆、小麦、鱼和树坚果的反应。所有已报道的儿童食物过敏中约有 80%是由花生、牛奶或鸡蛋引起的。大多数童年的食物过敏消失后，在成年很少能减轻对花生、树坚果和鱼的过敏。对于成年人来说，最常见的食物过敏是对花生、树坚果、鱼和甲壳的水生动物的过敏。IgE 介导的对某种特定粮食作物过敏的发生率正在增加，特别是在发展中国家，可能归因于蛋白质消费水平的增加。过敏反应一般是由蛋白质的一种特定亚群引起的，而这种蛋白质在食物中含量丰富。

## 致敏作用及其发生的免疫学基础

人类胃肠道将各种各样的食物消化为可被有效转运到上皮细胞中的营养物质，胃肠道黏膜也为许多有害生物提供了进入的通道。肠道淋巴组织是免疫系统的专门分支，负责区分潜在的病原体和膳食中无害的营养物质、自身蛋白质和肠道菌群。大部分的食物蛋白可激发起经口服的耐受性，同时病原体则被破坏和清除。对肠道病原体的免疫防御包括体液和细胞介导的免疫反应即黏液的分泌增加、IgA 的生成、呕吐和腹泻。另一方面，口服的耐受性是对摄入抗原的全身性无免疫应答性。值得注意的是免疫耐受性是一个主动过程，在免疫活化时开始，在免疫活性受到抑制、免疫无反应性或免疫缺失时结束。因此，可以认为食物过敏作用是免疫耐受性受到了破坏的结果。

大多数对食物蛋白的过敏反应是由 IgE 介导的。引起 IgE 产生的免疫过程依赖于 $CD_4^+$ T 细胞。现在普遍认为 $CD_4^+$ T 细胞分为 Th1 和 Th2 亚群。虽然起源于相同的祖先细胞，Th1 和 Th2 细胞呈现出相反的光谱末端。白细胞介素（IL）-4、IL-5、IL-10 和 IL-13 的分泌和它们对 γ-干扰素（IFN）的应答性以及对 IL-12 的无应答性可用来区分

Th2 细胞；通过它们对 γ-干扰素、IL-2 和肿瘤坏死因子（TNF-β）的分泌，对 IL-12 的应答性及对 γ-干扰素的无应答性可区分 Th1 细胞。Th1 细胞调节细胞介导的免疫反应。Th1 和 Th2 细胞不仅促进不同类型的免疫反应，还积极地抑制其他的免疫反应。免疫偏离过程是这一抑制作用的结果，在此过程中免疫系统通过 Th1 反应或 Th2 反应，而不是同时通过两者对病原体发生反应。Th2 细胞促进对细胞外病原体的体液免疫反应，在某些条件下会引起免疫系统产生特异性反应。

特异性免疫反应的最终结果是产生对过敏原有特异性的 IgE 抗体。IgE 抗体通过与肥大细胞和嗜碱性粒细胞上的高亲合力 IgE 受体（FcεRI）的结合介导食物的过敏过程。这些 FcεRI 受体通过过敏原发生交联，进而引起炎症介质的释放。这些介质的释放引起瘙痒、麻疹、呕吐、腹泻和过敏反应等过敏的反应症状，普通的诊断性过敏测试也需要特异性的 IgE 抗体，如皮肤穿刺试验和鉴定过敏原的来源的体外化验。即使循环中 IgE 的滴度相对较低，FcεRI 受体对 IgE 的高亲合力也确保大部分肥大细胞和嗜碱性粒细胞上的 FcεRI 受体分布于全身各处。

为了释放过敏症状的介质，IgE 必须交联在效应细胞上，这就决定了一种单一的过敏原至少有两个亲合力高的 IgE 结合的抗原决定簇。既然过敏原在通过胃肠道时会遇到多种蛋白酶，那么可以从理论上计算出不产生临床症状的肽的大小。为了估计可与肥大细胞上受体有效交联的最小肽的大小，必须多方估计抗原决定簇的大小和肽的结构。第一种估计是有关在食物过敏原中发现的典型结合 IgE 的抗原决定簇的大小的。据报道，大多数过敏原结合 IgE 的抗原决定簇的长度为 6～15 个氨基酸（Stanley and Bannon，1999），因此，一条肽的最小绝对长度为 12～30 个氨基酸含有两个结合 IgE 的抗原决定簇。但是，这种估计并未考虑到 Kane 等（1986，1988）的研究结果。Kane 等（1986，1988）认为，结合 IgE 的抗原决定簇之间必须至少相距 80～240Å 的距离才最适合脱粒。假定两个 IgE 结合的抗原决定簇之间的最小距离是 80Å，氨基酸如丙氨酸的直径是 5Å，则产生过敏反应临床症状的肽的最小长度为 29 个氨基酸或是肽的大小为 3190Da [29 个（氨基酸）×110（氨基酸的平均分子质量）]。这些计算并未考虑肽的二级结构。例如，肽的构象可能是呈 α 螺旋、β 折叠或随机卷曲，这取决于它的氨基酸序列。依赖于肽的二级结构，如果每个片段的末端代表一种很强的结合 IgE 的抗原决定簇，且如果肽呈 β 折叠构象，就可能使肥大细胞脱粒。依照此原理，一种耐蛋白酶水解的小于 3kDa 的片段似乎不可能引起肥大细胞脱粒。

## 临床表现

对某种食物的过敏反应常常几分钟后就变得明显，但也可能在摄入食物后的任何时间，直至 2h 后出现。症状取决于受影响的身体部位（Sampson，1997）。最初的症状是嘴唇、舌、上腭或喉的发痒或刺痛。这些感觉发展非常快，接着可能出现嘴唇和舌的肿大，喉咙感觉憋闷、干咳和嗓音嘶哑。当过敏原进入胃和肠时，会引起恶心、腹部绞痛、呕吐或腹泻。当过敏原被带入眼睛和鼻中时会产生发痒、流泪、鼻子阻塞，并伴有鼻结膜炎。当过敏原进入肺部时，会引起气喘，进而导致胸闷、喘气和呼吸变短促（Quirce *et al.*，1997）。通常，皮肤会受到影响，引起荨麻疹、血管性水肿、湿疹和特异性皮炎（Burks *et al.*，1998)。过敏反应的最危险的后果是全身性的过敏反应，在此过程中肥大细胞和嗜碱性粒细胞释放组胺、前列腺素和白三烯，这些物质可使血管扩张、增加血管通透性和引起平滑肌收缩。一种典型的过敏反应除了有低血压、心动过速、发绀、呼吸困难和休克症状外，还可能包括以上所有症状。全身性过敏反应可能发展很快，在吃了不良食物后的几分钟到几小时都可能发生。花生、树坚果、鱼和甲壳类水生动物是最易引起全身性过敏的食物。一般 50mg 食物即可引起过敏反应。但是，对于一些极度过敏的个体，少到 2mg 食物就能产生过敏症状（Hourihane *et al.*，1997)。

## 诊　断

确定某个人是否对某一特定食物有过敏反应的唯一可靠方法是进行双盲空白对照的食物攻击试验。在此试验中给患者食用可疑的过敏原或一种对照物，直至临床症状很明显地出现（Sicherer *et al.*，2000)。但是，因为时间长、费用高和对患者有危险性，用这些试验来确定大量的可疑性食物过敏原是不可行的。皮肤穿刺试验是测试某个人对某种可疑过敏原的反应性的最常用方法。皮肤穿刺试验对过敏症的阴性结果具有极好的预测率，但对阳性结果的出现只有 50%～60%的预测率（Hill *et al.*，2001)。大多数错误的阳性结果是由其他食物中相似蛋白质之间的交叉所引起的。鉴定过敏原的特异性 IgE 抗体的放射变应原吸附测定（radioallergosorbent test，RAST）和其他相似的体外化验在准确性方面与皮肤穿刺试验相似。因此，如果获得阴性结果，它们在消除可疑性食物方面是有用的。但是，它们对鉴定某些食物是否确实能使患者产生临床症状的预测值是非常有限的。

## 食物过敏原的一般特征

人们发现许多过敏原可刺激 IgE 的形成并引起人类的过敏性疾病。从对各种来源的食物过敏原的研究中得出的资料揭示了过敏原的一些一般特征。大多数过敏原是有酸性等电点的糖蛋白，这种糖蛋白在过敏原的来源物中含量丰富。此类分子通常对蛋白酶、热和变性剂具有耐受性，在食物加工和消化过程中可以使这些蛋白质免受降解（Metcalf，1985)。这些特征在使过敏原快速穿过肠黏膜、促进最初的致敏过程以及后来快速产生在过敏患者体内可观察到的症状方面可能起着重要作用。尽管我们有关过敏原的主要序列和结构方面的知识在增加，过敏的致敏过程所需的特征还未被确定（Anderson and Sogn，1984)。

现在普遍认为在食物过敏原中存在线性和构象性两种结合 IgE 的抗原决定簇，当过敏原需具有二级结构和三级结构 IgE 才能与其结合时，构象性抗原决定簇开始起作用。相反，线性抗原决定簇只需具有与 IgE 结合的主要氨基酸序列即可。构象性的结合 IgE 的抗原决定簇对需气过敏原介导的过敏反应的病原具有重要性，线性抗原决定簇对食物过敏原具有重要性，主要是因为只有它们部分变性或被人类胃肠道消化后免疫系统才会遭遇它们。因此，食物过敏原的线性结合 IgE 的抗原决定簇比相对不太普遍构象性抗原决定簇更能引起人们的注意。

# 从大多数常见的引起过敏症的食物中鉴别过敏原

## 花生过敏原

花生（*Arachis hypogaea* L.）属于豆科家族，是南美洲一年生的本地植物。弗吉尼亚（Virginia）、西班牙（Spanish）和蔓藤枝是最常见和最重要的商品代花生品种。前几年的各种研究已确定花生中过敏原的特性和存在部位（Gillespie *et al.*，1976；Bush *et al.*，1989）。这些研究证实花生过敏原的蛋白质存在于子叶中，在不同种类花生之间无明显差异（Taylor *et al.*，1981）。

许多花生蛋白由于能与来自花生过敏患者体内的 IgE 结合而被鉴定为过敏原。用此种方式鉴定出的第一种主要的花生过敏原是 Ara h 1（Burks *et al.*，1991）。Ara h 1 蛋白在十二烷基硫酸钠聚丙烯酰胺凝胶中的平均分子质量是 63.5kDa，等电点是 4.55。此种过敏原的基因最近已被克隆和测序，并发现和植物 7S 球蛋白的序列具有明显同源性（Burks *et al.*，1995）。已知 Ara h 1 是糖基化的，有一个与天冬氨酸相连的碳水化合物结合位点（Burks *et al.*，1995；van Rhee *et al*，2000）。对花生有阳性反应的患者中超过 90%的人有对应于 Ara h 1 的特异性的 IgE。基于 IgE 的识别，Ara h Ⅰ 被认为是花生中主要的过敏原之一。分离并已克隆出的第二种过敏原是 Ara h 2（Burks *et al.*，1992）。Ara h 2 的平均分子质量为 17.5kDa，等电点为 5.2，与植物羽扇豆球蛋白具有明显的同源序列（Stanley *et al.*，1997；Viquez *et al.*，2001）。使用大豆吸收的来自过敏患者的血清中的 IgE 鉴定出第三种花生过敏原（Ara h 3）（Eigenman *et al.*，1996），此种方法可用来识别花生过敏原的特异性蛋白。曾推断 Ara h 3 的氨基酸序列已被发现和种子储存蛋白质的 11S 家族具有同源性。此种过敏原的重组形式曾在一种细菌系统中表达，并被约 50%的对花生过敏的患者的血清 IgE 所识别（Rabjohn *et al.*，1999）。4 种其他的蛋白质也已被鉴定为花生过敏原，它们被称为 Ara h 4～7（Kleber-Janke *et al.*，1999）。除了 Ara h 5 外，它们均与 Ara h 1、2 和 3 具有明显的同源性（Kleber-Janke *et al.*，1999）。Ara h 5 是抑制蛋白家族的一个成员，但是只能被一小部分（13%）对花生过敏的人的 IgE 所识别（Kleber-Janke *et al.*，1999）。

## 鸡蛋过敏原

在美国的小儿科人群中，对鸡蛋的过敏是最普通的食物过敏症之一（Bock *et al.*，1988；Bock and Atkins，1989；Sampson *et al.*，1992）。在一项研究中，用双盲空白对照食物攻击试验对 470 个孩子可能出现的食物过敏反应进行测试，结果发现，2/3 的小孩对食物的阳性反应归因于鸡蛋蛋白（Sampson，1997）。卵类黏蛋白（Gal d 1）、卵清蛋白（Gal d 2）、卵转铁蛋白（Gal d 3）和溶菌酶（Gal d 4），这些都是鸡蛋白蛋白中最重要的过敏原（Langeland，1982；Holen and Elsyed，1990；Hansen *et al.*，1997）。这四种过敏原均为糖蛋白，并已被克隆和测序。卵类黏蛋白约占鸡蛋中总白蛋白的 10%，是主要的过敏原（Bernhisel-Broadbent *et al.*，1994）。

卵类黏蛋白的长度为 186 个氨基酸，分子质量为 28kDa，等电点为 4.1，与胰腺分泌的胰蛋白酶抑制因子具有明显的同源性。蛋白质分三个结构域，每个区域含 60 个氨基酸。结构域内有两个二硫键，确保每个结构域维持其空间结构。前两个结构域各含有两个糖基化位点，第三个结构域只含有一个位点。碳水化合物约占糖蛋白质量的 25％（Egge *et al.*，1983）。

对卵转铁蛋白（Gal d 3）和溶菌酶（Gal d 4）的研究程度不同于对卵类黏蛋白和卵清蛋白的研究。卵转铁蛋白是一种分子质量为 70kDa 的铁结合糖蛋白，在鸡蛋蛋白中具有杀菌活性，等电点为 5.6～6.2。溶菌酶是分子质量约 14.3kDa 的糖蛋白，等电点为 11（Holen and Elsayed，1990）。卵类黏蛋白、卵清蛋白、卵转铁蛋白和溶菌酶四种过敏原约占鸡蛋蛋白中蛋白质的 80％。

### 牛奶过敏原

牛奶除了是最易引起过敏的食物外，也是最早添加到婴儿膳食中的食物之一。已使用血清 IgE 抗体鉴定出 $\alpha$-酪蛋白（约 32.4kDa）、$\beta$-酪蛋白（约 26.6kDa）、$\alpha$-乳清蛋白（约 14.2kDa）、$\beta$-乳球蛋白（约 18kDa）、牛血清白蛋白（约 66kDa）和牛 $\gamma$-球蛋白（约 150kDa）为牛奶中的过敏原。但是，对于对牛奶过敏的个人来说，$\alpha$-酪蛋白（约 32.4kDa）和 $\beta$-乳球蛋白（约 26.6kDa）似乎是最常见的过敏原（Docena *et al.*，1996）。

### 鱼和甲壳类水生动物的过敏原

儿童和成年人食物过敏的最常见的原因是对鱼和甲壳类水生动物的过敏反应。鱼和甲壳类水生动物引起的过敏反应与花生引起的过敏反应相似，两者都不能随着年龄的增长而消除。研究较深入的两种过敏原是 Gad c1 和 Pen a1。Gad c1 是存在于鳕鱼（cod-fish）中的小白蛋白的同系物。它是一种分子质量约为 12.5kDa 的糖蛋白。已调查的大多数种类的鱼中含有小白蛋白，它是最大的 IgE 反应蛋白（Hansen *et al.*，1997；Janes *et al.*，1997）。到目前为止，Gad c1 的 IgE 结合的抗原决定簇的存在位置还未被测定。

已鉴定出虾中的原肌球蛋白（36kDa）是主要的过敏原。Pen a1 是褐色虾（*Penaeus aztecus*）的原肌球蛋白的同系物。Met e 1 是从脂背虾（*Metapenaeus ensis*）中分离出的肌球蛋白的同系物。最近，有人发现肌球蛋白是存在于龙虾中的一种过敏原（Leung *et al.*，1998）。来自带刺龙虾（*Panulirus stimpsoni*）的 Pan s 1 和来自美国龙虾（*Homarus americanus*）的 Hom a 1 均为原肌球蛋白的同系物。此两种过敏原已被克隆和测序，和虾中过敏原 Pen a 1 具有明显的同源性。抑制试验表明此两种龙虾过敏原将完全抑制彼此与来自对龙虾过敏的个体的 IgE 的结合。除此之外，重组的虾过敏原 Met e 1 能够阻止 IgE 与龙虾的过敏原结合。这提示原肌球蛋白是龙虾中主要的过敏原，且许多结合 IgE 的抗原决定簇是交叉反应的（Leung *et al.*，1998）。

## 普通食物过敏原中与 IgE 结合的抗原决定簇的分子特征

### 花生过敏原

人们用重叠肽和已证明对花生有超敏性的患者的血清 IgE，确定了 Ara h 1、Ara h

2 和 Ara h 3 三种主要的花生过敏原的线性的 IgE 结合的抗原决定簇的图谱。在 Ara h 1 的整个分子中，已鉴定出 21 个线性的结合 IgE 的抗原决定簇（Burks *et al.*，1997）。用同样的方法，已鉴定出 Ara h 2 和 Ara h 3 中的 10 个结合 IgE 的抗原决定簇（Stanley *et al.*，1997；Rabjohn *et al.*，1999）。这些抗原决定簇的长度为 6～15 个氨基酸，但是所有的肽并没有相同的序列基元。4 种 Ara h 1 抗原决定簇可能是具有免疫显性的结合 IgE 的抗原决定簇，因为它们可被大于 80％的受试患者的血清所识别，且它们比其他任何 Ara h 1 结合 IgE 的抗原决定簇能结合更多的 IgE。Ara h 2 的 3 种结合 IgE 的抗原决定簇和 Ara h 3 的一种抗原决定簇已被确定具有免疫显性。

对三种主要的花生过敏原的每个结合 IgE 的抗原决定簇进行定向位点诱变分析，结果显示这些肽内部的某一氨基酸变化对 IgE 结合的特点有很大的影响。现已发现每个抗原决定簇内部的一个或多个氨基酸对 IgE 的结合具有关键性作用。如果这些氨基酸中的一个被其他氨基酸所取代，会导致大多数 IgE 不能与之结合（Burks *et al.*，1997；Stanley *et al.*，1997；Shin *et al.*，1998）。对有这些效应的结合 IgE 的抗原决定簇内部的氨基酸的种类和位置进行分析，结果表明，位于抗原决定簇中心的疏水残基被替换更易引起 IgE 与之结合的失败（Shin *et al.*，1998）。人们已利用这些研究结果来开发这些过敏原的重组形式以便在免疫治疗中应用。

Ara h 1 形成一种高度稳定的同源三聚体，此三聚体通过疏水相互作用结合在一起。我们发现，组成三聚体的疏水氨基酸位于三维结构的末端，此处为单体与单体接触部位。同时，结合 IgE 的抗原决定簇也位于此区域。Ara h 1 三聚体的稳定性和结合 IgE 的抗原决定簇位于单体与单体接触的位置表明：这些特征在这一主要花生过敏原的整个致敏过程中可能起作用（Shin *et al.*，1998）。

## 鸡蛋过敏原

IgE 和 IgG 与卵类黏蛋白的结合对深度加热、蛋白水解的消化过程、糖基化和化学变性具有耐受性（Matsuda *et al.*，1982；Honma *et al.*，1996；Cooke and Sampson，1997）。这些结果提示线性抗原决定簇在卵类黏蛋白对抗体的识别方面起重要作用。IgE 和 IgG 抗体的线性和构象性抗原决定簇的重要性在人与人之间变化较大。一般而言，构象性抗原决定簇在新获得鸡蛋超敏性的一些个人中起更加重要的作用，而线性抗原决定簇对长期有过敏症的人来说更加重要。用代表全长蛋白质的重叠合成二肽来确定卵类黏蛋白的线性的结合 IgE 和 IgG 的抗原决定簇。从 7 个对鸡蛋过敏的人中收集的血清 IgE 可识别卵类黏蛋白的 5 个区域，而收集的血清 IgG 可结合 7 个区域。大多数 IgE 抗体结合位点也结合 IgG。

作为过敏原的卵清蛋白（Gal d 2）是分子质量为 42.8kDa 的糖蛋白，此糖蛋白代表了蛋清中 50％多的蛋白质。许多与卵清蛋白结合的 IgE 对酶或化学的降解和变性具有耐受性，表明线性 IgE 抗原决定簇的重要性（Kahlert *et al.*，1992；Honma *et al.*，1996）。根据已知的蛋白质序列，几个不同的研究室利用全长蛋白和合成肽的化学和酶的消化已鉴定出至少 7 个 IgE 结合位点。到目前为止，已鉴定出的 IgE 抗体决定都簇聚集在过敏原的 N 端和 C 端（Honma *et al.*，1994）。

# 牛奶过敏原

牛 $\alpha_{s1}$-酪蛋白是一种酸性的磷蛋白（pI4.1～4.5）。它的主要氨基酸序列的 3 个方位决定了它的三级结构。它的 199 个氨基酸中近半数的氨基酸是疏水的，其中 16 个丝氨酸残基可能是磷酸化的，有 17 个脯氨酸残基分布于整个分子上的各个部位。这些特点，加上缺乏二硫桥，使蛋白质呈现带有已退化的三级相互作用的线性构象（Kumosinski *et al.*，1991）。高度有序的结构的缺乏增加了重要的结合 IgE 的抗原决定簇是线性而不是构象性的可能性。两个独立研究小组最近确定了 $\alpha_{s1}$-酪蛋白的线性结合 IgE 和 IgG 的抗原决定簇的图谱，并且得出了不同的试验结果。Spuergin 等（1996）用重叠的合成肽和从 15 个对牛奶过敏的人采集的血清鉴定出 7 个结合 IgE 的抗原决定簇。这些抗原决定簇中的三种可能有免疫显性，可被 15 个人的血清 IgE 所识别。这三种免疫显性抗原决定簇可被一种高浓度的非极性芳香族氨基酸区分开。在 $\alpha_{s1}$-酪蛋白中，结合 IgG 的抗原决定簇和结合 IgE 的抗原决定簇必须存在于相同的部位。Nakajima-Adachi 等（1998）利用重叠的合成肽与消化碎片的结合确定了 IgE、$IgG_4$ 和 T 细胞的抗原决定簇的图谱，在过敏原的 C 端鉴定出一种单一的有免疫显性的结合 IgE 的抗原决定簇，此肽位于 $\alpha_{s1}$-酪蛋白的第 181～199 个氨基酸部位，可被 9 个对牛奶过敏的患者血清的 IgE 所识别。一个人的血清在第 106～125 个氨基酸位置识别了一种其他的肽。这些结果与 Spuergin 等（1996）的结果不同，Spuergin 等（1996）在此位置（第 172～183 个氨基酸位置和第 189～199 个氨基酸位置）只发现两个小的抗原决定簇，他认为主要的抗原决定簇存在于别的位置，而在此位置上 Nakajima-Adachi 等（1998）并未发现有 IgE 的结合。结果的不同可能归因于两个试验中肽的产生方法或对牛奶过敏的患者的人数不同。Spuergin 等（1996）使用 188 个长度为 10 个氨基酸的残基，其中有 9 个残基重叠的肽，Nakajima-Adachi 等（1998）使用 13 个含有 20 个残基和 5 个重叠残基的肽。两个研究小组通过酶联免疫吸附试验（enzyme linked immunosorbent assay，ELISA）检测了 IgE 与合成肽的结合，并使用不同的公式来确定什么样的结合水平才是显著的。Nakajima-Adachi 等（1998）所用的公式似乎更加严格，可以解释他们确定的结合 IgE 的位点为什么较少。而且，由于一个研究是在德国进行，而另一研究在日本进行，可能这两个试验中受试的过敏患者人数不同。Nakajima-Adachi 等（1998）报道 IgE 和 IgG4 与肽的结合没有差异，此结果与 Spuergin 等（1996）的研究结果相一致。

Nakajima-Adachi 等（1998）用从两个人体内分离出的 T 细胞系发现，在 $\alpha_{s1}$-酪蛋白上存在 6 个不同的 T 细胞抗原决定簇。通过对每个人识别的 T 细胞抗原决定簇进行比较，发现有两种不同的结合基元。从第一个人体内分离的 T 细胞系可识别 N 端含有谷氨酸残基、C 端含有亮氨酸残基且两端被 7 个氨基酸［-E-（X)$_7$-L-］隔开的肽。从第二个人体内分离的 T 细胞系可识别 N 端含有谷氨酸残基、C 端含有赖氨酸残基且两端被 6 个氨基酸［-E-（X)$_6$-K-］隔开的肽。有趣的是，已鉴定出有-E-（X）5-K-序列的 T 细胞活化基元是一种交叉反应的 T 细胞抗原决定簇，存在于许多不同的吸入性过敏原中。这一基元与 Nakajima-Adachi 等（1998）发现的-E-（X)$_6$-K-基元非常相似。而且，在 $\alpha_{s1}$-酪蛋白序列上的其他两个位置含有-E-（X)$_5$-K-序列。虽然在此试验中使用的是来自两个人的 7 个 T 细胞系均不能识别这种交叉反应的抗原决定簇，但它很可能在这些人和其他人体内起作用。

## 鱼和甲壳类水生动物的过敏原

原肌球蛋白也存在于脊椎动物来源的食物（即有骨的鱼肉、牛肉、猪肉和鸡肉）中，但在这些食物中很少能引起过敏症。这使得人们对可能解释过敏反应中存在的不同情况的分子特征产生了兴趣。Reese 等（1997）使用肽的表达文库和来自对虾过敏的患者血清的 IgE 从 Pen a 1 中确定了 IgE 反应肽的图谱，发现了 4 种肽（长度为 13～21 个氨基酸），并对它们进行了测序。这 4 种肽都存在于过敏原的 C 端。肽上存在肌原蛋白的保守和非保守区域。这一发现暗示，脊椎动物和虾的原肌球蛋白之间氨基酸的不同可能引起过敏反应的不同。但是，现在还未进行过用定点诱变来确定临界氨基酸残基的试验。Ayuso 等（2002）使用合成的重叠肽和来自对虾过敏的人的血清确定了虾的结合 IgE 的线性抗原决定簇。与 Reese 等（1997）的研究相反，Ayuso 等（2002）发现了 5 个主要的 IgE 结合位点，它们存在于分子中约 42 个氨基酸的有规则的间隙处。这一发现暗示原肌球蛋白的重复双螺旋结构和此分子的过敏性之间存在一定的关系。

# 改变食物过敏原的措施

传统过敏原的免疫治疗法包括增加过敏原提取物的注射剂量，直至达到维持水平。治疗是否有效，可能关键在于注射体内的相对较高的过敏原水平。不幸的是，传统过敏原的免疫治疗法对食物过敏并不是十分有效，现在，从饮食中消除过敏原已被证实是治疗食物过敏的唯一方法。即使不摄入某些食物，如小麦、牛奶、鸡蛋、大豆和花生，也很难保证不会偶然的摄入过敏原。而且，对于重要的食物如牛奶，会有对过敏患者的营养状态产生负面影响的风险。Madsen 和 Henderson（1997）估计了经诊断对牛奶过敏的 58 个患者的钙的摄入情况，他们发现，53％的患者的钙的摄入量低于每日推荐允许量（recommended daily allowance，RDA）。补充钙的患者中，有 21％的人的钙摄入量仍达不到他们的 RDA。人们可使过敏原变得无过敏性或低过敏性，这种能力给食物供应的安全性和降低食物过敏的发生率带来了希望。

## 物理化学和酶的加工过程

过敏临床症状的另一种治疗方法是修饰过敏原蛋白，使其不能再引起过敏反应。一些不稳定的过敏原不能耐受一般的加工过程。例如，许多水果过敏原是热敏感性的，在罐装过程中会受到破坏。毫不奇怪的是，许多食物过敏原可耐受极端的热、pH 和酶的降解，因为这一耐受性使得过敏原对消化系统的消化具有足够的耐受时间，从而可与免疫系统相互作用，且可以使某个人过敏或引起过敏反应。减少食物过敏原带来的危险的最有效和最常见的方法是食物蛋白的酶解。50 多年前，牛奶就可成功地水解为无过敏状态。对牛奶的过敏不仅是最常见的也是最悬而未决的食物过敏，因为对婴儿来说，牛奶的替代品很少。而且，从以牛奶为基础的处方到以大豆为基础的处方的变化，可能引起对大豆的过敏（Bishop *et al.*，1990）。因此，水解的牛奶是治疗过敏的婴儿处方中的一种常见的替代物。水解处方可以分为部分水解或深度水解处方。部分水解的牛奶仍含有能够结合 IgE 并启动过敏反应的一些全长的过敏原和大的过敏原碎片，因此不适合被对牛奶过敏的人食用（Halken *et al.*，1993）。有关深度水解处方的研究检测到过敏原生成肽仍能结合 IgE 和（或）引发皮肤穿刺试验中的阳性反应，但是深度水解处方在患者攻击试验中很少能引起过敏反应（Sampson and McCaskill，1985；Rugo *et al.*，1992；Halken *et al.*，1993；Isolauri *et al.*，1995）。酶的水解不能破坏食物的营养价值，因为它只专一地降解蛋白质，对食物中的脂类、碳水化合物或矿物质含量没有影响。酶解生成的肽和氨基酸更易被消化道吸收和利用。不幸的是，深度水解会产生苦味，游离氨基酸的增加也使渗透压增加。一些处方生产者通过在开始时只使用游离氨基酸来消除最终产品中接触过敏源的可能性（Sampson and McCaskill，1985；Isolauri *et al.*，1995）。酶的水解作用在大米（Watanabe *et al.*，1990）和小麦（Tanabe *et al.*，1996）中也被应用。

## 传统的植物育种方法和遗传工程

食物的化学或酶的水解是一种经济且可行的替代措施，通过育种或遗传工程从食物源中移去过敏原蛋白。腹腔疾病是对小麦醇溶蛋白的一种非 IgE 介导的食物超敏反应。已繁殖出缺乏不同醇溶蛋白基因的小麦品种，这些小麦品种缺乏大部分的醇溶蛋白。可使用同样的方法选育天然的低过敏原的水果和蔬菜品种。苹果是一种可引起过敏症的水果，其内引起过敏症的 Mal d 1 蛋白质的浓度在不同品种之间变化较大（Vieths *et al.*，1994），因此，可以开发苹果的低过敏原品种来替代在已加工的食物中更易引起过敏症的品种。花生品种中主要过敏原的浓度较一致。花生过敏原是存在于植物中的一种储存蛋白质，为花生胚芽的生长和发育提供氨基酸。用传统的育种方法培育低过敏原的花生很困难，因为此类引起过敏症的蛋白质含量丰富，且种子发芽也需要它们。基因工程可能提供一种处理过敏原（如在花生中发现的一些过敏原）的方法。已发现三种主要的花生过敏原结合 IgE 的抗原决定簇（Egge *et al.*，1983；Burks *et al.*，1997；Rabjohn *et al.*，1999）。用定点突变将重组过敏原中此类抗原决定簇突变为非 IgE 结合的形式，为防止对蛋白质的折叠产生干扰，在突变时应谨慎操作（Rabjohn *et al.*，2002）。这一方法可以在花生或其他引起过敏症的食物中使用，用来生成可替代内源基因的低过敏原的蛋白质基因。这种方法有许多潜在的优点。现在，避开过敏原是食物过敏的唯一有效的治疗措施。但是，花生超敏性的长期本性和作为其他食物添加剂的花生及其副产品使用的增加使得避开过敏原变得非常困难。Bock 和 Atkins（1989）发现，32 个患花生过敏症的人试图从他们的膳食中去除花生，24 个人在前 5 年中偶然地食入过花生，16 个人只在前一年食入过花生。花生的低过敏原品种不仅在偶然摄入时是安全的，而且也可减少新对花生产生过敏的人数。

许多设想和研究已经向试图阐明哪种蛋白质将会引起过敏症和哪种蛋白质是良性的机制的方向进行。人们推测对 pH、热和蛋白水解酶的耐受性可能是食物过敏原所需要的（Astwood *et al.*，1996）。但是，仅仅这些特征就足以使一种蛋白质引发过敏症的说法是值得怀疑的。过敏原只是免疫系统的攻击目标的可能性，以及第二个蛋白质或一些非蛋白化合物以一种佐剂起作用且启动遗传性过敏症的免疫反应的可能性是存在的。如果过敏原自身不能引发遗传性过敏反应，用不能结合 IgE 的突变物来代替原来的过敏原可能不会减少未来对花生产生过敏的人数。但是，为研究这一问题，可通过在折叠蛋白的暴露位置引入胃蛋白酶切割位点，从而在不影响蛋白质功能的情况下创造一种不稳定的蛋白质，如果食物过敏原需要对胃蛋白酶的耐受性，那么使过敏原对酶的切割变得不稳定将会干扰致敏过程的进行。

## 小结

食物中能引起过敏症的蛋白质只是摄入蛋白质的很小一部分，其功能在从储存到具有酶活性之间是可变的。食物过敏原的一般特征包括具有一个酸性等电点、在食物中含量丰富、对消化较稳定和含有许多线性的结合 IgE 的抗原决定簇。但是，这些能力也有例外的情况，使人们很难预测是否一个蛋白质就是一种潜在的过敏原。从摄入到与免疫系统相互作用，IgE 对一种过敏原的反应过程包括一系列的步骤。因为致敏过程的复杂性，继续在分子水平上研究有关已知过敏原抵制消化和刺激一种过敏免疫反应的能力是重要的。有了这些知识，人们就可能改进诊断方法、开发新的免疫治疗措施以及为了更安全的供应食物而开发低过敏原的植物。

## 参考文献


Anderson, J.A. and Sogn, D.D. (eds) (1984) *Adverse Reactions to Food.* American Academy of Allergy and Immunology Committee on Adverse Reactions to Foods, US National Institute of Allergy and Infectious Diseases, NIH publication no. 84-2442.

Astwood, J.D., Leach, J.N. and Fuchs, R.L. (1996) Stability of food allergens to digestion *in vitro*. *Nature Biotechnology* 14, 1269–1273.

Ayuso, R., Lehrer, S.B. and Reese, G. (2002) Identification of continuous, allergenic regions of the major shrimp allergen Pen a 1 (tropomyosin). *International Archives of Allergy and Immunology* 127, 27–37.

Bernhisel-Broadbent, J., Dintzis, H.M., Dintzis, R.Z. and Sampson, H.A. (1994) Allergenicity and antigenicity of chicken egg ovomucoid (Gal d III) compared with ovalbumin (Gal d I) in children with egg allergy and in mice. *Journal of Allergy and Clinical Immunology* 93, 1047–1059.

Bishop, J.M., Hill, D.J. and Hosking, C.S. (1990) Natural history of cow milk allergy: clinical outcome. *Journal of Pediatrics* 116, 862–867.

Bock, S.A. and Atkins, F.M. (1989) The natural history of peanut allergy. *Journal of Allergy and Clinical Immunology* 83, 900–904.

Bock, S.A., Sampson, H.A., Atkins, F.M., Zeiger, R.S., Lehrer, S., Sachs, M., Bush, R.K. and Metcalfe, D.D. (1988) Double-blind, placebo-controlled food challenge (DBPCFC) as an office procedure: a manual. *Journal of Allergy and Clinical Immunology* 82, 986–997.

Burks, A.W., and Sampson, H.A. (1993) Food allergies in children. *Current Problems in Pediatrics* 23, 230–252.

Burks, A.W., Williams, L.W., Helm, R.M., Connaughton, C., Cockrell, G. and O'Brien, T. (1991) Identification of a major peanut allergen, Ara h I, in patients with atopic dermatitis and positive peanut challenges. *Journal of Allergy and Clinical Immunology* 88, 172–179.

Burks, A.W., Williams, L.W., Connaughton, C., Cockrell, G., O'Brien, T.J. and Helm, R.M. (1992) Identification and characterization of a second major peanut allergen, Ara h II, with use of the sera of patients with atopic dermatitis and positive peanut challenge. *Journal of Allergy and Clinical Immunology* 90, 962–969.

Burks, A.W., Cockrell, G., Stanley, J.S., Helm, R.M. and Bannon, G.A. (1995) Recombinant peanut allergen Ara h I expression and IgE binding in patients with peanut hypersensitivity. *Journal of Clinical Investigation* 96, 1715–1721.

Burks, A.W., Shin, D., Cockrell, G., Stanley, J.S., Helm, R.M. and Bannon, G.A. (1997) Mapping and mutational analysis of the IgE-binding epitopes on Ara h 1, a legume vicilin protein and a major allergen in peanut hypersensitivity. *European Journal of Biochemistry* 245, 334–339.

Burks, A.W., James, J.M., Hiegel, A., Wilson, G., Wheeler, J.G., Jones, S.M. and Zuerlein, N. (1998) Atopic dermatitis and food hypersensitivity reactions. *Journal of Pediatrics* 132, 132–136.

Bush, R.K., Taylor, S.L. and Nordlee, J.A. (1989) Peanut sensitivity. *Allergy Proceedings* 10, 261–264.

Cooke, S.K. and Sampson, H.A. (1997) Allergenic properties of ovomucoid in man. *Journal of*

*Immunology* 159, 2026–2032.

Docena, G.H., Fernandez, R., Chirdo, F.G. and Fossati, C.A. (1996) Identification of casein as the major allergenic and antigenic protein of cow's milk. *Allergy* 51, 412–416.

Egge, H., Peter-Katalinic, J., Paz-Parente, J., Strecker, G., Montreuil, J. and Fournet, B. (1983) Carbohydrate structures of hen ovomucoid. A mass spectrometric analysis. *FEBS Letters* 156, 357–362.

Eigenmann, P.A., Burks, A.W., Bannon, G.A. and Sampson, H.A. (1996) Identification of unique peanut and soy allergens in sera adsorbed with cross-reacting antibodies. *Journal of Allergy and Clinical Immunology* 98, 969–978.

Gillespie, D.N., Nakajima, S. and Gleich, G.J. (1976) Detection of allergy to nuts by the radioallergosorbent test. *Journal of Allergy and Clinical Immunology* 57, 302–309.

Halken, S., Host, A., Hansen, L.G. and Osterballe, O. (1993) Safety of a new, ultrafiltrated whey hydrolysate formula in children with cow milk allergy: a clinical investigation. *Pediatric Allergy and Immunology* 4, 53–59.

Hansen, T.K., Bindslev-Jensen, C., Skov, P.S. and Poulsen, L.K. (1997) Codfish allergy in adults: IgE cross-reactivity among fish species. *Annals of Allergy, Asthma and Immunology* 78, 187–194.

Hanson, L. and Telemo, E. (1997) The growing allergy problem. *Acta Paediatrica* 86, 916–918.

Hill, D.J., Hosking, C.S. and Reyes-Benito, L.V. (2001) Reducing the need for food allergen challenges in young children: a comparison of *in vitro* with *in vivo* tests. *Clinical and Experimental Allergy* 31, 1031–1035.

Holen, E. and Elsayed, S. (1990) Characterization of four major allergens of hen egg-white by IEF/SDS-PAGE combined with electrophoretic transfer and IgE-immunoautoradiography. *International Archives of Allergy and Applied Immunology* 91, 136–141.

Honma, K., Kohno, Y., Saito, K., Shimojo, N., Tsunoo, H. and Niimi, H. (1994) Specificities of IgE, IgG and IgA antibodies to ovalbumin. Comparison of binding activities to denatured ovalbumin or ovalbumin fragments of IgE antibodies with those of IgG or IgA antibodies. *International Archives of Allergy and Immunology* 103, 28–35.

Honma, K., Kohno, Y., Saito, K., Shimojo, N., Horiuchi, T., Hayashi, H., Suzuki, N., Hosoya, T., Tsunoo, H. and Niimi, H. (1996) Allergenic epitopes of ovalbumin (OVA) in patients with hen's egg allergy: inhibition of basophil histamine release by haptenic ovalbumin peptide. *Clinical and Experimental Immunology* 103, 446–453.

Hourihane, J.O., Kilburn, S.A., Nordlee, J.A., Hefle, S.L., Taylor, S.L. and Warner, J.O. (1997) An evaluation of the sensitivity of subjects with peanut allergy to very low doses of peanut protein: a randomized, double-blind, placebo-controlled food challenge study. *Journal of Allergy and Clinical Immunology* 100, 596–600.

Isolauri, E., Sutas, Y., Makinen-Kiljunen, S., Oja, S.S., Isosomppi, R. and Turjanmaa, K. (1995) Efficacy and safety of hydrolyzed cow milk and amino acid-derived formulas in infants with cow milk allergy. *Journal of Pediatrics* 127, 550–557.

James, J.M., Helm, R.M., Burks, A.W. and Lehrer, S.B. (1997) Comparison of pediatric and adult IgE antibody binding to fish proteins. *Annals of Allergy, Asthma and Immunology* 79, 131–137.

Jansen, J., Kardinaal, A., Huijber, G., Vleig-Boestra, B. and Ockhuizen, T. (1994) Prevalence of food allergy and intolerance in the adult Dutch population. *Journal of Allergy and Clinical Immunology* 93, 446–456.

Kahlert, H., Petersen, A., Becker, W.M. and Schlaak, M. (1992) Epitope analysis of the allergen ovalbumin (Gal d II) with monoclonal antibodies and patients' IgE. *Molecular Immunology* 29, 1191–1201.

Kane, P., Erickson, J., Fewtrell, C., Baird, B. and Holowka, D. (1986) Cross-linking of IgE–receptor complexes at the cell surface: synthesis and characterization of a long bivalent hapten that is capable of triggering mast cells and rat basophilic leukemia cells. *Molecular Immunology* 23, 783–790.

Kane, P.M., Holowka, D. and Baird, B. (1988) Cross-linking of IgE–receptor complexes by rigid bivalent antigens greater than 200 Å in length triggers cellular degranulation. *Journal of Cell Biology* 107, 969–980.

Kleber-Janke, T., Crameri, R., Appenzeller, U., Schlaak, M. and Becker, W.M. (1999) Selective cloning of peanut allergens, including profilin and 2S albumins, by phage display technology. *International Archives of Allergy and Immunology* 119, 265–274.

Kumosinski, T.F., Brown, E.M. and Farrell, H.M. Jr (1991) Three-dimensional molecular modeling of bovine caseins: kappa-casein. *Journal of Dairy Science* 74, 2879–2887.

Langeland, T. (1982) A clinical and immunological study of allergy to hen's egg white. III. Allergens in hen's egg white studied by crossed radio-immunoelectrophoresis (CRIE). *Allergy* 37, 521–530.

Leung, P.S., Chen, Y.C., Mykles, D.L., Chow, W.K., Li, C.P. and Chu, K.H. (1998) Molecular identification of the lobster muscle protein tropomyosin as a seafood allergen. *Molecular Marine Biology and Biotechnology* 7, 12–20.

Madsen, C.D. and Henderson, R.C. (1997) Calcium intake in children with positive IgG RAST to cow's milk. *Journal of Paediatrics and Child Health* 33, 209–212.

Matsuda, T., Watanabe, K. and Nakamura, R. (1982) Immunochemical studies on thermal denaturation of ovomucoid. *Biochimica et Biophysica Acta* 707, 121–128.

Metcalfe, D.D. (1985) Food allergens. *Clinical Reviews in Allergy* 3, 331–349.
Nakajima-Adachi, H., Hachimura, S., Ise, W., Honma, K., Nishiwaki, S., Hirota, M., Shimojo, N., Katsuki, T., Ametani, A., Kohno, Y. and Kaminogawa, S. (1998) Determinant analysis of IgE and IgG4 antibodies and T cells specific for bovine alpha(s)1-casein from the same patients allergic to cow's milk: existence of alpha(s)1-casein-specific B cells and T cells characteristic in cow's-milk allergy. *Journal of Allergy and Clinical Immunology* 101, 660–671.
Quirce, S., Blanco, R., Diez-Gomez, M.L., Cuevas, M., Eiras, P. and Losada, E. (1997) Carrot-induced asthma: immunodetection of allergens. *Journal of Allergy and Clinical Immunology* 99, 718–719.
Rabjohn, P., Helm, E.M., Stanley, J.S., West, C.M., Sampson, H.A., Burks, A.W. and Bannon, G.A. (1999) Molecular cloning and epitope analysis of the peanut allergen Ara h 3. *Journal of Clinical Investigation* 103, 535–542.
Rabjohn, P., West, C.M., Connaughton, C., Sampson, H.A., Helm, R.M., Burks, A.W. and Bannon, G.A. (2002) Modification of peanut allergen Ara h 3: effects on IgE binding and T cell stimulation. *International Archives of Allergy and Immunology* 128, 15–23.
Reese, G., Jeoung, B.J., Daul, C.B. and Lehrer, S.B. (1997) Characterization of recombinant shrimp allergen Pen a 1 (tropomyosin). *International Archives of Allergy and Immunology* 113, 240–242.
Rugo, E., Wahl, R. and Wahn, U. (1992) How allergenic are hypoallergenic infant formulae? *Clinical and Experimental Allergy* 22, 635–639.
Sampson, H.A. (1997) Food allergy. *Journal of the American Medical Association* 278, 1888–1894.
Sampson, H.A. and McCaskill, C.C. (1985) Food hypersensitivity and atopic dermatitis: evaluation of 113 patients. *Journal of Pediatrics* 107, 669–675.
Sampson, H.A., James, J.M. and Bernhisel-Broadbent, J. (1992) Safety of an amino acid-derived infant formula in children allergic to cow milk. *Pediatrics* 90, 463–465.
Shin, D.S., Compadre, C.M., Maleki, S.J., Kopper, R.A., Sampson, H., Huang, S.K., Burks, A.W. and Bannon, G.A. (1998) Biochemical and structural analysis of the IgE binding sites on ara h1, an abundant and highly allergenic peanut protein. *Journal of Biological Chemistry* 273, 13753–13759.
Sicherer, S.H., Morrow, E.H. and Sampson, H.A. (2000) Dose-response in double-blind, placebo-controlled oral food challenges in children with atopic dermatitis. *Journal of Allergy and Clinical Immunology* 105, 582–586.
Spuergin, P., Mueller, H., Walter, M., Schiltz, E. and Forster, J. (1996) Allergenic epitopes of bovine alpha S1-casein recognized by human IgE and IgG. *Allergy* 51, 306–312.
Stanley, J.S. and Bannon, G.A. (1999) Biochemical aspects of food allergens. *Immunology and Allergy Clinics of North America* 19, 605–617.
Stanley, J.S., King, N., Burks, A.W., Huang, S.K., Sampson, H., Cockrell, G., Helm, R.M., West, C.M. and Bannon, G.A. (1997) Identification and mutational analysis of the immunodominant IgE binding epitopes of the major peanut allergen Ara h 2. *Archives of Biochemistry and Biophysics* 342, 244–253.
Tanabe, S., Arai, S. and Watanabe, M. (1996) Modification of wheat flour with bromelain and baking hypoallergenic bread with added ingredients. *Bioscience Biotechnology and Biochemistry* 60, 1269–1272.
Taylor, S.L., Busse, W.W., Sachs, M.I., Parker, J.L. and Yunginger, J.W. (1981) Peanut oil is not allergenic to peanut-sensitive individuals. *Journal of Allergy and Clinical Immunology* 68, 372–375.
van Rhee, R., Cabanes-Macheteau, M., Akkerdaas, J., Milazzo, J.P., Loutelier-Bourhis, C., Rayon, C., Villalba, M., Koppelman, S., Aalberse, R., Rodriguez, R., Faye, L. and Lerouge, P. (2000) Beta(1,2)-xylose and alpha(1,3)-fucose residues have a strong contribution in IgE binding to plant glycoallergens. *Journal of Biological Chemistry* 275, 11451–11458.
Vieths, S., Jankiewicz, A., Schoning, B. and Aulepp, H. (1994) Apple allergy: the IgE-binding potency of apple strains is related to the occurrence of the 18-kDa allergen. *Allergy* 49, 262–271.
Viquez, O.M., Summer, C.G., and Dodo, H.W. (2001) Isolation and molecular characterization of the first genomic clone of a major peanut allergen, Ara h 2. *Journal of Allergy and Clinical Immunology* 107, 713–717.
Watanabe, M., Miyakawa, J., Ikezawa, Z., Suzuki, Y., Hirao, T., Yashizawa, T. and Arai, S. (1990) Production of hypoallergenic rice by enzymatic decomposition of constituent proteins. *Journal of Food Science* 55, 781.

# 24　遗传修饰食物的安全性评价

Steve L. Taylor
（食品科学与技术系，内布拉斯加-林肯大学，林肯，内布拉斯加，美国）

## 引　言

用农业生物技术生产的食物正开始出现在消费者市场，在北美洲尤其如此。当这类产品在一些地区已被接受并获得成功时，在几个国家对通过这种技术生产的食物的存在出现了一些反对的声音。对于这样的食物进入消费者市场，反对者提出了许多反对意见，包括环境、伦理、经济及安全性关注。本章将仅论述与遗传修饰食物的安全性及安全性评价方法的特点和完善性有关的问题。

## 农业生物技术对农业生产的影响

农业生物技术已对生产农业产生了很大的影响，对某些农作物尤其如此。到目前为止，通过农业生物技术进行商业化生产的农作物的数量相对较少，包括玉米、马铃薯、低毒油菜、大豆、棉花、南瓜和番木瓜等（James，2001）。农业生物技术对两种主要农作物大豆和玉米的影响最大，在世界范围内，2000 年转基因大豆的种植面积是 2580 万 $hm^2$，而转基因玉米的种植面积是 1030 万 $hm^2$（James，2000）。这两种农作物的种植在北美洲特别成功，在那里，转基因大豆和玉米的种植面积正稳定地增加到总种植面积的 30％～50％。转基因棉花和低毒油菜也已在几个国家被成功引入，尽管这些农作物的含油部分仅作为食物供应产品而使用。虽然转基因马铃薯的种植面积还很小，但却已经商业化生产。转基因南瓜和番木瓜也已被少量引入市场。通过农业生物技术开发出抗病毒的番木瓜品种使被病毒感染所破坏的夏威夷番木瓜工业获得了新生。2000 年在美国、阿根廷和加拿大种植的转基因农作物占总转基因农作物的 97％，并且转基因农作物已在 15 个国家种植（James，2001）。

很明显，农民看到了遗传修饰农作物的优势，否则，这些品种的成功率就很小。目前仅一些有限的有益特性被导入通过农业生物技术而商业化生产的农作物中（IFT，2000a)。抗虫性提高或对除草剂耐受力增强的玉米、大豆、马铃薯、低毒油菜和棉花已被引入商业化生产中。最近，已开发出抗虫性和耐除草剂两方面都得到增加的农作物。引入南瓜和番木瓜的有益性状是对病毒的抗性。当前这代遗传修饰农作物具有的所有特性主要是带来农艺学益处。

因此，这些益处主要对农民有利。具有已改进农艺学特征的农作物在商业上的成功将很可能导致开发和引入具有这些优点和相似优点的其他转基因农作物。发展具有抗病、抗虫及抗逆条件如炎热、寒冷及干旱的其他农作物在将来是可能实现的。

除了注射重组牛生长激素的牛奶的外，遗传修饰动物产品的引入尚未广泛出现。但是，如果能得到管理部门的正式批准，那么生长性状增强的遗传修饰大马哈鱼就会平稳地进入美国市场。

随着世界范围的人口增长，农业生产将承担为养活这些人群而提供足够食物的日益增长的压力。仅靠常规的现代耕作技术将不足以解决当前的耕地约束问题。许多当今潜在的丰收每年都会因昆虫、与杂草的竞争、干旱条件和其他农艺方面的因素而丧失。传统的农业化学药品的应用能消除害虫和杂草，但是许多消费者都关注杀虫剂在其食物中残留的安全性和与这些技术有关的潜在负面环境影响。农业生物技术已使农作物抗虫性提高和耐除草剂的能力增强，从而使人们利用有限的农业化学应用技术就可控制害虫和杂草所带来的损失。正如前面所提到的，在可预见的未来，将很可能开发出对气候应激和贫瘠的土地状况耐受性增强的转基因农作物，而且，种子本身也含有这种改进的技术，以使不能接触先进和昂贵农业机械的农民可从中获益。例如，抗病毒的南瓜品种对非洲特别适合，因为在那里与病毒有关的农作物损失很大。希望随着抗病毒南瓜的开发，这种作物将成为一种更加可靠的食物来源，并能提高一些非洲国家的食品安全性。

## 农业生物技术对消费者的影响

目前，由于现有的主要益处与改进的农艺学性状有关，故大多数消费者都很可能感到由农业生物技术所带来的益处很少。已进入市场的与消费者利益直接有关的经农业生物技术开发的食品很少。已开发出成熟性状得到改进的番茄，这种番茄的风味品质得到了改善（Redenbaugh，1992），但这些番茄并未被成功引入市场。已开发出油酸（一种被认为对心血管健康有益的单不饱和脂肪酸）含量高的低毒油菜油，并可以很快获得（FAO/WHO，2000）。最近开发的β-胡萝卜素含量提高的所谓黄金稻已受到相当大的公众注意（Ye *et al.*，2000）。虽然黄金稻还未被商业化利用，但它有希望能帮助消除某些亚洲人群中普遍存在的维生素 A 缺乏和夜盲症（Underwood，1994）的现象。农业生物技术为开发更多为消费者带来直接利益的性状带来了潜力。通过农业生物技术能给消费者带来的益处包括提高营养物质和滋补药的含量、延长货架寿命、抗腐败、改善风味和外观，以及消除自然产生的毒性物质（包括过敏原）。

消费者，特别是北美的消费者，已广泛接触到源自遗传修饰农作物的食物和食物原料。玉米和大豆是特别重要的食物，用于生产许多食品原料包括玉米油、玉米淀粉、玉米浆、豆油、大豆蛋白分离物、大豆蛋白精、豆酱和大豆卵磷脂。此外，低毒油菜油和棉花籽油是来自那些特定的遗传修饰农作物的常见的食物成分。虽然接触来自通过农业生物技术生产的农作物的食品和食品成分的机会很多，但因缺乏很明显的消费者益处而仍看不到农业生物技术对消费者产生的影响。大多数的消费者很可能没意识到，他们正在摄入来自遗传修饰农作物的食品和食品成分。但是，表明遗传修饰食品引入市场后并未伴随任何已知的对健康产生的负面影响也很重要。显然，对通过农业生物技术生产的食物进行安全性评价对于确保这一安全记录的连续性是非常重要的。

# 农业生物技术的入门书

农业生物技术为增强由植物、动物和微生物生产的食物的有益特性提供了一种准确的方法。利用现代重组 DNA 方法学，几乎可从任何生物资源中选择基因，并移入一些其他品种的生物基因组中（IFT，2000a）。用这种方法，可选择特定有益特性的基因，范围可从改进了的农艺学特性缩小到改进了的营养组成。在重组 DNA 方法中，只有少数基因被嵌入易接受生物种类的基因组中，使其产生已选择的有益特性。在抗虫性玉米中，需要一种单独基因，此基因可导致产生已选择的昆虫毒性物质（Sanders *et al.*，1998）。在此过程中，也加入一些其他基因作为标记物或促进子（Sanders *et al.*，1998）。在黄金稻中，需要引入 4 个基因以提供促进 $\beta$-胡萝卜素合成的生物学能力（Ye *et al.*，2000）。对于如改善营养成分这样的更加复杂的特性，则需要加入更多的基因以获得预期结果。同样的道理，对于更加复杂的农艺学特性包括具有抗虫性和耐除草剂特性相结合的品种，也必须同时引入几种已经选择的基因才能发挥效应。

相比之下，用传统的育种方法，可将数百个基因从一种生物源移入另一种生物源。人们常常不知道所有这些基因的特性。在选择有益性状的过程中，其他不想要的基因也可能出现在新的品种中。用传统的育种方法，可能只能从亲缘关系相近的生物源中引入基因。相比之下，用重组 DNA 的方法，就可从非常广泛的生物源中选择基因。历史上，用传统的育种手段仅发生过几次关于已发现新品种含有潜在危险组分事件（Zitnak and Johnson，1970；IFT，2000b），但是，对用传统育种手段培育的新品种一般不进行全面的安全性评价。

无论通过传统的植物育种手段还是通过现代的 DNA 重组农业生物技术，都可能将新蛋白质引入新品种的可食部分。众所周知，历史上还未对使用传统植物育种手段产生的这些新蛋白质的安全性进行过评价。对使用传统植物育种手段生成的新蛋白质的安全性进行评价是极其困难的，因为一般不了解这些新蛋白质的数量和确切特性。但是，对于通过现代农业生物技术生产的食物，只有有限的通过特定嵌入基因生产的定义已明确的新蛋白质被引入新的植物品种中，因此，有机会对这些特定新蛋白质进行安全性评价。对新蛋白质的安全性评价只是对通过农业生物技术生产的食物的全面安全性评价过程的一部分。

# 对通过农业生物技术生产的食物的安全性评价过程

农业生物技术产品在商业化之前应进行认真而全面的安全性评价。在一种农作物的商业化得到批准以前，全球管理机构要求农业生物技术公司对该农作物进行评审中的全面安全性评价。这已引起人们的关注，因为在美国安全性评价过程是自愿的。但是，美国食品和医药管理局已宣布其对通过农业生物技术生产的食物进行必需安全性评价的意向。应该认识到，允许在市场上出现的遗传修饰食物已经受过一次事实上的必需安全性评价过程，因为政府管理机构在这些产品的商业化之前已对已有安全评价信息进行过审查。

到目前为止，所有引入市场的遗传修饰产品都已经过彻底的安全性评价。安全性评价的数据已经过全世界管理机构的审查。当前这代遗传修饰产品被不同国家的一个或多个管理机构认定对人和膳食动物消费是安全的。

在商品遗传修饰食物发展的各个阶段，都必须考虑安全性评价。遗传修饰食物的商业化过程可分成三个阶段：基因发现、品系选择和产品商业化推进。安全性评价在所有这三个阶段中都起作用。

如上所述，农业生物技术涉及赋予各种理想性状的新基因的引入。对于遗传修饰产品，安全性评价通常集中于引入的性状或基因产品上。

## 基因发现阶段

在基因发现阶段，战略选择一种产品概念，然后科学家筛选可实现此概念的基因。安全评价过程理想上应从最初基因发现阶段的最早期开始。如果提出任何问题，这些问题必须立即回答或随后在发展过程中回答。人们从有安全使用历史的来源中谨慎地选择基因，会使研发者避开许多更加困难的安全性评价问题。选择合适的基因应考虑各种因素，包括基因的来源、对这种源材料的前期消费者接触情况、这种源材料的安全使用历史、基因和基因产品。在这一早期阶段，通常也要考虑伦理问题。虽然在这种审查的范围之外，但在基因发现阶段的初期也会提出环境和生态关注问题。如果在这一早期阶段认识到不可接受或高度可疑的食物安全危险，那么很可能就应放弃这种产品概念，这在产品商业化之前如不能对所认识的风险完全和清晰地进行评价时尤其如此。

在基因发现阶段这种最初的安全性评价是重要的，因为它意在强调后期安全性评价过程中有效提出的关注和问题，除非在这点上就放弃这一概念。在其他问题中，在基因发现阶段人们可能关注的一些问题，如从中获取基因的源材料的过敏性，源材料中已知的自然形成的毒性物质，或与此源材料或基因产品有关的各种环境/生态问题。例如，如果从一种已知有过敏史的源材料如落花生、干果、鱼等，或豚草、白桦树和草（常见的花粉过敏源）中选择一种基因，那么必须确信，这一基因产品不是来自此源材料的过敏原之一。如果这种基因的源材料是常见过敏性的，那么这些关注就相对明显。但是，在一些情况下，这些关注并不总是很明显。例如，可选择甲壳质酶基因作为预防一些作物植物中常见的各种真菌疾病的潜在手段，因为霉菌菌丝主要由壳质组成。但是，来自一些源材料的壳质酶是过敏原（Breiteneder and Ebner，2000），因此，需要对与已知过敏性甲壳质酶产生的可能交叉反应进行安全性评价。

抗昆虫的农作物可作为在基因发现阶段需要考虑的一个很好的例子。通过从编码合成各种杀虫 Bt 蛋白的苏云金芽孢杆菌（*Bacillus thuringiensis*，Bt）中选择基因而研发抗昆虫的作物。通过商业种子公司将几种不同但相关的基因克隆进几种农作物中，以生产抗昆虫的品种。Bt 蛋白在 B 型苏云金菌中自然产生，已知存在许多不同但相关的 Bt 蛋白形式（Schnepf *et al.*，1998）。Bt 蛋白包含在商业农业喷雾剂中作为昆虫控制的一种选择已有几十年了。Bt 喷雾剂已广泛使用，特别是有组织的农民和家庭花农。作为 Bt 喷雾剂使用的微生物产品含有作为活性杀虫成分的 Bt 蛋白。Bt 蛋白对特定的昆虫具有选择性毒性，但对哺乳类动物基本是无毒性的。存在于这些市售农业喷雾剂中的 Bt 蛋白已经受到相当全面的毒物学评价，包括对实验动物进行的急性、亚慢性和慢性毒性试验，及甚至对人进行的经口用导管的强饲试验（McClintock *et al.*，1995）。对从这些不同试验中得出的数据进行管理评价后得出结论，即这些商业产品中的 Bt 蛋白是安全的。

用通过农业生物技术生产的 Bt 农作物进行的另外的研究也表明了它们的安全性（Sanders，1998）。各种 Bt 农作物包括玉米、棉花和马铃薯已获得美国和其他国家管理机构的批准，并自 20 世纪 90 年代中期以来就一直在市场上销售（James，2001）。

## 品系选择阶段

一旦发现合适且有前途的基因具有产生理想有益性状的潜力，人们就开始努力在受体植物中产生理想的转化，并研发遗传修饰农作物。开始时，大量的转化在实验室中产生。在品系选择阶段的末期，将发现一个或几个最有希望进入商业化阶段的品系。在品系选择阶段，转基因品系从实验室进展通过各种室度试验。最后，如果结果很有前途，则将进行田间试验。

进行温室和田间试验通常需要管理部门的批准（Taylor，2001）。在美国，在品系选择阶段期间必须扫清几个管理方面的障碍。首先，美国农业部（United States Department of Agriculture，USDA）必须审查和批准有关对植物进行研发和测试所用的温室和其他设施的计划。接着，研发者为了进行田间试验必须寻求获得 USDA 的批准。此外，USDA 必须批准研发者将种子从温室运到田间试验地点。温室和田间试验完成后，商业研发者必须将从试验中获得的全部数据上交给 USDA，并请求使遗传修饰农作物成为非管理状态。在批准过程的这一结合点，USDA 征求公众的意见。一旦 USDA 批准该农作物为非管理状态，那么就可种植、测试或使用这种农作物来进行传统作物育种，而无需寻求 USDA 的进一步批准。

虽然品系选择的主要焦点是从许多可选品系中识别出最好的品系，但做这些选择时也应考虑遗传修饰农作物的安全性。在品系选择阶段，在实验室、温室和田间进行的生长试验期间，要考虑许多农艺学特性。这些特性包括植物的高度、叶的取向、叶的颜色、早期植物的活力、根的强度和产量。在品系选择阶段期间，要将遗传修饰农作物的生物学和农艺学等价性与其传统的对应物进行比较。虽然不是专门为了安全性评价的目的，但在发展过程的这一阶段，应消除具有不寻常特性的潜在产品。这些具有不寻常农艺学或生物学特性的产品的消除很可能会促进选择安全产品的前景。如上所述，大多数有前途的新品种都会在品系选择阶段期间从研发中放弃。

在美国，在品系选择阶段，在新的植物品种具有杀虫特性（如 Bt 蛋白）的情况下，也必须另外经过政府的检查。对于具有这些特性的农作物，环境保护机构（Environmental Protection Agency，EPA）也参与管理批准过程。作为 EPA 管理过程的一部分，研发者具有杀虫蛋白或其他成分的任何农作物的面积超过 10 英亩（0.405 公顷）之前必须获得试验使用许可证。对这类 EPA 批准的请求必须征求公众意见。EPA 继续参与东虫成分含量超过其最初提供的田间许可量的遗传修饰农作物的管理过程。在商业化之前，EPA 对遗传修饰植物来源的食物中可允许的东虫成分的数量确定限量（容许量）。为达到此目的，EPA 审查并依赖于杀虫成分对人、动物和环境的安全性方面的数据。如果在杀虫成分的安全性方面已有相当多的数据，且有过安全使用历史，可能跟一些 Bt 蛋白的情况一样，可以请求免除容许量审查的需要，但是，在那些情况下，EPA 必须控制是否应该同意这样的免除请求。一般情况下，在此阶段 EPA 要进行对农作物及其杀虫成分的安全性的全面审查，并征求公众意见来确定其容许量。

## 产品的商品化阶段推进

在品系选择阶段之后和商业化之前，已选择的遗传修饰农作物品种应经过一个详细且更加正式的安全性评价过程。这种安全性评价应集中在与引入的基因相关的基因产品及与新基因源和引入作物有关的任何其他很可能的毒物学或抗营养因子上。在安全性评价过程中，必须考虑遗传修饰产品在包括食品和膳食应用方面的所有潜在使用情况。在美国，当有前途的遗传修饰农作物从温室和田间试验阶段过渡到商业化阶段时，食品和药品管理局（Food and Drug Administration，FDA）便参与管理批准过程。FDA 对用遗传修饰农作物产生的任何食品或食品成分的安全性方面的所有可利用数据进行审查。在现有的 FDA 使用的自愿的安全性评价方法中，该管理机构一般会在研发过程的早期与研发者会面，并对应进行的研究提供指导，以依照 FDA 的意见，确保遗传修饰食物在作为食品和动物饲料使用时的安全性。现在在美国还不了解悬而未决的必需安全评价方法的性质。安全性评价过程的细节很可能随遗传修饰食物的属性而发生变化。FDA 确曾发布了一系列它认为在评价遗传修饰食物和膳食的安全性方面合适的问题（FDA，1992）。尽管美国现存的管理安全评价方法具有自愿的性质，但现在美国市场上的所有遗传修饰食物都已经过 FDA 的详细审查。

当然，也必须从全世界范围来考虑遗传修饰食物的安全性。在国际上，有几个国际组织包括联合国食物和农业组织（Food and Agricultural Organization of United Nations，FAO）、世界卫生组织（World Healthy Organization，WHO）和经济合作与发展组织（Organization for Economic Cooperation and Development，OECD）都已建立了通过农业生物技术生产的食物的安全性评价的背景（FAO/WHO，2000；OECD，2000）。一般而言，在单个国家的管理机构已将其批准遗传修饰食物的方法建立在此背景上。这些报告的总体结论是，植物生物技术的产品安全性不是生来就小于那些通过传统育种方法研发的产品（FAO/WHO，1996）。这些组织也已得出结论，即食物安全性考虑基本上具有与那些来自传统育种方法的产品相同的特性。因此，传统的安全性评价方法适合于评价通过农业生物技术生产的食物和食物成分。对遗传修饰食物的已接受标准与美国现有标准及大多数有关所有食品的国际食品法相同，即在预期的消费条件下有意使用而不会产生有害影响的一种合理的保证。在这些情况下，通过与来自传统的育种技术的食物相比较，遗传修饰食物应提供同等或更大的安全性保证。

## 物质等价的概念

虽然有时被误解，从国际审议一开始，物质等价的概念就一直是遗传修饰食物安全性评价的一个主要特征。（OECD，1992；FAO/WHO，1996；IFT，2000b）物质等价的概念来自遗传修饰食物应与其传统的对应物一样安全的目的。实际上，当前得到批准的传统农作物的遗传修饰品种，如玉米和大豆，与它们传统的对应物相比，变化很小。在物质等价的概念中，安全性评价则集中于假设未改变的成分正好与传统的相应品种中的相同成分一样安全时出现的那些差异上。

在物质等价的概念中，遗传修饰食物（或食物成分）与其传统的对应物比较的特性包括新基因的来源、农艺学参数和主要营养物质、抗营养物质和过敏原这些成分以及消费模式。从通过物质等价的概念进行的比较中可得出如下三种可能的结果：①认为遗传修饰食物与其传统的对应物是物质等价的；②认为除了一个或多个确定的差异外，遗传修饰食物与其传统的对应物是物质等价的；③不认为遗传修饰食物与其传统的对应物是物质等价的（FAO/WHO，1996）。

在第一种情况下，可判断与其传统的对应物是物质等价的。在这种情况下，不需要进一步的安全测试。实际上，只是在很少见的情况下才会出现这一可能性。如果一种基因被去除或被沉默但没有新基因的引入，那么这一情况也可能出现。对于如棉籽油这样的成分也可发生这种情况，因为这种植物与特定新蛋白质的表达有一些确定的差异，但是油组分并不含有这些新蛋白质，且在成分上与来自棉籽的传统品种等价。围绕遗传修饰食物安全性评价的完善性问题出现的一些混淆是基于不正确的设想，即认为大多数遗传修饰食物与它们传统的对应物是物质等价的，因此而不进行安全性研究。事实上，人们并不认为大多数遗传修饰食物与其传统的对应物完全等同，且管理机构已要求对与传统对应物截然不同及不同的遗传修饰食物的那些特性进行安全性研究。

对于现在大多数的遗传修饰农作物，可判断除了确定的与引入性状有关的差异外，来自这些作物的食物或食物组分与其传统的对应物是物质等价的。在这种情况下，安全性测验则集中于引入性状或基因产品（通常是一种新蛋白质）的安全性上。即使在几种基因引入的情况下，除了来自特定引入基因和新基因外，此产品可认为与其传统的对应物是物质等价的。

最后，可判定遗传修饰食物与传统的食物或食物成分不是物质等价的。对这样的产品，很可能需要更加广泛的安全性评价。既然还没有产品进入商业市场或被世界管理机构批准，则还不能具体描绘对这些产品进行安全性评价的要求的性质。的确，安全性评价需要以一种灵活的方式进行，这种方式取决于这些新食品的特性。人们可能想对已研发的这种类型的任何产品进行更加严格的营养和毒物学评价。虽然还没有这样的产品进入商业市场，但农业生物技术对未来许多这样的产品提供了希望。

人们对遗传修饰食物进行全面的成分分析，以与其传统的对应物进行比较。当然，由于品种不同、气候条件和农艺学条件的差异，传统对应物的成分可能变化很大。为了进行比较，对传统对应物的选择是极其重要的。在这样的比较中，对农作物和从其中生产的食品或食品成分在蛋白质、碳水化合物、脂肪、脂肪酸组成、淀粉、氨基酸成分、纤维、灰分、矿物质、维生素和其他因素方面进行比较。如果已知抗营养物质存在于新基因的源材料或宿主植物中，则可测定这些抗营养物质的相对水平。类似地，如果已知新基因的生物来源或宿主植物是过敏性的，则要对转基因品种中过敏原的存在和水平进行测定，并与其传统的对应物进行比较。宿主植物的过敏性不太受关注，例如，对大豆过敏的消费者很可能会避免摄食所有的大豆，不管大豆是遗传修饰与否。

## DNA 的安全性

如上所述，除了某些确定的差异外，可认为现在市场上的绝大多数遗传修饰食物与它们传统的对应物是物质等价的。因此，安全性评价集中于引入的基因及由此基因而产生的性状或产品上。在遗传修饰食物中与嵌入基因有关的 DNA 并不是很受关注的安全性问题。不论来源如何，膳食中的 DNA 并不认为具有明显毒性（Beever and Kemp，

2000)。事实上，所有食物都含有DNA，且有大量的DNA被摄入。人类膳食中RNA和DNA的摄入量估计为0.1～1.0g/d（Doerfler and Schubbert，1997）。通过比较发现，从遗传修饰食物中摄入的新DNA很微小，且很可能低于总DNA摄入量的1/250 000（FAO/WHO，2000）。DNA也是高度易消化的（FAO/WHO，2000）。因此，遗传修饰食物中的新DNA极不可能造成任何安全问题。

人们也已测检了基因从遗传修饰食物转移到哺乳动物细胞中的可能性，并认为这样发生的可能性很低。在正常的饮食条件下，基因从遗传修饰植物转移入哺乳动物细胞需要满足以下所有条件（FAO/WHO，2000）：

- 植物DNA中的相关基因必须很可能作为线性片段被释放；
- 此基因在植物和胃肠道中必须能耐受核酸酶的降解；
- 此基因必须与膳食中来自传统源的DNA具有竞争性吸收作用；
- 受体哺乳动物细胞必须能转化，且该基因必须能抵抗其限制酶的作用；
- 该基因必须能几乎不经修复和重组就嵌入宿主DNA中。

针对评价植物DNA转移至微生物或哺乳动物细胞中的可能性已进行了试验。在这样的一个试验中，给小鼠口服高剂量的细菌来源的DNA，并获得一些标记，表明测试DNA明显合并入了小鼠细胞（Schubbert *et al.*，1998）。但是，这种观测已受到严重质疑（Beever and Kemp，2000）。FAO/WHO（2000）得出的结论是，没有现存的数据可表明植物DNA可被转移并稳定地存在于哺乳动物细胞中。此外，FAO/WHO（2000）认为，现在还没有证据可以证明，来自植物的完整基因可转移到并在哺乳动物细胞中表达。

人们也研究了来自遗传修饰食物的基因转移至胃肠道细菌中的可能性。没有证据提示，摄入遗传修饰食物的人体内发生了这样的转移情况。这样转移的可能性被认为是相当低的（FAO/WHO，2000）。除来自遗传修饰植物的基因与原核基因具有同源性的情况外，基因从植物细胞转入微生物细胞被认为是很不可能的（FAO/WHO，2000）。在实验室条件下已观测到基因的转移，但这仅在可能发生同源性重组的情况下才会出现（Nielsen *et al.*，1998）。在动物胃肠道进行的试验中尚未表明从植物细胞到微生物发生了基因转移。

## 抗生素抗性标记的安全性

抗生素抗性标记基因常与新的目标基因一起嵌入遗传修饰食物中。抗生素抗性标记目的是帮助追踪理想遗传材料与宿主基因组的成功结合。抗生素抗性标记的使用已引起一些人对安全性的关注。但是，没有证据表明，现在使用的抗生素抗性标记基因对人类有任何健康危害（FAO/WHO，2000）。由于前部分中提到的一些原因，抗生素抗性标记基因转移到哺乳动物细胞或胃肠道细菌细胞中的可能性很低，即使发生基因转移，对人类健康的影响也是很小的（FAO/WHO，2000）。最常见的抗生素抗性标记基因可编码新霉素磷酸转移酶Ⅱ。现已很好确定了常见抗生素抗性标记的安全性（WHO，1993；FAO/WHO，2000）。新霉素已限制了治疗价值。但是，如果使用新的标记，安全性评价这部分将变得更加重要，因为必须证明这些替代战略安全性。

## 新蛋白质的安全性

既然并不认为DNA的安全性是一项重要的考虑因素，那么安全性评价应集中于由嵌入基因产生的新蛋白质以及遗传修饰食物中由那种蛋白质是一种酶时的作用而可能产生的任何成分上。一般而言，新蛋白质的安全性评价最初涉及与已知毒素或过敏原进行氨基酸序列同源性方面的结构比较，通常也要评价新蛋白质的消化结果或对胃蛋白酶水解的抵抗性，因为如果新蛋白质被胃蛋白酶或在模拟的胃的消化性测试中迅速水解，那么蛋白质就不可能产生如毒性或过敏性这样的不良反应。新蛋白质安全性评价的另一种方法，是测验这种纯化的蛋白质在用鼠或另一种适合动物模型的急性口服畜性筛选中的毒性。当在毒物学测验中使用纯化蛋白质时，必须确保该纯化蛋白质实质上与遗传修饰食物中表达的蛋白质相同。植物中的蛋白质常被糖基化。糖基化和其他的翻译后修饰可潜在影响新蛋白质的毒性。新蛋白质安全性评价中使用的另一种方法是评定其热稳定性，虽然这一特性与蛋白质的安全性的相关性还存在争议。这些不同方法应依据已知的有关蛋白质和其可能的毒性方面的信息而灵活应用。但是，必须足以让世界管理机构信服的是，这种新蛋白质本身不是毒性的，或不可能是或成为过敏性的。

对新蛋白质过敏性的评价是安全性评价过程的另一重要部分。所有的食物过敏原都是蛋白质。因此，任何新引入的蛋白质会是过敏原或成为新过敏原的一些可能性是存在的。但是，食物中含成千上万种蛋白质，且只有几百种已知在典型的接触情况下是过敏原，因此，仅以概率为根据认为任何特定的新蛋白质会是一种过敏原的可能性是微小的。

尽管可能性很小，但一般认为，一种新蛋白质的潜在过敏性可能是作为遗传修饰食物安全性评价部分的最相关安全性评价问题。为了评价遗传修饰食物中新蛋白质的潜在过敏性，已积极提出了几种战略（Metcalfe *et al.*，1996；FAO/WHO，2000，2001）。所有这些战略都有几个共同的要素，包括基于新基因的来源和新基因的源材料的过敏性的考虑、新蛋白质与已知过敏原的氨基酸序列同源性的程度、新蛋白质与来自对源材料有已知过敏症的人的血清免疫球蛋白（IgE）的免疫反应性，以及新蛋白质对胃蛋白酶的抗性，而这种抗性可表明新蛋白质的消化稳定性。

过敏性的评价通常通过集中于引入宿主组织的 DNA 的来源开始。如果已知基因源是过敏性的，那么必须测定基因产品的过敏性。应考虑两种环境过敏原，即花粉和食物过敏原，因为一些已知的食物过敏原与花粉过敏原有交叉反应（Calkoven *et al.*，1987）。如果已知基因源是过敏性的，那么可使用一种特定的采用已知对该基因生物源过敏的人体的全血血清的血清筛选测验，就能较准确测定该新基因产品的潜在过敏性。这种假设应该是在这种情况下的基因源编码一种过敏原，除非所得数据反驳这种假设。在农业生物技术中，当从一种常见的过敏源中获取该基因时将会引起人们的最大关注。通常过敏的食物包括来自植物界的花生、大豆、坚果和小麦，以及来自动物界的奶、蛋、鱼和甲克纲动物（FAO，1995）。对这些少数食物或食物组的过敏很可能占世界范围内所有食物过敏的 90%以上（FAO，1995）。除了这 8 种食物或食物组之外，还有 160 多种食物和与食物有关的物质至少在某些情况下与个人的过敏反应有关（Hefle *et al.*，1996）。另一方面，如果从一种没有过敏史的来源中获得该种基因，那么就不可能有人血清存在，因而很明显就不可能进行特定的血清筛选试验。在大量的农业生物技术事例中，基因是从没有过敏史的来源中获取的。在这样的情况下，必须将可信度不高的方法，如氨基酸序列同源性与已知过敏原和新蛋白对胃蛋白酶的相对抗性结合起来，评价转基因产品中新蛋白质的潜在过敏性。

新引入的蛋白质与已知的食物及环境过敏原之间的氨基酸序列同源性程度的比较是迄今设计的所有方法中已被提倡的关键战略（Metcalfe *et al.*，1996；FAO/WHO，2000，2001)。许多食物和环境过敏原的氨基酸序列是已知的（Metcalfe 等，1996；Gendel，1998)。测定新蛋白质的过敏潜力的一种有用的方法，是将新蛋白质的氨基酸序列与已知过敏原的进行比较（Metcalfe *et al.*，1996；FAO/WHO，2001)。如果有足够的同源性，那么新蛋白质可能与已知的过敏原发生交叉反应，当被有那特定过敏症的人摄入后引起过敏症状。

在 IgE 介导的食物过敏症中，已知只有某些特定的蛋白质在典型暴露的环境下可诱导 IgE 的致敏过程。免疫系统只识别过敏原全部结构的一部分而不是整个结构。这些较小的区域被称为过敏原决定簇或抗原决定部位。抗原决定部位可能是连续的（一段线性的氨基酸序列），也可能是不连续的（取决于蛋白质的三维空间结构）（Taylor and Lehrer，1996)。线性抗原决定簇可能对食物过敏原更加重要，因为食物过敏原对热加工和消化很稳定，而这些条件经常导致蛋白质的三维结构出现解折叠（Taylor *et al.*，1987)。但是，最近获得的有关 Ara h 1（一种主要的花生过敏原）的证据提示，非连续构象的抗原决定簇可能在这种特定食物过敏原的情况下具有重要性（Shin *et al.*，1998)。尽管有此证据存在，但在对通过农业生物技术引入食物中的新蛋白质的过敏潜力评价中，线性抗原决定簇很可能仍很重要。IgE 介导的食物过敏症要经过两阶段致敏过程后才能开始诱发（Mekori，1996)。在致敏过程中，膳食中的蛋白质被抗原存在细胞中的水解作用加工，水解产生的肽然后与 T 细胞反应引起 B 细胞转而产生特定过敏原的 IgE 抗体。一旦特定过敏原的 IgE 抗体与肥大细胞和嗜碱性粒细胞结合后，这些相同蛋白质上的 IgE 结合抗原决定簇就与细胞结合的 IgE 抗体相互作用，使该过程进入诱发阶段，并从肥大细胞和嗜碱性粒细胞释放出组胺和过敏反应的其他介质。过敏原特异性 IgE 抗体能与线性还是构象性抗原决定簇结合，取决于此过程中所涉及的特定过敏原。如果通过农业生物技术引入的新蛋白质与现有的过敏原相同或与现有的过敏原发生交叉反应，那么被细胞结合的 IgE 识别的任何线性或构象性抗原决定簇可能都很重要。但是，对于真正新的蛋白质，不存在优先的致敏作用。在这些情况下，通过 T 细胞而产生的致敏作用很关键。T 细胞抗原决定簇唯一可能是线性的。

人们已建议用几个不同的指标来测定新蛋白质与已知过敏原之间的氨基酸序列同源性的显著程度。在一些方法（Metcalfe *et al.*，1996；FAO/WHO，2000）中，使用 8 个相邻的相同氨基酸的指标来确定一种阳性配比。在最近的 FAO/WHO 磋商中，则使用 6 个相邻的相同氨基酸的指标来确定一种阳性配比（FAO/WHO，2001)。使用 6 个相邻的相同氨基酸作为一种配比，预测到 Ara h 1 和 Ara h 2 的最小 IgE 结合抗原决定簇含有 6 个相邻氨基酸（Burks *et al.*，1997；Stanley *et al.*，1997)。T 细胞结合的抗原决定簇的最小肽的长度很可能是 8 个相邻的氨基酸（Metcalfe *et al.*，1996)。使用 6 个相邻的相同氨基酸的指标将很可能导致识别出大量的假阳性配比。因此，使用 7 个或 8 个相邻的相同氨基酸作为一种配比指标可能更加合理。既然这些方法可评价整个蛋白质序列，那么它们是以氨基酸序列的同一性为基础，而不只是以已知过敏原的已知 T 和 B 细胞结合的抗原决定簇为基础。因此，这些方法可以（且很可能会）识别与新蛋白质的过敏潜力不相关的配比序列。但是，可使用特定的血清筛选来消除临床不显著的配比。

FAO/WHO（2001）也建议，使用在新蛋白质的氨基酸序列与已知食物和环境过敏原的序列之间具有35%的整体结构同源性指标来识别具有相似功能且可能是交叉反应过敏原的蛋白质。许多常见的植物过敏原都在几个功能性种类之中（Breiteneder and Ebner，2000）。几种不同类型的与发病机制有关的蛋白质都明显参与其中（Breiteneder and Ebner，2000）。如果在通过农业生物技术研发的食物中引入的新蛋白质列入含有已知食物过敏原的功能性种类，那么35%的整体结构同源性指标将很可能识别出它们。应谨慎地评价这些特定蛋白质的潜在过敏性。

如果从已知过敏源中获得的基因，或寻找出序列同源性识别出一种与已知食物或环境过敏原的配比，那么就应评价新蛋白质与来自对源材料或过敏原过敏的人的血清IgE抗体的免疫反应性（FAO/WHO，2001）。既然还不知道来自所有过敏源的所有过敏原的结构，那么在从已知过敏源中获取基因的每种情况下都已提倡进行特异性血清筛选试验（FAO/WHO，2001）。在血清筛选试验中，用来自已知对特定过敏源过敏的单个人的全血血清进行测验，以测定血清中的过敏原特异性IgE是否与新蛋白质上的抗原决定簇发生反应。来自已熟悉其特征的有那特定过敏症的患者的血清的可用性是一个重要的且有时也富挑战性的问题。对一些类型过敏症的已熟悉其特征的人全血血清，可能是罕见的，且很难获取。阳性血清筛选测验结果肯定会引起人们对新蛋白质的可能过敏性的关注，且会导致得出新蛋白质很可能是致敏性的，因而，进一步的商业研发很可能会停止这样的结论。

当利用来自对新基因源过敏的单个人的血清或测定来自结构比较的同源序列是否相关而进行的血清筛选试验是过敏性评价中的一种公认方法时（Metcalfe *et al*.，1996；FAO/WHO，2000，2001），使用目的血清筛选则是更具争议的一种方法。最近的FAO/WHO磋商会议（FAO/WHO，2001）倡导使用目的血清筛选试验，在其中，使用来自对源料有广泛相关的材料有已知过敏症的人的血清IgE来评价新蛋白质的免疫反应性。此类例子包括：来自对草的花粉过敏的单个体人的血清（如果从单子叶植物源中获得基因）或来自对蟑螂过敏的个体人的血清（如果从昆虫源中获得基因）。在FAO/WHO的方法中，建议对目的血清筛选试验使用几种广泛的种类：单子叶植物、双子叶植物、无脊椎动物、脊椎动物和霉菌（FAO/WHO，2001）。如果基因源是细菌，那么就不能进行目的血清筛选试验，因为细菌蛋白质的暴露水平低且对这些蛋白质缺乏过敏性致敏作用，因而几乎不是过敏性的。对此很可能需要进一步的研究来测定目的血清筛选在评价通过农业生物技术生产的食物中的新蛋白质的潜在过敏性方面是否有用。当然，很可能出现假阳性结果，且获得已熟悉其特征的全血血清也是很关键的问题。

新蛋白质的水解稳定性是评价新蛋白质的过敏症潜力的另一个有用的指标。过敏原蛋白必须以一种足够完整的形式到达肠道，以激活免疫系统，快速消化的蛋白质将不很可能以免疫完整的形式到达肠道。已知的食物过敏原比已知的非过敏原食物蛋白在模拟的胃和肠消化模型中呈现出更大的蛋白水解稳定性（Astwood *et al*.，1996）。引入通过农业生物技术生产的食物中的某些新蛋白质在这些相同模型系统中也被快速消化（Astwood *et al*.，1996）。例如，转入大豆中使其具有耐除草剂草甘膦作用的酶可在体外被快速消化（Harrison *et al*.，1996）。这种新蛋白质因此不很可能诱导过敏性致敏过程。人们已倡导利用对胃蛋白酶的消化稳定性和（或）抵抗性来作为评价引入遗传修饰食物中的新蛋白质的潜在过敏性的方法（Metcalfe *et al.*，1996；FAO/WHO，2000，2001）。既然人的消化能力是个体变异的，那么消化稳定性试验就不能用来分析预测所有人与新蛋白质的消化稳定性。但是，对胃蛋白酶水解的比较抗性很可能是过敏性评价中的一种合理的比较方法。胃蛋白酶的抗性并不是过敏潜力的一个完善测定指标。已知新鲜水果和蔬菜中的一些过敏原对蛋白水解敏感（Moneret-Vautrin *et al.*，1997）。这些特殊的过敏原往往是那些与已知的花粉过敏原发生交叉反应的过敏原（Calkoven *et al.*，1987），因此很可能会在序列同源性测验中被发现。

在评价新蛋白质的过敏性方面使用动物模型还不成熟，因为现在还没有很有效化的动物模型可以达到此目的。但是，在研发合适的动物模型方面已取得了一些进步（Taylor，2002）。

当把来自巴西坚果的一种高蛋氨酸蛋白质引入大豆以纠正其固有的蛋氨酸缺乏症时，表明了这些方法用于评价通过农业生物技术引入的新蛋白质的潜在过敏性的适合性。在当时，已知巴西坚果是过敏性的（Arshad *et al.*，1991），但是还不知道巴西坚果中的过敏原的特性。根据用来自对巴西坚果过敏的个体人的血清进行的血清筛选试验，来自巴西坚果的新的高蛋氨酸蛋白质被认为是以前未发现的来自巴西坚果的主要过敏原（Nordlee *et al.*，1996）。因此，高蛋氨酸大豆的商业化还未进行。

### 用于动物饲养的安全性

安全性评价也应考虑遗传修饰农作物在动物饲养方面的可能应用。在遗传修饰农作物中，玉米、大豆、低毒菜子和棉籽对家畜饲养很重要。膳食安全性评价一般涉及用合适目标动物种类进行的饲养研究及对典型性能指标如生长速率的比较（Hammond *et al.*，1996）。

## 总　　结

在美国和其他国家，遗传修饰食物在早期的基因发现阶段与产品商业化阶段之间，要经过许多管理屏障。因此，现在全世界作为食品而出售的农业生物技术产品已经受严格的安全性评价。遗传修饰食物的安全性评价通常从将新食物与其传统的对应物进行比较开始。当前在大多数情况下，除了由于引入了特别的兴趣基因而导致的少数几种确定差异外，新食物与其传统的对应物是相似的。因此，安全性评价集中于引入基因且特别是由基因产生的新蛋白质的安全性上。已充分证实现在的遗传修饰食物在预期消费条件下的预期使用是安全的，因为它们已经受这样的安全性评价。随着未来与传统食物不同的遗传修饰食物的引入，安全性评价过程将变得更加困难。但是，还没有这样的遗传修饰农作物已被引入市场。世界范围内存在的管理系统可确保在其引入市场前要进行适当的测验，以保证消费者现在和未来所消费的新食物的安全性。主要的安全性问题是新蛋白质的潜在过敏性。但是，除非新基因是从一种已知的过敏源中获得或新蛋白质在遗传修饰食物中大量表达，否则，新蛋白质将成为一种新的和唯一的过敏原的可能性就很小。已经研发出评价通过农业生物技术引入的新蛋白质的潜在过敏性的方法。为更好地确保这些新蛋白质不会成为过敏原的改良方法的研究还在继续进行。

# 参 考 文 献


Arshad, S.H., Malmberg, E., Krapf, K. and Hide, D.W. (1991) Clinical and immunological characteristics of Brazil nut allergy. *Clinical and Experimental Allergy* 21, 373–376.

Astwood, J.D., Leach, J.N. and Fuchs, R.L. (1996) Stability of food allergens to digestion *in vitro*. *Nature Biotechnology* 14, 1269–1273.

Beever, D.E. and Kemp, C.F. (2000) Safety issues associated with the DNA in animal feed derived from genetically modified crops. A review of scientific and regulatory procedures. *Nutrition Abstracts and Reviews Series B: Livestock Feeds and Feeding* 70, 175–182.

Breiteneder, H. and Ebner, C. (2000) Molecular and biochemical classification of plant-derived food allergens. *Journal of Allergy and Clinical Immunology* 106, 27–36.

Burks, A.W., Shin, D., Cockrell, G., Stanley, J.S., Helm, R.M. and Bannon, G.A. (1997) Mapping and mutational analysis of the IgE-binding epitopes of Ara h 1, a legume vicilin protein and a major allergen in peanut hypersensitivity. *European Journal of Biochemistry* 245, 334–339.

Calkoven, P.G., Aalbers, M., Koshte, V.L., Pos, O., Oei, H.D. and Aalberse, R.C. (1987) Cross-reactivity among birch pollen, vegetables and fruits as detected by IgE antibodies is due to at least three distinct cross-reactive structures. *Allergy* 42, 382–390.

Doerfler, W. and Schubbert, R. (1997) Fremde DNA im Saugersystem. *Deutsches Arzteblatt* 94, 51–52.

FAO (1995) *Report of the FAO Technical Consultation on Food Allergies*. Food and Agriculture Organization of the United Nations. FAO, Rome, Italy.

FAO/WHO (1996) Biotechnology and food safety. *Report of a Joint FAO/WHO Expert Consultation*. Food and Agriculture Organization of the United Nations and World Health Organization. FAO, Rome, Italy.

FAO/WHO (2000) Safety aspects of genetically modified foods of plant origin. *Report of a Joint FAO/WHO Expert Consultation*. Food and Agriculture Organization of the United Nations and World Health Organization. WHO, Geneva, Switzerland.

FAO/WHO (2001) Evaluation of the allergenicity of genetically modified foods. *Report of a Joint FAO/WHO Expert Consultation*. Food and Agriculture Organization of the United Nations and World Health Organization. FAO, Rome, Italy.

FDA (1992) Statement of policy: foods derived from new plant varieties – Food and Drug Administration. *Federal Register* 57, 22984–23005.

Gendel, S.M. (1998) The use of amino acid sequence alignments to assess potential allergenicity of proteins used in genetically modified foods. *Advances in Food and Nutritrion Research* 42, 45–62.

Hammond, B.G., Vicini, J.L., Hartnell, G.F., Naylor, M.W., Knight, C.D., Robinson, E.H., Fuchs, R.L. and Padgette, S.R. (1996) The feeding value of soybeans fed to rats, chickens, catfish and dairy cattle is not altered by genetic incorporation of glyphosate tolerance. *Journal of Nutrition* 126, 717–727.

Harrison, L.A., Bailey, M.R., Naylor, M.W., Ream, J.E., Hammond, B.G., Nida, D.L., Burnette, B.L., Nickson, T.E., Mitsky, T.A., Taylor, M.L., Fuchs, R.L. and Padgette, S.R. (1996) The expressed protein in glyphosate-tolerant soybean, 5-enolpyruvylshikimate-3-phosphate synthase from *Agrobacterium* sp. strain CP4, is rapidly digested *in vitro* and is not toxic to acutely gavaged mice. *Journal of Nutrition* 126, 728–740.

Hefle, S.L., Nordlee, J.A. and Taylor, S.L. (1996) Allergenic foods. *Critical Reviews in Food Science and Nutrition* 36, S69–S89.

IFT (2000a) IFT expert report on biotechnology and foods. Introduction. *Food Technology* 54(8), 124–136.

IFT (2000b) IFT expert report on biotechnology and foods. Human food safety of rDNA biotechnology-derived foods. *Food Technology* 54(9), 53–61.

James, C. (2001) *Global Review of Commercialized Transgenic Crops, 2000. ISAAA Briefs No. 23*. International Service for the Acquisition of Agri-biotech Applications, Ithaca, New York.

McClintock, J.T., Schaffer, C.R. and Sjoblad, R.D. (1995) A comparative review of the mammalian toxicity of *Bacillus thuringiensis*-based pesticides. *Pesticide Science* 45, 95–105.

Mekori, Y.A. (1996) Introduction to allergic diseases. *Critical Reviews in Food Science andNutrition* 36, S1–S18.

Metcalfe, D.D., Astwood, J.D., Townsend, R., Sampson, H.A., Taylor, S.L. and Fuchs, R.L. (1996) Assessment of the allergenic potential of foods derived from genetically engineered crop plants. *Critical Reviews in Food Science and Nutrition* 36S, 165–186.

Moneret-Vautrin, D.A., Kanny, G., Rance, F. and Lemerdy, P. (1997) Les allergènes végétaux alimentaires. Allergies associées et réactions croisées. *Revue Francaise d'Allergologie et d'Immunologie Clinique* 37, 316–324.

Neilsen, K.M., Bones, A.M., Smalla, K. and van Elsas, J.D. (1998) Horizontal gene transfer from transgenic plants to terrestrial bacteria – a rare event? *FEMS Microbiology Reviews* 22, 79–103.

Nordlee, J.A., Taylor, S.L., Townsend, J.A., Thomas, L.A. and Bush, R.K. (1996) Identification of Brazil nut allergen in transgenic soybeans. *New England Journal of Medicine* 334, 688–692.

OECD (1992) *Safety Evaluation of Foods Derived by Modern*

*Biotechnology – Concepts and Principles*. Organisation for Economic Cooperation and Development, Paris.
OECD (2000) *Report of the Task Force for the Safety of Novel Foods and Feeds*. Organisation for Economic Cooperation and Development, Paris, 86/ADDI, May 17.
Redenbaugh, K., Hiatt, W., Martineau, B., Kramer, M., Sheehy, R., Sanders, R., Houck, C. and Emlay, D. (1992) *Safety Assessment of Genetically Engineered Fruits and Vegetables. A Case Study of the Flavr Savr Tomato*. CRC Press, Boca Raton, Florida.
Sanders, P.R., Lee, T.C., Groth, M.E., Astwood, J.D. and Fuchs, R.L. (1998) Safety assessment of insect-protected corn. In: Thomas, J.A. (ed.) *Biotechnology and Safety Assessment*, 2nd edn. Hemisphere Publishing, New York, pp. 241–256.
Schnepf, E., Crickmore, N., Van Rie, J., Lereclus, D., Baum, J., Feitelson, J., Zeigler, D.R. and Dean, D.H. (1998) *Bacillus thuringiensis* and its pesticidal crystal proteins. *Microbiology and Molecular Biology Reviews* 62, 775–806.
Schubbert, R., Hohlweg, U., Renz, D. and Doerfler, W. (1998) On the fate of orally ingested foreign DNA in mice: chromosomal association and placental transfer to the fetus. *Molecular and General Genetics* 259, 569–576.
Shin, D.S., Compadre, C.M., Maleki, S.J., Kopper, R.A., Sampson, H., Huang, S.K., Burks, A.W. and Bannon, G.A. (1998) Biochemical and structural analysis of the IgE binding sites on Ara h 1, an abundant and highly allergenic peanut protein. *Journal of Biological Chemistry* 273, 13753–13759.
Stanley, J.S., King, N., Burks, A.W., Huang, S.K., Sampson, H., Cockrell, G., Helm, R.M., West, C.M. and Bannon, G.A. (1997) Identification and mutational analysis of the immunodominant IgE binding epitopes of the major peanut allergen Ara h 2. *Archives of Biochemistry and Biophysics* 342, 244–253.
Taylor, S.L. (2001) Safety assessment of genetically modified foods. *Journal of Nematology* 33, 178–182.
Taylor, S.L. (2002) Protein allergenicity assessment of foods produced through agricultural biotechnology. *Annual Reviews of Pharmacology and Toxicology* 42, 99–112.
Taylor, S.L., and Lehrer, S.B. (1996) Principles and characteristics of food allergens. *Critical Reviews in Food Science and Nutrition* 36, S91–S118.
Taylor, S.L., Lemanske, R.F. Jr, Bush, R.K. and Busse, W.W. (1987) Food allergens: structure and immunologic properties. *Annals of Allergy* 59, 93–99.
Underwood, B.A. (1994) Vitamin A in human nutrition: public health considerations. In: Sporn, M.B., Roberts, A.B. and Goodman, D.S. (eds) *The Retinoids: Biology, Chemistry and Medicine*, 2nd edn. Raven Press, New York, pp. 217–227.
WHO (1993) *Health Aspects of Marker Genes in Genetically Modified Plants. Report of a WHO Workshop*. World Health Organization, Geneva, Switzerland.
Ye, X., Al-Babili, S., Kloti, A., Zhang, J., Lucca, P., Beyer, P. and Potrykus, I. (2000) Engineering the provitamin A (β-carotene) biosynthetic pathway into (carotenoid-free) rice endosperm. *Science* 287, 303–305.
Zitnak, A. and Johnston, G.R. (1970) Glycoalkaloid content of B5141-6 potatoes. *American Potato Journal* 47, 256–260.

# 索　　引